AF742922

AN INTRODUCTION TO THE STUDY OF MAMMALS LIVING AND EXTINCT

This is a volume in the Arno Press collection

BIOLOGISTS AND THEIR WORLD

Advisory Editor
Keir B. Sterling

Editorial Board
Kraig Adler
Armin Geus
David M. Knight
E. Gorton Linsley
Hans Querner
Ronald L. Stuckey

See last pages of this volume
for a complete list of titles

AN INTRODUCTION

TO THE STUDY OF

MAMMALS

LIVING AND EXTINCT

WILLIAM HENRY FLOWER

AND

RICHARD LYDEKKER

23 OCT 2006

ARNO PRESS
A New York Times Company
New York • 1978

599
FLO

Editorial Supervision: MARIE STARECK

Reprint Edition 1978 by Arno Press Inc.

BIOLOGISTS AND THEIR WORLD
ISBN for complete set: 0-405-10641-6
See last pages of this volume for titles.

Manufactured in the United States of America

Library of Congress Cataloging in Publication Data

Flower, William Henry, Sir, 1831-1899.
An introduction to the study of mammals living and extinct.

(Biologists and their world)
Reprint of the 1891 ed. published by A. and C. Black, London.
1. Mammals. I. Lydekker, Richard, 1849-1915, joint author. II. Title. III. Series.
QL703.F56 1978 599 77-81075
ISBN 0-405-10644-0

BAL774

AN INTRODUCTION

TO THE STUDY OF

MAMMALS

LIVING AND EXTINCT

BY

WILLIAM HENRY FLOWER

C.B., F.R.S, D.C.L., LL.D., P.Z.S, F.L.S., F.G.S., &c.

DIRECTOR OF THE NATURAL HISTORY DEPARTMENTS, BRITISH MUSEUM

AND

RICHARD LYDEKKER

B.A., F.G.S., F.Z.S., &c.

ILLUSTRATED

LONDON: ADAM AND CHARLES BLACK

MDCCCXCI

AN INTRODUCTION

TO THE

STUDY OF MAMMALS

PREFACE

One of the greatest difficulties experienced by all who undertake a work of this nature, not professing to be an exhaustive treatise on the subject with which it deals, is to determine the amount of detail desirable to be introduced to meet the requirements of the ordinary student, without rendering it too bulky or costly for general use. The experience of those who endeavour to profit by the book can alone decide how far the authors have succeeded in this respect. It will be observed that in many instances certain better-known or more interesting members of the class have been described at considerable length, while it has been necessary to treat others with much greater brevity.

With regard to the references to the literature of the various groups treated of, it has been the endeavour of the authors to make a selection of such memoirs and works as are likely to prove most valuable to the student for the amount of original information which they contain, and more especially of those giving full bibliographical data up to the time of their publication, the repetition of which has been considered unnecessary.

In a few instances new generic terms have been introduced to replace some which were already occupied; these have been proposed by Mr. Lydekker, and should be quoted as his.

The work is based largely upon the article "Mammalia," together with forty shorter articles, written by the senior of the two authors for the ninth edition of the Encyclopædia Britannica. The account of the orders Rodentia, Insectivora, and Chiroptera contributed to the article "Mammalia" by Dr. G. E. Dobson, F.R.S., as well as the articles "Mole," "Shrew," and "Vampyre," by the same writer, the articles "Marmot," "Mouse," "Opossum," "Phalanger," "Rat," "Squirrel," "Stoat," "Vole," and others, by Mr. Oldfield Thomas, and likewise the article "Ape," by Dr. St. G. Mivart, F.R.S., have also been made use of to a greater or less extent. The best thanks of the authors are due to these three gentlemen for freely permitting the incorporation of their own work in the present volume.

Mr. Lydekker undertook the task of arranging the various articles in their proper sequence, selecting from these such portions as seemed suitable, filling up the gaps, and adding new matter where necessary; a large amount of this new matter treating of the extinct forms, and also of the group Artiodactyla.

The subsequent revision, both before being sent to the printers, and also when passing through the press, has been made by both authors, who are thus jointly responsible for the whole work.

The illustrations are to a great extent those prepared for the various articles in the Encyclopædia, but many have been added —some drawn expressly for the work, and some borrowed from other publications. For most of the latter the authors take this opportunity of expressing their thanks to the Publication Committee of the Zoological Society of London, as well as to the individual writers in whose works they first appeared.

The authors have further much pleasure in acknowledging the ready and obliging way in which Mr. Oldfield Thomas has, throughout the progress of the work, placed his extensive knowledge of the group of animals of which it treats at their disposal.

LONDON, *March* 1891.

Corrigenda.

Page 280, *for* Chæropsis *read* Chœropsis.

Page 292, *for* Chæropotamidæ and Chæropotamus *read* Chœropotamidæ and Chœropotamus.

Page 590, *for* Pæcilogale *read* Pœcilogale.

CONTENTS

CHAPTER I

CHAPTER II

CHAPTER IX

CHAPTER X

CHAPTER XI

CHAPTER XII

CHAPTER XIII

CHAPTER XIV

AN INTRODUCTION

TO

THE STUDY OF MAMMALS

LIVING AND EXTINCT

CHAPTER I

INTRODUCTORY REMARKS

MAMMALIA (French, *Mammifères;* German, *Säugethiere*) is the name invented by Linnæus (from the Latin *mamma*), and now commonly used by zoologists, for one of the five great classes of vertebrated animals, which, though the best known and undoubtedly the most important group of the animal kingdom, has never received any generally accepted vernacular designation in our language. The unity of structure of the animals composing this class, and their definite demarcation from other vertebrates, were not recognised until comparatively modern times, and hence no word was thought of to designate what zoologists now term a mammal. The nearest equivalents in common use are "beast" and "quadruped," both of which, however, cover a different ground, since they are often used to include the larger four-footed reptiles, and to exclude certain undoubted mammals, as Man, Bats, and Whales.

The limits of the class as now understood by zoologists are perfectly well defined, and, although certain forms still existing on the earth (but not those mentioned above as excluded by the popular idea) are of exceedingly aberrant structure, and exhibit several well-marked characters connecting them with the lower vertebrated groups, common consent retains them in the class with which the great proportion of their characters ally them, and hitherto no traces of any species showing still more divergent or transitional characters have been discovered. There is thus an interval, not bridged over by any known forms, between mammals and other vertebrates; although recent discoveries have shown evidence of a more or less marked affinity between the most generalised mammals and a peculiar group of extinct reptiles known as the Anomodontia (or Theromora), which are themselves nearly related to the equally extinct Labyrinthodont amphibians of the Palæozoic and Mesozoic epochs.

In the gradual order of evolution of living beings, mammals, taken altogether, are certainly the highest in organisation, as, with the possible exception of birds, they were the last to appear on the earth's surface. But, as in speaking of all other large and greatly differentiated groups, this expression must not be understood in too limited a sense. The tendency to gradual perfection for their particular station in life, which all groups manifest, leads to various lines of specialisation, or divergence from the common or general type, which may or may not take the direction of elevation. A too complex and sensitive condition of organisation may in some circumstances of life be disadvantageous, and modification may then take place in a retrograde direction. Thus in mammals, as in other classes, there are low as well as high forms, but by any tests that can be applied—especially those based on the state of development of the central nervous system—it will be seen that the average exceeds that of any other class; that the class contains many species far excelling those of any other in perfection of structure, and especially one form which is unquestionably the culminating point yet arrived at amongst organised beings.

With regard to the time of the first appearance of mammals upon the earth, the geological record is provokingly imperfect. At the commencement of the Tertiary period they were abundant, and already modified into most of the leading types at present existing. It was at one time thought that they first came into being at this date, but the discovery of more or less fragmentary remains of numerous and generally small species has revealed the existence of some forms of the class at various periods throughout almost the whole of the age of the deposition of the Secondary or Mesozoic rocks. This subject will be reverted to later on.

It hardly need be said that mammals are vertebrated animals, and possess all the characteristics common to the members of that division of the animal kingdom. They are separated from the *Ichthyopsida* (fishes and amphibians), and agree with the *Sauropsida* (reptiles and birds) in the possession during their development of an amnion and allantois, and in never having external branchiæ or gills. They differ from reptiles and resemble birds in being warm-blooded, and having a heart with four cavities and a complete double circulation. They differ from both birds and reptiles in the red corpuscles of the blood being non-nucleated and, with very few exceptions, circular in outline; in the lungs being freely suspended in a thoracic cavity, separated from the abdomen by a complete muscular partition—the diaphragm—which is the principal agent in inflating the lungs in respiration; in having but one aortic arch, which curves over the left bronchus; in the skin being more or less clothed with hair; in the greater perfection of the commissural system of the cerebral hemispheres, which has either a complete corpus callosum, or an incomplete one associated with a very large anterior commissure; in having no syrinx or inferior vocal organ, but a complete larynx at the upper end of the trachea; in having a mandible of which each ramus (except in very early developmental conditions) consists of a single bone on each side, articulating to the squamosal without the intervention of a quadrate bone; in having a pair of laterally placed occipital condyles instead of one median one; and in the very obvious character of the female being provided with mammary glands, by the secretion of which the young (usually produced alive, although in the lowest forms by means of externally hatched eggs) are nourished for some time after birth.

In common with all vertebrated animals, mammals never have more than two pairs of limbs; as the larger number live ordinarily on the surface of the earth, in the great majority of the class both pairs are well-developed and functional, and adapted for terrestrial progression. Mammals are, however, by no means limited to this situation. Thus some species spend the greater part of their lives beneath the surface, their fore limbs being specially modified for burrowing; others, again, are habitually arboreal, their limbs being fitted for climbing or hanging to boughs of trees; some are as aerial as birds, the fore limbs being developed into wings of a special character; while in others which are as aquatic as fishes, the limbs assume the form of fins or paddles. In many of the latter the hinder extremities are either completely suppressed, or present only in a rudimentary state. In no known mammal are the fore limbs absent.

The hinder extremity of the axis of the body is usually prolonged into a tail, which may be a mere pendent appendage, or may be modified to perform various functions, as grasping boughs in climbing, or even gathering food, in the case of the prehensile-tailed Monkeys and Opossums, swimming in the Cetacea, and acting as a flap to drive away troublesome insects from the skin in the Ungulata.

The state of development of the young at the time of birth varies greatly in the different groups. Thus among the Marsupials where there is no connection during intra-uterine life between the circulatory systems of the parent and the fœtus, the young are born in an exceedingly imperfectly developed condition. For their protection the mother, in a large number of cases, has a special pouch enclosing the mammæ, into which the young are transferred at birth, and in which they remain till they are well developed. Among the higher, or Placental types, however, where a connection exists between the maternal and fœtal circulations previous to birth, the young are always born in a much more highly developed state than among the Marsupials, although we meet with great variations in this respect. In those forms which habitually live in holes, like many Rodents, the young are always very helpless at birth; and the same is also true of many of the Carnivora, which are well able to defend their young from attack. In the great order of Ungulate, or Hoofed Mammals, where in the majority of cases defence from foes depends upon fleetness of foot, or upon huge corporeal bulk, the young are born in a very highly developed condition, and are able almost at once to run by the side of the parent. This state of relative maturity at birth reaches its highest development in the Cetacea, where it is evidently associated with the peculiar conditions under which these animals pass their existence. In the Primates, however, we again find the young produced in a more or less helpless condition, and requiring a long period before they attain their full development, this being more especially the case with those higher forms which approximate in structure to man.

In point of size mammals vary to a greater extent than the existing members of any one class of animals, and include the largest living inhabitants of the earth. The extremes of size are marked on the one hand by the whale known as Sibbald's Rorqual, which attains a length of eighty feet and a weight of nearly as many tons, and on the other by the Pigmy-Shrew and the minute Harvest-mouse, which can climb a stem of wheat.

Of all the living creatures inhabiting our globe, mammals are by far the most important in their economic uses, since, in addition to being the only animals capable of labour for human benefit, they furnish the greater portion of the animal food of many races of man, and likewise a large amount of their clothing. In these respects the Ungulates hold the first place.

As regards employment for labour, with the exception of the Dogs used for sleighing by the Esquimaux, and those which among some European nations draw light carts, all the mammals in general use are Ungulates. Of the first importance are the Horses and Asses, which are employed as beasts of draught or burden over nearly the whole globe. Among many nations, however, cattle, as represented by the true Oxen, the Buffalos, and the Yaks of Tibet, occupy a still more important position, while in the highlands of Tibet Sheep are largely used for carrying burdens. In other regions, again, the place of the Horse and the Ass is taken by the Camels, which are peculiarly fitted for traversing parched and arid deserts, while in the Andes we find the Llamas serving the same office. In Lapland and other parts of the northern regions the Reindeer is the main agent employed in draught. Lastly, we must not omit to mention the Indian Elephant, which, from its vast strength, is so useful in transport through the wilder parts of its native country.

As regards food, we again find the Ungulates, and more especially the Artiodactyle division, taking the foremost place; and in this connection we have only to mention, among animals kept in a domestic condition, Swine, Cattle, Sheep, and Goats—the three latter affording not only their flesh, but also milk and its resulting cheese and butter. To many races, however, Mares and Camels are the chief milk producers, while the Laps make use of the milk of the Reindeer. The Rodents, as represented by Hares and Rabbits, occupy a minor position as furnishers of food.

In relation to clothing, the Ungulates are likewise of paramount importance, as exemplified by the wool of the Sheep, which is so valuable on account of its peculiar property of felting. Furs, however, are mostly yielded by mammals of other orders, among which the Fur-seals are perhaps the most important at the present day. Many other Carnivores yield valuable furs, among which may be mentioned Bears, Foxes, Racoons, Skunks, Minks, Otters, and Ermines. Of less importance are certain Rodents, such as the Squirrels, Rabbits, Hares, etc., while the hair of the Beaver was formerly much sought after for the manufacture of hats. Returning to the Ungulates, we may notice the importance of horse-hair, the employment of camel's hair for brushes, and the many uses of the bristles of the pig. Some of the Monkeys yield fur which has been extensively used. Leather, again, is almost exclusively supplied by mammals, and mainly by the Ungulates.

Three other important products, namely horn, buck's-horn, and ivory, are likewise obtained solely from the same great order. Horn, as we shall notice in the sequel, is the sheath covering the bony horn-cores of the Oxen, while buck's-horn is the commercial term applied to the antlers of the Deer, which are largely used for knife-handles and other purposes. True ivory is the product of the two species of Elephant; but other kinds of ivory are obtained from the teeth of the Sperm Whale and the tusks of the Walrus and Hippopotamus, the latter kind having been extensively employed some years ago for artificial teeth. For many purposes the place of ivory is taken by bone, this being mostly obtained from Ungulates. The bones of Camels are of an especially firm texture and good colour, and are largely employed in India for inlaying. Other important uses of bones are in the form of bone-dust as manure, and also as a source of phosphoric acid. The horns of the African Rhinoceros and the hide of the Hippopotamus are occasionally manufactured into small canes or whips. Horns and hoofs are also largely employed in the manufacture of glue.

Formerly the so-called whalebone, or more properly baleen, was much used, especially to form the ribs of umbrellas and in stiffening ladies' apparel, but the gradual destruction of the Right Whales, its only source of supply, has largely restricted its use of late years.

The Cetacea are also of great economical importance from the abundance of oil yielded by the thick layer of blubber underlying the skin. Large quantities of valuable oil are also furnished by the Walrus and the Seals. Spermaceti, which was at one time extensively used in the manufacture of candles, is obtained from a large cavity in the head of the Sperm Whale or Cachalot, and also from the *Hyperoödon* or Bottle-nosed Whale.

The nature of ambergris, a peculiar substance found floating on the surface of the sea and employed in perfumery, was long a matter of controversy; but it appears to be an intestinal concretion of the Sperm Whale. Other substances of more importance to the perfumer are musk, the product of the Musk-Deer of the Himalaya, and civet, which is obtained from the so-called Civet Cat and other allied Carnivores. A secretion of the Beaver has also been used in perfumery and in medicine.

CHAPTER II

GENERAL ANATOMICAL CHARACTERS

I. TEGUMENTARY STRUCTURES

Hair.—The external surface of the greater number of members of the class is thickly clothed with a peculiarly modified form of epidermis, commonly called hair. This consists of hard, elongated, slender, cylindrical or tapering, filiform, unbranched masses of epidermic material, growing from a short papilla sunk at the bottom of a follicle in the derm or true skin. Such hairs upon different parts of the same animal, or upon different animals, assume various forms, and are of various sizes and degrees of rigidity,—as seen in the delicate soft velvety fur of the Mole, the stiff bristles of the Pig, and the spines of the Hedgehog and Porcupine, all modifications of the same structures. Each hair is composed usually of a cellular pithy internal portion, containing much air, and a denser or more horny cortical part. In some animals, as Deer, the substance of the hair is almost entirely composed of the medullary or cellular substance, and it is consequently very easily broken; in others the horny part prevails almost exclusively, as in the bristles of the Wild Boar. In the Three-toed Sloth (*Bradypus*) the hairs have a central horny axis and a pithy exterior. Though generally nearly smooth, or but slightly scaly, the surface of some hairs is strongly imbricated, notably so in some Bats; while in the Two-toed Sloth (*Cholœpus*) the hairs are longitudinally grooved or fluted. Though usually more or less cylindrical or circular in section, hairs are often elliptical or flattened, as in the curly-haired races of men, the terminal portion of the hair of Moles and Shrews, and conspicuously in the spines of the Rodents *Xerus* and *Platacanthomys.* Hair having a property of mutual cohesion or "felting," which depends upon a roughened scaly surface and a tendency to curl, as in domestic Sheep (in which animal this property has been especially cultivated by selective breeding), is called "wool."

In a large number of mammals hairs of one kind only are scattered pretty evenly over the surface; but in many there are two kinds, one longer, stiffer, and alone appearing on the surface, and the other shorter, finer, and softer, constituting the under fur, analogous to the down of birds. This under fur, or *pashm* as it is called by the natives of Kashmir, is especially abundant in the mammals inhabiting the cold plateau of Tibet and the adjacent regions. In many cases hairs of a different character from those of the general surface grow in special regions, forming ridges or tufts on the median dorsal or ventral surface or elsewhere. The tail is very often completed in this way by variously disposed elongated hairs. The margins of the eyelids are almost always furnished with a special row of stiffish hairs, called *cilia* or eyelashes; and in most mammals specially modified hairs, constituting the *vibrissæ* or whiskers, and endowed, through the abundant nerve supply of their basal papillæ, with special tactile powers, grow from the lips and cheeks. In some mammals the hairy covering is partial and limited to particular regions; in others, as the Hippopotamus and the Sirenia, though scattered over the whole surface, it is extremely short and scanty; but in none is it reduced to so great an extent as in the Cetacea, in which it is limited to a few small bristles confined to the neighbourhood of the lips and nostrils, and often only present in the young or even fœtal condition.

Some kinds of hairs, as those of the mane and tail of the Horse, appear to persist throughout the life-time of the animal; but more generally, as in the case of the body hair of the same animal, they are shed and renewed periodically, generally annually. Many mammals have a longer hairy coat in winter, which is shed as summer comes on; and some few, which inhabit countries covered in winter with snow, as the Arctic Fox, Variable Hare, and Ermine, undergo a complete change of colour in the two seasons, being white in winter, and gray or brown in summer. The several species of Cape Mole (*Chrysochloris*), the Desmans or Water Moles (*Myogale*), and *Potamogale velox*, are remarkable as being the only mammals whose hair reflects those iridescent tints so common in the feathers of tropical birds.

The principal and most obvious purpose of the hairy covering is to protect the skin against external influences, especially cold and damp. Its function in the hairless Cetacea is supplied by the specially modified and thickened layer of adipose tissue beneath the skin, called "blubber."

Colour.—From the consideration of hair we are easily led to that of colour. As a general rule, bright and primary colours are absent in the class; but among the Baboons we find brilliant patches of scarlet or blue on some of the bare portions of the body, and one of the South American Monkeys (*Brachyurus*) has its whole face of a bright crimson. The most general colours are various shades of gray, brown, and tawny, with a frequent tendency to whiteness of the ventral surface of the body; but among the Squirrels, and more especially those provided with a parachute for flying, we find brilliant russets, passing into orange and red. Dark brown or black is also not very uncommon, as in the Bears and the Sable Antelope of South Africa. Entirely white mammals are rare, and mostly characteristic of the polar regions, or of countries having a long and snowy winter. An entirely white Bat (*Diclidurus albus*) occurs, however, in South America. In the large majority of mammals that exhibit a varied coloration, the upper and most exposed parts of the surface present the richest and darkest colours, the under parts being pale or often quite white. The Ratels, Gluttons, *Ælurus*, Hamsters, and some others are exceptions to this rule. A large number of mammals having a ground colour of gray, tawny, or dun are marked by stripes or spots, which are generally of a darker hue than the ground colour, as in many Carnivora, but more rarely are lighter, as in the Fallow and Axis Deer and several species of Antelope. These stripes very generally run transversely to the axis of the body, as in the Tasmanian Thylacine, the Tiger, and the Zebra; but they may be longitudinal, as in several of the Civet family. There has been considerable discussion as to whether the striped or the spotted is the more primitive type of coloration; but no very conclusive arguments have been brought forward in favour of either view. It is, however, manifest that in several groups of mammals there is a tendency to lose the spots, and more rarely the stripes, and to assume a uniform colour. Thus the young of nearly all the species of Deer are spotted, whereas the adults of only the Fallow and Axis Deer are so marked. The same is true of most of the Pigs; and the young of the Malayan and American Tapirs are marked by light-coloured stripes and spots on a dark ground. In like manner the young of the Lion and the Puma exhibit distinct spots which disappear with advancing age. In most of our domestic horses of various shades of bay and brown we may detect "dappling" on the under hair when the outer coat has been removed, which is not apparent on the surface of the latter. Many varieties of the Ass and the Horse also exhibit a tendency to the presence of stripes on the legs, which would seem to indicate a descent from a striped Zebra-like type.

A peculiar feature, which is, however, common to many other groups of animals, is the tendency to what is known as melanism, or the production of black or dark individuals or races of particular species, due to an excess of pigment in the skin and hair. Thus we may have black Leopards and Jaguars, black Wolves, and black Rabbits.

The opposite to melanism, and of more frequent occurrence, is albinism—a condition in which the pigment or colouring matter usually present in the tissues constituting the external coverings of the body, and which gives them their characteristic hue, is absent. When it occurs the hair is of an opaque white, the claws, hoofs, etc., of a pale horn-colour, and the skin and eyes pink, in consequence of the colour of the blood which circulates through them being no longer concealed by the stronger hues of the pigments. An animal in this condition is called an *albino*. In complete albinism there is a total absence of pigment throughout the system. This condition occurs occasionally as an individual peculiarity among wild animals of many kinds; but it has never been perpetuated among them in distinct races or species. The disadvantage of absence of pigment in the eye, causing a certain amount of intolerance of light, is probably sufficient to account for this. Several races of true albinos, as White Ferrets, Rabbits, Rats, and Mice, have, however, been established under the protection of man, and in them this abnormal condition is propagated from generation to generation.

Partial albinism—a condition in which the absence of pigment is limited to portions of the surface, or, at all events, does not extend to the eyes—is much more common as an individual variation both in domestic and in wild animals. It is possible that the artificial conditions incident to domestication increase the tendency to its occurrence; but, whether this be so or not, it certainly becomes perpetuated more frequently among domesticated than among wild animals. This may be accounted for partly by its proving of no disadvantage to them, and partly by the frequent selection by man of animals of such colour in preference to others. The result is that there is no completely domestic animal of which white races do not exist. On the other hand, to most wild animals even partial albinism seems to be a disadvantage in the struggle for existence, since, except in the case of species inhabiting lands continually covered with snow, it renders them more conspicuous objects both to their enemies and their prey, and hence it is rarely perpetuated. In northern regions, however, a large proportion of species are regularly and normally of a white colour, either, as the Polar Bear, all the year through, or, as the Ermine or Stoat, Arctic Fox, and Alpine Hare, during the winter season. The coloration in these cases is obviously protective, as it is also to a great extent in many other instances throughout the class.

Among conspicuously coloured mammals, it has been observed that the vertical black and tawny stripes of the Tiger harmonise so well with the brown and green grasses of its native jungle as to render the animal almost invisible when lying among them; while the dappled hide of the Giraffe is said to agree equally well with the chequered splashes of light and shade in the clumps of tall mimosas among which it feeds. The uniformly tawny hue of the

Lion accords well with the prevailing tint of its native desert; and any one who has seen an Elephant or Buffalo in the deep shades of an Indian forest will realise how perfectly adapted is their dull, slaty colour to concealment in such a spot. The dun colour of the Wild Ass of India is equally well suited to the sandy deserts of Kutch; it is also stated that the brilliant stripes of the Zebras of Africa are arranged in such proportion as exactly to match the pale tint which arid ground possesses when seen by moonlight.[1] The most remarkable instance of protective coloration is, however, to be found in the Sloths of South America, in which the coarse gray hairs so closely resemble a mass of lichenous growth that it is almost impossible to distinguish these animals when at rest from the gnarled and lichen-clad boughs from which they suspend themselves. This resemblance is increased by the fact that the hairs actually develop a growth of lichens upon themselves. That the sombre coloration of these animals has been produced to harmonise with their present surroundings seems to be evident by the circumstance that when the long hair is plucked off the under fur is seen to present a bold alternation of black and yellow stripes, which may probably be regarded as the original primitive coloration of this group.

Scales, etc.—True scales, or flat imbricated plates of horny material, covering the greater part of the body, so frequently occurring in reptiles, are found only in one family of mammals, the *Manidæ* or Pangolins; but these are also associated with hairs growing from the intervals between the scales, or on the parts of the skin not covered by them. Similarly, imbricated epidermic productions form the covering of the under surface of the tail of the flying Rodents of the genus *Anomalurus;* and flat scutes, with the edges in apposition, and not overlaid, clothe both surfaces of the tail of the Beaver, Rats, and others of the same order, and also of some Insectivores and Marsupials. The Armadillos alone have an ossified exoskeleton, composed of plates of true bony tissue, developed in the derm or corium, and covered with scutes of horny epidermis. Other epidermic appendages are the horns of Ruminants and Rhinoceroses,—the former being elongated, tapering, hollow caps of hardened epidermis of fibrillated structure, fitting on and growing from conical projections of the frontal bone, and always arranged in pairs, while the latter are of similar structure, but solid and without any internal bony support, and (in all existing species) situated in the median line. Callosities, or bare patches covered with hardened and thickened epidermis, are found covering the pads under the soles of the feet and undersurfaces of the toes of nearly all mammals, upon the ischial tuberosities of many Apes, the sternum of Camels, on the inner side of the limbs of the

[1] Galton's *South Africa*, p. 187.

Equidæ, the grasping under surface of the tail of the prehensile-tailed Monkeys, etc. The greater part of the skin of both species of one-horned Asiatic Rhinoceros is immensely thickened and stiffened by increase of the tissue both of the derm and epiderm, constituting the well-known jointed "armour-plated" hide of those animals.

Nails, Claws, and Hoofs.—With very few exceptions, the terminal extremities of the digits of both limbs are more or less protected or armed by epidermic plates or sheaths, constituting the various forms of nails, claws, or hoofs. These are wanting in the Cetacea alone. A perforated spur, with a special secreting gland in connection with it, is found attached to the hind leg of the males of the three genera of Monotremata, *Ornithorhynchus*, *Proechidna*, and *Echidna*.

Odour-secreting Glands.—Besides the universally distributed sebaceous glands connected with the pilose system, most mammals have special glands situated in modified portions of the integument, often involuted to form a shallow recess or a deep sac with a narrow opening, situated in various parts of the surface of the body, and secreting odorous substances, by the aid of which individuals appear to recognise one another, and probably affording the principal means by which wild animals are able to become aware of the presence of other members of the species, even at great distances. Although the commencement of the modifications of portions of the external covering for the formation of special secretions may be at present difficult to understand, the principle of natural selection will readily explain how such organs become fixed and gradually increase in development in any species, especially as there would probably be a corresponding modification and increased sensibility of the olfactory organs. Such individuals as by the intensity and peculiarity of their scent had greater power of attracting the opposite sex would certainly be those most likely to leave descendants to inherit and in their turn propagate the modification.

To this group of structures belong the suborbital gland or "crumen" of Antelopes and Deer, the frontal gland of the Muntjak and of Bats of the genus *Hipposiderus*, the submental gland of the Chevrotains and of *Taphozous* and some other Bats, the post-auditory follicle of the Chamois, the temporal gland of the Elephant, the lateral glands of the Musk-Shrew, the dorsal gland of the Peccary, the inguinal glands of Antelopes, the preputial glands of the Musk-Deer and Beaver (already alluded to in connection with the use made of their powerfully odorous secretion in medicine and perfumery) and also of the Swine and Hare, the anal glands of Carnivora, the perineal gland of the Civet (also of commercial value), the caudal glands of the Fox and Goat, the gland on the humeral membrane of Bats of the genus *Saccopteryx*, the post-digital gland of the Rhinoceros, the inter-digital glands of the Sheep and many Ruminants, and numerous others. In some of these cases the glands are peculiar to, or more largely developed in, the male; in others they are found equally developed in both sexes.

II. DENTAL SYSTEM

The dental system of mammals may be considered rather more in detail than space permits for some other portions of their structure, not only on account of the important part it plays in the economy of the animals of this class, but also for its interest to zoologists as an aid in the classification and identification of species. Owing to the imperishable nature of their tissues, teeth are preserved for an indefinite time, and in the case of extinct species frequently offer the only indications available from which to derive an idea of the characters, affinities, and habits of the animals to which they once belonged. Hence even their smallest modifications have received great attention from comparative anatomists, and they have formed the subject of many special monographs.[1]

Teeth are present in nearly all mammals, and are applied to various purposes. They are, however, mainly subservient to the function of alimentation, being used either in procuring food, by seizing and killing living prey or gathering and biting off portions of vegetable material, and more indirectly in tearing or cutting through the hard protective coverings of food substances, as the husks and shells of nuts, or in pounding, crushing, or otherwise mechanically dividing the solid materials before swallowing, so as to prepare them for digestion in the stomach. Certain teeth are also in many animals most efficient weapons of offence and defence, and for this purpose alone, quite irrespective of subserviency to the digestive process, are they developed in the male sex of many herbivorous animals, in the females of which they are absent or rudimentary.

Teeth belong essentially to the tegumentary or dermal system of organs, and, as is well seen in the lower vertebrates, pass by almost insensible gradations into the hardened spines and scutes formed upon the integument covering the outer surface of the body; but in mammals they are more specialised in structure and limited in locality. In this class they are developed only in the

[1] L. F. E. Rousseau, *Anatomie comparée du Système dentaire chez l'Homme et chez les principaux Animaux*, 2d ed., 1839; F. Cuvier, *Des Dents des Mammifères considérées comme caractères zoologiques*, 1822-25; R. Owen, *Odontography*, 1840-45; C. G. Giebel, *Odontographie*, 1855; C. S. Tomes, *Manual of Dental Anatomy, Human and Comparative*, 3d ed., 1889.

gums or fibro-mucous membrane covering the alveolar borders of the upper and lower jaws, or, in other words, the premaxillary and maxillary bones and the mandible. In the process of development, for the purpose of giving them that support which is needful for the performance of their functions, they almost always become implanted in the bone,—the osseous tissue growing up and moulding itself around the lengthening root of the tooth, so that ultimately they become apparently parts of the skeleton. In no mammal, however, does ankylosis or bony union between the tooth and jaw normally take place, as in many fishes and reptiles, —a vascular layer of connective tissue, the alveolo-dental membrane, always intervening.[1] The presence of two or more roots, frequently met with in the cheek-teeth of mammals, implanted in corresponding distinct sockets of the jaw, is now peculiar to animals of this class.[2]

Structure.—The greater number of mammalian teeth when fully formed are not simple and homogeneous in structure, but are composed of several distinct tissues, which are enumerated below.

The *pulp*, a soft substance, consisting of a very delicate gelatinous connective tissue, in which numerous cells are imbedded, and abundantly supplied with blood-vessels and nerves, constitutes the central axis of all the basal part of the tooth, and affords the means by which the vitality of the whole is preserved. The nerves which pass into the pulp and endow the tooth with sensibility are branches of the fifth pair of cranial nerves. The pulp occupies a larger relative space, and performs a more important purpose, in the young growing tooth than afterwards, as by the calcification and conversion of its outer layers the principal hard constituent of the tooth, the dentine, is formed. In teeth which have ceased to grow the pulp occupies a comparatively small space, which in the dried tooth is called the pulp-cavity. This communicates with the external surface of the tooth by a small aperture at the apex of the root, through which the branches of the blood-vessels and nerves, by which the tooth receives its nutrition and sensitiveness, pass in to be distributed in the pulp. In growing teeth the pulp-cavity is widely open, while in advanced age it often becomes obliterated, and the pulp itself entirely converted into bone-like material.

The *dentine* or *ivory* forms the principal constituent of the greater number of teeth. When developed in its most characteristic form, it is a very hard but elastic substance, white, with a yellowish tinge, and slightly translucent. It consists of an organic

[1] The lower incisors of some species of Shrews are, however, said to become ankylosed to the jaw in adult age.

[2] The teeth of the extinct Dinosaurian reptile *Triceratops* have two distinct roots, placed transversely to the axis of the jaws.

matrix, something like, but not identical with, that of bone, richly impregnated with calcareous salts (chiefly calcium phosphate), these constituting in a fresh human tooth 72 per cent of its weight. When subjected to microscopical examination it is seen to be everywhere permeated by nearly parallel branching tubes which run, in a slightly curving or wavy manner, in a general direction from the centre towards the free surface of the tooth. These tubes communicate by open mouths with the pulp-cavity, and usually terminate near the periphery of the dentine by closed ends or loops, though in Marsupials and certain other mammals they penetrate into the enamel. They are occupied in the living tooth by soft gelatinous fibrils connected with the cells of the pulp. A variety of dentine, permeated by canals containing blood-vessels, met with commonly in fishes and in some few mammals, as the *Megatherium*, is called vaso-dentine. Other modifications of this tissue occasionally met with are called osteo-dentine and secondary dentine,—the latter being a dentine of irregular structure which often fills up the pulp-cavity of old animals.

The *enamel* constitutes a thin investing layer, complete or partial, of the outer or exposed and working surface of the dentine of the crown of the teeth of most mammals. This is the hardest tissue met with in the animal body, containing from 95 to 97 per cent of mineral substance (chiefly calcium phosphate and some carbonate, with traces of fluoride). Its ultimate structure consists of prismatic fibres, placed generally with their long axes at right angles to the free surface of the tooth. Enamel is easily distinguished from dentine with the naked eye by its clear, bluish-white, translucent appearance.

The *cement* or *crusta petrosa* is always the most externally placed of the hard tissues of which teeth are composed, as will be understood when the mode of development of these organs is considered. It is often only found as a thin layer upon the surface of the root; but sometimes, as in the complex-crowned molar teeth of the Horse and Elephant, it is a structure which plays a very important part, covering and filling in the interstices between the folds of the enamel. In appearance, histological structure, and chemical composition it is closely allied to osseous tissue, containing lacunæ and canaliculi, though only when it is of considerable thickness are Haversian canals present in it.

Development.—The two principal constituents of the teeth, the dentine and the enamel, are developed from the two layers of the mucous membrane of the jaw—the dentine from the deeper or vascular, the enamel from the superficial or epithelial layer. The latter dips down into the substance of the gum, and forms the enamel-organ or germ, the first rudiment of the future tooth, which is constantly present even in those animals in which the enamel is not found as a constituent of the perfectly-formed tooth. Below the mass of epithelial cells thus embedded in the substance of the gum, and remaining connected by a narrow neck of similar structure with the epithelium of the surface, a portion of the vascular areolar tissue becomes gradually separated and defined from that which surrounds it, and assumes a distinct form, which is that of the crown of the future tooth,—a single cone in the case of simple teeth, or with two or more eminences in the complex forms. This is called the dental papilla or dentine germ, and by the gradual conversion of its tissue into dentine the bulk of the future tooth is formed, the uncalcified central portion remaining as the pulp. The conversion of the papilla into hard tissue commences at the outer surface of the apex, and gradually proceeds downwards and inwards, so that the form of the papilla exactly determines the form of the future dentine, and no alteration either in shape or size of this portion of the tooth, when once calcified, can take place by addition to its outer surface. In the meanwhile, calcification of a portion of the cells of the enamel-organ, which adapts itself like a cap round the top of the dentinal papilla, and has assumed a somewhat complex structure, results in the formation of the enamel-coating of the crown of the tooth. While these changes are taking place the tissues immediately surrounding the tooth-germ become condensed and differentiated into a capsule, which appears to grow up from the base of the dental papilla, and encloses both this and the enamel-germ, constituting the follicle or tooth-sac. By the ossification of the inner layer of this follicle the cement is formed. This substance, therefore, unlike the dentine, increases from within outwards, and its growth may accordingly be the cause of considerable modification of form and enlargement, especially of the roots, of certain teeth, as those of Seals and some Cetacea. The delicate homogeneous layer coating the enamel surface of newly-formed teeth, in which cement is not found in the adult state, and known as Nasmyth's membrane, is considered by Tomes as probably a film of this substance, too thin to exhibit its characteristic structure, though by others it is believed to be derived from the external layer of the enamel-organ. The homology of the teeth with the dermal appendages, hairs, scales, and claws, has already been alluded to, and it will now be seen that in both cases two of the primary embryonic layers are concerned in their development—the mesoblast and epiblast—although in very different proportions respectively. Thus in the hair or nail the part derived from the epiblast forms the principal bulk of the organ, the mesoblast only constituting the papilla or matrix. But in the tooth the epiblastic portion is limited to the enamel, and is always of relatively small bulk and often absent, while the dentine (the principal constituent of the tooth) and the cement are formed from the mesoblast.

When more than one set of teeth occur in mammals, those of

the second set are developed in a precisely similar manner to the first, but the enamel-germ, instead of being derived directly from an independent part of the oral epithelium, is formed from a budding out of the neck of the germ of the tooth succeeded. In the case of the true molars, which have no predecessors, the germ of the first has an independent origin, but that of the others is derived from the neck of the germ of the tooth preceding it in the series. The foundations of the permanent teeth are thus laid as it were almost simultaneously with those of their predecessors, although they remain in many cases for years before they are developed into functional activity.

Although the commencement of their formation takes place at an early period of embryonic life, teeth are in nearly all mammals still concealed beneath the gum at the time of birth. The period of eruption, or "cutting" of the teeth as it is called, that is, their piercing through and rising above the surface of the mucous membrane, varies much in different species. In some, as Seals, the whole series of teeth appears almost simultaneously; but more often there are considerable intervals between the appearance of the individual teeth, the front ones usually coming into place first, and those at the back of the mouth at a later period.

Forms of Teeth.—The simplest form of tooth may be exemplified on a large scale by the tusk of the Elephant (Fig. 1, I.) It is a hard mass almost entirely composed of dentine, of a conical shape at first, but during growth becoming more and more cylindrical or unifo m in width. The enamel-covering, present on the apex in its earliest condition, soon disappears, but a thin layer of cement covers the circumference of the tooth throughout life. In section it will be seen that the basal portion is hollow, and contains a large conical pulp, as broad at the base as the tooth itself, and deeply imbedded in the bottom of a recess, or socket, in the maxillary bone. This pulp continues to grow during the lifetime of the animal, and at the same time is converted at its surface into dentine. The tooth therefore continually elongates, but the use to which the animal subjects it in its natural state causes the apex to wear away, at a rate generally proportionate to the growth at the base, otherwise it would become of inconvenient length and weight. Such teeth of indefinite growth are said to be "rootless," or to have "persistent pulps."

One of the corresponding front teeth of man (Fig. 2, II. and III.) may be taken as an example of a very different condition. After its crown is fully formed by calcification of the germ, the pulp, though continuing to elongate, begins to contract in diameter; a neck or slight constriction is formed; and the remainder of the pulp is converted into the root (often, but incorrectly, called "fang"), a tapering conical process imbedded in the alveolar cavity of the bone, and having at its extremity a minute perforation, through which the vessels and nerves required to maintain the vitality of the tooth enter the pulp-cavity, which is very different from the widely open cavity at the base of the growing tooth. When the crown of the tooth is broad and complex in character, instead of having a single root, it may be supported by two or more roots, each of which is implanted in a distinct alveolar recess or socket, and to the apex of which a branch of the common pulp-cavity is continued (Fig. 1, IV.) Such teeth are called "rooted teeth." When they have once attained their position in the jaw, with the neck a little way above the level of the free margin of the alveolus, and embraced by the gum or tough fibro-vascular membrane covering the alveolar border, and having the root fully formed, they can never increase in length or alter their position; if they appear to do so in old age, it being only in consequence of absorption and retrocession of the surrounding alveolar margins. If, as often happens, their surface wears away in mastication, it is never renewed. The open cavity at the base of the imperfectly developed tooth (Fig. 1, II.) causes it to resemble the persistent condition of the rootless tooth. The latter is therefore a more primitive condition, the formation of the root being a completion of the process of tooth development. Functionally it is, however, difficult to say that the

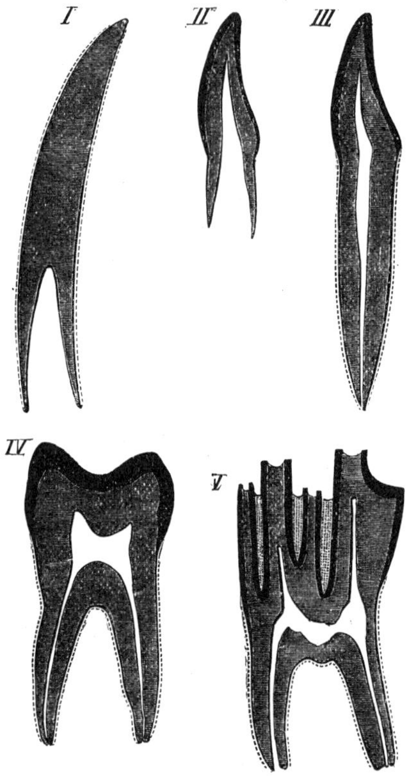

FIG. 1.—Diagrammatic Sections of various forms of Teeth. I. Incisor or tusk of Elephant, with pulp-cavity persistently open at base. II. Human incisor during development, with root imperfectly formed, and pulp-cavity widely open at base. III. Completely formed human incisor, with pulp-cavity contracted to a small aperture at the end of the root. IV. Human molar, with broad crown and two roots. V. Molar of the Ox, with the enamel covering the crown deeply folded, and the depressions filled up with cement. The surface is worn by use; otherwise the enamel coating would be continuous at the top of the ridges. In all the figures the enamel is black, the pulp white, the dentine represented by horizontal lines, and the cement by dots.

one is a higher form than the other, since they both serve important and different purposes in the animal economy.

As is almost always the case in nature, intermediate conditions between these two forms of teeth are met with. Thus some teeth, as the molars of the Horse, and of many Rodents, are for a time rootless, and have growing pulps producing very long crowns with parallel sides, the summits of which may be in use and beginning to wear away while the bases are still growing; but ultimately the pulp contracts, forms a neck and distinct roots, and ceases to grow. The canine tusks of the Musk Deer and of the Walrus have persistent pulps, and are open at their base until the animal is of advanced age, when they close, and the pulp ceases to be renewed. The same sometimes happens in the tusks of very old Boars.

The simplest form of the crown of a tooth is that of a cone; but this may be variously modified. Thus it may be flattened, with its edges sharp and cutting, and pointed at the apex, as in the laterally compressed premolars of most Carnivora; or it may be chisel- or awl-shaped, with a straight truncated edge, as in the human incisors; or it may be broad, with a flat or rounded upper surface. Very often there is a more or less prominent ridge encircling the whole or part of the base of the crown just above the neck, called the cingulum, which serves as a protection to the edge of the gum in masticating, and is most developed in flesh-eating and insectivorous animals, in which the gums are liable to be injured by splinters of bone or other hard fragments of their food. The form of the crown is frequently rendered complex by the development upon its surface of elevations or tubercules called cusps or cones, or by ridges usually transverse, but sometimes variously curved or folded. When the crown is broad and the ridges are greatly developed, as in the molars of the Elephant, Horse, and Ox (Fig. 1, V.), the interspaces between them are filled with cement, which supports them and makes a solid compact mass of the whole tooth. When such a tooth wears away at the surface by friction against the opposed tooth of the other jaw, the different density of the layers of the substances of which it is composed—enamel, dentine, and cement—arranged in characteristic patterns, causes them to wear unequally, the hard enamel ridges projecting beyond the others, and thus giving rise to a grinding surface of great mechanical advantage.

Succession.—The dentition of all mammals consists of a definite set of teeth, almost always of constant and determinate number, form, and situation, and, with few exceptions, persisting in a functional condition throughout the natural term of the animal's life. In many species these are the only teeth which the animal ever possesses,—the set which is first formed being permanent, or, if accidentally lost, or decaying in extreme old age, not being replaced by others. These animals are called Monophyodont. But in the larger number of mammals, certain of the teeth are preceded by others, which may be only of a very transient, rudimentary, and functionless character (being in the Seals, for example, shed either before or within a few days after birth), or may be considerably developed, and functionally occupy the place of the permanent teeth for a somewhat lengthened period, during the growth and development of the latter and of the jaws. In all cases these teeth disappear (by the absorption of their roots and shedding of the crowns) before the frame of the animal has acquired complete maturity, as evidenced by the coalescence of the epiphyses of the osseous system. As these teeth are, as a general rule, present during the period in which the animal is nourished by the milk of the mother, the name of "milk-teeth" (French *dents de lait*, German *milchzähne*) has been commonly accorded to them, although it must be understood that the epoch of their presence is by no means necessarily synchronous with that of lactation. Animals possessing such teeth are called Diphyodont. No mammal is known to have more than two sets of teeth; and the definite and orderly replacement of certain members of the series is a process of quite a different nature from the indefinite succession which takes place in all the teeth continuously throughout the lifetime of the lower vertebrates.

When the milk-teeth are well developed, and continue in place during the greater part of the animal's growth, as is especially the case with the Ungulata, and, though to a less degree, with the Primates and Carnivora, their use is obvious, since taken all together they form structurally a complete epitome on a small scale of the more numerous and larger permanent set (see Fig. 3), and, consequently, are able to perform the same functions, while time is allowed for the gradual maturation of the latter, and especially while the jaws of the growing animal are acquiring the size and strength sufficient to support the permanent teeth. Those animals, therefore, that have a well-developed and tolerably persistent set of milk-teeth may be considered to be in a higher state of development, as regards their dentition, than those that have the milk-teeth absent or rudimentary.

It is a very general rule that individual teeth of the milk and permanent set have a close relationship to one another, being originally formed, as mentioned above, in exceedingly near proximity, and with, at all events so far as the enamel-germ is concerned, a direct connection. Moreover, since the latter ultimately come to occupy the position in the alveolar border temporarily held by the former, they are spoken of respectively as the predecessors or successors of each other. But it must be understood that milk-teeth may be present which have no successors in the permanent series, and, what is far more general, permanent teeth may have no predecessors in the milk series.

The complete series of permanent teeth of most mammals forms a complex machine, with its several parts adapted for different functions,—the most obvious structural modification for this purpose being an increased complexity of the individual components of the series from the anterior towards the posterior extremity of such series. Since, as has just been said, the complete series of the milk teeth often presents structurally and functionally a similar machine, but composed of fewer individual members, and the anterior of which are as simple, and the posterior as complex as those occupying corresponding positions in the permanent series,—and since the milk-teeth are only developed in relation to the anterior or lateral, never to the most posterior of the permanent series,—it follows that the hinder milk-teeth are usually more complex than the teeth of which they are the predecessors in the permanent series, and represent functionally, not their immediate successors, but those more posterior permanent teeth which have no direct predecessors. This character is clearly seen in those animals in which the various members of the molar series are well differentiated from each other in form, as the Carnivora, and also in Man.

In animals which have two sets of teeth the number of those of the permanent series which are preceded by milk-teeth varies greatly, being sometimes, as in Marsupials and some Rodents, as few as one on each side of each jaw, and sometimes including the larger portion of the series.

Although there are difficulties in some cases in arriving at a satisfactory solution of the question, it is, on the whole, safest to assume that when only one set of teeth is present, this corresponds to the permanent teeth of the Diphyodonts. When this one set is completely developed, and remains in use throughout the animal's life, there can be no question on this subject. When, on the other hand, the teeth are rudimentary and transient, as in the Whalebone Whales, it is possible to consider them as representing the milk series; but there are weighty reasons in favour of the opposite conclusion.[1]

Arrangement, Homologies, and Notation of Teeth.—The teeth of the two sides of the jaws are always alike in number and character,

[1] This and other questions concerning the homologies, notation, and succession of the teeth of mammals are more fully developed in two memoirs by one of the present writers:—"Remarks on the Homologies and Notation of the Teeth of the Mammalia," in the *Journal of Anatomy and Physiology*, vol. iii. p. 262, 1869: and "Notes on the First or Milk Dentition of the Mammalia," in the *Trans. Odontological Society of Great Britain*, 1871. See also an important memoir by Oldfield Thomas on the "Homologies and Succession of the teeth in the Dasyuridæ," *Phil. Trans.* 1887, pp. 443-462.

except in cases of accidental or abnormal variation, and in the one remarkable instance of constant deviation from bilateral symmetry among mammals, the tusks of the Narwhal (*Monodon*), in which the left is of immense size, and the right rudimentary. In certain mammals, such as the Dolphins and some Armadillos, which have a very large series of similar teeth, not always constant in number in different individuals, there may be differences in the two sides; but, apart from these, in describing the dentition of any mammal, it is quite sufficient to give the number and characters of the teeth of one side only. Since the teeth of the upper and the lower jaws work against each other in masticating, there is a general correspondence or harmony between them, the projections of one series, when the mouth is closed, fitting into corresponding depressions of the other. There is also a general resemblance in the number, characters, and mode of succession of both series, so that, although individual teeth of the upper and lower jaws may not be in any strict sense of the term homologous parts, there is a great convenience in applying the same descriptive terms to the one as are used for the other.

The simplest dentition as a whole is that of many species of Dolphin (Fig. 2), in which the crowns are single-pointed, slightly

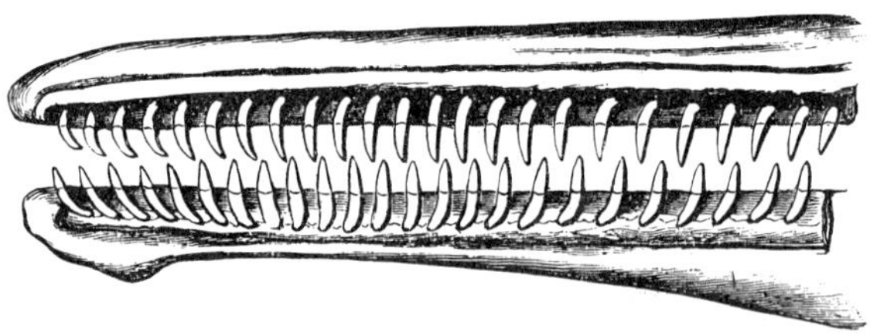

FIG. 2.—Upper and Lower Teeth of one side of the Mouth of a Dolphin (*Lagenorhynchus*) as an example of the homodont type of dentition. The bone covering the outer side of the roots of the teeth has been removed to show their simple character.

curved cones, and the roots also single and tapering, and all alike in form from the anterior to the posterior end of the series, though it may be with some slight difference in size, those at the two extremities of the series being rather smaller than the others. Such a dentition is called Homodont, and in the case cited, as the teeth are never changed, it is also Monophyodont. Such teeth are adapted only for catching slippery living prey, as fish.

In a very large number of mammals the teeth of different parts of the series are more or less differentiated in character, and have different functions to perform. The front teeth are simple and one-rooted, and are adapted for cutting and seizing. They are called "incisors." The back- or cheek-teeth have broader and more complex crowns, tuberculated or ridged, and are supported on two or more roots. They crush or grind the food, and are hence called "molars." Many animals have, between these two sets, a tooth at each corner of the mouth, longer and more pointed than the others, adapted for tearing or stabbing, or for fixing struggling prey. From the conspicuous development of such teeth in the Carnivora, especially the Dogs, they have received the name of "canines." A dentition with its component parts so differently formed that these distinctive terms are applicable to them is called Heterodont. In most cases, though by no means invariably, animals with Heterodont dentition are also Diphyodont.

This general arrangement is extremely obvious in a considerable number of mammals; and closer examination shows that, under very great modification in detail, there is a remarkable uniformity of essential characters in the dentition of a large number of members of the class belonging to different orders and not otherwise closely allied; so much so indeed that it has been possible (chiefly through the researches of Sir Richard Owen) to formulate a common plan of dentition from which the others have been derived by the alteration of some and suppression of other members of the series, and occasionally, but very rarely, by addition. The records of palæontology fully confirm this view, as by tracing back many groups now widely separated in dental characters we find a gradual approximation to a common type. In this generalised form of mammalian dentition (which is best exemplified in the genera *Anoplotherium* and *Homalodontotherium*) the entire number of teeth present is 44, or 11 above and 11 below on each side. Those of each jaw are placed in continuous series without intervals between them; and, although the anterior teeth are simple and single-rooted, and the posterior teeth complex and with several roots, the transition between the two kinds is gradual.

In dividing and grouping such teeth for the purpose of description and comparison, more definite characters are required than those derived merely from form or function. The first step towards a classification has been made by the observation that the upper jaw is composed of two bones, the premaxilla and the maxilla, and that the suture between these bones separates the three anterior teeth from the others. These three teeth, then, which are implanted by their roots in the premaxilla, form a distinct group, to which the name of "incisor" is applied. This distinction is, however, not so important as it appears at first sight, for, as mentioned when speaking of the development of the teeth, their connection with the bone is only of a secondary nature, and, although it happens conveniently for our purpose that in the great majority of cases the segmentation of the bone coincides with the interspace between the third and fourth tooth of the series, still, when it does not happen to do so, as in the case of the Mole, we must not give too much weight to this fact, if it contravenes other reasons for determining the homologies of the teeth. The eight remaining teeth of the upper jaw offer a natural division, inasmuch as the posterior three never have milk-predecessors; and, although some of the anterior teeth may be in the same case, the particular one preceding these three always has such a predecessor. These three then are grouped apart as the "molars," or, since some of the teeth in front of them often have a molariform character, "true molars." Of the five teeth between the incisors and molars the most anterior, or that which is usually situated close behind the premaxillary suture, almost always, as soon as any departure takes place from the simplest and most homogeneous type, assumes a lengthened and pointed form, and is the tooth so developed as to constitute the "canine" or "laniary" tooth of the Carnivora, the tusk of the Boar, etc. It is customary therefore to call this tooth, whatever its size or form, the "canine." The remaining four are the "premolars" or "false molars." This system of nomenclature has been objected to as being artificial, and in many cases not descriptive, the distinction between premolars and canine especially being sometimes not obvious; but the terms are now in such general use, and are so practically convenient—especially if, as it is best to do in all such cases, we forget their original signification and treat them as arbitrary signs—that it is not likely they will be superseded by any that have been proposed as substitutes for them.

With regard to the lower teeth the difficulties are greater, owing to the absence of any suture corresponding to that which defines the incisors above; but since the number of the teeth is the same, the corresponding teeth are preceded by milk-teeth, and in the large majority of cases it is the fourth tooth of the series which is modified in the same way as the canine (or fourth tooth) of the upper jaw, it is quite reasonable to adopt the same divisions as with the upper series, and to call the first three, which are implanted in the part of the mandible opposite to the premaxilla, the incisors, the next the canine, the next four the premolars, and the last three the molars. It may be observed that when the mouth is closed, especially when the opposed surfaces of the teeth present an irregular outline, the corresponding upper and lower teeth are not exactly opposite, otherwise the two series could not fit into one another; but as a rule the points of the lower teeth shut into the interspaces in front of the corresponding teeth of the upper jaw. This is seen very distinctly in the canine teeth of the Carnivora, and is a useful guide in determining the homologies of the teeth of the two jaws. Objections have certainly been made to this view, because, in certain rare cases, the tooth which, according to it, would be called the lower canine has the form and function of an incisor (as in Ruminants and Lemurs), and on the other hand (as in *Cotylops*, an extinct Ungulate from North America) the tooth that would thus be determined as the first premolar has the form of a canine; but it should not be forgotten that, as in all such cases, definitions derived from form and function alone are quite as open to objection as those derived from position and relation to surrounding parts, or still more so.

Dental formulæ.—For the sake of brevity the complete dentition, arranged according to these principles, is often described by the following formula, the numbers above the line representing the teeth of the upper, those below the line those of the lower jaw:—
incisors $\frac{3-3}{3-3}$, canines $\frac{1-1}{1-1}$, premolars $\frac{4-4}{4-4}$, molars $\frac{3-3}{3-3}=\frac{11-11}{11-11}$; total 44. Since, however, initial letters may be substituted for the names of each group, and it is quite unnecessary to give more than the numbers of the teeth on one side of the mouth, the formula may be conveniently abbreviated into—

$$i\,\tfrac{3}{3},\ c\,\tfrac{1}{1},\ p\,\tfrac{4}{4},\ m\,\tfrac{3}{3}=\tfrac{11}{11};\ \text{total } 44.$$

The individual teeth of each group are always enumerated from before backwards, and by such a formula as the following—

$$\frac{i\,1,\ i\,2,\ i\,3,\ c,\ p\,1,\ p\,2,\ p\,3,\ p\,4,\ m\,1,\ m\,2,\ m\,3}{i\,1,\ i\,2,\ i\,3,\ c,\ p\,1,\ p\,2,\ p\,3,\ p\,4,\ m\,1,\ m\,2,\ m\,3}$$

or more briefly—

$$i\,\frac{1,\ 2,\ 3}{1,\ 2,\ 3},\ c\,\frac{1}{1},\ p\,\frac{1,\ 2,\ 3,\ 4}{1,\ 2,\ 3,\ 4},\ m\,\frac{1,\ 2,\ 3}{1,\ 2,\ 3}.$$

A special numerical designation is thus given by which each one can be indicated. In mentioning any single tooth, such a sign as $\underline{m\,1}$ will mean the first upper molar, $\overline{m\,1}$ the first lower molar, and so on. The use of such signs saves much time and space in description.[1]

It was part of the view of the founder of this system of dental notation that, at least throughout the group of mammals whose dentition is derived from this general type, each tooth has its strict homologue in all species, and that in those cases in which fewer than the typical number are present (as in all existing mammals except the genera *Sus*, *Gymnura*, *Talpa*, and *Myogale*), the teeth that are missing can be accurately defined. According to this view, when the number of incisors falls short of three it is assumed that the absent ones are missing from the outer and posterior end of the series. Thus, when there is but one incisor present, it is *i* 1; when two, they are *i* 1 and *i* 2. Furthermore, when the premolars and the molars are below their typical number, the absent teeth are missing from the fore part of the premolar series, and from the back part of the molar series. If this were invariably so, the labours of those who describe teeth

[1] By many writers the letters indicating the different kinds of teeth are printed in capitals, as *I*, *C*, *P*, and *M*; while very frequently the symbol *Pm* is employed in place of *p*.

would be greatly simplified; but there are so many exceptions that a close scrutiny into the situation, relations, and development of a tooth is required before its nature can be determined, and in some cases the evidence at our disposal is scarcely sufficient for the purpose. In other instances, however, as among the Polyprotodont Marsupials, we have decisive evidence to show that the missing premolar teeth are not those at the extremity of the series.

The milk-dentition is expressed by a similar formula, *d* for deciduous or *m* for milk being commonly prefixed to the

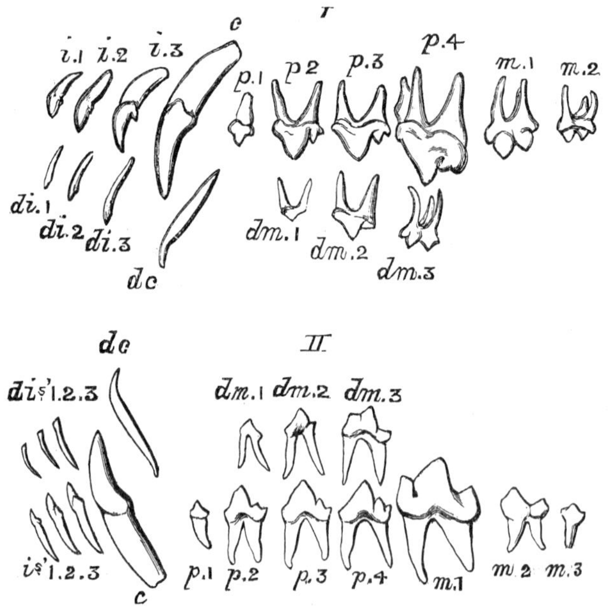

FIG. 3.—Milk and Permanent Dentition of Upper (I.) and Lower (II.) Jaw of the Dog (*Canis familiaris*), with the symbols by which the different teeth are commonly designated. The third upper molar (*m*.3) is the only tooth wanting in this animal to complete the typical heterodont mammalian dentition.

letter expressive of the nature of the tooth. Since the three molars, and almost invariably the first premolar of the permanent series, have no predecessors, the typical milk-dentition would be expressed as follows—*di* $\frac{3}{3}$, *dc* $\frac{1}{1}$, *dm* $\frac{3}{3}$, $= \frac{7}{7}$, total 28. In a few Ungulates, however, such as the Hyrax and Tapir, and in some instances the Rhinoceros and the extinct *Palæotherium*, the whole of the four premolars are preceded by milk-teeth; when we have the fullest development of cheek-teeth in the whole of the Eutheria. The teeth which precede the premolars of the permanent series are all called molars in the milk-dentition, although as a general rule, in form and function they represent in a condensed form the whole premolar and molar series of the adult. When there is a marked difference between the premolars and molars of the permanent dentition, the first milk-molar resembles a premolar, while the last has the characters of the posterior true molar.

The dentition of all the members of the orders Primates, Carnivora, Insectivora, Chiroptera, and Ungulata can clearly be derived from the above-described generalised type. The same may be said of the Rodents, and even the Proboscidea, though at least in the existing members of the order with greater modification. It is also apparent in certain extinct Cetacea, as *Zeuglodon* and *Squalodon*, but it is difficult to find any traces of it in existing Cetacea, Sirenia, or any of the so-called Edentata. All the Marsupials, different as they are in their general structure and mode of life, and variously modified as is their dentition, present in this system of organs some deep-lying common characters which show their unity of origin. The generalised type to which their dentition can be reduced presents considerable resemblance to that of the placental mammals, yet differing in details. It is markedly heterodont, and susceptible of division into incisors, canines, premolars, and molars upon the same principles. The whole number is, however, not limited to forty-four. The incisors may be as numerous as five on each side above, and they are almost always different in number in the upper and the lower jaw. The premolars and molars are commonly seven, as in the placental mammals, but their arrangement is reversed, as there are four true molars and three premolars.

The larger number of incisive and molar teeth among the Marsupials suggests that their additional teeth have disappeared in the Eutheria,[1] and Mr. O. Thomas has endeavoured to construct a generalised dental formula from which both the Marsupial and Eutherian modifications may have been derived by the suppression of particular teeth. Thus the hypothetical formula $i\,\frac{1,\,2,\,3,\,4,\,5}{1,\,2,\,3,\,4,\,5}$, $c\,\frac{1}{1}$, $p\,\frac{1,\,2,\,3,\,4}{1,\,2,\,3,\,4}$, $m\,\frac{1,\,2,\,3,\,4,\,5}{1,\,2,\,3,\,4,\,5}$, by the loss of the fifth lower incisor, and of the second premolars (which we know to be those which disappear in the Marsupials) and the fifth molars, will give $i\,\frac{1,\,2,\,3,\,4,\,5}{1,\,2,\,3,\,4,\,0}$, $c\,\frac{1}{1}$, $p\,\frac{1,\,0,\,3,\,4}{1,\,0,\,3,\,4}$, $m\,\frac{1,\,2,\,3,\,4}{1,\,2,\,3,\,4}$; or the formula of the Opossum (*Didelphys*), usually written $i\,\frac{5}{4}$, $c\,\frac{1}{1}$, $p\,\frac{3}{3}$, $m\,\frac{4}{4}$. Again, in the same formula the loss of the fourth and fifth incisors in both jaws, and also of the fourth molars, gives us $i\,\frac{1,\,2,\,3,\,0,\,0}{1,\,2,\,3,\,0,\,0}$, $c\,\frac{1}{1}$, $p\,\frac{1,\,2,\,3,\,4}{1,\,2,\,3,\,4}$, $m\,\frac{1,\,2,\,3}{1,\,2,\,3}$, or the formula of a typical Eutherian, like the

[1] According to Mr. G. E. Dobson there are four upper incisors in some of the *Soricidæ*.

Pig, which we generally write as $i\ \frac{3}{3}$, $c\ \frac{1}{1}$, $p\ \frac{4}{4}$, $m\ \frac{3}{3}$. Such a generalised formula will admit of modification into that of all existing, and a large number of fossil Marsupials, but it is possible that some of the Mesozoic types may have had more than four premolars, although there is no absolutely decisive evidence that such was the case. The presence of seven or eight true molars in some Mesozoic forms merely entails the addition of two or three additional figures to the ideal generalised formula.

The milk-dentition of all known Marsupials, existing or extinct, is (if not entirely absent) limited to a single tooth on either side of each jaw, this being the predecessor of the last permanent premolar. And if the view that the milk-dentition is an additional series grafted upon the original permanent series be correct, it is evident that we have in this single replacement the first stage of this additional development.

In very few mammals are teeth entirely absent. Even in the Whalebone Whales their germs are formed in the same manner and at the same period of life as in other mammals, and even become partially calcified, but they never rise above the gums, and completely disappear before the birth of the animal. In some species of the order Edentata, the true Anteaters and the Pangolins, no traces of teeth have been found at any age. The adult Monotremata are likewise devoid of teeth of the same structure as those of ordinary mammals; but well-developed molars occur in the young *Ornithorhynchus*, although no traces of teeth have hitherto been detected in *Echidna*.

Modifications of the Teeth in Relation to their Functions.—The principal functional modifications noticed in the dentition of mammalia may be roughly grouped as piscivorous, carnivorous, insectivorous, omnivorous, and herbivorous, each having, of course, numerous variations and transitional conditions.

The essential characters of a piscivorous dentition are best exemplified in the Dolphins, and also (as modifications of the carnivorous type) in the Seals. This type consists of an elongated, rather narrow mouth, wide gape, with numerous subequal, conical, sharp-pointed, recurved teeth, adapted simply to rapidly seize, but not to divide or masticate, active, slippery, but not powerful prey. All animals which feed on fish as a rule swallow and digest them entire, a process which the structure of prey of this nature, especially the intimate interblending of delicate, sharp-pointed bones with the muscles, renders very advantageous, and for which the above-described type of dentition is best adapted.

The carnivorous type of dentition is shown in its most specialised development among existing mammals in the *Felidæ*. The function being here to seize and kill struggling animals, often of large size and great muscular power, the canines are immensely developed, trenchant, and piercing, and are situated wide apart, so as to give the firmest hold when fixed in the victim's body. The jaws are as short as is consistent with the free action of the canines, so that no power may be lost. The incisors are very small, so as not to interfere with the penetrating action of the canines, and the crowns of the molar series are reduced to scissor-like blades, with which to pare off the soft tissues from the large bones, or to divide into small pieces the less dense portions of the bones for the sake of nutriment afforded by the blood and marrow they contain. The gradual modification between this and the two following types will be noticed in their appropriate places.

In the most typical insectivorous animals, as the Hedgehogs and Shrews, the central incisors are elongated, pointed, and project forwards, those of the upper and lower jaw meeting like the blades of a pair of forceps, so as readily to secure small active prey, quick to elude capture, but powerless to resist when once seized. The crowns of the molars are covered with numerous sharp edges and points, which, working against each other, rapidly cut up the hard-cased insects into little pieces fit for swallowing and digestion.

The omnivorous type, especially that adapted for the consumption of soft vegetable substances, such as fruits of various kinds, may be exemplified in the dentition of Man, of most Monkeys, and of the less modified Pigs. The incisors are moderate, subequal, and cutting. If the canines are enlarged, it is usually for other purposes than those connected with food, and only in the male sex. The molars have their crowns broad, flattened, and elevated into rounded tubercles. The name *Bunodont*, or hillock-toothed, has been proposed for molars of this type, and will frequently be found convenient.

In the most typically herbivorous forms of dentition, as seen in the Horse and Kangaroo, the incisors are well developed, trenchant, and adapted for cutting off the herbage on which the animals feed; the canines are rudimentary or suppressed; the molars are large, with broad crowns, which in the simplest forms have strong transverse ridges, but may become variously complicated in the higher degrees of modification which this type of tooth assumes.

Various forms of teeth of this type will be noticed among the Ungulates and Rodents.

The natural groups of mammals, or those which in our present state of knowledge we have reason to believe are truly related to each other, may each contain examples of more than one of these modifications. Thus the Primates have both omnivorous and insectivorous forms. The Carnivora show piscivorous, carnivorous, insectivorous, and omnivorous modifications of their common type of dentition. The Ungulata and the Rodentia have among them the omnivorous and various modifications, both simple and complex, of the herbivorous type. The Marsupialia exhibit examples of all forms, except the purely piscivorous. Other orders, more restricted in number or in habits, as the Proboscidea and Cetacea, naturally do not show so great a variety in the dental structure of their members.

Taxonomy.—In considering the taxonomic value to be assigned to the modifications of teeth of mammals, two principles, often opposed to each other, which have been at work in producing these modifications, must be held in view:—(1) the type, or ancestral form, as we generally now call it, characteristic of each group, which in most mammals is itself derived from the still more generalised type described above; and (2) variations which have taken place from this type, generally in accordance with special functions which the teeth are called upon to fulfil in particular cases. These variations are sometimes so great as completely to mask the primitive type, and in this way the dentition of many animals of widely different origin has come to present a remarkable superficial resemblance, as in the case of the Wombat (a Marsupial), the Aye-Aye (a Lemur), and the Rodents, or as in the case of the Thylacine and the Dog. In all these examples indications may generally be found of the true nature of the case by examining the earlier conditions of dentition; for the characters of the milk-teeth or the presence of rudimentary or deciduous members of the permanent set will generally indicate the route by which the specialised dentition of the adult has been derived. It is perhaps owing to the importance of the dental armature to the well-being of the animal in procuring its sustenance, and preserving its life from the attacks of enemies, that great changes appear to have taken place so readily, and with such comparative rapidity, in the forms of these organs—changes often accompanied with but little modification in the general structure of the animal. Of this proposition the Aye-Aye (*Chiromys*) among Lemurs, the Walrus among Seals, and the Narwhal among Dolphins form striking examples; since in all these forms the superficial characters of their dentition would entirely separate them from the animals with which all other evidence (even including the mode of development of their teeth) proves their close affinity.

Trituberculism.—Recent researches, and more especially those of Professors Cope and Osborn, tend to show that almost all of the extremely different forms of tooth-structure found among Mammals may be traced to one common type, in which the crown of each tooth carried three cusps, and hence termed the *tritubercular* type; these three cusps being arranged in a triangle, with the apex directed inwardly in the upper teeth (Fig. 4, 6), and outwardly in the lower ones (Fig. 4, 7). It is further probable that this tritubercular type was itself derived from a type of dentition in which the teeth were in the form of almost a quite simple cone; such a presumably primitive type of dentition being apparently retained among some existing Edentates, like the Armadillos, while it is possible that we should regard the dentition of the existing Cetacea (Fig. 2) as a reversion to the same primitive type. None of the Mesozoic mammals at present known exhibit this simple conical type of teeth, although we have an approximation to it in the extremely generalised genus *Dromatherium*. Starting then

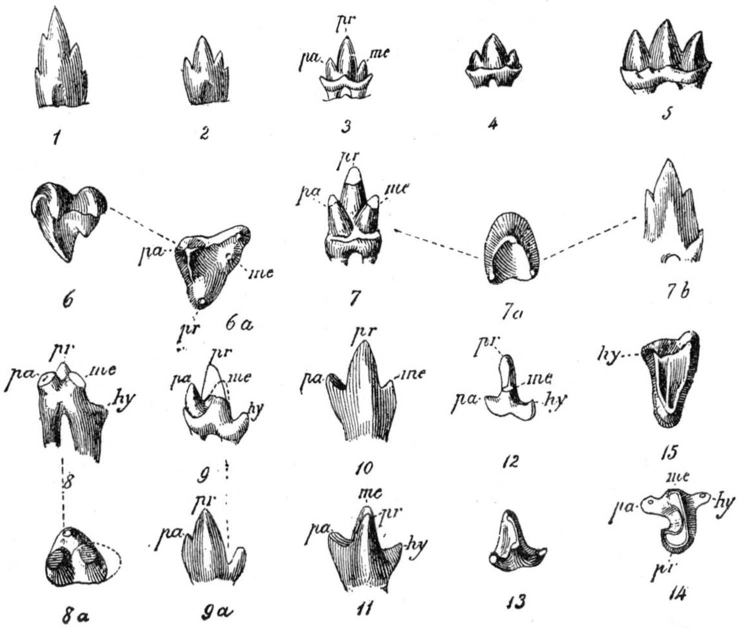

FIG. 4.—Molar teeth of Mesozoic Mammals (enlarged). Triconodont type—1, *Dromatherium*; 2, *Microconodon*; 3, *Amphilestes*; 4, *Phascolotherium*; 5, *Triconodon*. Tritubercular type—6, 7, *Spalacotherium*; 10, *Asthenodon*. Tubercular sectorial type—8, *Amphitherium*; 9, *Peramus*; 11-13, *Amblotherium*; 14 (?) *Amblotherium*. *pr*, Protocone; *hy*, hypocone; *pa*, paracone; *me*, metacone, in the upper teeth; and protoconid, hypoconid, paraconid, and metaconid in the lower. 6 and 15 are upper molars, and the rest lower molars. (After Osborn.)

from this presumed simple cone it appears that the teeth of *Dromatherium* (Fig. 4, 1) present the first stage towards trituberculism, the crown of each tooth having one main cone, with minute lateral cusps, and the root being grooved. In the next or true Triconodont stage (Fig. 4, 3-5) the crown has become elongated antero-posteriorly, and consists of one central and two lateral cones or cusps, while the root is divided. From this the transition is easy to the tritubercular type, in which the three cusps, instead of being placed in a line, are arranged in a triangle; the upper teeth (Fig.

4, 6) having one inner and two outer cusps, while the reverse condition obtains in those of the lower jaw (Fig. 4, 7). These three cusps of the simple tritubercular tooth are collectively designated as the primitive triangle; in the upper tooth the inner cusp is termed the protocone, the antero-external one the paracone, and the postero-external the metacone; the corresponding cusps of the lower tooth being named protoconid, paraconid, and metaconid—the protoconid being here on the outer side of the crown.

It is thus apparent that in the first, or haplodont type, as well as in the triconodont type, the upper and lower molars are alike; while in the simple tritubercular type they have a similar pattern, but with the arrangement of the cusps reversed. This simple tritubercular type occurs in the Mesozoic genus *Spalacotherium* (Fig. 4, 6 and 7), and apparently in the existing *Chrysochloris*; but in the majority of tritubercular forms, while this primitive triangle forms the main portion of the crown, other secondary cusps are added, the homologies of which in the upper and lower teeth are somewhat doubtful. At the same time that we have the addition of these secondary cusps we also find trituberculism differentiating into a secodont and a bunodont series, according as to whether the dentition becomes of a cutting or a crushing type.

Thus in the lower molars (Fig. 4, 8 and 9) we very frequently find the three cusps of the primitive triangle elevated and connected by cross crests, while there is an additional low posterior heel or talon, which may be termed the hypoconid. This tubercular-sectorial sub-type, as it is termed, is found in the lower molars of many Polyprotodont Marsupials and Insectivores, and it also occurs in the lower carnassial teeth of the true Carnivora. The presence of two cusps (inner and outer) to the talon converts this modification into a quinquetubercular form; while, by the suppression of one of the three primitive cusps, it develops into the quadritubercular type of the bunodont series.

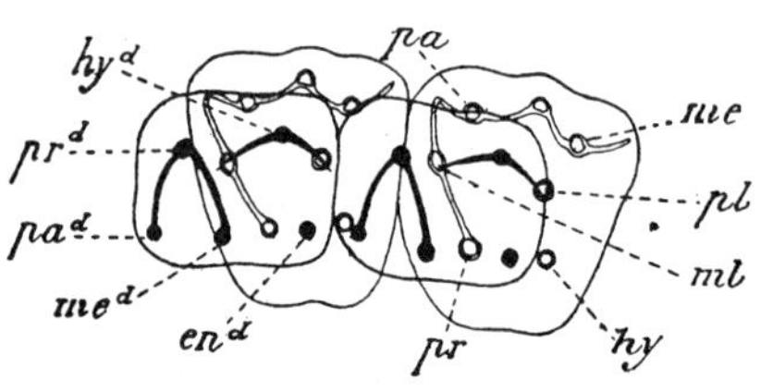

FIG. 5.—Diagram of two upper and two lower left quadritubercular molars in mutual apposition. The cusps and ridges of the upper molars in double lines, and those of the lower in black lines. The lower molars are looked at from below, as if transparent. *pr*, Protocone; *hy*, hypocone; *pa*, paracone; *me*, metacone; *ml*, protoconule; *pl*, metaconule; *prd*, protoconid; *hyd*, hypoconid; *pad*, paraconid; *med*, metaconid; *end*, entoconid. (After Osborn.)

In the upper molars the primitive triangle in the secodont series may remain purely tricuspid; but the addition of intermediate cusps, both in the secodont and bunodont series, may give rise to a quinquetubercular type; these intermediate cusps being respectively designated as the protoconule and metaconule (Fig. 5, *ml*, *pl*). Finally, in the bunodont series, the addition of a postero-internal cusp (Fig. 5, *hy*), termed the hypocone, forms the sextubercular molar.

The following table exhibits, in a collective form, the names and relations of all the above-mentioned cusps, and the letters by which they are indicated in the figures :—

Upper Molars.

Antero-internal cusp	= protocone	= *pr.*
Postero " or 6th cusp	= hypocone	= *hy.*
Antero-external cusp	= paracone	= *pa.*
Postero " "	= metacone	= *me.*
Anterior intermediate cusp	= protoconule	= *ml.*
Posterior " "	= metaconule	= *pl.*

Lower Molars.

Antero-external cusp	= protoconid	= *prd.*
Postero " "	= hypoconid	= *hyd.*
Antero-internal or 5th cusp	= paraconid	= *pad.*
Intermediate (or in quadritubercular molars antero-internal) cusp.	= metaconid	= *med.*
Postero-internal cusp	= entaconid	= *end.*

The common occurrence of trituberculism in the mammals of the earlier geological epochs is, as remarked by Osborn, very significant of the uniformity of molar origin. Thus, among the Mesozoic mammals (with the exception of the group known as Multituberculata, in which the molars are constructed on a different type), trituberculism occurs in the great majority of the genera; while out of 82 species, belonging to five different suborders from the Lowest or Puerco Eocene of the United States, all but four exhibit this feature; and the same holds good for the mammals of the corresponding European horizon. At the present day trituberculism persists in the Lemuroidea, Insectivora, Carnivora, and Marsupialia. In the Carnivora there is a tendency to lose the metaconid, while in the bunodont molars of the Ungulata it is the paraconid that disappears.

III. THE SKELETON.

Definition.—The skeleton is a system of hard parts, forming a framework which supports and protects the softer organs and tissues of the body. It consists of dense fibrous and cartilaginous tissues, portions of which remain through life in this state, but the greater part is transformed during the growth of the animal into bone or osseous tissue. This is characterised by a peculiar

3

histological structure and chemical composition, being formed mainly of a gelatinous basis, strongly impregnated with salts of calcium, chiefly phosphate, and disposed in a definite manner, containing numerous minute nucleated spaces or cavities called lacunæ, connected together by delicate channels or canaliculi, which radiate in all directions from the sides of the lacunæ. Parts composed of bone are, next to the teeth, the most imperishable of all the organs of the body, often retaining their exact form and internal structure for ages after every trace of all other portions of the organisation has completely disappeared, and thus, in the case of extinct animals, affording the only means of attaining a knowledge of their characters and affinities.[1]

In the Armadillos and their extinct allies alone is there an ossified exoskeleton, or bony covering developed in the skin. In all other mammals the skeleton is completely internal. It may be described as consisting of an axial portion belonging to the head and trunk, and an appendicular portion belonging to the limbs. There are also certain bones called splanchnic, being developed within the substance of some of the viscera. Such are the *os cordis* and *os penis* found in some mammals.

It is characteristic of all the larger bones of the mammalia that their ossification takes its origin from several distinct centres. One near the middle of the bone, and spreading throughout its greater portion, constitutes the *diaphysis*, or "shaft," in the case of the long bones. Others near the extremities, or in projecting parts, form the *epiphyses*, which remain distinct during growth, but ultimately coalesce with the rest of the bone.

Axial skeleton.—The axial skeleton consists of the skull, the vertebral column (prolonged at the posterior extremity into the tail), the sternum, and the ribs.

Skull.—In the *skull* of adult mammals, all the bones, except the lower jaw, the auditory ossicles, and the bones of the hyoid arch, are immovably articulated together, their edges being in close contact, and often interlocking by means of fine denticulations projecting from one bone and fitting into corresponding depressions of the other; they are also held together by the investing periosteum, or fibrous membrane, which passes directly from one to the other, and permits no motion, beyond perhaps a slight yielding to external pressure. In old animals there is a great tendency for the different bones to become actually united by the extension of ossification from one to the other, with consequent obliteration of the sutures.

[1] See for the principal modifications of the skeleton of the class, the large and beautifully illustrated *Ostéographie* of De Blainville, 1835-54; the section devoted to the subject in Bronn's *Klassen und Ordnungen des Thier-Reichs*, by Giebel, 1874-79; and *An Introduction to the Osteology of the Mammalia*, by W. H. Flower, 3d ed., 1885.

The cranium, thus formed of numerous originally independent ossifications, which may retain throughout life more or less of their individuality, or be all fused together, according to the species, the age, or even individual peculiarity, consists of a brain-case, or bony capsule for enclosing and protecting the brain, and a face for the support of the organs of sight, smell, and taste, and of those concerned in seizing and masticating the food. The brain-case articulates directly with the anterior cervical vertebra, by means of a pair of oval eminences, called condyles, placed on each side of the large median foramen which transmits the spinal cord. It consists of a basal axis, continuous serially with the axes or centra of the

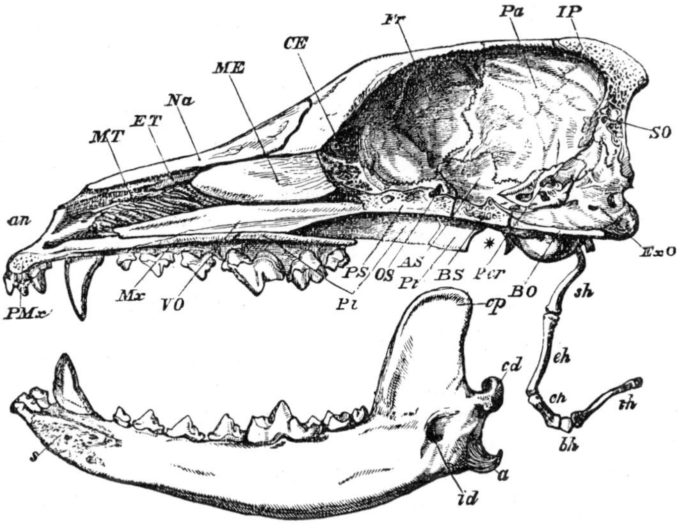

FIG. 6.—Longitudinal and vertical section of the skull of a Dog (*Canis familiaris*), with mandible and hyoid arch. *an*, Anterior narial aperture; *MT*, maxillo-turbinal bone; *ET*, ethmo-turbinal; *Na*, nasal; *ME*, ossified portion of the mesethmoid; *CE*, cribriform plate of the ethmo-turbinal: *Fr*, frontal; *Pa*, parietal; *IP*, interparietal; *SO*, supraoccipital; *ExO*, ex-occipital; *BO*, basioccipital; *Per*, periotic; *BS*, basisphenoid; *Pt*, pterygoid; *AS*, ali-sphenoid; *OS*, orbitosphenoid; *PS*, presphenoid; *Pl*, palatine; *VO*, vomer; *Mx*, maxilla; *PMx*, premaxilla; *sh*, stylohyal; *eh*, epihyal; *ch*, ceratohyal; *bh*, basihyal; *th*, thyrohyal; *s*, symphysis of mandible; *cp*, coronoid process; *cd*, condyle; *a*, angle; *id*, inferior dental canal. The mandible is displaced downwards, to show its entire form; the * indicates the part of the cranium to which the condyle is articulated.[1]

vertebræ, and of an arch above, roofing over and enclosing the cavity which contains the cephalic portion of the central nervous system (see Fig. 6). The base with its arch is composed of three segments placed one before the other, each of which is comparable to a vertebra with a greatly expanded neural arch. The hinder or

[1] This and many of the following figures in this chapter are taken from Flower's *Osteology of the Mammalia*.

occipital segment consists of the basioccipital, exoccipital, and supraoccipital bones; the middle segment of the basisphenoid, alisphenoid, and parietal bones; and the anterior segment of the presphenoid, orbitosphenoid, and frontal bones. The axis is continued forwards into the mesethmoid, or septum of the nose, around which the bones of the face are arranged in a manner so extremely modified for their special purposes that anatomists who have attempted to trace their serial homologies with the more simple portions of the axial skeleton have arrived at very diverse interpretations. The characteristic form and structure of the face of mammals is mainly dependent upon the size and shape of (1) the orbits, a pair of cup-shaped cavities for containing the eyeball and its muscles, which may be directed forwards or laterally, placed near together or wide apart, and may be completely or only partially encircled by bone; (2) the nasal fossæ, or cavities on each side of the median nasal septum, forming the passage for the air to pass between the external and the internal nares, and containing in their upper part the organ of smell; (3) the zygomatic arch, a bridge of bone for the purpose of muscular attachment, which extends from the side of the face to the skull, overarching the temporal fossa; (4) the roof of the mouth, with its alveolar margin for the implantation of the upper teeth. The face is completed by the mandible, or lower jaw, consisting of two lateral rami, articulated by a hinge joint with the squamosal (a cranial bone interposed between the posterior and penultimate segment of the brain-case, where also the bony capsule of the organ of hearing is placed), each being composed of a single solid piece of bone, and the two united together in the middle line in front, at the symphysis,—which union may be permanently ligamentous or become completely ossified. Into the upper border of the mandibular rami the lower teeth are implanted.

In addition to the bones already mentioned as entering into the formation of the cranium, there are many others, the most important of which may be briefly noticed. The anterior extremity of the skull is formed by the premaxillæ (Figs. 6, 7, *PMx*), which carry the incisors; behind them are the maxillæ, in which all the remaining upper teeth are implanted. Both the premaxillæ and maxillæ meet in a median suture on the palate, where they form a floor to the nasal passage; this floor being continued backwards by the plate-like palatines, at the hinder extremity of which the posterior nares are usually situated. In a few instances, however, as in certain Edentates and Cetaceans, the small pair of bones forming the posterior continuation of the lateral borders of the palatines, and known as the pterygoids (Fig. 6, *Pt*), likewise meet in the middle line below the nasal passage, and thus cause the aperture of the posterior nares to be situated near the occiput. On the upper, or frontal aspect of the cranium the paired nasals roof over the nasal passage and fill the interval left between the premaxilla and maxilla of either side. Behind the nasals and maxillæ, the anterior part of the brain-case is formed by the large paired frontals (Figs. 6, 7, *Fr*), behind which are the parietals, which may be of still larger size, and form the greater part of the brain-case. A median interparietal ossification (Fig. 6, *IP*) may divide the parietals posteriorly, and is itself articulated with the supra-occipital, to the lateral borders of which the parietals are also joined. The squamosal (Fig. 7, *Sq*) forms the lateral wall of the hinder part of the brain-case, and articulates superiorly with the parietal, and posteriorly with the exoccipital. The glenoid cavity (Fig. 8), for the reception of the articular condyle of the mandible, is formed by the inferior portion of the squamosal, at the point where it gives off the zygomatic process to form the hinder portion of the zygomatic arch. The middle portion of that arch is formed by the jugal, or malar bone (Fig. 7, *Ma*), which articulates posteriorly with the zygomatic process of the squamosal, and anteriorly with the maxilla. The jugal (as in Fig. 7) may also articulate with a small bone situated on the anterior border of the orbit known as the lachrymal. It is important to observe that the zygomatic or temporal arch is a squamoso-maxillary one, and that an arcade thus composed is found elsewhere only among the extinct Anomodont reptiles, which have already been mentioned as showing signs of mammalian affinity. The relative position occupied by the orbito- and alisphenoid is sufficiently indicated in Fig. 7.

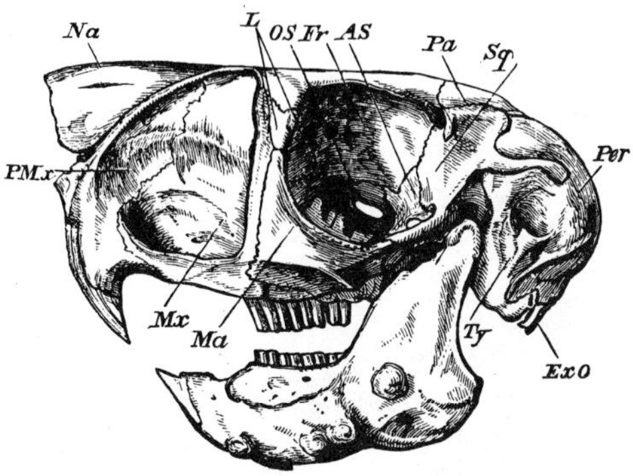

FIG. 7.—Side view of skull of Cape Jumping Hare (*Pedetes caffer*). × $\frac{3}{5}$. *PMx*, Premaxilla; *Mx*, maxilla; *Ma*, jugal or malar; *Fr*, frontal; *L*, lachrymal; *Pa*, parietal; *Na*, nasal; *Sq*, squamosal; *Ty*, tympanic; *ExO*, exoccipital; *AS*, alisphenoid; *OS*, orbitosphenoid; *Per*, mastoid bulla.

Wedged in between the squamosal and the bones of the occipital and basisphenoidal region are the bones connected with the organ of hearing, known as the periotic and tympanic. The position of the periotic, which encloses the labyrinth or essential organ of hearing, is shown in Fig. 6. The periotic is divided into a very dense antero-internal moiety known as the petrosal, and a postero-external or mastoid portion (Fig. 8), which appears on the outer wall of the brain-case. The tympanic is produced horizontally outwards to form the external auditory meatus or tube of the ear, while the inner and under surface is frequently dilated into a shell-like auditory bulla (Fig. 8). The small bones of the internal ear known as the malleus, incus, and stapes are contained in the membranous *tympanic cavity*, which is situated in a space left among this group of bones. Further mention of these bones is made below under the head of the sense organs.

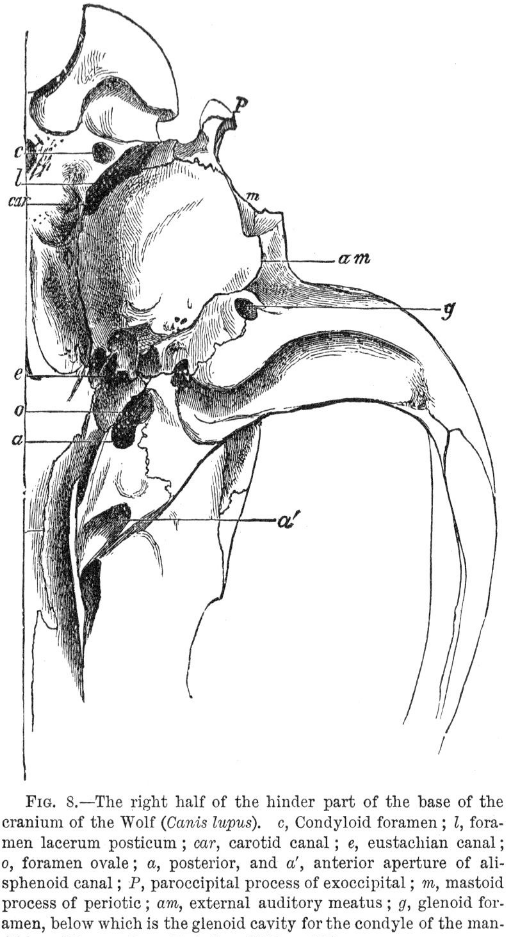

FIG. 8.—The right half of the hinder part of the base of the cranium of the Wolf (*Canis lupus*). *c*, Condyloid foramen; *l*, foramen lacerum posticum; *car*, carotid canal; *e*, eustachian canal; *o*, foramen ovale; *a*, posterior, and *a'*, anterior aperture of alisphenoid canal; *P*, paroccipital process of exoccipital; *m*, mastoid process of periotic; *am*, external auditory meatus; *g*, glenoid foramen, below which is the glenoid cavity for the condyle of the mandible. (Flower, *Proc. Zool. Soc.*, 1869, p. 25.)

In the Carnivora and some other groups the foramina on the base of the skull for the passage of blood-vessels and nerves are of considerable taxonomic importance. The position of the more important of these foramina is indicated in Fig. 8; but for details the reader may refer to the work on the *Osteology of the Mammalia* already mentioned. Attention may, however, be particularly directed to the so-called alisphenoid canal, the position of which is shown in Fig. 8, since this is a feature of some importance in the classification of the Carnivora. This canal is a short channel running horizontally forward from near the foramen ovale through the alisphenoid, and opening anteriorly with the foramen rotundum; it is traversed by the external carotid artery.

Only in those species, as Man and the smaller kinds of the Primates and some other orders, in which the brain holds a large relative proportion to the rest of the body, does the external form of the skull receive much impress from the real shape of the cavity containing the brain. The size and form of the mouth, and the modifications of the jaws for the support of teeth of various shape and number, the ridges and crests on the cranium for the attachment of the muscles necessary to put this apparatus in motion, and outgrowths of bone for the enlargement of the external surface required for the support of sense organs or of weapons, such as horns or antlers (which outgrowths, to prevent undue increase of weight, are filled with cells containing air), cause the principal variations in the general configuration of the skull. These variations are, however, only characteristically developed in perfectly adult animals, and are in many cases more strongly marked in the male than the female sex. Throughout all the later stages of growth up to maturity the size and form of the brain-case remain comparatively stationary, while the accessory parts of the skull rapidly increase and assume their distinctive development characteristic of the species.

The hyoidean apparatus in mammals (Fig. 6) supports the tongue and larynx, and consists of an inferior median portion termed the basihyal, from which two pairs of half arches, or cornua, extend upwards and outwards. The anterior is the more important, being connected with the periotic bone of the cranium. It may be almost entirely ligamentous, but more often has several ossifications, the largest of which is usually the stylohyal. The posterior cornu (thyrohyal) is united at its extremity with the thyroid cartilage of the larynx, which it suspends in position. The median portion, or basihyal, is sometimes, as in the Howling Monkeys, enormously enlarged and hollowed, admitting into its cavity an air-sac connected with the organ of voice.

Vertebral Column.—The *vertebral column* consists of a series of distinct bones called vertebræ, arranged in close connection with each other along the dorsal side of the neck and trunk, and in the median line.[1] It is generally prolonged posteriorly beyond the trunk, to form the axial support of the appendage called the tail. Anteriorly it is articulated with the occipital region of the skull. The number of distinct bones composing the vertebral column varies greatly among the Mammalia, the main variation being due to the degree of elongation of the tail. Apart from this, in most mammals the number is not far from thirty, though it may fall as low as twenty-six (as in some Bats), or rise as high as forty (*Hyrax* and *Cholœpus*). The different vertebræ, with some exceptions, remain through life quite distinct from each other, though closely connected by means of fibrous structures which allow of a certain, but limited, amount of motion between them. The exceptions are the following:—(1) near the posterior part

[1] For the sake of uniformity, in all the following descriptions of the vertebral column, the long axis of the body is supposed to be in the horizontal position.

of the trunk, in nearly all mammals which possess completely developed hinder limbs, two or more vertebræ become ankylosed together to form the "sacrum," or portion of the vertebral column to which the pelvic girdle is attached; (2) in some species of Whales and Armadillos there are constant ossific unions of certain vertebræ of the cervical region.

Although the vertebræ of different regions of the column of the same animal or of different animals present great diversities of form, yet there is a certain general resemblance among them, or a common plan on which they are constructed, which is more or less modified by alteration of form or proportions, or by the addition or suppression of parts to fit them to fulfil their special purpose in the economy. An ordinary or typical vertebra consists, in the first place, of a solid piece of bone, termed the body or centrum (Fig. 9, *c*), of the form of a disk or short cylinder. The bodies of con-

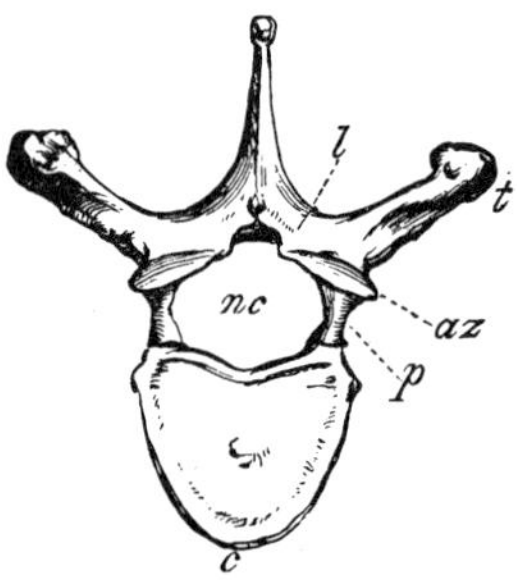

FIG. 9.—Anterior surface of Human thoracic vertebra (fourth). *c*, Body or centrum; *nc*, neural canal; *p*, pedicle, and *l*, lamina of the arch; *t*, transverse process; *az*, anterior zygapophysis.

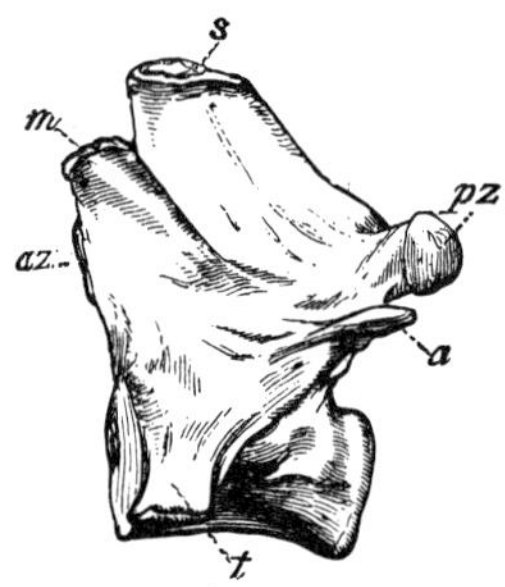

FIG. 10.—Side view of the first lumbar vertebra of a Dog (*Canis familiaris*). *s*, Spinous process; *az*, anterior zygapophysis; *pz*, posterior zygapophysis; *m*, metapophysis; *a*, anapophysis; *t*, transverse process.

tiguous vertebræ are connected together by a very dense, tough, and elastic material called the "intervertebral substance," of peculiar and complex arrangement. This substance forms the main, and in some cases the only, union between the vertebræ. Its elasticity provides for the vertebræ always returning to their normal relation to each other and to the column generally, when they have been disturbed therefrom by muscular action. A process (*p*) arises on each side from the dorsal surface of the body. These processes, meeting in the middle line above, form an arch, surmounting a space or short canal (*nc*). Since it contains the posterior prolongation of the great cerebro-spinal nervous axis, or spinal cord, this space is called the neural canal, and the arch the neural arch, in contradistinction to another arch on the ventral surface of the body of the vertebræ, called the hæmal arch. The latter is, however, never formed in mammals by any part of the vertebra itself, but by certain distinct bones placed more or less in apposition to it, namely the ribs in the thoracic, and the "chevron bones" in the caudal region. In most cases the arch of one vertebra is articulated with that of the next by distinct surfaces with synovial joints, placed one on each side, called "zygapophyses" (*az*, *pz*), but these are often entirely wanting when flexibility is more needed than strength, as in the greater part of the caudal region of long-tailed animals. In addition to the body and the arch, there are certain projecting parts called processes, chiefly serving for the attachment of the numerous muscles which move the vertebral column. Of these two are single and median, viz. the spinous process, neural spine, or neurapophysis (*s*), arising from the middle of the upper part of the arch, and the hypapophysis from the under surface of the body. The latter, however, is as frequently absent as the former is constant. The other processes are paired and lateral. They are the transverse processes (*t*), of which there may be two, an upper and a lower, in which case the former is called, in the language of Owen (to whom we are indebted for the terminology of the parts of vertebræ in common use), "diapophysis," and the latter "parapophysis." Other processes less constantly present are called respectively "metapophyses" (*m*) and "anapophyses" (*a*).

The vertebral column is divided for convenience of description into five regions—the cervical, thoracic or dorsal, lumbar, sacral, and caudal. This division is useful, especially as it is not entirely arbitrary, and in most cases is capable of ready definition; but at the contiguous extremities of the regions the characters of the vertebræ of one are apt to blend into those of the next region, either normally or as peculiarities of individual skeletons.

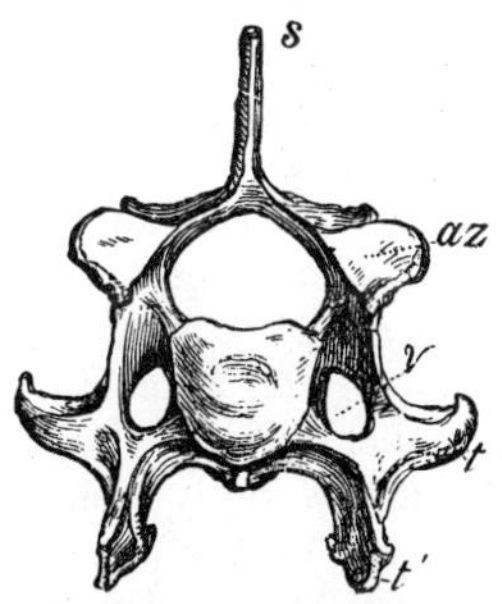

FIG. 11.—Anterior surface of sixth cervical vertebra of Dog. *s*, Spinous process; *az*, anterior zygapophysis; *v*, vertebrarterial canal; *t*, transverse process; *t′*, its inferior lamella.

Cervical Vertebræ.—The *cervical* region constitutes the most anterior portion of the column, or that which joins the cranium. The vertebræ which belong to it are either entirely destitute of movable ribs, or if they have any these are small, and do not join the sternum. As a general rule they have a considerable perforation through the base of the transverse process (the vertebrarterial canal, Fig. 11, *v*); or, as it is sometimes described, they have two transverse processes, superior and inferior, which meet at their extremities to enclose a canal. This, however, rarely applies to the last vertebra of the region, in which only the upper transverse process is usually developed. The transverse process, moreover, very often sends down near its extremity a more or less compressed plate (inferior lamella), which, being considered serially homologous with the ribs of the thoracic vertebræ (though not developed autogenously), is often called the "costal" or "pleurapophysial" plate. This is usually largest on the sixth, and altogether wanting on the seventh vertebra. The first and second cervical vertebræ, called respectively "atlas" and "axis," are specially modified for the function of supporting and permitting the free movements of the head. They are not united together by the intervertebral substance, but connected only by ordinary ligaments and synovial joints.

The cervical region in mammals presents the remarkable peculiarity that whatever the length or flexibility of the neck, the number of vertebræ is the same, viz. seven, with the exception of the Manatee and Hoffman's Two-toed Sloth (*Cholœpus hoffmanni*), which both have but six, and the Three-toed Sloth (*Bradypus tridactylus*), which has nine, though in this case the last two usually support movable ribs, which are not sufficiently developed to reach the sternum.

According to Parker there may occasionally be eight cervicals in the Pangolins (*Manis*).

Dorsal Vertebræ.—The *dorsal* (or, as it would be more correctly termed, *thoracic*) region consists of the vertebræ succeeding those of the neck, which have ribs movably articulated to them. These ribs arch round the thorax—the anterior one, and usually the greater number of those that follow, being attached below to the sternum.

Lumbar Vertebræ.—The *lumbar* region consists of those vertebræ of the trunk in front of the sacrum which bear no movable ribs. It may happen that, as the ribs decrease in size posteriorly (the last being sometimes more or less rudimentary), the step from the thoracic to the lumbar region may be gradual and rather undetermined in a given species; but most commonly this is not the case, and the distinction is as well defined here as in any other region. As a general rule there is a certain relation between the number of the thoracic and lumbar vertebræ, the whole number being tolerably constant in a given group of animals, and any increase of the one being at the expense of the other. Thus in all known Artiodactyle Ungulata there are 19 dorso-lumbar vertebræ; but these may consist of 12 dorsal and 7 lumbar vertebræ, or 13 dorsal and 6 lumbar, or 14 dorsal and 5 lumbar. The smallest number of dorso-lumbar vertebræ in mammals occurs in some Armadillos, which have but 14. The number found in Man, the higher Apes, and most Bats, viz. 17, is exceptionally low; 19 prevails in the Artiodactyla, nearly all Marsupials, and very many Rodents; 20 or 21 in Carnivora and most Insectivora;

and 23 in Perissodactyla. The highest and quite exceptional numbers are in the Two-toed Sloth (*Cholœpus*) 27, and the Hyrax 30. The prevailing number of rib-bearing vertebræ is 12 or 13, any variation being generally in excess of these numbers.

Sacral Vertebræ.—The *sacral* region offers more difficulties o. definition. Taking the human "os sacrum" as a guide for comparison, it is generally defined as consisting of those vertebræ between the lumbar and caudal regions which are ankylosed together to form a single bone. It happens, however, that the number of such vertebræ varies in different individuals of the same or nearly allied species, especially as age advances, when a certain number of the tail vertebræ generally become incorporated with the true sacrum. Other suggested tests—as those vertebræ which have a distinct additional (pleurapophysial) centre of ossification between the body and the ilium, those to which the ilium is directly articulated, or those in front of the insertion of the ischio-sacral ligaments—being equally unsatisfactory or unpractical, the old one of ankylosis, as it is found to prevail in the average condition of adults in each species, is used in the enumeration of the vertebræ in the following pages. The Cetacea, having no iliac bones, have no part of the vertebral column modified into a sacrum.

Caudal Vertebræ.—The *caudal* vertebræ are those placed behind the sacrum, and terminating the vertebral column. They vary in number greatly—being reduced to 5, 4, or even 3, in a most rudimentary condition, in Man and in some Apes and Bats, and being numerous and powerfully developed, with strong and complex processes, in many mammals, especially among the Edentata, Cetacea, and Marsupialia. The highest known number, 46, is possessed by the African Long-tailed Pangolin. Connected with the under surface of the caudal vertebræ of many mammals which have the tail well developed are certain bones formed more or less like an inverted arch, called chevron bones, or by the French *os en V*. These are always situated nearly opposite to an intervertebral space, and are generally articulated both to the vertebra in front and the vertebra behind, but sometimes chiefly or entirely either to one or the other.

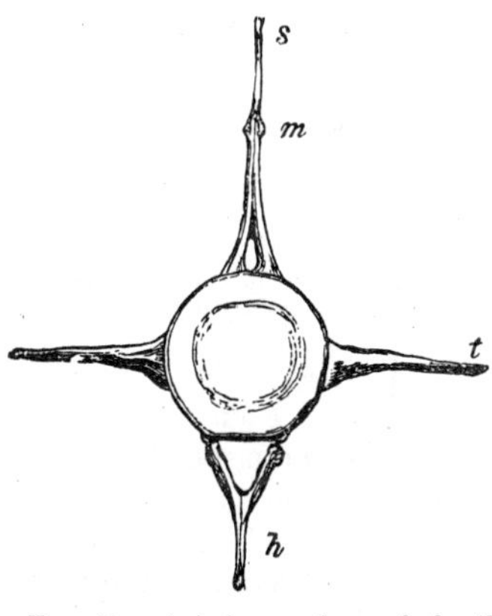

FIG. 12.—Anterior surface of fourth caudal vertebræ of Porpoise (*Phocœna communis*). *s*, Spinous process; *m*, metapophysis; *t*, transverse process; *h*, chevron bone.

In some of the Anomodont Reptiles and Labyrinthodont Amphibians these chevrons are attached to the intercentra—or imperfect disks alternating with the true centra—which suggests that they are primarily intercentral elements which have been transferred to the edges of the centra by the disappearance of the intercentra.

Sternum.—The *sternum* of mammals is a bone, or generally a series of bones, placed longitudinally in the mesial line, on the inferior or ventral aspect of the thorax, and connected on each side with the vertebral column by a series of more or less ossified bars called "ribs." It is present in all mammals, but varies much in character in the different groups. It usually consists of a series of distinct segments placed one before the other, the anterior being called the presternum or "manubrium sterni" of human anatomy, and the posterior the xiphisternum, or xiphoid or ensiform process, while the intermediate segments, whatever their number, constitute the mesosternum or "body." In the Whalebone Whales the presternum alone is developed, and but a single pair of ribs is attached to it.

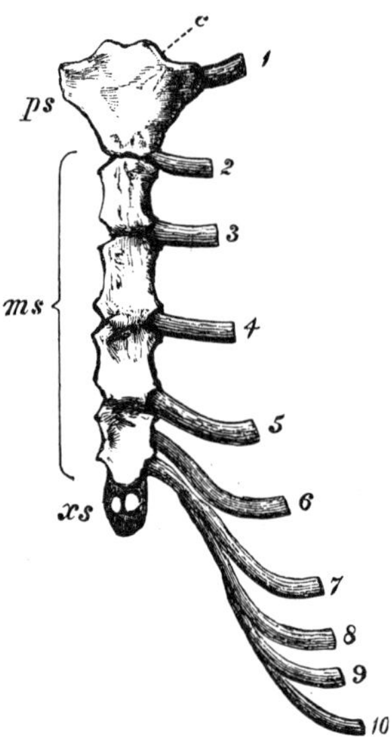

FIG. 13. Human sternum and sternal ribs. *ps*, Presternum; *ms*, mesosternum; *xs*, xiphisternum; *c*, point of attachment of clavicle; 1 to 10, the cartilaginous sternal ribs.

Ribs.—The *ribs* form a series of long, narrow, and more or less flattened bones, extending laterally from the sides of the vertebral column, curving downwards towards the median line of the body below, and mostly joining the sides of the sternum. The posterior ribs, however, do not directly articulate with that bone, but are either attached by their extremities to the edges of each rib in front of them, and thus only indirectly join the sternum, or else they are quite free below, meeting no part of the skeleton. These differences have given rise to the division into "true" and "false" ribs (by no means good expressions), signifying those that join the sternum directly and those that do not; and of the latter, those that are free below are called "floating" ribs. The portion of each rib nearest the vertebral column and that nearest the sternum differ in their characters, the latter being usually but imperfectly ossified, or remaining permanently cartilaginous. These are called "costal cartilages," or when ossified "sternal ribs."

In the anterior part of the thorax the vertebral extremity of each rib is divided into two parts, "head" or "capitulum," and "tubercle"; the former is attached to the side of the body of the vertebra, the latter to its transverse process; the former attachment corresponds to the interspace between the vertebræ, the head of the rib commonly articulating partly with the hinder edge of the body of the vertebra antecedent to that which bears its tubercle. Hence the body of the last cervical vertebra usually supports part of the head of the first rib. In the posterior part of the series the capitular and tubercular attachments commonly coalesce, and the rib is attached solely to its corresponding vertebra. The number of pairs of ribs is of course the same as that of the thoracic vertebræ.

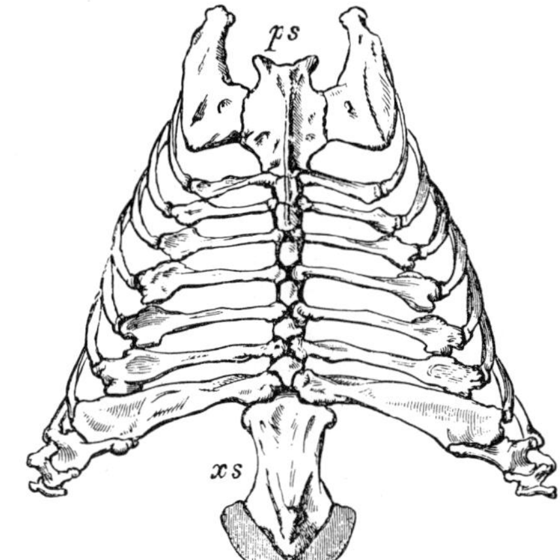

Fig. 14.—Sternum and strongly ossified sternal ribs of Great Armadillo (*Priodon gigas*). *ps*, Presternum; *xs*, xiphisternum.

The circumstance that in some of the Anomodont reptiles and

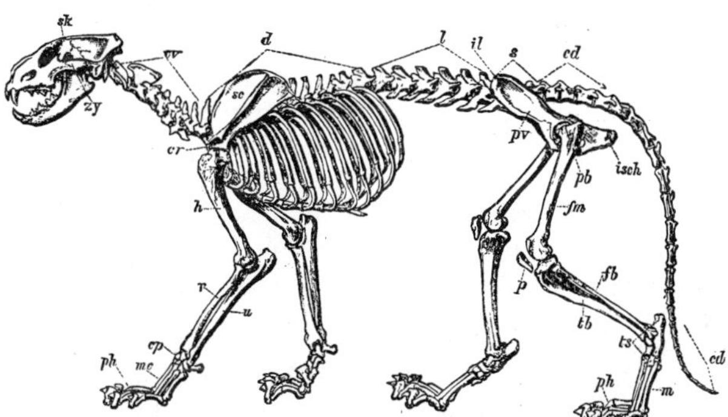

Fig. 15.—Skeleton of Lion (*Felis leo*). *cd*, Caudal vertebræ; *cp*, carpus; *cr*, coracoid process of scapula; *cv*, cervical vertebræ; *d*, dorsal vertebræ; *fb*, fibula; *fm*, femur; *h*, humerus; *il*, ilium; *isch*, ischium; *l*, lumbar vertebræ; *m*, metatarsus; *mc*, metacarpus; *p*, patella; *pb*, pubis; *ph*, phalanges; *pv*, pelvis; *r*, radius; *s*, sacral vertebræ; *sc*, scapula; *sk*, skull; *tb*, tibia; *ts*, tarsus; *u*, ulna; *zy*, zygomatic arch.

Labyrinthodonts the capitula of the ribs articulate with the intercentral elements of the vertebral column has suggested, as in the instance of the chevron bones, that the intercentral capitular articulation of the ribs of mammals is a feature directly inherited from those extinct types by the gradual disappearance of the intercentra.

Appendicular Skeleton.—The appendicular portion of the framework consists, when completely developed, of two pairs of limbs, anterior and posterior (Fig. 15).

Anterior Limb.—The anterior limb is present and fully developed in all mammals, being composed of a shoulder girdle and three segments belonging to the limb proper, viz. the upper arm or brachium, the fore-arm or antebrachium, and the hand or manus.

Shoulder-girdle.—The *shoulder* or *pectoral girdle* in the large majority of mammals is in a rudimentary or rather modified condition, compared with that in which it exists in other vertebrates. In the Monotremata (*Ornithorhynchus* and *Echidna*) alone is the ventral portion, or coracoid, complete and articulated with the sternum below, as in the Sauropsida; and in this group alone do we find an anterior ventral element, apparently corresponding with the precoracoid of the Anomodont reptiles, although generally known as the epicoracoid. In all other mammals the coracoid, though ossified from a distinct centre, forms only a process, sometimes a scarcely distinct tubercle, projecting from the anterior border of the glenoid cavity of the scapula. The last-named cavity, which in the Monotremes is formed jointly by the scapula and coracoid, receives the head of the humerus, or arm-bone. The scapula is always well developed, and generally broad and flat (whence its vernacular name "blade bone"), with a ridge called the "spine" on its outer surface, which usually ends in a free curved process, the "acromion." As the scapula affords attachment to many of the muscles which act upon the anterior limb, its form and the development of its processes are greatly modified according to the uses to which the member is put. Thus it is most reduced and simple in character in those animals whose limbs are mere organs of support, as the Ungulates; and most complex when the limbs are also used for grasping, climbing, or digging. The development or absence of the clavicle or "collar-bone," an accessory bar which connects the sternum with the scapula and steadies the shoulder-joint, has a somewhat similar relation, though its complete absence in the Bears shows that this is not an invariable rule. A complete clavicle is found in Man and all the Primates, in Chiroptera, all Insectivora (except *Potamogale*), in many Rodents, in most Edentates, and in all Marsupials, except *Perameles*. More or less rudimentary clavicles (generally suspended freely in the muscles) are found in the Cat, Dog, and most Carnivora, *Myrmecophaga*, and some Rodents. Clavicles are altogether absent in most of the *Ursidæ*, all the Pinnipedia, *Manis* among Edentates, the Cetacea, Sirenia, Ungulates, and some Rodents.

The Monotremes are peculiar in possessing a T-shaped interclavicle like that of many reptiles, lying upon the sternum, and articulating superiorly with the clavicles.

Brachium and Antebrachium.—The proximal segment of the anterior or pectoral limb proper contains a single bone, the humerus, and the second segment two bones, the radius and the ulna, placed side by side, and articulating with the humerus at their proximal, and with the carpus at their distal extremity (Fig. 15). In their primitive and unmodified condition these bones may be considered as placed one on each border of the limb, the radius being preaxial or anterior, and the ulna postaxial or posterior, when the distal or free end of the limb is directed outwards, or away from the trunk. This is their position in the earliest embryonic condition, and is best illustrated among adult mammals in the Cetacea, where the two bones are fixed side by side and parallel to each other. In the greater number of mammals the bones assume a very modified and adaptive position, usually crossing each other in the forearm, the radius in front of the ulna, so that the preaxial bone (radius), though external (in the ordinary position of the limb) at the upper end, is internal at the lower end; and the hand, being mainly fixed to the radius, also has its preaxial border internal. In the large majority of mammals the bones are fixed in this position, but in some few, as in Man, a free movement of crossing and uncrossing—or pronation and supination, as it is termed—is allowed between them, so that they can be placed in their primitive parallel condition, when the hand (which moves with the radius) is said to be supine, or they may be crossed, when the hand is said to be prone.

The humerus frequently has a foramen piercing the inner border of the distal extremity, known as the entepicondylar foramen, which corresponds with a similar one found in the Anomodont reptiles. The hollow in the head of the ulna for the reception of the head of the humerus is known as the greater sigmoid cavity, and that for the head of the radius as the lesser sigmoid cavity (Fig. 16). The term olecranon is applied to that process of the ulna which forms the prominence of the elbow.

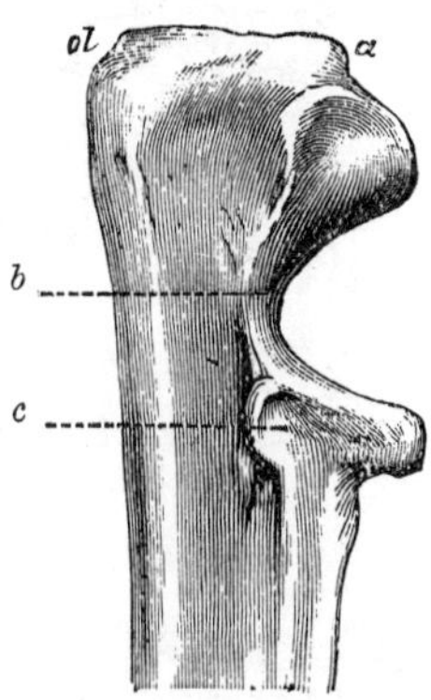

FIG. 16.—Outer aspect of the proximal extremity of the right ulna of a Bear (*Ursus*). *a*, Anterior tubercle; *ol*, olecranon; *b*, greater sigmoid cavity; *c*, lesser do.

In most mammals walking on four limbs, in which the hand is permanently prone, the ulna is much reduced in size, and the radius increased, especially at the upper end; so that the articular surface of the latter, instead of being confined to the external side of the trochlea of the humerus, extends all across its anterior surface, and the two bones, instead of being external and internal, are anterior and posterior. In many hoofed or Ungulate mammals, and in Bats, the ulna is reduced to little more than its upper articular extremity, and firmly ankylosed to the radius —stability of these parts being more essential than mobility.

Manus.—The terminal segment of the anterior limb is the hand or manus. Its skeleton consists of three divisions: (1) the "carpus," a group of small, more or less rounded or angular bones with flattened surfaces applied to one another, and, though articulating by synovial joints, having scarcely any motion between them; (2) the "metacarpus," a series of elongated bones placed side by side, with their proximal ends articulating by almost immovable joints with the carpus; (3) the "phalanges" or bones of the digits, usually three in number to each, articulating to one another by freely movable hinge-joints, the first being connected in like manner to the distal end of the metacarpal bone to which it corresponds.

Carpus.—To understand thoroughly the arrangement of the bones of the carpus in mammals, it is necessary to study their condition in some of the lower vertebrates. Fig. 17 represents the manus in one of its fullest and at the same time most generalised forms, as seen in one of the Water Tortoises (*Chelydra serpentina*). The carpus consists of two principal rows of bones. The upper or proximal row contains three bones, to which Gegenbaur has applied the terms *radiale* (*r*), *intermedium* (*i*), and *ulnare* (*u*), the first being on the radial or preaxial side of the limb.[1] The lower or distal row contains five bones, called *carpale* 1, 2, 3, 4, and 5 respectively, commencing on the radial side. Between these two rows, in the middle of the carpus, is a single bone, the *centrale* (*c*). In this very symmetrical carpus it will be observed that the *radiale* supports on its distal side two bones, *carpale* 1 and 2; the *intermedium* is in a line with the *centrale* and *carpale* 3, which together form a median axis of the hand, while the *ulnare* has also two bones articulating with its distal end, viz. *carpale* 4 and 5. Each of the carpals of the distal row supports a metacarpal.

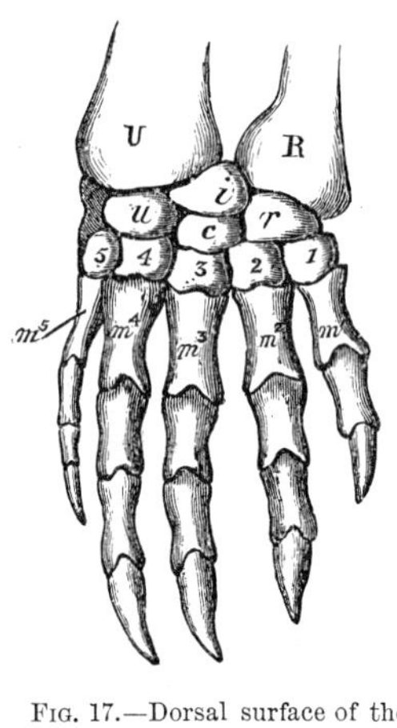

FIG. 17.—Dorsal surface of the right manus of a Water Tortoise (*Chelydra serpentina*). After Gegenbaur. U, Ulna; R, radius; *u*, ulnare; *i*, intermedium; *r*, radiale; *c*, centrale; 1-5, the five bones of the distal row of the carpus; m^1-m^5, the five metacarpals.

[1] The opinion has recently been expressed by Baur that the bone termed radiale in Fig. 17 is really a second centrale, and that the radiale is represented by a minute bone generally known as the radial sesamoid. The mammalian

In the carpus of the Mammalia there are usually two additional bones developed in the tendons of the flexor muscles, one on each side of the carpus, which may be called the radial and ulnar sesamoid bones ; the latter, which is the more constant and generally larger, is commonly known as the pisiform bone. The fourth and fifth carpals of the distal row are always united into a single bone, and the centrale is very often absent. As a general rule all the other bones are present and distinct, though it not unfrequently happens that two may have coalesced to form a single bone, or one or more may be altogether suppressed.

The following table shows the principal names in use for the various carpal bones,—those in the second column being the terms generally employed by English anatomists :—

Radiale	= Scaphoid	= *Naviculare.*
Intermedium	= Lunar	= *Semilunare, Lunatum.*
Ulnare	= Cuneiform	= *Triquetrum, Pyramidale.*
Centrale	= Central	= *Intermedium* (Cuvier).
Carpale 1	= Trapezium	= *Multangulum majus.*
Carpale 2	= Trapezoid	= *Multangulum minus.*
Carpale 3	= Magnum	= *Capitatum.*
Carpale 4, *Carpale* 5	= Unciform	= *Hamatum, Uncinatum.*

The radial and ulnar sesamoids are regarded by Bardeleben[1] as the rudiments of a prepollex and a postminimus digit ; the primitive number of digits being thus supposed to have been seven. These bones have been observed in all orders of mammals having five complete digits. Occasionally, as in *Pedetes caffer*, the so-called prepollex consists of two bones, of which the distal one bears a distinct nail-like horny covering. In *Bathyergus maritimus* the pisiform, or postminimus, is likewise double ; the two elements being regarded by their describer as representing the carpal and metacarpal of the presumed seventh digit.

Similarly in the posterior limb the tibial sesamoid, and a fibular ossification corresponding to the pisiform, are regarded as representing a prehallux and a postminimus.

Metacarpus and Phalanges.—The metacarpal bones, with the digits which they support, are never more than five in number, and are described numerically—first, second, etc., counting from the radial towards the ulnar side. The digits are also sometimes named (1) the pollex, (2) index, (3) medius, (4) annularis, (5) minimus.

scaphoid is accordingly also regarded as a second centrale. In the same communication, Dr. Baur expresses his disbelief in the existence of remnants of a prepollex and of a seventh digit in mammals and other vertebrates. (See *Anat Anzeiger*, vol. iv. pp. 49-52, 1889.)

[1] On the Præpollex and Præhallux, etc., *Proc. Zool. Soc.* 1889, pp. 259-262.

One or more may be in a rudimentary condition, or altogether suppressed. If one is absent, it is most commonly the first. Excepting the Cetacea, no mammals have more than three phalanges to each digit, but they may occasionally have fewer by suppression or ankylosis. The first or radial digit is an exception to the usual rule, one of its parts being constantly absent, since, while each of the other digits has commonly a metacarpal and three phalanges, it has only three bones altogether; whether the missing one is a metacarpal or one of the phalanges is a subject which has occasioned much discussion, and has not yet been satisfactorily decided. The terminal phalanges of the digits are usually specially modified to support the nail, claw, or hoof, and are called "ungual phalanges." In walking, some mammals (as the Bears) apply the whole of the lower surface of the carpus, metacarpus, and phalanges to the ground; to these the term "plantigrade" is applied. Many others (as nearly all the existing Ungulata) only rest on the last one or two phalanges of the toes, the first phalanx and the metacarpals being vertical and in a line with the fore-arm. These are called "digitigrade." Intermediate conditions exist between these two forms, to which the terms "phalangigrade" (as the Camel) and "subplantigrade" (as in most Carnivora), are applied. When the weight is borne entirely on the distal surface of the ungual phalanx, and the horny structures growing around it, as in the Horse, the mode of progression is called "unguligrade."

In the Chiroptera the digits are enormously elongated, and support a cutaneous expansion constituting the organ of flight. In the Cetacea the manus is formed into a paddle, being covered by continuous integument, which conceals all trace of division into separate digits, and shows no sign of nails or claws. In the Sloths the manus is long and very narrow, habitually curved, and terminating in two or three pointed curved claws in close apposition with each other, and incapable, in fact, of being divaricated; so that it is reduced to the condition of a hook, by which the animal suspends itself to the boughs of the trees among which it lives. These are only examples of the endless modifications to which the distal extremity of the limb is subjected in adaptation to the various purposes to which it is applied.

Posterior Limb.—The posterior limb is constructed upon a plan very similar to that of the anterior extremity. It consists of a pelvic girdle and three segments belonging to the limb proper, viz. the thigh, the leg, and the foot or pes (Fig. 15).

Pelvic Girdle.—The pelvic girdle is present in some form in all mammals, though in the Cetacea and the Sirenia it is in an exceedingly rudimentary condition. In all mammals except those belonging to the two orders just named, each lateral half of the pelvic girdle consists essentially, like the corresponding part of the anterior limb, of a flattened rod of bone crossing the long axis of the trunk, having an upper or dorsal and a lower or ventral end. The upper end diverges from that of the opposite side, but the lower end approaches, and, in most cases, meets it, forming a symphysis, without the intervention of any bone corresponding to the sternum. The pelvic girdle differs from the shoulder girdle in being firmly articulated to the vertebral column, thus giving greater power to the hinder limb in its function of supporting and propelling the body. Like the shoulder girdle, it bears on its outer side, near the middle, a cup-shaped articular cavity ("acetabulum"), into which the proximal end of the first bone of the limb proper is received. Each lateral half of the girdle is called the "os innominatum," or innominate bone, and consists originally of three bones which unite at the acetabulum. The "ilium" or upper bone is that which articulates with the sacral vertebræ. Of the two lower bones the anterior or "pubis" unites with its fellow of the other side at the symphysis; the posterior is the "ischium." These lower elements form two bars of bone, united above and below, but leaving a space between them in the middle, filled only by membrane, and called the "thyroid" or "obturator" foramen. The whole circle of bone formed by the two innominate bones and the sacrum is called the pelvis. In the Monotremata and Marsupialia, a pair of thin, flat, elongated ossifications called epipubic or marsupial bones are attached to the fore part of the pubis, and project forward into the muscular wall of the abdomen.

Thigh and Leg.—The first segment of the limb proper has one bone, the femur, corresponding with the humerus of the anterior limb. The second segment has two bones, the tibia and fibula, corresponding with the radius and ulna. These bones always lie in their primitive unmodified position, parallel to each other, the tibia on the preaxial and the fibula on the postaxial side, and are never either permanently crossed or capable of any considerable amount of rotation, as in the corresponding bones of the fore limb. In the ordinary walking position the tibia is internal, and the fibula external. In many mammals the fibula is in a more or less rudimentary condition, and it often ankyloses with the tibia at one or both extremities. The patella or "knee-cap," which is found in an ossified condition in all mammals, with the exception of some of the Marsupialia, is a large sesamoid bone developed in the tendon of the extensor muscles of the thigh, where the tendon passes over the front of the knee-joint, to which it serves as a protection. There are frequently smaller ossicles, one or two in number, situated behind the femoral condyles, called "fabellæ." The processes for the attachment of muscles near the upper end of the femur are termed trochanters; and the third trochanter, found on the hinder aspect of the shaft of this bone in many forms is of considerable taxonomic importance.

Pes.—The terminal segment of the hind limb is the foot or pes. Its skeleton presents in many particulars a close resemblance to that of the manus, being divisible into three parts: (1) a group of short, more or less rounded or square bones, constituting the tarsus; (2) a series of long bones placed side by side, forming the metatarsus; and (3) the phalanges of the digits or toes.

The bones of the tarsus of many of the lower Vertebrata closely resemble both in number and arrangement those of the carpus, as shown in Fig. 17. They have been described in their most generalised condition by Gegenbaur under the names expressed in the first column of the following table. The names in the second column are those by which they are generally known to English anatomists, while in the third column some synonyms occasionally employed are added.

Tibiale (?) / *Intermedium*	= Astragalus[1]	= *Talus.*
Fibulare	= Calcaneum	= *Os calcis.*
Centrale	= Navicular	= *Scaphoideum.*
Tarsale 1	= Internal cuneiform	= *Entocuneiforme.*
Tarsale 2	= Middle cuneiform	= *Mesocuneiforme.*
Tarsale 3	= External cuneiform	= *Ectocuneiforme.*
Tarsale 4 / *Tarsale* 5	= Cuboid.	

The bones of the tarsus of mammals present fewer diversities of number and arrangement than those of the carpus. The proximal row (see Fig. 18) always consists of two bones, namely the astragalus (*a*), which probably represents the coalesced scaphoid and lunar of the hand, and the calcaneum (*c*). The former is placed more to the dorsal side of the foot than the latter, and almost exclusively furnishes the tarsal part of the tibio-tarsal or ankle-joint. The calcaneum, placed more to the ventral or "plantar" side of the foot, is elongated backwards to form a more or less prominent tuberosity, the "tuber calcis," to which the tendon of the great extensor muscles of the foot is attached. The navicular bone (*n*) is interposed between the proximal and distal row on the inner or tibial side of the foot, but on the outer side the bones of the two rows come into contact. The distal row, when complete, consists of four bones, which, beginning on the inner side, are the three cuneiform bones, internal (c^1), middle (c^2), and external (c^3), articulated to the distal surface of the navicular, and the cuboid (*cb*), articulated with the calcaneum. Of these the middle cuneiform is usually the smallest in animals

[1] Cope and Baur consider that the astragalus corresponds only with the intermedium, and that the tibiale may exist as a distinct element.

in which all five digits are developed; but when the hallux is wanting the internal cuneiform may be rudimentary or altogether absent. The three cuneiform bones support respectively the first, second, and third metatarsals, and the cuboid supports the fourth and fifth; they thus exactly correspond with the four bones of the distal row of the carpus.

In addition to these constant tarsal bones, there may be supplemental or sesamoid bones: one situated near the middle of the tibial side of the tarsus, largely developed in many Carnivora and Rodentia; another, less frequent, on the fibular side; and a third, often developed in the tendons of the plantar surface of the tarsus, is especially large in Armadillos. There is also usually a pair of sesamoid bones on the plantar aspect of each metatarso-phalangeal articulation. In the young of the carnivorous genus *Cryptoprocta* there may be a second centrale, which usually coalesces with the ectocuneiform.

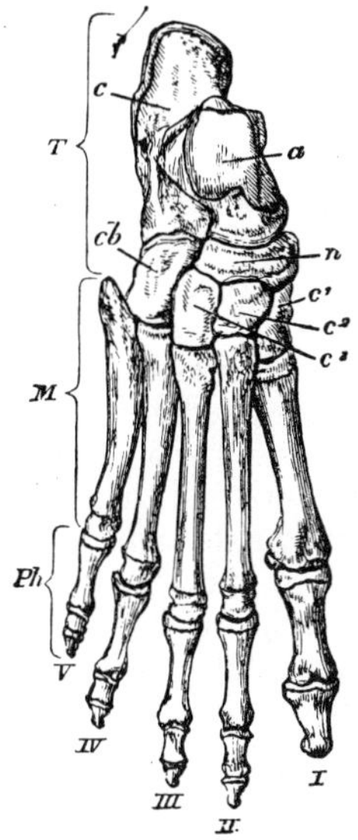

FIG. 18.—Bones of the right Human foot. *T*, Tarsus; *M*, metatarsus; *Ph*, phalanges; *c*, calcaneum; *a*, astragalus; *cb*, cuboid; *n*, navicular; c^1, internal cuneiform; c^2, middle cuneiform; c^3, external cuneiform. The digits are indicated by Roman numerals, counting from the tibial to the fibular side.

The metatarsal bones never exceed five in number, and the phalanges follow the same numerical rule as in the manus, never exceeding three in each digit. Moreover, the first digit, counting from the tibial side, or hallux, resembles the pollex of the hand in always having one segment less than the other digits. As the function of the hind foot is more restricted than that of the hand the modifications of its structure are less striking. In the Cetacea and the Sirenia it is entirely wanting, though in some existing members of the first-named order rudiments of the bones of both the first and second segments of the limb have been detected, and a femur is present in the Miocene Sirenian *Halitherium*.

IV. THE DIGESTIVE SYSTEM.

General Considerations.—The search after the purpose which every modification of structure subserves in the economy is always full of interest, and, if conducted with due caution and sufficient knowledge of all the attendant circumstances, may lead to important generalisations. It must always be borne in mind, however, that adaptation to its special function is not the only cause of the particular form or structure of an organ, but that this form, having in all probability been arrived at by the successive and gradual modification of some other different form from which it is now to a greater or less degree removed, has other factors besides use to be taken into account. In no case is this principle so well seen as in that of the organs of digestion. These may be considered as machines which have to operate upon alimentary substances in very different conditions of mechanical and chemical combination, and to reduce them in every case to the same or precisely similar materials; and we might well imagine that the apparatus required to produce flesh and blood out of coarse fibrous vegetable substances would be different from that which had to produce exactly the same results out of ready-made flesh or blood; and in a very broad sense we find that this is so. Thus, if we take a large number of carnivorous animals, belonging to different fundamental types, and a large number of herbivorous animals, and strike a kind of average of each, we shall find that there is, pervading the first group, a general style, if we may use the expression, of the alimentary organs, different from that of the others. That is to say, there is a specially carnivorous and a specially herbivorous modification of these parts. But, if function were the only element which has guided such modification, it might be inferred that, as one form must be supposed to be best adapted in its relation to a particular kind of diet, that form would be found in all the animals consuming such diet. But this is far from being the case. Thus the Horse and the Ox, for instance—two animals whose food in the natural state is precisely similar—are most different as regards the structure of their alimentary canal, and the processes involved in the preparation of that food. Again, the Seal and the Porpoise, both purely fish-eaters, which seize, swallow, and digest precisely the same kind of prey, in precisely the same manner, have a totally different arrangement of the alimentary canal. If the Seal's stomach is adapted in the best conceivable manner for the purpose it has to fulfil, why is not the Porpoise's stomach an exact facsimile of it, and *vice versâ?* We can only answer that the Seal and Porpoise belong to different natural groups of animals, formed either on different primitive types, or descended from differently constructed ancestors. On this principle only can we account for the fact that, whereas, owing to the comparatively small variety of the different alimentary substances met with in nature, few modifications would appear necessary in the organs of digestion, there is really endless variety in the parts devoted to this purpose.

Mouth.—The digestive apparatus of mammals, as in other vertebrates, consists mainly of a tube with an aperture placed at or near either extremity of the body,—the oral and the anal orifice,—

and furnished with muscular walls, the fibres of which are so arranged as by their regular alternate contraction and relaxation to drive onwards the contents of the tube from the first to the second of these apertures. The anterior or commencing portion of this tube and the parts around it are greatly and variously modified in relation to the functions assigned to them of selecting and seizing the food, and preparing it by various mechanical and chemical processes for the true digestion which it has afterwards to undergo before it can be assimilated into the system. For this end the tube is dilated into a chamber or cavity called the mouth, bordered externally by the lips, which are usually muscular and prehensile, and supported by a movable framework carrying the teeth; the structure and modifications of which have been already described. The roof of the mouth is formed by the palate, terminating behind by a muscular, contractile arch, having in Man and some few other species a median projection called the uvula, beneath which the mouth communicates with the pharynx. The anterior part of the palate is composed of mucous membrane tightly stretched over the flat or slightly concave bony lamina separating the mouth from the nasal passages, and is generally raised into a series of transverse ridges, which sometimes, as in Ruminants, attain a considerable development. In the floor of the mouth, between the rami of the mandible, and supported behind by the hyoidean apparatus, lies the tongue; an organ the free surface of which, especially in its posterior part, is devoted to the sense of taste, but which also, by its great mobility (being composed almost entirely of muscular fibres), performs important mechanical functions connected with masticating and procuring food. Its modifications of form in different mammals are very numerous. Between the long, extensile, vermiform tongue of the Anteaters, which is essential to the peculiar mode of feeding of those animals, and the short, sessile, and almost functionless tongue of the Porpoise, every intermediate condition is found. Whatever the form, the upper surface is always covered with numerous fine papillæ, in which the terminal filaments of the gustatory nerve are distributed.

Salivary Glands.—The fluid known as the saliva is secreted by an extensive and complex system of glands discharging into the cavity of the mouth (buccal cavity), the position and relation of some of which are exhibited in the woodcut on the next page (Fig. 19).

This apparatus consists of small glands embedded in the mucous membrane or submucous tissue lining the cavity of the mouth, which are of two kinds (the follicular and the racemose), and of others in which the secreting structure is aggregated in distinct masses removed some distance from the cavity; other tissues besides the lining membrane being usually interposed, and pouring their secretion into the cavity by a distinct tube or duct, which traverses the mucous membrane. To the latter alone the name of "salivary glands" is ordinarily appropriated, although the distinction between them and the smaller racemose glands is only one of convenience for descriptive purposes, their structure being more or less nearly identical; and, since the fluids secreted by all become mixed in the mouth, their functions are, at all events in great part, common. Under the name of salivary glands are commonly

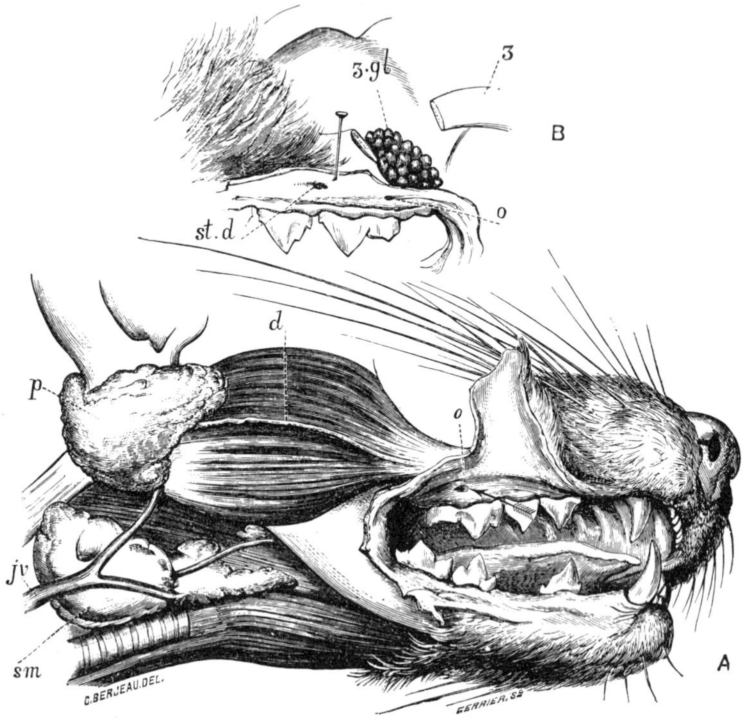

FIG. 19.—Salivary Glands of the Genet. *A*, Right side of the head dissected; *p*, parotid gland; *d*, Steno's duct; *sm*, submaxillary gland, traversed by the jugular veins (*jv*); *o*, aperture of Steno's duct. *B*, Part of the head with the lip drawn up to show (*st.d*) aperture of Steno's duct; *z.gl*, zygomatic gland; *o*, aperture of do.; *z*, zygomatic arch. (Mivart, *Proc. Zool. Soc.* 1882, p. 504.)

included—(1) the "parotid" (*p*), situated very superficially on the side of the head, below or around the cartilaginous external auditory meatus, and the secretion of which enters the mouth by a duct (often called Steno's or Stenson's) which crosses the masseter muscle and opens into the upper and back part of the cheek (Fig. 19); and (2) the "submaxillary" (*sm*), situated in the neck, near or below the angle of the mandible, and sending a long duct

(Wharton's) forwards to open on the fore-part of the floor of the cavity of the mouth, below the apex of the tongue. These are the most largely developed and constant of the salivary glands, being met with in various degrees of development in almost all animals of the class. Next in constancy are (3) "the sublingual," closely associated with the last-named, at all events in the locality in which the secretion is poured out; and (4) the "zygomatic" (*z.gl*), found only in some animals in the cheek, just under cover of the anterior part of the zygomatic arch, its duct entering the buccal cavity near that of the parotid.

The most obvious function common to the secretion of these various glands, and to that of the smaller ones placed in the mucous membrane of the lips, the cheeks, the tongue, the palate, and fauces, is the mechanical one of moistening and softening the food, to enable it the more readily to be tasted, masticated, and swallowed, though each kind of gland may contribute in different manner and different degree to perform this function. The saliva is, moreover, of the greatest importance in the first stage or introduction to the digestive process, as it dissolves or makes a watery extract of all soluble substances in the food, and so prepares them to be further acted on by the more potent digestive fluids met with subsequently in their progress through the alimentary canal. In addition to these functions it seems now well established by experiment that saliva serves in Man and many animals to aid directly in the digestive process, particularly by its power of inducing the saccharine transformation of amylaceous substances. As a general rule, in mammals the parotid saliva is more watery in its composition, while that of the submaxillaries, and still more the sublingual, contains more solid elements and is more viscid;—so much so that some anatomists consider the latter, together with the small racemose glands of the cheeks, lips, and tongue, as mucous glands, retaining the name of salivary only for the parotid. These peculiar properties are sometimes illustrated in a remarkable degree, as, for example, the great secretion of excessively viscid saliva which lubricates the tongue of the Anteaters and Armadillos, associated with enormously developed submaxillary glands; while, on the other hand, the parotids are of great size in those animals which habitually masticate dry and fibrous food.

Stomach.—After the preparation which the aliment has undergone in the mouth,—the extent of which varies immensely in different forms, being reduced almost to nothing in such animals as the Seals and Cetaceans, which, to use the familiar expression, "bolt" their food entire, and most fully carried out in the Ruminants, which "chew the cud,"—it is swallowed, and carried along the œsophagus by the action of its muscular coats into the stomach. In the greater number of mammals this organ is a simple saccular dilatation of the alimentary canal, as in Figs. 20, 21, but in others it undergoes remarkable modifications and complexities. The lining of the stomach is thickly beset with tubular glands, which are generally considered to belong to two different forms, recognisable by their structure, and different in their function—the most numerous and important secreting the gastric juice (the active agent in stomachic digestion), and hence called "peptic" glands, while the others are concerned only in the elaboration of mucus. The relative distribution of these glands in different regions of the walls of the stomach varies greatly in different animals, and in many species there are large tracts of the mucous membrane which do not secrete a fluid having the properties of gastric juice, but often constitute more or less distinct cavities devoted to storing

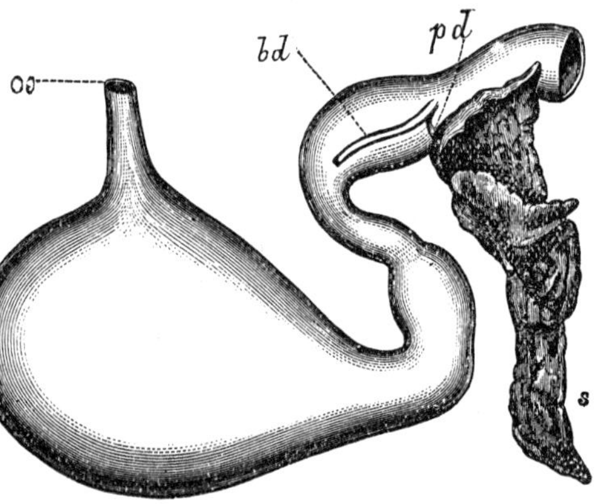

FIG. 20.—Stomach and pancreas of the Genet. Posterior or dorsal surface. *œ*, Œsophagus; *s*, pancreas; *pd*, pancreatic duct; *bd*, biliary duct from the liver. (From Mivart, *Proc. Zool. Soc.* 1882, p. 305.)

and perhaps softening or otherwise preparing the food for digestion. Sometimes there is a great aggregation of glands forming distinct thickened patches of the stomach wall, as in the Beaver and Koala, or even collected in pyriform pouches with a common narrow opening into the cavity, as in the Manatee and the curious African Rodent *Lophiomys*. The action of the gastric fluid is mainly exerted upon the nitrogenous elements of the food, which it dissolves and modifies so as to render them capable of undergoing absorption, effected partly by the blood-vessels of the stomach, although the greater part passes through the pylorus, an aperture surrounded by a circular muscular valve, into the intestinal canal. Here it comes in contact with the secretion of a vast number of small glands called the crypts of Lieberkuhn, somewhat similar to those of the stomach; and also of several special glands of a different character, namely, the small racemose, duodenal, or

Brunner's glands, the pancreas, and the liver; the position of the ducts of the two latter organs being indicated in Fig. 20.

Intestinal Canal.—The intestinal canal varies greatly in relative length and capacity in different animals, and it also offers manifold peculiarities of form, being sometimes a simple cylindrical tube of nearly uniform calibre throughout, but more often subject to alterations of form and capacity in different portions of its course,—the most characteristic and constant being the division into an upper and narrower, and lower and wider portion, called respectively the small and the large intestine, the former being divided quite arbitrarily and artificially into duodenum, jejunum, and ileum, and the latter into colon and rectum. One of the most striking peculiarities of this part of the alimentary canal is the frequent presence of a diverticulum or blind pouch, the *caput cæcum coli*, as it was first called, a name generally abbreviated into "cæcum," situated at the junction of the large and the small intestine, a structure presenting an immense variety of development, from the smallest bulging of a portion of the side wall of the tube to a huge and complex sac, greatly exceeding in capacity the whole of the remainder of the alimentary canal. It is only in herbivorous animals that the cæcum is developed to this great extent, and among these there is a curious complementary relationship between the size and complexity of this organ and that of the stomach. Where the latter is simple the cæcum is generally the largest, and *vice versâ*. Both the cæcum and colon are often sacculated, a disposition caused by the arrangement of the longitudinal bands of muscular tissue in their walls; but the small intestine is always smooth and simple-walled externally, though its lining membrane often exhibits various contrivances for increasing the absorbing surface without adding to the general bulk of the organ, such as the numerous small villi by which it is everywhere beset, and the more obvious transverse, longitudinal, or reticulating folds projecting into the interior, met with in many animals, of which the "valvulæ conniventes" of Man form well-known examples.

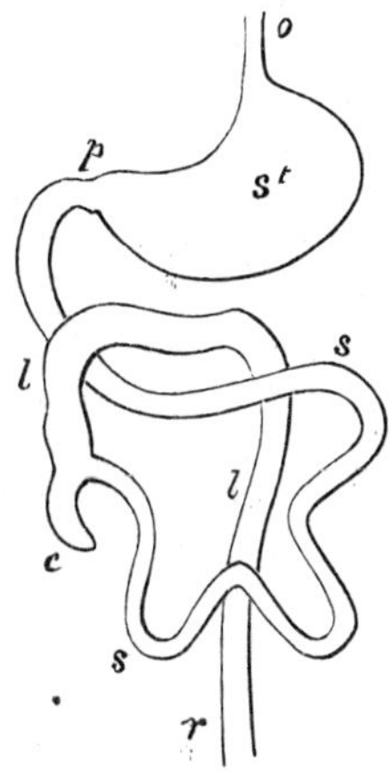

FIG. 21.—Diagrammatic plan of the general arrangement of the alimentary canal in a typical Mammal. *o*, Œsophagus; *st*, stomach; *p*, pylorus; *s*, *s*, small intestine (abbreviated); *c*, cæcum; *l*, *l*, large intestine or colon, ending in *r*, the rectum.

Besides the crypts of Lieberkuhn found throughout the intestinal canal, and the glands of Brunner confined to the duodenum, there are other structures in the mucous membrane, about the nature of which there is still much uncertainty, called "solitary" and "agminated" glands; the latter being more commonly known by the name of "Peyer's patches." These were formerly supposed to be secretory organs, which discharged some kind of fluid into the intestine, but are now more generally considered to belong to that group of structures of somewhat mysterious function of which the lymphatic and lacteal glands are members. The solitary glands are found scattered irregularly throughout the whole intestinal tract; the agminated, on the other hand, are always confined to the small intestine, and are most abundant in its lower part. They are subject to great variation in number and in size, and even in different individuals of the same species, and also differ in character at different periods of life, becoming atrophied in old age.

Liver.—The distinct glands situated outside the walls of the intestinal canal, but which pour their secretion into it, are the pancreas and the liver. The latter is the more important on account of its size, if not on account of the direct action of its secretion in the digestive process. This large gland, so complex in structure and function, is well developed in all mammals, and its secreting tube, the bile-duct, always opens into the duodenum, or that portion of the canal which immediately succeeds the stomach. It is situated on the right side of the abdomen in contact with the diaphragm and the stomach, but varies greatly in relative size, and also in form, in different groups of mammals. In most mammals a gall-bladder, consisting of a pyriform diverticulum from the bile-duct, is present, but in many this appendage is wanting, and it is difficult to find the rationale of its presence or absence in relation to use or any other circumstance in the animal economy.

The descriptions of the livers of various animals to be met with in treatises or memoirs on comparative anatomy are very difficult to understand for want of a uniform system of nomenclature. The difficulty usually met with arises from the circumstance that this organ is divided sometimes, as in Man, Ruminants, and the Cetacea, into two main lobes, which have been always called respectively right and left, and in other cases, as in the lower Monkeys, Carnivora, Insectivora, and several other orders, into a larger number of lobes. Among the latter the primary division usually appears at first sight tripartite, the whole organ consisting of a middle, called "cystic" or "suspensory" lobe, and two lateral lobes, called respectively right and left lobes. This introduces confusion in describing livers by the same terms throughout the whole series of mammals, since the right and left lobes of the Monkey or Dog, for instance, do not correspond with parts designated by the same names in Man and the Sheep. There are, moreover, conditions where neither the bipartite nor the tripartite system of nomenclature will answer, so that we should have considerable difficulty in describing them without some more general system. In order to arrive at such a system it appears desirable to consider the liver in all cases as primarily divided by the umbilical vein (see Fig. 22, *u*) into two segments, right and left. This corresponds with its development and with the condition characteristic of the organ in the inferior classes of vertebrates. The situation of this division can almost always be recognised in adult animals by the persistence of some traces of the umbilical vein in the form of the round ligament, and by the position of the suspensory ligament.

When the two main parts into which the liver is thus divided are entire, as in Man, the Ruminants, and Cetacea, they may be spoken of as the right and left lobes; when fissured, as the right and left segments of the liver, reserving the term lobe for the sub-

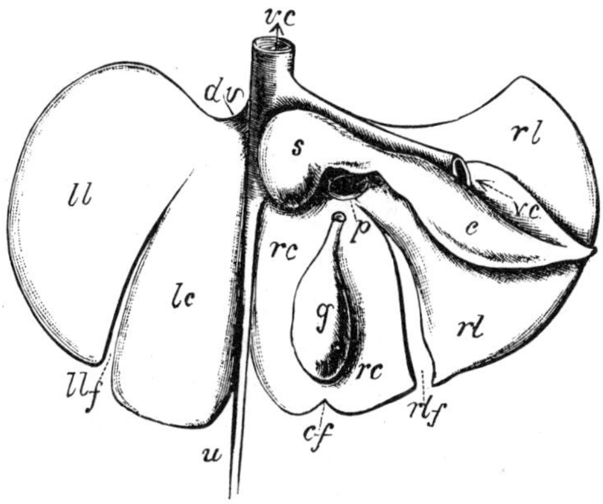

Fig. 22.—Diagrammatic plan of the inferior surface of a multilobed liver of a Mammal. The posterior or attached border is uppermost. *u*, Umbilical vein of the fœtus, represented by the round ligament in the adult, lying in the umbilical fissure; *dv*, the ductus venosus; *vc*, the inferior vena cava; *p*, the vena portæ entering the transverse fissure; *llf*, the left lateral fissure; *rlf*, the right lateral fissure; *cf*, the cystic fissure; *ll*, the left lateral lobe; *lc*, the left central lobe; *rc*, the right central lobe; *rl*, the right lateral lobe; *s*, the Spigelian lobe; *c*, the caudate lobe; *g*, the gall-bladder.

divisions. This will involve no ambiguity, for the terms right and left lobe will no longer be used for divisions of the more complex form of liver. In the large majority of mammals each segment is further divided by a fissure, more or less deep, extending from the free towards the attached border, which are called right and left lateral fissures (Fig. 22, *rlf* and *llf*). When these are more deeply cut than the umbilical fissure (*u*), the organ has that tripartite or trefoil-like form just spoken of, but it is easily seen that it is really divided into four regions or lobes, those included between the lateral fissures being the right and left central (*rc* and *lc*) separated by the umbilical fissure, and those beyond the lateral fissures on each side being the right and left lateral lobes (*rl* and *ll*).

The essentially bipartite character of the organ and its uniformity of construction throughout the class are thus not lost sight of, even in the most complex forms. The left segment of the liver is rarely complicated to any further extent, except in some cases by minor or secondary fissures marking off small lobules, generally inconstant and irregular, and never worthy of any special designation. On the other hand, the right segment is usually more complex. The gall-bladder, when present, is always attached to the under surface of the right central lobe, sometimes merely applied to it, in other cases deeply embedded in its substance. In many instances the fossa in which it is sunk is continued to the free margin of the liver as an indent, or even a tolerably deep fissure (*cf*). The portal fissure (*p*), through which the portal vein and hepatic artery enter and the bile-duct emerges from the liver, crosses the right central lobe transversely, near the attached border of the liver. The right lateral lobe always has the great vena cava (*vc*) either grooving its surface or tunnelling through its substance near the inner or left end of its attached border; and a prolongation of this lobe to the left, between the vein and the portal fissure, sometimes forming a mere flat track of hepatic substance, but more often a prominent tongue-shaped process, is the so-called "Spigelian lobe" (*s*). From the under surface of the right lateral lobe a portion is generally partially detached by a fissure, and called the "caudate lobe" (*c*). In Man this lobe is almost obsolete, but in most mammals it is of considerable magnitude, and has very constant and characteristic relations. It is connected by an isthmus at the left (narrowest or attached) end to the Spigelian lobe, behind which isthmus the vena cava is always in relation to it, channelling through or grooving its surface. It generally has a pointed apex, and is deeply hollowed to receive the right kidney, to the upper and inner side of which it is applied.

Considerations derived from the comparatively small and simple condition of the liver of the Ungulata, compared with its large size and complex form in the Carnivora, have led to the perhaps too hasty generalisation that the first type is related to a herbivorous and the latter to a carnivorous diet. The exceptions to such a proposition are very numerous. The fact of the great difference between the liver of the Cetacea and that of the Seals cannot be accounted for by difference of habits of life, though it perhaps may be by difference of origin.[1]

[1] For further details of these modifications, see Flower's "Lectures on the Comparative Anatomy of the Organs of Digestion of the Mammalia," *Medical Times and Gazette*, Feb.-Dec. 1872.

V. CIRCULATORY, ABSORBENT, RESPIRATORY, AND URINARY SYSTEMS.

Blood.—The blood of mammals is always red, and during the life of the animal hot, having a nearly uniform temperature, varying within a few degrees on each side of 100° Fahr. The corpuscles are, as usual in the vertebrates, of two kinds: (1) colourless, spheroidal, nucleated, and exhibiting amœboid movements; while (2) the more numerous, on which depends the characteristic hue of the fluid in which they are suspended, are coloured, non-nucleated, flattened, slightly biconcave discs, with circular outline in all known species except the Camels and Llamas, where they have the elliptical form characteristic of the red corpuscles of nearly all the other vertebrates, though adhering to the mammalian type in the absence of nucleus and relatively small size. As a rule they are smaller as well as more numerous than in other classes, but vary considerably in size in different species, and not always in relation to the magnitude of the animal; a Mouse, for instance, having as large corpuscles as a Horse. Within the limits of any natural group there is, however, very often some such relation, the largest corpuscles being found among the large species and the smallest corpuscles among the small species of the group, but even to this generalisation there are many exceptions. The transverse diameter of the red corpuscles in Man averages $\frac{1}{3200}$ of an inch, which is exceptionally large, and only exceeded by the Elephant ($\frac{1}{2745}$), and by some Cetacea and Edentata. They are also generally large in Apes, Rodents, and the Monotremata, and small in the Artiodactyles, least of all in the Chevrotains (*Tragulus*), being in *T. javanicus* and *meminna* not more than $\frac{1}{12325}$.[1]

Heart.—The heart of mammals consists of four distinct cavities, two auricles and two ventricles. Usually the ventricular portion is externally of conical form, with a simple apex, but in the Sirenia it is broad and flattened, and a deep notch separates the apical portion of each ventricle. A tendency to this form is seen in the Cetacea and the Seals. It is characteristic of mammals alone among vertebrates that the right auriculo-ventricular valve is tendinous like the left, consisting of flaps held in their place by fibrous ends (*chordæ tendiniæ*) and arising from projections of the muscular walls of the ventricular cavity (*musculi papillares*). In the Monotremata a transition between this condition and the simple muscular flap of the Sauropsida is observed. In most of the larger Ungulates a distinct but rather irregular ossification (*os cordis*) is developed in the central tendinous portion of the base of the heart.

Blood-vessels.—The orifices of the aorta and pulmonary artery are

[1] G. Gulliver, *Proc. Zool. Soc.*, 1862, p. 91.

each guarded by three semilunar valves. The aorta is single, and arches over the left bronchial tube. After supplying the tissues of the heart itself with blood by means of the coronary arteries, it gives off large vessels ("carotid") to the head and ("brachial") to the anterior extremities. The mode in which these vessels arise from the aorta varies much in different mammals, and the study of their disposition affords some guide to classification. In nearly all cases the right brachial and carotid have a common origin (called the "innominate artery" in anthropotomy). The other two vessels may come off from this, as is the rule in Ungulates, the common trunk constituting the "anterior aorta" of veterinary anatomy; or they may be detached in various degrees, both arising separately from the aorta, as in Man, or the left carotid from the innominate and the left brachial from the aorta, a very common arrangement; or the last two from a common second or left innominate, as in some Bats and Insectivores. The aorta, after giving off the intercostal arteries, passes through the diaphragm into the abdomen, and, after supplying the viscera of that cavity by means of the gastric, hepatic, splenic, mesenteric, renal, and spermatic vessels, gives off in the lumbar region a large branch (iliac) to each of the hinder extremities, which also supplies the pelvic viscera, and is continued onwards in the middle line, greatly diminished in size, along the under surface of the tail as the caudal artery. In certain mammals, arterial plexuses, called *retia mirabilia*, formed by the breaking up of the vessel into an immense number of small trunks, which may run in a straight course parallel to one another (as in the limbs of Sloths and Slow Lemurs), or form a closely packed network, as in the intracranial plexuses of Ruminants, or a sponge-like mass of convoluted vessels, as in the intercostals of Cetaceans, are peculiarities of the vascular system the meaning of which is not in all cases clearly understood. In the Cetacea they are obviously receptacles for containing a large quantity of oxygenated blood available during the prolonged immersion, with consequent absence of respiration, to which these animals are subject.

The vessels returning the blood to the heart from the head and upper extremities usually unite, as in Man, to form the single *vena cava superior* or precaval vein, but in some Insectivores, Chiroptera, and Rodents, in the Elephant, and all Marsupials and Monotremes, the two superior caval veins enter the right auricle without uniting, as in birds. In Seals and some other diving mammals there is a large venous sinus or dilatation of the inferior vena cava immediately below the diaphragm. In the Cetacea the purpose of this is supplied by the immense abdominal venous plexuses. As a rule the veins of mammals are furnished with valves, but these are said to be altogether wanting in the Cetacea, and in the superior and inferior cava, subclavian and iliac veins, the veins of the liver (both portal and hepatic), heart, lungs, kidneys, brain, and spinal cord of other mammals. Many of the veins within the cranium are included in spaces formed by the separation of the laminæ of the dura mater, and do not admit of being dilated beyond a certain size; these are termed sinuses. The portal circulation in mammals is limited to the liver, the portal vein being formed by the superior and inferior mesenteric, the splenic, the gastro-epiploic, and the pancreatic veins. The kidney is supplied solely by arterial blood, and its veins empty their contents only into the inferior cava.

Lymphatic Vessels.—The *absorbent* or *lymphatic* system of vessels is very fully developed in the Mammalia. Its ramifications extend through all the soft tissues of the body, and convey a colourless fluid called lymph, containing nucleated corpuscles, and also, during the process of digestion, the chyle, a milky fluid taken up by the lymphatics (here called lacteals) of the small intestine, and pour them into the general vascular system, where they mix with the venous blood. The lymphatic vessels of the hinder extremities, as well as those from the intestinal canal, unite in the abdomen to form the "thoracic duct," the hinder end or commencement of which has a dilatation called the *receptaculum chyli*. This duct, which is of irregular size and sometimes double, often dividing and uniting again in its course, or even becoming plexiform, passes forwards close to the bodies of the thoracic vertebræ, and empties itself, by an orifice guarded by a valve, into the great left brachio-cephalic vein, having previously received the lymphatics from the thorax and the left side of the head and left anterior extremity. The lymphatics from the right side of the head and right anterior limb usually enter by a small distinct trunk into the corresponding part of the right brachio-cephalic vein. The duct, and also the principal lymphatic vessels, are provided with valves.

Lymphatic glands, rarely met with in the Sauropsida, are usually present in mammals, both in the general and in the lacteal system; the latter being called "mesenteric glands." They are round or oval masses, situated upon the course of the vessels, which break up in them and assume a plexiform arrangement, and then reunite as they emerge. No structures corresponding to the pulsating "lymphatic hearts" of the lower vertebrates have been met with in mammals.

Ductless Glands.—Associated with the vascular and lymphatic systems are certain bodies (the functions of which are not properly understood), usually, on account of their general appearance, grouped together under the name of "ductless glands." The largest of these is the "spleen," which is single, and always placed in mammals in relation to the fundus or left end of the stomach, to which it is attached by a fold of peritoneum. It is dark-coloured and spongy in substance, and has a depression or "hilus" on one side, into which the splenic artery, a branch of the cœliac axis of the abdominal aorta, enters, and from which the vein joining the portal system emerges. The spleen varies much in size and form in different mammals, being relatively very small in the Cetacea. It is sometimes almost spherical, but more often flattened, oval, triangular, or elongated, and occasionally, as in Monotremes and most Marsupials, triradiate. The "suprarenal bodies" or "adrenals" are two in number, each situated either in contact with, or at a short distance in front of the anterior extremity of the kidney. They are abundantly supplied with nerves, and are much larger relatively in early than in adult life. The "thyroid bodies," of which there are generally two, though in Man and some other species they are connected by an isthmus passing across the middle line, are constant in mammals, though only met with in a rudimentary condition, if at all, in other vertebrates. They are situated in the neck, in contact with the sides of the anterior extremity of the trachea. The "thymus" lies in the anterior part of the thorax, between the sternum and the great vessels at the base of the heart, and differs from the thyroid in being median and single, and having a central cavity. It attains its greatest development during the period in which the animal is nourished by its mother's milk, and then it diminishes, and generally disappears before full growth is attained.

Nostrils.—Mammals breathe occasionally through the mouth, but usually, and in many cases exclusively, through the nostrils or *nares.* These are apertures, always paired (except in the toothed Cetacea, where they unite to form a single external opening), and situated at the fore part of the face, generally at or beneath the end of the muzzle, a median prominence above the mouth. This is sometimes elongated to form a proboscis, to the extremity of which the nostrils are carried, and which attains its maximum of development in the Elephant. In the Cetacea the nostrils are situated at a considerable distance behind the anterior end of the face, upon the highest part of the head, and are called "blow-holes," from the peculiar mode of respiration of those animals. The nostrils are kept open by means of cartilages surrounding their aperture, which many animals have the power of moving so as to cause partial dilatation or contraction. In diving animals, as Seals and Cetacea, they can be completely closed at will so as to prevent the entrance of water when beneath the surface. The passage to which the nostrils lead is in most mammals filled by a more or less complex sieve-like apparatus, formed of the convoluted turbinal bones and cartilages, over which a moist, vascular, ciliated mucous membrane is spread, which intercepts particles of dust, and also aids in warming the inspired air before it reaches the lungs. In the Proboscidea, in which these functions are performed by the walls of the long tubular proboscis, this apparatus is entirely wanting.

Trachea.—The narial passages have the organ of smell situated in their upper part, and communicate posteriorly with the pharynx, and through the glottis with the "trachea" or windpipe, a tube by which the air is conveyed to and from the lungs. The permanent patency of the trachea during the varied movements of the neck is provided for by its walls being stiffened by a series of cartilaginous rings or hoops, which in most mammals are incomplete behind. Having entered the thorax, the trachea bifurcates into the two bronchi, one of which enters, and, dividing dichotomously, ramifies through each lung. In some of the Cetacea and Artiodactyla a third bronchus is given off from the lower part of the trachea, above its bifurcation, and enters the right lung.

Larynx.—The upper end of the trachea is modified into the organ of voice or "larynx," the air passing through which to and from the lungs is made use of to set the edges of the "vocal cords," or fibrous bands stretched one on each side of the tube, into vibration. The larynx is composed of several cartilages, such as the "thyroid," the "cricoid," and the "arytenoid" which are moved upon one another by muscles, and suspended from the hyoidean arch. By alteration of the relative position of these cartilages the cords can be tightened or relaxed, approximated or divaricated, as required to modulate the tone and volume of the voice. A median tongue-shaped fibro-cartilage at the top of the larynx, the "epiglottis," protects the "glottis," or aperture by which the larynx communicates with the pharynx, from the entry of particles of food during deglutition. The form of the larynx and development of the vocal cords present many variations in different members of the class, the greatest modification from the ordinary type being met with in the Cetacea, where the arytenoid cartilages and epiglottis are united in a tubular manner, so as to project into the nasal passage, and, being grasped by the muscular posterior margin of the palate, provide a direct channel of communication from the lungs to the external surface. An approach to this condition is met with in the Hippopotamus and some other Ungulates; it is indeed so general as an abnormality, that Howes suggests that an internarial epiglottis may have been a primitive feature common throughout the class. Nearly all mammals have a voice, although sometimes it is only exercised at seasons of sexual excitement. Some Marsupials and Edentates appear to be quite mute. In no mammal is there an inferior larynx, or "syrinx," as in birds.

Diaphragm.—The thoracic cavity of mammals differs from that of the Sauropsida in being completely separated from the abdomen by a muscular partition, the "diaphragm," attached to the vertebral column, the ribs, and the sternum. This is much arched, with the convexity towards the thorax, so that when its fibres contract and it is flattened the cavity of the thorax is increased, and when they are relaxed the cavity is diminished.

Lungs.—The lungs are suspended freely in the thorax, one on each side of the heart, being attached only by the root, which consists of the bronchus or air-tube and pulmonary arteries and veins by which the blood is passed backwards and forwards between the heart and the lungs. The remaining part of the surface of each lung is covered by serous membrane, the "pleura"; and whatever the state of distension or contraction of the chest-wall, is accurately in contact with it. Inspiration is effected by the contraction of the diaphragm and by the intercostal and other muscles elevating or bringing forward the ribs, and thus throwing the sternum farther away from the vertebral column. As the surface of the lung must follow the chest-wall, the organ itself is expanded, and air rushes in through the trachea to fill all the minute cells in which the ultimate ramifications of the bronchi terminate. In ordinary expiration very little muscular power is expended, the elasticity of the lungs and surrounding parts being sufficient to cause a state of contraction and thus drive out at least a portion of the air contained in the cells, when the muscular stimulus is withdrawn. The lungs are sometimes simple externally, as in the Sirenia (where they are greatly elongated) and the Cetacea, but are more often divided by deep fissures into one or more lobes. The right lung is usually larger and more subdivided than the left. It often has a small distinct lobe behind, wanting on the left side, and hence called *lobulus azygos.*

Air-sacs.—Most mammals have in connection with the air passages certain diverticuli or pouches containing air, the use of which is not always easy to divine. The numerous air sinuses situated between the outer and inner tables of the bones of the head, represented in Man by the antrum of Highmore and the frontal and sphenoidal sinuses, and attaining their maximum of development in the Indian Elephant, are obviously for the mechanical purpose of allowing expansion of the osseous surface without increase of weight. They are connected with the nasal passages. The Eustachian tubes pass from the back of the pharynx to the cavity of the tympanum, into which and the mastoid cells they allow air to pass. In the *Equidæ* there are large post-pharyngeal air-sacs in connection with them. The Dolphins have an exceedingly complicated system of air-sacs in connection with the nasal passages just within the nostrils, and the Tapirs, Rhinoceroses, and Horses have blind sacs in the same situation. In the males of some Seals (*Cystophora* and *Macrorhinus*) large pouches, which the animal can inflate with air, and which are not developed in the young animal or the female, arise from the upper part of the nasal passages, and lie immediately under the skin of the face. These appear analogous, although not in the same situation, to the gular pouch of the male Bustard. The larynx frequently has membranous pouches in connection with it, into which air passes. These may be lateral and opening just above the vocal cords, when they constitute the *sacculi laryngis*, found in a rudimentary state in Man, and attaining an enormous development, so as to reach to the shoulders and axillæ, in some of the Anthropoid Apes; or they may be median, opening in front either above or below the thyroid and cricoid cartilages, as in the Howling and other Monkeys, and also in the Whalebone Whales and Great Anteater.

Urinary Organs.—The kidneys of mammals are more compact and definite in form than in other vertebrates, being usually more or less oval, with an indent on the side turned towards the middle line, from and into which the vessels and ducts pass. They are distinctly divided into a cortical secretory portion, composed mainly of convoluted tubes, and containing the so-called Malpighian bodies; and a medullary excreting portion, formed of straight tubes converging towards a papilla, embraced by the commencement of the ureter or duct of the organ. The kidneys of some mammals, as most Monkeys, Carnivores, Rodents, etc., are simple, with a single papilla into which all the renal tubuli enter. In others, as Man, there are many pyramids of the medullary portion, each with its papilla, opening into a division (calyx) of the upper end of the ureter. Such kidneys, either in the embryonic condition only, or throughout life, are lobulated on the surface. In some cases, as in Bears, Seals, and especially the Cetacea, the lobulation is carried further, the whole organ being composed of a mass of renules, loosely united by connective tissue, and with separate ducts, which soon join to form the common ureter.

Bladder.—In all mammals except the Monotremes the ureters terminate by slit-like valvular openings in the urinary bladder. This receptacle when filled discharges its contents through the single median urethra, which in the male is almost invariably included in the penis, and in the females of some species of Rodents, Insectivores, and Lemurs has a similar relation to the clitoris. In the Monotremes, though the bladder is present, the ureters do not enter into it, but join the urino-genital canal some distance below it, with the orifice of the genital duct intervening.

VI. NERVOUS SYSTEM AND ORGANS OF SENSE.

Brain.—The brain of mammals shows a higher condition of organisation than that of other vertebrates. The cerebral hemispheres have a greater preponderance compared with other parts, especially to the so-called optic lobes, or corpora quadrigemina, which are completely concealed by them. The commissural system of the hemispheres is much more complex, both fornix and corpus callosum being present in some form; and when the latter is rudimentary, as in Marsupials and Monotremes, its deficiency is made up for by the great size of the anterior commissure. The lateral lobes of the cerebellum, wanting in lower vertebrates, are well developed and connected by a transverse commissure, the pons Varolii. The whole brain, owing especially to the size of the cerebral hemispheres, is considerably larger relatively to the bulk of the animal than in other classes, but it must be recollected that the size of its brain depends upon many circumstances besides the degree of intelligence which an animal possesses, although this is certainly one. Man's brain is many times larger than that of all other known mammals of equal bulk, and even three times as large as that of the most nearly allied Ape. Equal bulk of body is here mentioned, because, in drawing any conclusions from the size of the brain compared with that of the entire animal, it is always necessary to take into consideration the fact that in every natural group of closely allied animals the larger species have much smaller brains relatively to their general size than the smaller species, so that, in making any effective comparison among animals belonging to different groups, species of the same size must be selected. It may be true that the brain of a Mouse is, as compared with the size of its body, larger than that of a Man, but, if it were possible to reduce an animal having the general organisation of a Man to the size of a Mouse, its brain would doubtless be very many times larger; and conversely, as shown by the rapid diminution of the relative size of the brain in all the large members of the Rodent order, a Mouse magnified to the size of a Man would, if the general rule were observed, have a brain exceedingly inferior in volume. Although the brain of the large species of Whales is, as commonly stated, the smallest in proportion to the bulk of the animal of any mammal, this does not invalidate the general proposition that the Cetacea have very large brains compared with terrestrial mammals, like the Ungulata, or even the aquatic Sirenia, as may be proved by placing the brain of a Dolphin by the side of that of a Sheep, a Pig, or a Manatee of equal general weight. It is only because the universally observed difference between the slower ratio of increase of the brain compared with that of the body becomes so enormous in these immense creatures that they are accredited with small brains.

The presence or absence of "sulci" or fissures on the surface of the hemisphere, dividing it into "convolutions" or "gyri," and thus increasing the superficies of the cortical gray matter, as well as allowing the pia mater with its nutrient blood-vessels to penetrate into the cerebral substance, follow somewhat similár rules. The sulci are related partly to the high or low condition of organisation of the species, but also in a great degree to the size of the cerebral hemispheres. In very small species of all groups, even the Primates, they are absent, and in the largest species of groups so low in the scale as the Marsupials and Edentates they are found. They reach their maximum of development in the Cetacea.

The accompanying woodcut (Fig. 23) shows the principal parts of a mammalian brain, as seen from the superior, lateral, and inner surfaces. The sylvian fissure (*sf*) is one of the most constant of the sulci found in the hemispheres.

The researches of Palæontologists, founded upon studies of casts of the interior of the cranial cavity of extinct forms, have shown that, in many natural groups of mammals, if not in all, the brain has increased in size, and also in complexity of surface foldings, with the advance of time,—indicating in this, as in so many other respects, a gradual progress from a lower to a higher type of development.

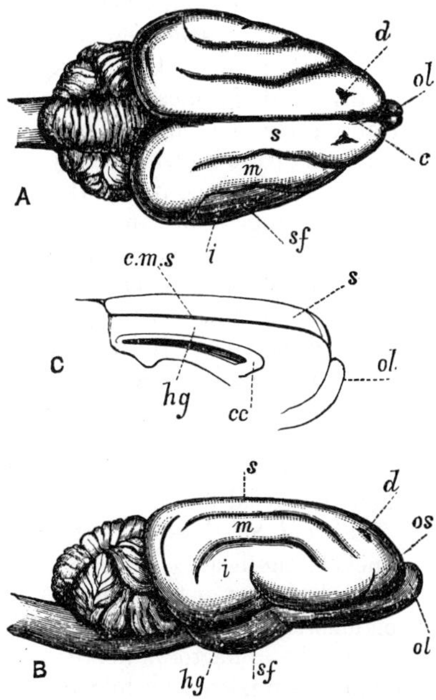

FIG. 23.—Brain of the Genet (*Genetta tigrina*). A, From above; B, from the right side; C, inner surface of right hemisphere; *cc*, corpus callosum; *c.m.s*, calloso-marginal sulcus; *c*, notch representing crucial sulcus of other forms; *d*, depression on superior lateral gyrus of hemisphere; *hg*, hippocampal gyrus; *i*, inferior lateral gyrus of hemisphere; *m*, middle lateral gyrus of do.; *s*, superior lateral gyrus of do.; *os*, supraorbital sulcus of do.; *sf*, sylvian fissure of do.; *ol*, olfactory lobes. The deeply convoluted part behind the cerebral hemisphere is the cerebellum, below which lies the medulla oblongata, or commencement of the spinal cord. (Mivart, *Proc. Zool. Soc.* 1882, p. 516.)

Nerves.—The twelve pairs of cranial nerves generally recognised in vertebrates are usually all found in mammals, though the olfactory nerves are excessively rudimentary, if not altogether absent, in the Toothed Whales. The spinal cord, or continuation of the central nervous axis, lies in the canal formed by the neural arches of the vertebræ, and gives off the compound double-rooted nerves of the trunk and the extremities, corresponding in number to the vertebræ, through the interspaces between which they pass out to their destination. The cord is somewhat enlarged at the two points where it gives off the great nerves to the anterior and the posterior extremities, which, from their interlacements soon after their origin, are called respectively the brachial and lumbar plexuses. The ganglionic or sympathetic portion of the nervous system is well developed, and presents few modifications.

Sense of Touch.—The sense of touch is situated in the skin generally, but is most acute in certain regions more or less specialised for the purpose by the presence of tactile papillæ, such as portions of the face, especially the lips and end of the snout, and the extremities of the limbs when these are used for other purposes than mere progression, and the under surface of the end of the tail in some Monkeys. The "vibrissæ" or long stiff bristles situated on the face of many mammals are rendered extremely sensitive to touch by the abundant supply of branches from the fifth nerve to their basal papillæ. In Bats the extended wing membranes, and probably also the large ears and the folds and prominences of skin about the face of some species, are so sensitive as to receive impressions even from the different degrees of resistance of the air, and so enable the animals to avoid coming in contact with obstacles to their nocturnal flight.

Taste and Smell.—The organs of the other special senses are confined to the head. Taste is situated in the papillæ scattered on the dorsal surface of the tongue. The organ of smell is present in all mammals except the Toothed Whales. It consists of a ramification of the olfactory nerves over a plicated, moist, mucous membrane, supported by folded plates of bone, placed on each side of the septum nasi in the roof, or often in a partially distinct upper chamber, of the nasal passage, so arranged that, of the air passing into the lungs in inspiration, some comes in contact with it, causing the perception of any odorous particles with which it may be charged. Many mammals possess intense powers of smelling certain odours which others are quite unable to appreciate, and the influence which this sense exercises over the well-being of many species is very great, especially in indicating the proximity of others of the same kind, and giving warning of the approach of enemies. The development and modification of the sense of smell is probably associated with that of the odorous secretion of the cutaneous glands.

Sight.—The organ of sight is quite rudimentary, and even concealed beneath the integument, in some burrowing Rodents and Insectivores, and is most imperfectly developed in the *Platanista*, or Freshwater Dolphin of the rivers of India. In all other mammals the eyeball has the structure characteristic of the organ in the higher Vertebrata, consisting of parts through which the rays of light are admitted, regulated, and concentrated upon the sensitive expansion of the optic nerve lining the posterior part of the ball. A portion of the fibro-vascular and highly pigmented layer, the choroid, which is interposed between the retina and the outer sclerotic coat, is in many mammals modified into a brilliantly-coloured light-reflecting surface, the *tapetum lucidum.* There is never a pecten or marsupium like that of the Sauropsida, nor is the sclerotic ever supported by a ring of flattened ossicles, as is so frequently the case in the lower vertebrated classes. The eyeball is moved in various directions by a series of muscles—the four straight, two oblique, and, except in the higher Primates, a posterior retractor muscle called choanoid. The superior oblique muscle passes through a tendinous pulley fastened to the roof of the orbit, which is a feature not found beyond the limits of the mammalian class. The eye is protected by the lids, generally distinctly separated into an upper and a lower movable flap, which, when closed, meet over the front of the eye in a more or less nearly horizontal line; but sometimes, as in the Sirenia, the lids are not distinct, and the aperture is circular, closing to a point. In almost all mammals below the Primates, except the Cetacea, a "nictitating membrane" or third eyelid is placed at the inner corner of the eyeball, and works horizontally across the front of the ball within the true lids. Its action is instantaneous, being apparently for the purpose of cleaning the front of the transparent cornea;—a function unnecessary in animals whose eyes are habitually bathed in water, and which in Man and his nearest allies is performed by winking the true eyelids. Except in Cetacea the surface of the eye is kept moist by the secretion of the lachrymal gland, placed under the upper lid at its outer side, and the lids are lubricated by the Harderian and Meibomian glands, the former being situated at the inner side of the orbit, and especially related to the nictitating membrane, the latter in the lining membrane of the lids.

Hearing.—The organ of hearing is inclosed in a bony capsule (periotic) situated in the side of the head, intercalated between the posterior (occipital) and the penultimate (parietal) segment of the skull. It has, in common with other vertebrates, three semicircular canals and a vestibule, but the cochlea is more fully developed than in the Sauropsida, and, except in the Monotremes, spirally convoluted. The tympanic cavity is often dilated below, forming a smooth rounded prominence on the base of the skull, the auditory bulla (Fig. 8). The three principal ossicles, the "malleus," "incus," and "stapes," are always present, but variable in characters. In the Sirenia, Cetacea, and Seals they are massive in form, being in the first-named order of larger size than in any other mammals. In the Cetacea the malleus is ankylosed to the tympanic; but in other mammals it is connected only with the membrana tympani. The stapes in the lower orders—Edentates, Marsupials, and Monotremes

—has a great tendency to assume the columnar form of the corresponding bone in Sauropsida, its two rami entirely or partially coalescing.[1] The tympanic membrane (drum of the ear) forms the outer wall of the cavity. In the fœtal state it is level with the external surface of the skull, and remains so permanently in a few mammals, as the American Monkeys; but commonly, by the growth of the squamosal bone, it becomes deeply buried at the bottom of a bony tube (*meatus auditorus externus*), which is continued to the surface of the skin in a fibrous or fibro-cartilaginous form. In Whales, owing to the thickness of the subcutaneous adipose tissue, this meatus is of great length, and is also extremely narrow. In most aquatic and burrowing animals it opens upon the surface by a simple aperture, but in the large majority of the class there is a projecting fold of skin, strengthened by fibro-cartilages, called the pinna, auricle, or "external ear," of very variable size and shape, generally movably articulated on the skull, and provided with muscles to vary its position; this pinna helping to collect and direct the vibrations of sound into the meatus.

VII. REPRODUCTIVE ORGANS.

Testes.—In the male the testes retain nearly their primitive or internal position throughout life in the Monotremata, Sirenia, Cetacea, most Edentata, Hyracoidea, Proboscidea, and Seals, but in other groups they either periodically (as in Rodentia, Insectivora, and Chiroptera) or permanently pass out of the abdominal cavity through the inguinal canal, forming a projection beneath the skin of the perineum, or becoming suspended in a distinct pouch of integument called the scrotum. All the Marsupials have a pedunculated scrotum, the position of which differs from that of other mammals, being in front of, instead of behind, the preputial orifice. As regards the presence, absence, or comparative size and number of the accessory generative glands—prostate, vesicular, and Cowper's glands, as they are called—there is much variation in different groups of mammals.

Penis.—The penis is almost always completely developed, consisting of two corpora cavernosa attached to the ischial bones, and of a median corpus spongiosum enclosing the urethra, and forming the glans at the distal portion of the organ. In Marsupials, Monotremes, and the Sloths and Anteaters, the corpora cavernosa are not attached directly to the ischia, and in the last-named the penis is otherwise of a very rudimentary character, the corpus

[1] The modifications of these bones are fully described by A. Doran, "Morphology of the Mammalian *Ossicula auditus*," *Trans. Linn. Soc.* ser. 2, vol. i. pp. 371-497, pl. lviii.-lxiv. (1878).

spongiosum not being present. In many Marsupials the glans penis is bifurcated. In most Primates, Carnivora, Rodentia, Insectivora, and Chiroptera, but in no other orders, an *os penis* is present.

Ovaries and Oviduct.—In the female, the ovaries permanently retain their original abdominal position, or only descend a short distance into the pelvis. They are of comparatively smaller size than in other vertebrates, have a definite flattened oval form, and are enclosed in a more or less firm "tunica albigenia." The oviduct has a trumpet-like, and usually fimbriated abdominal aperture, and is more or less differentiated into three portions:—(1) a contracted upper part, called in Man and the higher mammals the "Fallopian tube"; (2) an expanded part with muscular walls, in which the ovum undergoes the changes by which it is developed into the fœtus, called the "uterus"; (3) a canal, the "vagina," separated from the last by a valvular aperture, and terminating in the urino-genital canal, or common urinal and genital passage, which in higher mammals is so short as scarcely to be distinct from the vagina. The complete distinction of the oviducts of the two sides throughout their whole length, found in all lower vertebrates, only occurs in this class in Monotremes; a prevailing mammalian characteristic being their more or less perfect coalescence in the middle line to form a single median canal. In the Marsupials this union only includes the lower part of the vagina; but in most Placentals it extends to the whole vagina and a certain portion of the uterus, which cavity is then described as "bicornuate." In the higher mammals, as in Man, and also in some of the Edentates, the whole of the uterus is single, the contracted upper portion of the oviducts or Fallopian tubes, as they are then called, entering its upper lateral angles by small apertures. In certain lower forms the urino-genital canal opens with the termination of the rectum into a common cloaca, as in other vertebrates; but it is characteristic of the majority of the class that the two orifices are more or less distinct externally.

Mammary Glands.—Mammary glands secreting the milk by which the young are nourished during the first portion of their existence after birth, are present in both sexes in all mammals, though usually only functional in the female. In the Monotremes alone their orifices are mere scattered pores in the skin, but in all other forms they are situated upon the end of conical elevations, called mammillæ or teats, which, taken into the mouth of the young animal, facilitate the process of sucking. These are always placed in pairs upon some part of the ventral surface of the body, but vary greatly in number and position in different groups. In the Cetacea, where the prolonged action of sucking would be incompatible with their subaqueous life, the ducts of the glands are dilated into large reservoirs from which the contents are injected into the mouth of the young animal by the action of a compressor muscle.

Secondary Sexual Characters.—Secondary sexual characters, or modifications of structure peculiar to one sex, but not directly related to the reproductive function, are very general in mammals. They almost always consist of the acquisition or perfection of some character by the male as it attains maturity, which is not found in the female or the young in either sex. In a large number of cases these clearly relate to the combats in which the males of many species engage for the possession of the females during the breeding season; others are apparently ornamental, and of many it is still difficult to apprehend the meaning. Many suggestions on this subject will, however, be found in the chapters devoted to it in Darwin's work on *The Descent of Man and Selection in Relation to Sex*, where most of the best-known instances are collected. Superiority of size and strength in the male of many species is a well-marked secondary sexual character related to the purpose indicated above, being probably perpetuated by the survivors or victors in combats transmitting to their descendants those qualities which gave them advantages over others of their kind. To the same category belong the great development of the canine teeth of the males of many species which do not use these organs in procuring their food, as the Apes, Swine, Musk and some other Deer, the tusk of the male Narwhal, the antlers of Deer, which are present in most cases only in the males, and the usual superiority in size and strength of the horns of the *Bovidæ*. Other secondary sexual characters, the use of which is not so obvious, or which may only relate to ornament, are the presence of masses or tufts of long hair on different parts of the body, as the mane of the male Lion and Bison, the beards of some Ruminants and Bats (as *Taphozous melanopogon*), Monkeys, and of Man, and all the variations of coloration in the sexes, in which, as a general rule. the adult male is darker and more vividly coloured than the female. Here may also be mentioned the presence or the greater development of odoriferous glands in the male, as in the Musk Deer, and the remarkable perforated spur with its glands and duct, so like the poison-tooth of the venomous serpents, found in the males of both *Ornithorhynchus* and *Echidna*, the use of which is at present unknown.

Placenta.—The development of the mammalian ovum, and the changes which the various tissues and organs of the body undergo in the process of growth, are too intricate subjects to be explained without entering into details incompatible with the limits of this work, especially as they scarcely differ, excepting in their later stages, from those of other vertebrates, upon which, owing to the greater facilities these present for examination and study, the subject has been more fully worked out. There are, however, some points which require notice, as peculiar to the mammalian class, and as affording at least some hints upon the difficult subject of the affinities and classification of the members of the group.

The nourishment of the fœtus during intra-uterine life takes place through the medium of certain structures, partly belonging to the fœtus itself and partly belonging to the inner parietes of the uterus of the parent. These in their complete form constitute the complex organ called the "placenta," serving as the medium of communication between the mother and fœtus, and in which the physiological processes that are concerned in the nutrition of the latter take place; but, as we shall see, though a placenta, in the usual acceptation of the term, is peculiar to the mammalian class, it is not in all of its members that one is developed. The structures to which we shall have especially to refer are the outer tunic of the ovum, to which, however formed, the term "chorion" is commonly applied, and two sac-like organs connected with the body-cavity of the embryo, both formed from the splanchnic mesoblast, lined by a layer of the hypoblast. These are the "umbilical vesicle" or "yolk-sac" and the "allantois."

The umbilical vesicle is a thin membrane enclosing the yolk, which by the doubling in of the ventral walls of the embryo becomes gradually formed into a distinct sac external to the body, with a pedicle (the omphalo-enteric duct) by which for a time a communication is maintained between its cavity and the intestinal canal. In the walls of this sac blood-vessels (omphalo-meseraic or vitelline) are developed in connection with the vascular system of the embryo, through which, either by their contact with the outer surface of the walls of the ovum, or by the absorption through them of the contents of the yolk-sac, the nutrition of the embryo in the lower vertebrates chiefly takes place. In mammals the umbilical vesicle plays a comparatively subordinate part in the nourishment of the fœtus, its function being generally superseded by the allantois.

The last-named sac commences at a very early period as a diverticulum from the hinder end of the alimentary tract of the embryo. Its proximal portion afterwards becomes the urinary bladder, the contracted part between this and the cavity of the allantois proper constituting the urachus, which passes out of the body of the fœtus at the umbilicus together with the vitelline duct. The mesoblastic tissue of the walls of the allantois soon becomes vascular; its arteries are supplied with fœtal blood by the two hypogastric branches of the iliacs, or main divisions of the abdominal aorta, and the blood is returned by venous trunks uniting to form the single umbilical vein which runs to the under surface of the liver, where, part of it joining the portal vein and part entering the vena cava directly, it is brought to the heart. These are the vessels which, with their surrounding membranes, constitute the umbilical cord—the medium of communication between the fœtus and the placenta, when that organ is fully developed.

The egg membranes of the Monotremes present many points of agreement with those of the ovum of the Marsupials,[1] and differ from those of the Placental types. Thus Monotremes and Marsupials agree in having a vitelline membrane, which appears between the young ovum and the follicular epithelium, persisting in the one case until the time of hatching, and in the other till a late uterine stage. There are also several other common features fully described in Mr. Caldwell's memoir, but which cannot be detailed in this work.

In the Marsupialia the observations made many years ago by Sir R. Owen upon the development of the Kangaroo have been confirmed by those of Dr. H. C. Chapman,[2] while Dr. Selenka,[3] and Professor H. F. Osborn[4] have contributed important evidence as to the structure and relations of the fœtal membranes of the Opossums and others. It thus appears that up to the period of the very premature birth of these animals the outer covering of the ovum, or false chorion, is free from persistent villi, and not adherent to the epithelium of the uterine walls; for, although fitting into the folds of the latter, it is perfectly and readily separable in its entire extent from them. The umbilical vesicle or yolk-sac is large, vascular, and adherent to a considerable portion of the false chorion or subzonal membrane, while the allantois is relatively small, and although the usual blood-vessels can be traced into it, it does not appear to contract any connection with the false chorion, and, therefore, much less with the walls of the uterus, of such a nature as to constitute a placenta. In some forms, however, such as the Opossums, the umbilical vesicle or yolk-sac develops temporary villi, which unite with the subzonal membrane, or false chorion, to form a disc-like area closely attached to the cells covering the utricular glands of the uterine epithelium, and thus forming a so-called *yolk-sac placenta*. The function of this organ is considered to be the transmission of the secretions of the utricular glands to the embryo by means of the umbilical vesicle; the function of the allantois being either respiratory or the absorption of the fluid secreted in the uterine cavity by the utricular glands.

While in the uterus the nourishment of the fœtus seems, therefore, to be derived from the umbilical vesicle, as in reptiles and

[1] See B. H. Caldwell—"The Embryology of Monotremata and Marsupialia," *Phil. Trans.* for 1887, p. 463.

[2] *Proc. Acad. Nat. Sci. Philadelphia*, 1881, p. 468.

[3] "*Studien ueber Entwickelungeschichte der Thiere*," pt. 4, Wiesbaden, 1886.

[4] *Journal of Morphology*, vol. i. p. 373 (1887).

birds, rather than from the uterine walls by means of the allantoic vessels, as in the higher mammals. The latter vessels, in fact, play even a much less important part in the development of these animals, not only than in the placental mammals, but even than in the Sauropsida, for they can scarcely have the respiratory function assigned to them in that group: pulmonary respiration and the lacteal secretion of the mother very early superseding all other methods of providing the due supply both of oxygen and of food required for the development and growth of the young animal. In this sense the Marsupials may be looked upon as the most typically "mammalian" of the whole class. In no other group do the milk-secreting glands play such an important part in providing for the continuity of the race.

In the third primary division of the Mammalia, the so-called Placentalia, the umbilical vesicle generally does not quite unite with the chorion, and disappears as development proceeds, so that no trace of it can be seen in the membranes of an advanced embryo; but it may persist until the end of the intra-uterine life as a distinct sac in the umbilical cord, or lying between the allantois and amnion. The disappearance or persistence of the umbilical vesicle does not, according to our present knowledge, appear to be correlated with a higher or lower general grade of development, as might be presupposed. It is stated to have been found in Man even up to the end of intra-uterine life, and also in the Carnivora, while in the Ungulata and Cetacea it disappears at an earlier age. In many, if not all, of the Rodentia, Insectivora, and Chiroptera, it plays a more important part, becoming adherent to a considerable part of the inner surface of the chorion, to which it conveys blood-vessels, although villi do not appear to be developed from the surface of this part, as they are on the portion of the chorion supplied by the allantoic vessels. These orders thus present to a certain extent a transitional condition from the Marsupials, although essentially different, in possessing the structures next to be described.

The special characteristic of the whole of the placental mammals constituting the majority of the class, is that the allantois and its vessels become intimately blended with a smaller or greater part of the parietes of the ovum, forming a structure on the outer surface of which villi are developed, and which, penetrating into corresponding cavities of the "decidua," or soft, vascular, hypertrophied lining membrane of the uterus, constitutes the placenta. This organ may be regarded, as Sir William Turner says, both in its function and in the relative arrangement of its constituent textures, as a specially modified secreting gland, the ducts of which are represented by the extremities of the blood-vessels of the fœtal system. The passage of material from the maternal to the fœtal system of vessels is not a simple percolation or diffusion through their walls, but is occasioned by the action of a layer of cells derived from the maternal or uterine structures, and interposed between the blood-vessels of the maternal part of the placenta and those of the villi covering the chorion, in which the embryonic vessels ramify.

The numerous modifications in the details of the structure of this organ relate to augmenting the absorbing capacity of the vessels of the chorion, and are brought about either by increasing the complexity of the fœtal villi and maternal crypts over a limited area, or by increasing the area of the part of the chorion covered by the placental villi, or by various combinations of the two methods.

The first class of variations has given rise to a distinction into two principal kinds of placenta: (1) simple or non-deciduate, and (2) deciduate. In the former the fœtal villi are received into corresponding depressions of the maternal surface, from which at the period of parturition they are simply withdrawn. In the second, or more complex form, the relation is more intimate, a layer of greater or less thickness of the lining membrane of the uterus, called "decidua," becoming so intimately blended with the chorion as to form part of the placenta proper, or that structure which is cast off as a solid body at parturition. In other words, in the one case the line of separation between the placenta and uterus at birth takes place at the junction of the fœtal and maternal structures, in the other through the latter, so that a portion of them, often of considerable thickness, and containing highly organised structures, is cast off with the former. It was once thought that the distinction between these two forms of placentation is so important as to constitute a sufficiently valid basis for a primary division of the placental mammals into two groups. It has, however, been shown that the distinction is one rather of degree than of kind, as intermediate conditions may exist, and it is probable that in different primary groups the simpler, non-deciduate form may have become developed independently into one or other of the more complex kinds.

Apart from its intimate structure, the placenta may be met with of very varied general form. It may consist of villi scattered more or less regularly over the greater part of the surface of the chorion, the two extremities or poles being usually more or less bare. This form is called the "diffused placenta." It is probably a primitive condition, from which most of the others are derived, although its existence must presuppose the absence of the umbilical vesicle as a constituent of the chorionic wall. It is found at present in the Manis among Edentates, the Cetacea, the Perissodactyle Ungulates, and the Camels, Pigs, and Chevrotains among the Artiodactyles. Such placentæ are always non-deciduate. Recent observations by Sir W. Turner on the placentation of the Dugong show that the

Sirenia present the peculiarity of having a zonary placenta, which is either entirely or in great part non-deciduate, and is, therefore, transitional between the diffused and the true zonary type.

In the true Ruminants or Pecora, among the Artiodactyle Ungulates, the villi are aggregated in masses called cotyledons, with bare spaces between. Such a placentation is called "polycotyledonary." In another modification the villi are collected in a more or less broad band encircling the chorion, leaving a very large portion of the two poles bare, constituting the "zonary placenta," characteristic of the Carnivora, and also occurring in the Elephant, Hyrax, and Orycteropus. The fact of the form of the placenta of these three last-named animals agreeing together, and with that of the Carnivora, does not, however, necessitate the ascription of zoological affinities, as the same ultimate form may have been attained by different processes of development.

In another form one pole only of the chorion is non-vascular, the placenta assuming a dome or bell shape, as in the Lemurs and the Sloths. The transition from this, by the gradual restriction of the vascular area, is easy to the oval or discoidal form of placenta of the Anteaters, Armadillos, and higher Primates. The discoidal placenta of the Rodents, Insectivores, and Chiroptera, though showing so much superficial resemblance to that of the last-named order as to have led to the inclusion of all these forms in one primary group, is now known to be developed in another manner, not by the concentration of villi from a diffused to a limited area, but by retaining the area to which it was originally restricted in consequence of the large surface of the chorion occupied, as before mentioned, by the umbilical vesicle. To compensate for the smallness of area, the complex or deciduate structure has been developed. Among some Rodents there is evidence to show that the discoidal placenta has been derived from a zonary one, of which distinct vestiges have been detected in the Mouse. We may conclude that, although the characters and arrangement of the fœtal structures may not have that extreme importance which has been attributed to them by some zoologists, they will form, especially when more completely understood, valuable aids in the study of the natural affinities and evolution of the Mammalia.[1]

[1] For a full exposition of the present state of knowledge on this subject, see the various memoirs of Sir William Turner, also F. M. Balfour's *Treatise on Comparative Embryology*, vol. ii. (1881), and J. A. Ryder in *American Naturalist*, vol. xxi. p. 780 (1887).

CHAPTER III

ORIGIN AND CLASSIFICATION OF THE MAMMALIA

Origin.—Although, as stated in the first chapter, the mammalian class, as at present known either by existing or extinct forms, is completely isolated from all other groups of the animal kingdom, yet it is impossible to refrain from speculating as to its origin and nearest affinities. In arranging the classes of vertebrates in a linear series it is customary to place them in the following order—Pisces, Amphibia, Reptilia, Aves, Mammalia,—an order which probably indicates the relative degree of elevation to which the mos highly developed members of each class has attained. Such an arrangement appears to express the true relationship of the first four classes to one another, but it is quite clear that the Mammalia have no sort of affinity with the Aves. Writing in 1879, Professor Huxley [1] came to the conclusion that, in looking among vertebrates for the progenitors of the Mammalia, we must pass over all known forms of birds and reptiles, and go straight down to the Amphibia. In addition to the characters derived from the conformation of the pelvis upon which the argument was primarily based, the following reasons were given for this conclusion: "The Amphibia are the only air-breathing Vertebrata which, like mammals, have a dicondylian skull. It is only in them that the articular element of the mandibular arch remains cartilaginous, while the quadrate ossification is small, and the squamosal extends down over it to the osseous elements of the mandible, thus affording an easy transition to the mammalian condition of those parts. The pectoral arch [girdle] of the Monotremes is as much amphibian as it is sauropsidian; the carpus and the tarsus of all Sauropsida, except the Chelonia, are modified away from the Urodele type, while those of the mammal are directly reducible to it. Finally, the fact that in all Sauropsida it is a right aortic arch which is the main conduit of arterial blood leaving the heart, while in mammals it is a left aortic arch which

[1] *Proceedings of the Royal Society of London*, vol. xxviii. p. 395 (1879).

performs this office, is a great stumbling-block in the way of the derivation of the Mammalia from any of the Sauropsida. But, if we suppose the earliest forms of both the Mammalia and the Sauropsida to have had a common Amphibian origin, there is no difficulty in the supposition that, from the first, it was a left aortic arch in the one series, and the corresponding right aortic arch in the other, which became the predominant feeder of the arterial system." Subsequently Professor E. D. Cope[1] in a suggestive paper called attention to the remarkable resemblances to the Monotremes presented by the skeleton of that group of early secondary reptiles which he then designated the Theromorpha, but which may be included in the Anomodontia of Sir R. Owen, and came to the conclusion that in that group we have the true ancestors of the Mammalia. This conclusion was, however, disputed by Dr. Baur,[2] who considered that the Anomodontia were too specialised to have been the actual progenitors of the Mammalia, and that they should rather be regarded as a divergent branch of the stem which had given origin to the Mammalia. Since that date observations made on the structure of the South African Anomodonts have shown such an intimate connection between that group and the Labyrinthodont Amphibians, that there can be no hesitation in regarding the one as the direct descendant of the other; and we may probably regard the Mammalia as having originated from the same ancestral stock at the time the Amphibian type was passing into the Reptilian. From this point of view, some of the mammalian features found in the more specialised Anomodonts may probably be regarded as having been acquired during a parallel line of development.

Both the Anomodontia and the Mammalia differ from the Amphibians in the loss of the splint-like parasphenoid which underlies the basisphenoid axis of the skull, and by the ossification of that axis; but while the former have become monocondylic by the participation of the basioccipital in the support of the cranium, the latter retain the Amphibian dicondylic plan. The skull of the Anomodonts presents mammalian resemblances not found in any other Reptiles, this being especially noticeable in the region of the squamosal; and it is only in this group and mammals that the temporal or zygomatic arch is a squamoso-maxillary one (see p. 37). The resemblance between the pectoral and pelvic girdles of the Anomodonts and those of the Monotreme Mammals is noticed under the head of the latter, where reference is also made to the similarity in the structure of the humerus in the two groups.

1 "The Relations between the Theromorphous Reptiles and the Monotreme Mammalia," *Proceedings of the American Association for the Advancement of Science*, vol. xxxiii. p. 471 (1885).

2 "On the Phylogenetic Arrangement of the Sauropsida," *Journal of Morphology*, vol. i. pp. 93-104 (1887).

The pes of the Amphibia and Anomodontia agree in having a distinct intermedium, tibiale, fibulare, and centrale, whereas in other Reptiles these bones are not generally distinct; in Mammals the intermedium, fibulare, and centrale are distinct, and according to Cope's interpretation there may be a distinct tibiale.

Classification.—In the present condition of the world, mammals have become so broken up into distinct groups by the extinction of intermediate forms, that a systematic classification is perfectly practicable. Most of the associations of species, which we call "orders," and even the "suborders" and "families," are natural groups. In isolating, defining, and naming them, we are really dealing with facts of nature of a totally different order from the artificial and fanciful divisions formed in the infancy of zoological science.

When, however, we pass to the extinct world, all is changed. In many cases the boundaries of our groups become enlarged until they touch those of others. New forms are discovered which cannot be placed within any of the existing divisions. As the horizon of our vision is thus expanded, the principles upon which a scheme of classification is constructed must be altogether changed. Our present divisions and terminology are no longer sufficient for the purpose; and some other method will have to be invented to show the complex relationships existing between different animal forms when viewed as a whole. The present time, pre-eminently distinguished by the rapidly changing and advancing knowledge of extinct forms, is scarcely one in which this can be done with any satisfactory result; so that all attempts to form a classification embracing even the already known extinct species must be only of a provisional and temporary nature.

In systematic descriptions in books, in lists, and catalogues, and in arranging collections, the objects dealt with must be placed in a single linear series. But by no means whatever can such a series be made to coincide with natural affinities. The artificial character of such an arrangement, the constant violation of all true relationships, are the more painfully evident the greater the knowledge of the real structure and affinities. But the necessity is obvious; and all that can be done is to make such an arrangement as little as possible discordant with facts.

The following table contains a list of the orders, suborders, and families of existing mammals as recognised by the authors, and placed in the order in which they will be treated of in this work. The more important of the groups containing only extinct forms are added in a different type, being interpolated, as near as may be, among those that appear to be their existing relatives.

A few explanatory remarks upon the mutual relations of some of the principal groups mentioned in the table may be useful here, but the subject will be more fully developed in treating separately of each division.

One of the most certain and fundamental points in the classification of the Mammalia is, that all the animals now composing the class can be grouped primarily into three natural divisions, which, presenting very marked differential characters, and having no existing, or yet certainly demonstrated extinct, intermediate, or transitional forms, may be considered as subclasses of equal value, taxonomically speaking, though very different in the numbers and importance of the animals at present composing them. These three groups are often called by the names originally proposed for them by Blainville—(1) *Ornithodelphia*, (2) *Didelphia*, (3) *Monodelphia*—the first being equivalent to the order *Monotremata*, the second to the *Marsupialia*, and the third including all the remaining members of the class. Although actual palæontological proof is wanting, there is much reason to believe that each of these, as now existing, are survivors of distinct branches to which the earliest forms of mammals have successively given rise, and for which hypothetical branches Professor Huxley has proposed the names of *Prototheria*, *Metatheria*, and *Eutheria*, names which, being far less open to objection than those of Blainville, are here used as equivalents of the latter.

The only known existing PROTOTHERIA, although agreeing in many important characters, evidently represent two very divergent stocks, perhaps as far removed as are the members of some of the accepted orders of the Eutheria. It would, however, be merely encumbering zoological science with new names to give them any other than the ordinarily known family designations of *Ornithorhynchidæ* and *Echidnidæ*.

Similarly with regard to the METATHERIA, although the great diversity in external form, in anatomical characters, and in mode of life of the various animals of this section might lead to their division into groups equivalent to the orders of the Eutheria, we do not think it advisable to depart from the usual custom of treating them all as forming one order, called Marsupialia, the limits of which are equivalent to those of the subclass. The characters of the six families which compose the group are extremely well marked and easily defined; and since they form a regular gradation between two extreme types, they can be satisfactorily arranged in a serial order. A marked distinction in the dentition enables us to divide them into primary groups or suborders.

The remaining mammals are included in the EUTHERIA, PLACENTALIA, or MONODELPHIA. Their affinities with one another are so complex that it is impossible to arrange them serially with any regard to natural affinities. Indeed each order is now so isolated that it is almost impossible to say what its affinities are; and none of the hitherto proposed associations of the orders into larger groups stand the test of critical investigation. All serial arrangements of the orders are therefore perfectly arbitrary; and although it would be of very great convenience for reference in books and museums if some general sequence, such as that here proposed, were generally adopted, such a result can scarcely be expected, since equally good reasons might be given for almost any other combination of the various elements of which the series is composed. In fact, we have already seen reason to depart in some respects from that used in the "Encyclopædia."

The Edentata, Sirenia, and Cetacea stand apart from all the rest in the fact that their dentition does not conform to the general heterodont, diphyodont type to which that of all other Eutheria can be reduced, and which is such a close bond of union between them. In all three orders, however, some indications may be traced of relationship, however distant, with the general type.

With regard to the Edentata, reasons will be given for believing that both the Sloths and Anteaters are nearly related, and that the Armadillos, though much modified, belong to the same stock, but that the Pangolins and the Aard-varks represent very isolated forms.

There is no difficulty about the limits of the order Sirenia, comprising aquatic, vegetable-eating animals, with complete absence of hind limbs, and low cerebral organisation, represented in our present state of knowledge only by two existing genera, *Halicore* and *Manatus*, and a few extinct forms, which, though approaching a more generalised mammalian type, show no special characters allying them to any of the other orders. The few facts as yet collected relating to the former history of the Sirenia leave us as much in the dark as to the origin and affinities of this peculiar group of animals as we were when we only knew the living members. They lend no countenance to their association with the Cetacea; and, on the other hand, their supposed affinity with the Ungulata receives no very material support from them.

Another equally well-marked and equally isolated, though far more numerously represented and diversified order, is that of the Cetacea, placed simply for convenience next to the Sirenia; with which, except in their fish-like adaptation to aquatic life, they have little in common. The old association of these orders in one group can only be maintained either in ignorance of their structure or in an avowedly artificial system. Among the existing members of the order, there are two very distinct types, the toothed Whales or Odontoceti, and the Baleen Whales or Mystacoceti, which present as many marked distinguishing structural characters as are found between many other divisions of the Mammalia usually reckoned as orders. Since the extinct Zeuglodonts, so far as their characters are known, do not fall into either of these groups, but are in some respects annectant forms, we have placed them provisionally, at least, in a third group by themselves, named Archæoceti. There is nothing known at present to connect the Cetacea with any other order of Mammals; but it is quite as likely that they are offsets of a primitive Ungulate as of a Carnivorous type, or perhaps of a still more generalised mammalian stock.

The remaining Eutherian mammals are clearly united by the characters of their teeth, being all heterodont and diphyodont, with their dental system reducible to a common formula.

Although older views of the relationship of Ungulate mammals expressed by the terms *Pachydermata*, *Ruminantia*, and so forth, still linger in some corners of zoological literature, no single point in zoological classification can be considered so firmly established as the distinction between the Perissodactyle and Artiodactyle Ungulates; both being in the existing fauna of the world perfectly natural and distinctly circumscribed groups. The breaking-up of the latter into four equivalent sections, the Pecora, Tylopoda, Tragulina, and Suina, is equally in accordance with all known facts. Less certain, however, is the association of the Proboscidea and the Hyracoidea with the true Ungulates. By many zoologists they are each, although containing so very few existing species, made into distinct orders; and much is to be said in favour of this view. The discovery, however, of a vast number of extinct species of Ungulates which cannot be brought under the definition of either Perissodactyla or Artiodactyla, and yet are evidently allied to both, and to a certain extent bridge over the interval between them and the isolated groups just mentioned, make it necessary either to introduce a number of new and ill-defined ordinal divisions, or so to widen the scope of the original order as to embrace them all, considering the Elephants and the Hyraces as representing sub-orders equivalent to the great Perissodactyle and Artiodactyle groups. It is the latter alternative that we have adopted.

The Rodentia, although generally presenting a low grade of development, are a very specialised and distinct group. The position here assigned to them would accord with apparent relationships with the Ungulates, through the Elephant on the one hand and the extinct *Typotherium* on the other.

In the present state of the fauna of the earth, the Carnivora form a very distinct order, though naturally subdivided into two groups, the members of the one being more typical, while those of the other (the *Pinnipedia*) are aberrant, having the whole of their organisation specially modified for living habitually in the water.

The Insectivora comprise various lowly organised and generalised forms, exhibiting considerable divergence of character, and apparently connected through transitional extinct species with the

Carnivora. As no other order can claim the family *Galeopithecidæ*, it is placed here, but rather for convenience than for any other consideration, since it has but little if any relationship with any of the other members. Its isolated position is indicated by assigning it a distinct subordinal rank.

The Chiroptera have always been placed near the Insectivora; but they are really a highly specialised group, as much isolated from all other mammals by the modification of their anterior limbs in adaptation to aerial locomotion, as the Cetacea and the Sirenia, by the absence of hind limbs, are specially adapted for an aquatic life.

Lastly, the Primates, which in any natural system must be placed at the head of the series, are divisible into two very distinct groups—one containing the various forms of Lemurs (Lemuroidea), and the other the Monkeys and Man (Anthropoidea). Whether the Lemuroidea should form part of the Primates (according to the traditional view), or a distinct order altogether removed from it, is as yet an undetermined question, for both sides of which there is much to be said. There can, however, be no doubt that the Anthropoidea form a perfectly natural group, presenting a series of tolerably regular gradations from the Marmosets (*Hapale*) to Man. Certain breaks in the series, however, enable us to divide it into five distinct families:—*Hapalidæ* or Marmosets; *Cebidæ* or American Monkeys, with three premolar teeth on each side of each jaw; *Cercopithecidæ*, containing the majority of Old-world Monkeys; *Simiidæ*, consisting of the genera *Hylobates*, *Simia*, *Gorilla*, and *Anthropopithecus*, the true Man-like Apes; and, lastly, *Hominidæ*, containing the genus *Homo* alone.

Subclass I. PROTOTHERIA.

Order i. MONOTREMATA—Monotremes.

Fam. 1. *Ornithorhynchidæ*—Duck-bill.
2. *Echidnidæ*—Spiny Anteater.

Group. **MULTITUBERCULATA.**[1]

Fam. 1. **Plagiaulacidæ**—Plagiaulax.
2. **Polymastodontidæ**—Polymastodon.
3. **Tritylodontidæ**—Tritylodon.

Subclass II. METATHERIA.

Order ii. MARSUPIALIA—Marsupials.

Suborder 1. POLYPROTODONTIA—Polyprotodonts.

[1] The names of the groups containing only extinct forms are printed in heavier type than those which contain species still existing.

Fam. 1. **Dromatheriidæ**—Dromatherium
2. **Amphitheriidæ**—Amphitherium, etc.
3. **Spalacotheriidæ**—Spalacotherium.
4. **Tritylodontidæ**—Tritylodon.
5. *Didelphyidæ*—Opossums.
6. *Dasyuridæ*—Thylacine and Dasyures.
7. *Peramelidæ*—Bandicoots.

Suborder 2. Diprotodontia—Diprotodonts.

Fam. 8. *Phascolomyidæ*—Wombats.
9. *Phalangeridæ*—Phalangers.
10. **Diprotodontidæ**—Diprotodon.
11. **Nototheriidæ**—Notothere.
12. *Macropodidæ*—Kangaroos.

Subclass III. Eutheria.

Order iii. Edentata—Edentates.

Fam. 1. *Bradypodidæ*—Sloths.
2. **Megatheriidæ**—Ground Sloths.
3. *Myrmecophagidæ*—Anteaters.
4. *Dasypodidæ*—Armadillos.
5. **Glyptodontidæ**—Glyptodonts.
6. *Manidæ*—Pangolins.
7. *Orycteropodidæ*—Aard-varks.

Order iv. Sirenia—Sirenians.

Fam. 1. *Manatidæ*—Manatees.
2. **Rhytinidæ**—Rhytina.
3. *Halicoridæ*—Dugongs.
4. **Halitheriidæ**—Halithere.

Order v. Cetacea—Cetaceans.

Suborder 1. Mystacoceti—Baleen Whales.

Fam. 1. *Balænidæ*—Greenland Whale, etc.

Suborder 2. **ARCHÆOCETI.**

Fam. 2. **Zeuglodontidæ**—Zeuglodonts.

Suborder 3. Odontoceti—Toothed Whales.

Fam. 3. *Physeteridæ*—Sperm Whale.
4. *Platanistidæ*—Freshwater Dolphins.
5. *Delphinidæ*—Dolphins, Porpoises, etc.

Order vi. Ungulata—Hoofed Mammals.

Suborder 1. Artiodactyla—Artiodactyles.

Section A. Suina—Pig-like Artiodactyles.

Fam. 1. *Hippopotamidæ*—Hippopotamus.
2. *Suidæ*—Pigs and Peccaries.

Annectant types.
3. **Chœropotamidæ**—Chœropotamus.
4. **Anthracotheriidæ**—Anthracothere.
5. **Merycopotamidæ**—Merycopotamus.
6. **Cotylopidæ**—Oreodonts.
7. **Anoplotheriidæ**—Anoplothere.
8. **Dichodontidæ**—Dichodon.

Section B. TRAGULINA—Chevrotains.
9. *Tragulidæ*—Chevrotains.

Section C. TYLOPODA—Camels.
10. *Camelidæ*—Camels and Llamas.
11. **Poebrotheriidæ**—Poëbrotherium.

Section D. PECORA—True Ruminants.
12. *Cervidæ*—Deer.
13. *Giraffidæ*—Giraffe.
14. *Antilocapridæ*—Prong-buck.
15. *Bovidæ*—Sheep, Cattle, etc.

Suborder 2. PERISSODACTYLA—Perissodactyles.
Fam. 16. *Tapiridæ*—Tapirs.
17. **Lophiodontidæ**—Lophiodonts.
18. **Palæotheriidæ**—Palæotheres.
19. *Equidæ*—Horses.
20. *Rhinocerotidæ*—Rhinoceroses.
21. **Lambdotheriidæ**—Palæosyops.
22. **Chalicotheriidæ**—Chalicothere.
23. **Titanotheriidæ**—Titanothere.
24. **Macraucheniidæ**—Macrauchenia.

Suborder 3. **TOXODONTIA**—Toxodonts.
Fam. 25. **Toxodontidæ**—Toxodon.
26. **Typotheriidæ**—Typothere.

Suborder 4. **CONDYLARTHRA**.
Fam. 27. **Periptychidæ**—Periptychus.
28. **Phenacodontidæ**—Phenacodus.
29. **Meniscotheriidæ**—Meniscothere.

Suborder 5. HYRACOIDEA—Hyraces.
Fam. 30. *Hyracidæ*—Hyrax.

Suborder 6. **AMBLYPODA**.
Fam. 31. **Pantolambdidæ**—Pantolambda.
32. **Coryphodontidæ**—Coryphodon.
33. **Uintatheriidæ**—Uintathere.

Suborder 7. PROBOSCIDEA—Proboscideans.
Fam. 34. **Dinotheriidæ**—Dinothere.
35. *Elephantidæ*—Elephants.

Group. **TILLODONTIA**—Tillodonts.
Fam. **Anchippodontidæ**—Anchippodus.
Calamodontidæ—Calamodon.

Order vii. RODENTIA—Rodents.
Suborder 1. SIMPLICIDENTATA.
Fam. 1. *Anomaluridæ*—Anomalurus.
2. *Sciuridæ*—Squirrels and Marmots.
3. *Haplodontidæ*—Haplodon.
4. **Ischyromyidæ**—Ischyromys.
5. *Castoridæ*—Beavers.
6. *Myoxidæ*—Dormice.
7. *Lophiomyidæ*—Lophiomys.
8. *Muridæ*—Rats, Mice, and Voles.
9. *Spalacidæ*—Mole-rats.
10. *Geomyidæ*—Pouched Rats.
11. *Dipodidæ*—Jerboas.
12. **Theridomyidæ**—Theridomys.
13. *Octodontidæ*—Spiny Mice.
14. **Castoroididæ**—Castoroides.
15. *Hystricidæ*—Porcupines.
16. *Chinchillidæ*—Chinchillas.
17. *Dinomyidæ*—Dinomys.
18. *Caviidæ*—Cavies.
19. *Dasyproctidæ*—Agouties.
Suborder 2. DUPLICIDENTATA.
Fam. 20. *Lagomyidæ*—Picas.
21. *Leporidæ*—Hares and Rabbits.

Order viii. CARNIVORA—Carnivores.
Suborder 1. CARNIVORA VERA—Fissipedes.
Fam. 1. *Felidæ*—Cats.
2. *Hyænidæ*—Hyænas.
3. *Proteleidæ*—Earth-wolf.
4. *Viverridæ*—Civets and Ichneumons.
5. *Canidæ*—Wolves and Foxes.
6. *Ursidæ*—Bears.
7. *Mustelidæ*—Weasels and Otters.
8. *Procyonidæ*—Racoons and Cat-bear.
Suborder 2. PINNIPEDIA—Pinnipedes.
Fam. 9. *Otariidæ*—Eared Seals.
10. *Trichechidæ*—Walrus.
11. *Phocidæ*—Seals.
Suborder 3. **CREODONTA**—Creodonts.
Fam. 12. **Hyænodontidæ**—Hyænodon.
13. **Proviverridæ**—Proviverra.
14. **Arctocyonidæ**—Arctocyon.
15. **Mesonychidæ**—Mesonyx.

Order ix. INSECTIVORA—Insectivores.

Suborder 1. INSECTIVORA VERA.

Fam. 1. *Tupaiidæ*—Tupaias.
2. *Macroscelididæ*—Elephant-Shrews.
3. *Erinaceidæ*—Hedgehogs.
4. *Soricidæ*—Shrews.
5. *Talpidæ*—Moles.
6. *Potamogalidæ*—Potamogale.
7. *Solenodontidæ*—Solenodon.
8. *Centetidæ*—Centetes.
9. *Chrysochloridæ*—Golden Moles.

Suborder 2. DERMOPTERA.

Fam. 10. *Galeopithecidæ*—Galeopithecus.

Order x. CHIROPTERA—Bats.

Suborder 1. MEGACHIROPTERA—Frugivorous Bats.

Fam. 1. *Pteropodidæ*—Flying Foxes.

Suborder 2. MICROCHIROPTERA—Insectivorous Bats.

Fam. 2. *Vespertilionidæ*—Common Bats.
3. *Nycteridæ*—Nycteris.
4. *Rhinolophidæ*—Leaf-nosed Bats.
5. *Emballonuridæ*—Emballonura.
6. *Phyllostomatidæ*—Vampyres.

Order xi. PRIMATES.

Suborder 1. LEMUROIDEA—Lemuroids.

Fam. 1. **Hyopsodontidæ**—Hyopsodus.
2. *Chiromyidæ*—Aye-Aye.
3. *Tarsiidæ*—Tarsier.
4. *Lemuridæ*—Lemurs.

Suborder 2. ANTHROPOIDEA—Anthropoids.

Fam. 5. *Hapalidæ*—Marmosets.
6. *Cebidæ*—American Monkeys.
7. *Cercopithecidæ*—Old World Monkeys.
8. *Simiidæ*—Gibbons and Man-like Apes.
9. *Hominidæ.*—Man.

The distinctive character of these subclasses and orders, with an account of their subdivisions and the principal forms contained in each, will be given in subsequent chapters.

CHAPTER IV

GEOGRAPHICAL AND GEOLOGICAL DISTRIBUTION

I. GEOGRAPHICAL DISTRIBUTION.[1]

In considering the present distribution of mammals over the globe, we may, in the first place, direct our attention to terrestrial or land types, reserving the consideration of aerial types, like the Bats, and aquatic forms, as exemplified by the Cetaceans, Sirenians, and Seals, to separate sections.

Among terrestrial forms each species has a certain definite area of distribution in space, which may be of very wide extent, or may be confined to a restricted region. This distributional area is, however, always connected, or continuous; that is to say, that although we may have a single species inhabiting two continents, like the Lion in Asia and Africa, or dwelling both on a continent and adjacent continental islands, like the Javan Rhinoceros of India, Java, and Borneo, yet we shall always find that such areas, if not still connected, show evident signs of having been so connected in comparatively late geological epochs; and we never find instances of the same species inhabiting totally disconnected areas, such as India and South America. As examples of mammals with a wide distribution we may mention the Lion and the Leopard, which are now found throughout Africa, and also occur in India, as well as in the intervening areas of Arabia and Persia. In the case of the former species, palæontology further teaches us that its distribution in the last geological epoch was even more extensive, since we have good evidence to show that it formerly ranged over the greater part of Europe, including the British Isles. The Jackal affords another well-known instance of a species common

[1] On this subject see A. Murray, *Geographical Distribution of Mammals*, 1866; and especially A. R. Wallace, *The Geographical Distribution of Animals*, 2 vols., 1876, and *Island Life*, 1881; also A. Heilprin, *The Geographical and Geological Distribution of Animals*, 1887.

to India and Africa. The American Puma, again, may be cited as an example of a mammal having a very wide range in latitude, since it is found from Patagonia in the south to Canada in the north. As instances of wide range in the opposite direction we have only to mention the Reindeer and the Elk or Moose, found in the northern regions of both the Old and New Worlds, which are only separated from one another by the narrow channel of Behring Strait.

Of mammals with extremely restricted distributional areas, we may mention many of the Insectivora, such as the Desman of the Pyrenees, and some of the Madagascar types of this order, the Lemurs from the same island, some of the species of Marmots, the remarkable bear-like *Æluropus* of Eastern Tibet, one species of Zebra, and other Ungulates from Africa.

The distribution of a genus (except of course when the genus is represented only by a single form) is very generally more extensive than that of a species; and this may be markedly the case when there are only some two or three species in a genus. In genera, moreover, we meet with what is known as discontinuous distribution, that is, where the distributional area of one or more species is totally separated from that of others. The best instance of this occurs in the case of the Tapirs, where we find one species inhabiting the Malayan Peninsula, and no others anywhere in the world, with the exception of South America. The explanation of such an apparently anomalous feature in distribution is to be found in the past history of the globe, which shows us that Tapirs once existed in China, Europe, and North America, and, therefore, indicates that the existing isolated species are the sole survivors of a group once spread over a large portion of the earth's surface. In regard to generic distribution it must, however, be mentioned that this depends to a great extent on the limits which we are disposed to assign to genera themselves.

As the distributional area of a genus generally exceeds that of a species, so that of a family, or group of genera, is larger than that of a single genus; and similarly the distribution of an order, or assemblage of families, usually occupies a larger area than that of a single family. Thus, for instance, the genus *Thylacinus*, represented only by the so-called Tasmanian Wolf or Thylacine, is now entirely restricted to Tasmania; but the family *Dasyuridæ*, to which that genus belongs, ranges all over Australia, while the order Marsupialia, which includes the *Dasyuridæ*, is found both in Australia and America, and in past epochs was probably spread over the entire globe.

A remarkable feature in connection with the distribution of the terrestrial Mammalia is the circumstance that, with the exception of certain species introduced by human agency, and small forms which can easily have been transported on floating timber or other similar means, they are totally absent from what are known as oceanic islands—that is islands arising from great depths in the ocean, mainly composed of coral or volcanic rocks, and showing no signs of having ever been connected with the existing continents, or the larger and so-called continental islands. The obvious explanation of this feature is, that from their total isolation these islands have never been able to receive a mammalian fauna from the great continental areas on which mammalian life was probably first developed.

As an intermediate step between these islands which are practically void of mammalian life and the continents which teem with such a variety of forms, are certain larger islands and portions of continents containing a mammalian fauna more or less markedly distinct from that of the whole of the other regions of the globe. The best instance of this is Australia, which, with the exception of one dog—the Dingo—and certain *Muridæ* and Bats, has no mammals except Monotremes and Marsupials. The latter are, moreover, perfectly distinct from those of America, which, if we exclude the islands in the neighbourhood of Australia, is the only other region which now possesses any Marsupials at all. Here also we have a ready and full explanation which accords with all the facts; since it is evident that Australia has been isolated from the Asiatic continent from some very remote geological epoch, at which period it is probable that Monotremes and Marsupials were the dominant if not the sole representatives of the Mammalia then existing. Consequently Australia has never been able to receive an influx of the Eutherian orders, which have probably swept away all the Marsupials except the small American Opossums from the rest of the globe. Again, the large island of Madagascar, which has a fauna of an African type, but still very markedly different from that of the mainland, may be considered to have been connected with the latter at a time when the Eutheria had become the dominant forms, but has been separated for a sufficiently long period to have enabled a large number of its species and genera to have become distinct from those of the adjacent continent. Similarly, there is evidence to show that South America was probably cut off for a considerable period from the northern half of the American continent, in consequence of which its lowly organised fauna of Edentates were enabled to attain such a remarkable development in the later geological periods.

In contrast to the mammalian fauna of islands of the preceding type is, or rather was, that of the British Islands, which in the early historic and prehistoric periods was identical with that of the Continent. This leads to the inference that at a comparatively late epoch there was a direct land communication between Britain and the Continent, which is shown by geological evidence to have actually been the case.

The above instances are sufficient to show what an important influence the date of separation of islands from the adjacent continents has had upon their existing mammalian fauna, and how largely the present distribution of mammalian life is bound up with the past history of our globe. We must, however, not omit to mention another very important agency of past times which has likewise had great influence on the present distribution of the various faunas of the northern hemisphere. This is the so-called glacial epoch, which took place immediately before the establishment of the present condition of things, and appears to have been the cause of the extinction of many of the larger mammalian types which formerly inhabited Europe, and whose retreat to the warmer regions of the south was apparently cut off by the Mediterranean.

Zoological Regions.—Zoologists are now generally agreed in dividing the land surfaces of the globe into a number of zoological regions or provinces, characterised by a more or less distinctly marked general *facies* of their fauna as a whole. Some of these regions are much more distinctly defined than the others; and in the majority of cases there is a kind of neutral ground or No-man's-land at the junction between any two of these regions. It must also be remembered that in the Old World proper as we go back in time we find a gradual assimilation in the mammalian faunas of the different regions, indicating that originally there was one large fauna of a generally similar type occupying the greater portion of this area. Thus we find that Hippopotami, Giraffes, Kudus, Elands, and other types of Antelopes now restricted to Africa, formerly extended to Europe and India, while there is also evidence to show that the group of large anthropoid Apes, now found only in Africa and the Bornean region, were likewise spread over a large part of the south-western half of the Old World. Moreover, while at the present day there is a marked connection between the mammals of the northern regions of both the Old and New Worlds, in the Tertiary period it appears that the fauna of the whole of North America was much more nearly allied to that of the central regions of the Old World than is now the case. Thus in the Tertiary rocks of America we meet with remains of what we are accustomed to regard as such essentially Old World genera as Horses and Rhinoceroses. On the other hand there are no traces in America of the existence at any period of Apes, Giraffes, Hippopotami, or Hyænas, while that continent has yielded evidence of groups of Ungulates totally unrepresented in the eastern hemisphere.

The chief zoological regions of the globe, proposed by Mr. Sclater in 1857, and now recognised by the majority of authorities, are six in number, and are named as follows. Firstly, the Palæarctic region, embracing the whole of Europe, Persia, Northern Arabia, and all of Asia northward of the line of the Himalaya proper, Japan, that part of Africa lying northward of the Sahara Desert, and the oceanic islands of the North Atlantic. Secondly, the Ethiopian region, which comprises all Africa lying to the south of the Sahara, the southern part of Arabia, Madagascar, and the Mascarene Islands. Thirdly, the Oriental or Indian region, which is taken to include India south of the Himalaya, and to the north-west as far as Beluchistan, the Malay peninsula, southern China, Sumatra, Java, Borneo, and the Philippines. Fourthly, the Australasian region, which is usually defined as being bounded to the north-west by the deep sea channel lying between Borneo and Celebes known as Wallace's line, and is taken to include Celebes, Lumbok, New Guinea, Australia, Tasmania, New Zealand, and the host of oceanic islands in the South Pacific. Several writers, however, prefer to regard Celebes and some of the adjacent islands as representing a transitional Austro-Malayan region. Fifthly, the Nearctic region, comprising Greenland and North America as far south as the north of Mexico. And, sixthly, the Neotropical region, which embraces the remaining portion of the American continent and the West Indies.

Various minor modifications of this scheme have been proposed. Thus some writers are disposed to raise India to the rank of a distinct primary region, while others propose the same for New Zealand. The Palæarctic and Nearctic regions have a large number of common types, more especially among the mammals, and Dr. A. Heilprin[1] has expressed his opinion that they should be regarded as a single primary region under the name of the Holarctic. The same writer would also separate the South Pacific Islands as constituting a Polynesian region.

Minor divisions or sub-regions have also been marked out, but it will be unnecessary to indicate their limits in the present work. We may, however, mention the Mediterranean sub-region of the Palæarctic, which includes the peninsular portion of southern Europe, North Africa, Asia Minor, Persia, Afghanistan, Beluchistan, and Northern Arabia, as a good instance of the transition from one region to another, since its fauna has a mingling of Palæarctic, Ethiopian, and Oriental types, the former being, however, the predominant ones.

Of the chief mammalian types characteristic of these various regions only a brief sketch can be given in this work.

Palœarctic Region.—The Palæarctic region is of enormous extent, and includes countries varying greatly in their flora, climate, and elevation. Thus it embraces the Arctic plains of Siberia, the warm regions of Italy, Southern France, and Northern Africa, the forest-

[1] *Distribution of Animals.*

clad slopes of the outer Himalaya, and the lofty arid plains of Turkestan and Tibet, scorched by a burning sun in summer and chilled by a still more terrible cold in winter. Its extreme limits in the west are marked by the Canaries and Azores, and in the east by distant Japan; and yet throughout this vast expanse we find a great uniformity of life, as exemplified by the large number of British genera which occur also in Japan. The mammals which are on the whole the most characteristic of this region are the Sheep and Goats, forming a section of the great family of *Bovidæ*, nearly all the species of which are Palæarctic, although we meet with one Goat (*Capra*) in the Nilgherries of Southern India, and a Sheep (*Ovis*) in the Nearctic region. The Musk Ox (*Ovibos*) is characteristic of the Palæarctic and Nearctic regions. At least one species of Camel is characteristic of this region, and it is not improbable that the second may also have originated in it. There are a few characteristic types of Antelopes, such as the Alpine Chamois (*Rupicapra*), the Saiga of Tartary, and the Chiru (*Pantholops*) of Tibet, each of which is represented by only a single species; and we miss the host of Antelopes so characteristic of the Ethiopian region. Deer (*Cervus*) are abundant, although by no means confined to this region; and the Musk Deer (*Moschus*), the sole representative of the subfamily *Moschinæ*, is exclusively Palæarctic. Monkeys, as a rule, are absent, although we meet with one species of *Macacus* in Northern Africa and at Gibraltar, and some other types on the southern border of Tibet. The Moles (*Talpa*) are mainly Palæarctic, although one species enters Northern India, while the Desmans (*Myogale*) of the Pyrenees and Southern Russia are unknown beyond the limits of this region. The Water-shrew (*Nectogale*) is likewise a peculiar eastern Palæarctic type. Among the Rodents, the Picas or Tailless Hares (*Lagomys*) and the Dormice (*Myoxus*) are essentially Palæarctic forms, only one species of each being found beyond the limits of the region, and the one extra-Palæarctic species of *Lagomys* occurring in the cognate Nearctic region. The Mice and Rats are represented by the typical genus *Mus* and other types, and Hares (*Lepus*) and one species of Squirrel (*Sciurus*) are common. The Carnivora include two species of Bears (*Ursus*), Wolves and Foxes (*Canis*), a Lynx and a few species of Cats (*Felis*), as well as numerous weasels (*Mustela*), and some other types.

Ethiopian Region.—The Ethiopian region is of great interest to the student of mammals, since it is inhabited by a number of forms remarkable for their large size. A considerable portion of the area consists of desert, especially in the north; but there is also a wide extent of grassy plains (veltd), as well as vast tracts of equatorial forests of great density. Perhaps the most striking feature in the Ethiopian fauna is the number of Ungulates, both of the Artiodactyle and Perissodactyle sections. In the former section we have the Giraffes (*Giraffa*) represented by one species, which is the type of a family, and is unknown elsewhere. Equally characteristic are the Hippopotami, which likewise form the type of a family, while the Pigs are represented by the Wart-hogs (*Phacochœrus*) and the River-hogs, forming an aberrant group of the genus *Sus*. The Oxen (*Bos*) are represented by Buffaloes, but there are no species of true Oxen or Bison. The Antelopes attain an extraordinary development, the number of species being estimated at from eighty to ninety, which are referred to a large number of genera, although several of these are more or less ill defined. Most of these genera are peculiar to this region, but the Gazelles (*Gazella*) are also found in the desert regions of other parts of the Old World, and *Oryx* ranges into Arabia and Persia. In contrast to this abundance of Antelopes is the total absence of the Deer family, or *Cervidæ*, which are so characteristic of the Palæarctic and Oriental regions. The Chevrotains or *Tragulidæ* are, however, represented by *Dorcatherium*.[1] In the Perissodactyle section we may notice the presence of two species of *Rhinoceros*, both furnished with two horns, and distinguished from those of the Oriental region by the absence of incisor and canine teeth. The Horse family (*Equidæ*) is also represented by several species, and includes the peculiar group of Zebras, characterised by their beautifully striped skins. Of other Ungulates the Elephants, which, like the Rhinoceroses, are now peculiar to the Ethiopian and Oriental regions, have one species, which is widely different from its Indian congener. The Hyraces are mainly characteristic of this region, although one species occurs in Syria and Palestine. The Carnivora include some forms like the Lion, Leopard, and Jackal, common to the Oriental region, but likewise include certain peculiar types like the Earth-wolf (*Proteles*), which may be regarded as the type of a distinct family, and two species of Hyænas, which are referred by some authorities to a distinct genus (*Crocuta*). There is also the Hunting-dog (*Lycaon*), and the peculiar group of Foxes known as the Fennecs, together with *Otocyon*. Bears, Wolves, and true Foxes are absent; but Civets, etc., are abundant, although not characteristic of the region. The Primates yield several very characteristic types, such as the Gorilla and the Chimpanzee (*Anthropopithecus*) among the *Simiidæ*, which, with the exception of the Orangs of Borneo, are the only existing large man-like Apes, and the group of Dog-faced Baboons (*Cynocephalus*) in the *Cercopithecidæ*. The genus *Colobus* is also a group of the latter family, absolutely characteristic of the region. Lemurs, again, occur on the continent of Africa, but the great development of this group is in the adjacent island of Madagascar, where several peculiar genera occur, and where the larger Carnivora and Ungulata are

[1] Generally known as *Hyomoschus*, but first described as an extinct form under the above name.

absent. These peculiarities of the fauna of Madagascar apparently point, as previously mentioned, to its separation from the mainland before the latter was overrun by the larger types, and at a time when its chief mammals were Lemurs and Insectivores. There are two genera of Edentates, the Pangolins (*Manis*), and the Aard-vark (*Orycteropus*), the latter being peculiar.

Although the foregoing groups of mammals are now so characteristic of the Ethiopian region, it cannot be too strongly insisted that their restriction to this region is, so to speak, merely a feature of the present day, and that at a late geological epoch nearly all the peculiar genera were represented in India, and many of them also in Europe.

Oriental Region.—The third or Oriental region is likewise of very considerable extent, and is the only one, in addition to the Ethiopian, which is the home of huge Ungulates, like Elephants and Rhinoceroses, and the large man-like Apes. A large proportion of this extensive area is occupied by tropical and subtropical forests and swamps; these being especially abundant in Burma, Southern China, Siam, and the southern ridges of the Himalaya, collectively constituting the Indo-Chinese sub-region, and also in the Indo-Malayan sub-region of the Malay peninsula and adjacent islands. In the third or Indian sub-region, comprising peninsular India, with the exception of the Carnatic, there are large tracts of open country, including some of the hottest regions in the world, parts of which form plains more or less covered with vegetation during the cooler and rainy seasons, while others are barren rocky table-lands, as in the Deccan, or arid deserts like those of parts of the Punjab and Sind. Finally, in the fourth or Cingalese sub-region, represented by the Carnatic and the island of Ceylon, we find vast areas of luxuriant forest and jungle. In the north-western desert area of the Indian sub-region the fauna includes a mixture of Palæarctic and Ethiopian forms, with those characteristic of the Oriental region.

Among the chief features of the mammalian fauna of this region we may notice the absence of Hippopotami and Giraffes, the greatly diminished number of Antelopes, as compared with those of Africa, and the abundance of Deer and true Pigs. The Antelopes comprise the two peculiar genera *Boselaphus* (Nilghai) and the typical *Antilope* (Black-buck), each of which is represented by only a single species, while the Deer belong to the so-called Rusine group, which is markedly different from that to which the Palæarctic Red Deer belongs. True Chevrotains (*Tragulus*) are peculiar to this region. The Oxen include the true Buffalo, differing in many respects from the African species of the same group, and also certain species of true Oxen, such as the Gaour and Banting, belonging to the Bibovine group, which is confined to this region. In the Perissodactyla Horses (*Equus*) are represented only by a single species in the desert area of the Indian sub-region, while the two species of *Rhinoceros* differ from those of Africa in being furnished with canines and incisors. The Malayan Tapir is the only Old World species of its genus. The Indian Elephant differs, moreover, so markedly from its African ally that some writers regard the two as types of distinct genera. The Carnivora include the Lion, Leopard, Jackal, and Hunting-Leopard, which are common to Africa; but the Tiger is very characteristic of this region, although extending northwards into the Palæarctic. Civets are abundant, comprising some peculiar genera, of which it will suffice to mention the well known *Paradoxurus*. Wolves closely allied to the Palæarctic species occur in Northern India, and there are also Foxes related to the typical species. The Dog-like animals which hunt in packs, and are separated by some writers from *Canis* under the name of *Cyon*, occur in the present and the Palæarctic region. The striped Hyæna is the Indian representative of its genus. Ratels are common to this and the Ethiopian region, and constitute the genus *Mellivora*. The most striking feature in the Carnivorous fauna of this region, as distinguished from the Ethiopian, is, however, the presence of Bears, some of which belong to the typical genus *Ursus*, while one species is usually generically separated under the name of *Melursus*. Among the Rodents we may especially notice the abundance of the *Muridæ* and *Sciuridæ*. In the former family we have numbers of true Mice (*Mus*), and also the peculiar genus *Nesocia* (Bandicoot-Rat), while in the latter both the true Squirrels (*Sciurus*) and the Flying-Squirrels (*Pteromys*) attain great development. The genus (*Pteromys*) is, indeed, mainly characteristic of this region, although in Kashmir and Japan it enters the Palæarctic. The Bats are very numerous, being represented by all the families, with the exception of the *Phyllostomatidæ*, or Vampyres, of South America. Among the Insectivora the genera *Tupaia* and *Galeopithecus* (Flying Lemur) are peculiar to this region, although not found in India. Finally, in the Primates we have the genera *Macacus* and *Semnopithecus* very abundantly represented, although both also enter the Palæarctic region; but the Anthropoid types are confined to the south-eastern half of the region, and include the Orangs (*Simia*) of Borneo, and the smaller long-armed Gibbons (*Hylobates*), which are abundant in the Malay peninsula, both genera not being found beyond this region. The Lemurs are much less abundant than in the Ethiopian region, but they include the peculiar Tarsier of Sumatra, Borneo, and Celebes (Austro-Malayan region), which differs so markedly in dentition and structure of the feet from all other forms that it has been made the type of a separate family. The Edentates, so poorly represented in the Old World, include only Pangolins (*Manis*), which, as we have already seen, also occur in the Ethiopian region.

Australasian Region.—With the fourth or Australasian region we come to a mammalian fauna so peculiar that we have no difficulty whatever in defining it from all the other regions of the globe, although it should be observed that in the Austro-Malayan islands we have a partial mingling of the Australasian and Malayan faunas. If we exclude Celebes from this region we find that, with the exception of a Pig in New Guinea, of the Dingo in Australia, of numerous Mice and Rats (*Muridæ*), and Bats, there are no Eutherian mammals throughout the area. The mammals of this region are restricted to the Australian mainland, the island of Tasmania, New Guinea, and the Aru islands, the whole area of New Zealand having been totally devoid of mammalian life until introduced by man. The whole of the Monotremata, constituting the subclass Prototheria, and all the Marsupials, exclusive of the few outlying forms ranging into the transitional Austro-Malayan area, and with the exception of the American family of the Opossums (*Didelphyidæ*), are absolutely confined to this region.

Celebes.—The mammals of Celebes—the typical representative of the Austro-Malayan transitional region or sub-region—include the peculiar Ape known as *Cynopithecus*, *Tarsius* (also Oriental), the Anoa, and the single species of *Babirusa.* Several other types of placental mammals are found in this transitional area, while the Marsupials are represented by *Phalanger* and *Petaurus.*

Nearctic Region.—The two remaining regions we have to consider are comprised in the New World. The first of these is the Nearctic, which, as already mentioned, has a fauna showing such a strongly marked relationship to that of the Palæarctic region, that it has been proposed to unite the two regions. Among types common to these two regions we may mention closely allied species of true Deer (*Cervus*) as exemplified by the Red Deer and the Wapiti; the allied Bisons of the two regions; the Reindeer and Elk common to both; as well as nearly related, and in some cases identical, species of Cats, Lynxes, Bears, Wolves, Foxes, Beavers, Squirrels, Marmots, and Hares. The Glutton or Wolverene, and the Musk Ox is also common to the Arctic portions of the two regions. The Ungulates are very poorly represented, but we have, in addition to the forms already mentioned, one species of the Palæarctic genus *Ovis*, namely the Big-horn, and the Prong-buck (*Antilocapra*), which is quite peculiar. There are, however, no Perissodactyla. The Racoons and Coatis (*Procyonidæ*) constitute a family represented out of the New World only by the aberrant Cat-Bear (*Ælurus*) of Nipal. The characteristic American feline known as the Puma extends over this region; but there are no Edentates, and the Marsupials are represented only by a single species of Opossum. Rodents are extremely numerous, and comprise several characteristic types, which alone would tell us what part of the globe we were visiting. The most distinctive are the Pouched Rats (*Geomyidæ*), and the Beaver-like rodents known as the *Haplodontidæ*. True Rats and Mice (*Mus*), which are represented throughout the Old World, are totally wanting in the New, where they are replaced by the Vesper-mice, which may be included in the European genus *Cricetus*, although often separated as *Hesperomys*. This feature alone would seem to justify the distinction of the Nearctic from the Palæarctic region. The Musquash (*Fiber*) is a genus of Nearctic rodents unknown in the Old World. Among other characteristic genera we may mention, in the Carnivora, the Skunk (*Mephitis*) and the American Badger (*Taxidea*). Primates are absent from the entire region.

Neotropical Region.—The last of the six main regions is the Neotropical, including Mexico, South America, and the West Indies. A very large extent of this area is occupied by forests, which are described as being denser and more luxuriant than those of any other part of the globe. Alternating with these forest areas are the vast grassy plains known in different regions as llanos, savannas, and pampas. The back-bone of the region is formed by the great chain of the Andes. Next to the Australasian, this region is perhaps better characterised by its mammalian fauna than any of the others. Commencing with the Ungulates, we find a total absence of Antelopes, Sheep, and Oxen, and also of all Perissodactyles except Tapirs. Deer are, however, represented, although by peculiar forms (*Cariacus*) unknown beyond the New World. The Peccaries (*Dicotyles*), which are often made the type of a distinct family, take the place of the Old World Pigs, while the Llamas and Alpacas (*Auchenia*) are the substitutes for the Palæarctic Camels. The Carnivora include several Cats (*Felis*), among which the Puma and the Jaguar are the most noticeable; and there are also Racoons, Coatis, Foxes, and one species of Bear. Insectivora are totally wanting; but the Bats are characterised by the presence of the Vampyres (*Phyllostomatidæ*), which are almost restricted to this region. The Rodents likewise include three families unknown elsewhere, namely the Chinchillas and Viscacha (*Chinchillidæ*), the Agouties (*Dasyproctidæ*), and the Cavies (*Caviidæ*); while a large number of the *Octodontidæ* are Neotropical, all the other forms being Ethiopian. In the Primates, again, we have all the forms quite peculiar to this region, and constituting two families, viz. the *Cebidæ* or Prehensile-tailed Monkeys, and the *Hapalidæ*, or Marmosets, both of which differ decidedly in their dentition, as well as in other features, from the Old World Monkeys. Lemuroids are unknown. Perhaps, however, the mammals which may be considered as most characteristic of the Nearctic region are the numerous Edentates, which form three families, mostly confined to it. These comprise the *Bradypodidæ* or Sloths, which solely inhabit the forest region; the *Myrmecophagidæ* or Anteaters; and the *Dasypodidæ* or Armadillos, of which one species has crept northward as far as Texas. Almost equally characteristic are the numerous Opossums, the majority of which belong to the genus *Didelphys*. Finally, it should be observed that the West Indies are distinguished from the rest of the region by the absence of Primates, Carnivora, and Edentates.

Aquatic Mammals.—Many mammals grouped for the present purpose as terrestrial pass a great portion of their lives in brooks, lakes, or rivers, and, being dependent upon such waters for obtaining their subsistence, are necessarily confined to their vicinity; but the truly aquatic mammals, or those living constantly in the water, and unable to move their quarters from place to place by land, are the orders Cetacea and Sirenia, with which may also be grouped the Seals, forming the Pinniped division of the order Carnivora.

For the marine Cetacea, animals mostly of large size and endowed with powers of rapid locomotion, there are obviously no barriers to universal distribution over the surface of the earth covered by sea, except such as are interposed by uncongenial temperature or absence of suitable food. Nevertheless it was thought some years ago that the fact of a Whale or a Dolphin occurring in a sea distant from that in which it had usually been found was sufficient justification for considering it as a distinct species and imposing a new name upon it. There are now, however, so many cases known in which Cetaceans from the northern and southern seas, from the Atlantic and Pacific Oceans, present absolutely no distinguishing external or anatomical characters upon which specific determination can be based that the opposite view is gaining ground; and, since some species are undoubtedly very widely distributed, being in fact almost cosmopolitan, there seems little reason why many others should not be included in the same category. The evidence is satisfactory enough in those instances in which the intermediate regions are inhabited by the same forms;—the cases of "continuous areas" of distribution. In those in which the areas of distribution are apparently discontinuous, there may be more room for doubt; but it must not be forgotten that the negative evidence is here of much less value than in the case of land animals, since the existence of Cetaceans in any particular part of the ocean may be easily overlooked. The great Sperm Whale (*Physeter macrocephalus*) is known to be almost cosmopolitan, inhabiting or passing through all the tropical and temperate seas, although not found near either pole. At least three of the well-known species of Rorqual (*Balænoptera*) of the British coasts are represented in the North Pacific, on the South American shores, and near New Zealand, by species so closely allied that it is difficult to point out any valid distinctive characters, though it may perhaps be desirable to wait for a more exhaustive examination of a large series of individuals before absolutely pronouncing them to be specifically identical. There is nothing yet known by which we can separate the "Humpback Whales" (*Megaptera*) of Greenland, the Cape of Good Hope, and Japan. The same may be said of the common Dolphin of the European seas (*Delphinus delphis*) and the so-called *D. bairdi* of the North Pacific and *D. forsteri* of the Australian seas. The Pilot Whale (*Globicephalus melas*) and the *Pseudorca* of the North Atlantic and of New Zealand are also, so far as present knowledge enables us to judge, respectively alike. Many other similar cases might be given. Captain Maury collected much valuable evidence about the distribution of the larger Cetacea, and, finding Right Whales (*Balæna*) common in both northern and southern temperate seas, and absent in the intermediate region, laid down the axiom that "the torrid zone is to the Right Whale as a sea of fire, through which he cannot pass." Hence all cetologists have assumed that the Right Whale of the North Atlantic (*B. biscayensis*), that of the South Seas (*B. australis*), and that of the North Pacific (*B. japonica*), are necessarily distinct species. The anatomical structure and external appearance of all are, however, so far as yet known, marvellously alike, and, unless some distinguishing characters can be pointed out, it seems scarcely justifiable to separate them from geographical position alone; as, though the tropical seas may be usually avoided by them, it does not seem impossible, or even improbable, that some individuals of animals of such size and rapid powers of swimming may have at some time traversed so small a space of ocean as that which divides the present habitual localities of these supposed distinct species. If identity or diversity of structural characters is not to be allowed as a test of species in these cases, as it is usually admitted to be in others, the study of their geographical distribution becomes an impossibility.

Although many species are thus apparently of such wide distribution, others are certainly restricted; thus the Arctic Right Whale (*Balæna mysticetus*) has been conclusively shown to be limited in its range to the region of the northern circumpolar ice, and no corresponding species has been met with in the southern hemisphere. In this case, not only temperature, but also the peculiarity of its mode of feeding, may be the cause. The Narwhal and the Beluga have a very similar distribution, though the latter occasionally ranges farther south. The common Hyperoödon is restricted to the North Atlantic, never entering, so far as is yet known, the tropical seas. Other species are exclusively tropical or austral in their range. One of the true Whalebone Whales (*Neobalæna marginata*) has only been met with hitherto in the seas round Australia and New Zealand; and a large Ziphioid (*Berardius arnouxi*) only near the last-named islands.

The Cetacea are not limited to the ocean, or even to salt water, some entering large rivers for considerable distances, and others being exclusively fluviatile. One species of *Platanista* is extensively distributed throughout nearly the whole of the river systems of the Ganges, Brahmaputra, and Indus, ascending as high as there is water enough to swim in, but apparently never passing out to sea. The individuals inhabiting the Indus and the Ganges must therefore have been for long ages isolated without developing any definite distinguishing anatomical characters; for those by which the supposed *P. indi* was formerly separated from *P. gangetica* have been shown by Anderson to be of no constant value. *Orcella fluminalis* appears to be limited to the Irawaddy river, and at least two distinct species of Dolphin belonging to different genera are found in the waters of the upper Amazon. A *Neomeris* has been found in the great Chinese river, the Yang-tsi-Kiang, nearly a thousand miles from the sea. It is remarkable, however, that none of the great lakes or inland seas of the world are, according to our present knowledge, inhabited by Cetaceans. A regular seasonal migration has been observed in many of the oceanic Cetacea, especially those inhabiting the North Atlantic, but further observations upon this subject are still much needed.

The great difference in the manner of life of the Sirenia, as compared with that of the Cetacea, causes a corresponding difference in their geographical distribution. Slow in their movements, and feeding exclusively upon vegetable substances, water-grasses, or fuci, the Sirenia are confined to rivers, estuaries, or coasts where these grow, and are not denizens of the open sea, although of course there is a possibility of accidental transport by the assistance of oceanic currents across considerable distances. Of the three genera existing within historic times, one (*Manatus*) is exclusively confined to the shores of the tropical Atlantic and the rivers entering into it, individuals scarcely specifically distinguishable being found both on the American and the African side of the ocean. The Dugong (*Halicore*) is distributed in different colonies, at present isolated, throughout the Indian Ocean from Arabia to North Australia. The *Rhytina* or Northern Sea-Cow was, for some time before its extinction, limited to a single island in the extreme north of the Pacific Ocean.

The Pinnipeds, although capable of traversing long reaches of ocean, are less truly aquatic than the last two groups, always resorting to the land or to extensive ice-floes for the purpose of breeding. The geographical range of the various species is generally more or less restricted, usually according to climate, as they are mostly inhabitants either of the Arctic or Antarctic seas and adjacent temperate regions, very few being found within the tropics. For this reason the northern and the southern species are for the most part quite distinct. In fact, the only known exception is the case of a colony of the Sea-Elephant (*Macrorhinus leoninus*), the general range of which is in the southern hemisphere, inhabiting the coast of California. Even in this case a different specific name has been given to the northern form; but the characters by which it is distinguished are not of great importance, and probably, except for the abnormal geographical distribution, would never have been noticed. The most remarkable circumstance connected with the distribution of the Pinnipeds is the presence of members of the suborder in the three isolated great lakes or inland seas of Central Asia—the Caspian, Aral, and Baikal; these forms, notwithstanding their long isolation, having varied but slightly from species now inhabiting the Polar Seas.

II. GEOLOGICAL DISTRIBUTION.

Geological Sequence.—In order to understand the geological distribution, or in other words the distribution in time of mammals, it is necessary to be acquainted with the chief divisions, or time-periods, of the strata constituting the crust of the globe. These are shown in the following table, which commences with the uppermost or most recent beds and ends with the lowest and oldest.

I. Cainozoic or Tertiary—
  1. Pleistocene—River alluvia, etc.
  2. Pliocene—Suffolk Crag.
  3. Miocene—Hempstead Beds of Hampshire.
  4. Eocene—Paris Gypsum and London Clay.

II. Mesozoic or Secondary—
  1. Cretaceous—Chalk, Greensands, etc.
  2. Jurassic—Oolites and Lias.
  3. Triassic—Red Marls, Dolomites, etc.

III. Palæozoic or Primary—
  1. Permian—Beds overlying the Coal.
  2. Carboniferous—Coal-measures, etc.
  3. Devonian—Old Red Sandstone.
  4. Silurian—Wenlock Limestone, etc.
  5. Cambrian—Llanberis Slate, etc.
  6. Archæan—Gneiss and other schists.

The names in the first column indicate the primary divisions or life-periods, while those in the second column are the great systems, each of which is again divided into minor groups, the popular names of a few of these minor groups being given in the third column. There are at present no means of arriving at any satisfactory conclusion as to the absolute length of time indicated by either the primary or secondary divisions; but there is little doubt that the whole of the Tertiary period is only equal to a fraction of the Mesozoic as regards its duration, while it is probable that the duration of the Mesozoic epoch was largely exceeded by that of the Palæozoic.

Mesozoic Mammals.—The earliest date at which mammals are at present known is in the upper part of the Triassic period, which forms the base of the great Mesozoic epoch; and from this date they are represented more or less abundantly in various horizons of the Jurassic and Cretaceous.

The very rapid advances in our knowledge of these forms which have been made in the last few years, especially in consequence of the explorations of rich fossiliferous beds in North America, have not only completely changed the present aspect of the science, but give such promise for the future, that any sketch which we may now attempt of this branch of the subject can only be regarded as representing a transient phase of knowledge. It will be well, however, to gather together in this place the leading facts now ascertained with regard to the most ancient forms, as, owing to the uncertainty of their relationship with any of the existing orders, they will be most conveniently treated of separately, while the ascertained facts relating to the geological history of the forms more nearly allied to those now living will be more appropriately described under the account of the different groups into which the class may now be divided.

The remains of mammals which existed anterior to the Tertiary period hitherto discovered nearly all belong to creatures of very small size, many of the largest scarcely exceeding the common Polecat or Squirrel. Some are known only by a few isolated teeth, others by nearly complete sets of these organs, and the majority by more or less nearly perfect specimens of the rami of the lower jaw. It is a very curious circumstance that this part of the skeleton alone has been preserved in such a large number of instances. Only very rarely has a nearly complete cranium been found; and there is no satisfactory evidence of the structure of the vertebral column of any single individual, and only one known case of a complete limb.[1] The species already described from European strata are numerous, although the number of genera and species has lately been reduced. Of these by far the greater number have been found at a single spot near Swanage in Dorsetshire, in a bed of calcareous mud only forty feet long, ten feet wide, and averaging five inches in depth. The marvellous results obtained by the exploration by Mr. S. H. Beckles of this small fragment of the earth's surface show by what accidents, as it were, our knowledge of the past history of life

[1] The fore limb from S. Africa described as *Theriodesmus*, which appears to be mammalian, and may belong to *Tritylodon*.

has been gained, and what may still remain in store where little thought of at present. A bed, apparently equally rich, has been discovered in the Jurassic of Wyoming, North America, the contents of which have been made known by Professor Marsh, while another fertile source of these remains occurs in the Laramie beds of the Upper Cretaceous of the United States.[1]

The whole of the Mesozoic mammals at present known may be divided into two great groups, the one characterised by a type of dentition more or less clearly resembling that found among the existing Polyprotodont Marsupials, while the other presents an altogether peculiar modification, recalling in some respects that of the Diprotodont Marsupials, although differing so decidedly as to

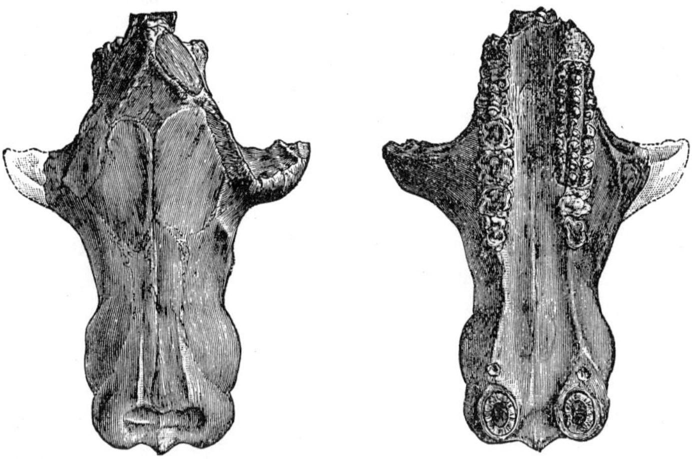

FIG. 24.—Frontal and oral aspects of the cranium of *Tritylodon longævus*; from the Karoo system of Basuto-land, South Africa. ⅔ natural size. (After Owen.)

show that the owners of this form of dentition cannot be included in that group.

Multituberculata.—The name Multituberculata has been proposed for the group exhibiting the type of dentition last mentioned, and is generally adopted, although the term Allotheria has been also suggested. The essential characteristic of the dentition of this group is the presence of a single scalpriform incisor on each side of the

[1] The subjects referred to under this heading are mostly described and figured in detail in Owen's "Monograph of the Fossil Mammalia of the Mesozoic Formations," *Palæontographical Society's Publications*, 1871; and in various papers by Marsh, in the *American Journal of Science and Arts*, 1878-89. Important contributions to our knowledge of these forms have also been made by Professors Cope and Osborn, and the reader should especially consult the memoir by the latter writer on the "Structure and Affinities of the Mesozoic Mammals," published in the *Journal of the Philadelphia Academy* (1888), vol. ix.

lower jaw (Fig. 25) and of one larger incisor, and in some instances of one or two smaller ones in each premaxilla (Fig. 24). These incisors are separated by an interval or diastema from the first of the premolars. The true molars, and in some instances the premolars (Fig. 24), are characterised by having longitudinal rows of tubercles separated by one or more grooves; there being either two or three of these rows in the upper molars of those forms in which these teeth are known, while there are, at least usually, only two in those of the lower jaw. In other cases the premolars are of a secant type, with a highly convex cutting-edge, and usually either serrated or obliquely grooved (Figs. 25, 26). From a certain resemblance between these secant premolars and those of some of the smaller *Macropodidæ* it was at one time considered that we had in these mammals representatives of Diprotodont Marsupials. The great difference in the structure of the molar teeth of these forms,

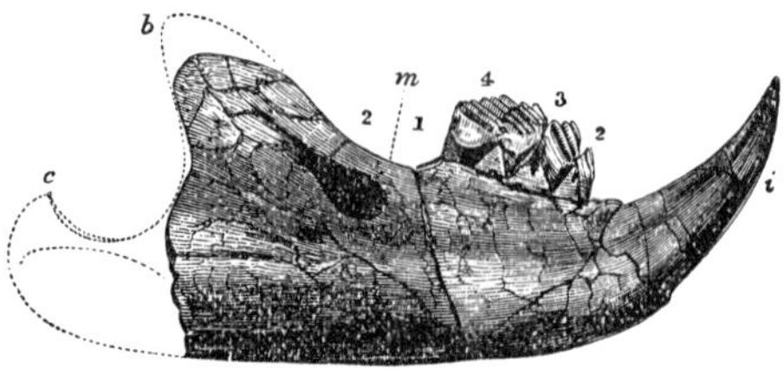

FIG. 25.—The right ramus of the mandible of *Plagiaulax beklesi*; from the Purbeck of Swanage. Twice natural size. *i*, Incisor; *m*, molar; *b*, coronoid process; *c*, condyle. (After Owen.)

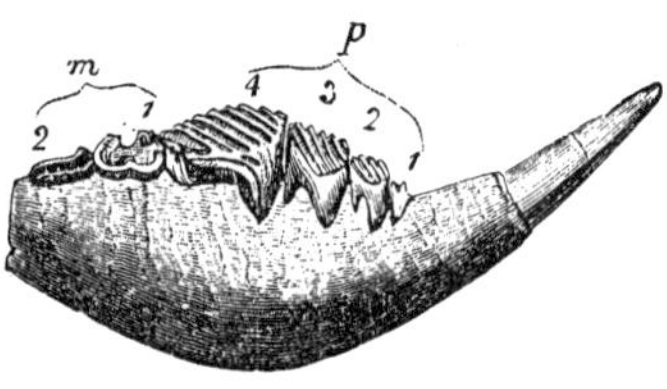

FIG. 26.—The imperfect right ramus of the mandible of *Plagiaulax minor*; from Swanage. Four times natural size. *p*, Premolars; *m*, molars. (After Lyall.)

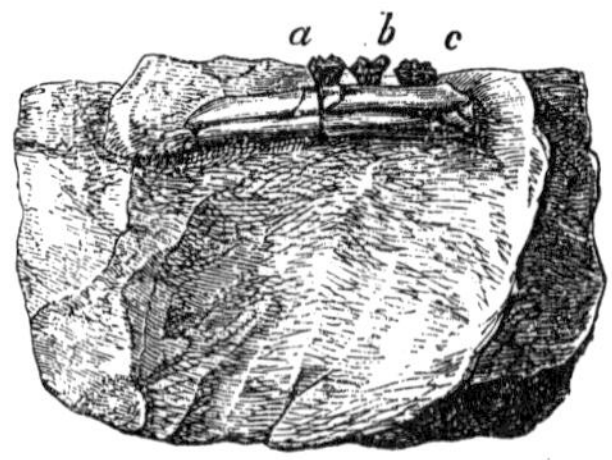

FIG. 27.—*Stereognathus oölithicus*. Fragment of jaw with three teeth (*a*, *b*, *c*), in matrix; from the Stonesfield Slate. Natural size. (After Owen.)

coupled with the circumstance that when the number of upper incisors is reduced below three it is the second in place of the first which becomes enlarged and opposed to the incisor of the lower jaw, seems to prevent the acceptation of this view. Moreover, in their peculiar structure the molars seem, on the whole, to make a nearer approximation to the teeth of *Ornithorhynchus* than to any other known mammal; and it has accordingly been suggested that the Multituberculata may really represent an order of Prototheria. Some support is afforded to this suggestion by certain fragmentary bones from the Cretaceous of the United States, which are regarded by Marsh as parts of a coracoid and interclavicle. The peculiar character of the whole dentition of these forms indicates that if they are really Prototherians they cannot be regarded as primitive and ancestral types.

It would be beyond the scope of the present work to describe in detail, or even to mention the names of all the members of this group, and it will therefore suffice to refer to a few of the principal types. Of the forms with tubercular premolars the best known is the genus *Tritylodon* (Fig. 24), which occurs typically in beds of Lower Mesozoic in South Africa, but is also known from the Trias of Stuttgart. In the Stonesfield Slate, near Oxford, which belongs to the lower part of the Jurassic system, and is separated from the Trias by the intervening Lias, a fragmentary jaw with three teeth (Fig. 27) appears to indicate an allied type, the teeth having three longitudinal ridges separated by grooves. In the Purbeck beds of Dorsetshire, forming the top of the Jurassic system, we find another member of this group, which has been described as *Bolodon*, closely allied to which is *Allodon* of the Upper Jurassic of the United States.

The first discovery of the remains of Mesozoic mammals was made in the Keuper or Upper Trias of the Rhætian Alps in Bavaria. In 1847 Professor Pleininger of Stuttgart, while sifting some sand from the Keuper of Diegerloch and Steinenbronn, found, among an immense mass of teeth, scales, and unrecognisable fragments of skeletons of fish and saurians, two minute teeth, each with well-defined, enamelled, tuberculated crowns and distinct roots, plainly showing their mammalian character. These were considered by their discoverer to indicate a predaceous and carnivorous animal of very small size, to which he gave the name of *Microlestes antiquus*. Subsequently Mr. C. Moore discovered in a bone bed of Rhætic (topmost Trias) age, filling a fissure in the Mountain Limestone at Holwell, near Frome in Somersetshire, various isolated teeth with their crowns much worn, but apparently including both upper and lower molars and a canine, which are assigned by Sir R. Owen to Pleininger's genus *Microlestes*, and described specifically as *M. moorei*. Under the name of *Hypsiprymnopsis rhæticus*, Professor Boyd Dawkins described a single tooth with two roots discovered in the Rhætic Marlstone at Watchet in Somersetshire. Sir R. Owen referred the latter tooth to *Microlestes*, and if its describer is right in regarding it as a much worn premolar of the type of those of *Plagiaulax* (Fig. 25) there would be evidence that *Microlestes* was closely allied to the latter, from the molars of which those of *Microlestes* are scarcely distinguishable.

Plagiaulax, of the Dorsetshire Purbeck (Figs. 24, 25), is at once distinguished from *Tritylodon* by its secant premolars, which, as already mentioned, recall those of some of the *Macropodidæ*, although readily distinguished by the convexity of the cutting edge and their oblique grooving. This remarkable and highly specialised type has been the occasion of one of the most interesting discussions on the inferences which may be drawn as to the affinities and habits of an otherwise unknown animal from the structure of a small portion of its organisation which occurs in the annals of natural history—a discussion carried on with great ability, ingenuity, and wealth of illustration on both sides. Dr. Falconer maintained that it was more nearly allied to the Rat-Kangaroo (*Potorous* or *Hypsiprymnus*) than to any other existing form, and that, as it is known that these animals feed upon grass and roots, "it may be inferred of *Plagiaulax* that the species were herbivorous or frugivorous. I can see nothing in the character of their teeth," he adds, "to indicate that they were either insectivorous or omnivorous." Sir R. Owen, on the other hand, from the same materials came to the conclusion that "the physiological deductions from the above-described characteristics of the lower jaw and teeth of *Plagiaulax* are that it was a carnivorous Marsupial. It probably found its prey in the contemporary small insectivorous mammals and Lizards, supposing no herbivorous form like *Stereognathus* to have co-existed during the Upper Oolitic period."

It is impossible here to give at any length the arguments by which these opposing views are respectively supported, but it may be indicated that the first-mentioned is strongly countenanced by the consideration of the following facts: (1) all existing Marsupials may be divided, so far as their dentition is concerned, into two groups—(*a*) those which have a pair of large more or less procumbent incisors close to the symphysis of the lower jaw, and rudimentary or no canines (diprotodont dentition), and (*b*) those which have numerous small incisors and large pointed canines (polyprotodont dentition); (2) the vast majority of the former group are purely vegetable feeders, and almost all of the latter are carnivorous or insectivorous; and (3) *Plagiaulax*, so far as its structure is known, shows an analogy with the former group; and, as we have no sure basis for inferences as to the habits of an unknown animal, but the knowledge of the habits of such as are known, we have no grounds for supposing that its habits differed from those forms having an analogous type of dental structure.[1]

Allied types, such as *Ctenacodon*, are also met with in the Upper

[1] The whole discussion is contained in the following memoirs: (1) H. Falconer, "Description of Two Species of the Fossil Mammalian genus *Plagiaulax*, from Purbeck," *Quart. Journ. Geol. Soc.* vol. xiv. 1857; (2) R. Owen, art. "Palæontology," *Encyclopædia Britannica*, 8th ed., 1859; (3) H. Falconer, "On the Disputed affinity of the Mammalian genus *Plagiaulax*," *Quart. Journ. Geol. Soc.* vol. xviii. 1862; (4) R. Owen, "Monograph of the Fossil Mammalia of the Mesozoic Formation," *Palæontographical Society*, 1871.

Jurassic of North America; and the *Plagiaulacidæ* also persisted into the lower part of the Eocene division of the Tertiary period; *Neoplagiaulax* being a Tertiary form common to Europe and the United States, while *Liotomus* and *Ptilodus* are at present known only from the latter country.

The present group is also represented in the upper Cretaceous of the United States by *Selenacodon* (*Meniscoëssus* in part), *Cimoliomys*, etc. *Polymastodon*, of the Lowest or Puerco Eocene of New Mexico is the largest known form, and is characterised by the presence of only one premolar and the elongated molars. The angle of the mandible is inflected after the Marsupial fashion.

Polyprotodont Types.—The second type of mammalian dentition found in the Mesozoic period resembles that occurring among recent Polyprotodont Marsupials—that is to say there are at least three lower incisors, the canines are well developed, and the premolars and molars are cuspidate, the number of the latter reaching in some cases to seven or eight. There has been much discussion as to the taxonomic position of these forms, and while the majority of writers admit the Marsupial affinities of at least a moiety, it has been contended that others indicate distinct ordinal groups more or less closely allied to the Insectivora. At present, however, there is no decisive evidence to support such a view. Important proof of the Marsupial affinity of one of these forms is afforded by the replacement of the teeth, which appears to be of the same nature as in the existing Marsupials, that is to say, the last premolar alone is preceded by a milk-tooth.

The most generalised forms appear to be *Dromatherium* and *Microconodon*, from Lower Mesozoic beds in the United States, of which enlarged views of the teeth are given in Fig. 4 (1, 2), p. 31. Professor Osborn points out the extremely simple character of these teeth, and it is quite possible that these forms may prove to be *Prototheria.* There are three premolars and seven molars in the lower jaw of *Dromatherium.*

A common form in the Purbeck of Dorsetshire is *Triconodon* (*Triacanthodon*), in which the formula of the lower teeth is *i* 3, *c* 1, *p* 4, *m* 3-4. A lower jaw is shown in Fig. 28, and an enlarged view of a molar tooth in Fig. 4 (5). The molar teeth consist of three flattened cones placed in the same antero-posterior line, those of the upper and lower jaw being alike. *Priacodon*, of the Jurassic of the United States, is probably inseparable from *Triconodon.* In the genus *Phascolotherium* (Fig. 29) of the Lower Jurassic Stonesfield Slate, the lower teeth may be classified as *i* 4, *c* 1, *p* 3, *m* 4, the premolars and molars being

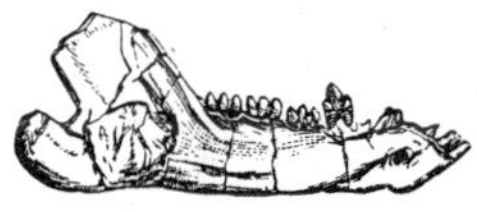

FIG. 28.—Reversed view of the left ramus of the mandible of *Triconodon mordax*; from the Purbeck of Swanage. Natural size. (After Owen.)

much alike. The molars approximate to the type of those of *Triconodon*, but the anterior and posterior cones are relatively smaller. Like that of the last-named genus, the mandible of

FIG. 29.—Inner view of the right ramus of the mandible of *Phascolotherium bucklandi*; from the Stonesfield Slate. The outline shows the natural size. *i*, Incisors (one missing); *c*, canine; *p*, premolars; *m*, molars. The mylohyoid groove is seen near the lower border. (After Owen.)

Phascolotherium is remarkable for the extremely low position of its articular condyle. In *Amphilestes* (Fig. 30) of the Stonesfield Slate the molars appear to be of the same general type as those of *Phascolotherium*, but are more numerous, although their exact number cannot be determined. A somewhat different type of lower molar is displayed by the genus *Amblotherium*, of the Dorsetshire Purbeck, to which *Amphitherium* of the Stonesfield Slate was probably allied. This type of tooth is shown in Fig. 4 (8, 9, 12) p. 31, and, as there stated, represents that modification of the tritubercular type known as the tubercular sectorial. The three primitive tritubercular cusps form what is known as the blade of the tooth, behind which there is the talon or hypocone. A similar form of molar occurs in the existing Opossums and Bandicoots. The number of lower teeth in *Amblotherium* is *i* 4, *c* 1, *p* 4, *m* 7-8. Numerous allied types, such as *Achyrodon* and *Dryolestes* occur in the Upper Jurassic of Europe or the United States, while from only one side of the jaw being exposed in each case so-called genera like *Stylodon* and *Stylacodon* have been formed upon specimens showing the opposite side to that which is exposed in the types of *Amblotherium* and *Amphitherium*. The

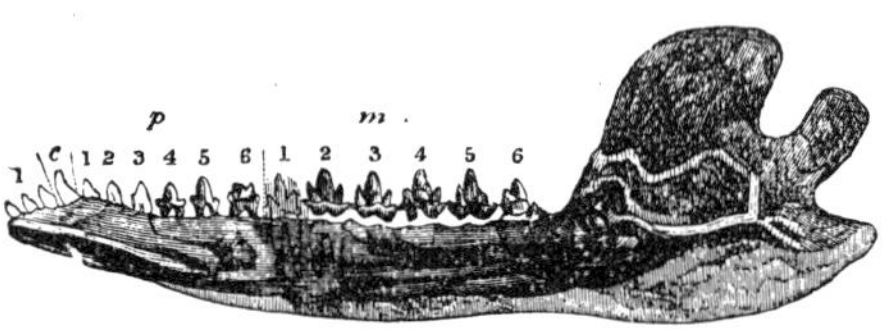

FIG. 30.—Reversed inner view of the left ramus of the mandible of *Amphilestes broderipi*; from the Stonesfield Slate. Twice natural size. The restoration of the anterior teeth is conjectural, and the condyle is placed too high. (After Owen.)

only parallel among existing forms to the excessive number of molar teeth found in these Mesozoic genera occurs in the Marsupial genus *Myrmecobius*, of which a description is given in a succeeding chapter. Jaws more or less closely resembling those described under the names mentioned above are also found in the uppermost Cretaceous of the United States. A feature common to these Mesozoic mammals and *Myrmecobius* and some other existing forms is the presence of a narrow channel on the inner side of the mandibular ramus known as the mylohyoid groove (Fig. 29).

The last type of molar dentition occurring among the Mesozoic Mammalia is that found in the lower jaws (Fig. 31), upon which the genus *Spalacotherium* was established, the upper jaws, described as *Peralestes*, being apparently referable to the same animal. Upper and lower teeth of this form are represented in Fig. 4 (6, 7), p. 31, where they are described as typical examples of the tritubercular type of molars, the upper teeth having one inner and two outer cusps, and the reverse condition obtaining in the lower ones. This type of molar presents a marked resemblance to that found in the existing Insectivorous genus *Chrysochloris*; the number of lower teeth in *Spalacotherium* is, however, *i* 3, *c* 1, *p* + *m* 10, by which it is widely distinguished from all the Insectivora. *Menacodon*, of the Upper Jurassic of the United States, appears to be allied to *Spalacotherium*.

FIG. 31.—Part of the left ramus of the mandible, viewed from the outer side, of *Spalacotherium tricuspidens*; from the Purbeck of Swanage. Twice natural size. (After Owen.)

Tertiary Mammals.—The more important types of Tertiary mammals will, as already mentioned, be noticed under the heads of the groups to which they are severally allied; but a few general remarks on this subject may be advantageously recorded in this chapter. In the first place, it may be observed that the comparatively scanty evidence of mammalian life hitherto yielded by the Cretaceous, coupled with the number and variety of forms approximating to the existing groups found even in the lowest Tertiary, indicates a great imperfection of the geological record. At present, indeed, we have no decisive evidence of the existence of any members of the Eutherian subclass previously to the Tertiary; but it can hardly be doubted that in some part of the world they had made their appearance before that epoch. The Eutherian mammals of the lowest Eocene, both in Europe and the United States, are of an extremely generalised type; and although many of them approximate to existing groups, they show such a combination of characters, now restricted to individual groups, as to indicate that several of the various orders into which the subclass is now divided were at that period very intimately connected. A marked feature of these early Eutherians is the prevalency of trituberculism in the dentition, not less noteworthy being the frequent occurrence of pentadactylism in the feet, while many of the individual bones were devoid of the grooves and ridges found in those of later types. By the time that we reach the upper division of the Eocene period, such as the horizon of the well-known gypsum of the Paris basin, nearly all the chief groups of mammals had become clearly differentiated from one another, although their representatives were usually more generalised than their existing allies. From this date to the later geological periods there is a gradual approximation to the types of mammalian life existing at the present day.

In addition to the features of trituberculism and pentadactylism so characteristic of the oldest known Eutherians, we may notice some other points in connection with the earlier types. Thus the older Tertiary mammals, as we have already stated, had relatively smaller and simpler brains than the later types, so that a gradual evolution in this respect may be traced from the Eocene to the Pleistocene. Again, there is a great tendency among the Eocene forms to a retention of the typical Eutherian dental formula noticed on page 25, and also to the absence of an interval, or diastema, in the dental series. Concomitantly with this feature we may notice the short crowns and simpler structure of the molar teeth of the earlier Ungulates as compared with those of to-day, of which details will be given in a later chapter. Another instance of the more generalised characters of the earlier mammals is afforded by the absence or slight development of horns, antlers, and tusks among the Ungulata. Thus the earlier Rhinoceroses were hornless, and the Deer either without antlers or with antlers of a very simple kind, while the male Swine were not furnished with the formidable tusks of the existing Wild Boars. Finally, all, or nearly all of the mammals, from the lowest Eocene of Rheims present the peculiarity of having a vertical perforation in the astragalus.

The intimate connection existing during the Middle Tertiary between many families of mammals now widely distinguished from one another may be more conveniently noted when we come to the consideration of the families in question.

CHAPTER V

THE SUBCLASS PROTOTHERIA OR ORNITHODELPHIA

General Characters.—The characters of the Prototheria can at present only be deduced from the two existing families, since hitherto no extinct animals which can be referred with certainty to other divisions of this remarkable and well-characterised group have been discovered. These two isolated forms, in many respects widely dissimilar, yet having numerous common characters which unite them together and distinguish them from the rest of the Mammalia, are the *Ornithorhynchidæ* and the *Echidnidæ*, both restricted in their geographical range to the Australian region of the globe. Taken altogether they represent the lowest type of evolution of the mammalian class, and most of the characters in which they differ from the other two subclasses tend to connect them with the inferior vertebrates, the Sauropsida and Amphibia; for, though the name Ornithodelphia owes its origin to the resemblance of the structure of the female reproductive organs to those of birds, there is nothing especially bird-like about them.

Their principal distinctive characters are these. The brain has a very large anterior commissure, and a very small corpus callosum, agreeing exactly in this respect with the Marsupials. The cerebral hemispheres, in *Echidna* at least, are well developed and convoluted on the surface. The auditory ossicles present a low grade of development, the malleus being very large, the incus small, and the stapes columelliform. The coracoid bone is complete, and articulates with the sternum, and there is a precoracoid (epicoracoid) in advance of the coracoid, while there is also a large "interclavicle" or episternum in front of the sternum, and connecting it with the clavicles. There are also "epipubic" bones. The oviducts (not differentiated into uterine and Fallopian portions) are completely distinct, and open, as in oviparous vertebrates, separately into a cloacal chamber, and there is no distinct vagina. The testes of the male are abdominal in position throughout life, and the vasa deferentia open into the cloaca, not into a distinct urethral passage. The penis, attached to the ventral wall of the cloaca, is perforated by a canal in the greater part of its length, and not merely grooved, as in reptiles and those birds which have such an organ. The canal is open at the base and brought only temporarily in contact with the termination of the vasa deferentia, so as to form a seminal urethra when required; but it never transmits the urinary secretion. This condition is a distinct advance on that of the Sauropsida in the direction of the more complex development of these parts in most of the other Mammalia. The ureters do not open into the bladder, but behind it into the dorsal wall of the genito-urinary passage. The mammary glands have no distinct nipple, but pour out their secretion through numerous apertures situated in a cup-shaped depression of the abdominal skin, forming a mammary marsupium, especially developed in the females during lactation. It should be mentioned that, according to the observations of Professor Gegenbaur, the mammary glands of the Monotremes are the simplest found in the entire class. The region of the glands is, indeed, distinguished from the rest of the abdomen merely by its thicker layers of muscles. The glands themselves are closely connected with the hair-follicles, and belong to the sudoriparous type, whereas the glands of all other mammals are of sebaceous origin.

The young are produced from eggs laid by the female parent, which are meroblastic, like those of birds; that is to say only a portion of the yolk segments and forms the embryo, the remainder serving for the nourishment of the latter.

The above are the principal distinguishing characters of the group, and apply not only to the subclass, but of course equally to the one order Monotremata, in which the two existing genera are included. In addition to these more important characters, the following minor features may also be mentioned.

The scapula differs from that of all other mammals in that the ridge corresponding to the spine of other forms is situated on the anterior border instead of in the middle of the outer or dorsal surface. The humerus is much expanded at its two extremities, and has a very prominent deltoid crest, and a well-marked entepicondylar foramen.

The dorso-thoracic vertebræ are nineteen in number, and have no terminal epiphyses to their bodies. The tranverse processes of the cervical vertebræ are of autogenous formation, and remain suturally connected with the remainder of the vertebra until the animal is full-grown. Though in this respect they present an approximation to the Sauropsida (Reptiles and Birds), they differ from these classes, inasmuch as there is not a gradual transition from these autogenous transverse processes of the neck (or cervical ribs, as they may be considered) into the thoracic ribs, for in the seventh vertebra the costal element is much smaller than in the others, indicative of a very marked separation of neck from thorax, not seen in the existing Sauropsida. The upper ends of the ribs are attached to the sides of the bodies of the dorsal vertebræ only, and not to the transverse processes. The sternal ribs are well ossified, and there are distinct partly ossified intermediate ribs. The cerebral cavity, unlike that of the lower Marsupials or the Reptiles, is large and hemispherical, flattened below and arched above, and about as broad as long. The cribriform plate of the ethmoid is nearly horizontal. The cranial walls are very thin, and smoothly rounded externally, and the sutures become completely obliterated in adult skulls, as in Birds. The broad occipital region slopes upwards and forwards, and the face is produced into a long and depressed rostrum. The bony palate is prolonged backwards, so that the posterior nares are nearly on a level with the glenoid fossæ. The mandible is without distinct ascending ramus; the coronoid process and angle are rudimentary, and the two halves are loosely connected at the symphysis. The fibula has a broad, flattened process, projecting upwards from its upper extremity above the articulation, like an olecranon. In the male there is an additional, flat, curved ossicle on the hinder and tibial side of the plantar aspect of the tarsus, articulating chiefly to the tibia, which supports in the adult a sharp-pointed perforated horny spur, with which is connected the duct of a gland situated beneath the skin of the back of the thigh, the function of which is not yet clearly understood. (A rudimentary spur is found in the young female *Ornithorhynchus*, but this disappears when the animal becomes adult.) The stomach is sub-globular and simple; the alimentary canal has no ileo-cæcal valve, or marked distinction between large and small intestine, but has a small, slender vermiform cæcum with glandular walls. The liver is divided into the usual number of lobes characteristic of the Mammalia, and is provided with a gall-bladder.

In the presence of three distinct bones developed from cartilage in the shoulder-girdle (viz. scapula, coracoid, and pre- or epi-coracoid) the Monotremes agree with the Anomodont reptiles (see p. 83), and with no other representatives of that class. The precoracoid of the Anomodonts is, however, distinguished by extending upwards to articulate with the acromial process of the scapula. The Monotreme humerus is, moreover, strikingly like the corresponding bone of many of the Anomodonts and of some of the allied Labyrinthodont Amphibians.

Family ORNITHORHYNCHIDÆ.

Ornithorhynchus.[1]—Cerebral hemispheres smooth. Premaxillæ and mandible expanded anteriorly and supporting a horny beak

[1] Blumenbach, *Voigts Magazin*, vol. ii. p. 205 (1800).

something like that of a duck, bordered by a naked and very sensitive membranous expansion. The place of teeth in the adult is supplied functionally by horny structures, elongated, narrow, and sharp-edged, along the anterior part of the sides of the mouth, and broad, flat-topped or molariform behind. Functional molar teeth present in the young and adolescent condition. Legs short, fitted for swimming; feet webbed, each with five well-developed toes armed with large claws, beyond which in the fore feet the interdigital membrane is extended. Vertebræ: C 7, D 17, L 2, S 2, Ca 21. Acetabulum not perforated. Tongue not extensile. Mucous membrane of small intestine covered with delicate, close-set transverse folds or ridges. Tail rather short, broad, and depressed. Eyes very small. Fur close and soft.

The Duck-billed Platypus (*Platypus anatinus*) was the name assigned to one of the most remarkable of known animals by Shaw, who had the good fortune to introduce it to the notice of the scientific world in the *Naturalist's Miscellany* (vol. x., 1799). In the following year it was independently described by Blumenbach (*Voigts Magazin*, ii. p. 205) under the name of *Ornithorhynchus paradoxus*. Shaw's generic name, although having priority to that of Blumenbach, could not be retained, as it had been used at a still earlier time (1793) by Herbst for a genus of Coleoptera. *Ornithorhynchus* is therefore now universally adopted as the scientific designation, although Duck-billed Platypus or Duck-bill may be conveniently retained as a vernacular appellation. By the colonists it is called "Water-Mole," but it need scarcely be said, its affinities with the true moles are of the slightest and most superficial description. Until the last few years the early stages of the development of the young were not fully known. It had, indeed, been repeatedly affirmed, in some cases by persons who have had actual opportunities of observation, that the Platypus lays eggs; but these statements were generally received with scepticism and even denial. This much-vexed question was, however, settled by the researches of Mr. W. H. Caldwell in 1884, who found that these animals, although undoubtedly mammals throughout the greater part of their structure, are oviparous, laying eggs, which in the manner of their development bear a close resemblance to the development of those of the Reptilia. Two eggs are produced at a time, each measuring about three-fourths of an inch in its long, and half an inch in its short, axis, and enclosed in a strong, flexible, white shell.

The Platypus is pretty generally distributed in situations suitable to its aquatic habits throughout the island of Tasmania and the southern and eastern portions of Australia. Slight variations in the colouring and size of different individuals have given rise to the idea that more than one species may exist; but all naturalists who have had the opportunity of investigating this question by the aid of a good series of specimens have come to the conclusion that there is but one, and no traces of any extinct allied forms have yet been discovered.

The length of the animal when full grown is from eighteen to twenty inches from the extremity of the beak to the end of the tail, the male being slightly larger than the female. The fur is short, dense, and rather soft to the touch, and composed of an extremely fine and close under-fur, and of longer hairs projecting beyond this, each of which is very slender at the base, and expanded,

Fig. 32.—Platypus or Duck-bill (*Ornithorhynchus anatinus*). From Gould's *Mammals of Australia.*

flattened, and glossy towards the free end. The general colour is deep brown, but paler on the under parts. The tail is short, broad, and depressed, and covered with coarse hairs, which in old animals generally become worn off from the under surface. The eyes are small and brown. There is no projecting pinna or ear-conch. The mouth, as is well known, bears a striking resemblance to the bill of a Duck. It is covered with a naked skin, a strong fold of which projects outwards around its base. The nostrils are situated near the extremity of the upper surface. There are no true teeth in the adult, but their purposes are served by horny prominences, or cornules, two on either side of each jaw—those in the front narrow, longitudinal, sharp-edged ridges, and those behind broad, flattened, and molariform. The upper surface of the lateral edges of the mandible has also a number of parallel fine transverse ridges, like those on the bill of a Duck. Until 1888 it was thought that true teeth were totally wanting throughout the life of this animal; but in the spring of that year Mr. E. B. Poulton [1] announced the discovery in an embryo of teeth which were regarded as quite functionless. In the following year, however, Mr. O. Thomas [2] was fortunate enough to find some young skulls with functional teeth *in situ*, and was thus enabled to give a detailed account of their structure and of their relations to the cornules. From these specimens it appears that the teeth are functional for a considerable part of the life of the animal, cutting the gum in the usual manner, and, after being worn down by friction with food and sand, are shed from the mouth in the same manner as are the milk-teeth of other mammals. The cornules are developed from the epithelium of the mouth under and around the teeth, and the hollows found in the middle of them are the vestiges of the alveoli from which the teeth have been shed. One of the skulls showed on either side, both above and below, two completely calcified teeth; but in another example there were three teeth on either side of the lower jaw. According to Mr. Thomas's account, "the teeth themselves are broad, flat, and low-crowned. The upper ones have each two high, conical, internal cusps, from which minute ridges run downwards and outwards to the outer borders of the crowns, where the edge is peculiarly crenulate rather than cuspidate, in the ordinary sense of the word. On the whole, the anterior and posterior upper teeth are essentially similar to one another, except that the former are narrower, and their outer edges are less markedly crenulated. In the lower jaw there is a greater difference between the two. The anterior is triangular in outline, its longest side is placed antero-externally, and its anterior and postero-external angles have each a high pointed cusp, ridged on its internal aspect, while the posterior and internal borders are indistinctly crenulated. The posterior tooth is broadly quadrangular in outline, with a projecting antero-internal angle. As in the corresponding tooth above, there are two cusps on one side, and a series of crenulations on the other, but they are of course reversed, the cusps being external and the crenulations internal. The cusps are high, and connected with transverse ridges running across towards the internal border."

In trying to find any teeth like those of the Duck-bill among other known mammals Mr. Thomas considers, as was first suggested by Professor Cope, that those of the Mesozoic Multituberculata (p. 109) make the nearest approximation. He adds, however, that "it must be insisted that the resemblance between the Multituberculate

[1] *Proceedings of the Royal Society of London*, vol. xliii. p. 353 (1888).

[2] *Ibid.* vol. xlvi. p. 126 (1889).

and the Ornithorhynchus teeth is of the most general character, and that the two are certainly widely separated generically, even if we do admit that they appear to possess a relationship nearer to each other than to any other known groups of mammals."

Reverting to the description of the Duck-bill, we find that in the cheeks are tolerably capacious pouches, which appear to be used as receptacles for food. The limbs are strong and very short, each with five well-developed toes provided with strong claws. In the fore feet the web not only fills the interspaces between the toes, but extends considerably beyond the ends of the long, broad, and somewhat flattened nails, giving great expanse to the foot when used for swimming, though capable of being folded back on the palm when the animal is burrowing or walking on the land. On the hind foot the nails are long, curved, and pointed, and the web extends only to their base. On the heel of the male is a strong, curved, sharply pointed, movable horny spur, directed upwards and backwards, attached by its expanded base to the accessory bone of the tarsus. This spur, which attains the length of nearly an inch, is traversed by a minute canal, terminating in a fine longitudinal slit near the point, and connected at its base with the duct of a large gland situated at the back part of the thigh. The whole apparatus is so exactly similar in structure to the poison-gland and tooth of a venomous snake as to suggest a similar function, but evidence that the Platypus ever employs its spur as an offensive weapon has, at all events until lately, been wanting. A case is, however, related by Mr. Spicer in the *Proceedings of the Royal Society of Tasmania for* 1876 (p. 162), of a captured Platypus inflicting a severe wound by a powerful lateral and inward movement of the hind legs, which wound was followed by symptoms of active local poisoning. It is not improbable that both the inclination to use the weapon and the activity of the secretion of the gland may be limited to the breeding season, and that their purpose may be, like that of the antlers of deer and many similar organs, for combat among the males. In the young female the spur is present in a rudimentary condition, but it disappears in the adult of that sex.

The Platypus is aquatic in its habits, passing most of its time in the water or close to the margin of lakes and streams, swimming and diving with the greatest ease, and forming for the purpose of sleeping and breeding deep burrows in the banks, which generally have two orifices—one just above the water level, concealed among long grasses and leaves, and the other below the surface. The passage at first runs obliquely upwards in the bank, sometimes to a distance of as much as fifty feet, and expands at its termination into a cavity, the floor of which is lined with dried grass and leaves, and in which the eggs are laid and the young brought up. The food consists of aquatic insects, small crustaceans, and worms, which are caught under water, the sand and small stones at the bottom being turned over with the bill. The creatures appear at first to deposit what they have thus collected in their cheek pouches, and when these are filled they rise to the surface and quietly triturate their meal with the horny plates before swallowing it. Swimming is effected chiefly by the action of the broad forepaws, the hind feet and tail taking little share in locomotion in the water. When asleep they roll themselves into a ball, as shown in the figure. In their native haunts they are extremely timid and wary, and very difficult to approach, being rarely seen out of their burrows in the daytime. Mr. A. B. Crowther, who has supplemented the often quoted observations of Dr. Bennett upon the habits of these animals in confinement, says, "They soon become very tame in captivity; in a few days the young ones appeared to recognise a call, swimming rapidly to the hand paddling the water; and it is curious to see their attempts to procure a worm enclosed in the hand, which they greedily take when offered to them. I have noticed that they appear to be able to smell whether or not a worm is contained in the closed hand to which they swim; for they desisted from their efforts if an empty fist was offered." When irritated they utter a soft low growl, resembling that of a puppy.

Family ECHIDNIDÆ.

Cerebral hemispheres larger and well convoluted. Facial portion of skull produced into a long, tapering, tubular rostrum, at the end of which the anterior nares are situated. Rami of mandible slender, styliform. Opening of mouth small, and placed below the extremity of the rostrum. No teeth or laterally placed horny plates, though the palate and tongue are furnished with spines. Tongue very long, vermiform, slender, and protractile. Lining membrane of small intestine villous, but without transverse folds. Feet not webbed, but with long strong claws fitted for scratching and burrowing. The hinder feet with the ends of the toes turned outwards and backwards in the ordinary position of the animal when on the ground. Tail very short. Acetabulum with a large perforation, as in Birds. Calcaneal spur and gland of the male much smaller than in *Ornithorhynchus*. Fur intermixed with strong, sharp-pointed spines. Terrestrial and fossorial in habits, feeding exclusively on ants, and recalling in the structure of the mouth and various other parts relating to their peculiar mode of life the true Anteaters of the order Edentata.

The Echidnas or Spiny Anteaters constitute a family which appears in some respects to be less specialised than the *Ornithorhynchidæ*. According to Mr. O. Thomas, all the living forms may be included in two species, which, with some hesitation, are referred to two genera—*Echidna* and *Proechidna* (*Acanthoglossus*).

Echidna.[1]—In *Echidna* there are five toes, all of which are provided with claws, those of the fore feet being broad, slightly curved, and directed forwards, while the posterior ones are slender, more curved, and inclined outwardly. The beak is about as long as the rest of the head, and either nearly straight, or slightly curved upwards, while the palate is comparatively wide, and but slightly vaulted. The number of the vertebræ is C 7, D 16, L 3, S 3, Ca 12. The one existing representative of the genus (*E. aculeata*) occurs in New Guinea, Tasmania, and Australia.

So much variation is displayed by this animal, that it has been divided into several species, but the latest researches tend to show that these variations cannot be regarded as indicating more than races, of which there are three well-marked types.

The first race, or variety, has been termed the Port Moresby Echidna, and is only known from that Papuan locality. It is distinguished from the typical form by its smaller size, by the shorter spines on the back, which admit of the fur being seen, and by the more spinous covering of the head, belly, and limbs, as well as by the lighter skull and relatively larger beak.

The typical variety is confined to the Australian mainland, and is of medium size. The spines of the back are very long and stout, often reaching a length of two inches, and almost completely concealing the hair. The colour of these spines varies from yellow at the roots to black at the tips, but some may be altogether yellow. The hair of the back is black or dark brown in colour, but it may be occasionally absent, or in the region of the loins may exceed the spines in length. The limbs and under surface of the body are covered with dark brown hair, thinly interspersed with short spines; and the hair of the face is of the same general hue as that of the body. The skull has a slender rostrum and a flat and narrow brain-case.

In the third or Tasmanian race, which is confined to Tasmania, the average size is somewhat larger than in the typical form. The most characteristic feature is, however, the shortness of the spines of the back, which in the greater part of that region are almost or quite concealed by the hairs. The hairs of the back are dark brown, those of the under surface and sides of the head being generally rather paler. There is often a white spot on the chest. Very frequently there is a difference in the proportionate lengths of the hinder claws from those of the typical race. In the skull the beak is comparatively short and stout, and the brain-case large and wide.

Echidnas are usually found in rocky districts, and more especially

[1] Cuvier, *Tableau Élémentaire d'Hist. Nat.* p. 143 (1798).

in the mountains. In a wild state they live mainly on ants. Specimens have been brought to this country and kept in the Zoological Society's Gardens; and in captivity they will readily eat eggs, and bread-and-milk. They are able, however, to endure long fasts, an individual having been known to go without food for upwards of a month.

These animals seem to be mainly of nocturnal habits, and if brought out during the day-time appear to be sluggish and stupid, crouching to the ground with the head between the legs, and thus presenting a mass of spines to an enemy. They burrow rapidly in soft ground, sinking directly downwards, and not going head forwards. A specimen placed on a large chest of earth containing plants reached the bottom in less than two minutes; and it is said that the muzzle assists in the work of burrowing.

Proechidna.[1]—The one known representative of the genus *Proechidna* (Fig. 33) attains dimensions about equal to those of

FIG. 33.—The Three-toed Echidna (*Proechidna bruijnii*). From Gervais.

the largest race of *Echidna aculeata*. The skull is less depressed than in the latter, with the anterior portion of the palate very concave, and the deflected beak nearly twice the length of the remainder of the skull. As a rule, there are only three claws to each foot; but the first and fifth digits are represented by several phalanges, and one instance is known where there are five complete claws on the anterior and four on the posterior feet. There are two more vertebræ in the dorsal and lumbar region than in *Echidna*.

The head and body are covered with a thick coat of hair, among which there are a number of short spines in the region of the back, which are much less numerous than in the typical race of the last species. The colour of the fur is generally dark brown or black, but the head may be almost white; and the spines are usually entirely white, although in certain cases they may be brown at the root.

[1] Gervais, *Ostéographie des Monotremes*, p. 43 (1877).

This species is known only from New Guinea, the recorded specimens being from the north-western regions of that country. It inhabits rocky ground, and dwells chiefly in the mountains, the specimens which were first described having been obtained at an elevation of about 3500 feet above the sea level. The Papuans capture it by digging trenches in the ground to a depth of about a yard, by which means they generally come upon its runs.

Fossil Species.—Remains of a species of Echidna of very much larger size than the existing forms have been obtained from the cave-deposits of New South Wales, which appear to be of Pleistocene age. This species was named *Echidna oweni* by the late Mr. Krefft, but was subsequently called *E. ramsayi* by Sir R. Owen. In referring this species to the genus *Echidna*, that term must be regarded as including *Proechidna.*

CHAPTER VI

THE SUBCLASS METATHERIA OR DIDELPHIA

General Characters.—The Metatheria or Didelphia are represented at present by numerous species, presenting great diversities of general appearance, structure, and habits, although all united by many essential anatomical and physiological characters, which, taken altogether, give them an intermediate position between the Prototheria and the Eutheria.

Although the striking differences in external form, in many anatomical characters, and in mode of life of various animals of this section might lead to their division into groups equivalent to the orders of the Eutheria, it is more convenient on the whole to adhere to the usual custom of treating them all as forming one order called MARSUPIALIA,[1] the limits of which are therefore equivalent to that of the subclass. The more essentially distinctive characters are as follows.

In the structure of the brain and the presence of epipubic bones they agree with the Prototheria, while in the structure of the ear-bones and the shoulder-girdle and the presence of teats on the mammary glands they resemble the Eutheria, the reproductive organs belonging to neither one nor the other type, but having a special character representing an intermediate grade of development. The ureters open into the base of the bladder. The oviducts are differentiated into uterine and Fallopian portions, and open into a long and distinct vagina, quite separate from the cystic urethra. The penis is large, but its crura are not directly attached to the ischia. The spongy body has a large bifurcated bulb. The young are born in an exceedingly rudimentary condition, and are never nourished by means of an allantoic placenta, but are transferred to the nipple of the mother, to which they remain firmly

[1] For the detailed characters of all the genera and species of Marsupials the reader should consult the British Museum *Catalogue of Marsupialia and Monotremata*, by Oldfield Thomas, 1888.

attached for a considerable time, nourished by the milk injected into the mouth by compression of the muscle covering the mammary gland. They are therefore the most typically mammalian of the whole class. The nipples are nearly always concealed in a fold of the abdominal integument or "pouch" (marsupium) which serves to support and protect the young in their early helpless condition.

Entering more fully into the characters of the subclass, which are also those of the order Marsupialia, it may be observed that the brain is generally small in proportion to the size of the animal, and the surface-folding of the cerebral hemispheres, though well marked in the larger species, is never very complex in character, and is absent in the medium-sized and smaller species. The arrangement of the folding of the inner wall of the cerebrum differs essentially from that of all known Eutheria, the hippocampal fissure being continued forward above the corpus callosum, which is of very small size. The anterior commissure is, on the other hand, greatly developed.

The teeth are always divisible, according to their position and form, into incisors, canines, premolars, and molars; but they vary much in number and character in the different families. Except in the genus *Phascolomys*, the number of incisors in the upper and lower jaws is never equal. The true molars are very generally four in number on either side of each jaw. The chief peculiarity in the dentition lies, however, in the mode of succession. Thus there is no vertical displacement and succession of the teeth, except in the case of a single tooth on either side of each jaw, which is always the hindermost of the premolar series, and is preceded by a tooth having more or less of the characters of a true molar (see Fig. 34); this deciduous tooth being the only one comparable to the "milk-teeth" of the diphyodont Eutheria. In some cases (as in *Potorous*) this tooth retains its place and function until the animal has nearly, if not quite, attained its full stature, and is not shed and replaced by its successor until after all the other teeth of the permanent series, including the posterior molars, are fully in place and use. In others, as the Thylacine, it is very rudimentary in form and size, being shed or absorbed before any

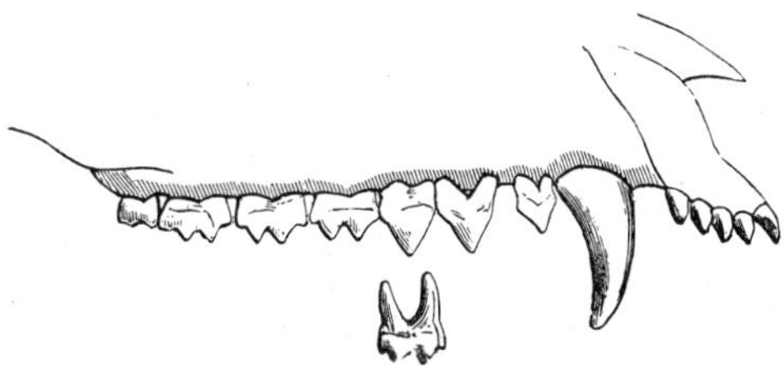

FIG. 34.—Teeth of upper jaw of Opossum (*Didelphys marsupialis*), all of which are unchanged, except the last premolar, the place of which is occupied in the young animal by a molariform tooth, represented in the figure below the line of the other teeth.

of the other teeth have cut the gum, and therefore quite functionless. It must further be noted that there are some Marsupials, as the Wombat, *Myrmecobius*, and the Dasyures, in which no such milk-tooth, even in a rudimentary state, has yet been discovered, possibly in some cases from want of materials for observation at the right stage of development.

Epipubic or marsupial bones are present in both sexes of nearly all species. In one genus alone, *Thylacinus*, they are not ossified. The number of dorso-lumbar vetebræ is always nineteen, although there are some apparent exceptions caused by the last lumbar being modified into a sacral vertebra. The number of pairs of ribs is nearly always thirteen. The tympanic bone remains permanently distinct. The carotid canal perforates the basisphenoid. The lachrymal foramen is situated upon or external to the anterior margin of the orbit, and there are generally large vacuities in the bony palate. The angle of the mandible is (except in *Tarsipes*) more or less inflected. The hyoid bones have always a peculiar form, consisting of a small, more or less lozenge-shaped basi-hyal, broad cerato-hyals, with the remainder of the anterior arch usually unossified, and stout, somewhat compressed thyro-hyals. There are two anterior venæ cavæ,[1] into each of which a "vena azygos" enters. In the male the testes are always contained in a scrotum, which is suspended by a narrow pedicle to the abdomen in front of the penis. The vasa deferentia open into a complete and continuous urethra, which is also the passage by which the urine escapes from the bladder, and is perfectly distinct from the passage for the fæces, although the anus and the termination of the urethro-sexual canal are embraced by the same sphincter muscle. The glans is often bifurcated anteriorly. In the female the oviducts never unite to form a common cavity or uterus, but open separately into the vagina, which at least for part of its course is double. The mammæ vary much in number, but are always abdominal in position, having long teats, and in most of the species are more or less enclosed in a fold of the integument forming a pouch or marsupium, though in some this is entirely wanting, and the newly-born, blind, naked, and helpless young, attached by their mouths to the teat, are merely concealed and protected by the hairy covering of the mother's abdomen. In this stage of their existence they are fed by milk injected into their stomach by the contraction of the muscles covering the mammary gland, the respiratory organs being modified temporarily, much as they are permanently in the Cetacea—the elongated upper part of the larynx projecting into the posterior nares, and so maintaining a free communication between the lungs and the external surface

[1] Except in *Petaurus* (*Belideus*) *breviceps* (Forbes, *Proc. Zool. Soc.* 1881, p. 188).

independently of the mouth and gullet, thus averting the danger of suffocation while the milk is passing down the latter passage.

Distribution.—The existing species of Marsupials are, with the

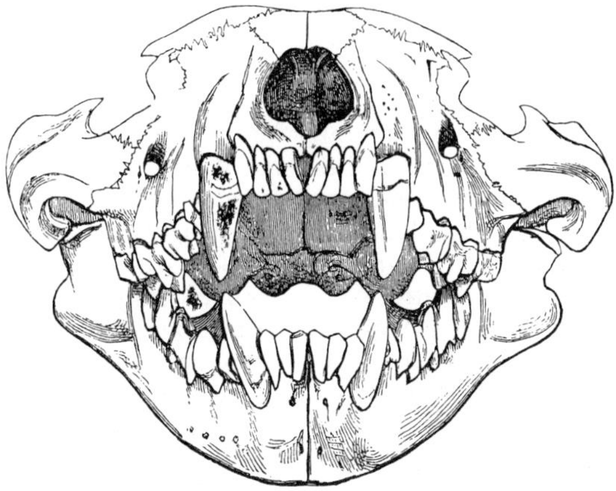

FIG. 35.—Front view of skull of *Sarcophilus ursinus*, showing polyprotodont and carnivorous dentition (*Quart. Journ. Geol. Soc.* vol. xxiv. p. 313).

exception of one family (the *Didelphyidæ*), limited in geographical distribution to the Australasian region,[1] forming the chief mammalian fauna of Australia, New Guinea, and some of the adjacent islands. The *Didelphyidæ* are almost purely Neotropical, one or two species ranging northwards into the Nearctic region. Fossil remains of members of this family have also been found in Europe and America in strata of the Eocene and early Miocene periods; and it is probable that at least many of the polyprotodont Mesozoic mammals noticed in Chapter IV. are referable to the Marsupialia.

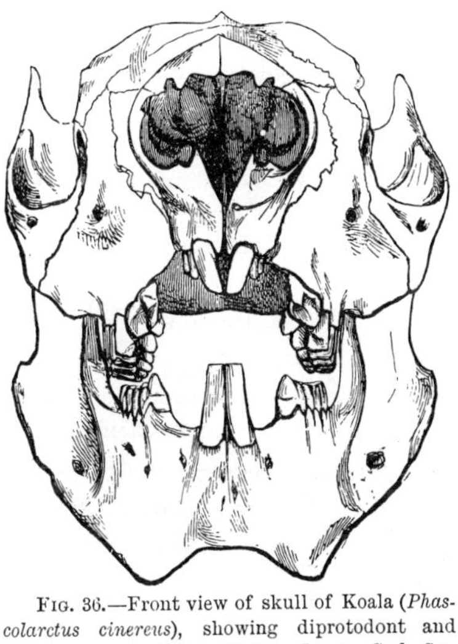

FIG. 36.—Front view of skull of Koala (*Phascolarctus cinereus*), showing diprotodont and herbivorous dentition (*Quart. Journ. Geol. Soc.* vol. xxiv. p. 313).

Classification.—In dividing the Marsupials into minor groups, it may be observed that one of the most obvious distinctive characters among them is derived from the form and arrangement of the teeth.

[1] Including the transitional Austro-Malayan region.

In certain species, as the Opossums, Dasyures, and Thylacine, the incisors are numerous, small, and subequal in size, and the canines large, as in the typical placental Carnivores (Fig. 35). To these the term "polyprotodont" is applied, and they are all more or less carnivorous in their habits. In others the central incisors are very prominent, and the lateral incisors and canines absent or subordinate in function (Fig. 36). These are called "diprotodont," and they are all wholly or in great part vegetable feeders. In one group of these, the Wombats, there are but two incisors above and the same number below; but all the others, including the Kangaroos, Koalas, and Phalangers, have two functional incisors below and as many as six above, three on each side, but of these the first or central pair is the most fully developed.

Some hesitation has frequently been expressed as to whether the Polyprotodont and Diprotodont types are entitled to constitute distinct primary groups, owing to the presence of syndactylism among the *Peramelidæ* in the former, as well as in the latter; but if Mr. O. Thomas is right in regarding this feature as acquired independently in the two groups we may safely adopt such a division. Taking various combinations into consideration, the existing Marsupials readily group themselves into six very natural families, the leading characters of which may be summarised as follows:—

Order MARSUPIALIA.

A. POLYPROTODONTIA.—Incisors numerous, small, subequal. Canines larger than the incisors. Molars with sharp cusps.

α. Incisors $\frac{5}{4}$. Hind feet with the four outer toes subequal, distinct, and a well-developed opposable hallux. *Didelphyidæ.*

β. Incisors $\frac{4}{3}$. Hind feet with four outer toes distinct. Hallux small or rudimentary, rarely opposable. *Dasyuridæ.*

γ. Incisors $\frac{(4-5)}{3}$. Hind feet long and narrow. Fourth toe larger than the others. Hallux rudimentary or absent. Second and third toes very slender, and united in a common integument (syndactylous). *Peramelidæ.*

B. DIPROTODONTIA.—Incisors not exceeding $\frac{3}{3}$, usually $\frac{3}{1}$, but occasionally $\frac{1}{1}$. Central (first) upper and lower incisors large and cutting. Upper canines generally, and lower invariably, absent or small. Molars with bluntly tuberculated or transversely ridged crowns.

α. Teeth with persistent pulps. Incisors $\frac{1}{1}$, large, scalpriform, with enamel on the outer surface only. No canines. Hind feet with four subequal outer toes, partially syndactylous, and with rudimentary hallux. *Phascolomyidæ.*

β. Teeth rooted. Three upper incisors and a canine. Hind limbs not disproportionately large. Feet syndactylous, broad, with four subequal outer toes, and a large opposable hallux. *Phalangeridæ.*

γ. Teeth rooted. Three upper incisors, and frequently a canine. Hind limbs disproportionately large, with syndactylous feet as in *Peramelidæ*. *Macropodidæ.*

Suborder POLYPROTODONTIA.

The leading characters of this group are given in the foregoing schedule. This group is the only one represented at the present day, and so far as we know also in past epochs, beyond the confines of the Australasian region and adjacent islands.

Family DIDELPHYIDÆ.

Dentition: $i\ \frac{5}{4}$, $c\ \frac{1}{1}$, $p\ \frac{3}{3}$, $m\ \frac{4}{4}$; total 50. Incisors very small and pointed. Canines large. Premolars with compressed pointed crowns. Molars with numerous sharp cusps. The last premolar preceded by a deciduous multicuspidate milk-molar, which remains in place until the animal is nearly adult (Fig. 34). Limbs of moderate development, each with five complete and distinct toes, all of which are provided with short, compressed, curved, sharp claws of nearly equal size, except the first toe of the hind foot or hallux (Fig. 37), which is large, widely separable from the others, to which it is opposed in climbing, and terminates in a dilated rounded extremity, without a nail. Tail generally long, partially naked and prehensile. Stomach simple. Cæcum of small or moderate size. Pouch generally absent, sometimes represented by two lateral folds of the abdominal integument, partially covering the teats, rarely complete. Vertebræ: C 7, D 13, L 6, S 2, C 19-35.

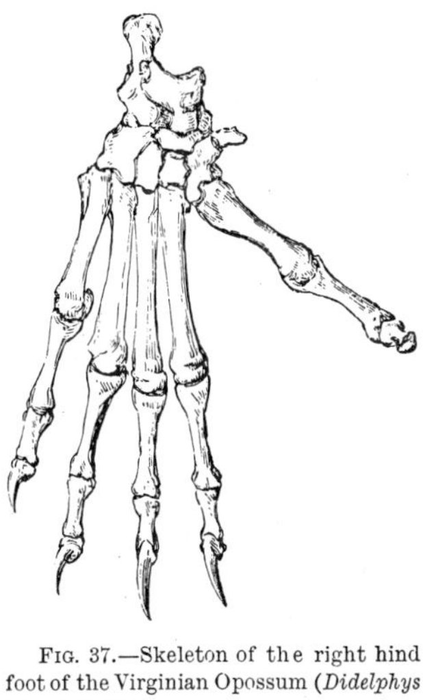

FIG. 37.—Skeleton of the right hind foot of the Virginian Opossum (*Didelphys marsupialis*).

The *Didelphyidæ*, or true Opossums, differ from all other existing Marsupials in their habitat, being peculiar to the American continent. They are mostly carnivorous or insectivorous in their diet, and arboreal in habits.

Opossums occur throughout the greater part of the American

continent, ranging from the United States to Patagonia, the greater number of species being found in the warmer regions. In South America the opossums take the place of the Eutherian Insectivora, and the sharp cusps on their teeth are admirably adapted for crushing the insects on which they mainly subsist.

Chironectes.[1]—The family comprises two genera only, namely *Didelphys*, containing all the species, with the exception of the curious Yapock, which forms by itself the genus *Chironectes*, and is distinguished from all other Opossums by its webbed feet, non-tuberculated soles, and peculiar coloration. Its ground colour is light gray, with four or five sharply-contrasted brown bands passing across its head and back, and thus giving it a very peculiar mottled appearance. It is almost wholly aquatic in its habits, living on small fish, crustaceans, and water insects. Its range extends from Guatemala to southern Brazil.

Didelphys.[2]—The type genus *Didelphys* is a very large one, containing, according to Mr. O. Thomas, twenty-three existing species. It may be divided into five groups, or sub-genera, all of which have received distinct names. The typical group is represented only by the common or Virginian Opossum (*D. marsupialis*), of which the numerous varieties have received separate specific names. This species is of large size, with a long, scaly, prehensile tail, and long bristle-like hairs mingled with the fur. The pouch is complete. It ranges over all temperate North America, and is also found in central and tropical South America, where it is commonly known as the Crab-eating Opossum. This animal is extremely common, being even found living in the towns, where it acts as a scavenger by night, retiring for shelter by day upon the roofs of the houses or into the sewers. The female produces in the spring from six to sixteen young ones, which are placed in her pouch immediately after birth, and remain there until able to take care of themselves.

The second or *Metachirine* group includes three species found all over the tropical parts of the New World. They are of medium size, with short close fur, very long, scaly, and naked tails, and less developed ridges on their skulls than in the type species. As a rule there is no pouch adapted to carry the young, which commonly ride on their mother's back, holding on by winding their prehensile tails round hers. The *Philanderine* group is closely allied to the preceding, but is readily distinguished by the woolly hair, and the brown streak down the middle of the face. The Woolly Opossum (*D. lanigera*), which is represented in the accompanying woodcut (Fig. 38) carrying its young in the fashion mentioned above, is one of the two species of this group. In the

[1] Illiger, *Prod. Syst. Mamm. et Aves*, p. 76 (1811).
[2] Linn. *Syst. Nat.* Ed. 12, vol. i. p. 71 (1766).

fourth or *Micoureine* group the numerous species are all smaller than in the preceding groups, and have short and close hair, and no dark streak down the face. The best known species is the Murine Opossum (*D. murina*), little larger than a House-Mouse, and of a bright red colour, which is found as far north as central Mexico, and extends thence right down to the south of Brazil. The last or *Peramyne* group contains several extremely shrew-like species, of very small size, with short, hairy, and usually non-prehensile tails, not half the length of the trunk, and with wholly unridged skulls. The most striking member of the group is the Three-striped Opossum (*D. americana*), from Brazil, which is of a reddish-gray colour, with three clearly-defined deep-black bands

FIG. 38.—The Woolly Opossum (*Didelphys lanigera*).

down its back, very much as in some of the striped mice of Africa.

The numerous fossil species of Opossum found in the Upper Eocene and Lower Miocene of Europe are of especial interest from a distributional point of view, since they indicate how the Opossums of America may have been connected with the Australian Marsupials. These forms were originally referred to *Didelphys*, but have been subsequently described as *Peratherium* and *Amphiperatherium*. The characters of the molar teeth on which these genera are based do not appear to be sufficiently important to justify their separation from *Didelphys*. Allied forms occur in the Tertiaries of North America, which were originally described under the name of *Herpetotherium*, but have been subsequently referred to *Peratherium*. Remains of many of the existing species of Opossum are found in a fossil condition in the Pleistocene cave-deposits of Brazil.

Family DASYURIDÆ

Dentition: $i\,\frac{4}{3}$, $c\,\frac{1}{1}$, p and m numerous, variable. Incisors small; canines well developed; molars with pointed cusps. Limbs equal. Fore feet with five subequal toes terminating in claws. Hind feet with the four outer toes well developed, and distinct from each other and bearing claws; the first (or hallux) clawless, generally rudimentary, sometimes entirely wanting. Stomach simple. No cæcum. Predatory carnivorous or insectivorous animals, inhabitants of Australia, Tasmania, and the southern parts of New Guinea and some of the adjacent islands. The aberrant genus *Myrmecobius*, though clearly a member of this family, is so sharply distinguished

FIG. 39.—The Thylacine (*Thylacinus cynocephalus*).

from all the others as to render a division into two subfamilies necessary.

Subfamily **Dasyurinæ.**—This comprises the more typical *Dasyuridæ*, in which the premolars and molars never exceed the normal number of seven on either side of each jaw, and in which the tongue is not specially extensile.

Thylacinus.[1]—Dentition: $i\,\frac{4}{3}$, $c\,\frac{1}{1}$, $p\,\frac{3}{3}$, $m\,\frac{4}{4} = 46$. Incisors small, vertical, the outer one in the upper jaw larger than the others. Summits of the lower incisors, before they are worn, with a deep transverse groove dividing them into an anterior and a posterior cusp. Canines long, strong, and conical. Premolars separated from one another by intervals, with compressed crowns, increasing in size from before backwards. True molars in general characters re-

[1] Temminck, *Monographies de Mammalogie*, vol. i. p. 60 (1827).

sembling those of *Dasyurus*, but of more simple form, the cusps being not so distinct nor sharply pointed. Milk-molar very small, and shed before the animal leaves the mother's pouch. Humerus with an entepicondylar foramen. General form very Dog-like. Head elongated. Muzzle pointed. Ears moderate, erect, triangular. Fur short and closely applied to the skin. Tail of moderate length, thick at the base and tapering towards the apex, clothed with short hair. Hallux (including the metacarpal bone) wanting. Vertebræ: C 7, D 13, L 6, S 2, C 23. Marsupial bones represented only by small unossified fibro-cartilages.

The only known existing species of this genus, *T. cynocephalus* (Fig. 39), though smaller than a common Wolf, is the largest predaceous Marsupial at present living. It is now entirely confined to the island of Tasmania, although fragments of bones and teeth found in caves afford evidence that a closely allied species once inhabited the Australian mainland. The general colour of the Thylacine is

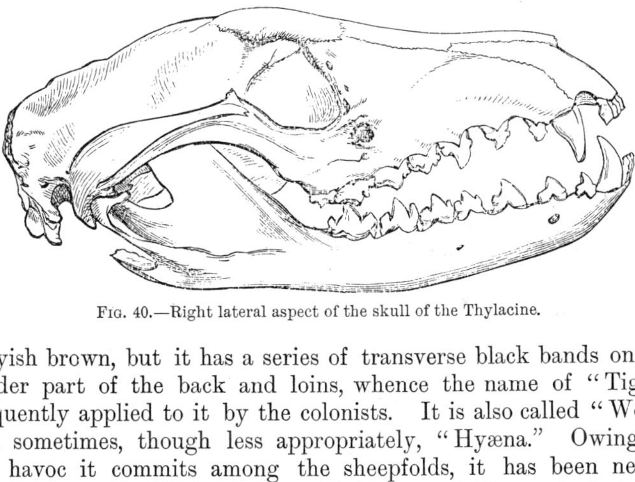

FIG. 40.—Right lateral aspect of the skull of the Thylacine.

grayish brown, but it has a series of transverse black bands on the hinder part of the back and loins, whence the name of "Tiger" frequently applied to it by the colonists. It is also called "Wolf," and sometimes, though less appropriately, "Hyæna." Owing to the havoc it commits among the sheepfolds, it has been nearly exterminated in all the more settled parts of Tasmania, but still finds shelter in the almost impenetrable rocky glens of the more mountainous regions of the island. The female produces four young at a time. The pouch opens backwardly, and there are four mammæ. The figure of the skull exhibits the peculiar Dog-like form so characteristic of the genus.

Sarcophilus.[1]—Dentition: $i\,\frac{4}{3}$, $c\,\frac{1}{1}$, $p\,\frac{2}{2}$, $m\,\frac{4}{4}$. Upper incisors nearly equal, and placed vertically, the first not differentiated from the rest. Premolars rounded and closely crowded between the canine and molars, with broad crowns; molars broad and heavy, the last one without a distinct hind talon. Form thick and powerful;

[1] F. Cuvier, *Hist. Nat. des Mammifères*, iv. (1837).

head disproportionately large for the body; muzzle short and broad; ears broad and rounded; tail of moderate length, and evenly hairy. Hallux wanting; soles of feet naked, without defined pads. Humerus with entepicondylar foramen.

This genus is now represented only by a single species (*S. ursinus*) found in Tasmania, where, from its ferocious and destructive habits, it is commonly known under the name of the "Devil." A front view of the skull is shown in Fig. 35.

The prevailing colour of this animal is black, and the size about equal to that of an English Badger; its habits are fossorial, and it is very destructive to sheep. On account of the similarity in the number of its teeth this genus has been generally included in the next one, but in the structure of the teeth it is much nearer to *Thylacinus*. An extinct species is found in the Pleistocene deposits of the mainland of Australia.

It may be observed that the two premolars missing from the typical series of four in this and the next genus are the second and the fourth; the fourth milk-molar being likewise absent. In *Thylacinus* and other Polyprotodonts with three premolars it is the second that is missing.

Dasyurus.[1]—Dentition: $i\ \frac{4}{3}$, $c\ \frac{1}{1}$, $p\ \frac{2}{2}$, $m\ \frac{4}{4}$; total 42. Upper incisors nearly equal, and placed vertically; first slightly longer, narrower, and separated from the rest. Lower incisors sloping forwards and upwards. Canines large and sharply pointed. Premolars with compressed and sharp-pointed crowns, and slightly developed anterior and posterior accessory basal cusps. True molars with numerous sharp-pointed cusps. In the upper jaw the first three with crowns having a triangular oral surface, the fourth small, simple, narrow, and placed transversely. In the lower jaw the molars more compressed, with longer cusps; the fourth not notably smaller than the others. Form viverrine. Ears long and narrow, prominent, and obtusely pointed. Hallux rudimentary, or absent; its metatarsal bone always present. Tail long and well clothed with hair. Humerus without an entepicondylar foramen. Vertebræ: C 7, D 13, L 6, S 2, C 18-20.

The Dasyures are small Civet-like animals with a gray or brown pellage profusely spotted with white; they are mostly inhabitants of the Australian continent and Tasmania, where in the economy of nature they take the place of the smaller predaceous Carnivora, the Cats, Civets, and Weasels of other parts of the world. They hide themselves in the daytime in holes among rocks or in hollow trees, but prowl about at night in search of the small living mammals and birds which constitute their prey. The species are not numerous, and include *D. maculatus*, about the size of a common Cat, inhabiting Tasmania and the southern part of Australia; *D. viver-*

[1] Geoffroy, *Bull. Soc. Philom.* vol. i. p. 106 (1796).

rinus, Tasmania and Victoria; *D. geoffroyi*, nearly all Australia; *D. hallucatus*, North Australia; *D. albopunctatus*, New Guinea.

Remains referred to *D. viverrinus* occur in the Australian Pleistocene deposits.

Phascologale.[1]—This genus comprises a considerable number of small Marsupials, none of them exceeding a common Rat in size, differing from the Dasyures in possessing an additional premolar—the dentition being $i\,\frac{4}{3}$, $c\,\frac{1}{1}$, $p\,\frac{3}{3}$, $m\,\frac{4}{4}$; total 46,—and having the teeth generally developed upon an insectivorous rather than a carnivorous pattern, the upper middle incisors being larger and inclined forwards, the canines relatively smaller, and the molars with broad crowns, armed with prickly tubercles. The muzzle is pointed. Ears moderately rounded and nearly naked. Feet broad and short. Fore feet with five subequal toes, having compressed, slightly curved, pointed claws. Hind feet with the four outer toes subequal, having claws similar to those in the fore feet; the hallux always distinct and partially opposable, though small and nailless. Tail long, very variable in its covering, being either bushy, crested, or nearly naked. Pouch represented merely by a few folds of skin. Mammæ varying from four to ten in number. The food of these animals is almost entirely insects; some species pursuing their prey among the branches of trees, while others are purely terrestrial. They are found throughout Australia, and also in New Guinea and the Aru and some of the adjacent islands.

P. cristicaudata, a species with a thick compressed tail ornamented upon its apical half with a crest of black hair, differs from the others by the very reduced size of the fourth premolar in the upper, and its complete absence in the lower jaw, thus forming an interesting transition in dentition towards *Dasyurus*. It constitutes the genus *Chætocercus* of Krefft, but is included by Mr. O. Thomas in *Phascologale*, the frequent absence of the fourth lower premolar in *P. thorbeckiana* indicating that the total absence of this tooth in the known specimens of this species cannot be regarded as of generic importance. All the members of this and the two following genera can be at once distinguished from *Dasyurus* by the absence of white spots on the fur.

Sminthopsis.[2]—The genus *Sminthopsis* includes several small species allied to *Phascologale* but characterised by the narrowness of the hind foot, and by the soles of the feet being either granulated or hairy, instead of naked.

Antechinomys.[3]—The last genus of the *Dasyurinæ* is *Antechinomys*, represented only by *A. laniger* of Queensland and New South Wales. This elegant little mouse-like creature, which has large oval ears and

[1] Temminck, *Monographies de Mammalogie*, vol. i. p. 56 (1827).

[2] Thomas, *Ann. Mus. Genov.* ser. 2, vol. iv. p. 503 (1887).

[3] Krefft, *Proc. Zool. Soc.* 1866, p. 434.

a long tail with the terminal part bushy, is distinguished from *Sminthopsis* by the absence of the hallux and the great elongation of the limbs. The tympanic bullæ of the skull are also unusually large, with the mastoid portion much swollen. A full account of the habits and anatomy of this animal, which appears to be of very rare occurrence, is given in the *Proc. Zool. Soc.* 1880, p. 454.

Subfamily **Myrmecobiinæ.**—Molars and premolars exceeding the normal number of seven on each side. Tongue, long cylindrical, and extensile.

Myrmecobius.[1]—Dentition : $i\ \frac{4}{3}$, $c\ \frac{1}{1}$, $p\ \frac{3}{3}$, $m\ \frac{5}{5}$ or $\frac{6}{6}$; total 52 or 56,

FIG. 41.—*Myrmecobius fasciatus.* From Gould.

being the largest number of teeth in any existing Marsupial. The distinction between the molars and premolars is founded not on a knowledge of the succession of the teeth, but on their form. The teeth are all small and (except the four posterior inferior molars) separated from each other by an interval. Head elongated, but broad behind. Muzzle long and pointed. Ears of moderate size, ovate, and rather pointed. Fore feet with five toes, all having strong, pointed, compressed claws, the second, third, and fourth nearly equal, the fifth somewhat, and the first considerably, shorter. Hind feet with no trace of hallux externally, but the metatarsal bone

[1] Waterhouse, *Proc. Zool. Soc.* 1836, p. 69.

present. Tail long, clothed with long hairs. Fur rather harsh and bristly. Female without any pouch, the young when attached to the nipples being concealed only by the long hair of the abdomen. Vertebræ: C 7, D 13, L 6, S 3, C 23. A gland on the under surface of the body just in advance of the sternum.

Of this singular genus but one species is known, *M. fasciatus* (Fig. 41), found in western and southern Australia. It is about the size of an English squirrel, to which animal its long bushy tail gives it some resemblance; but it lives entirely on the ground, especially in sterile, sandy districts, feeding on ants. Its prevailing colour is chestnut-red, but the hinder part of the back is elegantly marked with broad, white, transverse bands on a dark ground.

The special interest of this form lies in its apparent relationship to those Mesozoic mammals which possess a large number of true molars (see p. 114); and it is suggested by Thomas that it may eventually be found advisable to include some of the latter in the present subfamily.

Family PERAMELIDÆ.

Dentition: $i\,\frac{(4-5)}{3}$, $c\,\frac{1}{1}$, $p\,\frac{3}{3}$, $m\,\frac{4}{4}$; total 46 or 48. Upper incisors small, with short broad crowns. Lower incisors moderate, narrow, proclivous. Canines well developed. Premolars compressed, pointed. Molars with quadrate tuberculated crowns. Fourth premolar preceded by a small molariform tooth, which remains in place until the animal is nearly full grown. Fore feet with two or three of the middle toes of nearly equal size, and provided with strong, sharp, slightly curved claws; the other toes rudimentary. Hind feet long and narrow; the hallux rudimentary or absent; the second and third toes very slender, and united in a common integument; the fourth very large, with a stout elongated conical claw; the fifth smaller than the fourth (see Fig. 43). The ungual phalanges of the large toes of both feet cleft at their extremities (as in *Manis* among the Edentata, but in no other Marsupials). Head elongated. Muzzle long, narrow, and pointed. Stomach simple. Cæcum of moderate size. Pouch complete, opening backwards. Alone among Marsupials they have no clavicles.

The *Peramelidæ* form a very distinct family, in some respects intermediate between the sarcophagous *Dasyuridæ* and the phytophagous *Macropodidæ*. In dentition they resemble the former, but they agree with the latter in the peculiar structure of the hind feet. In the construction of the fore feet they differ from all other Marsupials.

The Bandicoots, as these Marsupials are popularly termed, are of fossorial habits, and subsist either on an insectivorous or omnivorous diet. It has been generally considered that their syndactylous feet indicate direct affinity with the Diprotodonts, but owing to the essentially Polyprotodont character of the organisation—which extends even to their carpal and tarsal bones—Thomas dissents from this view, and concludes that their syndactylism is an independently acquired character, and that they are really a direct offshoot from the *Dasyuridæ*. Some individuals are remarkable for the presence of a longitudinal groove in the root of the canines, by which feature they approximate to some of the Mesozoic Polyprotodont forms. They may be divided into three genera.

Perameles.[1]—Anterior and posterior extremities not differing greatly in development. Fore feet with the three middle toes well

FIG. 42.—*Perameles gunni*. From Gould.

developed, the third slightly larger than the second, the fourth somewhat shorter, provided with long, strong, slightly curved, pointed claws. First and fifth toes very short and without claws. Hind feet with hallux of one or two phalanges, forming a distinct tubercle visible externally; the second and third toes very slender, of equal length, joined as far as the ungual phalanges, but with distinct claws; the fifth intermediate in length between these and the largely developed fourth toe. Ears of moderate or small size, ovate, pointed. Tail rather short, clothed with short adpressed hairs. Fur short and harsh. Vertebræ; C 7, D 13, L 6, S 1, C 17. Skull long and narrow, with the bulla single, and its mastoid portion not inflated.

The animals of this genus are all small, and live entirely on the ground, making nests composed of dried leaves, grass, and sticks in

[1] Geoffroy, *Bull. Soc. Philom.* vol. iii. p. 249 (1803).

hollow places. They are rather mixed feeders; but insects, worms, roots, and bulbs constitute their ordinary diet. The various species are widely distributed over Australia, Tasmania, New Guinea, and several of the adjacent islands, as Aru, Kei, and New Ireland. The best known are—*P. gunni* (Fig. 42), *bougainvillei*, *nasuta*, *obesula*, and *macrura* from Australia, and *P. doreyana*, *raffrayana*, and *longicaudata* from New Guinea.

Remains apparently referable to existing species are found in the cave-deposits of New South Wales.

Peragale.[1]—Molar teeth curved, typically with longer crowns and shorter roots than in the last. Hinder extremities proportionally longer, and hallux without claw. Muzzle much elongated and narrow. Fur soft and silky. Ears very large, long, and pointed. Tail long, its apical half clothed on the dorsal surface with long hairs which form a crest. Vertebræ: C 7, D 13, L 6, S 2, C 23. Skull distinguished from that of *Perameles* by the large size and double structure of the auditory bulla, of which the mastoid portion is inflated. There is also an abrupt contraction of the muzzle at the third premolar.

The type species of Rabbit-Bandicoot (*P. lagotis*), as these animals are called, is found in Western Australia, and also occurs fossil in the cave-deposits of New South Wales. It is the largest member of the family, being about the size of the common Rabbit, to which animal it bears sufficient superficial resemblance to have acquired the name of "Native Rabbit" from the colonists. It burrows in the ground, but in other respects resembles the true Bandicoots in its habits.

The smaller *P. leucura* has short-crowned molars, with distinct cusps, which are almost obsolete in the type species.

Chœropus.[2]—Dentition generally resembling that of *Perameles*, but the canines are less developed, and in the upper jaw two-rooted. Limbs very slender; posterior nearly twice the length of the anterior. Fore feet with the functional toes reduced to two, the second and third, of equal length, with closely united metacarpals and short, sharp, slightly curved, compressed claws. First toe represented by a minute rudiment of a metacarpal bone; the fourth by a metacarpal and two small phalanges without a claw, and not reaching the middle of the metacarpal of the third; fifth entirely absent. Hind foot (Fig. 43) long and narrow, mainly composed of the strongly developed fourth toe, terminating in a conical pointed nail, with a strong pad behind it; the hallux absent or represented by a rudimentary metatarsal; the remaining toes completely developed, and with claws, but exceedingly slender; the united second and third reaching a little way beyond the metatarso-phalangeal articulation of

[1] Gray, in *Grey's Australia*, vol. ii. p. 401 (1841).
[2] Ogilby, *Proc. Zool. Soc.* 1838, p. 25.

the fourth; the fifth somewhat shorter. Tail not quite so long as the body, and covered with short hairs forming a slight crest. Ears large and pointed, and folded down when the animal is at rest. Fur soft and loose. Vertebræ: C 7, D 13, L 6, S 1, C 20. Skull short and wide, with a small and single bulla, and a contraction of the muzzle at the third premolar.

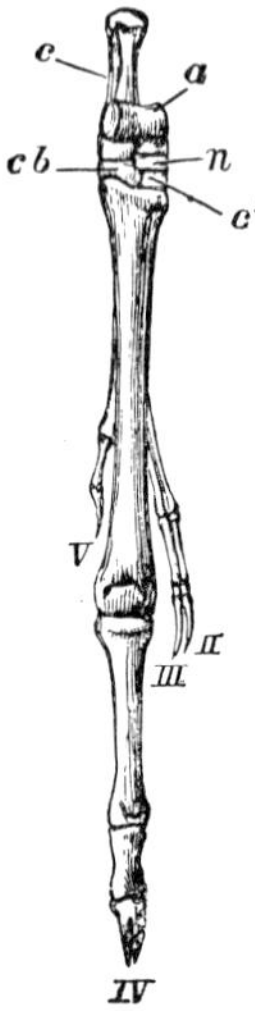

FIG. 43.—Skeleton of right hind foot of *Chœropus castanotis*. *c*, Calcaneum; *a*, astragalus; *cb*, cuboid; *n*, navicular; c^3, ectocuneiform; II and III, the conjoined second and third digits; IV, the large and only functional digit; V, the rudimentary fifth digit.

The only known species of this genus (Fig. 44), chiefly remarkable for the singular construction of its limbs, is an animal about the size of a small Rat, found in the interior of the Australian continent. Its general habits and food appear to resemble those of the other *Peramelidæ*. It was first described as *C. ecaudatus* by Ogilby from a mutilated specimen, but the specific name was afterwards changed, as being inappropriate, by Gray to *castanotis*.

Suborder DIPROTODONTIA.

For the leading characters of this group, see page 132.

Family PHASCOLOMYIDÆ

Dentition: $c\ \frac{1}{1}$, $i\ \frac{0}{0}$, $p\ \frac{1}{1}$, $m\ \frac{4}{4} = 24$. All the teeth with persistent pulps. The incisors large, scalpriform, with enamel only on the front surface, as in the Rodentia. The molars strongly curved, forming from the base to the summit about a quarter of a circle, the concavity being directed outwards in the upper and inwards in the lower teeth. The first of the series, or premolar, appears to have no milk-predecessor, and is single-lobed; the other four composed of two lobes, each subtriangular in section. Limbs equal, stout, and short. Fore feet with five distinct toes, each furnished with a long, strong, and slightly curved nail, the first and fifth considerably shorter than the other three. Hind feet with a very short nailless hallux, the second, third, and fourth toes partially united by integument, of nearly equal length, the fifth distinct and rather shorter; all four provided with long and curved nails. In the skeleton of the foot, the second and third toes are distinctly more slender than the fourth, showing a slight tendency towards the peculiar character so marked in the next two families. Tail rudimentary. Stomach simple, provided with a special gland situated near the cardiac orifice. Cæcum very short, wide, and with a peculiar vermiform appendage. Pouch present. The auditory bullæ of the skull are imperfect, open behind, with their anterior wall formed by a descending process of the squamosal, instead of the

FIG. 44.—*Chœropus castanotis.* From Gould.

alisphenoid. Masseteric fossa of mandible with a perforation and a deep pit.

FIG. 45.—Common Wombat (*Phascolomys ursinus*).

Phascolomys.[1]—The existing Wombats (Fig. 45) comprise three

[1] Geoffroy, *Ann. du Muséum*, vol. ii. p. 365 (1803).

species, all of which are included in the one genus *Phascolomys*, and all of which date from the Pleistocene.

In the typical group we find the following characters. Fur rough and coarse. Ears short and rounded. Muffle naked. Post-orbital process of the frontal bone obsolete. Ribs fifteen pairs. Vertebræ: C 7, D 15, L 4, S 4, C 10-12. The Wombat of Tasmania and the islands of Bass's Straits (*P. ursinus*) and the closely similar but larger animal of the southern portion of the mainland of Australia (*P. mitchelli*) belong to this group.

In the second group the characters are as follows. Fur smooth and silky. Ears large and more pointed. Muffle hairy. Frontal region of skull broader than in the other group, with well-marked postorbital processes. Ribs thirteen. Vertebræ: C 7, D 13, L 6, S 4, C 15-16. One species, *P. latifrons*, the Hairy-nosed Wombat of Southern Australia.

In their general form and actions the Wombats resemble small bears, having a somewhat similar shuffling manner of walking, but they are still shorter in the legs, and have broader, flatter backs than bears. They live entirely on the ground, or in burrows or holes among rocks, never climbing trees, and feed entirely on grass, roots, and other vegetable substances. They sleep during the day, and wander forth at night in search of food, and are shy and gentle in their habits generally, though they can bite strongly when provoked. The only noise the common Wombat makes is a low kind of hissing, but the Hairy-nosed Wombat is said to emit a short quick grunt when annoyed. The prevailing colour of the last-named species, as well as of *P. ursinus* of Tasmania, is a brownish gray. The large wombat of the mainland is very variable in colour, some individuals being found of a pale yellowish brown, others dark gray, and some quite black. The length of head and body is about three feet.

It is noteworthy that *P. mitchelli* was first described from the evidence of fossil remains, the living form subsequently described as *P. platyrhinus* being found to be indistinguishable. Other extinct species occur in the Pleistocene of Australia.

Phascolonus.[1]—Remains of a large extinct Wombat, which must have nearly equalled the dimensions of a Tapir, occur in the Pleistocene of Queensland, and have been described as *Phascolonus*. It is probable that the expanded and flattened upper incisors from the same deposits upon the evidence of which the presumed genus *Sceparnodon* was founded, are likewise referable to the same form. The characters of both the upper and lower incisors distinguish *Phascolonus* from *Phascolomys*.

[1] Owen, *Phil. Trans.* 1872, p. 257.

Family PHALANGERIDÆ.

Dentition extremely variable, owing to the presence of minute rudimental teeth not constant in the same species, or even in the two sides of the jaws of the same individual; exclusive, however, of *Tarsipes*, the formula $i\,\frac{3}{1}$, $c\,\frac{1}{0}$, $p\,\frac{(2-3)}{(0-2)}$, $m\,\frac{(3-4)}{(3-4)}$ represents fairly the general condition of the functional teeth. First incisors long and stout; the lower pair very large and pointed, but without the scissor-like action found in the existing *Macropodidæ;* second and third lower incisors minute and probably functionless. Fourth premolar generally secant; milk-molar generally minute and deciduous at an early period. Molars either with sharp cutting-crests or bluntly tuberculate; fourth sometimes absent. Mandible without pit, and at most a very minute perforation in the masseteric fossa. Limbs subequal. Fore feet with five distinct, subequal toes, furnished with claws. Hind feet short and broad, with five well-developed toes; the hallux large, nailless and opposable; the second and third slender, and united by a common integument as far as the claws. Tail generally long, and frequently more or less prehensile. Stomach simple. Cæcum present (except in *Tarsipes*), and usually large. Pouch complete. Animals of small or moderate size and arboreal habits, usually feeding on a vegetable or mixed diet, inhabiting Australia and the Papuan Islands.

The homologies of the lower functionless teeth between the first incisor and fourth premolar are very difficult to determine, but it is probable that one represents a canine only when the largest known number is present; this tooth, according to Mr. Thomas, being the first to disappear.

Phalangers are small woolly-coated animals, with long, powerful, and often prehensile tails, large claws, and, as in the American opossums, with opposable nailless great toes. Their expression seems in the day to be dull and sleepy, but by night they appear to decidedly greater advantage. They live mostly upon fruit, leaves, and blossoms, although some few feed habitually upon insects, and all relish, when in confinement, an occasional bird or other small animal. Several of the Phalangers possess flying membranes stretched between their fore and hind limbs (Fig. 48), by the help of which they can make long and sustained leaps through the air, like the Flying Squirrels, but it is interesting to notice that the possession of these flying membranes does not seem to be any indication of special affinity, the characters of the skull and teeth sharply dividing the flying forms, and uniting them with other species of the non-flying groups. Their skulls (Fig. 47) are as a rule broad and flattened, with the posterior part swollen out laterally, owing to the numerous air-cells situated in the substance of the squamosal.

The Phalangers are interesting from an historical point of view, since the Gray Cuscus (*Phalanger orientalis*) was the first of the Marsupials of the eastern hemisphere brought to the notice of Europeans, having been described in a work published at Leyden in 1611, from an account of a specimen seen at Amboyna during the third expedition of Admiral Van der Hagen.

The present family corresponds to the *Dasyuridæ* among the

FIG. 46.—*Tarsipes rostratus.* From Gould.

Polyprotodonts as presenting, on the whole, the most generalised types of the suborder. The existing forms may be divided into three subfamilies.

Subfamily **Tarsipedinæ.**—Cheek-teeth almost rudimentary and variable in number. Tongue long, slender, pointed, and very extensile. Tail long. Cæcum absent.

Tarsipes.[1]—So named from some supposed resemblance of its foot to that of the Lemurine genus *Tarsius*; but it must be remarked that it has none of the peculiar elongation of the calcaneum and navicular so characteristic of that genus. Head with elongated

[1] Gervais and Verraux, *Proc. Zool. Soc.* 1842, p. 1.

and slender muzzle. Mouth-opening small. The two lower incisors are long, very slender, sharp-pointed, and horizontally placed. All the other teeth are simple, conical, minute, and placed at considerable and irregular intervals apart in the jaws, the number appearing to vary in different individuals and even on different sides of the same individual. The formula in a specimen in the Museum of the Royal College of Surgeons is $i\ \frac{2}{1}$, $c\ \frac{1}{0}$, p and $m\ \frac{3}{2}$ on one side, and $\frac{4}{3}$ on the other; total 20. Rami of the mandible extremely slender, nearly straight, and without coronoid process or inflected angle. Fore feet with five well-developed toes, furnished with small, flat, scale-like nails, not reaching to the extremity of the digits. Hind feet rather long and slender compared with those of the *Phalangerinæ*, having a well-developed opposable and nailless hallux; second and third digits syndactylous, with sharp compressed curved claws; the fourth and fifth free, and with small flat nails. Ears of moderate size and rounded. Tail longer than the body and head, scantily clothed with short hairs, prehensile. Vertebræ: C 7, D 13, L 5, S 3, C 24.

Of this singular genus but one species, *T. rostratus* (Fig. 46), is known, about the size of a common Mouse. It inhabits Western Australia, lives in trees and bushes, uses its tail in climbing, and feeds on honey, which it procures by inserting its long tongue into the blossoms of *Melaleucæ*, etc. One kept in confinement by Mr. Gould was also observed to eat flies.

Subfamily **Phalangerinæ**.—Teeth normal. One or more rudimentary teeth between the upper canine and fourth premolar, and between the first lower incisor and fourth premolar. Tongue of ordinary structure. No cheek-pouches. Stomach and ascending colon simple. Cæcum long, simple. Tail well-developed, generally prehensile.

A numerous group of animals, varying from the size of a mouse to that of a large cat, arboreal in their habits, and abundantly distributed throughout the Australian region. The members of this group are the typical representatives of the family, and are commonly known to the colonists as Opossums.

Phalanger.[1]—The typical genus *Phalanger* (*Cuscus*) presents the following characters. No flying membrane; size large or medium, and build stout and clumsy; fur thick and woolly. Ears short or medium, hairy externally, and in some cases also internally. Toes of fore feet subequal, their relative lengths in the order 4, 3, 5, 2, 1. Claws long, stout, and curved. Soles of feet naked and striated, with large ill-defined pads. Tail stout and markedly prehensile, with the proximal half furred like the body, and the terminal portion entirely naked. Four mammæ. Skull (Fig. 47)

[1] Storr, *Prodromus Meth. Mamm.* p. 33 (1780). Syn. *Phalangista*, Geoffroy, *Bull. Soc. Philom.* vol. i. p. 106 (1796).

stout and strong, with large vacuities in the hinder half of the palate, and the auditory bullæ thick and inflated. Dentition usually $i\,\frac{3}{2}$, $c\,\frac{1}{0}$, $p\,\frac{3}{3}$, $m\,\frac{4}{4}$. First upper incisor with nearly circular section, or only slightly flattened in front; canine more or less closely approximated to third incisor (which is very small), and situated partly in front of the suture between the premaxilla and maxilla. Fourth premolar large, secant, and placed obliquely to line of molars. Molars four-cusped, with the inner cusps of the upper ones crescentoid, and imperfect transverse ridges connecting each pair of cusps.

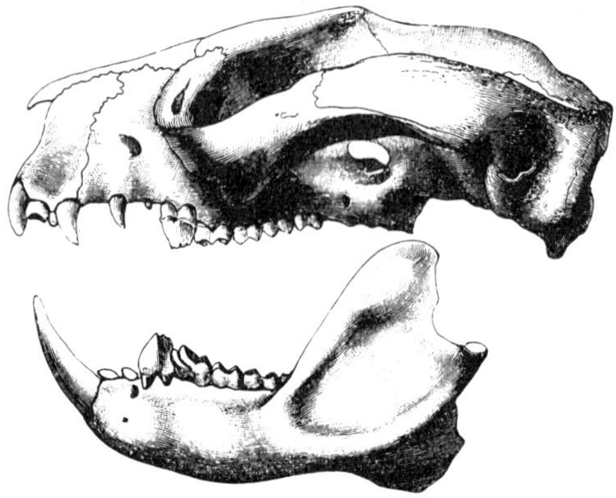

FIG 47.—Left lateral view of skull of Gray Cuscus (*Phalanger orientalis*). After Peters.

The Cuscuses are curious sleepy-looking animals, inhabiting the various islands of the East Indian Archipelago as far west as Celebes, and being the only Marsupials found west of New Guinea. As already noted, it was a member of this genus, the Gray Cuscus (*P. orientalis*), a native of Amboyna, Timor, and the neighbouring islands, which was the first Australasian Marsupial known to European naturalists. There are altogether five species known, all of about the size of a large cat; their habits resemble those of other Phalangers, except that they are said to be somewhat more carnivorous.

Trichosurus.[1]—The members of the genus *Trichosurus* are of relatively large size, and are distinguished from *Phalanger* by the following characters. Ears more or less hairy behind. Relative lengths of toes of fore feet in the order 4, 3, 2, 5, 1. Hair on the soles of the hind feet beneath the heel, but not elsewhere. Tail thick, not tapering, covered with bushy hair up to the extreme tip, which is naked, but with a naked strip on the inferior surface in the distal third or half. A gland on the chest. Dentition usually $i\,\frac{3}{2}$, $c\,\frac{1}{0}$, $p\,\frac{2}{2}$, $m\,\frac{4}{4}$. Upper incisors of nearly uniform length, the first much flattened in front. Canine situated some distance behind the third upper incisor, which it scarcely exceeds in size. Last premolar and molars very similar to those of *Phalanger*.

The true Phalangers comprise two species, of which the best known is the Vulpine Phalanger (*T. vulpecula*), so common in

[1] Lesson, *Dict. Class. d'Hist. Nat.* vol. xiii. p. 333 (1828).

zoological gardens, where, however, it is seldom seen, owing to its nocturnal habits. It is of about the size and general build of a small fox, whence its name. In the typical variety the colour is gray, with a yellowish white belly, white ears, and a black tail. This variety is a native of the greater part of the continent of Australia, but is replaced in Tasmania by the closely allied Brown Phalanger (*var. fuliginosa*). Its habits are very similar to those of the Yellow-bellied Flying-Phalanger (*Petaurus australis*) described below, except that it is unable to take the flying leaps of that animal. Like all the other phalangers, its flesh is freely eaten both by the natives and the lower class of settlers.

Pseudochirus.[1]—The genus *Pseudochirus* agrees with the preceding in the absence of a flying membrane, and presents the following leading characters. Size large or medium. Fur comparatively short and woolly. Ears medium or short, hairy behind, although seldom closely furred over all this aspect. Claws medium. Fore toes subequal, the first two distinctly opposable to the other three. Soles of feet naked, with large, striated, round pads, and hair beneath the heels. Tail tapering, markedly prehensile, with its distal third and the whole of the under surface short-haired; tip naked underneath for a short distance. Four mammæ. No gland on chest. Skull with larger nasals than in the preceding genera; the posterior part of the palate in most cases fully ossified, and the auditory bullæ generally somewhat inflated. Dentition (at most) $i\ \frac{(2-3)}{2},\ c\ \frac{(0-1)}{0},\ p\ \frac{3}{3},\ m\ \frac{4}{4}$. Upper teeth nearly uniform in length, but the first incisor distinctly longer than second. Upper premolars variable. Molars with both inner and outer cusps distinctly crescentoid, and recalling those of the Selenodont Artiodactyle Ungulates.

Range.—Tasmania, Australia, and New Guinea.

There are about ten species of this genus known, of which the commonest is Cook's Ring-tailed Phalanger (*Pseudochirus peregrinus*), an animal discovered by Captain Cook during his first voyage, at Endeavour river, North Queensland.

The complex and sub-selenodont character of the molars of this and the following genus readily distinguish them from the more typical Phalangers, and show an approximation to the type of dentition prevailing in *Phascolarctus;* according, however, to Mr. O. Thomas, a tendency towards the same structure is observable in unworn molars of young Cuscuses. The genus may be divided into three groups, of which the first, as typified by the common *P. peregrinus*, is restricted to Australia and Tasmania, while the third, as represented by *P. canescens*, is only found in New Guinea. *P. albertisi* may be taken as the type of the second group, which is

[1] Ogilby, *Proc. Zool. Soc.* 1836, p. 26.

represented by that species in New Guinea, and by *P. archeri* in Queensland. With the exception of *P. peregrinus*, the species have a more or less restricted range. Remains of *Pseudochirus*, probably referable to existing species, are found in the cave-deposits of New South Wales.

Petauroides.[1]—With the genus *Petauroides*, containing only the single species *P. volans*, we come to the first of the Flying-Phalangers, characterised by the possession of a flying membrane along the flanks. The characters of this genus are as follows. Size large. Fur very long and silky. Ears large and oval, thickly furred on the back, but naked internally. Flying-membrane reaching from wrist to ankle, but very narrow along the sides of the fore-arm and lower leg. Fore toes subequal, their relative lengths in the order 4, 3, 5, 2, 1. Claws long, curved, and sharp. Tail long, cylindrical, and bushy, except near its tip, where it is naked and prehensile. Skull short and broad, with the nasals short, and not extending nearly as far forwards as the premaxillæ. Large vacuities in hinder part of palate. Auditory bullæ inflated and smooth. Dentition usually $i\,\frac{3}{2}$, $c\,\frac{1}{0}$, $p\,\frac{2}{2}$, $m\,\frac{4}{4}$. General characters of teeth very similar to those of *Pseudochirus*, but the first upper incisor scarcely longer than the second.

The single species is found in Australia, from Queensland to Victoria, and is commonly known as the Taguan Flying-Phalanger. The structure of the skull and teeth indicates close affinity with *Pseudochirus*, although the external form is widely different in the two genera. This Phalanger seems, indeed, to be, so to speak, a very specialised *Pseudochirus*, in which the teeth have become somewhat further diminished and the flying membrane has been developed.

Dactylopsila.[2]—The genus *Dactylopsila* is one of the forms without any trace of a flying membrane, its characters being as follows. Size medium. Body striped black and white. Ears oval, nearly naked at the ends. Fore toes of very unequal length, the fourth being enormously elongated; fourth and fifth toes of pes also markedly elongated. Claws long, moderately curved. Tail long, cylindrical, and evenly bushy, with the extremity more or less naked below. Skull narrow, but with the zygomatic arches greatly expanded; palate fully ossified. Dentition: $i\,\frac{3}{3}$, $c\,\frac{1}{0}$, $p\,\frac{3}{2}$, $m\,\frac{4}{4}$. Upper incisors very large, the third being directed horizontally forwards; canine small and approximated to the third incisor, which it resembles. The fourth premolar of moderate size, with its longer axis placed obliquely. First lower incisor longer than in any other genus. Molars oblong, with four cusps.

The typical *D. trivirgata*, or Striped Phalanger, inhabits the

[1] Thomas, *Cat. Marsupials Brit. Mus.* p. 163 (1888).

[2] Gray, *Proc. Zool. Soc.* 1858, p. 109.

Papuan and North Australian sub-region; a second species (*D. palpator*), characterised by the still greater elongation of the fourth finger, occurring in South New Guinea. These animals are said to be of insectivorous habits, the elongated fourth finger, as in the analogous instance of the Lemuroid genus *Chiromys*, being apparently specially adapted for extracting insects and larvæ from their hiding places.

Petaurus.[1]—Size medium or small. Fur very soft and silky. A broad flying membrane extending from the outer side of the fifth digit of the manus to the ankle. Fore toes usually increasing regularly in length from the first to the fifth, but in some of the smaller species the fourth is the longest. Claws strong, sharp, and much curved. Tail long, evenly bushy to the extremity. Glands on the chest and between the ears. Skull short and wide, with the nasals expanded posteriorly, and usually two small palatal vacuities near the second molars. Auditory bullæ inflated, and variable in size. Dentition: $i\,\frac{3}{2}$, $c\,\frac{1}{0}$, $p\,\frac{3}{3}$, $m\,\frac{4}{4}$. First upper incisors very large, and taller than canine. Molars with square crowns rounded at the angles, and four cusps, except in the last, which is triangular.

This genus, which ranges from New Ireland to South Australia, but is not found in Tasmania, contains three species, the largest of which is the Yellow-bellied Flying-Phalanger (*P. australis*), whose habits are recorded by Mr. Gould as follows. "This animal is common in all the brushes of New South Wales, particularly those which stretch along the coast from Port Philip to Moreton Bay. In these vast forests trees of one kind or another are perpetually flowering, and thus offer a never-failing supply of the blossoms upon which it feeds; the flowers of the various kinds of gums, some of which are of great magnitude, are the principal favourites. Like the rest of the genus, it is nocturnal in its habits, dwelling in holes and in the spouts of the larger branches during the day, and displaying the greatest activity at night while running over the small leafy branches, frequently even to their very extremities, in search of insects and the honey of the newly-opened blossoms. Its structure being ill adapted for terrestrial habits, it seldom descends to the ground except for the purpose of passing to a tree too distant to be reached by flight. When chased or forced to flight it ascends to the highest branch and performs the most enormous leaps, sweeping from tree to tree with wonderful address; a slight elevation gives its body an impetus which, with the expansion of its membrane, enables it to pass to a considerable distance, always ascending a little at the extremity of the leap; by this ascent the animal is prevented from receiving the shock which it would otherwise sustain."

[1] Shaw, *Naturalist's Miscellany*, vol. ii. pl. lx. (1791).

A second species, *P. sciureus*, in some ways one of the most beautiful of all mammals, has been chosen for the accompanying woodcut.

Gymnobelideus.[1]—Like *Petaurus* in every respect, but without any trace of a flying membrane, and with the fifth digit of the manus slightly shorter than the third. This genus is represented only by *G. leadbeateri* of Victoria, and according to Mr. Thomas, may be regarded as the primitive form from which the specialised *Petaurus* has been developed.

FIG. 48.—Squirrel Flying-Phalanger (*Petaurus sciureus*).

Dromicia.[2]—Size small, and general appearance dormouse-like. Ears large and thin, almost naked, and without internal or basal tufts. No flying membrane. Digits of normal proportions, the relative lengths of those of the manus in the order 3, 4, 2, 5, 1; fore claws rudimentary, hind ones long and sharp. Tail mouse-like, cylindrical, furry at base, the remainder scaly, with fine hairs, except at the tip, which is naked and prehensile.

1 M'Coy, *Ann. Mag. N. H.* (3) xx. p. 287 (1867).
2 Gray, in *Grey's Australia*, appendix, vol. ii. p. 407 (1841).

Skull short and broad, with the hinder part of the palate incomplete, and the auditory bullæ large, much inflated, and transparent. Dentition: $i\,\frac{3}{2},\ c\,\frac{1}{0},\ p\,\frac{3}{3},\ m\,\frac{(3-4)}{(3-4)}$. First upper incisor spatulate, and much longer than either of the others. Canine large, placed at some distance behind the third incisor. Molars (except the last) with evenly rounded crowns, carrying four small smooth cusps.

This genus, which occurs in New Guinea, Western Australia, and Tasmania, is represented by four species. It seems to be intermediate between *Petaurus* and *Acrobates*, and it has apparently had to yield place to those more highly organised types in regions where they have come in contact with one another.

Distœchurus.[1]—Size small. Ears rather short, thinly covered with hair, but with small tufts at the base. No flying membrane. Digits of normal proportions, without expanded terminal pads. Claws curved and sharp. Tail, skull, and dentition as in *Acrobates*, with the exception that the fourth premolar is small in the upper, and absent in the lower jaw.

The one species of Feather-tailed Phalanger (*D. pennatus*) is found in New Guinea.

Acrobates.[2]—Size very small. Ears moderate, thinly covered with hair, but with small tufts round the base and on the internal prominences. A narrow flying membrane, fringed with long hairs, running from the elbow to the flank, and from the latter to the knee. Four mammæ. Digits furnished with expanded and striated terminal pads, the relative length of those of the manus being in the order 4, 3, 5, 2, 1. Claws sharp, although somewhat concealed by the terminal pads. Tail short-haired above and below, with a broad fringe on either side. Skull short, wide, and depressed. Posterior portion of palate very imperfectly ossified; anterior palatal vacuities almost confined to the maxillæ. Auditory bullæ low, rounded, and but slightly prominent. Dentition: $i\,\frac{3}{2},\ c\,\frac{1}{0},\ p\,\frac{3}{3},\ m\,\frac{3}{3}$. Teeth sharp, and of an insectivorous type. Upper canine long, and approximated to third incisor. The three upper premolars large, functional, and taller than the molars. Molars small and rounded, with smooth unridged cusps.

There is only one species in this genus, the beautiful little Pigmy Flying-Phalanger (*A. pygmæus*), not so big as a Mouse, which is found in Queensland, New South Wales, and Victoria, and feeds on the honey it abstracts from flowers, and on insects. Its agility and powers of leaping are exceedingly great, and it is said by Mr. Gould to make a most charming little pet.

Subfamily **Phascolarctinæ.**—Teeth large, normal; no rudimentary premolars before the last upper premolar, or any teeth

[1] Peters, *Ann. Mus. Genov.* vol. vi. p. 303 (1874).

[2] Desmarest, *Nouv. Dict. d'Hist. Nat.* sér. 2, vol. xxv. p. 405 (1817).

between the first lower incisor and fourth premolar. Tongue of ordinary structure. Distinct cheek-pouches. Stomach with a special gland near the cardiac orifice. Cæcum very long, and (with the upper portion of the colon) dilated and provided with numerous longitudinal folds of mucous membrane. In many anatomical characters, especially the possession of a special gastric gland, this group resembles the *Phascolomyidæ*.[1]

Phascolarctus.[2]—Dentition: $i\,\frac{3}{1}$, $c\,\frac{1}{0}$, $p\,\frac{1}{1}$, $m\,\frac{4}{4}$; total 30. Upper incisors crowded together, cylindroidal, the first much larger than the others, with a bevelled cutting edge (Fig. 36). Canine very small; a considerable interval between it and the premolar, which is as long from before backwards but not so broad as the true molars, and has a cutting edge, with a smaller parallel inner ridge. The molars slightly diminishing in size from the first to the fourth, with square crowns, each bearing four pyramidal cusps, with curved ridges radiating from them, and having a structure very similar to these of *Pseudochirus*. The lower incisors are semiproclivous, compressed and tapering, bevelled at the ends. Premolars and molars in continuous series, as in the upper jaw. Milk-tooth very minute, and almost functionless. Fore feet with the two inner toes slightly separated from and opposable to the remaining three, all with strong, curved, and much compressed claws. Hind foot (Fig. 49) with the hallux placed very far back, large and broad, the second and third (united) toes considerably smaller than the other two; the fourth the largest. No external tail. Fur dense and woolly. Ears of moderate size, thickly clothed with long hairs. Vertebræ: C 7, D 11, L 8, S 2, C 8. Ribs eleven pairs, a rare exception to the usual number (13) in the Marsupialia.

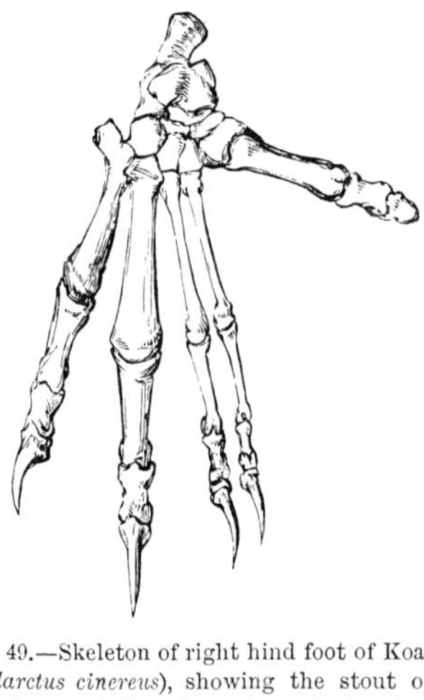

FIG. 49.—Skeleton of right hind foot of Koala (*Phascolarctus cinereus*), showing the stout opposable hallux, followed by two slender toes, which in the living animal are enclosed as far as the nails in a common integument.

There is but one species, the Koala or Native Bear of the Australian colonists (*P. cinereus*), an animal of comparatively large size and heavy build (Fig. 50), found in the south-eastern parts of the Australian continent. It is about two feet in length, and of an ash-gray colour, an excellent climber, and residing generally in lofty

[1] *Cf.* W. A. Forbes, "Anatomy of the Koala," *Proc. Zool. Soc.* 1881, p. 180.
[2] Blainville, *Bull. Soc. Philom.* 1816, p. 116.

Eucalyptus trees, on the buds and tender shoots of which it feeds, though occasionally descending to the ground in the night.

Extinct Phalangeroids.

Numerous imperfect remains recently described by De Vis are regarded as indicating large extinct types of *Phalangeridæ*, but further evidence is required before all these determinations can be definitely accepted. Thus part of an upper jaw is provisionally referred to a large species of *Pseudochirus*, while part of a scapula is made the type of a genus *Archizonurus* which appears to be

Fig. 50.—The Koala (*Phascolarctus cinereus*). From Sclater, *Proc. Zool. Soc.* 1880, p. 355.

allied to the former. Another fragmentary scapula is considered to indicate a large *Phalanger*. Finally, part of a fibula described under the name of *Koalemus* is regarded as affording evidence of the former existence of a large ancestral form allied to the Koala, and it is suggested that an upper jaw with teeth may belong to the same or an allied type.

Thylacoleo.[1]—Dentition of adult: $i\,\frac{3}{1}$, $c\,\frac{1}{0}$, $p\,\frac{3}{3}$, $m\,\frac{1}{2}$; total 28. First upper incisor much larger than the others; canine and first two premolars rudimentary. In the lower jaw the two small anterior premolars are functionless, and often deciduous; posterior premolars of both jaws formed on the same type as those of *Potorous*, but relatively much larger; true molars rudimentary, tubercular. One species, *T. carnifex*. This animal presents a most anomalous

[1] Owen, in *Gervais's Zool. et Pal. françaises*, 1st ed. pt. i. p. 192 (1849-52).

condition of dentition, the functional teeth being reduced to one pair of large cutting incisors situated close to the median line, and one great, trenchant, compressed premolar, on each side above and below. It was first described as a carnivorous Marsupial, and named, in accordance with its presumed habits, "as one of the fellest and most destructive of predatory beasts"; but, as its affinities are certainly with the *Phalangeridæ* and *Macropodidæ*, and its dentition completely unlike that of any known predaceous animal, this view has been called in question.

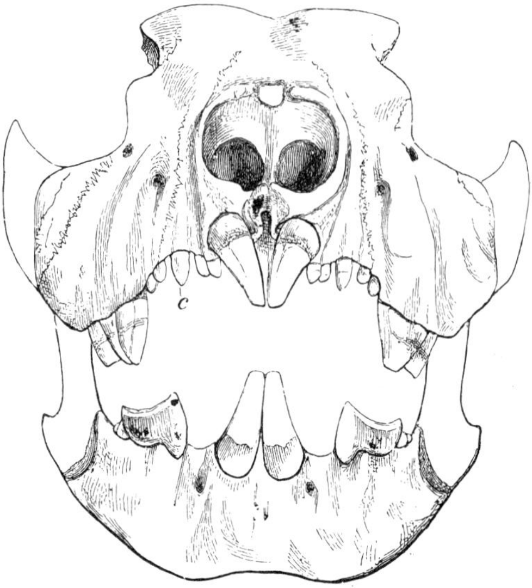

FIG. 51.—Front view of skull of *Thylacoleo carnifex*, restored. ⅓ natural size. From *Quart. Journ. Geol. Soc.* vol. xxiv. p. 312.

The dentition is nearer to that of the existing *Phalangeridæ* than to that of the *Macropodidæ*, and the genus may be provisionally regarded as the type of a distinct subfamily of the former.

Family MACROPODIDÆ.

Dentition $i\,\frac{3}{1}$, $c\,\frac{(0-1)}{0}$, $p\,\frac{2}{2}$, $m\,\frac{4}{4}$. Incisors sharp and cutting, those of the lower jaw frequently having a scissor-like action against one another; upper canine, if present, small. Penultimate premolar shed with the fourth milk-molar, which is molariform and long persistent. Molars wide, and either transversely ridged or bluntly tuberculate. Premolars and molars moving forwards in the skull as the age of the animal increases, this being most marked in the larger species. Masseteric fossa of mandible hollowed out below into a deep cavity walled in externally by a plate of bone, and communicating with the inferior dental canal by a large foramen. Hind limbs usually larger than the anterior ones, and progression generally saltatorial. Fore feet with five digits; hind feet syndactylous, the fourth digit being very large and strongly clawed; hallux usually absent. Tail generally long and hairy, occasionally prehensile; stomach sacculated. Pouch large and opening forwards.

The *Macropodidæ* or Kangaroos, taken as a whole, form a very well-marked family, easily distinguished from the other members of the suborder by their general conformation, and by peculiarities in the structure of their limbs, teeth, and other organs. They vary in size from that of a sheep down to a small rabbit. The head, especially in the larger species, is small, compared with the rest of the body, and tapers forward to the muzzle. The shoulders and fore limbs are feebly developed, and the hind limbs usually of disproportionate strength and magnitude, which gives them a peculiarly awkward appearance when moving about on all fours, as they occasionally do when feeding. Rapid progression is, however, performed only by the powerful hind limbs, the animal covering the ground by a series of immense bounds, during which the fore part of the body is inclined forwards, and balanced by the long, strong, and tapering tail, which is carried horizontally backwards. When not moving they often assume a perfectly upright position, the tail aiding the two hind legs to form a sort of supporting tripod, and the front limbs dangling by the side of the chest. This position gives full scope for the senses of sight, hearing, and smell to warn of the approach of enemies, from which these animals save themselves by their bounding flight. The fore paws have five distinct digits, each armed with a strong curved claw.

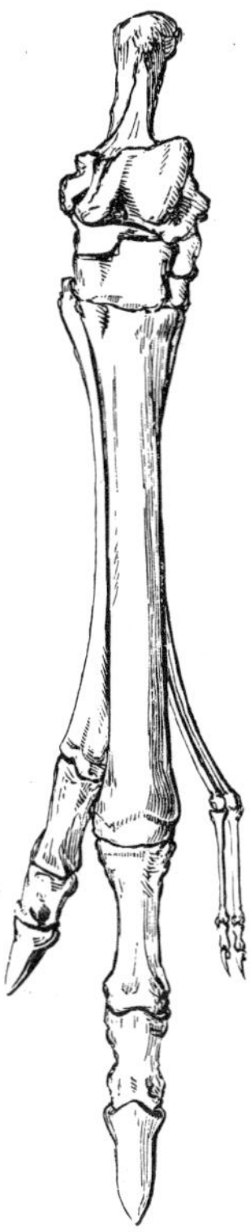

FIG. 52.—Skeleton of right hind foot of Kangaroo.

The hind foot (Fig. 52), as being a typical example of the syndactylous modification, may be noticed in some detail. It is extremely long and narrow, and (with only one exception) without any hallux or great toe. It consists mainly of one very large and strong toe, corresponding to the fourth of the human or other typically developed foot, ending in a strong, curved, and pointed claw. Close to the outer side of this lies a smaller fifth digit, and to the inner side two excessively slender toes (the second and third), bound together almost to the extremity in a common integument. The two little claws of these toes, projecting together from the skin, may be of use in scratching and cleaning the fur of the animal, but the toes themselves must have quite lost all connexion with the functions of support or progression.

The dentition of the Kangaroos, functionally considered,

consists of sharp-edged incisors, most fully developed near the median line of the mouth, for the purpose of cropping the various kinds of herbage on which they feed, and ridged and tuberculated molars for crushing it, there being no tusks or canines for offensive or defensive purposes.

The number of vertebræ is—in the cervical region 7, dorsal 13, lumbar 6, sacral 2, caudal varying according to the length of the tail, but generally from 21 to 25. In the fore limb the clavicle and the radius and ulna are well developed, allowing of considerable freedom of motion of the hand. The pelvis has large epipubic or "marsupial" bones. The femur is short, and the tibia and fibula

FIG. 53.—The Great Gray Kangaroo (*Macropus giganteus*).

are of great length, as is the foot, the whole of which is applied to the ground when the animal is at rest in the upright position.

The stomach is of large size, and very complex, its walls being puckered up by longitudinal muscular bands into a great number of sacculi, like those of the human colon. The alimentary canal is long, and the cæcum well developed. All the species have a marsupium or pouch formed by a fold of the skin of the abdomen, covering the mammary glands with their four nipples. In this pouch the young are placed as soon as they are born; there their growth and development proceeds; and to it they resort temporarily for the purpose of shelter, concealment, or transport, for some time after they are able to run and jump about the ground and feed upon the same herbage which forms the nourishment of the parent. During the early period of their sojourn in the pouch, the blind, naked, helpless young creatures (which in the Great Kangaroo (Fig. 53) scarcely exceed an inch in length) are attached by their mouths to the nipples of the mother, and are fed by milk injected into their stomach by the contraction of the muscle covering the mammary gland.

The Kangaroos are all vegetable feeders, browsing on grass and various kinds of herbage, the smaller species also eating roots. They are naturally timid, inoffensive creatures; but the larger ones when hard pressed will turn and defend themselves, sometimes killing a dog by grasping it in their fore paws, and inflicting terrible wounds with the sharp claws of their powerful hind legs, sustaining themselves meanwhile upon the tail. A few aberrant forms are arboreal. The great majority are inhabitants of Australia and Tasmania, forming one of the most prominent and characteristic features of the fauna of these lands, and in the scenery of the country, as well as the economy of nature, performing the part of the deer and antelopes of other parts of the world, which are entirely wanting in Australia. Kangaroos were very important sources of food-supply to the natives, and are hunted by the colonists, both for sport and with a view to their destruction, on account of the damage they naturally do in consuming the grass, now required for feeding cattle and sheep. Notwithstanding this, they have in some districts increased in numbers, owing to the suppression of their former enemies, the aborigines and the Dingo or native dog. A few species are found in New Guinea and the adjacent islands, which belong, in the zoological sense, to the Australian region.

Before noticing the various generic types of the *Macropodidæ*, a few words are necessary in respect of the tooth-change, and we may here quote the observations of Mr. O. Thomas on this subject. "The full dentition of the members of this family consists, in the upper jaw, first of three incisors, then of a small canine (often, however, suppressed, as in Fig. 55), and then of six cheek-teeth, of which the second in the series is the only one which has a milk or deciduous predecessor, and is therefore the one to be regarded as the last premolar of the typical mammalian dentition. The special characteristics that render the development and succession of the teeth in the *Macropodidæ*, and especially in the genus *Macropus*, so puzzling to systematic zoologists, are: firstly, a general progression forwards in the jaw of the whole tooth-row, comparable to that found elsewhere only in the Elephants and some Sirenians; and, secondly, the fact that before the tooth-change the first tooth of the series (*p* 3) and the single milk-tooth (*dm* 4) placed next to it, both of which fall out at the change, are respectively so very similar in shape and size to the first and second teeth of the permanent series, viz. the permanent premolar (*p* 4) and the first molar (m 1), as to be most naturally mistaken for, or compared with, them in specific descriptions. . . . The necessary knowledge as to the stage of dentition in which any skull may be, can often be gained only by cutting open the bone either above and behind the first tooth of the series to see if the true permanent p 4 be still buried there (in which case, of course, that first tooth is only p 3), or behind the last visible molar to see if there be yet another tooth behind it, showing it to be m 3 and not m 4. The first plan is, as a rule, the better, since p 4 is generally by far the most important tooth for diagnostic purposes, and its characters have, therefore, in any case to be taken into account."

The *Macropodidæ* are divided into three well-marked sections: (1) the true Kangaroos (*Macropodinæ*); (2) a group consisting of smaller animals, commonly called Rat Kangaroos, or (improperly) "Kangaroo Rats," or sometimes Potoroos; and (3) the *Hypsiprymnodontinæ*, now represented only by a single species.

Subfamily **Hypsiprymnodontinæ.**—Size very small. Claws small, feeble, and subequal. Hind feet with an opposable hallux. Tail naked and scaly. The fourth premolar twisted obliquely outwards, as in *Phalanger*. Other teeth as in the *Potoroinæ*.

This subfamily is now represented only by the genus *Hypsiprymnodon*,[1] which is a form of great interest, as showing a structure of foot connecting that of the Kangaroos with that of the Phalangers. The single known species, *H. moschatus*, was described by Ramsay from specimens discovered in north-east Australia. It was described almost simultaneously by Owen under the name of *Pleopus nudicaudatus*. From the resemblance in the structure of the foot and the obliquity of the premolars to the Phalangers Mr. Thomas has some hesitation as to which family should receive this genus, but the macropine characters of the mandible preponderate in favour of the *Macropodidæ*.

Triclis.[2]—A lower jaw of a much larger form from the Pleistocene deposits of Australia apparently indicates another member of this subfamily, having the outwardly directed and grooved premolar characteristic of *Hypsiprymnodon*. It differs, however, from that genus, and also from all other known *Macropodidæ*, in having a small tooth between the incisor and fourth premolar, which apparently represents a canine, or perhaps an anterior premolar. This form indicates, therefore, a closer connexion between the *Phalangeridæ* and *Macropodidæ* than any other.

Subfamily **Potoroinæ.**—The second section or subfamily, the *Potoroinæ*, have the first upper incisor narrow, curved, and much exceeding the others in length (Fig. 54). Upper canines always persistent, flattened, blunt, and slightly curved. Premolars of both

[1] Ramsay, *Proc. Linn. Soc. N. S. Wales*, vol. i. p. 33 (1876).

[2] De Vis, *Proc. Roy. Soc. Queensland*, ser. 2, vol. iii. p. 8 (1888).

jaws always having large, simple, compressed crowns, with a nearly straight or slightly concave free cutting edge, both outer and inner surfaces usually marked by a series of parallel, vertical grooves and ridges, these teeth being either set in the same line with the molars, or slightly bent outwards. Molars with quadrate crowns, having a blunt, conical cusp at each corner, the fourth notably smaller than the third, sometimes rudimentary, and appearing early. Fore feet narrow; three middle toes considerably exceeding the first and fifth in length; their claws long, compressed, and but slightly curved. Hind feet as in *Macropus*. Tail long and hairy, sometimes partially prehensile, being used for carrying bundles of grass with which these animals build their nests.

The Potoroos or Rat Kangaroos are all small animals, none of them exceeding a common rabbit in size. They inhabit Australia and Tasmania, are nocturnal, and feed on the leaves of various

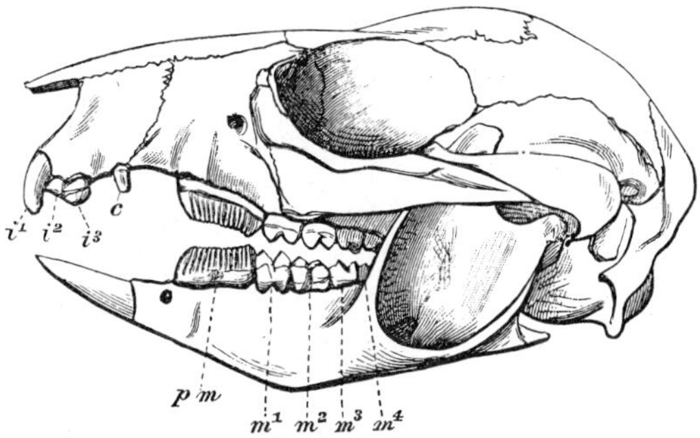

FIG. 54.—Skull and Teeth of Rat Kangaroo (*Bettongia lesueuiri*). c, Upper canine. The other letters as in Fig. 51.

kinds of grasses and other plants, as well as roots and bulbs, which they dig up with their fore paws. Nine species are known, presenting a considerable range of diversity in minor characters, and admitting of being grouped in four principal sections, which may be allowed the rank of genera. These are:

Potorous.[1]—Head long and slender. Auditory bullæ somewhat inflated. Ridges on premolars few and perpendicular. Large palatine foramina. Tarsus short. Muffle naked. Three species, viz. *P. tridactylus*, *P. gilberti*, and *P. platyops;* the last two being confined to West Australia.

Bettongia.[2]—Head comparatively short and broad. Ears short and rounded. Auditory bullæ generally much inflated. Large palatine foramina. Tarsus long. Ridges on premolars numerous

[1] Desmarest, *Nouv. Dict. d'Hist. Nat.* ser. 1, vol. xxiv. *Table Meth.* p. 20 (1804). Syn. *Hypsiprymnus*, Illiger, *Prodromus Syst. Mamm.* p. 79 (1811).

[2] Gray, *Charlesworth's Mag. Nat. Hist.* vol. i. p. 584 (1837).

and oblique. Tail more or less prehensile, thickly haired, and the hairs on the upper surface longer than those on the lower, and forming a crest. Muffle naked. Four species, viz. *B. penicillata*, *B. cuniculus*, *B. gaimardi*, *B. lesueuiri*.

Caloprymnus.[1]—Muffle naked, as in *Bettongia*, but the edge of the hairy part less emarginate backwards in the middle line. Ears short, rounded, and hairy. Auditory bullæ much inflated, and of large size. Nasals larger and wider behind than in the other genera. Very long anterior palatine foramina. Limbs as in *Bettongia*. Tail thin, cylindrical, evenly coated with short hair, without trace of a crest. Skull broad and flat, with a remarkably short and conical muzzle. The sole representative of this genus is *C. campestris* of South Australia, originally referred to *Bettongia*.

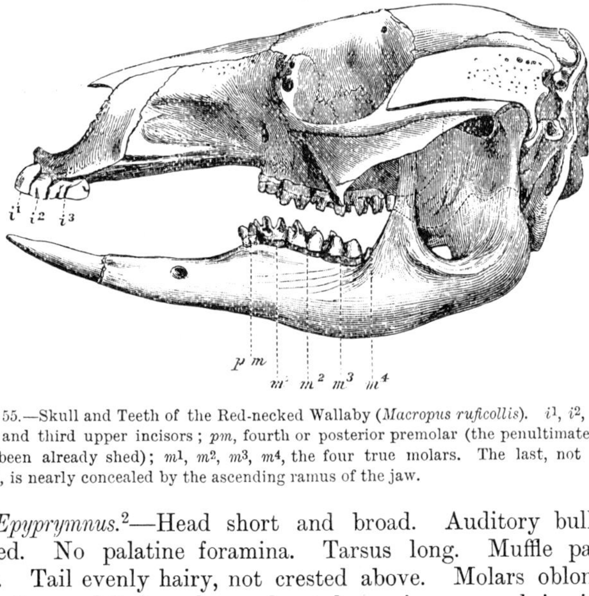

FIG. 55.—Skull and Teeth of the Red-necked Wallaby (*Macropus ruficollis*). i^1, i^2, i^3, First, second, and third upper incisors; *pm*, fourth or posterior premolar (the penultimate or third having been already shed); m^1, m^2, m^3, m^4, the four true molars. The last, not fully developed, is nearly concealed by the ascending ramus of the jaw.

Æpyprymnus.[2]—Head short and broad. Auditory bullæ not inflated. No palatine foramina. Tarsus long. Muffle partially hairy. Tail evenly hairy, not crested above. Molars oblong, less distinctly quadritubercular, and not decreasing so much in size posteriorly as in the other genera. Represented only by *Æ. rufescens*.

Remains of *Æ. rufescens* occur in the Pleistocene cave-deposits of New South Wales.

Subfamily **Macropodinæ**.—This subfamily includes the largest forms. The cutting edges of the upper incisors are nearly level, or the first pair but slightly longer than the others (Fig. 55). The canines are rudimentary and often wanting. The premolars are usually not longer (from before backwards) than the true molars

[1] Thomas, *Cat. Marsup. Brit. Mus.* p. 114 (1888).
[2] Garrod, *Proc. Zool. Soc.* 1875, p. 59.

and less compressed than in the last subfamily; they are placed in precisely the same line with the molars. The crowns of the molars always have two prominent transverse ridges; and these teeth increase in size from before backwards, the fourth molar appearing very late. The fore limbs are small, with subequal toes armed with strong, moderately long, curved claws. Hind limbs very long and strongly made. Head small, with more or less elongated muzzle. Ears generally rather long and ovate.

Upwards of forty-four existing species of this group have been described, and many attempts have been made to subdivide them into smaller groups or genera for the convenience of arrangement and description, but these have generally been based upon such trivial characters that it is preferable to speak of many of them as sections of the genus *Macropus*, reserving generic rank only to forms somewhat aberrant in structure. According to this arrangement the genera will be as follows:

Lagostrophus.[1]—Represented only by the Banded Wallaby (*L. fasciatus*) of Western Australia, which presents the following distinctive features. Size small. Muffle naked. Hind feet covered with long bristly hairs, concealing the claws. Lower part of back marked by dark cross-bands. Skull with a narrow pointed muzzle and inflated auditory bullæ; symphysis of mandible firmly united. No canine. Upper incisive series meeting at a sharp angle, and diverging but slightly behind. First incisor smaller in section than either of the others and scarcely longer, bluntly pointed; second with a flattened oral surface; third smaller, similarly flattened, but with a groove on oral surface forming a notch at its postero-external angle. Fourth premolar short, with a distinct inner ledge. Molars as in *Macropus*.

Dendrolagus.[2]—General proportions of limbs and body normal and unlike those of other members of the family. Muffle broad and only partly naked. Fur on nape, and sometimes on back, directed forwards. Fore limbs nearly as large as the hind; hind feet with the syndactylous second and third digits relatively large; claws of fourth and fifth hind digits curved like those of the manus. Tail very long, and thickly furred. Skull stout, with a short and wide muzzle; the posterior part of the palate fully ossified, and the auditory bullæ not inflated. A small canine. Fourth premolar large, but much shorter antero-posteriorly than in the next genus; molars as in the latter.

This genus includes four species of Tree-Kangaroos, three of which occur in New Guinea, while *D. lumholtzi* is found in North Queensland. They differ greatly from all the other forms in being chiefly arboreal in their habits, climbing with facility among the

[1] Thomas, *Proc. Zool. Soc.* 1886, p. 544.

[2] Schlegel and Müller, *Verh. Nat. Ges. Nederland*, p. 138 (1839-44).

branches of large trees, and feeding on the bark, leaves, and fruit. They are confined to the tropical forests of the regions mentioned; and it would appear that we must regard their resemblance in the proportions of the limbs and habits to the Phalangers as having been independently acquired.

Dorcopsis.[1]—Hind limbs relatively less large than in *Macropus*. Muffle large, broad, and naked. Ears small. Fur on nape directed wholly or partially forwards. Hind claws not concealed by hair. Tail with a nearly naked tip. Skull long and narrow, with the auditory bullæ not inflated. A well-developed canine. First upper incisor somewhat short; second and third nearly equal, notched externally. Fourth premolar greatly elongated antero-posteriorly, its length generally exceeding the united lengths of the first and second molars; a distinct inner ledge, and vertical grooves on both sides. Molars low and rounded, with the median longitudinal bridge between the ridges almost or quite aborted, and the talon in front of the first transverse ridge very narrow, and not extending to the inner side. The two series of cheek-teeth parallel, or nearly so, instead of converging at the extremities.

Three species of this genus are known, all of which are from New Guinea; the type being *D. muelleri*. In the characters of the dentition, the forward inclination of the fur on the nape, and other points, this genus is allied to *Dendrolagus*; but *Dorcopsis macleayi* connects the other species with *Macropus*.

Lagorchestes.[2]—Muffle entirely or partially covered with hair. Fourth hind digit with a long claw, not concealed by hair. Tail rather short, evenly furred, without a spur. Skull with short muzzle and diastema, and inflated auditory bulla. Canine present, sometimes very small. Fourth premolar large, not constricted in the middle, with a continuous inner ledge.

This genus includes the Hare-Kangaroos, a group of small hare-like animals, great leapers and swift runners, which mostly affect the open grassy ridges, particularly those of a stony character, sleeping in forms or seats like the common hare. Their limbs are comparatively small, their claws sharp and slender, and their muffle is clothed with velvet-like hairs. Three species—*M. leporoides*, *M. hirsutus*, *M. conspicillatus*.

The range extends over the whole of Australia, but does not embrace Tasmania.

Onychogale.[3]—Muffle hairy. Fourth hind claw long, narrow, compressed, and sharp. Tail long and tapering, covered with short hair, and furnished at the tip with a horny spur. Skull nearly as in *Macropus*, with the auditory bullæ more or less inflated. Canine

[1] Schlegel and Müller, *Verh. Nat. Ges. Nederland*, p. 130 (1839-44).

[2] Gould, *Monograph of Macropodidæ*, pl. xiii. (1841).

[3] Gray, in *Grey's Australia*, vol. ii. appendix, p. 402 (1841).

small or wanting. Upper incisors small, decreasing in size from first to third. Fourth premolar small, hour-glass shaped, and without inner ledge. Molars as in *Macropus*.

This genus contains three species, having the same distribution as *Lagorchestes*. Mr. O. Thomas observes: "The spur-tailed Wallabies form a natural little group, distinguished both by the shape of the incisors and the peculiar horny excrescence at the tip of the tail. The latter character is altogether unique among Marsupials, and is only found among other mammals in the Lion, which occasionally has a somewhat similar horny spur at the end of its tail. In the case of the Wallabies it is difficult to conceive what can be the use of this spur; and observations on the living animal are much needed with regard to this interesting point."

Petrogale.[1]—Muffle naked. Fur of nape directed backwards. Claw of fourth hind digit very short. Tail long, cylindrical, thinner than in *Macropus*, and thickly haired and pencilled at the extremity. Skull as in the smaller species of *Macropus*, with large posterior palatal vacuities, and the bullæ sometimes inflated. No canine. Upper incisors small, the third resembling that of *Macropus*. Fourth premolar large and stout, as in some of the Wallabies, with a continuous inner ledge, and two or three indistinct vertical ridges externally. Molars as in the Wallabies.

This genus is represented by six species, of which *P. penicillata* is a well-known example, ranging over the whole of the mainland of Australia. The Rock-Wallabies, as its members may be called, are very closely allied to some of the true Wallabies; and some hesitation may be expressed as to the advisability of accepting their generic separation from *Macropus*. They inhabit rocky regions, making their retreats in caverns and crevices, leaping with surprising agility from one narrow ledge to another, and browsing upon the scanty herbage that the neighbourhood of such situations affords. The species are *P. xanthopus*, *P. penicillata*, *P. lateralis*, *P. concinna*, *P. brachyotis*, *P. inornata*.

Remains of *P. penicillata* are found in a fossil state in the Pleistocene cave-deposits of New South Wales.

Macropus.[2]—Muffle generally completely naked. Ears large. Fur on nape (with an occasional exception in two species) directed backwards. Claw of fourth hind digit very long. Tail thick, tapering, and evenly furred. Four mammæ. Skull (Fig. 55) long, smooth, and rounded; the nasals expanded behind; generally large palatal vacuities; and the auditory bullæ not inflated. Canine minute, and shed at an early period. Incisor series forming an open curve; the first the tallest, and the third nearly always the longest antero-posteriorly, and generally with an infolding of enamel

[1] Gray, *Charlesworth's Mag. Nat. Hist.* vol. i. p. 583 (1837).

[2] Shaw, *Naturalist's Miscellany*, vol. i. pl. xxxiii. (1790).

near its postero-external angle. Fourth upper premolar with a secant edge, and an inner basal ledge or tubercle; corresponding lower tooth secant; both may be longer or shorter than first molar. Molars (except very occasionally) with a distinct longitudinal bridge connecting transverse ridges. Lower incisors long and scalpriform, with inner secant edges opposable, owing to the loose articulation of the mandibular symphysis.

This genus includes the true Kangaroos and Wallabies, the size of the individual existing species varying from that of a Rabbit to that of a Man. There are no less than twenty-three existing species, which may be divided into three groups, as well as many extinct ones. The genus is found in Australia and New Guinea, as well as in the eastern half of the Austro-Malayan transitional region.

The first group, or true Kangaroos, comprises the largest existing forms, which are generally of a uniform and sombre colour.

The skull is of a large and massive type, with the palate more or less well ossified posteriorly, while the molars frequently have a median longitudinal bridge connecting the first transverse ridge with the anterior talon, and no antero-external bridge between the same ridge and talon. The history of the discovery of the typical representative of this group, as being of considerable interest, may be given at some length. When Captain Cook, during his first memorable voyage of discovery, was detained for the purpose of refitting his ship at Endeavour river on the north-east coast of Australia, a strange-looking animal, entirely unknown to them, was frequently seen by the ship's company; and it is recorded in the annals of the voyage that, on the 14th of July 1770, "Mr. Gore, who went out this day with his gun, had the good fortune to kill one of the animals which had been so much the subject of our speculation, . . . and which is called by the natives kanguroo," a name which, though it does not appear to be now known to any of the aboriginal tribes of the country, has been adopted for this animal in all European languages, with only slight modifications of spelling. With the exception of a passing glimpse in the beginning of the same century by the Dutch traveller Bruyn of some living examples of an allied species, this was the first introduction to the civilised world of any member of a group of animals now so familiar. The affinities of the species, skins of which were brought home by Captain Cook and subsequent voyagers, were recognised by Schreber as nearer to the American opossums (then the only known Marsupials) than to any other mammals with which zoologists were acquainted, and consequently it was placed by him, in his great work on the Mammalia, then in the course of publication, in the genus *Didelphys*, with *gigantea* for a specific designation,—the latter having been bestowed upon it by Zimmermann under the impression that it was a huge species of jerboa. Soon afterwards (1791) Dr. Shaw very properly formed a new genus for its reception, which he named *Macropus*, in allusion to the peculiar length of its hind foot. By the name thus formed, *Macrŏpus giganteus*, this kind of Kangaroo has ever since been known in zoological literature. It is the common Gray Kangaroo, called "boomer," "forrester," or "old man" by the colonists, and frequents the open grassy plains of the greater part of eastern Australia and Tasmania; a figure being given in the woodcut on p. 160. The muffle is partly covered with hair, and the fourth premolar very short. Several varieties are known.

A sub-group, distinguished from the above by the naked muffle, includes some very large and handsome species, which principally dwell in rocky mountain ranges, as *M. rufus*, the great Red Kangaroo, *M. antilopinus*, and *M. robustus*. The fourth premolar is of large or medium size in these forms. Remains of *M. giganteus* occur fossil in the Pleistocene of Australia, where we also find the allied extinct *M. titan*, which attains somewhat larger dimensions. *M. robustus* also dates from the same geological epoch, where it was accompanied by two allied types known as *M. altus* and *M. cooperi*.

The second group includes the larger Wallabies, which are smaller than the true Kangaroos, with a brighter and more variegated coloration. The palate is generally more incomplete than in the typical group; and in the molars the anterior talon is connected with the first transverse ridge by an external instead of a median longitudinal bridge. The members of this group are frequenters of forests and dense impenetrable brushes and scrubs, and hence are often called Brush Kangaroos, though a native name, "Wallaby," is now generally applied to them. There are several species, of which *M. ruficollis*, *M. ualabatus*, *M. parryi*, and *M. agilis* are the best known.

M. ualabatus and *M. parryi* are found fossil in the Pleistocene deposits of Australia. In those beds we also meet with remains of several very large extinct species, which appear to be allied to those Wallabies in which the fourth premolar is large and elongated, all of them agreeing with the Wallabies in the absence of the median bridge between the first ridge and talon of the molars. These fossil forms comprise *M. brehus*, in which the skull was probably about one foot in length, and *M. rœchus*, and *M. anak*, which were of somewhat inferior dimensions. In the last-named species the length of the fourth upper premolar is equal to that of the first and half of the second molar.[1]

The third and last group of the genus includes the small

[1] For the characters of these species and the undermentioned distinct genera, see Owen's *Extinct Mammals of Australia* (1877), and Lydekker's *Catalogue of Fossil Mammalia in the British Museum*, pt. v. (1887).

Wallabies, which are small and lightly-built animals, in some instances not larger than a Rabbit. Their muffles are always naked, and in the skull the anterior palatine foramina are small and the posterior vacuities very large, while the posterior expansion of the nasals is very marked. The third upper incisor is smaller than in the last group. This group extends farther into the tropics than either of the others, being found in the New Britain and Aru islands, as well as in New Guinea. *M. brachyurus* is remarkable for its comparatively short and slender tail and small ears. The earliest known species of Kangaroo, referred to before, *M. bruni*, belongs to this section. Several examples were seen by Bruyn in 1711 living in captivity in the garden of the Dutch governor of Batavia, and described and figured in the account of his travels (*Reizen over Moskovie*, etc.) under the name of "Filander." It was quite lost sight of, and its name even transferred by S. Müller to another species (*Dorcopsis muelleri*), until rediscovered in 1865 by Rosenberg, who sent a series of specimens to the Leyden Museum from the islands of Aru and Great Key, thus determining its true habitat. *M. thetidis* is a well-known Australian representative of this group.

Extinct genera.—In addition to the fossil forms already mentioned which can be referred to existing genera, there are others from the Australian Pleistocene indicating extinct generic types of *Macropodidæ*, to which brief reference may now be made. The first of these is *Sthenurus*,[1] represented by a single large species (*S. atlas*), and characterised by the presence of a complete inner lobe to the fourth upper premolar, and of an outer one in the opposing lower tooth, so that these teeth present a flat and oval grinding surface when worn. The median longitudinal bridge connecting the transverse ridges of the molars is very imperfect; and in the upper molars there is no bridge between the first ridge and talon. In *Procoptodon*[2] the premolars resemble those of *Sthenurus*, but the molars are elongated, and usually have their enamel thrown into numerous vertical foldings. The most distinctive feature is, however, the complete ankylosis of the mandibular symphysis; the mandibular rami being deep, and the diastema in the dental series short. The lower incisors are nearly cylindrical, and the palate has large vacuities. Three species are known. The largest representation of the whole family is the type of the genus *Palorchestes*[3] (*P. azael*), in which the length of the skull is estimated at sixteen inches. It is distinguished from *Procoptodon* by the longer mandibular symphysis and diastema, and the spatulate lower incisors. The true molars have no distinct anterior talon, and are not grooved, while the palate was fully ossified.

[1] Owen, *Phil. Trans.* 1874, p. 264.

[2] Owen, *op. cit.* p. 788.

[3] Owen, *op. cit.* p. 797.

EXTINCT FAMILIES.

Here may be noticed two genera of extinct Marsupials, the remains of which have been found in the Pleistocene deposits of Australia, which agree with the *Macropodidæ* and the *Phalangeridæ* in having $\frac{3}{1}$ incisors, those of the lower jaw being very large and proclivous. As the whole of their structure, especially that of the hind feet, is not yet known, their precise affinities cannot be determined.

Diprotodon.[1]—Dentition: $i\ \frac{3}{1}$, $c\ \frac{0}{0}$, $p\ \frac{1}{1}$, $m\ \frac{4}{4}$; total 28. The first upper incisor very large and scalpriform (Fig. 56). True molars with prominent transverse ridges, as in *Macropus*, but wanting the longitudinal connecting bridge. Anterior and posterior limbs less disproportionate than in the Kangaroos. Humerus elongated, and differing from that of nearly all Marsupials in the absence of an

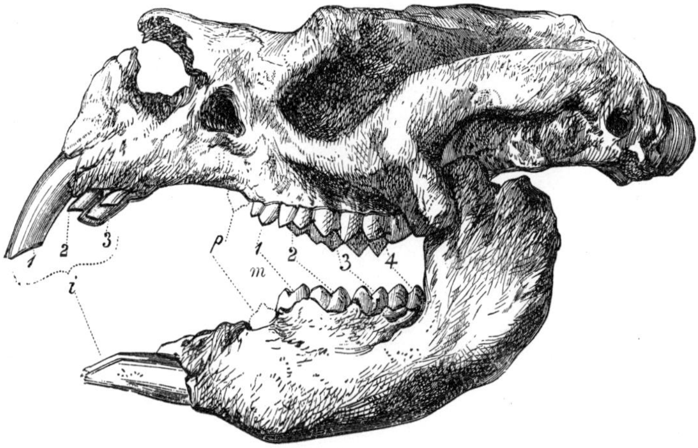

FIG. 56.—Left lateral aspect of the skull of *Diprotodon australis*; from the Pleistocene of Australia. $\frac{1}{10}$ natural size. *i*, Incisors; *p*, premolar; *m*, molars. (After Owen.)

entepicondylar foramen. The palate is fully ossified, and there is no pit or perforation in the masseteric fossa of the mandible. *D. australis* is the largest known Marsupial, being fully equal in bulk to a Rhinoceros. It may be regarded as the type of a family—*Diprotodontidæ*—having affinity on the one hand with the Phalangers and on the other with the Kangaroos.

Nototherium.[2]—Represented by a species of somewhat smaller size than the type of *Diprotodon*, with a shorter skull, in which the zygomatic arches are very wide and the nasals curiously expanded at their extremities. The mandibular symphysis is ankylosed;

[1] Owen, in *Mitchell's Eastern Australia*, 2d ed. vol. ii. p. 362 (1838).
[2] Owen, *Cat. Mamm. and Aves, Mus. R. Coll. Surgeons*, p. 314 (1845).

and, as in *Diprotodon*, there appears to have been no tooth-change. The humerus probably referable to *Nototherium* is of a short and widely expanded type, with a large entepicondylar foramen, and coming nearer to that of the Wombat than to that of any other existing form. The *Nototheriidæ* may apparently be regarded as a distinct family connecting the *Diprotodontidæ* with the *Phascolomyidæ* and *Phalangeridæ*.

Bibliography of Marsupialia.—G. R. Waterhouse, *Nat. Hist. of the Mammalia*, vol. i. "Marsupiata," 1846 ; J. Gould, *Mammals of Australia*, 1863 ; R. Owen, article "Marsupialia," in *Cyclop. of Anatomy and Physiology*, and various memoirs "On Extinct Mammals of Australia" in *Philosophical Transactions ;* W. H. Flower, "On the Development and Succession of the Teeth in the Marsupialia," *Phil. Trans.* 1867 ; O. Thomas, "On the Homologies and Succession of the Teeth in the Dasyuridæ," *Phil. Trans.* 1887 ; and "Catalogue of Marsupialia and Monotremata in the British Museum," 1888.

CHAPTER VII

THE SUBCLASS EUTHERIA AND THE ORDER EDENTATA

THE whole of the remaining groups of mammals are included in a single subclass, known by the names Eutheria, Monodelphia, or Placentalia.[1] The one distinctive feature they have in common (from which the last-mentioned name is derived) is the presence of an allantoic placenta by means of which the fœtus is nourished within the uterus of the mother. Throughout the entire subclass, as a general rule, the urino-genital organs open quite independently of the rectum; the corpus callosum of the brain is well developed; the mandible does not show a marked inflection of its angle; and distinct epipubic bones are not attached to the anterior margin of the pubic symphysis. In those cases where there is a heterodont and diphyodont dentition the dental formula can be reduced to some modification of the one given on p. 25, there being only one known genus where four true molars occur, and even that not invariably. As in the Metatheria, the coracoid is reduced to a mere appendage of the scapula, and the acetabular cavity of the pelvis is imperforate. While the survivors of the other subclasses have probably been for a long time in a stationary condition, these have, as there is already good evidence to show throughout all the Tertiary geological age, and by inference for some time before, been multiplying in numbers and variations of form, and attaining higher stages of development and specialisation in various directions. They consequently exhibit far greater diversity of external or adaptive modification than is met with in either of the other subclasses,—some being fitted to live as exclusively in the water as fishes, and others to emulate the aerial flight of birds.

To facilitate the study of the different component members of this large group, it is usual to separate them into certain

[1] The characters of the chief groups of the Eutheria here given are, in some measure, a fuller recapitulation of those already detailed in Chapter III., pp. 83-88.

divisions which are called "orders." In the main zoologists are now of accord as to the general number and limits of these divisions among the existing forms, but the affinities and relationships of the orders to one another are far from being understood, and there are very many extinct forms already discovered which do not fit at all satisfactorily into any of the orders as commonly defined.

Commencing with the most easily distinguished, we may first separate a group called Edentata, composed of several very distinct forms, the Sloths, Anteaters, and Armadillos, which under great modifications of characters of limbs and digestive organs, as well as habits of life, have just enough in common to make it probable that they are the very specialised survivors of an ancient group, most of the members of which are extinct, although the researches of palæontology have not yet revealed them to us. The characters of their cerebral, dental, and in many cases of their reproductive organs show an inferior grade of organisation to that of the generality of the subclass. The next order, about the limits of which there is no difficulty, is the Sirenia,—aquatic vegetable-eating animals, with complete absence of hind limbs, and low cerebral organisation,—represented in our present state of knowledge by but two existing genera, the Dugongs and Manatees, and by a few extinct forms, which, though approaching a more generalised mammalian type, show no special characters allying them to any of the other orders. Another equally well-marked and equally isolated, though far more numerously represented and diversified order, is that of the Cetacea, composed of the various forms of Whales, Dolphins, and Porpoises. In aquatic habits, external fish-like form, and absence of hind limbs, they resemble the last, though in all other characters they are as widely removed as are any two orders among the Eutheria.

All the remaining orders are more nearly allied together, the steps by which they have become modified from one general type being in most cases not difficult to realise. Their dentition especially, however diversified in detail, always responds to the formula already alluded to, and, although the existing forms are broken up into groups in most cases easy of definition, the discoveries already made in palæontology have in great measure filled up the gaps between them.

Very isolated among existing Eutheria are the two species of Elephant constituting the group called Proboscidea. These, however, are now known to be the survivors of a large series of similar animals, Mammoths, Mastodons, and Dinotheres, which as we pass backwards in time gradually assume a more ordinary or generalised type; and the interval which was lately supposed to exist between even these and the rest of the class is partially bridged over by the discovery in American Eocene and early Miocene formations of the gigantic Dinocerata, evidently offshoots of the great group of hoofed animals, or Ungulata, represented in the actual fauna by the Horses, Rhinoceroses, Tapirs, Swine, and Ruminants. Almost as isolated as the Proboscidea among existing mammals are the few small species constituting the family *Hyracidæ*, and in their case palæontology affords no help at present, and therefore, pending further discoveries, it has been thought advisable in most recent systems to give them the honour of an order to themselves, under the name of Hyracoidea. But the number of extinct forms already known allied to the Ungulata, though not coming under the definition of either of the two groups (Artiodactyla and Perissodactyla) under which all existing species range themselves, is so great that either many new orders must be made for their reception or the definition of the old order Ungulata so far extended as to receive them all, in which case both Proboscidea and Hyracoidea may be included within it. Again, the Rodentia or gnawing animals—Rabbits, Rats, Squirrels, Porcupines, Beavers, etc.—are, if we look only at the present state of the class, most isolated. No one can doubt what is meant by a Rodent animal, or have any difficulty about defining it clearly, at least by its dental characters; yet our definitions break down before the extinct South American *Typotherium*, half Rodent and half Ungulate, which leads by an easy transition to the still more truly Ungulate *Toxodon*, for the reception of which a distinct order (Toxodontia) has been proposed. It has also been suggested that the Rodents are connected by some of the extinct Tillodontia (or Tæniodontia) with the Edentates. The Insectivora and the Carnivora again are at present quite distinct orders, but they merge into one another through fossil forms, and are especially connected by the large group of primitive Carnivora, so abundantly represented in the Eocene deposits both of America and Europe, to which Cope has given the name of Creodonta. The Carnivora also appear to have been closely connected with the primitive Ungulates as represented by the extinct group called Condylarthra. In another direction the step from the Insectivores to the Lemurs is not great, and in past times the transition was probably complete. The Bats or Chiroptera are allied to the Insectivora in all characters except the extraordinary modification of their anterior extremities into wings; but this, like the want of the hind limbs in the Cetacea and Sirenia, makes such a clear distinction between them and all other mammals that, in the absence of any knowledge of any completely intermediate or transitional forms, they can be perfectly separated, and constitute as well-defined an order as any in the class. We have, however, an inkling of the mode in which the Insectivora were modified into Chiroptera shown us by the so-called Flying Lemur (*Galeopithecus*). Finally, we have the important and well-characterised group called Primates, including all the Monkeys and Man; and the question is not yet solved as to how and through what forms this is linked on to the other groups. It is commonly assumed that the Lemurs are nothing more than inferior Primates, but the interval between them in the actual fauna of the world is very great, and our knowledge of numerous extinct types recently discovered in America, said to be intermediate in characters, is not yet sufficient to enable us to form a definite opinion upon the subject.

The Edentata may be taken first as standing in some respects apart from all the others; and the Primates must be placed at the head of the series. The position of the others is quite arbitrary, as none of the hitherto proposed associations of the orders into larger groups stand the test of critical investigation, and palæontological researches have already gone far to show that they are all modifications of a common heterodont, diphyodont, pentadactylate form.

Order EDENTATA.

The name assigned to this group (which some zoologists think ought rather to be ranked as a subclass[1] than an order) by Cuvier is often objected to as inappropriate—for although some of the members are edentulous, others have very numerous teeth—and the Linnæan name Bruta is occasionally substituted. But that term is quite as objectionable, especially since the group to which Linnæus applied it is by no means equivalent to the order as now understood, as the names of the genera contained in it, viz. *Elephas*, *Trichechus*, *Bradypus*, *Myrmecophaga*, *Manis* and *Dasypus*, indicate. It contained, in fact, all the animals then known which are comprised in the modern groups of Proboscidea, Sirenia and Edentata together with the Walrus, one of the Carnivora. If retained at all, it should rather belong to the Proboscidea, as *Elephas* stands first in the list of genera in the *Systema Naturæ*. Cuvier's order included the *Ornithorhynchus* and *Echidna*, the structure of which was then imperfectly known, and which are now by common consent removed to an altogether different section of the class; but otherwise its limits are those now adopted. The name Edentata is so generally used, and its meaning so well understood, that it would be undesirable to change it now; in fact similar reasons might be assigned for ceasing to use nearly all the other current ordinal designations, for it might be equally well objected that all Carnivora are not flesheaters, many of the Marsupialia have not pouches, and so forth.

If the teeth are not always absent, they invariably exhibit certain imperfections, which are indeed almost the only common characters binding together the various extinct and existing members of the order. These are—that they are homodont and, with the

[1] The name Paratheria has been suggested for this proposed subclass.

remarkable exceptions of *Tatusia* and *Orycteropus*, monophyodont; they are never rooted, but have persistent pulps; except in some fossil forms, they are always deficient in one of the constituents which enter into the formation of the complete mammalian tooth, the enamel; and, at least among living forms, are never present either in the upper or lower jaw in the fore part of the mouth, the situation occupied by the incisors of other mammals.[1]

The peculiar nature of the dentition in the aberrant *Orycteropus* will be noticed under the heading of that genus. As a rule, the coracoid process of the scapula of the Edentates is more developed than in other Eutheria.

The degree of development of the brain varies considerably in the different families, the hemispheres being in some cases almost or quite smooth (Fig. 57), with a small corpus callosum, and large anterior commissure; while in other instances the hemispheres are convoluted, and the corpus callosum is larger.

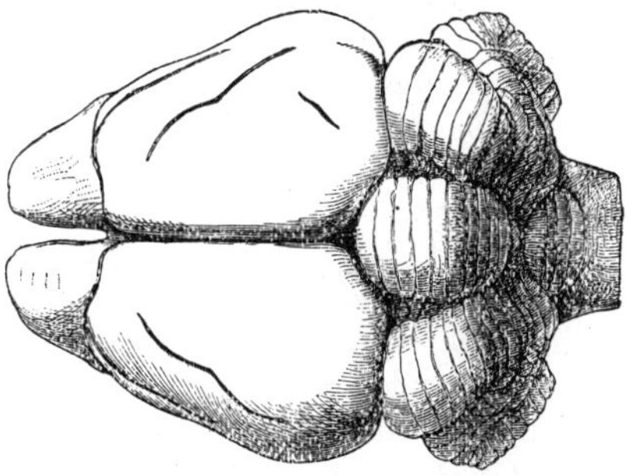

FIG. 57.—Upper surface of the brain of the Broad-banded Armadillo (*Xenurus unicinctus*). The large olfactory lobes are seen at the anterior extremity (left of figure); the hemispheres have only three sulci. (From Garrod, *Proc. Zool. Soc.* 1878, p. 230.)

There is so great a difference in structure and habits between some of the existing animals assigned to this order that, beyond the negative characters just mentioned, there seems little to connect them. The Sloths and Anteaters, for instance, in mode of life, general conformation of limbs, structure of digestive organs, etc., appear at first sight almost as widely separated as any mammals. Palæontology has, however, thrown great light upon their relations, and proved their real affinities. Perfectly intermediate forms have been discovered in the great Ground Sloths of America, which have the dentition and general form of the head of the Sloths, combined with the limbs and trunk of the Anteaters. It is, indeed, highly probable that the existing members of this order are very much differentiated representatives of a large group, the greater number of which are now extinct, and have become so without ever attaining a high grade of organisation. The great diversity of structure in the existing families, the high degree of specialisation to which many have attained, the paucity of species and even of individuals, their

[1] In some few Armadillos the suture between the premaxilla and maxilla passes behind the first upper tooth; but in all other known members of the order all the teeth are implanted in the maxilla.

limited area of distribution, and their small size compared with known ancestral forms, all show that this is an ancient and a waning group, the members of which seem still to hold their own either by the remoteness and seclusion of their dwelling-places, by their remarkable adaptation of structure to special conditions of life, or by aid of the peculiar defensive armature with which they are invested. Their former history can, however, only be thus surmised, rather than read, at present; for, though we have ample evidence of the abundance and superior magnitude of certain forms in the most recent or Pleistocene geological age, yet we have at present no definite evidence as to their origin, or relationship to other orders of mammals.

The existing members of the order readily group themselves into five distinct families, the limits of which are perfectly clear. These are (1) *Bradypodidæ*, or Sloths; (2) *Myrmecophagidæ*, or Anteaters; (3) *Dasypodidæ*, or Armadillos; (4) *Manidæ*, Pangolins or Scaly Anteaters; and (5) *Orycteropodidæ*, Aard-varks or African Anteaters. The geographical distribution of these families coincides with their structural distinction, the first three being inhabitants of the New and the last two of the Old World. It has been usual to arrange these families into two large groups or suborders: (1) the Phyllophaga, leaf-eaters, also called Tardigrada, containing the *Bradypodidæ* alone; and (2) the Entomophaga, insect-eaters, or Vermilingua, containing all the other families, from which sometimes the *Orycteropodidæ* are separated as a third suborder under the name of Effodientia, or Tubulidentata. Such an arrangement is, however, an artificial one, founded on superficial resemblance. The bonds which unite the *Manidæ* to the *Myrmecophagidæ* are mainly to be found in the structure of the mouth, especially the extensile character of the tongue, the great development of the submaxillary glands, and the absence of teeth. These characters are exactly analogous to those found in the Echidna among Monotremes, the Woodpeckers among Birds, and the Chamæleon among Reptiles,—the fact probably being that in countries where Termites and similar insects flourish various distinct forms of vertebrates have become modified in special relation to this abundance of nutritious food, which could only be made available by a peculiar structure of the alimentary organs. A close study of the more essential portions of the anatomy of these animals[1] leads to the belief that all the American Edentates at present known, however diversified in form and habits, belong to a common stock. Thus the *Bradypodidæ*, *Megatheriidæ*, and *Myrmecophagidæ* are certainly allied, the modifications seen in the existing families relating only to food and manner of life. The ancestral forms may have been omni-

[1] See Flower, "On the Mutual Affinities of the Animals composing the Order Edentata," *Proceedings of the Zoological Society*, 1882, p. 358.

vorous, and gradually separated into the purely vegetable and purely animal feeders; from the former are developed the modern Sloths, from the latter the Anteaters. The Armadillos (*Dasypodidæ*) are another modification of the same type, retaining some generalised characters, as those of the alimentary organs, but in other respects, as in their defensive armature, remarkably specialised. The two Old World families *Manidæ* and *Orycteropodidæ* are so essentially distinct, both from the American families and from each other, that it may even be considered doubtful whether they are derived from the same primary branch of mammals, or whether they may not be offsets of some other branch, the remaining members of which have been lost to knowledge. Further remarks on this point are recorded under the description of the *Orycteropodidæ*.[1]

Family BRADYPODIDÆ.

Externally clothed with long, coarse, crisp hair. Head short and rounded. External ears inconspicuous. Teeth $\frac{5}{4}$ in each jaw, subcylindrical, of persistent growth, consisting of a central axis of vaso-dentine, with a thin investment of hard dentine, and a thick outer coating of cement; without (so far as is yet known) any succession. Clavicles present. Fore limbs greatly longer than the hind limbs. All the extremities terminating in narrow, curved feet; the digits never exceeding three in number, encased for nearly their whole length in a common integument, and armed with long strong claws. Tail rudimentary. Stomach complex. No cæcum. Uterus simple and globular. Placenta deciduate, dome-like, composed of an aggregation of numerous discoidal lobes. Strictly

[1] An attempt has been made to represent these views by the following classification:

Order EDENTATA.
- Suborder PILOSA.
  - *Bradypodidæ.*
  - *Megatheriidæ.*
  - *Myrmecophagidæ.*
- Suborder LORICATA.
  - *Dasypodidæ.*
- Suborder SQUAMATA.
  - *Manidæ.*
- Suborder TUBULIDENTATA.
  - *Orycteropodidæ.*

It may be objected to this arrangement that the *present* divergence between the Sloths and Anteaters is hardly sufficiently indicated by their association in one suborder.—Flower, "On the Arrangement of the Orders and Families of Mammals," *Proc. Zool. Soc.* 1883, p. 178.

arboreal in habits, vegetable feeders, and limited geographically to the forest regions of South and Central America.

The Sloths, as the animals of this family are called on account of the habitual sluggishness of their movements, are the most strictly arboreal of all mammals, living entirely among the branches of trees, usually hanging under them, with their backs downwards (Fig. 58), and clinging to them with the simple hook-like organs to which the terminations of all their limbs are reduced. When they are obliged from any cause to descend to the ground, which they rarely, if ever, do voluntarily, their limbs, owing to their unequal length and the peculiar conformation of the feet—which allows the animals to rest only on the outer edge—are most inefficient

FIG. 58.—Two-toed Sloth (*Cholœpus hoffmanni*).

for terrestrial progression, and they crawl along a level surface with considerable difficulty. Though generally slow and inactive, even when in their natural haunts, Sloths can on occasions travel with considerable rapidity along the branches; and, as they do not leap, like most other arboreal creatures, they avail themselves of the swaying of the boughs by the wind to pass from tree to tree. They feed entirely on leaves and young shoots and fruits, which they gather in their mouth, the fore limbs aiding in dragging boughs within reach, but not being used like hands, as they are by monkeys, squirrels, etc. When sleeping they roll themselves up in a ball, and, owing to the dry shaggy character of their hair, are very inconspicuous among the mosses and lichens with which the trees of their native forests abound; the concealment thus afforded being heightened in some species by the peculiar greenish tint of the outer covering—very uncommon in mammals. This is not due to the colour of the hair itself, but to the presence upon its surface of an alga, the lodgment of which is facilitated by the fluted or rough surface of the exterior of the hair, and the growth of which is promoted by the dampness of the atmosphere in the gloomy tropical forests, as it soon disappears from the hair of animals kept in captivity in England. Sloths are nocturnal, silent, inoffensive, and solitary animals, and usually produce but one young at birth. They appear to show an almost reptilian tenacity of life, surviving the most severe injuries and large doses of poisons, and exhibiting longer persistence of irritability of muscular tissue after death than other mammals.

In the *Bradypodidæ*, as well as in the *Myrmecophagidæ*, the testes are placed close to each other, lying on the rectum between it and the bladder; the penis is quite rudimentary, consisting of a pair of small corpora cavernosa, not directly attached by their crura to the rami of the ischia, and having a glans scarcely larger than that of the clitoris of most mammals, and, as in birds and reptiles, without any true corpus spongiosum. In the females of both families the uterus is simple and globular; and the vagina, at least in the virgin state, is divided into two channels by a strong median partition. The deciduate placenta of *Cholœpus* is composed of a number of lobes aggregated into a dome-like mass; and it does not appear that the placenta of the Anteaters departs in any important characters from this type. According to the late Professor W. K. Parker, the embryos of the Sloths, Anteaters, and Pangolins have the stapes of the middle ear in the form of a rod, thus showing affinities with a very primitive type of mammalian organisation.

The Sloths were all included in the Linnæan genus *Bradypus*, but Illiger very properly separated the species with but two claws on the fore feet, under the name of *Cholœpus*, leaving *Bradypus* for those with three.

Bradypus.[1]—Three-toed Sloths. Teeth usually $\frac{5}{4}$ on each side; no tooth projecting greatly beyond the others; the first in the upper jaw much smaller than any of the rest; the first in the lower jaw broad and compressed; the grinding surfaces of all much cupped. Vertebræ: C 9, D and L 20 (of which 15 to 17 bear ribs), S 6, C 11. All the known species present the remarkable peculiarity of possessing nine cervical vertebræ, *i.e.* nine vertebræ in front of the one which bears the first thoracic rib (or first rib connected with the sternum, and corresponding in its general relations with the first rib of other mammals); but the ninth,

[1] Linn. *Syst. Nat.* 12th ed. vol. i. p. 50 (1766).

and sometimes the eighth, bears a pair of short movable ribs. The arms or fore limbs are considerably longer than the hind legs. The bones of the fore arm are complete, free, and capable of pronation and supination. The hand is long, very narrow, habitually curved, and terminates in three pointed curved claws, in close apposition with each other. The claws are, in fact, incapable of being divaricated, so that the hand is reduced to the condition of a triple hook, fit only for the function of suspension from the boughs of trees. The foot closely resembles the hand in its general structure and mode of use; the sole being habitually turned inwards, so that it cannot be applied to the ground in walking. The tongue is short and soft, and the stomach large and complex, bearing some resemblance to that of the ruminating Ungulates. The windpipe or trachea has the remarkable peculiarity among mammals—not unfrequent among birds and reptiles—of being folded on itself before it reaches the lungs. The mammæ are two, and pectoral in position.

"Ai" is the common name given in books to the Three-toed Sloths. They were all comprised by Linnæus under the species *Bradypus tridactylus*. More recently Dr. Gray described as many as eleven species, ranged in two genera, *Bradypus* and *Arctopithecus*; but the distinctions which he assigned both to species and genera do not bear close examination. Some are covered uniformly with a gray or grayish-brown coat; others have a dark collar of elongated hairs around the shoulders (*B. torquatus*); some have the hair of the face very much shorter than that of the rest of the head and neck; and others have a remarkable-looking patch of soft short hair on the back between the shoulders, consisting, when best marked, of a median stripe of glossy black, bordered on each side by bright orange, yellow, or white. There are also structural differences in the skulls, as in the amount of inflation of the pterygoid bones, which indicate real differences of species; but the materials in our museums are not yet sufficient to correlate these with external characters and geographical distribution. The habits of all are apparently alike. They are natives of Guiana, Brazil, and Peru, and one if not two species (*B. infuscatus* and *B. castaneiceps*) extend north of the Isthmus of Panama as far as Nicaragua. Of the former of these Dr. Seeman says that, though generally silent, a specimen in captivity uttered a shrill sound like a monkey when forcibly pulled away from the tree to which it was holding.

Cholœpus.[1]—Teeth $\frac{5}{4}$; the most anterior in both jaws separated by an interval from the others, very large, caniniform, wearing to a sharp, bevelled edge against the opposing tooth, the upper shutting in front of the lower when the mouth is closed (Fig. 59),

[1] Illiger, *Prodromus Syst. Mamm. et Avium*, p. 108 (1811).

unlike the true canines of heterodont mammals. Vertebræ: C 6 or 7, D 23-24, L 3, S 7-8, C 4-6. One species (*C. didactylus*) has the ordinary number of vertebræ in the neck; but an otherwise closely allied form (*C. hoffmanni*) has but six. The tail is very rudimentary. The hand generally resembles that of *Bradypus*; but there are only two functional digits with claws—those answering to the second and third of the typical pentadactylate manus. The structure of the hind limb generally resembles that of *Bradypus*, the appellation "two-toed" referring only to the anterior limb, for in the foot the three middle toes are functionally developed and of nearly equal size. *C. didactylus*, which has been longest known, is commonly called by the native name of *Unau*. It inhabits the forests of Brazil. *C. hoffmanni* (Fig. 58) has a more northern geographical range, extending from Ecuador through Panama to Costa Rica. Its voice, which is seldom heard, is like the bleat of a sheep, and if the animal is seized it snorts violently. Both species are very variable in external coloration.

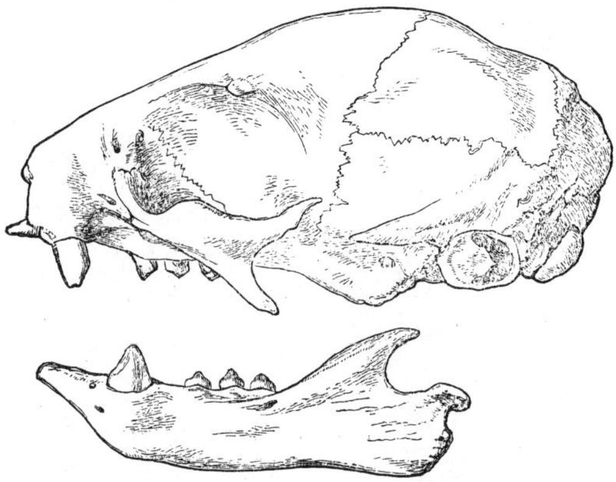

Fig. 59.—Skull of Two-toed Sloth (*Cholœpus didactylus*). From *Proc. Zool. Soc.* 1871, p. 432.

Nothropus.[1]—The only fossil form which has been referred to this family is indicated by a lower jaw, described by Dr. Burmeister, from the Pleistocene of Argentina, which appears to have belonged to an animal of about double the dimensions of *Cholœpus didactylus*. Professor Cope states, however, that this jaw really belongs to a Glyptodont; while it is referred by Dr. Ameghino to the next family.

Family MEGATHERIIDÆ.

The members of this family are all extinct. Their characters, so far as is known from the well-preserved remains of many species found abundantly in deposits of Pleistocene age in both North and South America, were intermediate between those of the existing *Bradypodidæ* and the *Myrmecophagidæ*, combining the head and

[1] Burmeister, *Sitzb. Ak. Berlin*, vol. xxviii. p. 613 (1882).

dentition of the former with the structure of the vertebral column, limbs, and tail of the latter. Almost all the known species are of comparatively gigantic size, the smallest, *Nothrotherium escrivanense*, exceeding the largest existing Anteater, and the Megatherium being larger than a Rhinoceros. The femur has no third trochanter, and the odontoid process of the axis vertebra has a peculiar facet on the ventral surface. The dentition is usually $\frac{5}{4}$ on each side, as in the Sloths, but $\frac{4}{3}$ in *Nothrotherium*.[1] This genus, and in a still more marked degree *Megatherium*, differ from all the others in the details of the structure of the teeth. They are very deeply implanted, of prismatic form (quadrate in transverse section), and the component tissues—hard dentine (Fig. 60, *d*), softer vaso-dentine

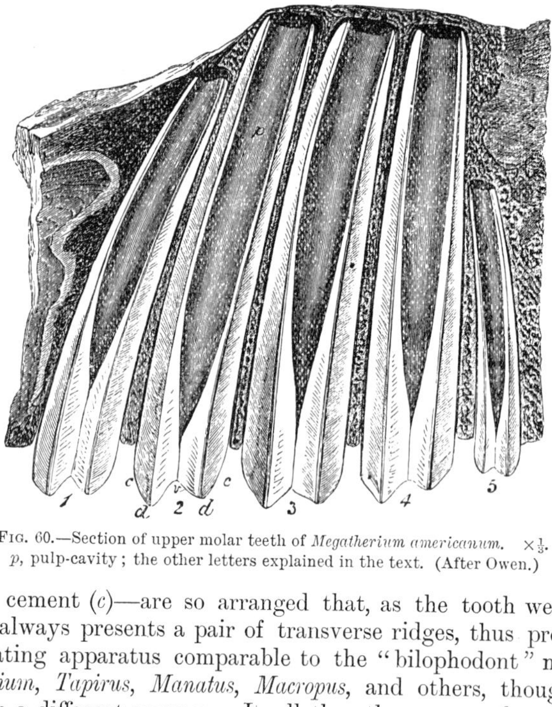

FIG. 60.—Section of upper molar teeth of *Megatherium americanum*. ×⅓. *p*, pulp-cavity; the other letters explained in the text. (After Owen.)

(*v*), and cement (*c*)—are so arranged that, as the tooth wears, the surface always presents a pair of transverse ridges, thus producing a triturating apparatus comparable to the "bilophodont" molar of *Dinotherium*, *Tapirus*, *Manatus*, *Macropus*, and others, though produced in a different manner. In all the other genera the teeth are more or less cylindrical, though sometimes laterally compressed or even longitudinally grooved on the sides, and on the grinding surface the prominent ridge of hard dentine follows the external contour, and is surrounded only by a thin layer of cement, as in the existing Sloths. The Ground Sloths, as the members

[1] Lydekker, in Nicholson and Lydekker's *Manual of Palæontology*, vol. ii. p. 1299 (1889). Originally described under the preoccupied name *Cœlodon*.

of this family may be conveniently designated, agree with the Sloths and Anteaters, and thereby differ from all other mammals, in that the coracoid process of the scapula and the coracoidal border of the same unite over the coraco-scapular notch, which is thus converted into a foramen. Large clavicles are present.

Megatherium.[1]—The typical genus *Megatherium*, as being the longest known representative of the family, may be noticed in some detail. A nearly complete skeleton, found on the banks of the River Luxan, near Buenos Ayres, and sent in 1789 to the Royal Museum at Madrid, long remained the principal if not the only source of information with regard to the species to which it belonged, and furnished the materials for many descriptions, notably that of Cuvier, who determined its affinities with the Sloths.[2] In 1832 an important collection of bones of the Megatherium was discovered near the Rio Salado, and secured for the Museum of the College

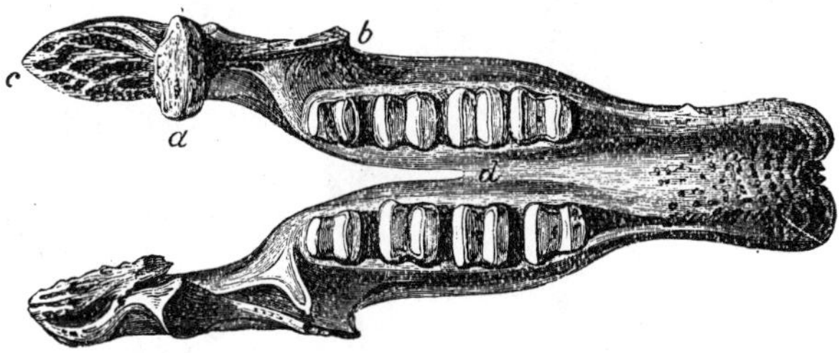

FIG. 61.—Oral surface of mandible of *Megatherium americanum*.
a, Condyle; *b*, masseteric process; *c*, angle; *d*, symphysis. (After Owen.)

of Surgeons of England; and these, with another collection found at Luxan in 1837, and now in the British Museum, supplied the materials for the complete description of the skeleton published by Sir R. Owen in 1861. Other skeletons have subsequently been received by several of the Continental museums, as Milan and Paris, and also by those in South America; and consequently our knowledge of the organisation of the Megatherium, so far as it can be deduced from the bones and teeth, is as complete as that of any other animal, recent or extinct.

The remains hitherto spoken of are all referred to one species, *Megatherium americanum* of Blumenbach (*M. cuvieri* of Desmarest), and are all from the newest or Pleistocene geological formations of the Argentine Republic and Paraguay, or the lands forming the

[1] Cuvier, *Tableau Elém. d'Hist. Nat. des Animaux*, p. 146 (1798).

[2] An excellent figure of this skeleton, which unfortunately was incorrectly articulated, and wanted the greater part of the tail, was published by Pander and D'Alton in 1821, and has been frequently reproduced in subsequent works.

basin of the Rio de la Plata. Dr. Leidy has described, from similar formations in Georgia and South Carolina, bones of a closely allied species, about one-fourth smaller, which he has named *M. mirabile*. Three other South American species have been described; but *M. laurillardi*, of Lund, founded upon remains found in Brazil, has been made the type of the genus *Ocnopus*.

The following description will apply especially to the best-known South American form, *Megatherium americanum*. In size it exceeded any existing land animal except the elephant, to which it was inferior only in consequence of the comparative shortness of its limbs; for in length and bulk of body it was its equal, if not

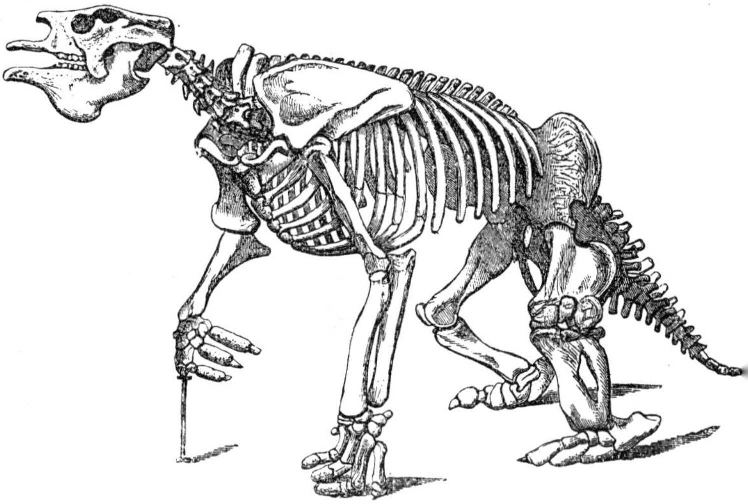

FIG. 62.—Skeleton of *Megatherium*, from the specimen in the Museum of the Royal College of Surgeons. $\times \frac{1}{25}$.

superior. The full length of a mounted skeleton (Fig. 62), from the fore part of the head to the end of the tail, is 18 feet, of which the tail occupies 5 feet. The head, which is small for the size of the animal, presents a general resemblance to that of the Sloth; the anterior part of the mouth is, however, more elongated, and the jugal bone, though branched posteriorly in the same way as that of the Sloth, meets the zygomatic process of the squamosal, thus completing the arch. The lower jaw has the middle part of its horizontal ramus curiously deepened, so as to admit of implantation of the very long-rooted teeth, the peculiar structure of which has been already described. A skull recently discovered shows that, instead of the wide gap between the extremity of the nasals and the premaxillæ exhibited in Fig. 62, there was a prenasal bone, towards which a process extended upwards and backwards from the extremity of the upper surface of the pre-maxillæ.

The vertebral column consists of seven cervical, sixteen dorsal, three lumbar, five sacral, and eighteen caudal vertebræ. The spinous processes are much better developed than in the Sloths, and are all directed backwards, there being no reversing of the inclination near the posterior end of the dorsal series, as in most active-bodied mammals. In the lumbar region, the accessory zyga-pophyses, rudimentary in Sloths, are fully developed, as in the Anteaters.

The tail is large, and its basal vertebræ have strong lateral and spinous processes and chevron bones, indicating great muscular development. The scapula resembles that of the Sloths in the union of the acromion with the coracoid, and in the bridging over of the supra-scapular notch. The clavicle is complete and very large, much resembling that of man on a large scale. The fore limbs are longer than the hind limbs. The humerus has no ent-epicondylar foramen. The radius and ulna are both well developed, and have a considerable amount of freedom of movement. The hand is singularly modified. The pollex is represented only by a rudimentary metacarpal, but the next three digits are large, and terminate in phalanges adapted for the support of immense claws, the middle one being especially large. The outer or fifth digit has no claw, and it may be considered as certain that the weight of the foot was, in standing and walking, chiefly thrown upon this one, which was protected by a callous pad below, as in the existing great Anteater, while the other toes were curved inwards towards the palm, and only came in contact with the ground by their outer surfaces. The mechanical arrangements by which the weight of the body was thrown entirely upon the outer side of the foot are very curious, and are fully described in Owen's memoir. The pelvis is remarkably wide, even more so than that of the Elephant, but it is formed on the same principle as in the Sloths. The femur is extremely broad and flattened; the tibia and fibula are short and strong, and united together at each end. The hind foot, contrary to the usual rule in the Edentata, is even more singularly modified than the hand. Thus the ankle-joint is formed upon a peculiar plan, quite unlike that of the Sloths, or of any other mammal, except the Megatherium's nearest allies; and the calcaneum projects nearly as far backwards as the fore part of the foot does forwards. There is no trace of great toe or hallux, or of its corresponding cuneiform bone; the second toe is rudimentary; while the third has an enor-mous ungual phalanx, which, as in those of the hand, is remarkable for the immense development of the bony sheath reflected from its proximal end around the base of the claw. The two outer toes have large and very peculiarly-shaped metatarsals, but only small phalanges, and no claws. The creature probably walked upon the outer edge of the sole, so that the great falcate claw of the third toe did not come into contact with the ground, and so was kept in a state of sharpness ready for use. The foot was therefore formed upon quite a different principle from that of the Anteaters or Sloths, though somewhat like the latter in having two of the toes aborted.

Taking all the various points of its structure together, they clearly indicate affinities both with the existing Sloths and with the Anteaters, the skull and teeth more resembling those of the former, and the vertebral column and limbs the latter. It is also not difficult to infer the food and habits of this enormous creature. That it was a leaf-eater there can be little doubt; but the greater size and more complex structure of its teeth might have enabled it to crush the smaller branches as well as the leaves and succulent shoots which form the food of the existing Sloths. It is, however, very improbable that it climbed into the branches of the trees like its diminutive congeners, and it is far more likely that it obtained its subsistence by tearing them down with the great hook-like claws of its powerful prehensile fore limbs, being easily enabled to reach them by raising itself up upon the massive tripod formed by the two hind feet, firmly fixed to the ground by the one huge falcate claw, and the stout, muscular tail. The whole conformation of the hinder part of the animal is strongly suggestive of such an action. There can also be little doubt but that all its movements were as slow and deliberate as those of its modern representatives.

An idea at one time prevailed that the Megatherium was covered externally with a coat of bony armour like that of the Armadillos; but this originated in dermal plates belonging to the Glyptodon having been accidentally associated with bones of the Megatherium. Similar plates, on a smaller scale, have indeed been found in connection with the skeleton of the Mylodon, but never yet with the Megatherium, which we may therefore imagine with a covering of coarse hair like that of its nearest living allies, the Sloths and Anteaters.

Scelidotherium, *Mylodon*, etc.—Of the more important remaining genera of this family a briefer notice will suffice. *Scelidotherium* (in which *Platyonyx* may be included) comprises several species of considerably smaller dimensions than the Megatherium, and is in some respects intermediate between that genus and *Mylodon*. The teeth have an oval cross-section, like those of the Sloths, while the skull, in which the length of the nasals is subject to great variation in the different species, approximates more or less closely to that of the *Myrmecophagidæ*. The humerus generally has an entepicondylar foramen; and the form and relations of the bones of the feet differ considerably from those obtaining in the type genus. *S. leptocephalum*, the type of the genus, occurs in Patagonia and Argentina but other species are found in Brazil and Chili. The genus *Mylodon*, in its widest sense, may be taken to include a number of comparatively large Edentates, some of which have been described under the names of *Grypotherium*, *Lestodon*, and *Pseudolestodon*. The teeth of the upper jaw are generally of an oval or subtriangular section; and in the more typical forms the first and second teeth are separated by a short interval, the former being horizontally worn. In other species, however, like *M.* (*Lestodon*) *armatus*, there is a considerable space between the first and second teeth, and the first is worn obliquely. The skull is exceedingly like that of the Sloths in general contour; and there is not the descending process at the angle of the mandible found in *Megatherium*. The humerus has no entepicondylar foramen. The species represented in Fig. 63 is from the Pleistocene of South America; but the type of the genus is *M. harlani*, from beds of corresponding age in Kentucky. The Patagonian *M.* (*Grypotherium*) *darwini* is a remarkable form, characterised by the presence of a bony arch connecting the premaxillæ with the nasals, of which, as already mentioned, there is an incomplete development in *Megatherium*. *Megalonyx*, from the Pleistocene of Kentucky, differs from *Mylodon* by the long interval between the first and second teeth, and also by the presence of an entepicondylar foramen in the humerus. *Nothrotherium* is a smaller form, occurring in the deposits of the Brazilian caves, of which the dental features have been already mentioned. The osteological characters of these and other allied genera have been fully described in the works of Cuvier, Owen, Burmeister, Leidy, Ameghino, Gervais, Reinhardt, and others.

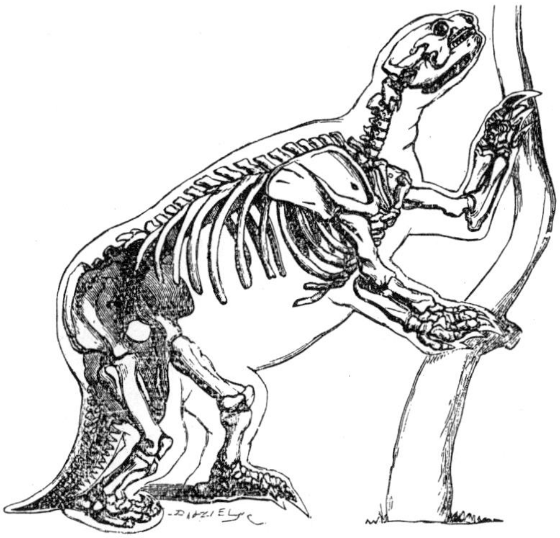

FIG. 63.—Skeleton of *Mylodon robustus* (Pleistocene, South America). From Owen.

Promegatherium.—Two genera from the infra-Pampean beds

of Argentina, described as *Promegatherium* and *Promylodon*, are respectively distinguished from *Megatherium* and *Mylodon* by the presence of bands of enamel on the teeth, which points to the descent of the Edentates from mammals with enamelled teeth.

The Tertiary North American forms described as *Moropus* and *Morotherium*,[1] and originally regarded as Edentates, would appear to be aberrant Ungulates.

Family MYRMECOPHAGIDÆ.

Externally clothed with hair. No teeth. Head elongated. Mouth tubular, with a small terminal aperture, through which the long, vermiform tongue, covered with the viscid secretion of the enormous submaxillary glands, is rapidly protruded in feeding, and withdrawn again with the adhering particles of aliment, which are then sucked into the pharynx. Clavicles rudimentary. In the manus, the third toe is greatly developed, and has a long falcate claw; the others are reduced or suppressed. The pes has four or five subequal digits with claws. Posterior dorsal and lumbar vertebræ, with additional interlocking zygapophyses. Tail long, sometimes prehensile. Uterus simple. Placenta dome-like or discoidal. Brain fairly convoluted, and with a large corpus callosum and anterior commissure. The animals of this family are the "Anteaters" *par excellence.* They feed exclusively on animal substances, mostly insects. One species is terrestrial, the others arboreal; none burrow in the ground. They are all inhabitants of the Neotropical region.

The reproductive organs, as noticed on p. 181, are of the same general type as in the *Bradypodidæ.*

Myrmecophaga.[2]—Skull greatly elongated and narrow, its upper surface smooth and cylindriform. Anteriorly the face is produced into a long, tubular rostrum, rounded above and flattened below, with terminal nares, and composed of the mesethmoid ossified for more than half its length, the vomer, the maxillæ, and the long and narrow nasal bones, the premaxillæ being extremely short and confined to the margin of the anterior nares. The zygomatic arch is incomplete, the styliform jugal only articulating with the maxilla in front, and not reaching to the very short zygomatic process of the squamosal. The lachrymal foramen is in front of the margin of the orbit. There are no postorbital processes to the frontals, or any other demarcation between the orbits and the temporal fossæ. Palate extremely elongated, and produced backwards as far as the level of

[1] See E. D. Cape, *Amer. Naturalist*, vol. xxiii. p. 152 (1889).

[2] Linn. *Syst. Nat.* 12th ed. vol. i. p. 51 (1766).

the external auditory meatus by the meeting in the middle line of the largely developed pterygoids. The glenoid fossa a shallow oval facet, with its long diameter from before backwards. Mandible very long and slender, with an exceedingly short symphysis, no distinct coronoid process, and a slightly elevated, elongated, flattened, condylar articular surface. Vertebræ: C 7, D 15-16, L 3-2, S 6, C 31. Clavicles rudimentary. In the manus the first digit is very slender, the second also slender, with compressed phalanges of nearly equal length. The third digit is immensely developed; though its proximal phalanx is extremely short, its ungual phalanx is so long that the entire length of the digit exceeds that of the second. The fourth has a long and rather slender metacarpal, and three phalanges diminishing in size, the ungual phalanx being very small. The fifth has the metacarpal nearly as long, but not so stout, as the fourth, and followed by two small phalanges, the last rudimentary and conical. Claws are developed upon all but the fifth. In walking the toes are kept strongly flexed, and have their points turned upwards and inwards, the weight being supported upon a callous pad over the end of the fifth digit, and by the dorsal surfaces of the third and fourth digits. The hind feet are short and rather broad, with five subequal claws, the fourth the longest, the first shortest; the whole sole is placed on the ground in walking. Body rather compressed, clothed with long, coarse hair. Tail about as long as the body, and covered with very long hair; not prehensile. Ears small, oval, erect. Eyes very small. Stomach consisting of a subglobular, thin-walled, cardiac portion, and a muscular pyloric gizzard with dense epithelial lining. No ileocolic valve, and a short wide ill-defined cæcum. Mammæ two, pectoral.

There is one species,[1] *M. jubata*, the Great Anteater, or Ant Bear (Fig. 64), measuring 4 feet in length without the tail, and upwards of 2 feet in height at the shoulder. Its prevailing colour is gray, with a broad black band, bordered with white, commencing on the chest, and passing obliquely over the shoulder, diminishing gradually in breadth as it approaches the loins, where it ends in a point. It is extensively distributed in the tropical parts of South and Central America, frequenting low swampy savannas along the banks of rivers, and the depths of the humid forests, but is nowhere abundant. Its food consists mainly of termites, to obtain which it opens their nests with its powerful sharp anterior claws, and as the insects swarm to the damaged part of their dwelling, it draws them into its mouth by means of its long, flexible, rapidly-moving tongue covered with glutinous saliva. The Great Anteater is quite terrestrial in its habits, being never known to climb trees, nor does it

[1] Professor Cope has recently come to the conclusion that there are three species; but further evidence is required in support of this view.

burrow underground like the Armadillos. Though generally an inoffensive animal, when attacked it can defend itself vigorously and effectively with its sabre-like anterior claws. The female bears but a single young at a birth.

The union of the pterygoids in the middle line to prolong the narial passage is a character found elsewhere among existing mammals only in the next genus, in one Armadillo (*Tatusia*), and in certain Cetacea. The contrast in length between the skull of the Great Anteater and that of the Sloth is, as Professor Parker observes, very marked indeed; the one being relatively the longest and the

FIG. 64.—The Great Anteater (*Myrmecophaga jubata*). (From Sclater, *List of Animals in Zoological Society's Gardens*, 1883, p. 190.)

other almost the shortest in the whole class. The small size and incomplete development of the jugal bone in the zygomatic arch affords another striking contrast to the Sloths (Fig. 59).

Tamandua.[1]—This genus closely resembles the last in anatomical structure, but the head is much less elongated, the fur is short and bristly, the tail tapering, prehensile, with the under side throughout and the whole of the terminal portion naked and scaly. The stomach is similar to that of *Myrmecophaga*, but with the muscular pyloric gizzard not quite so strongly developed. There is a distinct ileo-colic valve and a short globular cæcum. The fore foot has a very large claw on the third toe, moderate-sized claws on the second and

[1] Gray, *Annals of Philosophy*, new series, vol. x. p. 343 (1825).

fourth, a very minute one on the first, and none on the fifth, which is entirely concealed within the skin. The hind foot has five subequal claws. Vertebræ: C 7, D 17, L 2, S 5, C 37. There are very rudimentary clavicles.

The Tamandua (Fig. 65) is much smaller than the Great Anteater, and differs essentially from it in its habits, being mainly

FIG. 65.—Tamandua Anteater (*Tamandua tetradactyla*). From *Proc. Zool. Soc.* 1871, pl. xliii.

arboreal. It is an inhabitant of the dense primeval forests of South and Central America. As different individuals vary much in their coloration, it is possible that there may be more than one species. The usual colour is yellowish-white, with a broad black lateral band, covering nearly the whole of the side of the body.

Cycloturus.[1]—The skull is much shorter even than in *Tamandua*, and is arched considerably in the longitudinal direction. It differs from that of the other members of the family mainly in the long canal for the posterior nares not being closed by bone below, as the greater part of the palatines and the pterygoids do not meet in the middle line. The mandible has a prominent, narrow, recurved coronoid, and a well-developed angular process; it is strongly decurved in front. Vertebræ: C 7, D 16, L 2, S 4, C 40. Ribs remarkably broad and flat. Clavicles well developed. Manus remarkably modified, the third digit being greatly developed at the expense of all the others, and having a stout short metacarpal and but two phalanges, of which the most distal is large, compressed, pointed, and much curved, and bears a very strong hook-like claw. The second digit has the same number of phalanges, and bears a claw, but is very much more slender than the third. The fourth is represented only by the metacarpal and one nailless phalanx, the first and fifth only by very rudimentary metacarpals. The pes

[1] Gray, *Annals of Philosophy*, new series, vol. x. p. 343 (1825).

is also completely modified into a climbing organ. The hallux is rudimentary, consisting of a metatarsal and one phalanx, concealed beneath the skin; but the other four toes are subequal and much curved, with long pointed compressed claws. The tuber calcanei is directed towards the plantar surface, and parallel with it and extending to about double its length is a greatly elongated sesamoid ossicle. These together support a prominent calcarine cushion, to which the nails are opposed in climbing. Stomach pyriform, with muscular walls, but no distinct gizzard-like portion, as in the foregoing genera. Commencement of the colon provided with two small cæca (Fig. 66), resembling those of many birds, narrow at the base, and rather dilated at their terminal blind ends, and communicating with the general cavity by very minute apertures. Tail longer than the body, tapering, bare on the under surface, and very prehensile. Fur soft and silky.

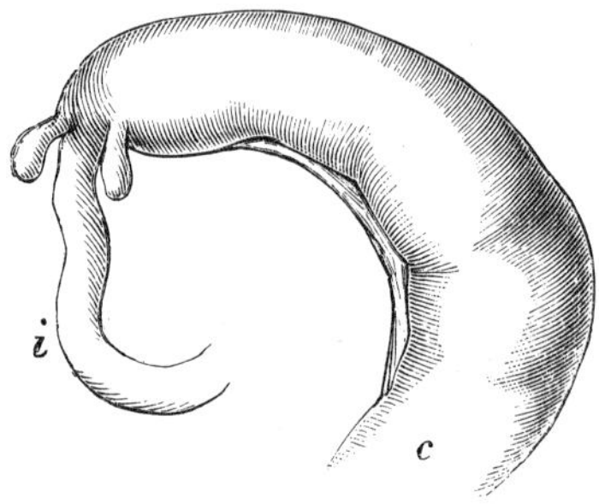

FIG. 66.—Cæca of the Two-toed Anteater (*Cycloturus didactylus*). *i*, Ileum; *c*, colon.

This genus has also but one species certainly known, the Little or Two-toed Anteater (*C. didactylus*), an animal not larger than a Rat, of a general yellowish-colour, and exclusively arboreal in its habits. It is a native of the hottest parts of South and Central America.

Family DASYPODIDÆ.

The greater part of the skin strongly ossified. On the back and sides the union of numerous quadrate or polygonal scutes forms a hard shield, usually consisting of an anterior (scapular) and posterior (pelvic) solid portion (which overhang on each side the parts of the body they respectively cover, forming chambers into which the limbs are withdrawn), and a variable number of rings between, connected by soft flexible skin so as to allow of curvature of the body. The top of the head has also a similar shield (cephalic), and the tail is usually encased in bony rings or plates. The outer or exposed surfaces of the limbs are protected by irregular bony scutes, not united at their margins; but the skin of the inner surface of the limbs and under side of the body is soft, and more or less clothed with hair. Hairs also in many species project through apertures between the bony scutes of the back. The ossified dermal scutes are everywhere covered by a layer of horny epidermis. Teeth numerous, simple, of persistent growth, and usually monophyodont, but in one genus (*Tatusia*) a succession of teeth has been observed. Zygomatic arch of skull complete. Cervical vertebræ with extremely short, broad, and depressed bodies. The atlas free, but the second and third, and often several of the others, ankylosed together both by their bodies and arches. Lumbar vertebræ with accessory zygomatic processes, and very large metapophyses, supporting the bony carapace. Clavicles well developed. A third trochanter on the femur. Tibia and fibula ankylosed at their distal extremities. Fore feet with strongly developed, curved claws, adapted for digging and scratching—three, four, or five in number. Hind feet plantigrade, with five toes, all provided with nails. Tongue long, pointed, and extensile, though to a less degree than in the Anteaters. Submaxillary glands largely developed. Stomach simple. Uterus simple. Placenta discoidal, deciduate. The brain is generally characterised by the large size of the olfactory lobes (Fig. 57), and the slight development of sulci on the hemispheres; the sylvian fissure being represented only by a very open and shallow angle. From the earliest stage of development the stapes is stirrup-shaped, thus showing a nearer affinity to the higher mammals than is presented by the Sloths.

The animals of this family are commonly called Armadillos, a word of Spanish origin, having reference to their armour-like covering. The existing species are all of small or moderate size. They are mostly, though not universally, nocturnal in their habits, and are all omnivorous, feeding on roots, insects, worms, reptiles, and carrion. Armadillos are harmless and inoffensive creatures, offering no resistance when caught, their principal means of escape from their enemies being the extraordinary rapidity with which they can burrow in the ground, and the tenacity with which they retain their hold in their subterranean retreats. Notwithstanding the shortness of their limbs they can run with great rapidity. Most of the species are esteemed good eating by the natives of the countries in which they live. They are all inhabitants of the open plains or the forests of the tropical and temperate parts of South America, with the exception of one species (*Tatusia novem-cincta*), which ranges as far north as Texas. Of the existing genera, *Chlamydophorus* stands apart from the rest in the formation of its external covering; but in all other respects *Tatusia* is the most aberrant form, exhibiting a peculiar type of structure of the fore feet, which in all the others show modifications, though in very varying degrees, of a single and different type.

The reproductive organs of the *Dasypodidæ* differ from those of the Sloths and Armadillos in the presence of a largely developed copulating organ in the male, and of a simple vagina of corresponding length in the female. The testes are still abdominal, although not in the same position; and the penis still wants both the glans and bulb. The uterus is nearly or quite as simple as in the Sloths and Anteaters.; and there is no reason to believe that the placentation is essentially different from that obtaining in the other groups.

Subfamily **Chlamydophorinæ.**—In most anatomical characters, especially the structure of the fore foot, this little group resembles the *Dasypodinæ;* but it differs remarkably from all other known Armadillos, living or extinct, in the peculiar modification of the dermal armour.

Chlamydophorus.[1]—Teeth $\frac{8}{8-9}$, subcylindrical, somewhat compressed, moderate in size, smaller at each end (especially in front) than at the middle of the series. Skull broad and rounded behind, pointed in front. Muzzle subcylindrical and depressed. A conspicuous rounded, rough prominence on the frontal bone, just before each orbit. Tympanic prolonged into a tubular auditory meatus, curving upwards round the base of the zygoma. Vertebræ: C 7, D 11, L 3, S 10, C 15. Upper part of head and trunk covered with four-sided horny plates (with very small thin ossifications beneath), forming a shield, free, and overhanging the sides of the trunk, and attached only along the middle line of the back. The plates are arranged in a series of distinct transverse bands, about twenty in number between the occiput and the posterior truncated end, and not divided into solid thoracic and pelvic shields with movable bands between. The hinder end of the body is abruptly truncated and covered by a vertically-placed, strong, solid, bony shield, of an oval (transversely extended) form, covered by thin epidermic plates. This shield is firmly ankylosed by five bony processes to the hinder part of the pelvis. Through a notch in the middle of its lower border the tail passes out. The latter is rather short, cylindrical in its proximal half, and expanded and depressed or spatulate in its terminal portion, and covered with horny plates. The dorsal surfaces of the fore and hind feet are also covered with horny plates. The remainder of the limbs and under surface and sides of the body beneath the overlapping lateral parts of the dorsal shield are clothed with rather long, very soft, silky hair. Eyes and ears very small, and concealed by the hair. Extremities short. Feet large, each with five well-developed claws, those on the fore feet very long, stout, and subcompressed, the structure of the digits being essentially the same as those of *Xenurus* and *Priodon*. Nipples two, pectoral. Visceral anatomy closely resembling that of *Dasypus*, the cæcum being broad, short, and bifid.

The Pichiciago (*C. truncatus*), a small burrowing animal, about 5 inches long, inhabits the sandy plains of the western part of the Argentine Republic, especially the vicinity of Mendoza. Its

[1] Harlan, *Ann. New York Lyceum Nat. Hist.* vol. i. p. 237 (1824).—Amended from *Chiamyphorus*.

horny covering is of a pinkish colour, and its silky hair snow white. It is rare, and its habits are but little known. A second species, *C. retusa*, from Bolivia, has been described by Burmeister. It is of rather larger size, and has the dorsal shield attached to the skin of the back as far as its edge, instead of only along the median line.

Subfamily **Dasypodinæ.**—Fore feet usually with all five digits developed and with nails, though the first and fifth may be suppressed. The first and second long and slender, with the normal number and relative length of phalanges. The others stout, with short broad metacarpals, and the phalanges greatly reduced in length and generally in number by coalescence. The ungual phalanx of the third very large, that of the others gradually diminishing to the fifth. *Dasypus*, as now restricted, has the most normal form of manus, but the modifications so markedly developed in all the others (and culminating in *Tolypeutes*) are foreshadowed, as it were, in it. Ears wide apart. Mammæ one pair, pectoral.

Dasypus.[1]—Teeth $\frac{9}{10}$ or $\frac{8}{9}$, of which the anterior in the upper jaw is usually implanted in the premaxillary bone. The series of teeth extends posteriorly some distance behind the anterior root of the zygoma, almost level with the hinder edge of the palate. They are large, subcylindrical, slightly compressed, diminishing in size towards each end of the series; the anterior two in the mandible much smaller, and more compressed than the others. Cranial portion of the skull broad and depressed. Facial portion triangular, broad in front and much depressed. Auditory bulla completely ossified, perforated on the inner side by the carotid canal, and continued externally into an elongated bony meatus auditorius, with its aperture directed upwards and backwards. (In all the remaining genera of *Dasypodinæ* the tympanic bone is a mere half ring, loosely attached to the cranium.) Mandible with a high ascending ramus, broad transversely-placed condyle, and high slender coronoid process. Vertebræ: C 7, D 11-12, L 3, S 8, C 17-19. Head broad and flat above. Muzzle obtusely pointed. Ears of moderate size or rather small, placed laterally, far apart. Body broad and depressed. Carapace with six or seven movable bands between the scapular and pelvic shields, each plate, or scute, being marked by a regular ellipse formed of widely separated punctures. Tail shorter than the body, tapering, covered with plates forming distinct rings near the base. Fore feet with five toes; the first much more slender than the others, and with a smaller ungual phalanx and nail; the second, though the longest, also slender. The third, fourth, and fifth gradually diminishing in length, all armed with very strong, slightly curved, compressed claws, sloping away from an elevated

[1] Linn. *Syst. Nat.*, 12th ed. vol. i. p. 54 (1766).

rounded inner border to a sharp, outer, and inferior edge. The hind foot rather short, with all five toes armed with stout, compressed, slightly curved, obtusely pointed claws—the third the longest, the second nearly equal to it, the fourth the next, the first and fifth shorter, and nearly equal.

To this genus belongs one of the best-known species of the group, the Six-banded Armadillo or Encoubert (*D. sexcinctus*) of Brazil and Paraguay. A very similar species, *D. villosus*, the Hairy Armadillo, replaces it south of the Rio Plata. There are also two very small species—*D. vellerosus*, from the Argentine Republic and North Patagonia, and *D. minutus* from La Plata. The latter differs from the other three in having no tooth implanted in the premaxillary bone. Remains apparently referable to *D. villosus* occur in the Pleistocene cavern-deposits of Brazil.

Xenurus.[1]—Teeth $\frac{9}{9}$ or $\frac{8}{8}$, of moderate size and subcylindrical. The most posterior placed a little way behind the anterior root of the zygoma, but far from the hinder margin of the palate. Cranium somewhat elongated, much constricted behind the orbits, and immediately in front of the constriction considerably dilated. Mandible slender; coronoid process very small and sharp-pointed, sometimes obsolete. Vertebræ: C 7, D 12-13, L 3, S 10, C 18. Head broad behind. Ears rather large and rounded, wide apart. Movable bands of carapace 12-13; the scutes being marked by an obscurely granular sculpture. Tail considerably shorter than the body, slender, and covered with nearly naked skin, with but a few small, scattered, dermal bony plates, chiefly on the under surface and near the apex. On the fore feet the first and second toes are long and slender, with small claws and the normal number of phalanges; the other toes have but two phalanges; the third has an immense falcate claw; the fourth and fifth similar but smaller claws. The hind feet are comparatively small, with five toes, bearing small, triangular, blunt nails; the third longest, the first shortest. The best known species of this genus, the Tatouay or Cabassou, *X. unicinctus*, is, after *Priodon gigas*, the largest of the group. It is found, though not abundantly, in Surinam, Brazil, and Paraguay, its remains occurring in the Pleistocene cavern-deposits of Brazil. Others, *X. hispidus* and *lugubris*, have been described, but little is as yet known of them.

Priodon.[2]—Teeth variable in number, and generally differing on the two sides of each jaw, usually from 20 to 25 on each side above and below, so that as many as 100 may be present altogether; but as life advances the anterior teeth fall out, and all traces of their alveoli disappear. The series extends as far back as the hinder edge of the anterior root of the zygoma. The teeth are

[1] Wagler, *Syst. Amphibien*, etc., p. 36 (1830).

[2] F. Cuvier, *Hist. Nat. des Mammifères* (1822).—*Priodontes*.

all very small; those in the anterior half of each series being strongly compressed, with flat sides and a straight free edge; the posterior ones are more nearly cylindrical, with flat truncated, free surfaces. Vertebræ: C 7, D 12, L 3, S 10, C 23. Head small, elongated, conical. Ears moderate, ovate. Carapace with 12-13 movable bands. Tail nearly equal to the body in length, gradually tapering, closely covered with quadrangular scales, arranged in a quincunx pattern. Fore feet with five toes, formed on the same plan as those of *Xenurus*, but with the claw of the third of still greater size, and that of each of the others, especially the fifth, proportionately reduced. Hind foot short and rounded, with five very short toes, with short, broad, flat, obtuse nails. The only known species, the Great Armadillo (*P. gigas*), is by far the largest of existing members of the family, measuring rather more than 3 feet from the tip of the nose to the root of the tail, the tail being about 20 inches long. It inhabits the forests of Surinam and Brazil. The powerful falcate claws of its fore feet enable it to dig with great facility. Its food consists chiefly of termites and other insects, but it is said to attack and uproot newly-made graves for the purpose of devouring the flesh of the bodies contained in them.

Tolypeutes.[1]—Teeth $\frac{9}{9}$ or $\frac{8}{9}$, rather large in proportion to the size of the skull, the hinder end of the series reaching nearly to the posterior margin of the palate. Vertebræ: C 7, D 11, L 3, S 12, C 13. Ears placed low on the sides of the head, rather large, broadly ovate. Carapace with its scapular and pelvic shields very free at the sides of the body, forming large chambers into which the limbs can be readily withdrawn. Only three movable bands; sculpture of scutes in the form of subconcentrically arranged granules. Tail short, conical, covered with large bony tubercles. The fore feet formed on the same type as in the last genus, but the peculiarities carried out to a still greater extent. The claw of the third toe is very long and falcate, the first and fifth greatly reduced and sometimes wanting. On the hind foot the three middle toes have broad, flat, subequal nails, forming together a kind of tripartite hoof; the first and fifth much shorter, with more compressed nails.

The Armadillos of this genus have the power of rolling themselves up into a perfect ball, the shield on the top of the head and the tuberculated dorsal surface of the tail exactly fitting into and filling up the apertures left by the notches at either end of the carapace. This appears to be their usual means of defence when frightened or surprised, as they do not burrow like the other species. They run very quickly, with a very peculiar gait, only the tips of the claws of the fore feet touching the ground. Three species are described:—*T. tricinctus*, the Apar; *T. conurus*, the

[1] Illiger, *Prodromus Syst. Mamm. et Avium*, p. 111 (1811).

Matico; and *T. muriei*. Remains apparently referable to *T. conurus* are of not uncommon occurrence in the Brazilian cavern-deposits.

Subfamily **Tatusiinæ.**—This group contains but one genus, *Tatusia*.[1] Teeth $\frac{8}{8}$ or $\frac{7}{7}$, very small subcylindrical. The first and second subcompressed, the last considerably smaller than the others. They present the remarkable peculiarity (elsewhere found among Edentates, so far as is yet known, only in *Orycteropus*) of all being, with the exception of the last, preceded by two-rooted milk teeth, which are not changed until the animal has nearly attained its full size. Vertebræ: C 7, D 9-11, L 5, S 8, C 20-27. Head narrow, with a long, narrow, subcylindrical, obliquely-truncated snout; pterygoids meeting in the middle line below the nasal passage. Ears rather large, ovate, and erect, placed close together on the occiput.

FIG. 67.—The Peba Armadillo (*Tatusia novemcincta*).

Carapace with seven to nine distinct movable bands; sculpture on scutes consisting of pits arranged in a V-shape. Body generally elongated and narrow. Tail moderate or long, gradually tapering; its dermal scutes forming very distinct rings for the greater part of its length. Fore feet with four visible toes, and a concealed clawless rudiment of the fifth. Claws all long, slightly curved, and very slender, the third and fourth subequal and alike, the first and fourth much shorter. Hind feet with five toes, all armed with strong, slightly curved, conical, obtusely-pointed nails. The third longest, then the second and fourth; the first and fifth much shorter than the others.

This genus differs from all the other Armadillos in having a pair of inguinal mammæ, in addition to the usual pectoral pair, and in

[1] Lesson, *Man. de Mammalogie*, p. 309 (1827); *ex.* F. Cuvier, *Tatusie*.

producing a large number (four to ten) of young at a birth, all the others having usually but one or two.

The Peba Armadillo, *T. novemcincta* (Fig. 67), is a well-known species, having an extensive range from Texas to Paraguay. It is replaced in the more southern regions of South America by a smaller species, with shorter tail, the Mulita (*T. hybrida*), so called from the resemblance of its head and ears to those of a mule. *T. kappleri* is a large species from Surinam.

A rare Armadillo from Peru described under the names of *Cryptophractus pilosus* and *Praopus hirsutus*, but which evidently belongs to *Tatusia*, is of some interest owing to the thick coat of hair with which it is covered. This animal appears to be closely allied to *T. novemcincta*, from which it mainly differs by having the whole of the carapace covered with a thick coating of light brown, fine, but rather stiff hair, about an inch and a half in length. Similar hair is found on the cheeks, the proximal portions of the limbs, and (although less abundantly and shorter) on the under surface of the body. The cephalic shield, snout, feet, and the tail, with the exception of the root, are bare. The coating of hair on the back and sides completely conceals the carapace, except near the margin of the scapular region; but by separating the hairs the bands and scutes are rendered visible.[1]

In the Pleistocene cavern-deposits of Brazil have been found remains of *T. novemcincta*, and also of *T. punctata*, which appears to be an extinct form nearly allied to *T. kappleri*, but of somewhat larger size.

Extinct genera.—In addition to remains referable to existing genera, the above-mentioned deposits have also yielded evidence of the former existence of extinct generic types of Armadillos, some of which attained very large dimensions. Of these *Eutatus* was a large form distinguished from all existing genera by the circumstance that the whole of the carapace was composed of movable bands, which were thirty-three in number. *Dasypotherium* was a still larger form, furnished with eight teeth, of which the second seems to have been larger than the others; this genus is regarded as connecting the modern Armadillos with the next one. The gigantic *Chlamydotherium*, the scutes of which are common in the Brazilian caves, is considered to have been as large as a Rhinoceros; the carapace has several movable bands, but the teeth

[1] A single imperfect skin, brought from the province of Ceara in Brazil, indicates a very remarkable form of Armadillo, named by A. Milne-Edwards *Scleropleura brunetti* (*Ann. Sc. Nat.* xvi. p. 8, 1872). The dermal scutes are said to be much less developed than in other members of the family, and confined to the sides, all the median portion of the back being clothed with a flexible hairy skin. The head is broad and short, the ears small and far apart. The tail is long, and almost entirely devoid of scutes. The feet are unknown.

approximate in structure to those of the next family, so that the genus tends to connect the Armadillos with the Glyptodonts.

Family GLYPTODONTIDÆ.

In the Pleistocene cavern-deposits of Brazil, but still more abundantly in the fluviatile deposits which cover the country in the neighbourhood of Buenos Ayres, are found the remains of some of the most remarkable forms of mammals yet discovered, the Glyptodonts, which may be regarded as forming a separate extinct family. They differ from the existing *Dasypodidæ* in their large size, and in having the carapace composed of a solid piece (formed by the union of a multitude of bony dermal scutes) without any movable rings, and in usually having also a ventral piece or plastron. The facial portion of the skull is very short. A long process of the maxillary bone descends from the anterior part of the zygomatic arch. The ascending ramus of the mandible is remarkably high. The teeth are $\frac{8}{8}$ in the known species, all much alike, having two deep grooves or flutings on each side, so as to divide them into three nearly distinct lobes (Fig. 68). The vertebral column is almost entirely ankylosed into a solid tube, and there is a complex joint at the base of the neck, to allow of the head being retracted within the carapace. The limbs are very strong, and the feet short and broad, resembling externally those of an elephant or tortoise. This family is mainly characteristic of the southern half of the American continent, but some species of the type genus ranged into Texas and Mexico. Many species of the family have been described and figured, especially by Burmeister (in the *Annales del Museo publico de Buenos Aires*), among which the following may be noticed. *Hoplophorus* is characterised by the sculptured and frequently thin scutes of the carapace, those of the periphery being flat, and not raised into prominences. The caudal sheath has several overlapping movable rings at the base, and ends in a long subcylindrical terminal tube similar to the one represented with the carapace of *Glyptodon* in Fig. 69, which in all probability really belongs to the genus under consideration. Each foot has four complete digits, and the humerus has an entepicondylar foramen. Most of the

FIG. 68.—Tooth of *Glyptodon* from the side, and from the grinding surface. (After Owen.)

species are of medium size. Part of a caudal tube from Uruguay described as *Eleutherocercus* indicates, however, a much larger allied form, in which the tail appears to have had a number of stout bristles protruding from the joints between the scutes. *Panochthus* comprises very large Glyptodonts, distinguished by the great thickness of the scutes of the carapace, which are ornamented with tubercles. The termination of the caudal sheath forms a tube bearing large radiated tubercles. *Euryurus* is distinguished by the radiate sculpture of the scutes of the carapace *Dœdicurus*, of which one species was about twelve feet in length, also has a rugose sculpture on the carapace; but the termination of the caudal tube is expanded into a club-like shape, flattened from above downwards,

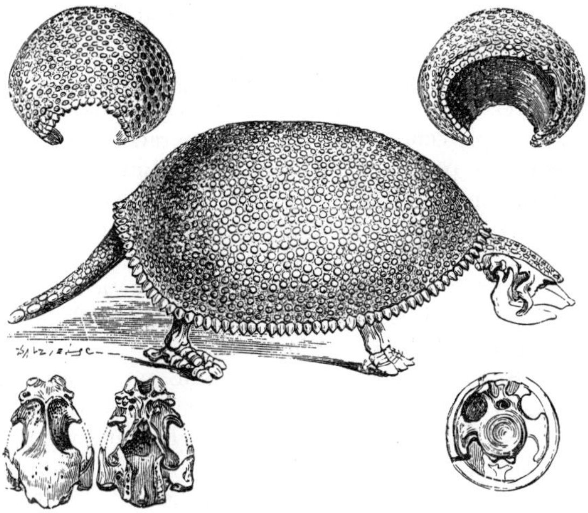

FIG. 69.—*Glyptodon clavipes* (Pleistocene, South America). From Owen. The tail is incorrectly restored, and it is probable that the figured portion belongs to *Hoplophorus*. The left lower corner shows an upper and a lower view of the skull, and the right a section of the caudal sheath.

and covered with tubercles mingled with a few large radiate discs, which, as in *Panochthus*, probably carried horny spines in the living condition. The typical genus *Glyptodon* has each scute of the carapace ornamented with a rosette-like sculpture, the peripheral scutes being raised into conical prominences (Fig. 69). The caudal sheath, instead of being like the one represented in the figure, was entirely composed of a series of movable rings, ornamented with large tubercles. The manus had five digits, and the pes four; and there was an entepicondylar foramen to the humerus. A species of this genus, which attained very large dimensions, was made the type of *Schistopleurum*, on the supposition that the tail of *Glyptodon* was of the type represented in Fig. 69. The genus *Thoracophorus*, of the Pleistocene of South America, as well as *Cariodermа*, of the Pliocene of Texas, differ from all the preceding in having the scutes of the carapace in the form of disconnected nodules. Glyptodonts also occur in South American beds of earlier age than the Pleistocene, some of these forms having enamel bands on the teeth. "Why such a form as the Glyptodon should have failed to keep his ground is," as the late Professor W. K. Parker remarks, "a great mystery; nature seems to have built him, as Rome was built, for eternity."

Family MANIDÆ.

Covered externally (except the under surface of the body and inside of the limbs) with large imbricated horny scales, and scattered hairs growing in the intervals. No teeth. Tongue long, vermiform, and protractile. No accessory articular processes to the lumbar vertebræ, but the anterior zygapophyses largely developed and deeply concave, completely embracing the semicylindrical surfaces of the posterior zygapophyses. Limbs short, with five complete digits on each foot. Scaphoid and lunar bones of carpus united. Uterus bicornuate. Placenta diffused and non-deciduate. All the existing forms belong to the Ethiopian and Oriental regions of the Old World. The absence of additional articular processes to the lumbar vertebræ is a character in which this and the following family differ from all the preceding forms.

Manis.[1]—Skull somewhat of the form of an elongated cone, with the small end turned forwards; very smooth and free from crests and ridges. No distinction between the orbits and temporal fossæ. The zygomatic arch usually incomplete, owing to the absence of the jugal bone. No distinct lachrymal bone. Palate long and narrow. The pterygoids extend backwards as far as the tympanics, but do not meet in the middle line below. Tympanic ankylosed to the surrounding bones, and more or less bullate, but not produced into a tubular auditory meatus. Rami of mandible very slender and straight, without any angle or coronoid process. From near the anterior extremity of the upper edge a sharp, conical, tooth-like process projects upwards and outwards. No clavicles. No third trochanter to the femur. Ungual phalanges bifid at their terminations. Caudal vertebræ with very long, strong transverse processes and numerous chevron bones. Tongue long, vermiform, flattened towards the tip; its retractor or sterno-glossal muscles arising from the hinder extremity of the immensely prolonged ensiform cartilage of the sternum. Stomach with thick lining membrane and muscular walls, and a special gland near the middle of the great curvature, consisting of a mass of complex

[1] Linn. *Syst. Nat.* 12th ed. vol. i. p. 52 (1766).

secreting follicles, the ducts of which terminate in a common orifice. No cæcum. A gall-bladder. Head small, depressed, narrow, pointed in front, with a very small mouth-opening. Eyes and pinna of ear very small. Body elongated, narrow. Tail more or less elongated, convex above, flat underneath. The whole of the upper surface of the head, the upper surface and sides of the body, the whole of the tail, and the outer sides of the extremities covered with large, overlapping, horny scales, but usually with a few stiff hairs growing between and projecting beyond them. The sides and under surface of the head, the under surface of the body, and the inner sides of the limbs without scales, but with a rather scanty covering of hair. Limbs short. In walking the dorsal surface and outer sides of the phalanges of the two outer digits of the front feet alone rest on the ground, the points of the nails turning upwards and inwards. The third toe the longest, with a powerful compressed curved claw; the second and fourth with similar but smaller claws, that of the pollex often almost rudimentary. Hind feet plantigrade, with the hallux very short, and the four other toes subequal, with moderate, curved, subcompressed nails.

The reproductive organs of *Manis* are of a totally different type from those of the families already noticed. The testes lie in the inguinal canal; and the penis is external and well developed. The uterus is truly bicornuate, the vagina not divided, and the placenta diffused and non-deciduate. All the organs and fœtal membranes are, indeed, formed very much on the plan of those of the Ungulates, without any trace of the special peculiarities obtaining in the typical American Edentates.

The animals of this genus, which includes all the existing forms, are called Pangolins or Scaly Anteaters, and are all of small or moderate size, terrestial and burrowing, and feed mainly on termites. Several of them can climb trees. Their length varies from 1 to 5 feet. They can roll themselves up in a ball when in danger. Their peculiar elongated form, short limbs, long, gradually-tapering tail, and scaly covering give them on a superficial inspection more the appearance of reptiles than of mammals. The species are not numerous, and may be divided into two groups distinguished by a few not very important external characters; these groups also coinciding with the present geographical distribution of the genus. These two groups, according to Mr. O. Thomas, may be distinguished as follows.

The Asiatic pangolins are characterised by having the central series of body-scales continued quite to the extreme end of the tail, by having many isolated hairs growing up between the scales of the back, and by their small external ears. They all have a small naked spot beneath the tip of the tail, which is said to be of service as an organ of touch. There are three species, viz. *Manis javanica*, ranging from Burma, through Malacca and Java, to Borneo; *M. aurita*, found in China, Formosa, and Nipal; and the common Indian Pangolin, *M. pentadactyla*, distributed over the whole of India and Ceylon. The African species have the central series of scales suddenly interrupted and breaking into two at a point about 2 or 3 inches from the tip of the tail; they have no hair between the scales, and no external ear-conch. The following are the four species belonging to this group:— the Long-tailed Pangolin (*M. macrura*), which has a tail nearly twice as long as its body, and containing as many as forty-nine caudal vertebræ, being the largest number known among mammals; the White-bellied Pangolin (*M. tricuspis*), Fig. 70, closely allied to the last, but with longer and tricuspid scales, and white belly hairs. These two, like the Indian species, have a naked spot beneath the tail tip, a character probably correlated with the power of climbing, and they are, moreover, peculiar in having the outer sides of their fore legs clothed with hair, all the other species being scaly there as elsewhere. Lastly, the Short-tailed and the Giant Pangolins (*M. temmincki* and *gigantea*), both of which have their tails covered entirely with scales, and evidently never take to arboreal habits. All the four species of the second group are found in the West African region, one only, *M. temmincki*, extending also into south and eastern equatorial Africa.

FIG. 70.—The White-bellied Pangolin (*Manis tricuspis*).

According to Professor W. K. Parker,[1] who remarks upon the peculiarly aberrant nature of the group, the horny scales of the Pangolins really consist of cemented hairs. This writer states that "in the early embryo lozenge-shaped tracts of skin are seen all over

[1] *Mammalian Descent*, p. 95.

its body, with lines of thinner cuticle between. Under the microscope, sections of these thicker tracts show that they are composed of fine hairs, cemented together by a copious growth of epidermic cells; here and there larger hairs are seen, but these fail to reach the surface, turning again towards the inside, like nails driven into wood that is too hard for their points."

The same author also observes[1] that there are occasional instances of the presence of eight cervical vertebræ in the Pangolins —a feature which has been considered to indicate some former genetic connection between this family and the Sloths.

The following account of the habits of *Manis tricuspis* is given by Mr. L. Fraser in his *Zoologia Typica*:—

"During my short residence at Fernando Po I succeeded in procuring two living specimens of this animal. The individuals, judging from the bones, were evidently not adult; the largest measured 30 inches in length, of which the head and body were 12 inches and the tail 18 inches. I kept them alive for about a week at Fernando Po, and allowed them the range of a room, where they fed upon a small black ant, which is very abundant and troublesome in the houses and elsewhere. Even when first procured they displayed little or no fear, but continued to climb about the room without noticing my occasional entrance. They would climb up the somewhat roughly-hewn square posts which supported the building with great facility, and upon reaching the ceiling would return head foremost; sometimes they would roll themselves up into a ball and throw themselves down, and apparently without experiencing any inconvenience from the fall, which was in a measure broken upon reaching the ground by the semi-yielding scales, which were thrown into an erect position by the curve of the body of the animal. In climbing, the tail, with its strongly pointed scales beneath, was used to assist the feet; and the grasp of the hind feet, assisted by the tail, was so powerful that the animal would throw the body back (when on the post) into a horizontal position, and sway itself to and fro, apparently taking pleasure in this kind of exercise. It always slept with the body rolled up; and when in this position in a corner of the building, owing to the position and strength of the scales, and the power of the limbs combined, I found it impossible to remove the animal against its will, the points of the scales being inserted into every little notch and hollow of the surrounding objects. The eyes are very dark hazel, and very prominent. The colonial name for this species of *Manis* is 'Attadillo,' and it is called by the Boobies, the natives of the island, 'Gahlah.' The flesh is said to be excedingly good eating, and is in great request among the natives."

[1] *Mammalian Descent*, p. 99.

The Indian species is said to live in pairs, and to give birth to one or two young at a time in the spring. Their burrow reaches a depth of some twelve feet, and terminates in a large chamber, which may be as much as six feet in diameter. A faint hiss appears to be the only sound emitted by these animals.

Remains of a large species of *Manis*, which are indistinguishable from the corresponding bones of the existing West African *M. gigantea*, are found fossil in cave-deposits in the Karnul district of Madras. This is one among several instances of the close connection between the Pleistocene and Pliocene mammalian fauna of India with the existing African fauna.

Palæomanis.[1]—The lower Pliocene deposits of the Isle of Samos, in the Turkish Archipelago, have yielded remains of a Pangolin fully three times the dimensions of *M. gigantea*, upon the evidence of which the genus *Palæomanis* has been established.

Family ORYCTEROPODIDÆ

External surface scantily covered with bristle-like hairs. Teeth numerous, apparently heterodont, diphyodont, and of peculiar and complex structure, being traversed by a number of parallel vertical pulp-canals. Lumbar vertebræ with no accessory zygapophyses. Femur with a third trochanter. Fore feet without pollex, but all the other digits well developed, with strong moderate-sized nails, suited to digging, the plantar surfaces of which rest on the ground in walking. Hind feet with five subequal toes. Mouth elongated and tubular. Tongue subvermiform. Uterus bicornuate. Placenta broadly zonular. Feeding on animal substances. Terrestrial and fossorial in habits. Now mainly limited to the Ethiopian region.

Orycteropus.[2]—The total number of permanent teeth appears to be from eight to ten in each side of the upper, and eight in the lower jaw; but they are never all in place at one time, as the small anterior teeth are shed before the series is completed behind. In the adult they number usually five on each side above and below, of which the first two are simple and compressed, the next two larger and longitudinally grooved at the sides, the most posterior simple and cylindrical. The last three in either jaw having no milk-predecessors, may be regarded as true molars. The structure of all these teeth is quite peculiar among mammals, though resembling that of some fishes. Their summits are rounded before they are worn; their bases do not taper to a root, but are evenly truncated and continually growing. Each tooth is made up of an aggregation of parallel dental systems, having a slender pulp-cavity

[1] Forsyth-Major, *Comptes Rendus*, vol. cvii. p. 1180 (1888).

[2] Geoffroy, *Décade Philosophique*, 1795 (*teste* Agassiz).

in the centre, from which the dentinal tubes radiate outwards, and being closely packed together each system assumes a polygonal outline as seen in transverse section. The small anterior teeth have milk-predecessors which are fully noticed below. Skull moderately elongated. The facial portion subcylindrical and slightly tapering. The zygoma complete and slender. The palate ends posteriorly in the thickened transverse border of the palatines, and is not continued back by the pterygoids. The tympanic is annular, and not ankylosed to the surrounding bones. The mandible is slender anteriorly, but rises high posteriorly, with a slender recurved coronoid, and an ascending pointed process on the hinder edge below the condyle, which is small, oval, and looks as much forwards as upwards. Vertebræ: C 7, D 13, L 8, S 6, C 27. The large number of lumbar vertebræ is peculiar among Edentates. Tongue less vermiform than in *Myrmecophaga*, being thick and fleshy at the base, and gradually tapering to the apex. The salivary apparatus is developed much in the same manner as in that genus, but the duct of the submaxillary gland has no reservoir. The stomach consists of a large subglobular cardiac portion, with a very thick, soft, and corrugated lining membrane, and a smaller muscular, pyloric part, with a comparatively thin and smooth lining. There is a very distinct ileo-cæcal valve, and a considerable-sized cæcum; also a gall-bladder. Head elongated, with a tubular snout, terminal nostrils, and small mouth-opening. Ears large, pointed, erect. Tail nearly as long as the body, cylindrical, very thick at the base, tapering to the extremity.

The reproductive organs and placentation of *Orycteropus* are formed upon a principle unknown in the more typical Edentates, or, in combination, in any other mammals. Thus the testes, in the one described example, were inguinal, but appeared to descend, at all events temporarily, into a scrotum; but the penis is scarcely larger than that of the Great Anteater. The uterus is still more fully bicornuate than in *Manis*, with its two lateral chambers opening separately into the vagina, as in certain Rodents. The placenta is broadly zonary, but it is not known whether it is deciduate or not. It might readily be derived from the diffused placenta of *Manis* by the abortion of the fœtal villi at the two poles of the ovum.

The *Orycteropodidæ* have long been regarded as widely different from other Edentates, their presumed affinity with the *Manidæ* being more or less problematical; but the discovery recently made by Mr. O. Thomas[1] that they have a milk-dentition still further emphasises their aberrant nature. According to this observer, it appears that there are normally no less than seven milk-teeth in the upper jaw, the hindmost of which is far larger than the others,

[1] *Proceedings of the Royal Society*, vol. xlvii. p. 246 (1890).

having a rudimentary crown, and a distinct anterior and posterior root. The other milk-teeth are styliform, the four anterior ones being very minute, and separated from one another by equal intervals; the foremost of all is situated immediately behind the premaxillo-maxillary suture. In the mandible only four milk-teeth have hitherto been detected, of which the hindmost has the comparatively complex form found in the corresponding upper tooth. None of these milk-teeth appear, however, to cut the gum, so that the whole set is entirely functionless. Under the microscope these milk-teeth show signs of possessing a commencement of the remarkable histological structure found in the permanent teeth.

Mr. Thomas remarks that since "the three large posterior teeth of *Orycteropus*, already distinguished by their more molariform shape, do not have milk-predecessors, while all the small teeth anterior to them do, and in addition the last milk-tooth is markedly different from those in front of it, we ought apparently no longer to look upon this animal as an homodont, but instead to consider it as an originally heterodont form in which the incisors and canines have been suppressed to allow free play to the mobile vermiform tongue.

"But important as a knowledge of the presence of a milk-dentition in *Orycteropus* is, it does not at present render any easier the difficult questions as to the phylogeny and systematic position of that animal. Although called an Edentate, it has always been recognised as possessing many characters exceedingly different from those of the typical American members of the order. It has in fact been placed with them rather on account of the inconvenience of forming a special order for its reception than because of its real relationship to them. Now, as they are either altogether toothless, or else homodont and monophyodont (apart from the remarkable exception of *Tatusia*), it seems more than ever incorrect to unite with them the solitary member of the Tubulidentata, toothed, heterodont, and diphyodont, and differing from them in addition by its placentation, the anatomy of its reproductive organs, the minute structure of its teeth, and the general characters of its skeleton.

"But if *Orycteropus* is not genetically a near relation of the Edentates, we are wholly in the dark as to what other mammals it is allied to, and I think it would be premature to hazard a guess on the subject. Whether even it has any special connection with *Manis* is a point about which there is the greatest doubt, and unfortunately we are as yet absolutely without any palæontological knowledge of the extinct allies of either. *Macrotherium* even, usually supposed from the structure of its phalangeal bones, to be related to *Manis*, has lately proved to have the teeth and vertebræ of a perissodactyle Ungulate, and one could not dare to suggest that ancestors of *Manis*, or *Orycteropus* were to be sought in that direction. Lastly, as the numerous fossil American Edentates do not show the slightest tendency to an approximation towards the Old World forms, we are furnished with an additional reason for insisting on the radical distinctness of the latter, whose phylogeny must therefore for the present remain one of the many unsolved zoological problems."

The Aard-Varks (Earth-Pigs) as these creatures are commonly termed, from the name bestowed on them by the Dutch Boers of the Cape, are of nocturnal habits, sleeping during the day in their burrows, which are usually found in the neighbourhood of the tall hills or mounds made by termites. Indeed, wherever these hills are abundant it is stated there is a good chance of finding an Aard-Vark, the food of these animals consisting almost exclusively of termites and ants.

Two existing species are recognised, namely the Cape Aard-Vark (*O. afra*) from South Africa, and another (*O. æthiopicus*) from the north-eastern parts of Africa, ranging into Egypt. An extinct species has been described from the Lower Pliocene of the Isle of Samos, in the Turkish Archipelago, differing from the existing forms by the larger proportionate size of the lateral metatarsals.

Bibliography of Edentata.—No general work on the order has been published since that of Rapp (*Anat. Untersuchungen über die Edentaten*, 2d ed. 1852). Among numerous memoirs on special groups the following may be cited:—*Myrmecophagidæ*:—R. Owen, "Anatomy of Great Anteater," *Trans. Zool. Soc.* vol. iv.; G. Pouchet, *Mém. sur le Grand Fourmilier*, 1874; W. A. Forbes, "Anat. of Great Anteater," *Proc. Zool. Soc.* 1882, p. 287. *Megatheriidæ*:—R. Owen, *Extinct Gigantic Sloth* (*Mylodon Robustus*), 1842; Id., "On the Megatherium," *Phil. Trans.* 1851-56; J. Leidy, "Extinct Sloth-tribe of North America," *Smithsonian Contrib. to Knowledge*, vii. 1855; H. Burmeister, *Description de la République Argentine*, t. iii. Mammifères, 1879,—which contains full references to various memoirs by Owen, Gervais, Reinhardt, and others. *Glyptodontidæ*:—Owen, *Catalogue of Fossil Mammals, Mus. Roy. Coll. Surgeons*, 1845; T. H. Huxley, "Osteol. of Glyptodon," *Phil. Trans.* 1865; H. Burmeister, *Annales del Museo Publico de Buenos Aires*, and *Descript. de la République Argentine*, 1879; H. Gervais and F. Ameghino, *Les Mammifères Fossiles de l'Amérique Méridionale*, Paris, 1880,—which also contains a list of all the S. American Edentates described at that date. *Dasypodidæ*:—J. Murie, "Anatomy of *Tolypeutes*," *Trans. Linn. Soc.* vol. xxx. 1874; A. H. Garrod, *Proc. Zool. Soc.* 1878. For Placentation of Edentates see W. Turner, *Trans. Roy. Soc. Edin.* xxvii. (1873) p. 72, and *Journ. Anat. and Physiol.* vols. viii. and x.; A. Milne-Edwards, *Ann. Sciences Nat.* [6] viii. p. 1; and for brain, P. Gervais, "Formes cérébrales des Edentés," *Nouv. Arch. du Muséum*, tom. v.; W. Turner, *Jour. Anatomy*, i. 313 (1867). For the dentition of *Orycteropus* see O. Thomas, "A Milk Dentition in *Orycteropus*," *Proc. Roy. Soc.* vol. xlvii. p. 246 (1890). Fuller observations on the mutual relations of the various families are given by W. H. Flower, "On the Mutual Affinities of the Animals composing the Order Edentata," *Proc. Zool. Soc.* 1882, p. 358.

CHAPTER VIII

THE ORDERS SIRENIA AND CETACEA

Order SIRENIA.

THE purely aquatic habits and fish-like form of the animals of this order caused them to be formerly confounded with the Cetacea, but a more intimate knowledge of their structure has shown that they really belong to a widely different type of the mammalian class.

The head is rounded and not disproportionate in size as compared with the trunk, from which it is scarcely separated by any externally visible constriction or neck. Nostrils valvular, separate, and placed above the fore part of the obtuse truncated muzzle. Eyes very small, with imperfectly formed eyelids, capable, however, of contraction, and with a well-developed nictitating membrane. Ear without any pinna. Mouth of small or moderate size, with tumid lips beset with stiff bristles. General form of the body depressed, fusiform. No dorsal fin. Tail flattened and horizontally expanded. Fore limbs paddle-shaped, the digits being enveloped in a common cutaneous covering, on which rudiments of nails are sometimes present. No trace of hind limbs in existing forms. External surface covered with a tough, finely wrinkled, or very rugose skin, naked, or with fine hairs sparsely scattered over it.

The skeleton is remarkable for the massiveness and density of most of the bones of which it is composed, especially the skull and ribs, which must add to the specific gravity of these slow-moving animals, and aid in keeping them to the bottom of the shallow waters in which they dwell, while feeding on aquatic vegetables. The skull presents many peculiarities, among which may be indicated the large size and backward position of the anterior narial aperture, a further modification of that met with in the Tapirs among Ungulates, and presenting some approach to that so characteristic of the Cetacea. The nasal bones are generally absent in the recent forms, or are only found in a most rudimentary condition, attached to the edge of the frontals, far away from the middle line; but in some at least of the extinct species these bones, though small in size, are normal in situation and relations. In very few other respects does the skull present any resemblance to that of the Cetacea. In the spinal column of existing forms none of the vertebræ are united together to form a sacrum, and the flat ends of the bodies do not ossify separately, so as to form disc-like epiphyses in the young state, as in nearly all other mammals; traces of epiphyses have, however, been recently detected in *Manatus*, and they were fully developed in *Halitherium* and other fossil forms. The anterior caudal vertebræ have well-developed chevron bones. In one genus (*Manatus*) there are only six cervical vertebræ. There are no clavicles. The humerus has a small but distinct trochlear articulation at the elbow-joint. The two bones of the forearm are about equally developed, and generally ankylosed together at both extremities. The carpus is short and broad, and the digits five in number, with moderately elongated and flattened phalanges, which are never increased in number beyond the limit usual in the Mammalia. The pelvis is extremely rudimentary, consisting of a pair of bones suspended at some distance from the vertebral column. In no existing species is there any trace of a hind limb, but in the extinct *Halitherium* an acetabular depression and rudimentary femur have been discovered.

Two kinds of teeth, incisors and molars, separated by a wide interval, are generally present. The former may be developed into tusks in the upper jaw, or may be quite rudimentary. The molars vary much in character. In one genus (*Rhytina*) no teeth of any kind are present, at least in the adult. Some fossil forms show a more decidedly heterodont dentition, while *Halitherium* has milk-teeth, of which no traces have been observed in the recent genera. In all recent types the anterior part of the palate, and a corresponding surface on the prolonged symphysis of the lower jaw, are covered with rough horny plates of peculiar structure, which doubtless assist in mastication. The tongue is small and fixed in position, with a surface resembling that of the plates just spoken of. The salivary glands are largely developed. The stomach is compound, being divided by a valvular constriction into two principal cavities, the first of which is provided with a singular glandular pouch near the cardiac end, and the second usually with a pair of elongated, conical, cæcal sacs or diverticula. The intestinal canal is long, and has very muscular walls. There is a cæcum, either simple, conical, and with extremely thick walls, as in *Halicore*, or bifid, as in *Manatus*. The heart is broad and flat, with its apex deeply cleft between the ventricles. The principal arteries form very extensive and complex retia mirabilia. The lungs are remarkably long and narrow, as, owing to the very oblique position of the diaphragm, the thoracic cavity extends far back over the abdomen. The epiglottis and arytenoid cartilages of the larynx do not form a tubular prolongation as in the Cetacea, so that the epiglottis is not intranarial. The brain is of comparatively small size, and the convolutions on the surface of the cerebrum are few and shallow. The kidneys are simple. The testes abdominal. The uterus is bicornuate. The placenta (in the Dugong) is non-deciduate and zonary. The umbilical vesicle disappears early. The mammæ are two, and pectoral, or rather post-axillary in position.

The Sirenia pass their whole life in the water, being denizens of shallow bays, estuaries, lagoons, and large rivers, but, unlike the Cetacea, are not met with in the high seas, far away from the shore. Their food consists entirely of aquatic plants, either marine algæ or freshwater grasses, upon which they browse beneath the surface, as the terrestrial herbivorous mammals do upon the green pastures on shore. They are generally gregarious, slow and inactive in their movements, mild, inoffensive, and apparently unintelligent in disposition. Though occasionally found stranded by the tide or waves, there is no satisfactory evidence that they voluntarily leave the water to bask or feed on the shore. The habit of the Dugong of raising its round head out of the water, and carrying its young under the fore fin, seems to have given rise, among the imaginative early voyagers in the Indian Ocean, to the legendary beings, half human and half fish, in allusion to which the name Sirenia was bestowed by Illiger on the order, though certainly the face of a Dugong, when closely inspected, does not bear the slightest resemblance to that of the mermaid of romance. The species now existing are very few, and there is reason to believe that the time is not far distant when they will all become extinct. One species, *Rhytina stelleri*, of the North Pacific, was totally exterminated through the agency of man during the last century; and the others, being valuable for their flesh as food, for their hides, and especially for the oil obtained from the thick layer of fat which lies immediately beneath their skin, rapidly diminish in numbers as civilised populations occupy the regions forming their natural habitat. The surviving species are confined to the tropical regions of the shores of both sides of the Atlantic and the great rivers which empty themselves into that ocean, and to the coasts of the Indian Ocean from the Red Sea to North Australia. In the Miocene and early Pliocene epoch Sirenians abounded in the seas of Europe, and their remains have been found in deposits of corresponding periods in North America. Evidence has also been discovered of the existence of an animal of this group in the seas at the bottom of which the Eocene nummulitic limestone mountain ranges of Egypt were deposited.

The existing genera present such well-marked distinguishing characters that it is on the whole convenient to place them in separate families, although, as in so many similar cases, our knowledge of the extinct forms, imperfect as it is, goes far to bridge over the distinction between them.

Family MANATIDÆ.

The characters of this and the two following families may be conveniently included under the heading of the single genus by which they are respectively represented.

Manatus.[1]—Incisors $\frac{2}{2}$, rudimentary, concealed beneath the horny oral plates, and disappearing before maturity. Molars $\frac{11}{11}$, but rarely more than $\frac{6}{6}$ present at one time, the anterior teeth falling before the posterior come into use; similar in characters from beginning to end of the series; with square, enamelled crowns, the grinding surface raised into tuberculated transverse ridges. The upper teeth with two ridges and three roots, the lower teeth with an additional (posterior) ridge, or talon, and two roots. The cervical vertebræ present the remarkable anomaly of being reduced to six in number, the usual vertebral formula being C 6, D 17, L 2, and C 23-25. Rostrum of the skull, formed by the union of the premaxillæ in front of the anterior narial aperture, shorter than the length of the aperture and scarcely deflected from the basi-cranial axis; premaxillæ and mandibular symphysis not markedly deflected (Fig. 72). Tail entire, rounded, or shovel-shaped. Rudimentary nails on the fore limbs. Cæcum bifid. Habitat the shores of, and the great rivers which empty themselves into, the Atlantic within the tropics. These animals are rather fluviatile than marine, ascending large rivers almost to their sources.

The Manatee may be selected for a somewhat full description, as being one of the best known representatives of this very remarkable order.

The name *Manati* was apparently first applied to this animal by the early Spanish colonists of the West Indies, in allusion to the hand-like use which it frequently makes of its fore limbs; by English writers from the time of Dampier (who gives a good account of its habits) downwards it has been generally spelt "Manatee." It was placed by Linnæus in his heterogeneous genus *Trichechus*, but Storr's name *Manatus* is now generally accepted for it by zoologists. The question of the specific distinction of the African and American Manatees will be treated of further on, but it will be chiefly to the latter and better known form that the following description applies.

The size of the Manatee has been much exaggerated, but

[1] Storr, *Prodromus Meth. Mamm.* p. 41 (1780).

there is no trustworthy evidence of its attaining a greater length than 8 feet. Its general external form may be seen in Fig. 71, taken from a living example in the Brighton Aquarium. The body is somewhat fish-like, but depressed and ending posteriorly in a broad, flat, shovel-like, horizontal tail, with rounded edges. The head is of moderate size, oblong, with a blunt, truncated muzzle, and divided from the body by a very slight constriction or neck. The fore limbs are flattened oval paddles, placed rather low on the sides of the body, and showing externally no signs of division into fingers, but with a tolerably free motion at the shoulder, elbow, and wrist joints, and with three diminutive flat nails near their extremities. No traces of hind limbs are discernible either externally or internally; and there is no dorsal fin. The mouth is very peculiar, the tumid upper lip being cleft in the middle line into two lobes, each of which is separately movable, as will be described in speaking of its manner of feeding. The nostrils are two semilunar

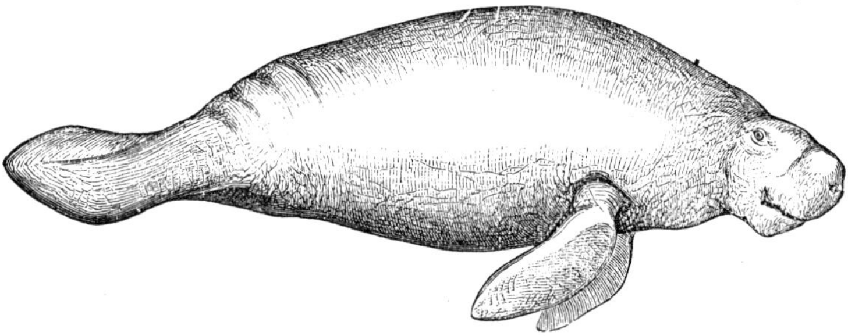

FIG. 71.—American Manatee (*Manatus americanus*), from life. *Proc. Zool. Soc.* 1881, p. 457.

valve-like slits, at the apex of the muzzle. The eyes are very minute, placed at the sides of the head, and with a nearly circular aperture with wrinkled margins. The external ear is a minute orifice situated behind the eye, without any trace of pinna. The skin generally is of a dark grayish colour, not smooth and glistening, like that of the Cetacea, but finely wrinkled. At a little distance it appears naked, but a close inspection, at all events in young animals, shows a scanty covering of very delicate hairs, and both upper and under lips are well supplied with short stiff bristles.

The general form of the skull is seen in Fig. 72. The cerebral cavity is rather small as compared with the size of the animal, and of oblong form; its roof is formed of the parietal bones as in ordinary mammals. The squamosal has an extremely large and massive zygomatic process, which joins the largely developed jugal bone in front. The orbit is small, but prominent and nearly surrounded by bone. The anterior nares taken together form a lozenge-shaped aperture, which looks upwards and extends backwards considerably behind the orbits. Their sides are formed by the ascending processes of the premaxillæ below, and by the supraorbital processes of the frontals above, no traces of nasals being found in most skulls, though these bones are occasionally present in a most rudimentary condition, attached to the edges of the frontals, far away from the middle line, in a position quite unique among the Mammalia. In front of the narial aperture the face is prolonged into a narrow rostrum, formed by the premaxillæ, supported below and at the sides by the maxillæ. The under surface of this is very rugose, and in life covered by a horny plate. The rami of the mandible are firmly united together at the symphysis, which is compressed laterally, slightly deflected, and has a rugose upper surface; to this another horny plate is attached, which, with that of the upper jaw, functionally supplies the

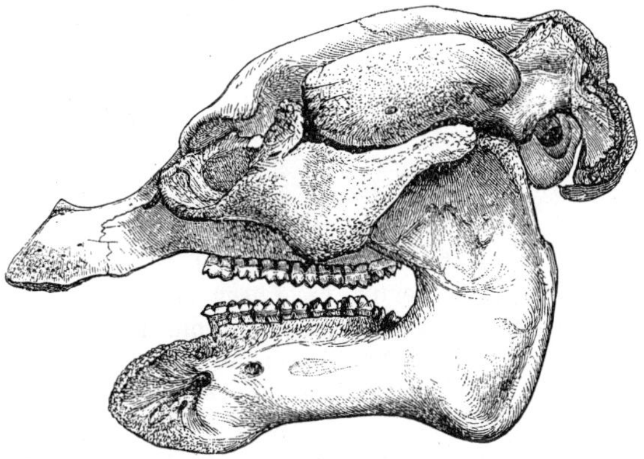

FIG. 72.—Skull of African Manatee (*Manatus senegalensis*). ⅛ natural size. From Mus. Roy. Coll. Surgeons.

place of teeth in the anterior part of the mouth. In the young state there are rudimentary teeth concealed beneath these horny plates, which never penetrate through them, and must therefore be quite functionless, and altogether disappear before the animal is full-grown. There is besides on each side of the hinder part of both upper and lower jaws, a parallel row of molar teeth, similar in characters from the beginning to the end of the series, with square enamelled crowns raised into tuberculated transverse ridges, something like those of the Tapir and Kangaroo. The upper teeth have two ridges and three roots; the lower teeth have an additional posterior small ridge or talon, and but two roots. These teeth succeed each other from before backwards, as in the Proboscidea, those at the front of the mouth being worn out and shed before those at the back are fully developed. There are altogether about eleven on either side of each jaw, but rarely more than six are present at one time. The brain is remarkably simple in structure, its hemispheres exhibiting none of the richness of convolution so characteristic of the Cetacea. The mammary glands of the female are situated just behind and to the inner side of the origin of the pectoral limb. The red corpuscles of the blood are among the largest of those of any members of the class, averaging in diameter, according to Gulliver, $\frac{1}{2400}$ of an inch.

Manatees pass the whole of their life in the water, inhabiting bays, lagoons, estuaries, and large rivers; but the open sea, so congenial to the Cetacea, is quite unsuited to their peculiar mode of life. As a general rule they prefer shallow water, in which, when not feeding, they lie near the bottom, supporting themselves on the extremity of the tail, or slowly moving about by the assistance of the fore limbs, the tips of which are just allowed to touch the ground, and only raising the top of the head above the surface for the purpose of breathing at intervals of two or three minutes. In deeper water they often float, with the body much arched, the rounded back close to the surface, and the head, limbs, and tail hanging downwards. The air in the lungs obviously assists them to maintain this position, acting in the same manner as that in the air-sac of fishes. Their food consists exclusively of aquatic plants, on which they browse beneath the water. They are extremely slow and inactive in their movements, and perfectly harmless and inoffensive. Frequent attempts have been made to keep specimens alive in captivity, and sometimes with considerable success, one having lived in the Brighton Aquarium for upwards of sixteen months. It was fed chiefly on lettuce and endive, but would also eat leaves of the dandelion, sow-thistle, cabbage, turnip, and carrot. From this and other captive specimens some interesting observations upon the mode of life of the animal have been made. One of these is the free use it makes of its fore limbs. From the shoulder-joint they can be moved in all directions, and the elbow and wrist permit of free extension and flexion. In feeding these creatures push the food towards their mouths by means of one of the hands, or both used simultaneously, and any one who has seen these members thus employed can readily believe the stories of their carrying their young about under their arms. Still more interesting and quite unique among mammals is the action of the peculiar lateral pads formed by the divided upper lip, thus described by the late Professor Garrod: "These pads have the power of transversely approaching towards and receding from one another simultaneously (see Fig. 73, A and B). When the animal is on the point of seizing (say) a leaf of lettuce, the pads are diverged transversely in such a way as to make a median gap of considerable breadth. Directly the leaf is within grasp the lip-pads are approximated, the leaf is firmly seized between their contiguous bristly surfaces, and then drawn inwards by a backward movement of the lower margin of the lip as a whole." The animal is thus enabled by the unaided means of the upper lip to introduce food placed before it without the assistance of the comparatively insignificant lower lip, the action greatly recalling to the observer that of the mouth of the silkworm and other caterpillars, in which the mandibles diverge and converge laterally during mastication. When out of water the Manatee is an extremely helpless animal; and, although statements are frequently met with in books of its voluntarily leaving the water for the purpose of basking or feeding on shore, all trustworthy observations of those acquainted with it, either in a state of nature

A B

FIG. 73.—Front view of head of American Manatee, showing the eyes, nostrils, and mouth. A, With the lobes of the upper lip divaricated; B, with the lip contracted. From Murie, *Trans. Zool. Soc.* vol. xi.

or in captivity, indicate that it has not the power of doing so. None of the specimens in confinement have been observed to emit any sound.

Manatees, though much less numerous than formerly, are still occasionally found in creeks, lagoons, and estuaries in some of the West India Islands, and at various spots on the Atlantic coast of America from Florida as far south as about 20° S. lat., and in the great rivers of Brazil, almost as high as their sources. They are also met with in similar situations on the opposite African coast, from about 16° N. to 10° S. lat., and as far into the interior as Lake Tchad. Their range may even extend, if native reports obtained by Schweinfurth are correctly interpreted, to the river Keebaly, 27° E. long.

A considerable number of specific names have been applied to the existing Manatees, but according to the researches of Dr. Hartlaub [1] they may be reduced to three species, distinguished from one another, among other features, by the characters of the skull, and more especially the relations of the nasals to the adjacent

[1] *Zool. Jahrbuch*, vol. i. p. 1 (1886).

bones. Of these the American Manatee may be known as *M. americanus*, although it has been described under the names of *M. latirostris*, and *M. australis*. The African Manatee (*M. senegalensis*) differs from the American species in the following cranial characters: the anterior part of the rostrum is shorter, shallower, and altogether smaller; the orbit is smaller; the zygomatic process is more deep and massive; the jugal bone is deeper from above downwards; the upper margin of the anterior nares is narrower and with a smooth and rounded, instead of a thin and serrated, edge; the upper surface of the frontal is flat, instead of concave; the foramen magnum and occipital condyles are narrower from side to side, and the symphysis of the mandible is smaller and shallower.

Finally, *M. inunguis* is a fluviatile species confined to the Amazon and Orinoco, which has been but recently fully brought under the notice of zoologists.

Family HALICORIDÆ.

Halicore.[1]—In the upper jaw a pair of large, nearly straight, tusk-like incisors, directed downwards and forwards, partially coated with enamel. In the male they have persistent pulps, and bevelled cutting edges, which project a short distance from the mouth, but in the female, though they remain through life in the alveolar cavity, they are not exserted, and, the pulp-cavity being filled with osteodentine, they soon cease to grow (as in the female Narwhal). In the young there is also a second small deciduous incisor on each side above. At this age there are also beneath the horny plate which covers the anterior portion of the mandible four pairs of slender conical teeth lodged in wide alveolar depressions; these become absorbed before the animal reaches maturity. The molars are usually $\frac{5}{5}$, sometimes $\frac{6}{6}$, altogether, but not all in place at once, as the first falls before the last rises above the gum; they are more or less nearly cylindrical in section (except the last, which is compressed and grooved laterally), without distinction into crown and root, increasing in size from before backwards, with persistent pulps and no enamel. The summits of the crowns are tuberculated before wearing, afterwards flattened or slightly concave. Skull with rostrum formed by the union of the premaxillæ in front of the narial aperture, longer than the aperture itself, bending downwards at a right angle with the basi-cranial axis, and enclosing the sockets of the large incisor tusks. Anterior part of the lower jaw bent down in a corresponding manner. Vertebræ: C 7, D 18-19, L and C 30. Tail broadly notched in the middle line, and with two pointed lateral lobes. No nails on the fore limbs. Cæcum single.

[1] Illiger, *Prodromus Syst. Mamm. et Avium*, p. 140 (1811).

The Dugongs are more distinctly marine in their habits than the Manatees, feeding chiefly on sea-water algæ. They inhabit the shallow bays and creeks of the Red Sea, east coast of Africa, Ceylon, islands of the Bay of Bengal and the Indo-Malayan Archipelago (including the Philippines), and the north coast of Australia, ranging from Barrow Reefs on the west to Moreton Bay on the east. Although the distinctive characters are not very obvious, they have been divided into three species, according to the localities which they respectively inhabit:—*H. tabernaculi* from the Red Sea, *H. dugong* from the Indian seas, and *H. australis* from Australia. The last-named has lately been the object of a regular "fishery," chiefly on account of its oil, which is peculiarly clear, limpid, and free from disagreeable smell, and is said to have the same medicinal properties as cod-liver oil. Although often stated in books to attain the length of 20 feet when adult, there does not appear to be any evidence from actual specimens in museums that Dugongs ever reach half that size, 8 feet being the common length of adult animals.

The placentation of this genus has been recently described by Sir W. Turner, who first indicated its zonary form.

Family RHYTINIDÆ.

Rhytina.[1]—No teeth, their place being supplied functionally by the dense, strongly-ridged, horny oral plates. Premaxillary rostrum about as long as the anterior narial aperture, and moderately deflected. Vertebræ: C 7, D 19, L and C 34-37. Head very small in proportion to the body. Tail with two lateral pointed lobes. Pectoral limbs small and truncated. Skin naked and covered with a very thick, hard, rugged, bark-like epidermis. Stomach without cæcal appendages to the pyloric cavity. Cæcum simple.

Only one species of this genus is known, *R. stelleri*, the Northern Sea-cow, by far the largest animal of the order, attaining the length of 20 to 25 feet. It was formerly an inhabitant of the shores of two small islands in the North Pacific, Behring and the adjacent Copper Island, on the former of which it was discovered by the ill-fated navigator whose name the island bears, when, with his accomplished companion, the German naturalist Steller, he was wrecked upon it in 1741. Twenty-seven years afterwards (1768), as is commonly supposed, the last of the race was killed,[2] and its

[1] Illiger, *Prodromus Syst. Mamm. et Avium*, p. 141 (1811).—Amended from *Rytina*.

[2] Nordenskiöld, during his voyage in the *Vega*, obtained some information from the natives of Behring Island which led him to believe that a few individuals may have survived to a much later date, even to 1854; but this conclusion is disputed by later writers.

very existence would have been unknown to science but for the interesting account of its anatomy and habits left by Steller, and the few more or less imperfect skeletons which have recently rewarded the researches carried on in the frozen soil of the islands around which it dwelt. There is no evidence at present of its having inhabited any other coasts than those of the islands just named, although it can hardly be supposed that its range was always so restricted. When first discovered it was extremely numerous in the shallow bays round Behring Island, finding abundant nutriment in the large laminariæ growing in the sea. Its extirpation is entirely due to the Russian hunters and traders who followed upon the track of the explorers, and, upon Steller's suggestion, lived upon the flesh of the great Sea-cows. Its restricted distribution, large size, inactive habits, fearlessness of man, and even its affectionate disposition towards its own kind when wounded or in distress, all contributed to accelerate its final extinction.

According to Steller's account, the Rhytina had a skin of a dark brown colour, sometimes spotted or streaked with white. The fore limb was covered with short brush-like hairs.

EXTINCT SIRENIANS.

Halitherium.[1]—The Miocene and early Pliocene seas of Europe abounded in Sirenians, to which the generic name of *Halitherium* was given by Kaup, but which have also received other names. They had large tusk-like incisors in the upper jaw, as in the existing Dugongs, though not so greatly developed. Their molar teeth were $\frac{5}{5}$ or $\frac{6}{6}$, anteriorly simple and single-rooted, posteriorly those above with three and those below with two roots, and with enamelled and tuberculated or ridged crowns, in all which respects they more resemble those of the Manatee than of the Dugong. The anterior molars were deciduous; and there is evidence of the presence of milk-teeth. Germs of inferior incisors were also present. Some species at least had nasal bones, short, broad, but normal in position, whereas in all the existing genera these bones are quite rudimentary. Another and still more important evidence of conformity to the general mammalian type is the better development of the pelvic bone, and the presence of a small styliform femur articulated to the acetabulum, although no traces of any other part of the limb have been discovered. These ancient Sirenians, which may be regarded as representing a distinct family —*Halitheriidæ*—were thus, in dental, cranial, and other osteological characters, less specialised than are either of the existing species,

[1] Kaup, *Neues Jahrbuch*, 1838, pp. 319 and 536.

and if the intermediate links could be discovered might well be looked upon as the ancestral forms from which the latter have been derived, but at present the transitional conditions have not been detected. So far as is yet known, when changes in the physical conditions of the European seas rendered them unfitted to be the habitation of Sirenians, the *Halitherium* type still prevailed. If the existing Dugongs and Manatees are descended from it, their evolution must have taken place during the Pliocene and Pleistocene epochs, the one in seas to the east, the other to the west of the African continent, which has long formed a barrier to their intercommunication. *Halitherium* remains have been found in many parts of Germany, especially near Darmstadt, also in France, Italy Belgium, Malta, etc. Until a few years ago none were known from England, probably owing to the absence of beds of an age corresponding to those in which they are found on the European continent; but a skull and several teeth have been detected among the rolled debris of which the Red Crag of Suffolk is partially composed. The species are not yet satisfactorily characterised. Some of them appear to have attained a larger size than the existing Manatee or Dugong. One of these, from the Pliocene of Italy and France, having but $\frac{5}{5}$ molar teeth, has been separated generically under the name of *Felsinotherium* by Capellini, by whom it has been fully described; but the difference in the number of the teeth does not afford sufficient grounds for separation from *Halitherium*. *Miosiren* of the Belgian Miocene, differs in that the last upper molar is the smallest, in place of the largest of the whole series of teeth.

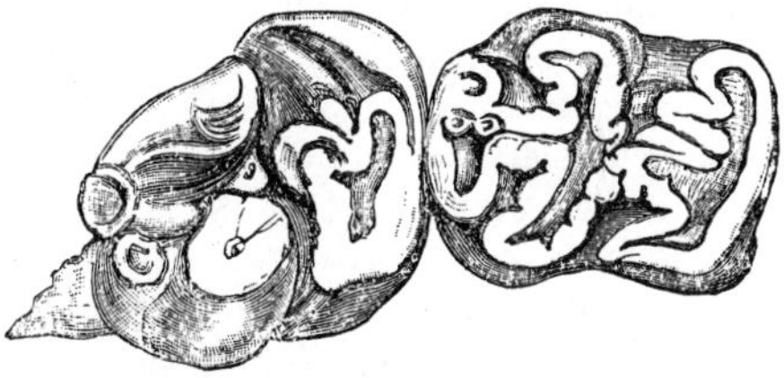

FIG. 74.—The penultimate and last right lower molars of *Halitherium fossile;* from the Miocene of the Continent. (After De Blainville.)

Other forms.—Remains from the Pliocene of France described as *Prohalicore* are regarded as indicating a Sirenian closely allied to *Halicore;* while a molar from the Tertiary of California has been made the type of *Desmotylus*, which is likewise referred to the *Halicoridæ*. *Dioplotherium*, from the Phosphorites of South Carolina, has been considered to connect *Halicore* with *Halitherium*, but even its ordinal position is uncertain.

A portion of a skull found in the Pliocene of Belgium has been described as *Crassitherium* by Van Beneden; and some compressed teeth, somewhat similar to but larger than those of the Dugong, discovered in the Miocene of the department of Lot-et-Garonne, France, gave origin to the genus *Rytiodus* of E. Lartet. Of this genus, which may be identical with *Trachytherium* of the French Miocene, better preserved remains have subsequently been described by Delfortrie. These show that the rostrum is more elongated than in *Halitherium*, but the skull is otherwise very similar, as are the molar teeth. The incisors are very large, exserted, strongly compressed, almost sabre-like, rounded on the upper or anterior surface, sharp below, concave on the external and convex on the inner side, and transversely striated.

Pachyacanthus from the Miocene of the Vienna basin is also, according to Van Beneden, another form of Sirenian, of which, however, the skull is not known. In various Miocene marine formations of the United States of America other remains of Sirenians have been found, but mostly in such a fragmentary condition that they afford at present little evidence of the early history of the group in that country. A more satisfactory discovery is that of a nearly complete skull and some bones from a Tertiary limestone formation in Jamaica. It is of smaller size than the Manatee, and, so far as the teeth are concerned, of a still more generalised character than *Halitherium*, the dentition being apparently $i\ \frac{3}{3}$, $c\ \frac{1}{1}$, $p + m\ \left(\frac{?8}{?8}\right) = 48$. The incisors are small, not developed into tusks; the canines (wanting in all existing Sirenians) are rather larger than the incisors, judging by the sockets; and the molars are bilophodont, and covered with enamel. It has been described by Sir R. Owen under the name of *Prorastomus sirenoides*. Some writers regard this genus as the type of a distinct family—the *Prorastomatidæ*. Unfortunately we have no knowledge of the geological antiquity of the formation in which it was embedded. Lastly must be mentioned the *Eotherium egyptiacum*, Owen, founded on the cast of a brain, with a small quantity of surrounding bone, discovered in the nummulitic limestone of Eocene age in the Mokattam Hills, near Cairo. The brain is narrower than in *Manatus*, and resembles that of *Halitherium*. This is of interest as the most ancient known evidence of any Sirenian whose age has been geologically determined. Teeth from the same deposits referred to *Manatus* not improbably belong really to *Eotherium*.

The few facts as yet collected relating to the former history of the Sirenia leave us as much in the dark as to the origin and affinities of this peculiar group of animals as we were when we only knew the living members. They lend no countenance to their association with the Cetacea, and on the other hand their supposed affinity with the Ungulata, so much favoured by modern zoologists, receives no very material support from them.

Bibliography of Sirenia.—J. F. Brandt, *Symbolæ Sirenologicæ*, St. Petersburg, 3 fasciculi, 1846-61-68—an exhaustive account of the anatomy, affinities, and literature of the group, with copious illustrations of the osteology of *Rhytina*.

Anatomy of Dugong:—Everard Home, *Phil. Trans.* 1820, p. 315; Owen, *Proc. Zool. Soc.* 1838, p. 29. *Placenta of do.*:—W. Turner, *Trans. Roy. Soc. Edin.* vol. xxxv. (1889). *Manatee*:—W. Vrolik, *Bijdragen tot de Dierkunde*, 1851; J. Murie, "On the Form and Structure of the Manatee," *Trans. Zool. Soc. Lond.* vol. viii. p. 127, 1872, and "Further Observations on the Manatee," *Ibid.* vol. xi. p. 19, 1880; A. H. Garrod, "Notes on the Manatee recently living in the Zoological Society's Gardens," *Ibid.* vol. x. p. 137, 1875; H. C. Chapman, "Observations on the Structure of the Manatee," *Proc. Acad. Nat. Sciences of Philadelphia*, 1875, p. 452; A. Crane, "Notes on the Habits of the Manatees in Captivity in the Brighton Aquarium," *Proc. Zool. Soc. Lond.* 1881, p. 456. *Extinct Sirenia*:—Gervais, *Journal de Zoologie*, tom. i. p. 332, 1872. R. Lydekker, *Catalogue of Fossil Mammalia in the British Museum*, pt. v.

Order CETACEA.

This is perhaps the most distinctly circumscribed and natural of all the larger groups into which the class is divided.

The external form is fish-like, the body being fusiform, passing anteriorly into the head without any distinct constriction or neck, and posteriorly tapering off gradually towards the extremity of the tail, which is provided with a pair of lateral, pointed expansions of skin supported by dense fibrous tissue, called "flukes," forming together a horizontally-placed triangular propelling organ, notched in the middle line behind.

The head is generally large, in some species attaining to even more than one-third of the entire length of the animal, and the aperture of the mouth is always wide, and bounded by stiff immobile lips. The fore limbs are reduced to the condition of flattened ovoid paddles, encased in a continuous integument, showing no external sign of division into arm, fore arm, and manus, or of separate digits, and without any trace of nails. There are no traces of hind limbs visible externally. The general surface of the skin is smooth and glistening, and devoid of hair, although in many species there are a few fine bristles in the neighbourhood of the mouth, which may either persist through life, or be present only in the young state. Immediately beneath the skin, and intimately connected with it, is a thick layer of fat, held together by a dense mesh of areolar tissue, constituting the "blubber," which serves the purpose of the hairy covering of other mammals in retaining the heat of the body. In nearly all species a compressed median dorsal tegumentary fin is present. The eye is small, and is not provided with a nictitating membrane or true lachrymal apparatus. The external auditory meatus is a very minute aperture in the skin situated at a short distance behind the eye, and there is no vestige of a pinna. The nostrils open either separately or by a single crescentic valvular aperture, not at the extremity of the snout, but near the vertex of the head.

The bones generally are spongy in texture, the cavities being filled with oil. In the vertebral column the cervical region is remarkably short and immobile, and the vertebræ, originally always seven in number, are in many species more or less fused together into a solid mass. The odontoid process of the axis, when that bone is free, is usually very obtuse, or even obsolete. None of the vertebræ are united together to form a sacrum. The lumbar and caudal vertebræ are numerous and large, and, as their arches are not connected by any articular processes (zygapophyses), they are capable of a very free motion in all directions. The epiphyses at the ends of the vertebral bodies are very distinct flattened disks, not uniting until after the animal has attained its full dimensions.[1] There are largely developed chevron bones, the presence of which indicates the distinction between the caudal and lumbar vertebræ.

The skull (Fig. 75) is modified in a very peculiar manner. The brain-case is short, broad, and high, in fact almost spherical. The supraoccipital bone rises upwards and forwards from the foramen magnum, to meet the frontals at the vertex, thus completely excluding the parietals from the upper region of the cranium. The frontals are expanded laterally to form the roof of the orbits. The anterior narial aperture opens upwards, and has in front of it a more or less horizontally prolonged rostrum, formed of the maxillæ, premaxillæ, vomer, and mesethmoid cartilage, extending forwards to form the upper jaw or roof of the mouth.

There are no clavicles. The humerus is freely movable on the scapula at the shoulder-joint, but beyond this the articulations of the limb are imperfect, the flattened ends of the bones coming in contact with each other, with fibrous tissue interposed, allowing of scarcely any motion. The radius and ulna are distinct, about equally developed, and much flattened, as are also all the bones of the manus. There are four, or more commonly five digits, and the number of the phalanges of the second and third digits always exceeds the normal number in mammals, sometimes very considerably (hyperphalangism); they present the exceptional character of having epiphyses at both ends.[2] The pelvis is represented by a pair of small styliform bones placed longitudinally, suspended below and at some distance from the vertebral column at the commencement of the caudal region. These appear to represent the ischia, as the crura of the corpora cavernosa are attached to them. In some species, to the outer surface of these are fixed other small bones or cartilages, the rudiments of the hind limb.

[1] This is an important distinction from the Sirenia, but a character common to nearly all other mammals. It is doubtful whether there is any foundation for the statement that these epiphyses remain ununited for an exceptionally long period in the Cetacea.

[2] A character repeated in some of the Seals.

Teeth are generally present, but exceedingly variable in number. In the existing species they are of simple, uniform character, all having conical or compressed crowns and single roots, and are never preceded by milk-teeth. They are therefore homodont and monophyodont. In one group, the Mystacocetes, the teeth are absent (except in the fœtal condition), and the palate is provided with numerous transversely placed horny laminæ or "baleen." The salivary glands are rudimentary or absent. The stomach is multilocular, its structure being fully noticed under the genus

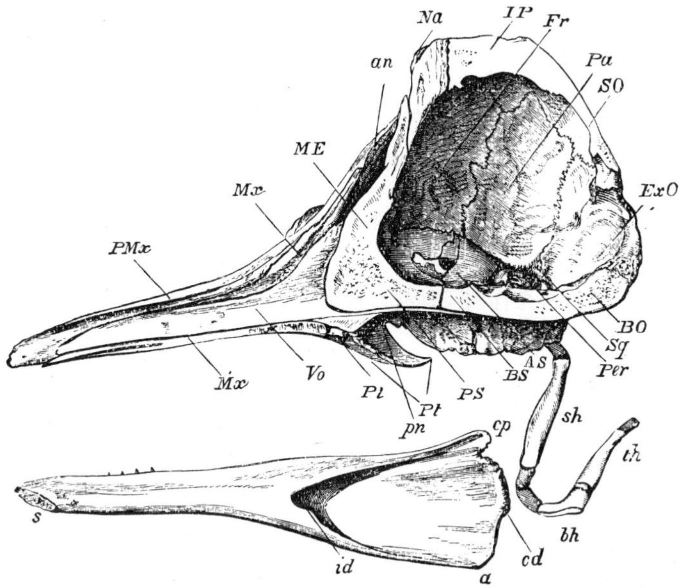

FIG. 75.—A section of the skull of a young Dolphin (*Globicephalus melas*). ×⅕. *PMx*, Premaxilla; *Mx*, maxilla; *ME*, ossified portion of the mesethmoid; *an*, anterior nares; *Na*, nasal; *IP*, inter-parietal; *Fr*, frontal; *Pa*, parietal; *SO*, supraoccipital; *ExO*, exoccipital; *BO*, basioccipital; *Sq*, squamosal; *Per*, periotic; *AS*, alisphenoid; *PS*, presphenoid; *Pt*, pterygoid; *pn*, posterior nares; *Pl*, palatine; *Vo*, vomer; *s*, symphysis of mandible; *id*, inferior dental canal; *cp*, coronoid process of mandible; *cd*, condyle; *a*, angle; *sh*, stylo-hyal; *bh*, basi-hyal; *th*, thyro-hyal. (From Flower's *Osteology of Mammalia*.)

Phocæna. The intestinal canal is simple, and only in some species provided with a small cæcum. The liver is very little fissured, and there is no gall-bladder. The vascular system is greatly complicated by arterial and venous plexuses, or *retia mirabilia*. The larynx is of peculiar shape, the arytenoid cartilages and the epiglottis being much elongated, and together forming a tubular prolongation, which projects into the posterior nares, and when embraced by the soft palate produces a continuous passage between the nostrils and the trachea, as in Ungulates, but in a more perfect manner. The brain is large relatively to the size of the animal, very round in form, and with its surface divided by sulci into very numerous and complex convolutions. The kidneys are deeply lobulated. The testes are abdominal. There are no vesiculæ seminales, nor os penis. The uterus is bicornuate, and the placenta nondeciduate and diffuse. The mammæ are two in number, and the nipples placed in depressions on each side of the vulva. The principal ducts of the gland are dilated during lactation into large reservoirs, into which the milk collects, and from which it is injected by the action of a compressor muscle into the mouth of the young animal, by which means the process of sucking under water is greatly facilitated and expedited.

The animals of the order Cetacea abound in all known seas, and some species are inhabitants of the larger rivers of South America and Asia. Their organisation necessitates passing their life entirely in the water, as on land they are absolutely helpless. They have, however, to rise very frequently to the surface for the purpose of respiration; and, in relation to the constant upward and downward movement in the water thus necessitated, their principal instrument of motion, the tail, is expanded horizontally, quite unlike that of a fish, whose movements are mainly in straightforward or lateral directions. The position of the respiratory orifice or nostril on the highest part of the head is very important for this mode of life, since it is the only part of the body of which the exposure above the surface is absolutely necessary. Of the numerous erroneous ideas connected with natural history, few are so wide spread and still so firmly believed, notwithstanding repeated expositions of its falsity, as that the Cetacea spout out through their blowholes water taken in at the mouth. The fact is, the "spouting," or more properly "blowing," of the Whale is nothing more than the ordinary act of expiration, which, taking place at longer intervals than in land animals, is performed with a greater amount of emphasis. The moment the animal rises to the surface it forcibly expels from its lungs the air taken in at the last inspiration, which of course is highly charged with watery vapour in consequence of the natural respiratory changes. This, rapidly condensing in the cold atmosphere in which the phenomenon is generally observed, forms a column of steam or spray, which has been erroneously taken for water. It also often happens, especially when the surface of the ocean is agitated into waves, that the animal commences its expiratory puff before the orifice has quite cleared the top of the water, some of which may thus be driven upwards with the blast, tending to complete the illusion. In hunting Whales the harpoon often pierces the lungs or air passages of the unfortunate victim, and then fountains of blood may be forced high in the air through the blowholes, as commonly depicted in scenes of Arctic adventure; but this is nothing more (allowance being made for the Whale's peculiar mode of breathing) than what always follows severe wounds of the respiratory organs of other mammals.

All the Cetacea are predaceous, subsisting on living animal food of some kind. One genus alone (*Orca*) eats other warm-blooded animals, as Seals, and even members of its own order, both large and small. Some feed on fish, others on small floating crustaceans, pteropods, and medusæ, while the principal staple of the food of many is constituted by the various species of cephalopods, *Loligo* and other *Teuthidæ*, which must abound in certain seas in vast numbers, as they form almost the entire support of some of the largest members of the order. In size the Cetacea vary much, some of the smaller Dolphins scarcely exceeding 4 feet in length, while others are the most colossal of all animals. It is true that most statements of their bulk found in general and even zoological literature are greatly exaggerated, but even when reduced to their actual dimensions (which will be stated under the respective genera) some of the existing Whales exceed in size any animal living either at present or in former times of which we have any certain evidence. With some exceptions, the Cetacea generally are timid inoffensive animals, active in their movements, and very affectionate in their disposition towards one another, especially the mother towards the young, of which there is usually but one, or at most two at a time. They are generally gregarious, swimming in herds or "schools" (so termed by the whalers) sometimes amounting to many thousands in number; though some species have hitherto only been met with either singly or in pairs.

Although by their mode of life so far removed from close observation that it is impossible to become as familiar with them in their natural condition as with many other animals, Whales are in many respects the most interesting and wonderful of all creatures; and there is much in their structure and habits well worthy of study, much that is difficult to understand, and much that leads to great generalisations and throws light upon far-reaching philosophical speculations. One of the first lessons which a study of these animals affords is that, in the endeavour to discover what a creature really is, from what others it is descended, and to what it is related, the general outward appearance affords little clue, and we must go deep below the surface to find out the essential characteristics of its nature. There was once, and may be still in many places, a common idea that a Whale is a fish. To realise the fallacy of this notion we have only to consider what a fish really is, what under all the diversities of form, size, and colour known among fishes there is common to them all, and we see that in everything which characterises a true fish and separates it from other classes, as reptiles, birds, and mammals, the Whale resembles the last-named and differs from the fish. It is as essentially a mammal as a Cow or a Horse, and simply resembles a fish externally because it is adapted to inhabit the same element; but it is no more on that account a fish than is a bat, because adapted to pass a great part of its existence on the wing in the air, nearly related to a bird. The whole structure of a whale is a most instructive instance of a type of organisation which is common to and characteristic of the class Mammalia, but specially modified or adapted to a peculiar mode of life. We see in every part the result of two great principles acting and reacting upon each other—on the one hand, adherence to type, or rather to fundamental inherited structural conditions, and, on the other, adaptation to the peculiar circumstances under which it lives, and to which in all probability it has become gradually more and more fitted. The external fish-like form is perfectly suited for swimming through the water; the tail, however, is not placed vertically as in fishes, but horizontally, a position which accords better with the constant necessity for rising to the surface for the purpose of breathing. The hairy covering characteristic of all mammals, which if present might interfere with rapidity of movement through the water, is reduced to the merest rudiments—a few short bristles about the chin or upper lip—which are often only present in very young animals; and the function of keeping the body warm is supplied by the "blubber." The fore-limbs, though functionally reduced to mere paddles, with no power of motion except at the shoulder-joint, have beneath their smooth and continuous external covering all the bones, joints, and even most of the muscles, nerves, and arteries of the human arm and hand; and the rudiments of hind legs found buried deep in the interior of the animal apparently subserve no useful purpose, but point an instructive lesson to those who are able to read it.

As before said, the Cetacea form a perfectly well-defined group, sharply separated from all other mammals, and with no outlying or doubtful forms at present known. Among the existing members of the order, there are two very distinct types, the Toothed Whales or Odontoceti and the Baleen Whales or Mystacoceti, which present as many marked distinguishing structural characters as are found between many other divisions of the Mammalia which are reckoned as orders. The extinct *Zeuglodon*, so far as its characters are known, does not fall into either of these groups, but is in some respects an annectant form, and therefore must be placed, provisionally at least, in a third group by itself.

The Mystacocetes appear at first sight to be the most specialised and aberrant of the existing Cetacea, as indicated by the absence of teeth, the presence of baleen, and the form and size of the mouth; but, as we see in other groups, dental characters, and all such as relate to the prehension of food generally, are essentially adaptive and consequently plastic or prone to variation, and hence cannot well be relied upon as tests of affinity. In another character, also adaptive, the laxity of the connection of the ribs with the vertebral column and with the sternum, and the reduction of that bone in size, allowing great freedom of expansion of the thoracic cavity for prolonged immersion beneath the water, the Mystacocetes have passed beyond the Odontocetes in specialisation. On the other hand, the greater symmetry of the skull, the more anterior position of the external nostrils and their double external orifice, the form of the nasal bones, the presence of a distinctly developed olfactory organ, the mode of attachment of the periotic bone to the cranium, the presence of a cæcum and the regular arrangement of the alimentary canal, the more normal characters of the manus and the better development of the muscles attached to it, and the presence, in many species at least, of parts representing not only the bones but also the ligaments and muscles of a hind limb,[1] all show less deviation from the ordinary mammalian type than is presented by the Odontocetes. Taking all these characters into consideration, it does not appear reasonable to suppose that either type has been derived from the other, at all events in the form in which we see it now, but rather that they are parallel groups, both modified in different fashions from common ancestors.

Among the Mystacocetes, in the especially distinguishing characters of the division, the Rorquals are less specialised than the Right Whales, which in the greater size of the head, the length and compression of the rostrum, the development of the baleen, and shortness of the cervical region, are exaggerated forms of the type, and yet they retain more fully some primitive characters, as the better development of the hind limb, the pentadactylous manus, and the absence of a dorsal fin. Both types are found distinct in a fossil state at least as far back as the early Pliocene age, but generally represented by smaller species than those now existing. Some of the Pliocene Rorquals (*Cetotherium*) were, in the elongated flattened form of the nasal bones, the greater distance between the occipital and frontal bone at the top of the head, and the greater length of the cervical vertebræ, more generalised than those now existing. In the shape of the mandible also, Van Beneden, to whose researches we are much indebted for a knowledge of these forms, discerns some approximation to the Odontocetes.

Among the last-named group there are several distinct types, of which that represented by *Platanista*, although in some respects singularly modified, has been considered to present on the whole approximations towards the more normal and general type of

[1] These have been described in detail by Professor Struthers in the *Journal of Anatomy and Physiology*, 1881.

mammalian structure. It is therefore interesting to find an apparently allied form well represented among the earliest fossil remains of Cetaceans in Europe. Almost all the other members of the suborder range themselves under the two principal heads of Ziphioids (or Physeteroids) and Delphinoids. The former is an ancient and once abounding type, of which the Sperm Whale (*Physeter*) is a highly specialised form. Among the latter, *Globicephalus* is a modified form as regards the structure of its anterior extremity, and *Monodon* as regards its dentition, while *Delphinus*, with the various allied genera, may be regarded as the dominating type of Cetaceans at the present day, abundant in slightly differentiated species and also in individuals. They are in this respect to the rest of the order much as the hollow-horned Ruminants are to the other Ungulates.

The earliest Cetaceans of whose organisation we have anything like complete evidence are the Zeuglodonts of the Eocene period.[1] which approach in the structure of the skull and teeth to a much more generalised mammalian type than either of the existing suborders. The smallness of the cerebral cavity compared with the jaws and the rest of the skull they share with the primitive forms of many other types. The forward position of the narial aperture and the length and flatness of the nasal bones, which distinguish them from all existing forms, we must also suppose to be a character at one time common to all Cetaceans, though now retained (but to a less degree) only by the Mystacocetes. Even *Squalodon*, which in its heterodont dentition so much resembles *Zeuglodon* as to have been placed by some zoologists in the same genus, entirely differs from it, and conforms with the ordinary Dolphins in its essential cranial characters.

The origin of the Cetacea is at present involved in much obscurity. They present no signs of closer affinity to any of the lower classes of vertebrates than do many other members of their own class. Indeed in all that essentially distinguishes a mammal from the oviparous vertebrates, whether in the osseous, nervous, reproductive, or any other system, they are as truly mammalian as any other group. Any supposed marks of inferiority, as absence of limb structure, of hairy covering, of lachrymal apparatus, etc., are obviously modifications (or degradations, as they may be termed) in adaptation to their special mode of life. The characters of the teeth of *Zeuglodon* and other extinct forms, and also of the fœtal Mystacocetes, clearly indicate that they have been derived from mammals in which the heterodont type of dentition was fully

[1] The ankylosed mass of cervical vertebræ, on which the genus *Palæocetus* was established, was regarded by its describer as having probably come from the Kimeridge Clay, but the mineral condition of the specimen points to the Red Crag as the place of origin.

established. The steps by which a land mammal may have been modified into a purely aquatic one are indicated by the .stages which still survive among the Carnivora in the *Otariidæ* and in the true Seals. A further change in the same direction would produce an animal somewhat resembling a Dolphin; and it has been thought that this may have been the route by which the Cetacean form has been developed. There are, however, great difficulties in the way of this view. Thus if the hind limbs had ever been developed into the very efficient aquatic propelling organs they present in the Seals, it is not easy to imagine how they could have become completely atrophied and their function transferred to the tail. So that from this point of view it is more likely that Whales were derived from animals with long tails, which were used in swimming, eventually with such effect that the hind limbs became no longer necessary. The powerful tail, with its lateral cutaneous flanges, of an American species of Otter (*Lutra brasiliensis*) may give an idea of this member in the primitive Cetaceans. But the structure of the Cetacea is, in so many essential characters, so unlike that of the Carnivora that the probabilities are against these orders being nearly related. Even in the skull of the Zeuglodon, which has been cited as presenting a great resemblance to that of a Seal, quite as many likenesses may be traced to one of the primitive Pig-like Ungulates (except in the purely adaptive character of the form of the teeth), while the elongated larynx,[1] complex stomach, simple liver, reproductive organs both male and female, and fœtal membranes of the existing Cetacea are far more like those of that group than of the Carnivora. Indeed it appears probable that the old popular idea which affixed the name of "Sea-Hog"[2] to the Porpoise contains a larger element of truth than the speculations of many accomplished zoologists of modern times. The fact that *Platanista*, which, as mentioned above, appears to retain more of the primitive characteristics of the group than any other existing form, and also the somewhat related *Inia* from South America, are both at the present day exclusively fluviatile, may point to the fresh-water origin of the whole group, in which case their otherwise rather inexplicable absence from the seas of the Cretaceous period would be accounted for.

On the other hand, it should be observed that the teeth of the Zeuglodonts approximate more to a carnivorous than to an ungulate type. It is scarcely necessary to allude to the hypothesis started by some Continental writers to the effect that the Whales are the most primitive type of mammals with which we are acquainted,

[1] There is much resemblance in the larynx of the Hippopotamus, but none in that of the Seal, to the same organ in the Cetacea.

[2] German *Meerschwein*, whence the French *Marsouin*. "Porpoise" is said to be derived from "*Porc-poisson*."

and that they are the descendants of the Mesozoic reptilian order Ichthyopterygia, from which their hyperphalangism is a direct inheritance. The Ichthyopterygia have been shown, on very strong evidence, to have been derived from land reptiles, and to have gradually acquired their hyperphalangism as an adaptive character suitable to their peculiar mode of life, and there can be but little doubt that a similar adaptation has taken place in the case of the Whales.

Suborder MYSTACOCETI,

the BALÆNOIDEA, *Whalebone, or True Whales.*[1]

Family BALÆNIDÆ.

Teeth never functionally developed, but always disappearing before the close of intra-uterine life. Palate provided with plates of baleen or "whalebone." Skull symmetrical. Nasal bones forming a roof to the anterior nasal passages, which are directed upwards and forwards. Maxilla produced in front of, but not over, the orbital process of the frontal. Lachrymal bones small and distinct from the jugal. Tympanic bone involuted (Fig. 76), and ankylosed with the periotic, which is attached to the base of the cranium by two strong diverging processes. Olfactory organ distinctly developed. Rami of mandible arched outwards, their anterior ends meeting at an angle, and connected by fibrous tissue without any true symphysis. All the ribs at their upper extremities articulating only with the transverse processes of the vertebræ; their capitular processes, when present, not articulating directly with the bodies of the vertebræ. Sternum composed of a single piece, and articulating only with a single pair of ribs. No ossified sternal ribs. External openings of nostrils distinct from each other, longitudinal. A short conical cæcum.

These animals have, when in the fœtal state, numerous minute calcified teeth lying in the dental groove of both upper and lower jaws. They are best developed about the middle of fœtal life, after which period they are absorbed, and no trace of them remains at the time of birth.[2] The baleen or whalebone does not make its appearance until after birth. It consists of a series of flattened horny plates, between three and four hundred in number, on each side of

[1] Icel. *hvalr;* Dan. and Swed. *hval;* Anglo-Saxon *hwæl;* Germ. *wal, walfisch.* The meaning apparently is "roller," the word being closely allied to "wheel" (Skeat).

[2] These were discovered in the Greenland Whale by Geoffroy St. Hilaire, whose observations were confirmed and extended to other genera by Eschricht. They have been very fully described in *Balænoptera rostrata* by Julin (*Archivs de Biologie*, i. 1880).

the palate, with a bare interval along the middle line. These plates are placed transversely to the long axis of the palate, with very short intervals between them. Each plate or blade is somewhat triangular in form, with the base attached to the palate and the apex hanging downwards. The outer edge of the blade is hard and smooth; but the inner edge and apex fray out into long bristly fibres, so that the roof of the Whale's mouth looks as if covered with hair, as described by Aristotle. At the inner edge of each principal blade are two or three much smaller or subsidiary blades. The principal blades are longest near the middle of the series, and gradually diminish towards the front and back of the mouth. The horny plates grow from a dense fibrous and highly vascular matrix, covering the palatal surface of the maxillæ, and sending out lamellar processes, one of which penetrates the base of each blade. Moreover, the free edge of these processes is covered with very long vascular thread-like papillæ, one of which forms the central axis of each of the hair-like epidermic fibres of which the blade is mainly composed. A transverse section of fresh whalebone shows that it is made up of numbers of these soft vascular papillæ, circular in outline, each surrounded by concentrically arranged epidermic cells, and the whole bound together by other epidermic cells, that constitute the smooth cortical (so-called "enamel") surface of the blade, which, disintegrating at the free edge, allows the individual fibres to become loose and assume the hair-like appearance before spoken of. These fibres differ from hairs in not being formed in depressed follicles in the enderon, but rather resemble the fibres composing the horn of the Rhinoceros. The whalebone in fact consists of nothing more than modified papillæ of the buccal mucous membrane, with an excessive and cornified epithelial development. The blades are supported and bound together for a certain distance from their base by a mass of less hardened epithelium, secreted by the surface of the palatal membrane or matrix of the whalebone in the intervals of the lamellar processes. This is the "intermediate substance" of Hunter, the "gum" of the whalers. Baleen varies much in colour in different species. In some it is almost jet black, in others slate-colour, horn-colour, yellow, or even creamy-white. In some the blades are variegated with longitudinal strips of different hues. Baleen differs also greatly in other respects, being short, thick, coarse, and stiff in some, and greatly elongated and highly elastic in those species in which it has attained its fullest development. Its function is to strain the water from the small marine molluscs, crustaceans, or fish upon which the Whales subsist. In feeding the immense mouth is filled with water containing shoals of these small creatures, and then, on the Whale closing the jaws and raising the tongue, so as to diminish the cavity of the mouth, the water streams out through the narrow intervals between the hairy fringe of the whalebone blades, and escapes through the lips, leaving the living prey to be swallowed.[1]

Our knowledge of the different structural modifications attained by members of this important group of mammals, though largely increased of late years, is still imperfect. Formerly they were all divided into Right Whales (*Balæna*) and Rorquals or Fin-Whales (*Balænoptera*), the latter distinguished by their smaller heads, elongated and slender form, free cervical vertebræ, tetradactylous manus, and the presence of very conspicuous longitudinal furrows or folds in the skin of the throat and chest, and of a small adipose dorsal fin. Recent discoveries have, however, brought to light several forms holding a somewhat intermediate position, and presenting combinations of characters not found in either of the longer known sections. According to our present knowledge the group is naturally divided into five very distinct genera, of which the leading characters are given below.

Balæna.[2]—Skin of throat smooth, not furrowed. No dorsal fin. Cervical vertebræ united into a single mass. Pectoral limb short, broad, and pentadactylous. Head very large. Baleen very long and narrow, highly elastic, and black. Scapula high, with a distinct coracoid and acromion process. Tympanic (Fig. 78) deep and angular, its inflation comparatively slight, and the involuted portion not fig-shaped, and frequently without a well-marked depression at the anterior extremity of the superior border of the inner surface for the Eustachian canal.

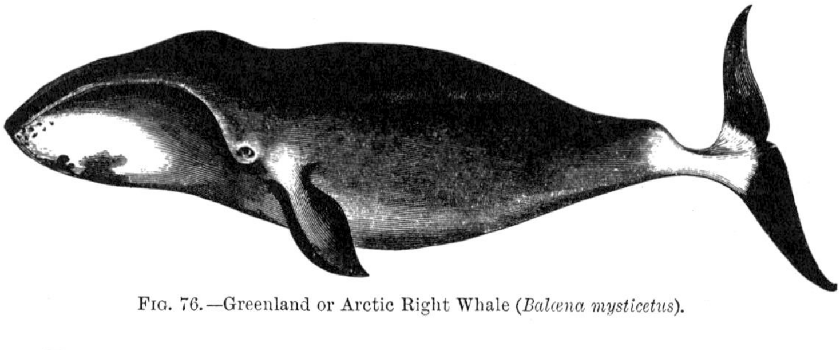

FIG. 76.—Greenland or Arctic Right Whale (*Balæna mysticetus*).

The Greenland, or more properly Arctic, Right Whale (*Balæna mysticetus*) attains, when full grown, a length of from 45 to 50 feet. Its usual vertebral formula is C 7, D 12, L 14, C 22. The external form is shown in Fig. 76 from a careful drawing by

[1] For the structure of whalebone see Hunter, "Observations on the Structure and Economy of Whales," *Phil. Trans.* 1787; Eschricht and Reinhardt, *On the Greenland Right Whale*, English translation by the Ray Society, 1866, pp. 67-78; and Sir W. Turner, in *Trans. Roy. Soc. Edin.* 1870.

[2] Linn. *Syst. Nat.* 12th ed. vol. i. p. 105 (1766).

Mr. Robert Gray. In this species all the peculiarities which distinguish the head and mouth of the Whales from those of other mammals have attained their greatest development. The head is of enormous size, exceeding one-third of the whole length of the creature. The cavity of the mouth is actually larger than that of the body, thorax and abdomen together. The upper jaw is very narrow, but greatly arched from before backwards, to increase the height of the cavity and allow for the great length of the baleen blades; the rami of the mandible are widely separated posteriorly, and have a still further outward sweep before they meet at the symphysis in front, giving the floor of the mouth the shape of an immense spoon. The baleen blades attain the number of 380 or more on each side, those in the middle of the series having a length of 10 or sometimes 12 feet. They are black in colour, fine and highly elastic in texture, and fray out at the inner edge and ends into long, delicate, soft, almost silky, but very tough, hairs. The remarkable development of the mouth and the structures in connection with it, which distinguishes the Right Whale among all its allies, is entirely in relation to the nature of its food. It is by this apparatus that the animal is enabled to avail itself of the minute but highly nutritious crustaceans and pteropods which swarm in immense shoals in the seas it frequents. The large mouth enables it to take in at one time a sufficient quantity of water filled with these small organisms, and the length and delicate structure of the baleen provide an efficient strainer or hair-sieve by which the water can be drained off. If the baleen were rigid, and only as long as is the aperture between the upper and lower jaws when the mouth is shut, a space would be left beneath it when the jaws were separated, through which the water and the minute particles of food would escape together. But instead of this the long, slender, brush-like, elastic ends of the whalebone blades fold back when the mouth is closed, the front ones passing below the hinder ones in a channel lying between the tongue and the lower jaw. When the mouth is opened, their elasticity causes them to straighten out like a bow unbent, so that at whatever distance the jaws are separated the strainer remains in perfect action, filling the whole of the interval. The mechanical perfection of the arrangement is completed by the great development of the lower lip, which rises stiffly above the jaw-bone and prevents the long, slender, flexible ends of the baleen from being carried outwards by the rush of water from the mouth, when its cavity is being diminished by the closure of the jaws and raising of the tongue.

If, as appears highly probable, the "bowhead" of the Okhotsk Sea and Behring Strait belongs to this species, its range is circumpolar. Though found in the seas on both sides of Greenland, and passing freely from one to the other, it is never seen so far south as Cape Farewell; but on the Labrador coast, where a cold stream sets down from the north, its range is somewhat farther. In the Behring Sea, according to Scammon, "it is seldom seen south of the fifty-fifth parallel, which is about the farthest southern extent of the winter ice, while on the Sea of Okhotsk its southern limit is about the latitude of 54°." As has been abundantly shown by Eschricht and Reinhardt in the case of the Greenland seas, "everything tends to prove," Scammon says, "that the *Balæna mysticetus* is truly an 'ice whale,' for among the scattered floes, or about the borders of the ice-fields or barriers, is its home and feeding-ground. It is true that these animals are pursued in the open water during the summer months; but in no instance have we learned of their being captured south of where winter ice-fields are occasionally met with." The occurrence of this species, therefore, on the British or any European coast is exceedingly unlikely, as when alive and in health the southern limit of its range in the North Sea has been ascertained to be from the east coast of Greenland at 64° N. lat. along the north of Iceland towards Spitzbergen, and a glance at a physical chart will show that there are no currents setting southwards which could bear a disabled animal or a floating carcase to British shores. To this *à priori* improbability may be added the fact that no authentic instance has been recorded of the capture or stranding of this species upon any European coast; for the cases in which it has been reported as seen in British waters may be explained by the supposition of one of the other species of the genus being mistaken for it. Still, as two other essentially Arctic Cetaceans, the Narwhal and the Beluga, have in a few undoubted instances found their way to British shores, it would be rash absolutely to deny the possibility of the Greenland Right Whale doing the same.

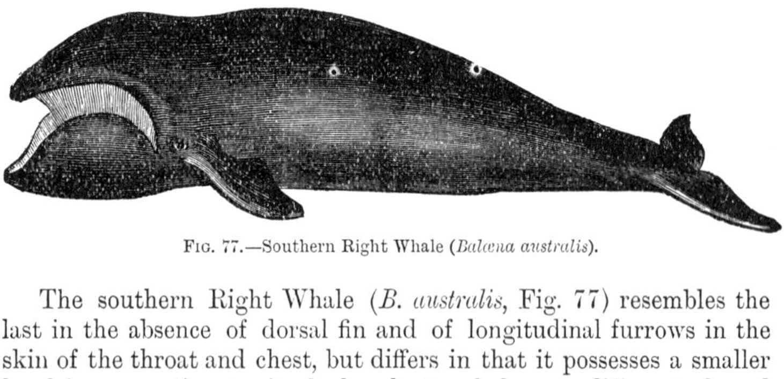

FIG. 77.—Southern Right Whale (*Balæna australis*).

The southern Right Whale (*B. australis*, Fig. 77) resembles the last in the absence of dorsal fin and of longitudinal furrows in the skin of the throat and chest, but differs in that it possesses a smaller head in proportion to its body, shorter baleen, a different shaped contour of the upper margin of the lower lip, and a greater number (fifteen) of ribs and dorsal vertebræ. This form inhabits the temperate seas of both northern and southern hemispheres, and is divided into several so-called species, according to their geographical distribution:—*B. biscayensis* of the North Atlantic, *B. japonica* of the North Pacific, *B. australis* of the South Atlantic, and *B. antipodarum* and *B. novæ-zealandiæ* of the South Pacific. The differential characters by which they have been separated, external as well as anatomical, are, however, slight and subject to individual variation; and the number of specimens available for comparison in museums is not yet sufficient to afford the necessary data to determine whether these characters can be regarded as specific or not. The most interesting of these is the Atlantic Right Whale, which was formerly abundant in the North Atlantic, but is now so scarce as to appear verging on extinction. This was the Whale the pursuit of which gave occupation to a numerous population on the shores of the Basque provinces of France and Spain in the Middle Ages. From the tenth to the sixteenth centuries Bayonne, Biarritz, St. Jean de Luz, and San Sebastian, as well as numerous other towns on the north coast of Spain, were the centres of an active Whale "fishery," which supplied Europe with oil and whalebone. In later times these Whales were pursued as far as the coast of Newfoundland. They were, however, already getting scarce when the voyages undertaken towards the close of the sixteenth century for the discovery of the north-eastern route to China and the East Indies opened out the seas around Spitzbergen; then for the first time the existence of the Greenland Whale became known, and henceforth the energies of the European whale-fishers were concentrated upon that animal. It is a singular fact that the existence of the Atlantic Right Whale was quite overlooked by naturalists till lately, all accounts referring to it being attributed to the Greenland Whale, supposed once to have had a wider distribution than now, and to have been driven by the persecution of man to its present circumpolar haunts. To the two Danish cetologists Eschricht and Reinhardt is due the credit of having proved its existence as a distinct species, from a careful collation of numerous historical notices of its structure, distribution, and habits; and their restoration of the animal, founded upon these documents, has been abundantly confirmed by the capture of various specimens in recent times, showing that it still lingers in some of the localities where it formerly was so abundant. The only known instances of its occurrence on the coasts of Europe in modern times are in the harbour of San Sebastian in January 1854, in the Gulf of Taranto, in the Mediterranean, in February 1877, and on the Spanish coast between Guetaria and Zarauz (Guipuzcoa) in February 1878. The skeletons of these three whales are preserved in the museums of Copenhagen, Naples, and San Sebastian respectively. On the coast of the United States several Whales of this species have been taken within the last few years. In the North Pacific a very similar if not identical species is regularly hunted by the Japanese, who tow the carcases ashore for the purposes of flensing and extracting the whalebone. In the tropical seas, however, according to Captain Maury's whale charts, Right Whales are never or rarely seen; but the southern temperate ocean, especially the neighbourhood of the Cape of Good Hope, Kerguelen's Island, Australia, and New Zealand, is inhabited by "Black Whales," once abundant, but now nearly exterminated through the wanton destruction of the females as they visit the bays and inlets round the coast, their constant habit in the breeding time. The range of these Whales southward has not been accurately determined; but no species corresponding with the Arctic Right Whale has as yet been met with in the Antarctic icy seas

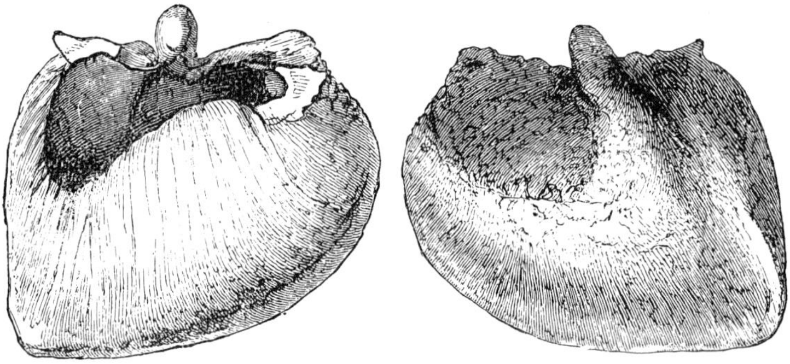

A B

FIG. 78.—The right tympanic bone of an immature individual of the Greenland Whale (*Balæna mysticetus*), from the inner (*A*) and outer (*B*) aspects. ⅓ natural size. (From the *Proc. Zool. Soc.*)

Remains of Right Whales are of not uncommon occurrence in the Pliocene Crag deposits of England and Belgium. The tympanics of *B. affinis* from these deposits appear to indicate a species closely allied to *B. mysticetus*, in which this bone is long and angulated anteriorly (Fig. 78); while the tympanics from the same deposits described as *B. primigenia* are shorter and more rounded at the antero-inferior angle, thus resembling those of *B. australis*. A smaller species, having an estimated length of about 20 feet, has been described as *Balænula balænopsis*, the generic distinction being made on account of the free condition of the atlas and seventh cervical vertebræ; but it seems scarcely advisable to regard such a feature as indicating more than a less specialised species. *Balæna* (*Balænotus*) *insignis* is a whale of somewhat larger dimensions, in which the atlas is generally, and the seventh cervical vertebra always, free, while in young individuals the axis vertebra may likewise be separate.

Neobalæna.[1]—Head about one-fourth the total length. Skin of the throat not plicated. A small falcate dorsal fin. Vertebræ, C 7, D 17, L 3, C 16 = 43 The cervical vertebræ are united. The manus small, narrow, and tetradactylous, wanting the pollex. The ribs remarkably expanded and flattened. The scapula very low and broad, with completely developed acromion and coracoid processes. Tympanic approximating to that of *Balæna*, but with certain very characteristic peculiarities of shape. Baleen very long, slender, elastic, and white. A single species, at present very rare, *N. marginata*, from the Australian and New Zealand seas is the smallest of the Whalebone Whales, being not more than 20 feet in length.

Rhachianectes.[2]—This combines the small head, elongated form, and narrow pectoral fin of *Balænoptera* with the smooth skin of the throat and absence of the dorsal fin of *Balæna.* The baleen is the shortest and coarsest of any of the group. Its osteology is imperfectly known. One species, *R. glaucus*, the Gray Whale of the North Pacific.

Megaptera.[3]—Head of moderate size. Baleen plates short and broad. Vertebræ, C 7, D 14, L 11, C 21 = 53. Cervical vertebræ free. Scapula with acromion and coracoid process absent or rudimentary. Skin of throat plicated. Dorsal fin low. Pectoral limb tetradactylous, very long and narrow, attaining about one-fourth of the length of the entire animal, the metacarpus and phalanges being greatly developed, and the latter very numerous. Tympanic still more inflated than in *Balænoptera*, with the involuted portion more distinctly pyriform, the Eustachian part of the aperture well defined, and two well-marked longitudinal ridges on the lower surface of adult specimens.

The Whale commonly called "Humpback" (*Megaptera boops*) by whalers, perhaps on account of the low hump-like form of the dorsal fin, is very distinctly characterised from all others of the group, especially by the immense length of the pectoral fins or flippers, which are indented or scalloped along their margins, and are, except at their base, of a white colour, nearly all the rest of the body being black. The baleen plates are of a deep black colour. Though common in the North Atlantic between Norway and Greenland, this Whale does not frequently appear on the coasts of the British Isles. One came ashore at Newcastle in 1839; another, a young one, was taken in the estuary of the Dee in 1863, and its skeleton is preserved in the Liverpool museum; and a nearly full-grown animal was captured in the mouth of the Tay in

[1] Gray, *Suppl. Cat. Seals and Whales in Brit. Mus.* p. 39 (1871).

[2] Cope, *Proc. Ac. Nat. Sci. Philad.* 1869, p. 15.

[3] Gray, *Zoology of Erebus and Terror*, p. 16 (1846).

the winter of 1883-84.[1] The usual length of the adult ranges from 45 to 50 feet, the female being larger than the male. Whales of

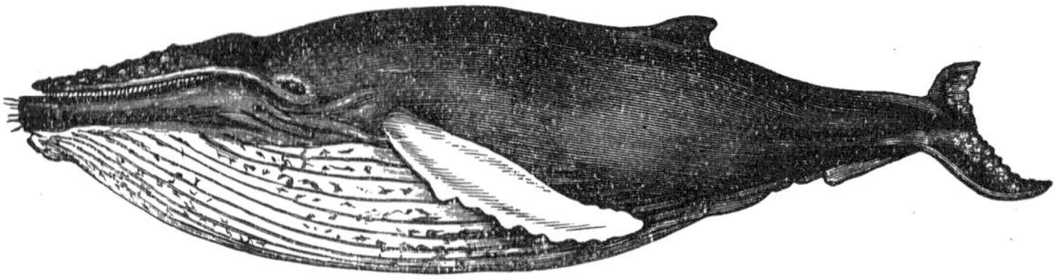

FIG. 79.—Humpbacked Whale (*Megaptera boops*).

the genus *Megaptera* are found in the South Atlantic and in both the North and the South Pacific. They resemble those of British seas so closely that it is doubtful whether the differences which have been observed, and upon which several species have been founded, may not be individual peculiarities; but zoologists have not yet had the opportunity of examining and comparing such a series of specimens of different ages and sexes from different localities as would be necessary to determine these points satisfactorily.

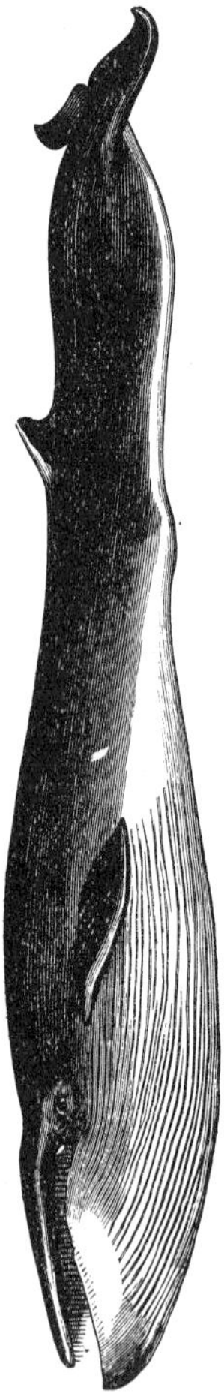

FIG. 80.—The Common Rorqual (*Balænoptera musculus*).

Tympanic bones of *Megaptera* occur in the English and Belgian Crags, although somewhat less commonly than those of *Balæna* and *Balænoptera;* they have been described under the names of *Megapteropsis* and *Burtinopsis.*

Balænoptera.[2]—Head small and flat, and pointed in front. Body long and slender. Skin of throat plicated. A small falcate dorsal fin. Baleen short and coarse. Cervical vertebræ free. Scapula low and broad, with a large acromion and coracoid process. Pectoral limb tetradactylous, small, narrow, and pointed. Tympanic (Fig. 81) long, much inflated, and rounded, with the involuted portion thickened and pyriform, and the notch for the Eustachian canal sharply defined; inner surface flattened, without the vertical groove found in *Megaptera.*

The Rorquals, Fin-Whales, Fin-backs, Finners, or Razor-backs, as they are variously called,

[1] See J. Struthers, "On the Anatomy of *Megaptera longimana,*" *Journ. Anatomy and Physiology*, 1887-89.

[2] Lacépède, "Table des Ordres," *Hist. Nat. des Cétacés*, p. xxxvi. (1804).

have the plicated skin of the throat like that of *Megaptera*, the furrows being more numerous and close set; but the pectoral fin is comparatively small, the dorsal fin distinct and falcate, and the tail very much compressed before it expands into the "flukes." The Rorquals are perhaps the most abundant and widely distributed of all the whales, being found in some of their modifications in all seas, except the extreme Arctic, and probably Antarctic regions. Owing to the small quantity and inferior quality of their whalebone, the comparatively limited amount of blubber, and their great activity and the difficulty of capturing them by the old methods, these Whales were not until recently an object of pursuit by whale-fishers; but, since the introduction of steam-vessels, and especially of explosive harpoons fired from guns in the place of those hurled by the human hand, a regular fishery has been established on the coast of Finmark. There are four distinct species of this genus in British seas. (1) *Balænoptera sibbaldi*, the "Blue Whale," the largest of all known animals, attains a length of 80 or even sometimes 85 feet. Its colour is dark bluish gray, with small whitish spots on the breast; the baleen is black; the flippers are larger proportionally than in other Rorquals, measuring one-seventh of the total length of the body; and the dorsal fin is small and placed very far back. This Whale has usually 64 vertebræ, of which 16 bear ribs. Like the others of the genus, this species seems to pass the winter in the open seas, and approaches the coast of Norway at the end of April or beginning of May. At this time its sole food is a small crustacean (*Euphausia inermis*) which

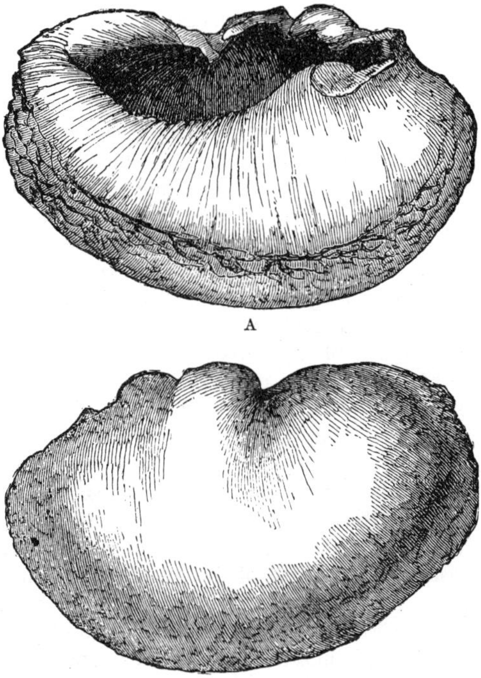

A

B

FIG. 81.—The right tympanic of *Balænoptera musculus* from the inner (A) and outer (B) aspects. ½ natural size. (From the *Proc. Zool. Soc.*)

swarms in the fjords. Several specimens have been taken on the British coasts, two fine skeletons from the Firth of Forth being preserved in the Edinburgh museums. (2) *Balænoptera musculus*, the Common Rorqual, has a length of 65 to 70 feet, is of a grayish slate colour above and white underneath, and the baleen is slate colour variegated with yellow or brown. It has usually 62 vertebræ, of which 15 bear ribs. This is the commonest of all the large Whales on the British coasts, scarcely a winter passing without the body of one being somewhere washed ashore, usually after stormy weather, and more frequently on the south coast, as this species has a more southern range than the last, and frequently enters the Mediterranean. It feeds largely on fish, and is frequently seen feasting among shoals of herring. (3) *Balænoptera borealis*, often called Rudolphi's Whale from its first describer, is a smaller species, scarcely attaining a length of 50 feet. It is bluish-black above, with oblong, light-coloured spots, whilst the under parts are more or less white; the whole of the tail and both sides of the flippers are black; the baleen is black, and the bristly ends fine, curling, and white; the flippers are very small, measuring one-eleventh of the total length of the body. There are 56 vertebræ, with 14 pairs of ribs. This species, according to Collett, feeds chiefly on minute crustaceans, mainly *Calanus finmarchicus* and *Euphausia inermis*, and not on fish. Until lately it was considered the rarest of the Whales of European seas, and was only known to science from a few individuals stranded on the coasts of northern Europe at long intervals, the skeletons of which have been preserved in museums. The most southern point at which it has been met with hitherto is Biarritz in France. Since the establishment of the whaling station near the North Cape it has been shown to be a regular summer visitor, and in 1885, 771 individuals were captured on the coast of Finmark. (4) *Balænoptera rostrata*, the lesser Fin-Whale or Rorqual, is the smallest species found in the northern seas, rarely exceeding 30 feet in length. Its colour is grayish-black above, whilst the under side is white, including the whole of the lower side of the tail; the inner side of the flippers is white; and there is a broad white band across the outer side, which is a very characteristic mark of the species; the baleen is yellowish-white. The dorsal fin in this and the last species is comparatively high, and placed far forwards on the body. This Whale has usually 48 vertebræ, 11 of which bear ribs. It is common in summer in the fjords of Norway, and is often seen around the British Isles. It has been taken, though rarely, in the Mediterranean; and ranges as far north as Davis's Straits.

Rorquals are met with in almost all seas throughout the world, but further and more accurate observations are required before their specific characters and geographical distribution can be made out. Nearly all the individuals hitherto examined with any care, whether from the North Pacific, the Australian seas, or the Indian Ocean, come very near in structure to one or the other of the Atlantic forms described above, so much so that some zoologists have been induced to believe that there are but four species, each of which has a wide, almost cosmopolitan range, while others have described and named almost every individual specimen captured as belonging to a different species.[1]

Tympanics, vertebræ, and other bones of Rorquals are among the commonest cetacean remains found in the Pliocene Crags of England and Belgium. Several species, varying in dimensions, are known from these deposits, *B. definita* (*sibbaldina*) being apparently nearly related to the existing *B. sibbaldi*. A caudal vertebra from the Upper Eocene of Hampshire has been referred to *Balænoptera*, but does not afford sufficient evidence to prove the existence of the genus at that date.

Extinct Genera.—The extinct genus *Cetotherium* of the European Pliocene may be taken to include a number of fossil Whalebone Whales allied to the Balænopterine group, several of which have been described under other names, such as *Plesiocetus*, *Heterocetus*, and *Amphicetus*. They are readily characterised by the form of the tympanic bone, which is much narrower in front than behind, the roughened inferior surface being in the shape of an isosceles triangle, and the notch for the Eustachian canal being smaller, and descending nearer to the inferior border of the inner wall than in *Balænoptera*. The skull is longer than the latter, with a greater interval between the occiput and the frontal, and with longer and more flattened nasals. The relative thickness of the cervical vertebræ is also greater. In the typical forms (*e.g. C. brialmonti* and *C. dubium*) the mandibular condyle is simple; but in *C.* (*Heterocetus*) *brevifrons* it is furnished with a projecting posterior talon, as in the Sperm Whale.

Herpetocetus is known by a comparatively small species from the Belgian and English Crags, characterised by the extreme inflation of the egg-shaped tympanic bone, which approximates to that of *Megaptera*, but has the greater part of the cavity filled by bone. There is a talon to the condyle of the mandible.

Palæocetus, as already mentioned (p. 232), is founded upon the ankylosed cervical vertebræ of a small Whale originally considered as having been derived from the Kimeridge Clay, but which doubtless came from the Suffolk Crag; if it belongs to the *Balænidæ* it indicates a Right Whale.

[1] See P. J. Van Beneden, "Histoire Naturelles des Balénoptères," *Mém. Acad. Belgique*, xli. 1887.

Suborder ARCHÆOCETI.

Family ZEUGLODONTIDÆ.

This group is formed to include certain extinct Cetacean-like animals at present only known by more or less fragmentary portions of their skeleton and teeth, and whose position and affinities are, therefore, still subject to doubt.[1]

In the anterior part of both jaws the teeth are simple, conical, or slightly compressed, and sharp pointed. The first three in the upper jaw are distinctly implanted in the premaxillary bone, and so may be reckoned as incisors. The tooth which succeeds, or the canine, is also simple and conical, but it does not exceed the others in size. This is followed by five teeth having two distinct roots and compressed pointed crowns, with denticulated cutting-edges. The dentition is therefore $i\ \frac{3}{3}$, $c\ \frac{1}{1}$, p and $m\ \frac{5}{5} = 36$, resembling that of some Seals.[2] General form of the skull elongated and much depressed. Brain-cavity very small, and the skull between it and the orbits elongated and narrow. Temporal fossæ very large. A strong sagittal crest. Rostrum long and narrow, differing from that of other Cetaceans in the large extent to which the premaxillæ form the sides of the anterior extremity. Nasal bones elongated, flat, and narrow, the opening of the anterior nares being over the middle of the elongated compressed rostrum. All the cervical vertebræ free. The characters of the dorsal vertebræ and mode of articulation of the ribs appear to have resembled those of *Platanista* rather than *Balæna*, *Physeter*, or *Delphinus*. Lumbar vertebræ with elongated bodies, low neural spines, and the transverse processes placed low down on the bodies. Characters of the limbs not known with certainty.[3]

All the known fossil remains belonging to the animals of this group may be referred, provisionally at least, to the genus *Zeuglodon*, so named because the first section of a molar tooth examined was taken from the base of the crown, where it was beginning to divide into the two roots, and looked like two single teeth "linked or

[1] In a recent memoir Professor D'Arcy Thompson has brought forward some arguments to show that the Zeuglodonts have no direct affinities with the Cetacea, but have on the other hand the strongest possible relation with the Pinnipede Carnivora. "On the Systematic position of Zeuglodon," *Studies from the Museum of Zoology, Dundee*, vol. i. No. 9, 1890.

[2] An appearance in one specimen has been described by C. G. Carus as indicating a vertical succession of the teeth, but the evidence upon which this rests is by no means satisfactory, and appears to admit of another explanation.

[3] A mutilated humerus of *Zeuglodon cetoides* has given rise to many conjectures, appearing to some anatomists to indicate seal-like freedom of motion at the elbow-joint, while to others its characters appear to be truly Cetacean.

yoked together." This name was substituted by Owen for the earlier one *Basilosaurus* of Harlan, with the consent of that author, on the mammalian nature of the animal being demonstrated.[1] The latter name is, however, still generally retained by American zoologists. The remains have hitherto been found chiefly in the Eocene formations of the States of Alabama, Louisiana, Mississippi, and Arkansas, and have been assigned to several species. A portion of a skull is recorded from the Barton Clay (Eocene) of Hampshire, England.

Suborder ODONTOCETI,

the DELPHINOIDEA, *or Toothed Whales.*

Calcified teeth always present after birth; generally numerous, but sometimes a very limited number (in a few cases none) are functionally developed. No baleen. Upper surface of the skull more or less asymmetrical. Nasal bones in the form of nodules or flattened plates, applied closely to the frontals, and not forming any part of the roof to the narial passage, which is directed upwards and backwards. Olfactory organ rudimentary or absent. Hinder end of the maxilla expanded and covering the greater part of the orbital plate of the frontal bone. Lachrymal bone either inseparable from the jugal, or, when distinct, very large, and forming part of the roof of the orbit. Tympanic bone not ankylosed with the periotic, which is usually only attached to the rest of the skull by ligament. Rami of mandible nearly straight, much expanded in height posteriorly, with a wide funnel-shaped aperture to the dental canal, and coming in contact in front by a flat surface of variable length, but always constituting a true symphysis. Several of the anterior ribs with well-developed capitular processes, articulating with the bodies of the vertebræ. Sternum almost always composed of several pieces, placed one behind the other, with which several pairs of ribs are always connected by the intervention of well-developed cartilaginous or ossified sternal ribs. External respiratory aperture single, the two nostrils uniting before they reach the surface, usually in the form of a transverse subcrescentic valvular aperture, situated on the top of the head. Manus always pentadactylous, though the first and fifth digits are usually very little developed. No cæcum, except in *Platanista*.

Family PHYSETERIDÆ.

No functional teeth in the upper jaw. Mandibular teeth various, often much reduced in number. Bones of the cranium raised so as

[1] See *Trans. Geol. Soc.* ser. 2, vol. vi. p. 67.

to form an elevated prominence or crest behind the nares. Pterygoid bones thick, produced backwards, meeting in the middle line, and not involuted to form the outer wall of the post-palatine air-sinuses, but simply hollowed on their outer side. Anterior facet of periotic bone (Fig. 87) for articulation with the tympanic quite smooth; and the posterior tympanic surface of the former broad, with a median longitudinal ridge. Transverse processes of the arches of the dorsal vertebræ, to which the tubercles of the ribs are attached, ceasing abruptly near the end of the series, and replaced by processes on the body at a much lower level, and not on a line or serially homologous with them, but serially homologous anteriorly with the heads of the ribs, and posteriorly with the transverse processes of the lumbar vertebræ. (In some genera, as *Physeter*, the two processes, upper and lower on each side, are both present and well developed in the same vertebra in the region of transition. In others, as *Ziphius* and *Berardius*, they are not both developed on any single vertebra.) Costal cartilages not ossified.

Subfamily **Physeterinæ.**—Numerous teeth in the mandible, which are not set in distinct bony alveoli, but in a long groove imperfectly divided by partial septa, and held in place by the strong, fibrous gum surrounding them. No distinct lachrymal bone. Cranium strikingly asymmetrical in the region of the narial apertures, in consequence of the left opening greatly exceeding the right in size.

Physeter.[1]—Upper teeth apparently of uncertain number, rudimentary, and functionless, being embedded in the gum. Lower jaw

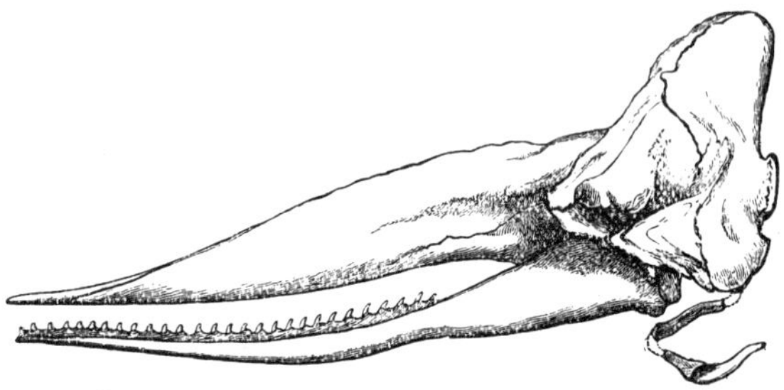

FIG. 82.—Skull of Sperm Whale (*Physeter macrocephalus*).

with from 20 to 25 teeth on each side, stout, conical, recurved, and pointed at the apex until they are worn, without enamel. Upper surface of the cranium concave; its posterior and lateral edges raised into a very high and greatly compressed semicircular crest or wall. Zygomatic processes of jugal bones thick and massive. Rostrum greatly elongated, broad at the base, and gradually tapering

[1] Linn. *Syst. Nat.* 12th ed. vol. i. p. 107 (1766).

to the apex. Upper edge of the mesethmoid forming a roughened irregular. projection between the narial apertures, inclining to the left side. Mandible exceedingly long and narrow, the symphysis being more than half the length of the ramus. Vertebræ: C 7, D 11, L 8, C 24; total 50. Atlas free; all the other cervical vertebræ united by their bodies and spines into a single mass. Eleventh pair of ribs rudimentary. Head about one-third the length of the body; very massive, high and truncated, and rather compressed in front; owing its huge size and remarkable form mainly to the accumulation of an oily substance secreted by the lining membranes of great cells surrounding the narial passage and filling the large hollow on the upper surface of the cranium and overlying the rostrum. The single blowhole is longitudinal, slightly sigmoid, and placed at the upper and anterior extremity of the head to the left side of the middle line. The opening of the mouth is on the under side of the head, considerably behind the end of the snout. Pectoral fin short, broad, and obliquely truncated. Dorsal fin a mere low protuberance.

The only representative of this genus is the Cachalot or Sperm

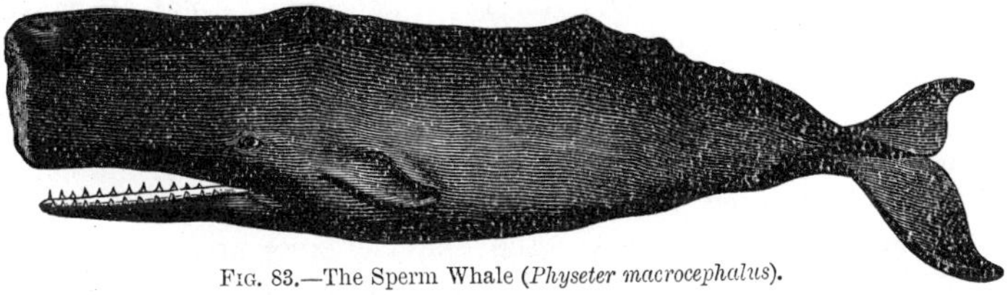

FIG. 83.—The Sperm Whale (*Physeter macrocephalus*).

Whale (*P. macrocephalus*, Fig. 83), one of the most colossal of animals, quite equalling, if not exceeding, the Greenland Whale in bulk. The length of the full-grown male is from 55 to 60 feet, but the female is stated not to reach more than half that size. The general colour of the surface is black above and gray below, the colours gradually shading into each other. The Sperm Whale is one of the most widely distributed of animals, being met with usually in herds or "schools" in almost all tropical and subtropical seas, but not occurring, except accidentally, in the Polar regions. Not unfrequently specimens appear on the coasts of the British Isles, but only as solitary stragglers, or as dead carcases, floated northwards by the Gulf Stream. It is remarkable that every one of these of which we have an accurate record has been an old male. The food of this Whale consists mainly of various species of cephalopods (squid and cuttle-fish), but fish of considerable size are also eaten. The substance called "ambergris," formerly used in medicine and now in perfumery, is a concretion formed in the intestine of this Whale, and is found floating on the surface of the seas it inhabits. Its genuineness is proved by the presence of the horny beaks of the cephalopods on which the Whale feeds.

The oil contained in the great cavity above the skull, when refined, yields "spermaceti," and the thick covering of blubber which everywhere envelops the body produces the valuable "sperm-oil" of commerce; hence this animal has long been the subject of a regular chase, by which its numbers have been greatly diminished.

Cogia.[1]—Teeth of the upper jaw absent, or reduced to a rudimentary pair in front; in the lower jaw 9 to 12 on each side, rather long, slender, pointed, and curved, with a coating of enamel. Upper surface of the cranium concave, with thick, raised posterior and lateral margins, massive and rounded at their anterior terminations above the orbits. Upper edge of the mesethmoid forming a prominent sinuous ridge, constituting a kind of longitudinal septum to the base of the great supra-cranial cavity. Rostrum not longer than the cranial portion of the skull, broad at the base, and rapidly tapering to the apex. Zygomatic process of the jugal styliform. Mandible with the symphysis less than half the length of the entire ramus. Vertebræ: C 7, D 13 or 14, L and C 30; total 50 or 51. All the cervical vertebræ united by their bodies and arches. External characters not well known, but, judging by the somewhat conflicting accounts of those that have had an opportunity of observing them, the head is about one-sixth of the length of the body, and obtusely pointed in front; the mouth small, and placed far below the apex of the snout; the spiracle crescentic, and placed obliquely on the top of the head anteriorly to the eyes, and to the left of the middle line; the pectoral fins are obtusely falcate; and there is a triangular dorsal fin.

The history of this genus is a good illustration of the difficulties in which the study of the Cetacea has been involved by the superficial manner in which it has been investigated. The first known example, a skull from the Cape of Good Hope in the Paris Museum, was described by De Blainville under the name of *Physeter breviceps.* This was afterwards with good reason generically separated by Gray. Until within a very few years ago only five other individuals had been met with, each of which had been described under a different specific name (viz. *grayi*, *macleayi*, *simus*, *floweri*, and *potsii*), and which are arranged by Gray in two distinct genera. The most careful examination of the description given of these specimens, or of the now numerous osteological remains available, fails to detect any differences beyond those which may be attributed to age or sex, and hence, according to our present knowledge, these six supposed species must all be included under one name, *C. breviceps*. This animal appears to attain the length of 10 feet when adult, and has

[1] Gray, *Zoology of Erebus and Terror*, p. 22 (1846). Usually spelt *Kogia*.

been met with at various distant localities in the Southern Ocean, and also off the coast of Madras and in the North Pacific.

Extinct Physeteroids.—Teeth of Physeteroids are of very common occurrence in the Belgian and English Crags, and evidently indicate the former existence of Whales more or less closely allied to the Sperm Whale, but often distinguished by the presence of an enamel-cap on the crowns of the teeth. The generic determination of these teeth is, however, exceedingly difficult, owing to the water-worn condition in which they are frequently found, and also on account of the impossibility of knowing whether small and large teeth may not be referable to different parts of the jaws of the same species or to individuals of different ages. Moreover, in the cases of isolated teeth it is impossible to know how many were contained in the jaws, and therefore to distinguish Physeteroid from Ziphioid teeth. *Physeterula* is a small form about one-third the dimensions of the Sperm Whale, and distinguished by the length of the mandibular symphysis being only about one-third that of the entire ramus; it is identified by Professor Cope with *Cogia*. *Eucetus* (*Dinoziphius*) is founded on teeth which are regarded as closely resembling those of *Physeter*, but distinguished by their subcylindrical form and the small size of the aperture of the pulp-cavity. It does not appear, however, to be certain that these teeth are not worn specimens of those described as *Scaldicetus*. *Physetodon*, from the Pliocene of Australia, is founded upon the evidence of similar teeth. The teeth from the Belgian Crag described as *Scaldicetus* are somewhat smaller than those of the Sperm Whale, and are readily characterised by their cap of grooved enamel. Other teeth with enamel-caps have been described as *Physodon* and *Hoplocetus*. The genus *Balænodon* is founded upon a very imperfect large tooth from the English Crag, which is not sufficiently well preserved to admit of exact comparison with the other types.

Subfamily **Ziphiinæ.**—Teeth of the mandible (at least in existing forms) quite rudimentary and concealed in the gum, except one, or very rarely two, pairs which may be largely developed, especially in the male sex. A distinct lachrymal bone. Externally the mouth is produced into a slender rostrum or beak, from above which the rounded eminence formed by a cushion of fat resting on the cranium in front of the blowhole rises somewhat abruptly. Spiracle or blowhole single, crescentic, median, as in the *Delphinidæ*. Pectoral fin small, ovate, the five digits all moderately well developed. A small obtusely falcate dorsal fin situated considerably behind the middle of the back. Longitudinal grooves on each side of the skin of the throat, diverging posteriorly, and nearly meeting in front. In external characters and habits the animals of this group closely resemble each other. They appear to be almost exclusively feeders on various species of cephalopods, and occur either singly, in pairs, or in small herds. By their dental and osteological characters they are easily separated into four distinct genera.

Hyperoödon.[1]—A small conical pointed tooth at the apex of each ramus of the mandible, concealed by the gum during life. Skull with the upper ends of the premaxillæ rising suddenly behind the nares to the vertex and expanded laterally, their outer edges curving backwards and their anterior surfaces arching forwards and overhanging the nares; the right larger than the left. Nasal bones lying in the hollow between the upper extremities of the premaxillæ, strongly concave in the middle line and in front; their outer edges, especially on the right side, expanded over the front of the inner border of the maxilla. Very high longitudinal crests on the maxillæ at the base of the rostrum, extending backwards almost to the nares, approaching each other in the middle line above; sometimes so massive that their inner edges come almost in contact. Anteorbital notch distinct. Mesethmoid but slightly ossified. Vertebræ: C 7, D 9, L 10, C 19; total 45. All the cervical vertebræ united. Upper surface of the head in front of the blow-

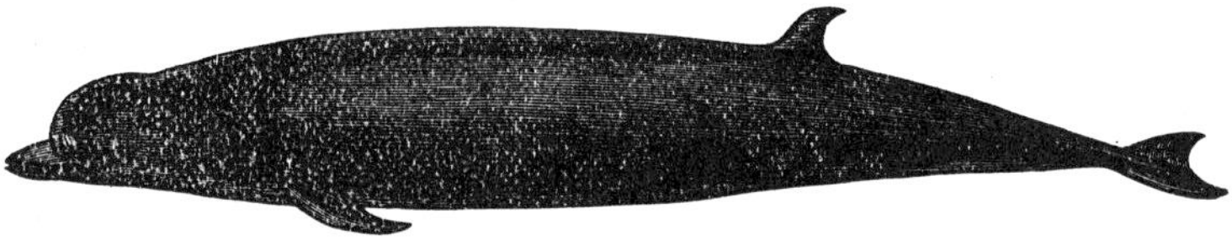

FIG. 84.—*Hyperoödon rostratus.* From a female specimen taken off the coast of Scotland, 1882.

hole very prominent and rounded, rising abruptly from above the small, distinct snout.

The genus is known typically by *H. rostratus* (Fig. 84), but an imperfect skull has been made the type of *H. planifrons*—a species differing considerably in cranial characters from the typical one. The females and young males of the first-named species have the contour of the head of the same general form as in Fig. 84; the premaxillary crests of the cranium being widely separated from one another, and terminating in comparatively sharp edges. In the males, however, as age advances the summits of these crests become gradually expanded and flattened, till they are almost or quite in contact in the middle line. This development of the maxillary crests produces a corresponding elevation and flattening of the front of the head, so that in very old males this aspect presents a flattened disc-like surface rising abruptly from the beak (which thus becomes almost buried) and situated in a plane nearly at right angles to the line of the back.[2] So different, indeed, is the appearance of the skull of an old male from that of a female individual that

[1] Lacépède, "Table des Ordres," *Hist. Nat. des Cétacés,* p. xliv. (1804).

[2] See the figures in the *Proc. Zool. Soc.* 1882, pp. 728, 729.

it was long considered that they belonged to different species—the male form having been described as *H. latifrons*. The length of an adult male reaches 30 feet, while that of the female does not exceed 24 feet.

The Hyperoödon, sometimes called "Bottlenose," a name also vaguely given to several species of Dolphin, is a regular inhabitant of the North Atlantic, passing the summer in the Spitzbergen seas and going farther south in winter. It resembles the Sperm Whale in possessing a large store of oil in the upper part of the head, which yields spermaceti when refined; on this account, and also for the sake of the blubber, which supplies an oil almost indistinguishable from sperm-oil, this Whale has been the object of a regular chase in recent years.

The following account of its habits is taken from a paper by Captain D. Gray, published in the *Zoological Society's Proceedings* for 1882:—

"These Whales are occasionally met with immediately after leaving the Shetland Isles in March, and north across the ocean until the ice is reached, near the margin of which they are found in the greatest numbers; but they are seldom seen amongst it. Although it is not in their nature to keep in amongst the ice, they like to frequent the open bays for the shelter it gives them from the sea. Sometimes a point of ice overlaps them; it is then only that they are seen going out again towards the ocean. They are also to be met with from the entrance of Hudson's Straits and up Davis's Straits, as far as 70° N. lat., and down the east side round Cape Farewell, all round Iceland, north along the Greenland ice to 77° N. lat.; also along the west coast of Spitzbergen, and east to Cherry Island in lat. 72° N. and long. 19° E. Beyond these limits I have never seen them; but doubtless they are to be found as far as the Straits of Belle Isle on the west, and east to Nova Zembla. From the fact that they are not seen in summer farther south than a day's sail from the ice, it would appear that they migrate south in the autumn, and north again in the spring. They are gregarious in their habits, going in herds of from four to ten. It is rare to see more than the latter number together, although many different herds are frequently in sight at the same time. The adult males very often go by themselves; but young bulls, cows, and calves, with an old male as a leader, are sometimes seen together. They are very unsuspicious, coming close alongside the ship, round about underneath the boats, until their curiosity is satisfied. . . . They vary in colour from black in the young to light brown in the older animals. The very old turn almost yellow, the beak and front of the head being quite white, with a white band round their necks; all of them are grayish-white on the belly. They can leap many feet out of the water, even having time while in the air to turn round their heads and look about them, taking the water head first, and not falling helplessly into it sideways like the larger whales. The full-grown whale is 30 feet long by 20 feet in circumference, and yields two tons of oil besides two hundredweight of spermaceti. . . . Their ordinary food consists of a bluish-white cuttle-fish, six inches long by three inches in circumference, and pointed towards the tail. . . . They evidently have a great depth to go to find them, judging from the length of time that they remain away, and from the long heavy blasts they make on coming to the surface again."

Periotic bones of *Hyperoödon* are found in the Red Crag of Suffolk, presenting no character by which they can be specifically distinguished from those of the common existing species.

Ziphius.[1]—A single conical tooth of moderate size on each side of the mandible close to the anterior extremity, and directed forwards and upwards. Skull with the premaxillæ immediately in front of, and at the sides of the nares expanded, hollowed, and with elevated lateral margins, the posterior ends rising to the vertex and curving forwards, the right being considerably more developed than the left; the conjoint nasals forming a strongly pronounced symmetrical eminence at the top of the cranium, projecting forwards over the nares, flat above, most prominent and rounded in the middle line in front, and separated by a notch on each side from the premaxillæ. Anteorbital notch not distinct. Rostrum (seen from above) triangular, gradually tapering from the base to the apex; upper and outer edges of maxillæ at base of rostrum raised into low roughened tuberosities. Mesethmoid cartilage densely ossified in adult age, and coalescing with the surrounding bones of the rostrum. Vertebræ: C 7, D 10, L 10, C 22; total 49. The three anterior cervical vertebræ united, the rest free.

The type of this genus is *Z. cavirostris* of Cuvier, founded upon an imperfect skull picked up in 1804 on the Mediterranean coast of France, and described and figured in the *Ossemens Fossiles* under the impression that it was that of an extinct species. Many other individuals have, however, been subsequently met with in various parts of the world, from the Shetland Islands to New Zealand, all referable to the same genus, if not to the same species; although, as is usual in such cases, they have mostly been described under different names, the so-called genera *Petrorhynchus* and *Epiodon* being probably referable to the type species.

It is quite probable that some of the Physeteroid teeth from the Crag deposits mentioned on p. 251 may be referable to *Ziphius*.

Mesoplodon.[2]—A much compressed and pointed tooth in each

[1] Cuvier, *Ossemens Fossiles*, 2d ed. vol. v. p. 352 (1823).

[2] Gervais, *Ann. Sci. Nat.* sér. 3, vol. xiv. p. 16 (1850). For the very complicated synonymy of this genus, see *Trans. Zool. Soc.* vol. viii. p. 208.

ramus of the mandible, variously situated, but generally at some distance behind the apex (Fig. 86); its point directed upwards, and often somewhat backwards, occasionally developed to a great size.

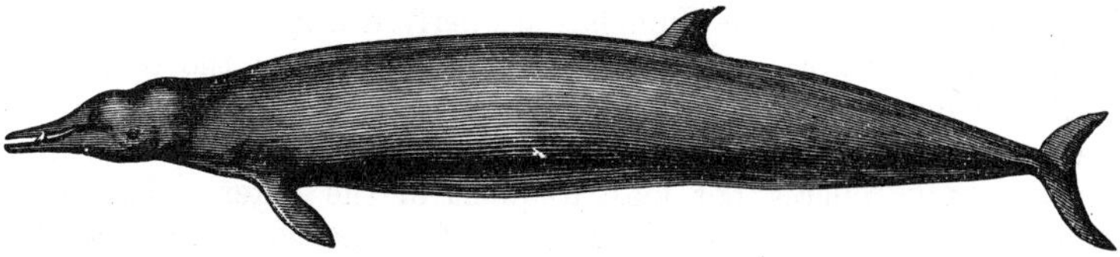

FIG. 85.—*Mesoplodon bidens.* From Reinhardt.

Skull with the region around the nares as in *Hyperoödon*, except that the nasals are narrow and more sunk between the upper ends of the premaxillæ; like those of *Hyperoödon*, they are concave in the middle line in front and above. No maxillary tuberosities. Anteorbital notch not very distinct. Rostrum long, narrow, and solid throughout. Mesethmoid in adult age ossified in its entire

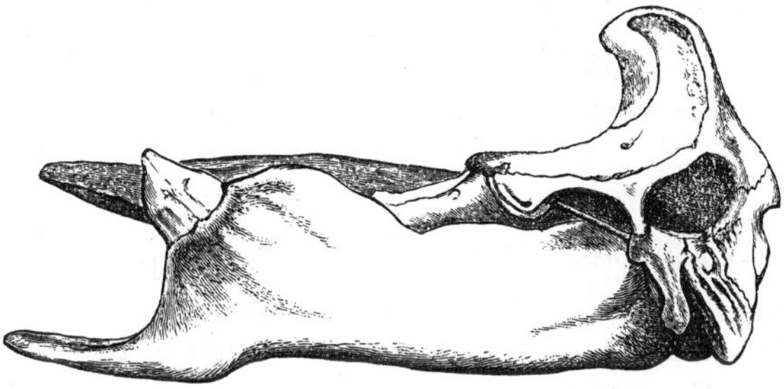

FIG. 86.—Left lateral view of skull of *Mesoplodon densirostris.*

length, coalescing with the surrounding bones, and showing as a narrow band on the upper surface of the rostrum. Vertebræ: C 7, D 10, L 10 or 11, C 19 or 20; total 46 to 48. Two or three anterior cervicals united, the rest usually free.

Though varying in form, the mandibular teeth of the different members of this genus agree in their essential structure, having a small and pointed enamel-covered crown, composed of true dentine, which, instead of surmounting a root of the ordinary character, is raised upon a solid mass of osteodentine. The continuous growth of this greatly alters the form and general appearance of the organ as age advances, as seen most strikingly in the case of *M. layardi*, where the long, narrow, flat, strap-like teeth, curving inwards at their extremities, actually meet over the rostrum, and must greatly interfere with the movements of the jaw. In one species (*M. grayi*) a row of minute, conical, pointed teeth, like those of ordinary Dolphins, 17 to 19 in number, are present even in the adults, on each side of the middle part of the upper jaw, but embedded by their roots only in the gum, and not in bony alveoli. This fact, with the frequent presence of rudimentary teeth in other species of this and the last genus in both upper and lower jaws, suggests the idea that the Ziphioids are derived from ancestral forms which had teeth of normal character in both jaws; the dentition of the living forms having become greatly specialised. The existing species of this genus are widely distributed in both northern and southern hemispheres, but most frequent in the latter. The best established are *M. bidens*, *M. europæus*, *M. densirostris*, *M. layardi*, *M. grayi*, and *M. hectori*; but there is still much to be learned with regard to their distinctive characters and geographical distribution. They were abundant in the Pliocene age, as attested by the frequency with which the most imperishable and easily recognised portion of their structure, the long, cylindrical rostrum of the skull, of more than ivory denseness, is found among the rolled and water-worn fragments of animal remains which compose the well-known "bone-bed" at the base of the Red Crag of Suffolk. Several species have been founded upon the evidence of these rostra. Periotic bones of this genus (Fig. 87) are of less common occurrence in the Crag; the figure is given to illustrate the characteristic features of this bone in the present family.

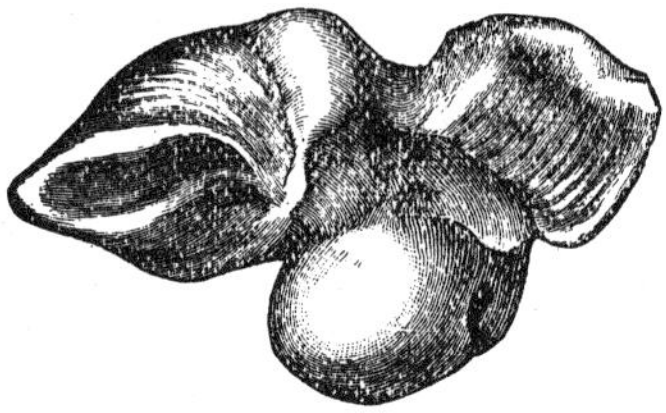

FIG. 87.—The left periotic bone of *Mesoplodon*; from the Red Crag of Suffolk. The smooth concave surface in the right upper corner of the figure forms the anterior articulation with the tympanic. (From the *Cat. Foss. Mamm. Brit. Mus.* pt. v. p. 70.)

Berardius.[1]—Two moderate-sized, compressed, pointed teeth on each side of the symphysis of the mandible, with their apices directed forwards, the anterior being the larger of the two and close to the apex. Upper ends of the premaxillæ nearly symmetrical, moderately elevated, very slightly expanded, and not curved forward over the nares. Nasals broad, massive, and rounded, of nearly equal size, forming the vertex of the skull, flattened in front, most prominent in the middle line. Anteorbital notch distinct. Rostrum long and narrow. Mesethmoid only partially ossified. Small rugous eminences on the outer edge of the upper surface of the maxillæ at base of rostrum. Vertebræ: C 7, D 10, L 12, C 19; total 48. The three anterior cervicals ankylosed, the rest free and well developed.

The only known species, *B. arnuxi*, attains the length of 30 feet, and has hitherto only been met with in the seas around New Zealand.

[1] Duvernoy, *Ann. Sci. Nat.-Zoologie*, sér. 3, vol. xv. p. 41 (1851).

Choneziphius.[1]—The rostral portions of crania from the Antwerp and Suffolk Crags, on the evidence of which this genus has been established, agree with those of *Mesoplodon* in having the premaxillæ in contact with the intervening bones throughout the length of their inner surfaces, and also in showing only a very small portion of the vomer on the inferior surface; they differ, however, in that the mesethmoid cartilage remains unossified, whereby a fistular vacuity remains. In some species the soldering of the inner surfaces of the premaxillæ is incomplete. The interorbital region of the skull is flat; and there are two pits in the nasal region, of which the right is the larger.

Family SQUALODONTIDÆ.

Numerous extinct forms, chiefly known by teeth and fragments of crania, may be provisionally placed here, until more of their osteological characters shall be brought to light. They differ from all existing Cetaceans in having the teeth distinctly differentiated into groups, as in the Archæoceti, the posterior molars being two-rooted. The cranium has, however, none of the distinguishing characteristics of the Zeuglodonts, but essentially resembles that of the Odontoceti, especially in the position of the anterior nares and form of the nasal bones.

Squalodon.[2]—All the forms may be included in this genus, the so-called *Rhizoprion* not being distinct. Dentition: $i\ \frac{3}{3}$, $c\ \frac{1}{1}$, simple teeth of the molar series (premolars ?) $\frac{4}{4}$, two-rooted molars $\frac{7}{7} = \frac{15}{15}$; total 60. The double-rooted molars differ from those of *Zeuglodon* in having the denticulations of the crown confined to the posterior border, or at all events much less developed on the front edge. Very little is known of the structure of these animals beyond the skull and teeth, fragments of which have been found widely distributed throughout the marine Miocene and early Pliocene formations of Europe, especially in the Vienna basin, many parts of France, and the Antwerp and Suffolk Crags. They have also been found in formations of corresponding age in North America and South Australia. A few isolated teeth have been met with in the cave-deposits of Italy, which, if contemporaneous with the beds in which they occur, indicate the survival of the genus into the Pleistocene period.

Family PLATANISTIDÆ.

Under this heading may be placed three very singular genera, which, though differing considerably from each other, have several

[1] Duvernoy, *op. cit.* p. 61.

[2] Grateloup, *Act. Ac. R. Sci. Bordeaux*, 1840, p. 208.

points in common, and do not altogether come under the definition either of the *Physeteridæ* or the *Delphinidæ*, especially in the important character of the mode of articulation of the ribs with the dorsal vertebræ, the tubercular and capitular articulations, distinct at the commencement of the series, gradually blending together, as they do in most ordinary mammals. The cervical vertebræ are all free. The lachrymal bone is not distinct from the jugal. The jaws are long and narrow, with numerous teeth in both. The symphysis of the mandible exceeds half the length of the whole ramus. Externally the head is divided from the body by a slightly constricted neck. Pectoral limbs broad and truncated. Dorsal fin small or obsolete. Fluviatile or estuarine in habits. There are three distinct genera, which might almost be made the types of families, but it is probably more convenient to keep them together, only regarding them as representing three subfamilies.

Platanista.[1]—Teeth about $\frac{30}{30}$ on each side, set near together, rather large, cylindrical, and sharp-pointed in the young; in old animals acquiring a large laterally compressed base, which in the posterior part of the series becomes irregularly divided into roots. As the conical enamel-covered crown wears away, the teeth of the young and old animals have a totally different appearance. The rostrum and dentigerous portion of the mandible are so narrow that the teeth of the two sides are almost in contact. Maxillæ supporting very large, incurved, compressed bony crests, which overarch the nares and base of the rostrum, and almost meet in the middle line above. Orbits very small and eyes rudimentary, without crystalline lens. External respiratory aperture longitudinal, linear. Vertebræ: C 7, D 10, L 9, C 26; total 52. A small cæcum. No pelvic bones. Dorsal fin represented by a low ridge.

One species, *P. gangetica*, entirely fluviatile, being extensively distributed throughout nearly the whole of the river systems, not

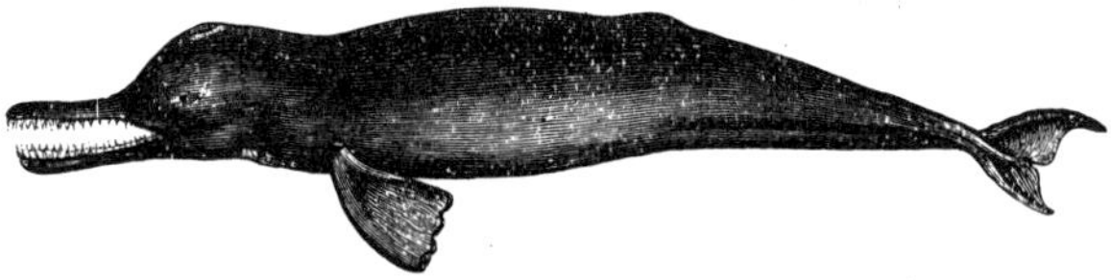

FIG. 88.—*Platanista gangetica.* (From Anderson.)

only of the Ganges, but of the Brahmaputra and Indus, ascending as high as there is water enough to swim in, but never passing out to sea. It is quite blind, and feeds on small fish and crustaceans, groping for them with its long snout in the muddy water at the bottom of the rivers. It attains the length of 8 feet.[2]

[1] Wagler, *Syst. Amphib.* etc., p. 35 (1830).

[2] The anatomy of *Platanista* is fully described by J. Anderson, *Zoological Results of Two Expeditions to Western Yunnan*, 1878.

Inia.[1]—Teeth variable, from 26 to 33 on either side of each jaw; those at the posterior part with a distinct tubercle at the inner side of the base of the crown. Vertebræ: C 7, D 13, L 3, C 18; total 41. Transverse processes of lumbar vertebræ very broad. Sternum short and broad, and consisting of a single segment only. Dorsal fin a mere ridge. The long cylindrical rostrum externally furnished with scattered, stout, and crisp hairs. One species only is known, *I. geoffroyensis*, about 7 feet in length, inhabiting the upper Amazon and its tributary streams.

Pontoporia.[2]—Teeth 50 to 60 on either side of each jaw, with a cingulum at the base of the crown. Jaws very long and slender. Vertebræ: C 7, D 10, L 5, C 19; total 41. Transverse processes of the lumbar vertebræ extremely broad. Sternum elongated, composed of two segments, with four sternal ribs attached. Dorsal fin rather small, triangular, pointed. External respiratory aperture

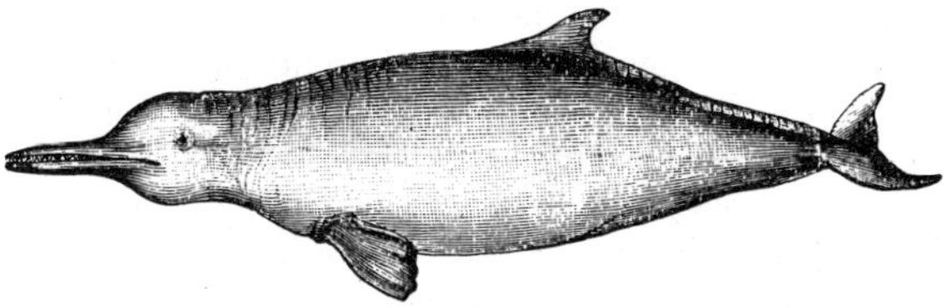

FIG. 89.—*Pontoporia blainvillei.* (From Burmeister.)

transverse, crescentic. This genus connects the last two forms with the true *Delphinidæ*. The only species, *P. blainvillei*, is one of the smallest members of the whole order, not exceeding 5 feet in length. It has only been met with at the mouth of the Rio de la Plata, near Buenos Ayres, and there is at present no evidence that it ascends into the fresh waters of the river.

Fossil forms.—Remains of a Cetacean from the Pleistocene of South America were referred by Bravard to *Pontoporia*, but they have been regarded by other writers as indicating a distinct genus, for which the names *Palæopontoporia* and *Pontistes* have been proposed. The Upper Tertiary European genera *Champsodelphis* and *Schizodelphis* are generally referred to the present family. The former has wide transverse processes to the lumbar vertebræ, as in *Inia*, while the teeth also resemble those of that genus. In *Schizodelphis* the form of the rostrum presents a great resemblance to that of the Delphinoid genus *Steno*, but the symphysis of the mandible is relatively longer. A number of fossil Cetaceans from the Miocene of the United States, such as *Priscodelphinus*, *Lophocetus*, *Ixacanthus*, *Rhabdosteus*, etc., are referred by Professor E. D. Cope to

[1] D'Orbigny, *Nouv. Ann. Mus. Paris*, vol. iii. p. 31 (1834).

[2] Gray, *Zoology of Erebus and Terror*, p. 46 (1846).

this family. *Agabelus*, from the same deposits, is an apparently allied, but toothless type.

Family DELPHINIDÆ.

Teeth usually numerous in both jaws. Pterygoid bones short, thin, each involuted to form with a process of the palate bone the outer wall of the post-palatine air-sinus. Symphysis of mandible short, or moderate, never exceeding one-third of the length of the ramus. Lachrymal bone not distinct from the jugal. The anterior facet on the periotic (Fig. 96) for articulation with the tympanic deeply grooved; and the posterior tympanic surface of the same bone comparatively narrow, with its ridge for articulation with the free border of the tympanic ill-defined, and situated close to one edge. Transverse processes of the dorsal vertebræ gradually transferred from the arches to the bodies of the vertebræ without any sudden break, and becoming posteriorly continuous serially with the transverse processes of the lumbar vertebræ. Anterior ribs attached to the transverse process by the tubercle, and to the body of the vertebra by the head; the latter attachment lost in the posterior ribs. Sternal ribs firmly ossified. External respiratory aperture transverse, crescentic, with the horns of the crescent pointing forwards.

A very large group, closely united in essential characters but presenting great modifications in details. The different types are mostly so connected by intermediate or osculant forms that there are great difficulties in grouping them into natural subfamilies. Even the formation of well-defined genera is by no means satisfactory in all cases. They may, however, be divided, perhaps artificially, into two groups.

Group A.—Head rounded, without distinct rostrum or beak. Rostrum of skull about as long as cranial portion.

Monodon.[1]—Besides some irregular rudimentary teeth, the entire dentition is reduced to a single pair of teeth which lie horizontally in the maxilla, and in the female remain permanently concealed within the alveolus, so that this sex is practically toothless, while in the male (see Fig. 90) the right tooth usually remains similarly concealed and abortive, and the left is immensely developed, attaining a length equal to more than half that of the entire animal, projecting horizontally from the head in the form of a cylindrical, or slightly tapering, pointed tusk, without enamel, and with the surface marked by spiral grooves and ridges, running in a sinistral direction. (When, as occasionally happens, both tusks are developed, the spiral grooves have the same direction in each.) Pterygoids very

[1] Linn. *Syst. Nat.* 12th ed. vol. i. p. 105 (1766).

small, not meeting in the middle line, but approximating posteriorly. Vertebræ: C 7, D 11, L 6, C 26; total 50. Cervical region comparatively long, and all the vertebræ distinct, or with irregular unions towards the middle of the series, the atlas and axis being usually free. Manus small, short, and broad; second and third digits nearly equal, fourth slightly shorter. No dorsal fin.

This genus is now represented only by the well-known Narwhal (*M. monoceros*), in which the horn-like tusk of the male often grows to a length of 7 or 8 feet. In very young animals several small additional teeth, irregular in number and position, are present, but these usually disappear soon after birth.

The head is rather short and rounded; the fore limbs or paddles are small and broad compared with those of most Dolphins; and (as in the Beluga) the median dorsal fin, found in nearly all other members of the group, is wanting or replaced by a low ridge. The general colour of the surface is dark gray above and white below, but variously marbled and spotted with different shades of gray. In the general contour of the body the Narwhal resembles the White Whale or Beluga.

The Narwhal is essentially an Arctic animal, frequenting the icy circumpolar seas, and but rarely seen south of 65° N. lat. Three instances have, however, been recorded of its occurrence on the British coasts, one in the Firth of Forth in 1648, one near Boston in Lincolnshire in 1800, while a third, which entangled itself among the rocks in the Sound of Weesdale, Shetland, in September 1808, is described by Fleming in the *Memoirs of the Wernerian Society*, vol. i. Like most other Cetaceans, it is gregarious in its habits, being usually met with in "schools" or herds of fifteen or twenty individuals. Its food appears to be various species of cephalopods, small fishes, and crustaceans. The purpose served in the animal's economy by the wonderfully developed asymmetrical tusk—or "horn," as it is commonly but erroneously called—is not known. As it is present only

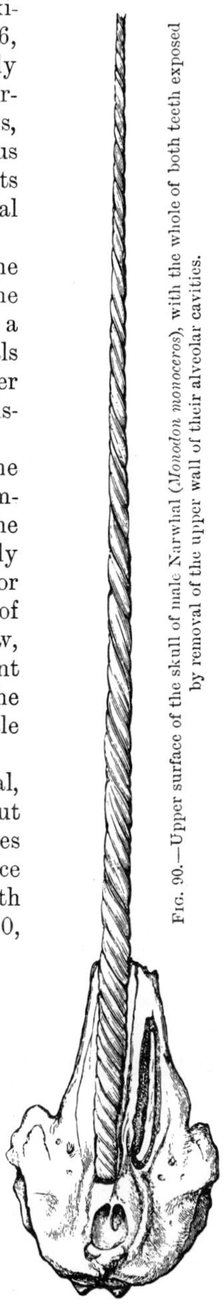

FIG. 90.—Upper surface of the skull of male Narwhal (*Monodon monoceros*), with the whole of both teeth exposed by removal of the upper wall of their alveolar cavities.

in the male sex, no function essential to the well-being of the individual, such as the procuring of sustenance, can be assigned to it, but it must be looked upon as belonging to the same category of organs as the antlers of deer, and perhaps may be applied to similar purposes. Very little is, however, known of the habits of Narwhals. Scoresby describes them as "extremely playful, frequently elevating their horns and crossing them with each other as in fencing." They have never been known to charge and pierce the bottom of ships with their weapons, as the swordfish often does. The name "Sea Unicorn," sometimes applied to the Narwhal, refers to the resemblance of its tusk to the horn represented as projecting from the forehead of the fabled unicorn. The ivory of which the tusk is composed is of very good quality, but, owing to the central cavity, which extends the greater part of its length, is only fitted for the manufacture of objects of small size. The entire tusks are sometimes used for decorative purposes, and are of considerable, though very fluctuating, commercial value.

Delphinapterus.[1]—This genus is closely allied to the last in external form, as well as anatomical structure, differing mainly in the very distinct character of the dentition. Teeth from $\frac{8}{8}$ to $\frac{10}{10}$, occupying the anterior three-fourths of the rostrum and corresponding portion of the mandible, rather small, conical, and pointed when unworn, but usually becoming obliquely truncated, separated by intervals considerably wider than the diameter of the tooth, and implanted obliquely, the crowns inclining forwards, especially in the upper jaw. Skull rather narrow and elongated, depressed. Premaxillæ convex in front of the nares. Rostrum about equal in length to the cranial portion of the skull, triangular, broad at the base, and gradually contracting towards the apex, where it is somewhat curved downwards. Vertebræ: C 7, D 11, L 9, C 23; total 50. Cervical vertebræ free. Manus broad, short, and rounded, all the digits being tolerably well developed, except the first. No dorsal fin, but a low ridge in its place.

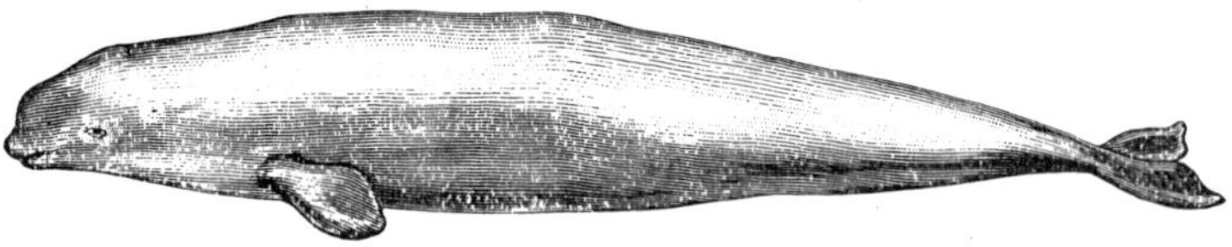

FIG. 91.—Beluga or White Whale (*Delphinapterus leucas*). From a specimen taken in the river St. Lawrence, and exhibited in London, 1877.

One existing species, *D. leucas* (Fig. 91), the Beluga or White Whale, so called from its pure white colour, about 12 feet long, abundant in the Arctic seas, and extending as far south on the

[1] Lacépède, *Hist. Nat. des Cétacés*, p. xli. (1804).

American coast as the river St. Lawrence, which it ascends for a considerable distance. On rare occasions it has been seen on the coast of Scotland.

Remains of a Cetacean from the Lower Pliocene of Tuscany have been referred by Brandt to this genus under the name *D. brocchii.*

In all the remaining genera of *Delphinidæ* the cervical region of the vertebral column is very short, and the first two, and usually more, of the vertebræ are firmly united.

Phocæna.[1]—Teeth $\frac{25}{25}$, small, occupying nearly the whole length of the rostrum, with compressed, spade-shaped crowns, separated from the root by a constricted neck (Fig. 92). Rostrum rather shorter than the cranium proper, broad at the base and tapering towards the apex. Premaxillæ raised into tuberosities in front of the nares. The frontal bones forming a somewhat square, elevated protuberance in the middle line of the skull behind the nares, rising altogether above the flattened nasals. Pterygoids very small, and widely separated in the middle line. Symphysis of mandible very short. Vertebræ: C 7, D 13, L 14, C 31; total 65 (subject to slight individual variations). First to sixth cervical vertebræ, and sometimes the seventh also, coalesced. Manus of moderate size, oval, slightly falcate; second and third digits nearly equal in length; fourth and fifth well developed, but shorter. Dorsal fin near the middle of the back, triangular; its height considerably less than the length of the base; its anterior edge frequently furnished with one or more rows of conical horny tubercles.

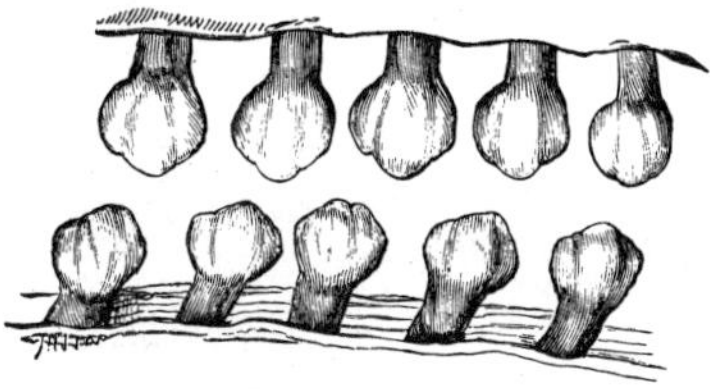

FIG. 92.—Teeth of Porpoise. Twice natural size.

The common Porpoise (Fig. 93), *P. communis*, is the best known of British Cetaceans. The word Porpoise (sometimes spelled Porpus and Porpesse) is apparently derived from the French *porc* and *poisson*, or the Italian *porco* and *pesce*, and thus corresponds with some of the English vernacular appellations, "hog-fish," "sea-hog," "herring-hog," and the German *Meerschwein*, whence the usual modern French name of the animal, *marsouin*. "Porpoise" is commonly used by sailors to designate all the smaller Cetaceans, especially those numerous species which naturalists call "Dolphins"; but in scientific language it is restricted to the genus *Phocæna* of Cuvier, of which the Porpoise of the British seas, *Phocæna communis*, Cuvier (*Delphinus phocæna*, Linnæus), is the type.

The Common Porpoise, when full grown, attains a length of 5 feet or a little more. The dimensions of an adult female specimen from the English Channel were as follows:—length in straight line

[1] Cuvier, *Règne Animal*, vol. i. p. 279 (1817).

from nose to median notch between the flukes of the tail, $62\frac{1}{2}$ inches; from the nose to the anterior edge of the dorsal fin, 29 inches; height of dorsal fin, $4\frac{1}{2}$ inches; length of base of dorsal fin, 8 inches; length of pectoral fin, $9\frac{1}{4}$ inches; breadth of pectoral fin, $3\frac{1}{2}$ inches; breadth of tail flukes, 13 inches. The under jaw projects about half an inch beyond the upper one. The aperture of the mouth is tolerably wide, and is bounded by stiff immobile lips, and curves slightly upwards at the hinder end. The eye is small, and the external ear represented by a minute aperture in the skin, scarcely larger than would be made by the puncture of a pin, situated about 2 inches behind the eye. The pectoral fins are of

FIG. 93.—The Common Porpoise (*Phocæna communis*).

moderate size, and slightly falcate. The upper parts are dark gray, or nearly black, according to the light in which they are viewed, and the state of moisture or otherwise of the skin; the under parts are pure white. The line of demarcation between these colours is not distinct (washes or splashes of gray encroaching upon the white on the sides), and varies somewhat in different individuals. Usually it passes from the throat (the anterior part of which, with the whole of the under jaw, is dark) above the origin of the pectoral fin, along the middle of the flank, and descends again to the middle line before reaching the tail. Both sides of the pectoral and caudal fins are black.

The Porpoise is sociable and gregarious in its habits, being usually seen in small herds, and frequenting coasts, bays, and estuaries rather than the open ocean. It is the commonest Cetacean in the seas around the British Isles, and not unfrequently ascends the river Thames, having been seen as high up as Richmond; it has also been observed in the Seine at Neuilly, near Paris. It frequents the Scandinavian coasts, entering the Baltic in the summer; and is found as far north as Baffin's Bay, and as far west as the coasts of the United States. Southward its range is more limited than that of the Common Dolphin, as, though very common on the Atlantic coasts of France, it rarely enters the Mediterranean.

It feeds on fish, such as mackerel, pilchards, and herrings, of which it devours large quantities, and, following the shoals, is often caught by fishermen in the nets along with its prey. In former times it was a common and esteemed article of food in England and in France, but is now rarely if ever eaten, being commercially valuable when caught only for the oil obtained from its blubber. Its skin is sometimes used for leather and boot-thongs, but the so-called "porpoise hides" are generally obtained from the Beluga.

A closely similar if not identical species from the American coast of the North Pacific has been described under the name of *Phocœna vomerina*, and another from the mouth of the Rio de la Plata as *P. spinipennis*.

The stomach of the Porpoise (Fig. 94) may be taken as a typical example of this organ in the Cetacea. The first and by far the largest compartment (*b*) may be regarded as a kind of crop, or dilatation of the large œsophagus (*a*). It is lined by a thick white epithelium, which ceases abruptly at the entrance into the next cavity. It corresponds to the cardiac compartment of the stomach in the Ungulates and certain Rodents; but, although its walls do not appear to contain peptic glands, its contents undergo partial digestion—probably caused by the regurgitation into it

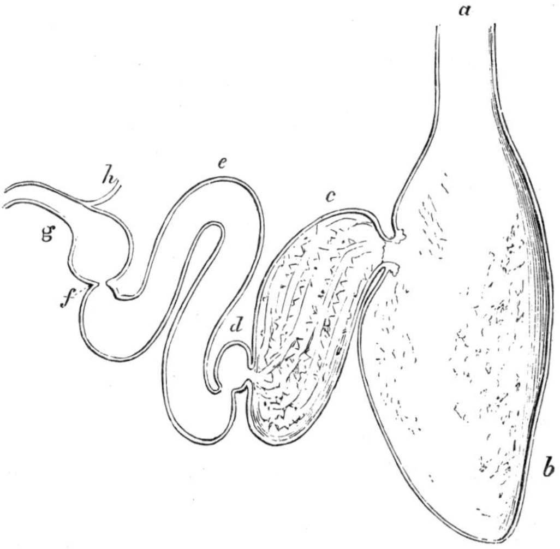

FIG. 94.—Diagrammatic section of the stomach of the Porpoise. *a*, Œsophagus; *b*, left, or cardiac, compartment; *c*, middle compartment; *d* and *e*, the two divisions of the right, or pyloric, compartment; *f*, pylorus; *g*, duodenum, dilated at its commencement; *h*, biliary duct.

of the secretions of the second, or true digestive compartment (*c*). This, which is much smaller than the first, has very thick walls, the mucous membrane being filled with numerous tubular glands. The surface of this membrane is smooth and soft, being thrown into numerous folds, which in this genus are arranged in a very peculiar and characteristic manner, so as to form a series of prominent longitudinal ridges, each of which sends off short lateral ridges at right angles to itself, which interdigitate with those proceeding from the next longitudinal ridge. The remainder of the stomach (*d* to *f*) may be compared to the pyloric antrum of the stomach of ordinary mammals. It is elongated, cylindrical, and intestiniform, with a smooth lining membrane, sharply bent upon itself, and terminating in a very small circular pyloric aperture (*f*). In the Porpoise the commencement of this cavity is constricted off from the remainder, so as to form a small globular sac. In most Dolphins (as *Tursiops*, *Globicephalus*, and *Grampus*) there are two such small sacs of very similar size and form, communicating by circular pylorus-like apertures; and in *Hyperoödon* the whole compartment is divided by a series of constrictions into as many as seven separate cavities, which have been regarded as distinct stomachs. Immediately beyond the pylorus the duodenum has a globular dilatation, as in the camels and some other Ungulates, into the lower end of which the biliary duct (*h*) enters.

An allied species, differing mainly in the absence of dorsal fin, and in the teeth (with the same form of crown) being fewer in number and of larger size, called *Delphinus phocænoides* by Cuvier, *D. melas* by Schlegel, forms the type of Gray's genus *Neomeris*.[1] It is rather smaller than the Common Porpoise, and almost entirely black in colour. Common off the coast of Bombay, it has been met with in other parts of the Indian Ocean, and near Japan. The British Museum recently received a specimen taken in the Chinese river Yang-tse-kiang nearly a thousand miles from the sea, which only differs from others from India in wanting a patch of small horny tubercles on the back. As such tubercles are present or absent in otherwise similar individuals of *P. communis*, it is doubtful whether they can be regarded as constituting a specific character.

Cephalorhynchus.[2]—Rostrum as long and sometimes slightly longer than the cranial portion of the skull. Pterygoids widely separated from one another. Teeth small (less than 3 mm. in diameter), $\frac{25}{25}$ to $\frac{30}{30}$. Vertebræ: C 7, D 13, L 15, C 30; total 65. Dorsal fin low, obtusely triangular or rounded. Pectoral fins rather

[1] *Zoology of Erebus and Terror*, p. 30 (1846). The name is preoccupied by Lamarck for a genus of Polyzoa (1816).

[2] Gray, *Cat. Cetacea Brit. Mus.* p. 106 (1850).

small, narrow, and ovate. Typified by *C. heavisidei*, from the southern seas. *C. eutropia* is a very distinct form from the same seas, known only by the skull, and referred provisionally to this genus.

Orcella.[1]—Teeth $\frac{12}{12}$ to $\frac{14}{14}$, small, conical, pointed, rather closely set, and occupying nearly the whole length of the rostrum. Skull subglobular, high. Rostrum nearly equal in length to the cranial portion of the skull, tapering. Pterygoids widely separated from one another. Manus of moderate size, not elongated, but somewhat pointed. All the bones of the digits broader than long, except the proximal phalanges of the index and third fingers. Dorsal fin rather small, placed behind the middle of the body. Two species, both of small size—*O. brevirostris*, from the Bay of Bengal, and *O. fluminalis*, from the Irawadi river, from 300 to 900 miles from the sea. Our present knowledge of the anatomy, geographical distribution, and habits of these interesting Cetaceans is almost entirely due to the researches of Dr. J. Anderson.[2]

Orca.[3]—Teeth about $\frac{12}{12}$, occupying nearly the whole length of the rostrum, very large and stout, with conical recurved crowns, and large roots, expanded laterally and flattened, or rather hollowed, on the anterior and posterior surfaces. Rostrum about equal in length to the cranial part of the skull, broad and flattened above, rounded in front; premaxillæ broad and rather concave in front of the nares, contracted at the middle of the rostrum, and expanding again towards the apex. Pterygoids of normal form, but not quite meeting in the middle line. Vertebræ: C 7, D 11-12, L 10, C 23; total 51 or 52. Bodies of the first and second and sometimes the third cervical vertebræ united; the rest free. Pectoral fin very large, ovate, nearly as broad as long. All the phalanges and metacarpals broader than long. General form of body robust. Dorsal fin near the middle of the back, very high and pointed. Anterior part of the head broad and depressed.

The animals composing this genus are met with in almost all seas from Greenland to Tasmania, but the number of species is still uncertain, and possibly they may be all reduced to one. They are readily known, when swimming in the water, by the high, erect, falcate dorsal fin, whence their common German name of *Schwertfisch* (Sword-fish). By English sailors they are generally known as "Grampuses" or "Killers." They are distinguished from all their allies by their great strength and ferocity, being the only Cetaceans which habitually prey on warm-blooded animals, for, though fish form part of their food, they also attack and devour Seals, and

[1] Gray, *Cat. Seals and Whales in Brit. Mus.* p. 285 (1866).

[2] *Anatomical and Zoological Researches, comprising an Account of the Zoological Results of the two Expeditions to Western Yunnan, in* 1868 *and* 1875 (1878).

[3] Gray, *Zoology of Erebus and Terror*, p. 33 (1846).

various species of their own order, not only the smaller Porpoises and Dolphins, but even full-sized Whales, which last they combine in packs to hunt down and destroy, as Wolves do the larger Ruminants.

FIG. 95.—The Killer Whale, or Grampus (*Orca gladiator*). From Hunter.

Orca citoniensis, of the Italian Pliocene, was of smaller size than the existing Killer. Teeth and periotic bones from the Suffolk Crag not improbably belong to the same species.

Pseudorca.[1]—Teeth about $\frac{10}{10}$. Cranial and dental characters generally like those of *Orca*, except that the roots of the teeth are cylindrical. Vertebræ: C 7, D 10, L 9, C 24; total 50. First to sixth or seventh cervical vertebræ united. Bodies of the lumbar vertebræ distinguished from those of the preceding genera by being more elongated, the length being to the width as 3 to 2. Pectoral fin of moderate size, narrow, and pointed. Dorsal fin situated near the middle of the back, of moderate size, falcate. Head in front of the blowhole high, and compressed anteriorly, the snout truncated.

This genus was first known by the discovery of a skull in a sub-fossil state in a fen in Lincolnshire, named by Sir R. Owen *Phocæna crassidens*. Animals of apparently the same species were afterwards met with in small herds on the Danish coast, and fully described by Reinhardt. Others subsequently received from Tasmania were supposed at first to indicate a different species, but comparison of a larger series of specimens from these extremely distant localities fails to establish any characteristic difference, and indicates an immense range of distribution for a species apparently so rare. The length of this Cetacean is about 14 feet, and its colour entirely black.

Globicephalus.[2]—Teeth $\frac{8-12}{8-12}$, confined to the anterior half of the rostrum and corresponding part of the mandible, small, conical, curved, sharp-pointed when unworn, sometimes deciduous in old age. Skull broad and depressed. Rostrum and cranial portion about equal in length. Upper surface of rostrum broad and flat.

[1] Reinhardt, *Overs. Dan. Sezsk. Forh.* 1862, p. 151.

[2] Lesson, *N. Tab. d. Règne Animal—Mamm.* p. 200 (1842).

Premaxillæ strongly concave in front of the nares, as wide at the middle of the rostrum as at the base, or wider, and very nearly or completely concealing the maxillæ in the anterior half of this region. Pterygoids of normal form, meeting, or very nearly so, in the middle line. Vertebræ: C 7, D 11, L 12-14, C 28-29; total 58 or 59. Bodies of the anterior five or six cervical vertebræ united. Length of the bodies of the lumbar and anterior caudal vertebræ about equal to their width. Pectoral limb very long and narrow, the second digit the longest, and having as many as 12 or 13 phalanges, the third shorter (with 9 phalanges), the first, fourth, and fifth very short. Fore part of the head very round, in consequence of the great development of a cushion of fat, placed on the rostrum of the skull in front of the blowhole. Dorsal fin low and triangular, the length of its base considerably exceeding its vertical height.

The type of this well-marked genus is *G. melas*, the Pilot Whale, Ca'ing Whale, or Grindhval of the Faroe islanders, which attains the length of 20 feet, and is of nearly uniform black colour, except the middle of the under surface, which is lighter. These animals are extremely gregarious, and, unlike the Killers, are mild and inoffensive in disposition, feeding principally on cephalopods. Their eminently sociable character constantly leads to their destruction, since when attacked they instinctively rush together and blindly follow the leaders of the herd. When they are seen in the neighbourhood of land, the fishermen endeavour to get to seaward of them in their boats, and with shouting and firing of guns to drive them into a bay or fjord, pursuing them until they run themselves on shore in their alarm. In this way many hundreds at a time are frequently driven ashore and killed, when a herd enters one of the bays or fjords of the Faroe Islands or north of Scotland. Animals of this well-marked genus are found in nearly all seas, and their specific distinctions are not yet made out. Specimens from the Australian coasts, where they are generally called "Blackfish," are quite indistinguishable, either by external or osteological characters, from those of the North Atlantic.

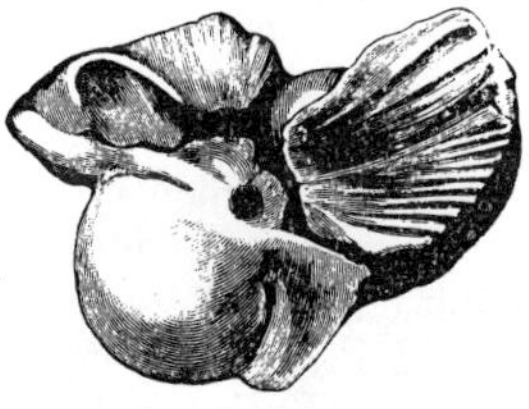

FIG. 96.—The left periotic bone of *Globicephalus uncidens*; from the Suffolk Crag. Natural size. The grooved surface on the right is the anterior facet for articulation with the tympanic; the posterior tympanic articulation being on the opposite side of the figure. (From the *Cat. Foss. Mamm. Brit. Mus.* pt. v.)

Teeth, periotic (Fig. 96) and tympanic bones from the Suffolk Crag, described as *G. uncidens*, indicate a form apparently closely allied to the existing species. The periotic is figured in order to illustrate the distinctive characters of that bone in the *Delphinidæ*.

Grampus.[1]—Teeth none in the upper jaw; in the mandible few (3 to 7 on each side), and confined to the region of the symphysis. Vertebræ: C 7, D 12, L 19, C 30; total 68. General external characters much as in *Globicephalus*, but the fore part of the head less rounded, and the pectoral fin less elongated.

But one species, *G. griseus*, is certainly known, about 13 feet long, and remarkable for its great variability of colour. It has been found, though rarely, in the North Atlantic and Mediterranean. A skull from the Cape of Good Hope, which differs slightly from that of the above, has been described under the name of *G. richardsoni*.

Feresia.[2]—This genus, known at present only by two skulls, may be provisionally placed here. These appear to indicate a form connecting *Globicephalus*, *Grampus*, and *Lagenorhynchus*. From the latter they differ chiefly in the smaller number (about $\frac{12}{12}$) and much larger size (6-7 mm. in diameter at base of crown) of the teeth.

Lagenorhynchus.[3]—Rostrum scarcely exceeding the length of the cranium, broad at the base and gradually tapering towards the apex, depressed. Pterygoids normal, meeting in the middle line. Teeth small (not exceeding 4 mm. in diameter), $\frac{23}{23}$ to $\frac{33}{33}$. Vertebræ very numerous, 80 to 90. Spines and transverse processes of the lumbar vertebræ very long and slender; centra short. Externally, head with a short but not very distinct beak. Two species, *L. albirostris* and *L. acutus*, are occasionally captured on the British coasts. Other species occur elsewhere.

Group B.—Head with distinctly elongated rostrum, or beak, generally marked off from the prenarial adipose elevation by a V-shaped groove. Rostrum of skull considerably longer than the cranial portion. Atlas and axis firmly united; all the other cervical vertebræ free.

If we add to it the above-mentioned genus, *Lagenorhynchus*, this group will include all the true Dolphins, Bottle-noses, or, as they are more commonly called by seafaring people, "Porpoises," which are found in considerable abundance in all seas, some species being habitually inhabitants of large rivers, as the Amazon. They are all among the smaller members of the order, none exceeding 10 feet in length. Their food is chiefly fish, for the capture of which their long narrow beaks, armed with numerous sharp-pointed teeth, are well adapted, but some appear also to devour crustaceans and molluscs. They are mostly gregarious, and the agility and grace of their movements in the water are constant themes of admiration to the spectators of the scene when a "school of Porpoises" is observed playing round the bows of a vessel at sea.

1 Gray, *Zoology of Erebus and Terror*, p. 30 (1846).

2 Gray, *Proc. Zool. Soc.* 1870, p. 77.

3 Gray, *Zoology of Erebus and Terror*, p. 35 (1846).

Delphinus.[1]—Teeth very numerous in both jaws, $\frac{40}{40}$ to $\frac{60}{60}$, occupying nearly the whole length of the rostrum, small, close-set, conical, pointed, slightly curved. Rostrum elongated, usually about double the length of the cranial portion of the skull. Pterygoids of normal form, meeting in the middle line throughout their length. Palate with deep lateral grooves. Vertebræ 73 to 75. Pectoral fin of moderate size, narrow, pointed, somewhat falcate. Second and third digits well developed; the rest rudimental.

The type of the genus is the Common Dolphin of the Mediterranean (*D. delphis*, Fig. 97), also found in the Atlantic, and of

FIG. 97.—The Common Dolphin (*Delphinus delphis*). From Reinhardt.

which a closely allied if not identical form is met with in the Australian seas (*D. forsteri*) and in the North Pacific (*D. bairdi*). Other species are *D. janira*, *D. major*, etc.

Tursiops.[2]—Rostrum tapering moderately from base to apex; palate not grooved; symphysis of mandible short; other cranial characters as in *Delphinus*. Teeth $\frac{21}{21}$ to $\frac{25}{25}$, stout (6 to 7 mm. in antero-posterior diameter). Vertebræ: C 7, D 13, L 17, C 27; total 64. Limbs as in *Delphinus*. Represented by the widely distributed *T. tursio*; *T. catalania* being a second form. Fossil remains of this genus from the Italian Pliocene have been recently described.

Prodelphinus.[3]—Rostrum somewhat variable; mandibular symphysis short (less than one-fifth the length of the ramus); other cranial characters as in the preceding genus. Teeth $\frac{30}{30}$ to $\frac{50}{50}$, small, not exceeding 3 mm. in diameter. Vertebræ 73 to 78. Limbs as in *Delphinus*. Four leading types of this genus are recognised (all of which have numerous synonyms) viz. *P. obscurus*, *P. euphrosyne*, *P. doris*, and *P. longirostris*.

Péron's Dolphin (*Delphinus leucorhamphus*, Péron, or *Leucorhamphus peroni*, Lilljeborg) resembles some forms of *Prodelphinus* in its cranial characters; but having no dorsal fin, it has been separated generically by some writers. It is not improbable that *Delphinus borealis*, Peale, from the North Pacific, in which there is likewise no dorsal fin, may be an allied form.

Steno.[4]—Rostrum long, narrow, and compressed, very distinct from the cranium; mandibular symphysis as long as, or longer than

[1] Linn. *Syst. Nat.* 12th ed. vol. i. p. 108 (1766).

[2] Gervais, *Hist. Nat. des Mammifères*, vol. ii. p. 323 (1855).

[3] Gervais, *Ostéographie des Cétacés*, p. 604 (1880).

[4] Gray, *Zoology of Erebus and Terror*, p. 43 (1846).

one-fourth the length of the ramus; other cranial characters as in the preceding genus. Teeth $\frac{21}{21}$ to $\frac{25}{25}$, of comparatively large size (5-6 mm. in diameter); surface of their crowns finely grooved. Vertebræ: C 7, D 12, L 15, C 32; total 66. Represented by *S. rostratus*, from which the forms which have received other names are probably not specifically separable.

Sotalia.[1]—Pterygoids narrow, not meeting in the middle line, and in their inner borders diverging posteriorly, instead of being parallel as in the preceding genera; other cranial characters much as in *Steno.* Teeth tolerably large (4-5 mm. in diameter), $\frac{30}{30}$ to $\frac{35}{35}$, with smooth enamelled surface. Vertebræ: C 7, D 12, L 10-14, C 22; total 51-55. Pectoral fin broad at base, the breadth being caused by the considerable development and position of the two outer digits. Six species are provisionally recognised as distinct, including the Chinese White Dolphin (*S. sinensis*) and *S. pallidus* from the river Amazon.

Bibliography of Cetacea.—D. F. Eschricht, *Untersuchungen über die Nordischen Wallthiere*, 1849, contains a copious bibliography of the group up to the date of publication. Since that time numerous monographs on special families and genera have been published, and a large illustrated general work, *Ostéographie des Cétacés*, by P. J. Van Beneden and P. Gervais, 1869-80. Besides those already referred to in the footnotes, the following may be mentioned; viz. J. F. Brandt, "Untersuchungen über die Fossilen und Subfossilen Cetaceen Europa's," in *Mém. de l'Acad. Imp. de St. Pétersbourg*, 7ième sér. vol. xx. 1873; C. M. Scammon, *Marine Mammals of the N. W. Coast of North America*, 1874; W. H. Flower, "On the characters and Divisions of the Families of the *Delphinidæ*," *Proc. Zool. Soc.* 1883, p. 466, and *List of the Specimens of Cetacea in the British Museum*, 1885; F. W. True, "Review of the Family Delphinidæ," *Bull. U.S. Nat. Museum*, No. 36, 1889; P. J. Van Beneden, *Histoire Naturelle des Cétacés des Mers d'Europe*, 1889.

For fossil forms, in addition to the works of Van Beneden, Gervais, and Brandt, already cited, the reader may refer to various memoirs published by the former writer in the *Bull. Ac. R. Belgique* and *Ann. Mus. R. Hist. Nat. Belg.* See also R. Lydekker, "The Cetacea of the Suffolk Crag," *Quart. Journ. Geol. Soc.* vol. xlii. p. 7 (1887), and *Catalogue of the Fossil Mammalia in the British Museum*, pt. v. (1887).

[1] Gray, *Cat. Seals and Whales Brit. Mus.* 2d ed. p. 393 (1866).

CHAPTER IX

THE ORDER UNGULATA

UNDER this term may be included provisionally a large and rather heterogeneous group of mammals, the existing members of which form the Pecora and Belluæ of Linnæus, the Ruminantia and Pachydermata of Cuvier. A few years ago it was found convenient to restrict the order to a well-marked and distinctly circumscribed group, comprising the two sections known as Perissodactyla and Artiodactyla, and to leave out such isolated forms as the Elephant and Hyrax; but the discovery of a vast number of extinct species, which could not be brought under the definition of either perissodactyle or artiodactyle Ungulates, and yet are evidently allied to both, and to a certain extent bridge over the interval between these and the isolated groups just mentioned, makes it necessary either to introduce a number of new and ill-defined ordinal divisions, or to widen the scope of the original order so as to embrace them all.

The existing forms are all animals eminently adapted for a terrestrial life, and in the main for a vegetable diet. Though a few are more or less omnivorous, and may under some circumstances kill living creatures smaller and weaker than themselves for food, none are distinctly and habitually predaceous. Their teeth are markedly heterodont and diphyodont,—the milk set being well developed and not completely changed until the animal attains its full stature. The molars have broad crowns with tuberculated or ridged surfaces. There are no clavicles.[1] Their toes are provided with blunt, broad nails, or in the majority of cases with hoofs, more or less enclosing the ungual phalanges. The scaphoid and lunar bones of the carpus are always distinct. The humerus has no entepicondylar foramen. The number of digits varies from five to one; and the radius and ulna may be united together.

[1] Since this was in type the discovery of transient rudimentary clavicles in the embryo of the Sheep has been announced by Wincza (*Morpholog. Jahrb.* xvi. p. 647).

The more generalised of the fossil forms do not conform in all respects to the above-mentioned characters; clavicles being present in *Typotherium*, and perhaps in some of the Condylarthra, while in the latter group the humerus may have an entepicondylar foramen, and thus approximate to the corresponding bone of the Carnivora. Wide as is the gap between existing Carnivores and Ungulates, there are indeed more or less strongly marked evidences of affinity between the earlier members of the two orders, as will be again noticed under the head of the suborder Condylarthra. A departure from the normal type of foot-structure is exhibited by the extinct *Macrotherium*, provisionally included in the Perissodactyla, where the digits terminated in long and curved claws.

As a general rule, the cheek-teeth have distinct roots, and in those of the existing suborders a gradual increase in the height of the crowns of these teeth may be noticed in passing from the more generalised to the more specialised types. Those teeth in which the crowns are low, and their whole structure visible from the grinding surface, are termed *brachydont* (Fig. 122); while those with higher crowns, in which the bases of the infoldings of enamel are invisible from the grinding surface, are known as *hypsodont* (Fig. 123). Again, when the tubercles on the crowns of the molars are more or less cone-like in form the tooth is said to be *bunodont;* but when they are expanded in an antero-posterior direction and curved into a crescent shape the tooth is described as *selenodont.*

FIG. 98.—Right fore foot of Indian Elephant. ×⅛. U, ulna; R, radius; *c*, cuneiform; *l*, lunar; *sc*, scaphoid; *u*, unciform; *m*, magnum; *td*, trapezoid; *tm*, trapezium; I to V, first to fifth digit.

The whole order may be divided into the Ungulata Vera, containing the suborders Perissodactyla and Artiodactyla, and a somewhat heterogeneous assemblage of animals which may be called Subungulata or Ungulata Polydactyla. Cope has pointed out a character in the structure of the carpus by which the latter are differentiated from the former. Thus in all the Subungulata the bones of the proximal and distal row retain the primitive or more typical relation to each other (see Fig. 98); the os magnum of the second row articulating mainly with the lunar of the first, or with the cuneiform, but not with the scaphoid. But in the group to which the vast majority of modern Ungulates belong the second or distal row has been shifted altogether towards the inner side of the limb (see Fig. 99), so that the magnum is brought considerably into relation with the scaphoid, and is entirely removed from the cuneiform, as in the great majority of existing mammals.

It will be on the whole more convenient to commence our survey of the members of this suborder with the more specialised group of the Ungulata Vera, in which the Artiodactyla will be taken first.

UNGULATA VERA.[1]

In the typical Ungulata the feet are never plantigrade, and the functional toes do not exceed four—the inner digit being suppressed, at all events in all forms which have existed since the Upper Eocene period.[2] The os magnum of the carpus articulates freely with the scaphoid. The allantois is largely developed, and the placenta, so far as is known, is non-deciduate; the chorionic villi being either evenly diffused or collected in groups or cotyledons (in Pecora). The testes descend into a scrotum. There is never an os penis. The uterus is bicornuate. The mammæ are usually few and inguinal, or may be numerous and abdominal (as in Suina), but are never solely pectoral. The cerebral hemispheres in existing Ungulates are well convoluted.

The group is now, and has been throughout almost the whole of the Tertiary period, composed of two perfectly distinct sections, differing from each other, not only in the obvious characters of the structure of the limbs, but in so many other parts of their organisation that they must be considered as of the rank at least of suborders. The characters of these divisions, first indicated by Cuvier, were thoroughly established by Owen, by whom the names whereby they are now generally known were proposed.

Suborder ARTIODACTYLA.

This is a well-defined group, traceable from the Eocene period, though then apparently by no means so numerous as the Perissodactyles. Some of its types, as that represented in the existing Swine, have retained to the present time much of the primitive character of the group; but others have been gradually becoming more specialised and perfected in structure, and its latest modification, the Cavicorn Ruminants or *Bovidæ* (Antelopes, Sheep, and Oxen), are now the dominating members of the great Ungulate order, widespread in geographical range, rich in generic and specific variation, and numerous in individuals—forming in all these

[1] Also known as Diplarthra.

[2] The pollex is present in the manus of the extinct *Cotylops*.

respects a great contrast to such decadent types as those represented by the Tapirs and Rhinoceroses.

The principal anatomical characters by which the Artiodactyles are distinguished from the Perissodactyles are as follows. The premolar and molar teeth usually not alike, the former being single and the latter two-lobed. The last lower molar of both first and second dentition almost invariably three-lobed; and the first tooth of the upper cheek series always without a milk-predecessor. Nasal bones not expanded posteriorly. No alisphenoid canal.

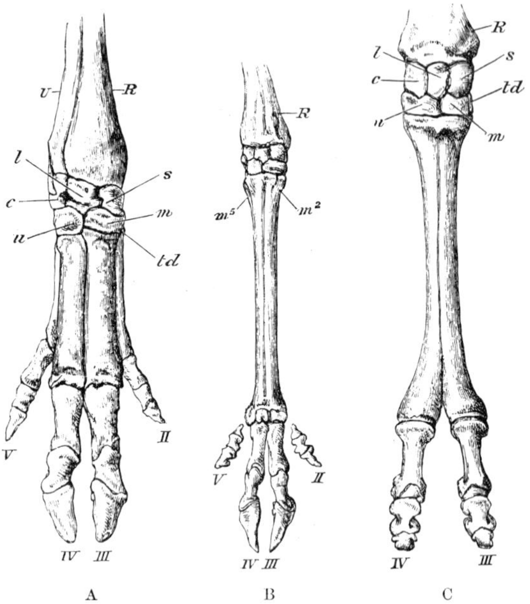

Fig. 99.—Bones of right fore foot of existing Artiodactyles. A, Pig (*Sus scrofa*), $\times \frac{1}{3}$; B, Red Deer (*Cervus elaphus*), $\times \frac{1}{7}$; C, Camel (*Camelus bactrianus*), $\times \frac{1}{8}$. *U*, Ulna; *R*, radius; *c*, cuneiform; *l*, lunar; *s*, scaphoid; *u*, unciform; *m*, magnum; *td*, trapezoid; *tm*, trapezium. From Flower's *Osteology of Mammalia*.

Dorsal and lumbar vertebræ together always nineteen, though the former may vary from twelve to fifteen. Femur without third trochanter. Third and fourth digits of both feet almost equally developed, and their ungual phalanges flattened on their inner or contiguous surfaces, so that each is not symmetrical in itself, but when the two are placed together they form a figure symmetrically disposed to a line drawn between them. Or, in other words, the axis or median line of the whole foot is a line drawn between the third and fourth digits, while in the Perissodactyles it is a line drawn down the centre of the third digit. Distal articular surface of the astragalus divided into two nearly equal facets, one for the navicular and the other for the cuboid bone. The calcaneum with an articular facet for the lower end of the fibula. Stomach almost always more or less complex. Colon convoluted. Cæcum small. Placenta diffused or cotyledonary. Mammæ few and inguinal, or numerous and abdominal.

In treating of many sections of mammals, it is only from the existing species that our characters and classification can be derived, and to these chiefly our observations upon the group must be directed, many of the extinct forms being so little known that they can only be referred to incidentally. With the Ungulata, however, it is quite otherwise. The history of the Artiodactyla throughout the Tertiary period is now well known, and throws great light upon the position and relations of the existing groups.

The principal modifications which have taken place in the type from its earliest known and most generalised manifestation have been the following:—

1. As regards the teeth. Assumption by the grinding surfaces of the molar teeth either of a bunodont or of a selenodont form. Modification of the latter from a brachydont to a hypsodont type. Loss of upper incisors. Development of canines into projecting tusks. Loss of anterior premolars.

2. As regards the limbs. Reduction of the ulna from a complete and distinct bone to a comparatively rudimentary state, in which it coalesces more or less firmly with the radius. Reduction of the fibula till nothing but its lower extremity remains. Reduction and final loss of external pair of digits (second and fifth), with coalescence of the metapodial bones of the two middle digits. Union of the navicular and cuboid, and sometimes the ectocuneiform, bones of the tarsus.

3. Change of form of the odontoid process of the axis vertebra from a cone to a hollow half-cylinder.

4. Development of horns or antlers on the frontal bones, and gradual complication of form of antlers.

5. By inference only, increasing complication of stomach with ruminating function superadded. Modification of placenta from simple diffused to cotyledonary form.

The primitive Artiodactyles, with the typical number (44) of incisor, canine, and molar teeth, brachydont molars, conical odontoid process, four distinct toes on each foot, with metapodium and all carpal bones distinct, no frontal appendages, and (in all probability) simple stomach and diffused placenta, were separated at a very early period into Bunodonts and Selenodonts, although there is evidence of intermediate forms showing a complete transition from the one modification to the other. These and other fossil forms so completely connect the four groups—Suina, Tylopoda,

Tragulina, and Pecora—into which the existing members of the suborder have become divided, that in a general classification embracing both living and extinct forms these divisions cannot be maintained. In the present work, however, it will be convenient to retain them, mention being made of some of the chief annectant forms in separate sections.

SUINA.

The existing members of this group are characterised by their bunodont molars, and the absence of a complete fusion of the third and fourth metapodials to form a "cannon-bone." The full Eutherian dentition is very frequently present.

Remains of very generalised swine-like animals have been abundantly found in Tertiary formations both in America and Europe. In the former continent they never (so far as present evidence indicates) underwent any great diversity of modification, but gradually dwindled away and almost died out, being only represented in the actual fauna by the two closely allied species of Peccary, among the smallest and most insignificant members of the group, which have existed almost unchanged since the Miocene age at least, if the evidence of teeth alone can be trusted. In the Old World, on the other hand, the swine have played a more important part in recent times, having become widely distributed, and throwing off some curiously specialised forms. At the present time, though not very numerous in species, they range through the greater part of the Old World, except within or near the Arctic Circle, although, in common with all the other members of the great Ungulate order, they were completely absent from the whole of the Australian region, until introduced by man in very recent times.

The existing swine-like animals may be divided naturally into three families:—I. *Hippopotamidæ*; II. *Suidæ*, or true Pigs; III. *Dicotylidæ*, or Peccaries.[1]

Family HIPPOPOTAMIDÆ.

Muzzle very broad and rounded. Feet short and broad, having four subequal toes, with short rounded hoofs, all reaching the ground in walking. Incisors not rooted, but continuously growing; those of the upper jaw curved and directed downwards; those of the lower straight and procumbent. Canines very large, curved, continuously growing; those of the upper jaw directed downwards. Stomach complex. No cæcum.

Hippopotamus.[2]—This genus may be taken to include all the known members of the family; it appears to have been always

[1] In the table on p. 89 the Peccaries are included in the *Suidæ*.

[2] Linn. *Syst. Nat.* 12th ed. vol. i. p. 101 (1766).

confined to the Old World. The dentition may be expressed by the formula $i\,\frac{2-3}{1-3}$, $c\,\frac{1}{1}$, $p\,\frac{4}{4}$, $m\,\frac{3}{3}$. The crowns of the molars (Fig. 100) when worn present trefoil-shaped surfaces of dentine; and those of the premolars are sharp. The facial portion of the skull is much elongated, the orbits are tubular and very prominent, and the mandible has a large rounded descending flange at its angle. The ears are small, the tail is short, and the legs are likewise so short that the belly is raised but a little distance above the ground. The brain is not richly convoluted, and differs very considerably from that of the Pigs, approximating in some respects to that of the Camel and Giraffe, but on the whole standing very much by itself. The stomach of the common species is of enormous dimensions, having an axial length of 11 feet, and measuring upwards of 15 feet along the greater curvature. Its axis is longitudinal, the pylorus being situated almost in the pelvis, and it is divided into three distinct compartments, of which the third is cylindrical. The liver of the adult is of extremely simple form, elongated transversely, and narrow from above downwards. With the exception of a few tufts of

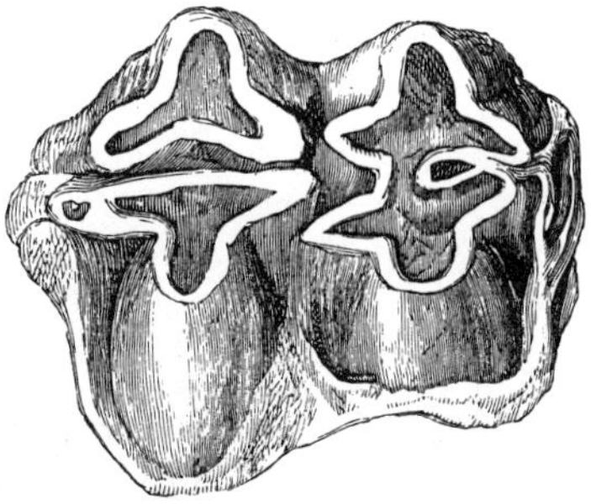

FIG. 100.—Grinding surface of a worn molar of *Hippopotamus amphibius*. (From Owen.)

FIG. 101.—The Hippopotamus (*Hippopotamus amphibius*).

hair on the lips, on the sides of the head and neck, and at the extremity of the short compressed tail, the skin of the hippopotamus, some portions of which are two inches in thickness, is entirely destitute of covering.

The common Hippopotamus (*H. amphibius*), widely distributed in the rivers and lakes of the African continent, is a huge bulky animal, characterised by having only two incisors on either side of each jaw; the central lower pair being very much larger than the outer ones. A male from the Upper Nile which lived for nearly thirty years in the gardens of the Zoological Society of London measured 12 feet along the back from the nose to the root of the tail.

The Hippopotamus lives in herds of from twenty to forty individuals on the banks and in the beds of rivers, in the neighbourhood of which it finds its food. This consists chiefly of grass and aquatic plants, of which it consumes enormous quantities, the stomach being capable of containing from 5 to 6 bushels. These animals feed principally by night, remaining in the water during the day, although in districts where they are undisturbed by man they are less exclusively aquatic. In such regions they put their heads boldly out of the water to blow, but when rendered suspicious by persecution, they become exceedingly cautious, only exposing their eyes and nostrils above the water, and even this they prefer doing amid the shelter of water plants. In spite of their enormous size and uncouth form, they are expert swimmers and divers, and can remain under the water from five to eight minutes. They are said to walk with considerable rapidity on the bottoms of rivers, beneath at least a foot of water. At nightfall they come on land to feed; and when, as often happens on the banks of the Nile, they reach cultivated ground, they do immense damage to growing crops, destroying by their ponderous tread even more than they devour.

A much smaller species, known as the Pigmy Hippopotamus (*H. liberiensis*), inhabits some of the rivers of Western Africa, and is characterised by having only a single pair of lower incisors. Mainly on this account, it has been proposed to regard this species as representing a distinct genus, under the name of *Chœropsis*; but since it agrees so essentially in other characters with the common form, and sometimes has two incisors on one side of the lower jaw, it appears preferable to include it in the type genus. The greater relative size of the brain-cavity as compared with the facial portion of the skull renders, indeed, the contour of the skull decidedly different from that of *H. amphibius*; but this is a feature generally found in young individuals of larger species, and also in the adults of allied smaller forms.

Both the existing species are now exclusively confined to Africa, but in the Pleistocene and Pliocene periods the genus was widely spread over the Old World. Thus in the Upper Pliocene of the Continent and the Pleistocene of England we meet with remains of a very large fossil Hippopotamus which cannot be specifically distinguished from *H. amphibius*. In the Pleistocene and Pliocene of India there are two species having three pairs of incisors in both jaws. Of these *H. palæindicus* has the second pair in the lower jaw very minute, and evidently just about to disappear; from which we learn that it is this pair which is missing in *H. amphibius*. In the still more generalised *H. sivalensis* the three incisors in the lower jaw are of equal size. Hexaprotodont species also occur in the Upper Tertiaries of Burma and Algeria. Small tetraprotodont species (*H. pentlandi* and *H. minutus*) have left their remains in enormous quantities in the caves and fissures of Sicily and Malta.

Family SUIDÆ.

An elongated mobile snout, with an expanded, truncated, nearly naked, flat, oval terminal surface in which the nostrils are placed. Feet narrow; four completely developed toes on each. Hoofs of the two middle toes with their contiguous surfaces flattened. The outer (second and fifth) digits of existing forms not reaching to the ground in the ordinary walking position. Teeth variable in number, owing to the suppression in some forms of an upper incisor and one or more premolars. Incisors rooted. Upper canines curving more or less outwards or upwards. Stomach simple, except for a more or less developed pouch near the cardiac orifice. A cæcum. Colon spirally coiled. Confined to the Old World.

The mandible has no descending flange at the angle. The crowns of the molars do not wear into such distinct trefoils as in the Hippopotamus, and are oblong in shape. The last molar of both the upper and lower jaw (Fig. 102) has an additional hinder lobe or talon, varying in size in the different species. The upper premolars are simpler than the true molars.

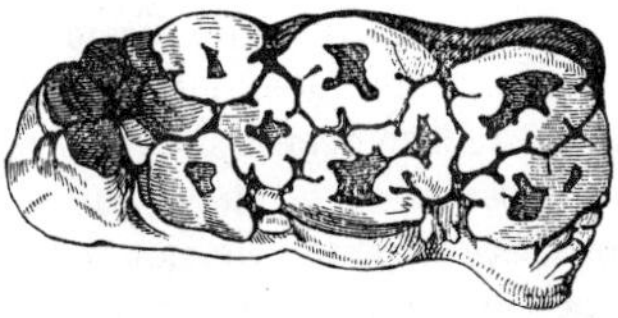

FIG. 102.—Grinding surface of a worn third right lower molar of the Wild Boar (*Sus scrofa*). After Owen.

Sus.[1]—Dentition: $i\,\frac{3}{3}$, $c\,\frac{1}{1}$, $p\,\frac{4}{4}$, $m\,\frac{3}{3}$; total 44. Upper incisors diminishing rapidly in size from the first to the third. Lower incisors long, narrow, closely approximated, and almost horizontal in position, their apices inclining towards the middle line; the second slightly larger than the first, the third much smaller. Canines strongly developed and with persistent roots and partial enamel-covering, those of the upper jaw not having the usual downward direction,

[1] Linn. *Syst. Nat.* 12th ed. vol. i. p. 102 (1766).

but curving strongly outwards, upwards, and finally inwards, while those of the lower jaw are directed upwards and outwards with a gentle backward curve, their hinder edges working and wearing against the front edges of the upper canines.[1] They appear externally to the mouth as tusks, the form of the upper lip being modified to allow of their protrusion, but are much less developed in the females than in the males. The teeth of the molar series gradually increase in size and complexity from first to last, and are arranged in contiguous series, except that the first lower premolar is separated by an interval from the second. First and second upper premolars with compressed crowns and two roots. The third and fourth have an inner lobe developed on the crown, and an additional pair of roots. The first and second true molars have quadrate crowns, with four principal obtuse conical cusps, around which numerous accessory cusps are clustered. The length of the third molar is nearly equal (antero-posteriorly) to that of the first and second together, its crown having, in addition to the four principal cusps, a large posterior talon or heel, composed of numerous clustered conical cusps, and supported by several additional roots. The lower molar teeth resemble generally those of the upper jaw, but are narrower. Milk dentition: $i\,\frac{3}{3}$, $c\,\frac{1}{1}$, $m\,\frac{3}{3}$; total 28,—the first permanent premolar having no predecessor in this series. The third incisor, in both upper and lower jaws, is large, developed before the others, and has much the size, form, and direction of the canine. Vertebræ: C 7, D 13-14, L 6, S 4, C 20-24. The hairy covering of the body varies much under different conditions of climate, but when best developed, as in the European Wild Boar, consists of long stiff bristles, mostly abundant on the back and sides, and of a close softer curling under-coat.

The skull of the Pigs (Figs. 103-105) has the axis of the face bent down upon the basicranial axis, as is also the case with the Sheep. Its most striking feature is the elevation and backward slope of the occipital crest formed by the union of the supraoccipital and parietals. The broad and flat frontals have small postorbital processes, which do not join the zygomata, so that the orbits are open behind. The nasals are very long and narrow; and the premaxillæ send up long nasal processes, stopping short of the frontals. A peculiar prenasal bone is developed at the anterior extremity of the mesethmoid, which serves to strengthen the cartilaginous snout. The palate is long and narrow, and extends behind the last molar

[1] If from any accidental circumstances these teeth are not constantly worn down by friction, they grow into a complete circle, the point penetrating the bone of the jaw close to the root of the tooth. The natives of the Fiji Islands avail themselves of this circumstance to produce one of their most valued ornaments—a circular boar's tusk: the upper canines being extracted, the lower ones are allowed to grow to the desired form.

tooth. In most species the occipital crest is more nearly vertical than in the skull represented in Fig. 104.

This genus occurs at present under three principal modifications or subgenera.

A.—*Sus* proper comprises a number of animals found in a wild state throughout the greater part of Europe (except where extermin-ated by human agency), the north of Africa, southern continental Asia, and the great islands of the Malayan archipelago, Formosa, and Japan. The following among others have been admitted by many zoologists as distinct species:—*Sus scrofa*, the Wild Boar of Europe, Asia Minor, and North Africa, once common throughout the British Isles; *S. sennaarensis*, North-East Africa; *S. cristatus*, India; *S. vittatus*, Java, Borneo, Amboyna, Batchian; *S. papuensis*, New Guinea; *S. timorensis*, Timor and Rotti; *S. andamanensis*, Anda-

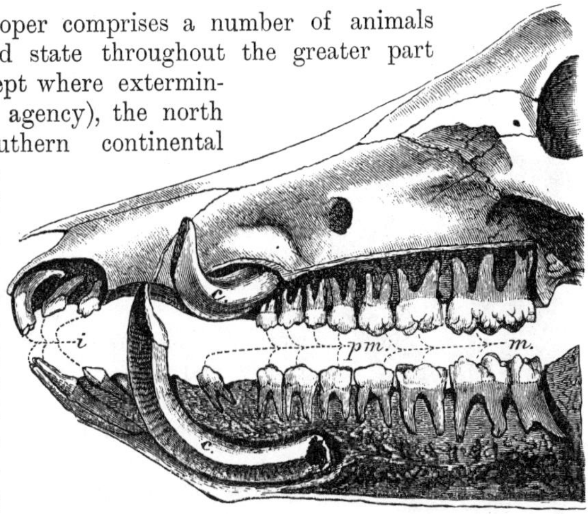

FIG. 103.—Left lateral view of the dentition of the Boar (*Sus scrofa*), the roots of the teeth being exposed by removing the external lamina of bone.

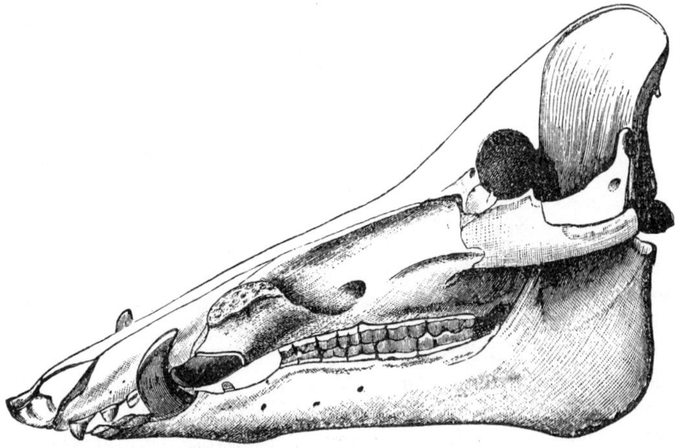

FIG. 104.—Left lateral view of the skull of *Sus longirostris*. ⅓ natural size. (From Nehring.)

man Islands; *S. taëvanus*, Formosa; *S. leucomystax*, Japan; *S. verrucosus*, Java, Borneo, Ceram; *S. barbatus*, Borneo; *S. celebensis*, Celebes, Philippines, and Moluccas; *S. longirostris*, Borneo and Java. The last four species form an allied group in which the facial portion of the skull may be greatly elongated; *S. barbatus* and *S. celebensis* being characterised by the small size and simple structure of the talon of the third molars. The skull of *S. longirostris* is shown in Figs. 104 and 105. The small *S. andamanensis* also has very simple third molars. *S. vittatus*, *S. leucomystax*, *S. cristatus*, *S. taëvanus*, and *S. papuensis* form another group, in which the third molar is generally of very complex structure, more or less closely allied to the Wild Boar; and Dr. Nehring is inclined to think that the whole five might be included under a single specific name. This list will give some idea of the geographical distribution of wild Pigs, but it must be borne in mind that through the whole of this region, and in fact now throughout the greater part of the habitable world, Pigs are kept by man in a domesticated state, and it is still an open question whether some of the wild Pigs of the islands named above may not be local races derived originally from, or crossed with, imported domestic specimens. In New Zealand a wild or rather "feral" race is already established, the origin of which is of course quite recent, since it is well ascertained that no animal of the kind ever lived upon the island until after its settlement by Europeans.

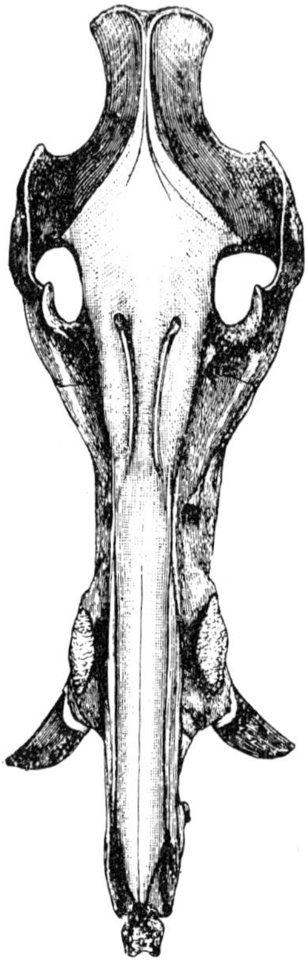

FIG. 105.—Frontal aspect of the cranium of *Sus longirostris*. ⅓ natural size. (From Nehring.)

Whether the various breeds of domestic Pigs have been derived from one or several sources is still unknown. As in so many similar cases, there is no historic evidence upon the subject, and the researches of naturalists, as Nathusius, Rütimeyer, Rolleston, Nehring, and others, who have endeavoured to settle the question on anatomical evidence, have not led to any satisfactory conclusions. It is, however, tolerably certain that all the species or forms of wild Pigs enumerated above and all the domestic races are closely allied, and it is probable (though of this there has been no opportunity of proof) will breed freely together. It is a curious circumstance that the young of all the wild kinds of Pigs (so far as yet is known) present a uniform coloration, being dark brown with longitudinal stripes of a paler colour, a character which completely disappears after the first few months. On the other hand, this peculiar marking is rarely seen in domestic Pigs in any part of the world, although it has been occasionally observed. It is stated by Darwin that the Pigs which have run wild in Jamaica and the semiferal Pigs of New Granada have resumed this aboriginal character, and produce longitudinally striped young; these must of course be the descendants of domestic animals introduced from Europe since the Spanish

FIG. 106.—Wild Boar and Young.

conquest, as before that time there were no true Pigs in the New World. Another character by which the European domestic Pig differs from any of the wild species is the concave outline of the frontal region of the skull, a form still retained by the feral Pigs in New Zealand.

B.—The diminutive Pig of the Nipal, Terai, and Bhutan, *Sus salvanius*, has been separated from the rest by Hodgson under the generic name of *Porcula*, but all the alleged distinctive characters prove on more careful investigation to have little real value. Owing to its retired habits and power of concealment under bushes and long grass in the depths of the great Sal Forest, which is its principal home, very little has been known of this curious little animal, scarcely larger than a hare. The acquisition of living specimens in the London Zoological Gardens has, however, afforded opportunities for careful anatomical observation.[1]

C.—Two well-marked species of African Swine have been with more reason separated under the name of *Potamochœrus*. The dentition differs from that of the true *Sus*, inasmuch as the anterior premolars have a tendency to disappear; sometimes in adult specimens the first upper premolar is retained, but it is usually absent, as well as the first and often the second lower premolars. The molar teeth are also less complex; the last especially having a

FIG. 107.—The Red River-Hog (*Sus porcus*). From Sclater, *Guide to Animals in Zoological Society's Gardens*, 1883, p. 183.

much less developed talon. There are likewise characteristic cranial differences. The two species are very distinct in outward appearance and coloration. One is *S. africanus*, the South African River-Hog, or Bosch-Vark, of a gray colour, and the other *S. porcus*, the West African Red River-Hog (Fig. 107), remarkable for its vivid colouring and long pencilled ears. It should be noted that the young of both these species, as well as of the pigmy *S. salvanius*, present the striped character of the true *Sus*, a strong indication of close affinities, whereas in all the following forms this is absent.

The genus *Sus*, in the above extended sense, is well represented in the Tertiaries of the Old World from the period of the Lower Pliocene upwards. In the Pliocene and Pleistocene of India

[1] See Garson, *Proc. Zool. Soc. Lond.* 1883, p. 413.

S. falconeri and *S. karnuliensis* are characterised by the extremely complex structure of the molars, in which they show decided signs of approximation to the Wart-Hogs; the same feature being exhibited by *S. phacochœroides* of the Algerian Pliocene. *S. titan* and *S. giganteus*, of the Indian Pliocene, together with *S. antiquus* and *S. erymanthius*, of the corresponding European deposits, are very large species characterised by their comparatively simple molars; *S. titan* being fully as large as a Tapir. *S. hysudricus* of the Pliocene of India, and *S. palæochœrus* of that of Europe, are smaller allied species not improbably related to *S. andamanensis*, with which they agree in molar structure. *S. arvernensis*, of the Upper Pliocene of France, appears to be allied to *S. africanus;* while in the diminutive *S. punjabiensis* of the Pliocene of North-Western India we probably have the direct ancestor of *S. salvanius*.

FIG. 108.—Head of Babirusa (*Babirusa alfurus*).

Babirusa.[1]—Dentition: $i\,\frac{2}{3}$, $c\,\frac{1}{1}$, $p\,\frac{2}{2}$, $m\,\frac{3}{3}$; total 34. The total number of teeth is therefore considerably reduced, the outer upper incisor and the two anterior premolars of both jaws being absent. The molars, especially the last, are smaller and simpler than in *Sus;* but the great peculiarity of this genus is the extraordinary development of the canines of the male. These teeth (Fig. 108) are ever-growing, long, slender, and curved, and entirely without enamel covering. Those of the upper jaw are directed upwards from their base, so that they never enter the mouth, but piercing the skin of the face, resemble horns rather than teeth, and curve backwards, downwards, and finally often forwards again, almost or quite touching the skin of the forehead. Vertebræ: C 7, D 13, L 16, S 4. There is but one species (*B. alfurus*), found only in the islands of Celebes and Buru. Its external surface is almost

[1] Lesson, *Man. d. Mamm.*, p. 337 (1827), "Babirusa."

entirely devoid of hair. With regard to the curiously modified dentition, Wallace (*Malay Archipelago*, vol. i. p. 435) makes the following observations:—"It is difficult to understand what can be the use of these horn-like teeth. Some of the old writers supposed that they served as hooks by which the creature could rest its head on a branch. But the way in which they usually diverge just over and in front of the eye has suggested the more probable idea, that they serve to guard these organs from thorns and spines while hunting for fallen fruits among the tangled thickets of rattans and other spiny plants. Even this, however, is not satisfactory, for the female, who must seek her food in the same way, does not possess them. I should be inclined to believe rather that these tusks were once useful, and were then worn down as fast as they grew, but that changed conditions of life have rendered them unnecessary, and they now develop into a monstrous form, just as the incisors of the Beaver and Rabbit will go on growing if the opposite teeth do not wear them away. In old animals they reach an enormous size, and are generally broken off as if by fighting."

Phacochœrus.[1]—The Wart-Hogs, so called from the large cutaneous lobes projecting from each side of the face, have the teeth still more remarkably modified than in *Babirusa*. The milk-dentition, and even the early condition of the permanent dentition, is formed on the same general type as that of *Sus*, except that certain of the typical teeth are absent, the formula being $i\,\frac{1}{3}$, $c\,\frac{1}{1}$, $p\,\frac{3}{2}$, $m\,\frac{3}{3}$, total 34; but as age advances all the teeth have a tendency to disappear, except the canines and the posterior molars, which in some cases are the only teeth left in the jaws, and attain an extraordinary development. The upper canines especially are of great size, and curve outwards, forwards, and upwards. Their enamel covering is confined to the apex, and soon wears away. The lower canines are much more slender, but follow the same curve; except on the posterior surface, their crowns are covered with enamel. Unlike those of the Babirusa, the canines of the Wart-Hog are large in both sexes. The third molar tooth of both jaws is of great size, and presents a structure at first sight unlike that of any other mammal, being composed of numerous (22-25) parallel cylinders or columns, each with pulp-cavity, dentine, and enamel covering, and packed together with cement. Careful examination will, however, show that a similar modification to that which has transformed the comparatively simple molar tooth of the Mastodon into the extremely complex grinder of the Indian Elephant has served to change the tooth of the common Pig into that of *Phacochœrus*; and, as already mentioned, some of the fossil Indian and African species of *Sus* indicate the mode in which this

[1] Cuvier, *Règne-Animal*, vol. i. p. 236 (1817).

transition came about. The tubercles which cluster over the surface of the crown of the molars of the common Pig are elongated and drawn out into columns in the Wart-Hog, as the low transverse ridges of the Mastodon's tooth become the leaf-like plates of the Elephant's.

Two species of this genus are commonly but rather doubtfully distinguished :—*P. africanus*, Ælian's Wart-Hog, widely distributed over the continent; and *P. æthiopicus*, Pallas's Wart-Hog, confined to South-Eastern Africa. In specimens attributed to the latter species the dentition reaches its most complete reduction, as in adult animals the upper incisors are absent and the lower ones worn down to the roots.

Family DICOTYLIDÆ.

Snout as in *Suidæ*. Dentition: $i\ \frac{2}{3}$, $c\ \frac{1}{1}$, $p\ \frac{3}{3}$, $m\ \frac{3}{3}$; total 38. Incisors rooted; upper canines directed downwards, with sharp cutting hinder edges. Toes, four on the fore feet and three on the hind feet (the fifth wanting). Stomach complex. A cæcum. Confined to the New World.

Dicotyles.[1]—The teeth of the Peccaries (*Dicotyles*) differ from those of the true Pigs (*Sus*) numerically in wanting the upper outer incisor and the anterior premolar on either side of each jaw, and also in the circumstance that the last premolar is nearly as complex as the molars. The upper canines have their points directed downwards, not outwards or upwards as in the Boars, and are very sharp, with cutting hinder edges, and completely covered with enamel until worn. The lower canines are large, directed upwards and outwards, and slightly curved backwards. The premolar and molar teeth form a continuous series, gradually increasing in size from the first to the last. The true molars have square quadricuspidate crowns. The stomach is much more complex than in the true Pigs, almost approaching that of the ruminants. In the feet the two middle (third and fourth) metapodial bones, which are completely separate in the Pigs, are united at their upper ends, as in the ruminants. On the fore foot the two (second and fifth) outer toes are equally developed as in Pigs, but on the hind foot, although the inner (or second) is present, the outer (or fifth) toe is entirely wanting, giving an unsymmetrical appearance of the member, very unusual in Artiodactyles. Vertebræ: C 7, D 14, L 5, S 4, C 7. As in the Pigs, the snout is truncated, and the nostrils are situated in its flat, expanded, disc-like termination. The ears are rather small, ovate, and erect; and there is no external appearance of a tail. The surface of the body is well covered with thick bristly hair, and rather behind the middle of the back is a large and

[1] Cuvier, *Règne Animal*, vol. i. p. 237 (1817).

peculiar gland, which secretes an oleaginous substance with a powerful musky odour. This was mistaken by the old travellers for a second navel, a popular error which suggested to Cuvier the name of *Dicotyles*. When the animal is killed for food, it is necessary speedily to remove this gland, otherwise it will taint the whole flesh so as to render it uneatable.

There are two species,[1] so nearly allied that they will breed together freely in captivity. Unlike the true Pigs, they never appear to produce more than two young ones at a birth. The Collared Peccary (*D. tajacu*, Linn., *torquatus*, Cuvier), Fig. 109, ranges from the Red River of Arkansas through the forest districts of

FIG. 109.—The Collared Peccary (*Dicotyles tajacu*).

Central and South America as far as the Rio Negro of Patagonia. Generally it is found singly or in pairs, or at most in small herds of from eight to ten, and is a comparatively harmless creature, not being inclined to attack other animals or human beings. Its colour is dark gray, with a white or whitish band passing across the chest from shoulder to shoulder. The length of the head and body is about 36 inches. The White-lipped Peccary or Warree (*D. labiatus*, Cuvier) is rather larger, being about 40 inches in length, of a blackish colour, with the lips and lower jaw white. Its range is less extensive, since it is not found farther north than British Honduras or south of Paraguay. It is generally met with in large herds of from fifty to a hundred or more individuals, and is of a more pugnacious disposition than the former species, and capable of

[1] Professor Cope considers that there is a third species, for which he has proposed the name *D. angularis*.

inflicting severe wounds with its sharp tusks. A hunter who encounters a herd of them in a forest has often to climb a tree as his only chance of safety. Both species are omnivorous, living on roots, fallen fruits, worms, and carrion; and when they approach the neighbourhood of villages and cultivated lands they often inflict great devastation upon the crops of the inhabitants.

Remains of the two existing species of Peccary, as well as of one much larger extinct form, are found in the cavern-deposits of Brazil; while large Peccaries also occur in the Pleistocene of the United States, which, although they have been referred to a distinct genus, *Platygonus*, on account of their relatively smaller incisors and somewhat simpler premolars, may well be included in *Dicotyles*.

Allied Extinct Genera.—In the Tertiary deposits of both the Old and New World occur remains of Pig-like animals which, so far as we can judge, appear to connect the Peccaries so closely with the true Pigs as to render the *Dicotylidæ* really inseparable from the *Suidæ*. Of these the American genus *Chænohyus* has the lower canine with a triangular cross section and received into a notch in the upper jaw, as in the Peccaries, but the fourth upper premolar is simpler than the molars, as in the under-mentioned genus *Hyotherium*. The typical forms have only three premolars, but in others, which it has been proposed to separate generically as *Bothriolabis*, there are four of these teeth. *Hyotherium*, of the Pliocene and Miocene of the Old World, is a generalised form allied both to *Sus* and *Dicotyles* as well as to certain extinct genera. The upper molars (Fig. 110) are characterised by their square crowns, the last having no distinct third lobe, and coming into use before the first is much worn, while the last premolar is simpler than the true molars. The canines, which have an oval section and are scarcely larger than the incisors, are not received into a notch in the upper jaw. In the Pliocene of India there occurs an apparently allied genus known as *Hippohyus*, in which the crowns of the molars are much taller, and have lateral infoldings of the enamel, producing a very complex pattern on the worn crowns. The European Miocene genus *Listriodon*, with the dental formula $i\ \frac{3}{3}$, $c\ \frac{1}{1}$, $p\ \frac{3}{3}$, $m\ \frac{3}{3}$, differs from all the preceding in having the anterior and posterior pairs of tubercles of the molars united into ridges running across their crowns, so that these teeth resemble the lower molars of the Tapir. The genus is also found in the Lower Pliocene of India.

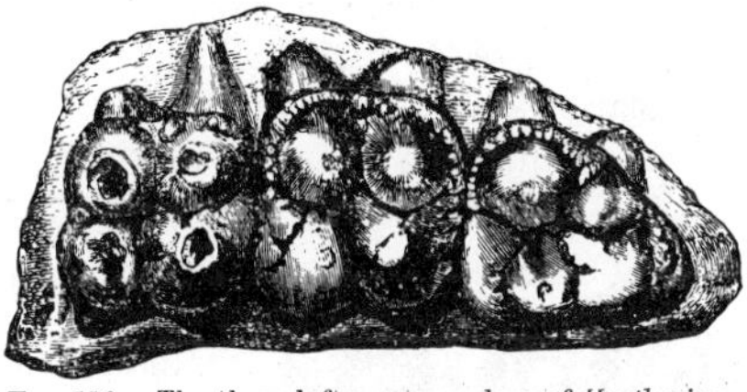

FIG. 110.—The three left upper molars of *Hyotherium perimense*, from the Pliocene of India.

EXTINCT TRANSITIONAL ARTIODACTYLES.

In this place it will be convenient to notice briefly a few of the extinct types of Tertiary Artiodactyles which connect the existing bunodont Suina with the more specialised selenodont groups mentioned below so closely as to show that in a strictly palæontological classification such groups cannot be maintained. It should be mentioned that while some of these extinct forms were in all probability actual ancestral links between the bunodonts and selenodonts, others, like the Anoplotheres, died out entirely without giving rise to any more specialised descendants.

Chæropotamidæ.—In this family the molars are intermediate in structure between those of the *Suidæ* and the next family. The upper ones have very broad crowns, with the five columns arranged as in *Anthracotherium;* while the premolars are not secant, and may be very large. The best known forms are the small *Cebochœrus* of the Phosphorites of Central France; *Chæropotamus* of the Upper Eocene, the type species of which was of the size of a large Pig, with the dental formula $i\ \frac{3}{3}$, $c\ \frac{1}{1}$, $p\ \frac{4}{3}$, $m\ \frac{3}{3}$, and no distinctly selenodont structure in the molars; the much larger *Elotherium*, from the Upper Eocene and Lower Miocene of both the Old and New Worlds, which presents the very rare feature of the absence of a third lobe to the last lower molar; and the equally large *Tetraconodon* of the Pliocene of India, in which this third lobe was present and the premolars were of enormous size. The remarkable North American Eocene genus *Achænodon* should perhaps also be placed here.

Anthracotheriidæ.—The genera *Anthracotherium* and *Hyopotamus*, of the upper Eocene and Miocene, have the typical Eutherian dental formula; the upper molars (Fig. 111) carrying three columns on the anterior and two on the posterior half of the crown, all of which are of a more or less decidedly selenodont structure. The mandible has a descending flange at the angle. The figured tooth (in which the antero-internal and antero-median columns are imperfect) may be compared with the diagram given in Fig. 5, p. 32, when the homology of the columns or tubercles will be at once apparent, the broken antero-median column representing the protoconule. Some of the species are of large size, while others are comparatively small.

FIG. 111.—The imperfect third left upper molar of *Hyopotamus giganteus*, Miocene, India. (From the *Palæontologia Indica*.)

Merycopotamus.—The genus *Merycopotamus* of the lower Pliocene of India may be regarded as an Anthracotheroid which has lost the antero-median column to the upper molars (Fig. 112), so that these teeth are consequently quadrituberculate; and may thus be regarded as typical examples of the brachy-selenodont modification of molar structure.

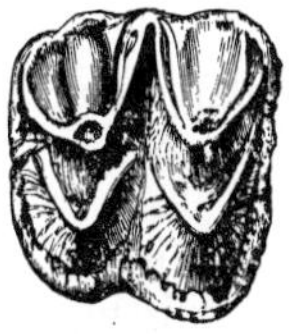

FIG. 112.—A right upper molar of *Merycopotamus pusillus*, Pliocene, India. (From the *Palæontologia Indica.*)

Cotylopidæ.—The Miocene genus *Cotylops* (*Oreodon*[1]) is the type of a large American family in which the upper molars are selenodont and usually have four columns, while the lower canine is approximated to the incisors and its form and function assumed by the first premolar. The last upper premolar is simpler than the molars. There is no flange to the angle of the mandible; and the feet have four digits. The affinities of this peculiar family are probably widely spread, but they may have been derived from the *Anthracotheriidæ.* The type genus has the full Eutherian dentition, but in some of the more specialised forms (*Cyclopidius*) the upper incisors may be wanting, and large vacuities occur in the lachrymal region. The generalised genus *Protoreodon*, of the Upper or Uinta Eocene, has five cusps on the upper molars, arranged as in the *Anthracotheriidæ.* The pollex is retained in the manus of the type genus.

The family may be divided into subfamilies as follows:—

I. Upper molars with four columns.
  1. Orbits open, no lachrymal fossa, a diastema, the last upper premolar with two outer columns, outer wall of upper molars concave and inclined inwards.—*Agriochœrinæ* (*Agriochœrus*).
  2. Orbits closed, a lachrymal fossa, no diastema, the last upper premolar with one outer column; outer wall of upper molars flattened.—*Cotylopinæ* (*Cotylops*, *Eporeodon*, *Merycochœrus*, *Cyclopidius*, etc.)

II. Upper molars with five columns.—*Protoreontinæ* (*Protoreodon*).

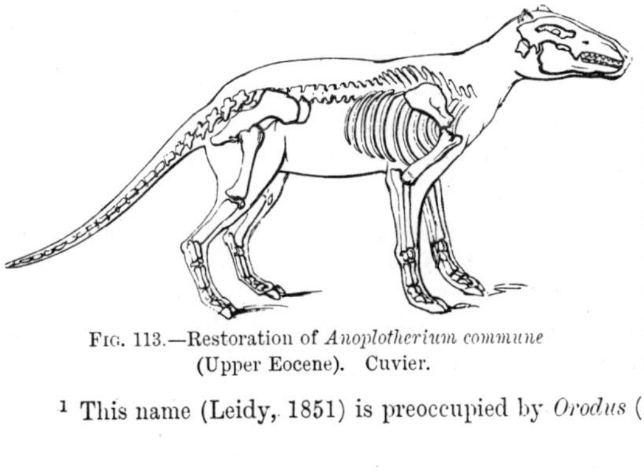

FIG. 113.—Restoration of *Anoplotherium commune* (Upper Eocene). Cuvier.

Anoplotheriidæ.—This family includes several Upper Eocene European genera, with selenodont upper molars, carrying five columns arranged as in *Anthracotherium.* One of

[1] This name (Leidy, 1851) is preoccupied by *Orodus* (Agassiz, 1838).

the earliest known, *Anoplotherium*, was fully described by Cuvier from remains found in the Paris gypsum-beds (Upper Eocene). Its forty-four teeth formed a series unbroken by a gap or diastema, and were of uniform height (as in Man alone of existing mammals). Its tail was long, with large chevron bones underneath, not usually found in Ungulates, and there were either three or two toes on each foot. It was in many respects a much-specialised form, apparently not on the line of descent of any of the existing groups.

Dacrytherium is an allied genus whose dentition leads on to that of the smaller *Xiphodon*. The latter genus is characterised by the compressed and elongated form of the crowns of the first three premolars, which thus approximate to those of the Chevrotains. There were only two functional digits to the feet. The so-called *Hyopotamus picteti*, of the Swiss Eocene, is a species of *Dacrytherium*.

Cœnotheriidæ.—The typical representatives of this family are small animals not larger than the Chevrotains, with the full complement of teeth, generally no marked gap in the series, and the crowns of the upper molars carrying two columns on the anterior and three on the posterior half of the crown—precisely the reverse of the arrangement obtaining in the *Anthracotheriidæ*. The known forms are from the Upper Eocene and Lower Miocene of Europe. In *Cœnotherium* the molars are selenodont, while they are bunodont in *Dichobunus*. *Homacodon*, of the Bridger Eocene of the United States, is closely allied to the the latter. The first lower premolar of *Dichobunus* assumes the form and function of a canine. *Spaniotherium* (*Metriotherium*) is a much larger form, in which the molars are not unlike those of *Anthracotherium*, if the arrangement of the cusps were reversed; it occurs in the Eocene Phosphorites of France. It is suggested that the *Tylopoda* may have originated from this group.

Tapirulus is a small Eocene Artiodactyle with the columns of the upper molars, which are somewhat like those of *Hyopotamus*, tending to form transverse ridges; its family position is uncertain.

Dichodontidæ.—The European genera included in this family all have quadritubercular selenodont molars, and show signs of approximating more or less closely to existing types. *Dichodon*, from the Upper Eocene and Lower Miocene, has the full complement of teeth, which show no diastema, and have low crowns. The fourth upper premolar has four columns, like the true molars, and the corresponding lower tooth three complete lobes; these features being unknown in any other Selenodonts. In *Lophiomeryx*, of the same beds, the somewhat higher crowns of the molars approximate to those of the *Cervidæ*, but the hinder lobes of the upper ones are imperfectly developed; the genus may be allied to the *Tragulidæ*. In the small *Gelocus*, of the Lower Miocene, the molars are not unlike those of *Dichodon;* but the navicular and cuboid bones of the tarsus were fused together, and the metatarsals had united to form a "cannon-bone," although the metacarpals still remained distinct. It is not improbable that upper incisors were wanting; and it has been suggested that we have in this genus the ancestral type of the *Tragulidæ* and *Cervidæ.*

TYLOPODA.

Family CAMELIDÆ.

This group is represented at the present day by the two species of Camels of the Old World and the Llamas of South America, collectively constituting the family *Camelidæ.* The special characters which the Llamas and Camels have in common, and the combination of which distinguishes them from the rest of the Artiodactyles, are as follows. The premaxillæ have the full number of incisor teeth in the young state, and the outermost is persistent through life as an isolated laniariform tooth. The canines are present in both jaws, and those of the mandible are differentiated from the long, procumbent, and spatulate incisors, being suberect and pointed. The crowns of the true molars belong to the crescentic or selenodont type, and are very hypsodont; but one or more of the anterior premolars is usually detached from the series, and is of simple pointed form. The auditory bulla is filled with cancellous tissue. The hinder part of the body is much contracted, and the femur long and vertically placed, so that the knee-joint is lower in position, and the thigh altogether more detached from the abdomen than in most quadrupedal mammals. The limbs are long, but with only the third and fourth digits developed; no traces of any of the others being present. The trapezoid and magnum of the carpus, and the cuboid and navicular of the tarsus are distinct. The two metapodial bones of each limb are confluent for the greater part of their length, though separated for a considerable distance at the lower end. Their distal articular surfaces, instead of being pulley-like, with deep ridges and grooves, as in other recent Artiodactyles, are simple, rounded, and smooth. The proximal phalanges are expanded at their distal ends, and the wide, depressed middle phalanges are embedded in a broad cutaneous pad, forming the sole of the foot, on which the animal rests in walking, instead of on the hoofs. The ungual phalanges are very small and nodular, not flattened on their inner or opposed surfaces, and not completely encased in hoofs, but bearing nails on their upper surface only. The cervical region is long and flexuous, and the vertebræ of which it is composed are remarkable for the position of the canal for the transmission of the vertebral artery, which does not perforate the transverse process, but passes obliquely through the anterior part of the pedicle of the arch (a condition only found in two other genera of mammals, *Macrauchenia* and *Myrmecophaga*). There are no horns or antlers. Though these animals ruminate, the stomach differs considerably in the details of its construction from that of the Pecora. The interior of the rumen or paunch has no villi on its surface, and there is no distinct psalterium or maniplies. Both the first and second compartments are remarkable for the presence of a number of pouches or cells in their walls, with muscular septa, and a sphincter-like arrangement of their orifices, by which they can be shut off from the rest of the cavity, and into which the fluid portion only of the contents of the stomach is allowed to enter.[1] The placenta is diffuse, as in the Suina and Tragulina, not cotyledonary, as in the Pecora. Finally, the *Camelidæ* differ not only from other Ungulates, but from all other mammals, in the fact that the red corpuscles of the blood, instead of being circular in outline, are oval, as in the inferior vertebrated classes.

Camelus.[2]—Dentition of adult: $i\,\frac{1}{3}$, $c\,\frac{1}{1}$, $p\,\frac{3}{2}$, $m\,\frac{3}{3}$; total 34. First upper premolar simple, placed immediately behind the premaxillæ, and separated by a long diastema from the penultimate tooth of that series. Lower incisors somewhat proclivous, the outermost the largest. Skull elongated, with an overhanging occiput, orbits completely surrounded by bone, and the premaxillæ not articulating with the arched and somewhat elongated nasals. Vertebræ: C 7, D 12, L 7, S 4, C 13-15. Ears comparatively short and rounded. One or two dorsal adipose humps. Feet broad, with the toes very imperfectly separated. Tail well developed, tufted at the end. Hair nearly straight, and not woolly. Size very large and bulky.

The genus is now represented by two species, viz. the single-humped Arabian Camel (*Camelus dromedarius*), and the double-humped Bactrian Camel (*C. bactrianus*, Fig. 114).[3] The former

[1] The stomach of the Camel inhabiting the Arabian desert is commonly looked upon as a striking example of specialised structure, adapted or modified in direct accordance with a highly specialised mode of life; it is therefore very remarkable to find an organ exactly similar, except in some unessential details, in the Llamas of the Peruvian Andes and the Guanacos of the Pampas. No hypothesis except that of a common origin will satisfactorily account for this, and, granting that this view is correct, it becomes extremely interesting to find for how long a time two genera may be isolated and yet retain such close similarities in parts which in other groups appear readily subject to adaptive modifications.

[2] Linn. *Syst. Nat.* 12th ed. vol. i. p. 90 (1766).

[3] There is much confusion as to the proper use of the names Camel and Dromedary. It is now generally accepted that the former is the common term for all the members of the genus, and that Dromedary should be confined to the lighter and swifter breeds of the one-humped species. One of the oldest pictures of

is quite unknown in a wild state, but it is reported that wild Bactrian Camels occur in the more remote parts of Turkestan. The latter species is found in a domesticated state throughout a large portion of Turkestan and the neighbouring region, extending as far as the Crimea in the west and to Lake Baikal and Pekin in the east. It is a heavier and more clumsy animal than the Arabian Camel, with thicker hair, shorter legs, and the feet more callous and better adapted to a hard ground. The hair is most developed upon the top of the head, neck, humps, arm, and wrist. Bactrian Camels are occasionally brought over the stupendous mountain

FIG. 114.—The Bactrian Camel (*Camelus bactrianus*).

passes south of Yarkand to within a few days' journey of Leh, in Kashmir territory.

The Arabian Camel is commonly employed as a beast of burden in Africa and India, and has of late years been introduced into Australia for the same purpose; it is especially valuable in crossing long stretches of arid desert from its power of existing for a considerable period of time without water. The female goes fully eleven months with young, and produces but a single calf at a birth, which is suckled for a whole year. In disposition the Camel is surly and subject to furious outbursts of temper, especially during the rutting season. At such periods the male utters a peculiar and highly disagreeable bubbling noise in its throat, well known to all who have travelled in India with Camels as their transport. It has been said that the Camel is docile, but Palgrave observes:—

the two-humped Camel extant, painted on the wall of the Chapter House of Westminster Abbey, has, however, "Dromedary" inscribed under it.

"If docile means stupid, well and good; in such a case the Camel is the very model of docility. But if the epithet is intended to designate an animal that takes an interest in its rider so far as a beast can, that in some way understands his intentions, or shares them in a subordinate fashion, that obeys from a sort of submissive or half-fellow-feeling with his master, like the horse or elephant, then I say that the camel is by no means docile—very much the contrary. He takes no heed of his rider, pays no attention whether he be on his back or not, walks straight on when once set agoing, merely because he is too stupid to turn aside, and then should some tempting thorn or green branch allure him out of the path, continues to walk on in the new direction simply because he is too dull to turn back into the right road. In a word, he is from first to last an undomesticated and savage animal, rendered serviceable by stupidity alone, without much skill on his master's part, or any co-operation on his own save that of an extreme passiveness. Neither attachment nor even habit impress him; never tame, though not wide-awake enough to be exactly wild." The two species breed together freely, and among the Yourouks of Asia Minor, hybrids, or mules, the produce generally of a male Bactrian and a female Arabian camel are preferred to either of the pure breeds.

Fossil remains of Camels are found in the Pliocene of the Siwalik Hills in Northern India. These differ from the existing representatives of the genus in having a vertical ridge at the antero-external angle of the lower molars, whereby they resemble *Auchenia;* their cervical vertebræ are also intermediate in structure between those of the latter and the existing Camels. A fossil Camel is also found in the Pleistocene of Algeria.

Auchenia.[1]—Dentition of adults normally: $i\ \frac{1}{3}$, $c\ \frac{1}{1}$, $p\ \frac{2}{2}$, $m\ \frac{3}{3}$; total 32—one of the lower premolars may, however, be wanting. In the upper jaw there is a compressed, sharp, pointed laniariform incisor near the hinder edge of the premaxilla, followed, in the male at least, by a moderate-sized, pointed, curved true canine in the anterior part of the maxilla. The isolated canine-like premolar which follows in the Camels is not present. The teeth of the molar series, which are in contact with each other, consist of two very small premolars (the first almost rudimentary) and three broad molars, constructed generally like those of *Camelus.* In the lower jaw the three incisors are long, spatulate, and procumbent; the outer ones being the smallest. Next to these is a curved, suberect canine, followed after an interval by an isolated, minute, and often deciduous simple conical premolar; then a contiguous series of one premolar and three molars, which differ from those of existing species of *Camelus* in having a small accessory column at the anterior outer edge. The skull generally resembles that of *Camelus,* the relatively larger brain-cavity and

[1] Illiger, *Prodromus Syst. Mamm.* p. 103 (1811).

orbits and less developed cranial ridges being due to its smaller size. The nasal bones are shorter and broader, and are joined by the premaxillæ. Vertebræ: C 7, D 12, L 7, S 4, C 15-20. Ears rather long and pointed. No dorsal hump. Feet narrow, the toes being more separated than in the camels, each having a distinct plantar pad. Tail short. Hairy covering long and woolly. Size (in existing forms) smaller, and general form lighter than in the Camels. At present and within historic times the

FIG. 115.—Llama (*Auchenia glama*), from an animal living in the Gardens of the Zoological Society of London.

genus is entirely confined to the western side and southernmost parts of South America, but fossil remains have been found in the caves of Brazil, in the pampas of the Argentine republic, and in Central and North America.

The word Llama, sometimes spelt Lama, is the name by which the Peruvians designated one of a small group of closely allied animals, which, before the Spanish conquest of America, were the only domesticated hoofed mammals of the country, being kept, not only for their value as beasts of burden, but also for their flesh, hides, and wool,—in fact, supplying in the domestic economy of the people the place of the horse, the ox, the goat, and the sheep of the Old World. The word is now sometimes restricted to one particular species or variety of the group, and sometimes used in a generic sense to cover the whole. Although they were often compared by early writers to sheep, and spoken of as such, their affinity to the camel was very soon perceived, and they were included in the genus *Camelus* in the *Systema Naturæ* of Linnæus. They were, however, separated by Cuvier in 1800 under the name of *Lama*, changed by Illiger in 1811 to *Auchenia* (in allusion to the great length of neck, *αὐχήν*), a term afterwards adopted by Cuvier, and almost universally accepted by systematic zoologists, although there has been of late a disposition to revive the earlier name.

In essential structural characters, as well as in general appearance and habits, all the animals of this genus very closely resemble each other, so that the question as to whether they should be considered as belonging to one, two, or more species has been one which has led to a large amount of controversy among naturalists. The question has been much complicated by the circumstances of the great majority of individuals which have come under observation being either in a completely or partially domesticated state, and descended from ancestors which from time immemorial have been in like condition, one which always tends to produce a certain amount of variation from the original type. It has, however, lost much of its importance since the doctrine of the distinct origin of species has been generally abandoned.

The four forms commonly distinguished by the inhabitants of South America are recognised by some naturalists as distinct species, and have had specific designations attached to them, though usually with expressions of doubt, and with great difficulties in defining their distinctive characteristics. These are (1) the Llama, *Auchenia glama* (Linn.), or *Lama peruana* (Tiedemann); (2) the Alpaca, *A. pacos* Linn.); (3) the Guanaco or Huanaco, *A. huanacus* (Molina); and (4) the Vicugna, *A. vicugna* (Molina), or *A. vicunna*, (Cuv.) The first and second are only known in the domestic state, and are variable in size and colour, being often white, black, or piebald. The third and fourth are wild, and of a nearly uniform light-brown colour, passing into white below. They certainly differ from each other,

Fig. 116.—Head of Vicugna, from an animal living in the Gardens of the Zoological Society of London.

the Vicugna being smaller, more slender in its proportions, and having a shorter head (Fig. 116) than the Guanaco (Fig. 117). It may therefore, according to the usual view of species, be considered distinct. It lives in herds on the bleak and elevated parts of the mountain range bordering the region of perpetual snow, amidst rocks and precipices, occurring in various suitable localities throughout Peru, in the southern part of Ecuador, and as far south as the middle of Bolivia. Its manners very much resemble those of the Chamois of the European Alps; and it is as vigilant, wild, and timid. The wool is extremely delicate and soft, and highly valued for the purposes of weaving, but the quantity which each animal produces is not great.

FIG. 117.—Head of Guanaco, from an animal living in the Gardens of the Zoological Society of London.

The Guanaco has an extensive geographical range, from the highlands of the Andean region of Ecuador and Peru to the open plains of Patagonia, and even the wooded islands of Tierra del Fuego. It constitutes the principal food of the Patagonian Indians, and its skin is invaluable to them, as furnishing the material out of which their long robes are constructed. It is about the size of a European Red Deer, and is an elegant animal, being possessed of a long, slender, gracefully curved neck and fine legs. Dr. Cunningham,[1] speaking from observation on wild animals, says:—

"It is not easy to describe its general appearance, which combines some of the characters of a camel, a deer, and a goat. The body, deep at the breast but very small at the loins, is covered with long, soft, very fine hair, which on the upper parts is of a kind of fawn-colour, and beneath varies from a very pale yellow to the most beautiful snow-white. The head is provided with large ears, in general carried well back, and is covered with short grayish hair, which is darkest on the forehead. Occasionally the face is nearly black. As a rule it lives in flocks of from half a dozen to several hundreds, but solitary individuals are now and then to be met with. They are very difficult to approach sufficiently near to admit of an easy shot, as they are extremely wary, but, on being disturbed, canter off at a pace which soon puts a safe distance between them

[1] *Natural History of the Strait of Magellan*, 1871.

and the sportsman, even though he should be mounted. Despite their timidity, however, they are possessed of great curiosity, and will sometimes advance within a comparatively short distance of an unknown object, at which they will gaze fixedly till they take alarm, when they effect a speedy retreat. Their cry is very peculiar, being something between the belling of a deer and the neigh of a horse. It would be difficult to overestimate their numbers upon the Patagonian plains; for in whatever direction we walked we always came upon numbers of portions of their skeletons and detached bones."

Darwin, who has given an interesting account of the habits of the Guanaco in his *Naturalist's Voyage*, says that they readily take to the water, and were seen several times at Port Valdes swimming from island to island.

The Llama is only known as a domestic animal, and is chiefly met with in the southern part of Peru. Burmeister, a very competent writer on the subject, says that he is perfectly satisfied that it is the descendant of the wild Guanaco, an opinion opposed to that of Tschudi. It generally attains a larger size than the Guanaco, and is usually white or spotted with brown or black, and sometimes altogether black. The earliest and often-quoted account of this animal by Agustin de Zarate, treasurer-general of Peru in 1544, will bear repeating as an excellent summary of the general character and uses to which it was put by the Peruvians at the time of the Spanish conquest. He speaks of the Llama as a sheep, observing, however, that it is camel-like in shape though destitute of a hump:—

"In places where there is no snow the natives want water, and to supply this they fill the skins of sheep with water and make other living sheep carry them; for, it must be remarked, these sheep of Peru are large enough to serve as beasts of burden. They can carry about one hundred pounds or more, and the Spaniards used to ride them, and they would go four or five leagues a day. When they are weary they lie down upon the ground; and as there are no means of making them get up, either by beating or assisting them, the load must of necessity be taken off. When there is a man on one of them, if the beast is tired and urged to go on, he turns his head round and discharges his saliva, which has an unpleasant odour, into the rider's face. These animals are of great use and profit to their masters, for their wool is very good and fine, particularly that of the species called Pacas, which have very long fleeces; and the expense of their food is trifling, as a handful of maize suffices them, and they can go four or five days without water. Their flesh is as good as that of the fat sheep of Castile. There are now public shambles for the sale of their flesh in all parts of Peru, which was not the case when the Spaniards came first; for when one Indian had killed a sheep his neighbours came and took what they wanted, and then another Indian killed a sheep in his turn."

The disagreeable habit here noticed of spitting in the face of persons whose presence is obnoxious is common to all the group, as may be daily witnessed in specimens in confinement in the menageries of Europe. One of the principal labours to which the Llamas were subjected at the time of the Spanish conquest was that of bringing down ore from the mines in the mountains. Gregory de Bolivar estimated that in his day as many as three hundred thousand were employed in the transport of the produce of the mines of Potosi alone; but since the introduction of horses, mules, and donkeys the importance of the Llama as a beast of burden has greatly diminished.

The Alpaca, though believed by many naturalists to be a variety of the Vicugna, is more probably, like the Llama, derived from the Guanaco, having the naked callosities on the hind limbs, and the relatively large skull of the latter. It is usually found in a domesticated or semi-domesticated state, being kept in large flocks which graze on the level heights of the Andes of southern Peru and northern Bolivia at an elevation of from 14,000 to 16,000 feet above the sea-level, throughout the year. It is smaller than the Llama, and, unlike that animal, is not used as a beast of burden, but is valued only for its wool, of which the Indian blankets and ponchas are made. Its colour is usually dark brown or black.

Mention has already been made of the occurrence of fossil Llamas in America, but some diversity of view obtains as to the generic position of some of these forms, owing to variations in their dental formula. Remains apparently referable to the existing species occur in the cavern-deposits of Brazil. In the Pleistocene of Mexico we meet with *A.* (*Palauchenia*) *magna*, which attained the size of a Camel, and had always two, and occasionally three, lower premolars; while in one South American Pleistocene species, which has been generically separated as *Hemiauchenia*, there were invariably three premolars in each jaw. In *A.* (*Holomeniscus*) *hesterna*, from the Pleistocene of North America, which was equal in size to *A. magna*, the premolars were reduced to one in each jaw; and the same condition obtains in *A.* (*Eschatius*) *vitakeriana*, where, however, the upper one is of simpler structure.

Extinct Cameloids.—Until within the last few years the existence of two genera having so very much in common as the Camels and the Llamas, and yet so completely isolated geographically, had not received any satisfactory explanation; for the old idea that they in some way "represented" each other in the two hemispheres of the world was a mere fancy without philosophical basis. The discoveries made mostly within the past twenty years of a vast and previously unsuspected extinct fauna in the American continent of the Tertiary period, as interpreted by Leidy, Cope, Marsh, and others, has thrown a flood of light upon the early history of this family, and upon its relations to other mammals.

There have been found in these regions many Camel-like animals exhibiting different generic modifications; and, what is more interesting, a gradual series of changes, coinciding with the antiquity of the deposits in which they are found, have been traced from the thoroughly differentiated species of the modern epoch down through the Pliocene to the early Miocene beds, where, their characters having become by degrees more generalised, they have lost all that specially distinguishes them as *Camelidæ*, and are merged into forms common to the ancestral type of all the other sections of the Artiodactyles. Hitherto none of these annectant forms have been found in any of the fossiliferous strata of the Old World; and it may therefore be fairly surmised (according to the evidence at present before us) that America was the original home of the Tylopoda, and that the true Camels have passed over into the Old World, probably by way of the north of Asia, where we have every reason to believe there was formerly a free communication between the continents, and then, gradually driven southward, perhaps by changes of climate, having become isolated, have undergone some further special modifications; while those members of the family that remained in their original birthplace have become, through causes not clearly understood, restricted solely to the southern or most distant part of the continent. The occurrence in the dentition of the fossil Siwalik Camels of a feature now found only in *Auchenia* is especially interesting from this point of view.

Briefly referring to some of these fossil types, we may note that *Pliauchenia*, of the Loup Fork beds (Lower Pliocene) of the United States, has three lower premolars, while in *Procamelus* there were four of these teeth. In *Protolabis* of the Miocene we have a more generalised form, in which the dental formula is $i\ \frac{3}{3},\ c\ \frac{1}{1},\ p\ \frac{4}{4},\ m\ \frac{3}{3}$; and from this type a transition may be traced to *Poëbrotherium*, which, while having the same dental formula, was no larger than a Fox, and had the third and fourth metacarpals separate, with rudiments of the fourth and fifth. The earliest undoubted representative of the group is *Leptotragulus*, of the Uinta Eocene, which appears to have been closely allied to *Poëbrotherium*. It is, however, probable that the first lower premolar was wanting; while the other premolars of the mandible were much shorter antero-posteriorly than in the last-named genus. The manus, moreover, appears to have been less reduced, the second metacarpal retaining its connection with the magnum. It is suggested that *Leptotragulus* may have been derived from the

Bunodont genus *Homacodon* of the Bridger Eocene, mentioned among the *Cœnotheriidæ*.

TRAGULINA.

Family TRAGULIDÆ.

No teeth in premaxillæ. Upper canines well developed, especially in the males; narrow and pointed. Lower canines incisiform. No caniniform premolars in either jaw, all the premolars except the last in the upper jaw being secant. Molariform teeth in a continuous series, consisting of $p\ \frac{3}{3}$, $m\ \frac{3}{3}$. Odontoid process of axis vertebra conical. Fibula complete. Four complete toes on each foot. The middle metapodials generally confluent, the outer ones (second and fifth) very slender but complete, *i.e.* extending from the carpus or tarsus to the digit. Navicular, cuboid, and ectocuneiform bones of tarsus united. Tympanic bullæ of skull filled with cancellar tissue. No frontal appendages. Ruminating, but the stomach with only three distinct compartments, the maniplies or third cavity of the stomach of the Pecora being rudimentary. Placenta diffused.

This section is represented only by the single family *Tragulidæ*, containing a few animals of small size, commonly known as Chevrotains, intermediate in their structure between the Deer, the Camels, and the Pigs. The large size of the canines of the male and the absence of horns caused them to be associated formerly with *Moschus*, one of the *Cervidæ;* hence they are often spoken of as "Pigmy Musk-Deer," although they have no musk-secreting gland, or, except in the above-named trivial external characters, no special affinities with the true Musk-Deer. There has scarcely been a more troublesome and obdurate error in zoology than in this association of animals so really distinct. It has been troublesome, not only in preventing a just conception of the relations of existing Artiodactyles, but also in causing great confusion and hindrance in palæontological researches among allied forms; and most obdurate, inasmuch as all that has been recently done in advancing our knowledge of both groups has not succeeded in eradicating it, not only from nearly every one of our zoological text-books, whether British or Continental, but even from works of the highest scientific pretensions.

The family is now generally divided into two genera.

Tragulus,[1] containing the smallest of the existing Ungulates, animals having more of the general aspects and habits of some Rodents, as the Agoutis, than of the rest of their own order. The best-known species are *T. javanicus*, *T. napu*, *T. stanleyanus*, and

[1] Pallas, *Spicilegia Zoologica*, vol. xiii. p. 27 (1779).

T. memmina. The first three are from the Malay Peninsula, or the islands of the Indo-Malayan Archipelago, the last from Ceylon and India. A fossil species occurs in the Pliocene of the latter country.

Dorcatherium[1] is distinguished chiefly by the feet being stouter and shorter, the outer toes better developed, and the two middle metacarpals not ankylosed together. Its dental formula (as that of *Tragulus*) is usually $i\,\frac{0}{3}$, $c\,\frac{1}{1}$, $p\,\frac{3}{3}$, $m\,\frac{3}{3} = 34$. Vertebræ: C 7, D 13, L 6, S 5, C 12-13. The only existing species, *D. aquaticum* (Fig. 118), from the west coast of Africa, is rather larger than any

FIG. 118.—The African Water-Chevrotain (*Dorcatherium aquaticum*).

of the Asiatic Chevrotains, which it otherwise much resembles, but it is said to frequent the banks of streams, and have much the habits of Pigs. It is of a rich brown colour, with back and sides spotted and striped with white. It is evidently the survivor of a very ancient form, as remains of the type species (*D. naui*), only differing in size, occur in the lower Pliocene and Miocene of Europe; fossil species are also found in the Indian Pliocene. In *D. naui* there are, at least frequently, four lower premolars, while the existing species has but three of these teeth.

Extinct Traguloids.—A number of small selenodont Artiodactyles

[1] Kaup, *Ossemens Fossiles de Darmstadt*, pt. 5, p. 92 (1836). This name, which was proposed for a fossil species, antedates *Hyomoschus*, Gray, applied to the living form.

from various Miocene and Pliocene deposits appear to connect the modern Tragulina so closely with *Gelocus* (p. 294), and thus with the ancestral *Cervidæ*, that their classification is almost an impossibility. Thus *Leptomeryx*, from the Miocene of the United States, is regarded as a Traguloid, having four premolars in each jaw and with the metatarsals fused into a cannon-bone. *Prodremotherium*, of the Upper Eocene Phosphorites of France, differs in that the metacarpals also form a cannon-bone; while in the American *Hypertragulus*, both metacarpals and metatarsals remain separate. *Bachitherium*, of the French Phosphorites, apparently presents affinity with *Gelocus*, *Prodremotherium*, and *Dorcatherium*. In this genus the first of the four lower premolars assumes the character and function of a canine, the true canine being incisor-like, and there are traces of minute upper incisors.

Pecora, or Cotylophora.

No premaxillary teeth or caniniform premolars. Upper canines generally absent, though sometimes largely developed. Inferior incisors, three on each side with an incisiform canine in contact with them. Molariform teeth consisting of *p* $\frac{3}{3}$, *m* $\frac{3}{3}$, in continuous series. Auditory bullæ simple and hollow within. Odontoid process in the form of a crescent, hollow above. Distal extremity of the fibula represented by a distinct malleolar bone of peculiar shape, articulating with the outer surface of the lower end of the tibia. Third and fourth metacarpals and metatarsals confluent. Outer or lateral toes small and rudimentary, or in some cases entirely suppressed; their metapodial bones never complete in existing forms. Navicular and cuboid bones of tarsus united. Horns or antlers usually present, at least in the male sex. Left brachial artery arising from a common innominate trunk, instead of coming off separately from the aortic arch as in the preceding sections. Stomach with four complete cavities. Placenta cotyledonous.[1]

The Pecora or true Ruminants form at the present time an extremely homogeneous group, one of the best-defined and most closely united of any of the Mammalia. But, though the original or common type has never been departed from in essentials, variation has been very active among them within certain limits; and the great difficulty which all zoologists have felt in subdividing them into natural minor groups arises from the fact that the changes in different organs (feet, skull, frontal appendages, teeth, cutaneous glands, etc.) have proceeded with such apparent irregularity and absence of correlation that the different modifica-

[1] For the anatomy of this group see A. H. Garrod, *Proc. Zool. Soc.* 1877, p. 2.

tions of these parts are most variously combined in different members of the group. It appears, however, extremely probable that they soon branched into two main types, represented in the present day by the *Cervidæ* and the *Bovidæ*,—otherwise the antlered and horned Ruminants. Intermediate smaller branches produced the existing Musk-Deer and Giraffe, as well as the extinct *Helladotherium* inclining to the first-named group, and the extinct *Sivatherium*, *Brahmatherium*, *Hydaspitherium*, and others more allied to the latter, although upon the true relationship of these forms there is a difference of opinion.

The earliest forms of true Pecora, as *Palæomeryx*, generally had no frontal appendages, and some few forms continue to the present day in a similar case. In the very large majority, however, either in both sexes or in the male only, a pair or occasionally two pairs (*Tetraceros* and the extinct *Sivatherium*) of processes are developed from the frontal bones as weapons of offence and defence, these being almost always formed on one or other of two types.

FIG. 119.—A shed right antler of the Red Deer (*Cervus elaphus*), found in an Irish lake. *a*, Brow tine; *b*, bez tine; *c*, tres tine; *d*, crown or royal tine. (After Owen.)

1. "Antlers" are outgrowths of true bone, covered during their growth with vascular, sensitive integument coated with short hair. When the growth of the antler is complete, the supply of blood to it ceases, the skin dies and peels off, leaving the bone bare and insensible, and after a time, by a process of absorption near the base, it becomes detached from the skull and is "shed" (Fig. 119). A more or less elongated portion or "pedicle" always remains on the skull, from the summit of which a new antler is developed. In the greater number of existing species of Deer this process is repeated with great regularity at the same period of each year. The antler may be simple, straight, subcylindrical, tapering and pointed, but more often it sends off one or more branches called "tines" or "snags" (Fig. 119). In this case the main stem is termed the "beam." Commonly all the branches of the antler are cylindrical and gradually tapering. Sometimes they are more or less expanded and flattened, the

antler being then said to be "palmated." In young animals the antlers are always small and simple, and in those species in which they are variously branched or palmated, this condition is only gradually acquired in several successive annual growths. An interesting parallel has been observed here, as in so many other cases, between the development of the race and that of the individual. Thus the earliest known forms of Deer, those of the Lower Miocene, generally have no antlers, as in the young of the existing species. The Deer of the Middle Miocene have simple antlers, with not more than two branches, as in existing Deer of the second year; but it is not until the Pliocene and Pleistocene times that Deer occur with antlers developed with that luxuriance of growth and beauty of form characteristic of some of the existing species in a perfectly adult state. Among recent *Cervidæ*, antlers are wanting in the genera *Moschus* and *Hydropotes;* they are present in both sexes in *Tarandus* (the Reindeer), and in the male sex only in all others.

FIG. 120.—Head of Deer (*Cervus schomburgki*), showing antlers. From Sclater, *Proc. Zool. Soc.* 1877, p. 682.

In those forms with the most complex antlers (Figs. 119, 120) the tine immediately over the forehead is termed the *brow tine*, the next one the *bez tine*, and the third one the *tres tine;* the mass of points at the summit of the antler being termed either the *royal* and *surroyal tines*, or collectively the *crown*. The nodulated bony ring at the base of the antler, just above the point at which it separates from the pedicle when it is shed, is termed the *burr*.

2. The horns of the *Bovidæ* consist of permanent, conical, usually curved bony processes, into which air-cells continued from the frontal sinuses often extend, called "horn-cores," ensheathed in a case of true horn, an epidermic development of fibrous structure, which grows continuously, though slowly, from the base, and wears away at the apex, but is very rarely shed entire. The only existing species in which the latter process occurs regularly and periodically is the American Prong-Buck (*Antilocapra*), in which the horns also differ from those of all others in being bifurcated. Horns are not present at birth, but begin to grow very soon afterwards. The males of all existing *Bovidæ* possess them, and they are also present (though usually not so fully developed) in the females of all except the genera *Boselaphus*, *Strepsiceros*, *Tragelaphus*, *Antilope*, *Æpyceros*, *Saiga*, *Cobus*, *Cervicapra*, *Pelea*, *Nanotragus*, *Neotragus*, *Cephalophus*, and *Tetraceros*; as well as in some species of *Gazella*, such as *G. picticaudata* and *G. walleri*.

FIG. 121.—Head of Antelope (*Gazella granti*), showing horns. From Sir V. Brooke, *Proc. Zool. Soc.* 1878, p. 724.

Another character by which different members of the *Pecora* can be distinguished among themselves is derived from the nature of the molar teeth. Although there is nothing in the general mode and arrangement of the enamel-folds, or in the accessory columns, absolutely distinctive between the two principal families, existing species may generally be distinguished, inasmuch as the true molars of the *Cervidæ* are more or less brachydont, and those of the *Bovidæ* generally hypsodont, *i.e.* the teeth of the former have comparatively short crowns (Fig. 122), which, as in most mammals, take their place at once with the neck (or point where the crown and root join) on a level with or a little above the alveolar border, and remain in this position throughout the animal's life; whereas in the other forms (Fig. 123), the crown being lengthened and the root small, the neck does not come up to the alveolar level until a considerable part of the surface has worn away, and the crown of the tooth thus appears for the greater part of the animal's life partially buried in the socket. In this form of tooth (which is almost always most developed in the posterior molars of the permanent series) the constituent columns of the crown are necessarily nearly parallel, whereas

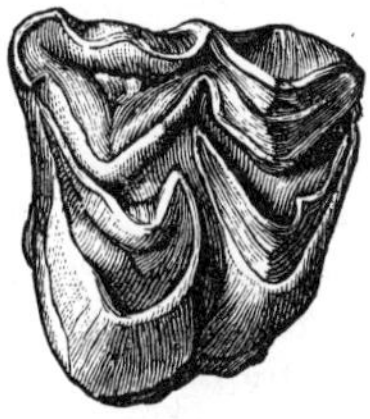

FIG. 122.—Crown surface of a worn left upper molar of *Palæomeryx sivalensis*, to show brachydont type. (From the *Palæontologia Indica.*)

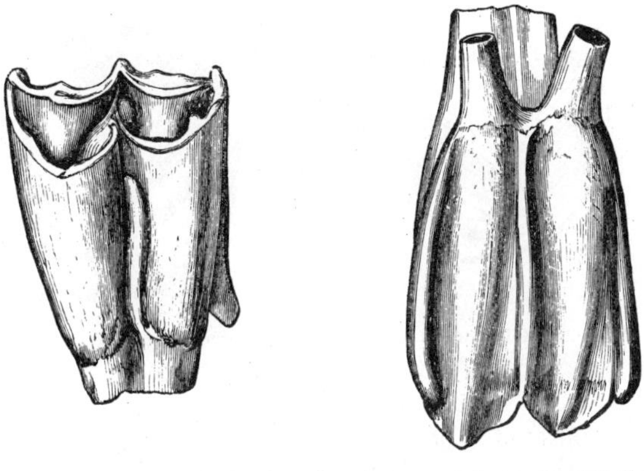

FIG. 123.—Inner and outer aspects of an almost unworn left upper molar of the Nilghai (*Boselaphus tragocamelus*), to show hypsodont type. (From the *Palæontologia Indica.*)

in the first-described they diverge from the neck towards the free or grinding surface of the tooth. In the completely hypsodont form the interstices of the lengthened columnar folds of enamel and dentine are filled up with cement, which gives stability to the whole organ, and is entirely or nearly wanting in the short-crowned teeth. The same modification from low to high crowns without essential alteration of pattern is seen in an even still more marked manner in some of the Perissodactyle Ungulates, the tooth of the Horse bearing to that of *Anchitherium* the same relation as that of an Ox does to the early selenodont Artiodactyles.

A parallel modification has also taken place in the molar teeth of the Proboscidea.

As the hypsodont tooth is essentially a modification of, and, as it were, an improvement upon, the brachydont, it is but natural to expect that all intermediate forms may be met with. Even among the Deer themselves, as pointed out by Lartet, the most ancient have very short molars, and the depressions on the grinding surface are so shallow that the bottom is always visible; while in the *Cervidæ* of the more recent Tertiary periods, and especially the Pleistocene and living species, these same cavities are so deep that whatever be the state of the dentition the bottom cannot be seen. Some existing Deer, as the Axis, are far more hypsodont than the majority of the family; and, on the other hand, many of the Antelopes (as *Tragelaphus*) retain much of the brachydont character, which is, however, completely lost in the more modern and highly specialised Sheep and Oxen.

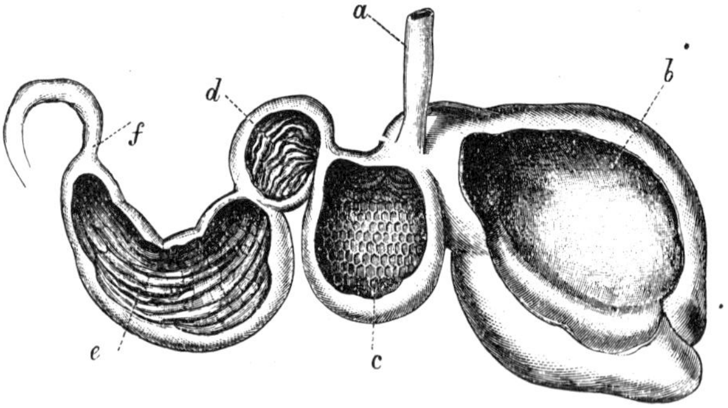

FIG. 124.—Stomach of Ruminant opened to show internal structure. *a*, Œsophagus; *b*, rumen or paunch; *c*, reticulum or honey-comb bag; *d*, psalterium or manyplies; *e*, abomasum or reed; *f*, duodenum.

The complicated stomach of the Pecora (Fig. 124), which is necessary for the performance of the peculiar function known as "chewing the cud"—a function common also to the Tragulina and Tylopoda—is divided into four well-defined compartments, known as (1) the Rumen or Paunch, (2) the Reticulum or Honey-comb Bag, (3) the Psalterium or Manyplies, (4) the Abomasum or Reed. The paunch is a very capacious receptacle, shaped like a blunted cone bent partly upon itself. Into its broader base opens the œsophagus or gullet at a spot not far removed from its wide orifice of communication with the second stomach or honey-comb bag. Its inner walls are nearly uniformly covered with a pale mucous membrane, which is beset with innumerable close-set, short, and slender villi, resembling very much the "pile" on velvet The honey-comb bag is very much smaller than the paunch.

It is nearly globose in shape, and receives its name on account of the peculiar arrangement of its mucous membrane which forms shallow hexagonal cells all over its inner surface. Running along its upper wall there is a deep groove, coursing from the first to the third stomach. This groove plays an important part in the act of rumination. Its walls are muscular, like those of the viscus with which it is associated, which allows its calibre to be altered. Sometimes it completely closes round so as to become converted into a tube by the opposition of its edges. At others it forms an open canal. The manyplies is globular in form, and its lining membrane is raised into longitudinal folds or laminæ arranged very much like the leaves of a book, and very close together. Their surfaces are roughened by the presence of small projections or papillæ. The reed is the proper digestive stomach, corresponding with the same organ in man. Its shape is somewhat pyriform, and its walls are formed of a smooth mucous membrane, which secretes the gastric juice.

When the food is first swallowed it is conveyed into the paunch, and after undergoing a softening process there it is regurgitated into the mouth, and undergoes a further trituration by the molar teeth and mixture with the secretion of the salivary and buccal glands. It is then swallowed again, but now passes directly through the before-mentioned groove into the manyplies, and, after filtering through the numerous folds of the lining membrane of this cavity, finally reaches the fourth or digestive stomach.

The placenta of the Pecora is characterised by the fœtal villi being collected into groups or cotyledons, which may present either a convex or a concave surface to the uterus. These cotyledons are received into permanent elevations in the mucous membrane of the uterus, the surfaces of which present a curvature which is the reverse of the cotyledons.

Family CERVIDÆ.

Frontal appendages, when present, in the form of antlers. First molar, at least, in both jaws brachydont. Two orifices to the lachrymal duct, situated on or inside the rim of the orbit. An antorbital or lachrymal vacuity of such dimensions as to exclude the lachrymal bone from articulation with the nasal. Upper canines usually present in both sexes, and sometimes attaining a very great size in the male (see Fig. 134). Lateral digits of both fore and hind feet almost always present, and frequently the distal ends of the metapodials. Placenta with few cotyledons. Gall-bladder absent (except in *Moschus*). This family contains numerous species, having a wide geographical distribution, ranging in the New World from the Arctic

Circle as far south as Chili, and in the Old World throughout the whole of Europe and Asia, though absent in the Ethiopian and Australian regions.

It may be divided into two subfamilies.

Subfamily **Moschinæ.**—This subfamily is represented solely by the Musk-Deer, which differs so remarkably from the true Deer that it is considered by several writers as the representative of a separate family. The late Professor Garrod even suggested that it should be regarded as an extremely aberrant member of the *Bovidæ.*

Moschus.[1]—The Musk-Deer (Fig. 125) in many respects stands by

FIG. 125.—The Musk-Deer (*Moschus moschiferus*).

itself as an isolated zoological form, retaining characters belonging to the older and more generalised types of ruminants before they were distinctly separated into the horned and the antlered sections now dominant upon the earth. One of these characters is that both sexes are entirely devoid of any sort of frontal appendage. In this, however, it agrees with one existing genus of true Deer (*Hydropotes*); and, as in that animal, the upper canine teeth of the males are remarkably developed, long, slender, sharp pointed, and gently curved, projecting downwards out of the mouth with the ends turned somewhat backwards. Vertebræ: C 7, D 14, L 5, S 5, C 6. Among the anatomical peculiarities in which it differs from all true Deer and agrees with the *Bovidæ* is the presence of a gall-

[1] Linn. *Syst. Nat.* 12th ed. vol. i. p. 91 (1766).

bladder. The hemispheres of the brain are but slightly convoluted, and the cotyledons of the placenta are arranged in a peculiar linear manner.[1]

Although, owing to variations of colour presented by different individuals in different localities and seasons, several nominal species have been described, zoologists are now generally agreed that there is but one, the *Moschus moschiferus* of Linnæus. In size it is rather less than the European Roe Deer, being about 20 inches high at the shoulder. Its limbs, especially the hinder ones, are long. The feet are remarkable for the great development of the lateral pair of hoofs, and for the freedom of motion they all present, so that they appear to have the power of grasping projecting rocky points,—a power which must be of great assistance to the animal in steadying it in its agile bounds among the crags of its native haunts. The ears are large, and the tail quite rudimentary. The hair covering the body is long, coarse, and of a peculiarly brittle and pith-like character, breaking with the application of an extremely slight force; it is generally of a grayish-brown colour, sometimes inclined to yellowish-red, and often variegated with lighter patches. The Musk-Deer has a wide distribution over the highlands of central and eastern Asia, including the greater part of southern Siberia, and extends to Kashmir on the south-west and Cochin-China on the south-east, always, however, at considerable elevations,—being rarely found in summer below 7000 feet above the sea-level, and ranging as high as the limits of the thickets of birch or pines, among which it mostly conceals itself in the day-time. It is a hardy, solitary, and retiring animal, chiefly nocturnal in its habits, and almost always found alone, rarely in pairs, and never in herds. It is exceedingly active and sure-footed, having few equals in traversing rocky and precipitous ground; and it feeds on moss, grass, and leaves of the plants which grow on the mountains among which it makes its home.

Most of the animals of the group to which the Musk-Deer belongs, in fact the large majority of mammals, have some portion of the cutaneous surface peculiarly modified and provided with glands secreting some odorous and oleaginous substance specially characteristic of the species. This, correlated with the extraordinary development of the olfactory organs, appears to offer the principal means by which animals in a state of nature become aware of the presence of other individuals of their own species, or of those inimical to them, even at very great distances, and hence it is of extreme importance both to the well-being of the individual and to the continuance of the race. The situation of this specially modified portion of skin is extremely various, sometimes between the toes, as in Sheep, sometimes on the face in front of the eyes, as in many

[1] For the anatomy of *Moschus* see Flower, *Proc. Zool. Soc.* 1875, p. 159; and Garrod, *ibid.* 1877, p. 287.

Deer and Antelopes. Sometimes it is in the form of a simple depression or shallow recess, often very deeply involuted, and in its fullest state of development it forms a distinct pouch or sac with a narrow tubular orifice. In this sac a considerable quantity of the secretion can accumulate until discharged by the action of a compressor muscle which surrounds it. This is the form taken by the special gland of the Musk-Deer, which has made the animal so well known, and has proved the cause of an unremitting persecution to its possessor. It is found in the male only, and is a sac about the size of a very small orange, situated beneath the skin of the abdomen, the orifice being immediately in front of the preputial aperture. The secretion with which the sac is filled is of dark-brown or chocolate colour, and when fresh described as being of the consistence of "moist gingerbread," but becoming dry and granular after keeping. It has a peculiar and very powerful scent, which when properly diluted and treated forms the basis of many of our most admired perfumes. When the animal is killed the whole gland or "pod" is cut out and dried, and in this form reaches the market of the Western World, chiefly through China.

Subfamily **Cervinæ**.—This subfamily includes all the true Deer. According to the arrangement proposed by Sir V. Brooke [1] the existing *Cervinæ* may be divided into the sections Plesiometacarpalia and Telemetacarpalia.

Plesiometacarpalia.—In this section, which is mainly characteristic of the Old World, the proximal portions of the lateral (second and fifth) metacarpals persist, and the vomer is never so ossified as to divide the posterior osseous nares into two distinct passages. The premaxillæ nearly always articulate with the nasals.

Cervulus.[2]—Antlers half the length of the head, placed on pedicles nearly equal to them in length. Brow tine short, inclined inwards and upwards; terminal extremity of beam unbranched, and curved downwards and inwards. Lachrymal fossa of skull very large, and extending into facial part of jugal; lachrymal (antorbital) vacuity moderate. Ascending portion of premaxillæ at least as long as nasals. A permanent ridge extending from each pedicle over the orbit, lachrymal fossa and vacuity. Auditory bulla much inflated. Upper canines of males very large. Ectocuneiform united with naviculo-cuboid of tarsus. No traces of the phalanges of the lateral digits.

The native name Muntjac has been generally adopted in European languages for a small group of Deer indigenous to the southern and eastern parts of Asia and the adjacent islands, which are separated by very marked characters from all their allies. They are also called "Kijang" or "Kidjang," and constitute the genus

[1] *Proc. Zool. Soc.* 1878, p. 889.

[2] De Blainville, *Bull. Soc. Philom.* 1816, p. 74.

Cervulus of Blainville and most zoologists ;—*Styloceros* of Hamilton-Smith, and *Prox* of Ogilby. They are all of small size compared with the majority of Deer, and have long bodies and rather short limbs and neck. The antlers, which as in most Deer are present in the male only, are small and simple, and the main stem or beam, after giving off a very short brow tine, inclines backwards and upwards, is unbranched and pointed, and when fully developed curves inwards and somewhat downwards at the tip. These small antlers are supported upon pedicles or permanent processes of the frontal bones, longer than in any other Deer, and the front edges of which are continued downwards as strong ridges passing along the sides of the face above the orbits, and serving to protect the large supra-orbital glands lying on their inner sides. The lachrymal fossa of the skull, in which is lodged the large suborbital gland or crumen, is of great depth and extent. The upper canine teeth of the males are strongly developed and sharp, curving downwards, backwards, and outwards, projecting visibly outside the mouth as tusks, and loosely implanted in their sockets. In the females they are very much smaller. The limbs exhibit several structural peculiarities not found in other Deer. The lateral digits of both fore and hind feet are very little developed, the hoofs alone being present and their bony supports (found in all other Deer) wanting. There is a tufted gland on the outer side of the metatarsus.

The Muntjacs are solitary animals, very rarely even two being seen together. They are fond of hilly ground covered with forests, in the dense thickets of which they pass most of their time, only coming to the skirts of the woods at morning and evening to graze. They carry the head and neck low and the hind-quarters high, their action in running being peculiar and not very elegant, somewhat resembling the pace of a sheep. Though with no power of sustained speed or extensive leap, they are remarkable for flexibility of body and facility of creeping through tangled underwood. They are often called by Indian sportsmen "Barking Deer," a name given on account of their alarm cry, a kind of short shrill bark, like that of a fox but louder, which may often be heard in the jungles they frequent both by day and by night. When attacked by dogs the males use their sharp canine teeth with great vigour, inflicting upon their opponents deep and even dangerous wounds.

There is some difference of opinion among zoologists as to the number of species of the genus *Cervulus*. Sir Victor Brooke, who investigated this question in 1878 (see *Proceedings of the Zoological Society of London* for that year, p. 898), came to the conclusion that there are certainly three which are quite well marked, viz.—

C. muntjac (Fig. 126), found in British India, Burma, the Malay Peninsula, Sumatra, Java, Hainan, Banca, and Borneo. The general colour is a bright yellowish-red, darker in the upper parts of the back; the fore legs from the shoulder downwards and the lower part of the hind legs, dark bluish-brown; anterior parts of the face from the muzzle to between the eyes, brown—a blackish line running up the inside of each frontal ridge; chin, throat, inside of hind legs, and under surface of tail white. The female has a black bristly tuft of hair on the spot from which the pedicles of the antlers of the male grow. The average length of the male, according to Jerdon, is 3½ feet, tail 7 inches, height 26 to 28 inches. The female is a little smaller. The specimens from Java, Sumatra, and Borneo are

FIG. 126.—The Muntjac (*Cervulus muntjac*).

of larger size than those from the mainland, and may possibly be of distinct species or race.

C. lacrymans of Milne-Edwards, or Sclater's Muntjac of Swinhoe, from Moupin, and near Hangchow, China.

C. reevesi, a very small species from southern China.

Subsequently the name *C. crinifrons* has been applied to a Muntjac from Ningpo, China, readily distinguished from all other species by its bushy forehead and long tail. Another species from Tenasserim has been described as *C. feæ*.

Small Deer from the European Pliocene have been provisionally referred to *Cervulus*, but the so-called *Prox furcatus*, of the Miocene, is now included in *Palæomeryx*.

Elaphodus.[1]—Antlers very small, unbranched, supported on long,

[1] Milne-Edwards, *Nouv. Arch. du Muséum*, vol. vii. Bull. p. 93 (1872).

slender, converging pedicles. Ascending rami of premaxillæ shorter than nasals. No supraorbital ridges or frontal glands. Upper canines of male long, but not everted. A distinct frontal tuft of hair. Other characters as in *Cervulus*.

This genus (which has also received the name of *Lophotragus*) is represented by a small Deer (Fig. 127) from China of about the same size as the Indian Muntjac. The male has minute simple antlers and very large canine teeth. There are no supraorbital

FIG. 127.—Male of *Elaphodus michianus*. From Sclater *Proc. Zool. Soc.* 1876, p. 273.

glands, nor is there a tufted gland on the metatarsus. The limbs have the same peculiarities as in *Cervulus*, but the mesocuneiform may also ankylose with the ectocuneiform, and traces of the metacarpals may remain. The hair is coarse and somewhat quill-like.

Cervus.[1]—The great majority of the Deer of the Old World may be included in this large genus, which is one not easy of definition. The antlers of the male are, however, large, and two or three times the length of the head, and may be either rounded or palmate; the canines are never large; the ectocuneiform of the tarsus remains distinct from the naviculo-cuboid; the lateral digits are represented by their phalanges; and the skull does not carry prominent frontal ridges. Vertebræ: C 7, D 13, L 6, S 4, C 11-14. The size of the

[1] Linn. *Syst. Nat.* 12th ed. vol. i. p. 92 (1766).

lachrymal fossa and vacuity, and the degree of inflation of the auditory bulla, are subject to variation in the different groups into which the genus may be divided.

The *Rusine* group is characteristic of the Oriental region, where it is typically represented by the Sambur (*C. aristotelis*) of India, Burma, and China. The antlers are rounded, and often strongly grooved, without a bez tine, and with the beam simply forked at the extremity, upright, and but slightly curved; the angle formed by the brow tine, which rises close to the burr, being acute. The molars are markedly hypsodont, with small accessory columns. The lachrymal fossa is deep and the vacuity large; the auditory bulla is slightly inflated and rugose. Tail moderate; neck maned.

The Sambur, which is abundant in hilly districts, is a fine animal, standing nearly 5 feet in height, and of massive build; the general colour being deep brown. *C. equinus*, of Borneo, Sumatra, and Singapore, *C. swinhoei*, of Formosa, *C. philippinus*, and *C. alfredi* of the Philippines, are closely allied species, of which the two latter are of smaller dimensions. The Indian Hog Deer (*C. porcinus*) is a still smaller form, not larger than the Roe. *C. hippelaphus* of Java, *C. timoriensis*, and *C. moluccensis* are distinguished by the posterior branch of the beam of the antler being considerably larger than the anterior.

The *Rucervine* group is another strictly Oriental one, and is represented by the Swamp Deer (*C. duvaucelli*) of India, the closely allied *C. schomburgki* of Siam, of which the antlers are shown in Fig. 119 (p. 309), and *C. eldi* of Burma and Hainan. The beam of the antler is somewhat flattened, and more curved than in the Rusine group; the large brow tine is given off from the beam at an obtuse angle and curves upwards; the beam bifurcates into two branches, which again divide. Skull as in the Rusine group, but relatively narrower. Tail short; neck maned.

The Swamp Deer is somewhat smaller than the Sambur, and of a full yellowish colour. Fossil representatives of this group occur in the Pliocene of India.

The *Elaphurine* group is represented only by the very aberrant *C. davidianus* of Northern China. In size and proportions this species approximates to the Swamp Deer, but the antlers are peculiar in rising straight from the brow and then giving off a long and straight back tine (correlated by Sir V. Brooke with the posterior branch of the Rusine antler); the summit of the beam is forked, and in old individuals the two tines of the fork may again branch. Nasals long, and much expanded between the lachrymal vacuities, of which they form the inner border; lachrymal fossa large and deep. Tail long; neck maned.

The *Axine* group includes only the well-known Axis of India, readily distinguished by the white spots with which the body is marked. Antlers of a Rusine type, the beam being much curved, and the brow tine usually given off at an acute or right angle. Molars very hypsodont. The coloration of the Axis is more brilliant than that of any other member of the family.

Here may be noticed a group of Deer mainly characteristic of the eastern Palæarctic region, frequently known as the *Pseudaxine* group, which appears to connect the Axine with the Elaphine type. Well-known representatives of this group are *C. sika* (Fig. 128) of Japan, *C. mantchuricus* of China, and *C. taëvanus* of Formosa.

FIG. 128.—The Japanese Deer (*Cervus sika*). From Lord Powerscourt, *Proc. Zool. Soc.* 1884, p. 209.

The antlers have a brow and tres tine, and then a forked beam, of which the posterior tine is the smaller. The lachrymal vacuity and fossa are of moderate size; and the auditory bulla is only moderately inflated, and quite smooth externally. Tail moderate; neck maned. In summer the coat is spotted, but is plain in winter. A herd of *C. sika* have been acclimatised in Ireland by Viscount Powerscourt, at Powerscourt, County Wicklow. A number of Deer from the Pliocene of Europe, such as *C. perrieri* and *C. etueriarum*, appear to be allied both to the Pseudaxine and Axine groups.

The *Elaphine* or typical group is at once characterised by the presence of a bez tine to the antlers (Fig. 129), in which the beam is rounded, and splits up near the summit into a larger or smaller number of snags, often arranged in a cup-like manner. Skull as in the preceding group. All the species large. The Red Deer, *C. elaphus*, which is dark brown in colour, with a light patch on the rump, inhabits Europe, Western Asia, and Northern Africa—the so-called Barbary Deer not being specifically distinct. A full-grown Scotch Stag is fully 4 feet in height at the withers. The antlers are shed between the end of February and the early part of April; old animals shedding earlier than younger ones. The young, which (as in all the members of the genus except some of the Rusine species) are spotted, are born at the end of May or the beginning

FIG. 129.—Head of the Wapiti (*Cervus canadensis*).

of June. The points on the antlers increase in number with the age of the creature, and when twelve are present it is known in Scotland as a "royal stag." This number, however, is sometimes exceeded, as in the case of a pair of antlers, weighing 74 lbs., from a stag killed in Transylvania, which had forty-five points. The antlers during the second year consist of a simple unbranched stem, to which a tine or branch is added in each succeeding year, until the normal development is attained, after which their growth is somewhat irregular. Many of the antlers dug up in British peat-beds (as Fig. 118) are larger than those of living individuals, and in the cave-deposits of England and the Continent antlers are met with rivalling those of the Wapiti in size; these large fossil antlers probably indicating the ancestral form from which the Red Deer and several of the allied species are descended.

The North American Wapiti (*Cervus canadensis*, Fig. 129), the Persian Maral (*C. maral*), the Kashmir Stag (*C. cashmeerianus*), as well as *C. affinis* of Tibet, are all closely allied to the Red Deer, but are of larger size, this being especially the case with the first two. A fine example of the antlers of the Wapiti is shown in the accompanying woodcut, and exhibits the absence of a cup at the surroyals, by which this species is distinguished from the Red Deer.

The last, or *Damine* group of existing Deer includes the Common and the Persian Fallow Deer. These are readily characterised by the palmation of the antlers in the region of the surroyals and the spotted coat. The Common Fallow Deer (*C. dama*) stands about three feet in height. The Persian Fallow Deer (*C. mesopotamicus*) is very closely allied, differing only in its slightly larger size and the form of the antlers, the two breeding together. The common species, although now kept in English parks, does not appear to be a native of this country, having probably been introduced from the regions bordering the Mediterranean. The fur is of a yellowish-brown colour (whence the name "fallow"), marked with white spots; there is, however, a uniformly dark brown variety found in Britain. The bucks and does live apart, except during the pairing season; and the doe produces one or two, and sometimes three fawns at a birth. The Fallow Deer from the Pleistocene and Pliocene deposits of the East Coast described under the names of *C. browni* and *C. falconeri* appear to have been closely allied to the existing species. The remarkable *C. verticornis*, of the Norfolk Forest-bed, is regarded as an aberrant member of this group, in which the antlers are very short and thick, with the brow tine cylindrical and downwardly curved, and the beam expanded above the tres tine into a crown with two points.

The extinct Irish Deer (*Cervus giganteus*), of which the skeleton is shown in the woodcut (Fig. 130), is the only representative of the *Megacerotine* group. The antlers, which may have a span of over 11 feet, are enormously palmated, and have a bifurcated brow tine, a small bez tine, and a third posterior tine. The skeleton measures upwards of 6 feet at the withers. Remains of this species are especially common in the peat-bogs of Ireland, but are also met with in Pleistocene deposits over a large part of Europe. In addition to the forms already mentioned there are many other fossil species of *Cervus*, some of which, like the English Pleistocene *C. sedgewicki*, cannot be included in any of the existing groups. There is no conclusive evidence of the existence of any species of *Cervus* before the Lower Pliocene period.

Telemetacarpalia.—This section includes all the Deer of the New World, together with some Old World forms, and is characterised by retaining the distal extremities of the lateral (second and fifth) metacarpals. With the exception of *Alces*, *Capreolus*, and *Hydropotes* (which are either partly or entirely Old World types), the vomer is so much ossified as to divide the posterior bony nares into two distinct orifices (Fig. 132).

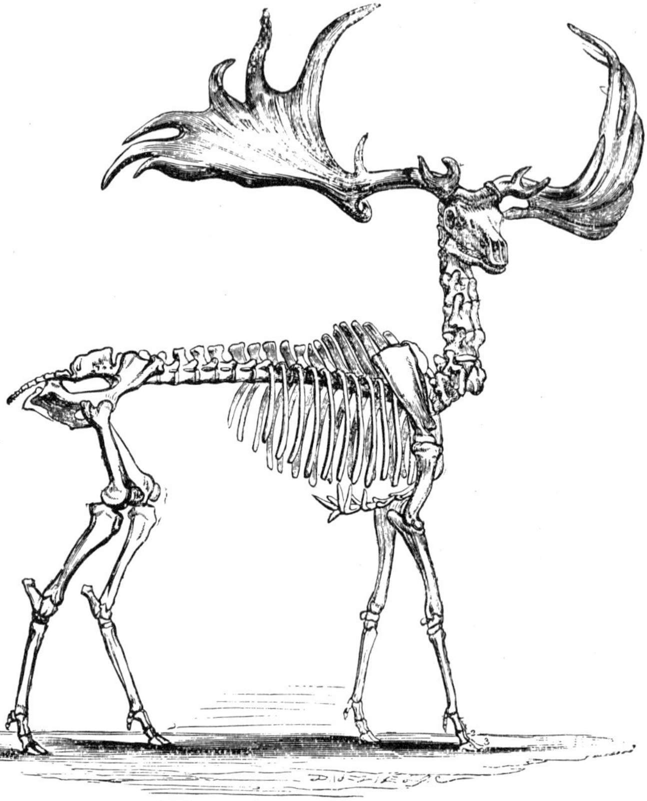

FIG. 130.—Skeleton of the Gigantic Irish Deer (*Cervus giganteus*). After Owen.

Rangifer.[1]—The Reindeer, or Caribou as it is termed in North America, is the sole representative of the genus *Rangifer*, which is sufficiently distinguished from all its allies by the presence of antlers in both sexes. The lachrymal vacuity is small. This animal is distributed over the northern parts of Europe, Asia, and America; the differences which may be observable in specimens from different regions not being sufficient to allow of specific distinction. The Reindeer is a heavily built animal, with short limbs, in which

[1] Hamilton-Smith, in *Griffith's Animal Kingdom*, vol. v. p. 304 (1827).

the lateral hoofs are well developed, and the cleft between the two main hoofs is very deep, so that these hoofs spread out as the animal traverses the snow-clad regions in which it dwells. The antlers (Fig. 131) are of very large relative size. There is a bez as well as a brow tine, which are peculiar in being either branched or palmated. In the American race (Caribou), as well as in some of the specimens found fossil in the English Pleistocene (Fig. 131), one of the brow tines is generally aborted to allow of the great development of the other. The dentition of the Reindeer is frequently remarkable for the very small size of the posterior lobe of the last lower molar. Vertebræ: C 7, D 14, L 5, S 5, C 11.

FIG. 131.—Skull and antlers of the Reindeer (*Rangifer tarandus*), from an English Pleistocene deposit. *br*, Brow tine; *bz*, bez tine. (After Owen.)

The Reindeer has long been domesticated in Scandinavia, and is of especial value to the Laplanders, whom it serves as a substitute for the Horse, Cow, Sheep, and Goat. It is capable of drawing a weight of 300 lbs., and its fleetness and endurance are remarkable. Harnessed to a sledge it will travel without difficulty 100 miles a day over the frozen snow, on which its broad and deeply cleft hoofs are admirably adapted for travelling. During the summer the Lapland Reindeer feeds chiefly on the young shoots of the willow and birch; and since at this season migration to the coast seems necessary to the well-being of this animal, the Laplander, with his herds, sojourns for several months in the neighbourhood of the sea. In winter its food consists chiefly of the so-called reindeer-moss and other lichens which the animal makes use of its hoofs in seeking for beneath the snow. The wild Reindeer grows to a much greater size than the tame breed; but in Northern Europe the former are being gradually reduced through the natives entrapping and domesticating them. The tame breed found in Northern Asia is much larger than the Lapland form, and is there used to ride on. Remains referable to the existing species are found in the cavern and other Pleistocene deposits of Europe.

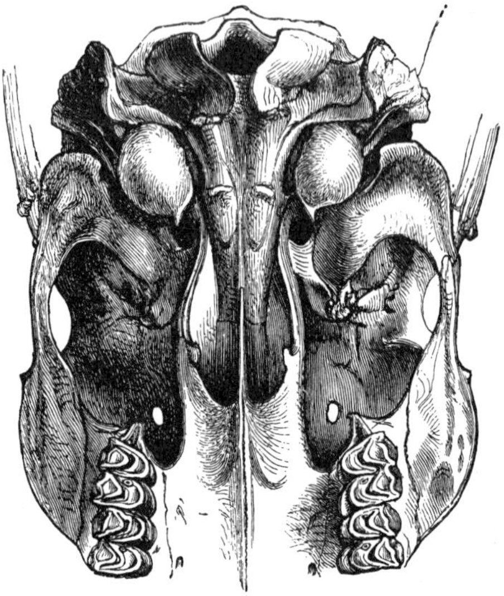

FIG. 132.—Hinder part of the base of the cranium of the Virginian Deer (*Cariacus virginianus*). From Garrod, *Proc. Zool. Soc.* 1877, p. 13.

Alces.[1]—The Elk or Moose (*Alces machlis*) has the same general distribution as the Reindeer, and is likewise the single existing representative of its genus. It is the largest existing member of the family, attaining sometimes a height of 8 feet at the withers. The antlers (Fig. 133) have neither brow nor bez tine, but form an enormous basin-shaped palmation, primarily composed of an anterior and a posterior branch; their weight may be as much as 60 lbs. The nasal bones are very short, and the narial aperture of great size. The Elk is covered with a thick coarse fur of a brownish colour, longest on the neck and throat. Its legs are long and its neck short, and as it is thus unable to feed close to the ground, it browses on the tops of low plants, the leaves of trees, and the tender shoots of the willow and birch. Its antlers attain their full length by the fifth year, but in after years they increase in breadth and in the number of snags, until fourteen of these are produced. Although spending a large part of their lives in forests, Elks do not suffer much inconvenience from the great expanse of their antlers, as in making their way among trees they are carried horizontally to prevent entanglement with the branches. Their usual pace is a shambling trot, but when frightened they break into a gallop. The natural timidity of the Elk forsakes the male at the rutting season, and he will then attack whatever animal comes in his way. The antlers and hoofs are his

[1] Hamilton-Smith, in *Griffith's Animal Kingdom*, vol. v. p. 303 (1827).

principal weapons, and with a single blow from the latter he has been known to kill a wolf. The female often gives birth to two fawns, and with these she retires into the deepest recesses of the forest, the young remaining with her till their third year. The Elk ranges, but in scanty numbers, over the whole of Northern Europe and Asia, as far south as East Prussia, the Caucasus, and North China, and over North America from the New England States westward to British Columbia. Fossil species are found in the Pleistocene deposits of Europe.

Cervalces.[1]—A remarkable extinct Deer from the Pleistocene of North America, described as *Cervalces*, appears in some respects

FIG. 133.—Head of Elk (*Alces machlis*).

(although a true Telemetacarpalian) to connect *Alces* with *Cervus*. Thus the palmated antlers are divided into anterior and posterior branches, but below this division there are two tines apparently corresponding to the bez and posterior tines of *Cervus giganteus* (Fig. 130).

Capreolus.[2]—Antlers (in the existing species) less than twice the length of the head, usually with three tines on each. Brow tine developed from the anterior surface of the upper half of the antler, and directed upwards. Lachrymal vacuity small. Premaxillæ not always articulating with nasals. Auditory bullæ slightly inflated, rugose externally. Vertebrae: C 7, D 13, L 6, S 6, C 8. Tail very short. Glands in fore feet rudimentary; large in hind feet.

The Roe, or Roe Deer (*Capreolus caprea*), is a small form dis-

[1] Scott, *Proc. Ac. Nat. Sci. Philad.* 1885, p. 181.

[2] Hamilton-Smith, in *Griffith's Animal Kingdom*, vol. v. p. 313 (1827).

tributed over Europe and Western Asia, being one of the species found in the British Isles. The male is somewhat over two feet in height at the withers, of a dark reddish-brown colour in summer, with a white patch on the rump. The small antlers are approximated at their bases, and consist of a rugged beam rising vertically for some distance, then bifurcating, and the posterior branch again dividing. The Roe dates from the Pleistocene period. Extinct Deer from the Continental Pliocene have been provisionally referred to *Capreolus*.

Hydropotes.[1]—No antlers in either sex. Lachrymal fossa deep and short (Fig. 134); lachrymal vacuity of moderate size. Orbits

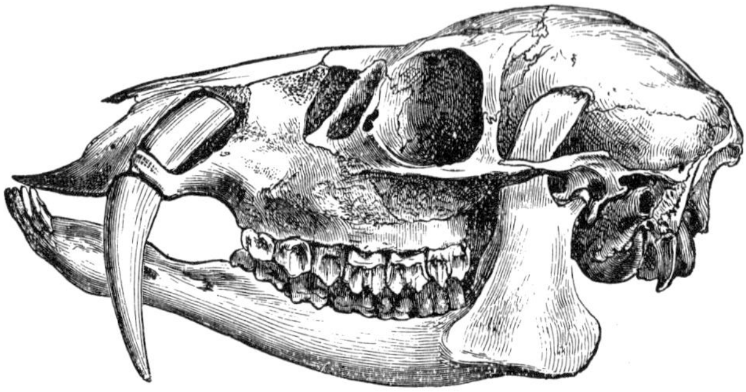

FIG. 134.—The left lateral view of the skull of a male Chinese Water Deer (*Hydropotes inermis*), with the wall of the maxilla cut away to show the root of the canine. ½ natural size. (From Sir V. Brooke, *Proc. Zool. Soc.* 1872, p. 524.)

small and but slightly prominent. Auditory bulla much inflated. Angle of mandible much produced backwardly (Fig. 134); alveolar margins of mandible in diastema sharp and everted. Canines of male very large, and slightly convergent. Vertebræ: C 7, D 12, L 6, S 4, C 10. No tufts on metatarsals. Foot glands small in fore feet, deep in hind ones.

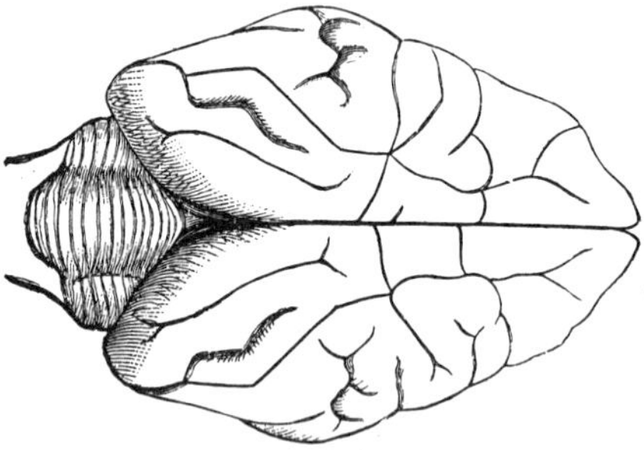

FIG. 135.—Upper surface of the brain of *Hydropotes inermis*. (From Garrod, *Proc. Zool. Soc.* 1877, p. 792.)

The Chinese Water Deer (*H. inermis*) is the sole representative of this genus. In the absence of antlers and the large canines of the male it resembles *Moschus*, although very different in other respects. Thus the brain (Fig. 135) has the hemispheres much convoluted, as in other *Cervinæ*, and approximates to that of *Pudua*;

[1] Swinhoe, *Proc. Zool. Soc.* 1870, p. 90.

while the placenta and viscera likewise agree with those of the true Deer. In the total absence of any ossification of the vomer to divide the posterior nares *Hydropotes* resembles *Capreolus* and differs from all the following genera. The Chinese Water-Deer is nearly of the same size as the Indian Muntjac. It has short legs and a long body, the hair covering the latter being of a light reddish-brown. It is a remarkably prolific animal, differing from all other Deer in producing five or six young at a time.

The mandible of a ruminant from the Middle Miocene of Gers in France, described under the name of *Platyprosopus*, presents such a marked remblance to *Hydropotes* in the form of the angle as to suggest a more or less intimate affinity.

Cariacus.[1]—Skull (Fig. 132) with the vomer dividing the posterior nares into two distinct chambers; premaxillæ not reaching nasals. Antlers never greatly exceeding the length of the head. Lachrymal vacuity very large, and lachrymal fossa small. Auditory bullæ slightly inflated. Vertebræ: C 7, D 13, L 6, S 4, C 13. Tail long or short. Colour uniform in adult.

This genus, which agrees with the Reindeer in the division of the posterior nares by the ossified vomer, comprises a number of species confined to the New World, none of which attain very large dimensions, and the antlers of which are relatively smaller than in the existing species of *Cervus*. The genus may be divided into groups.

The typical *Cariacine* group, as represented by *C. virginianus*, has well-developed antlers, with a short brow tine rising from the inner side of the beam, and directed upwards, and several branches; a long tail; and no upper canines. In this species, as well as in *C. mexicanus* and other forms, the antlers do not divide dichotomously, and the lachrymal fossa is of moderate depth. The Mule Deer (*C. macrotis*) of North America is distinguished by the dichotomous branching of the antlers and the deeper lachrymal fossa. The Virginian Deer is somewhat smaller than the Fallow Deer, and of a uniform reddish-yellow colour in summer, and light gray in winter.

The *Blastocerine* group of South America is represented by *C. paludosus* and *C. campestris*, and has dichotomous antlers, with no brow tine, and the posterior branch the larger, a short tail, and no upper canines. The *Furciferine* group includes *C. chilensis* and *C. antisiensis*, confined to western South America. The antlers are not longer than the head, with a large anterior tine curving forwards at right angles to the simple posterior one. Auditory bullæ slightly inflated, and rugose. Upper canines may be present. The species are of medium size. *C. clavatus*, of Central America, while resembling this group in the characters of the skull and the arrangement

[1] Gray, *Proc. Zool. Soc.* 1850, p. 237.

of the hair on the face, agrees with the next one in having simple spike-like antlers.

The South American *Coassine* group comprises the small forms known as Brockets, in which the antlers form simple spikes not exceeding half the length of the head. Some six species are known.

Remains of *Cariacus*, mostly or entirely referable to existing species, are of common occurrence in the Brazilian cave-deposits. *Blastomeryx*, of the Pliocene of North America, is believed to be an allied type.

Pudua.[1]—Antlers in the form of minute simple spikes. Distinguished from the Coassine group of *Cariacus* by the articulation of the premaxillæ with the nasals (as in the *Furciferine* group), and the coalescence of the ectocuneiform with the naviculo-cuboid. as well as by various external characters. No upper canines. Represented only by the very small *P. humilis* of the Chilian Andes.

Extinct Genera.—In the European and other Tertiary deposits several genera of extinct *Cervidæ* occur, of which the more important may be briefly mentioned. *Amphitragulus*, of the Lower Miocene of the Continent, has four lower premolars, brachydont molars, and no antlers; the largest species being somewhat bigger than the Musk-Deer. The closely allied *Palæomeryx* (*Dremotherium* or *Micromeryx*) generally has but three lower premolars, and the brachydont upper molars (Fig. 122), like those of *Amphitragulus*, want the small accessory inner column[2] found in modern Deer. In *P. feignouxi*, of the Lower Miocene, the lateral metacarpals, although slender, were complete, and the males had large canines, but no antlers. *P. furcatus*, of the Middle Miocene, had small antlers, and the canines appear to have been reduced in size. This genus, besides being represented in the European Miocene, also occurs in the Pliocene of India and China; some of the species being as large as the Red Deer.

Family GIRAFFIDÆ.

In the existing genus the frontal appendages consist of a pair of short, erect, permanent bony processes placed over the union of the frontal and the parietal bones, ossified from distinct centres, though afterwards ankylosed to the skull, covered externally with a hairy skin, present in both sexes, and even in the new-born animal. Anterior to these is a median protuberance on the frontal and contiguous parts of the nasal bones, which increases with age, and is sometimes spoken of as a third horn. Skull with a lachrymal vacuity. No upper canines. Molars brachydont, with rugose

[1] Gray, *Proc. Zool. Soc.* 1850, p. 242.

[2] This accessory column is shown in the figure of the molar of *Boselaphus* on p. 311.

enamel; the upper ones having no inner accessory column. Lateral digits entirely absent on both fore and hind feet, even the hoofs not developed. Humerus with double bicipital groove. Vertebræ:

FIG. 186.—The Giraffe (*Giraffa camelopardalis*).

C 7, D 14, L 5, S 3, C 20. Gall-bladder generally absent. Male reproductive organs and placenta of a Bovine type. Dentition: $i\,\frac{0}{3}$, $c\,\frac{0}{1}$, $p\,\frac{3}{3}$, $m\,\frac{3}{3}$.

Giraffa.[1]—The Giraffe (*G. camelopardalis*) is the sole existing representative of the genus, now confined to the Ethiopian region.

[1] Zimmermann, *Geograph. Geschichte*, vol. ii. p. 125 (1780).

In addition to the characters noticed above, the Giraffe is characterised by its great size and peculiar proportions; the neck and limbs being of great length, and the back inclining upwards from the loins to the withers.

To produce the extremely elongated neck the seven cervical vertebræ are proportionately long, which gives a somewhat stiff and awkward motion to the neck. The ears are large, the lips long and thin, the nostrils closable at the will of the animal, the tongue very long and extensile, and the tail of considerable length, with a large terminal tuft. An adult male may have a total height of 16 feet. The coloration consists of large blotches of darker or lighter chestnut-brown on a paler ground, the lower limbs and under parts being of a uniform pale colour. The Giraffe feeds almost exclusively on the foliage of trees, showing a preference for certain varieties of mimosa, and for the young shoots of the prickly acacia, for browsing on which its prehensile tongue and large free lips are specially adapted. It is gregarious in its habits, living in small herds of about twenty individuals, although Sir S. Baker, who hunted it in Abyssinia, states that he has seen as many as a hundred together.

Fossil species of *Giraffa* occur in Pliocene deposits over Greece, Persia, India, and China, thus affording one of many striking instances of the former wide distribution of the generic types now confined to the Ethiopian region.

Allied Extinct Types.—The Pliocene deposits of many parts of the Old World yield remains of a number of large Ruminants which show such evident signs of affinity with the Giraffe that it is difficult to draw up a definition by which they can be separated in characters of family value from that genus. On the other hand, some of these forms approximate in the characters of the skull to some of the brachydont members of the *Bovidæ*, although it is quite clear from the nature of the cranial appendages that they cannot be included in that family. All these forms have brachydont molars, with rugose enamel, like those of the Giraffe; while several of them have limb-bones approximating to those of the latter—the humerus, when known, having a double bicipital groove. The nature of the cranial appendages (when present) is not fully understood, but it appears that in some cases these approximated more to the type of an antler than to that of a horn; although, from the absence of a "burr," they appear never to have been shed. A gradual diminution in the length of the limbs and neck can be traced from the more Giraffoid to the more Bovoid forms of this extinct group; and it is manifest that if these animals be included in the *Giraffidæ* the definition of that family as given above must be somewhat modified. Only brief mention can be made of the more important genera.

The imperfectly known *Vishnutherium*, of the Pliocene of India and Burma, seems to make the nearest approach to the Giraffe, but the limbs and cervical vertebræ were decidedly shorter, although of a similar slender type. *Helladotherium*, of the Pliocene of Greece and India, is represented by a species of considerably larger size than the Giraffe, with no appendages or lachrymal vacuity to the skull, and with shorter and stouter limbs and neck.

Hydaspitherium, *Bramatherium*, and *Sivatherium* are Indian genera, characterised by the presence of large palmated and antler-like cranial appendages, varying considerably in arrangement. The former genus has a large lachrymal vacuity which is absent in the two latter. In the first and second genera all the appendages rise from a common base; but in *Sivatherium* there is a pair of simple horn-like projections on the orbits in addition to the posterior palmated antlers. *Sivatherium* was an animal of huge bulk, being the largest known representative of the Pecora.

Another apparently allied type is *Samotherium*, of the Pliocene of the Isle of Samos, which appears also to have some affinity with the Antelopes. The skull is nearly as large as that of the Giraffe, and is of the same elongated shape, although depressed between the conical horn-cores, which rise vertically above the orbits, and without a median bony prominence on the frontals. The horn-cores form mere processes of the frontals. The diastema and the mandibular symphysis are shorter than in the Giraffe, and the latter is less deflected. The teeth, although larger, are almost indistinguishable from those of the Giraffe, the only well-marked difference being that the last lower premolar has a double in place of a single postero-internal column.

Family ANTILOCAPRIDÆ.

Closely allied to the *Bovidæ*, but the horns deciduous and branched.

Antilocapra.[1]—The Prongbuck, or Prong-horned Antelope (*Antilocapra americana*), as the single existing member of this family is called, is an animal of nearly the same size as the Fallow Deer, but of a lighter and more graceful build. It is an inhabitant of the prairies of North America, where it is one of the few representatives of the Cavicorn Pecora. The bony horn-cores are unbranched, and form vertical, blade-like projections immediately above the orbit. The horns themselves are compressed, and nearly one foot in length, having a gentle backward curvature, the short branch arising somewhat above the middle of its height, and inclining forwards. When the horn is about to be cast off it becomes loosened, and a new one is formed upon the bony core beneath it. The ears are long and pointed, and the tail is short. The neck has a thick mane of long chestnut-coloured hair, and there is a white patch on the rump.

[1] Ord. *Journ. de Physique*, vol. lxxxvii. p. 149 (1818).

Family BOVIDÆ.

Frontal appendages, when present, in the form of non-deciduous horns. Molars frequently hypsodont. Usually only one orifice to the lachrymal canal, situated inside the rim of the orbit. Lachrymal bone almost always articulating with the nasal. Canines absent in both sexes. The lateral toes may be completely absent, but more often they are represented by the hoofs alone, supported sometimes by a very rudimentary skeleton, consisting of mere irregular nodules of bone. Distal ends of the lateral metapodials never present. Gall-bladder almost always present. The number of cotyledons in the placenta generally varies from 60 to 100; whereas in the *Cervidæ* the number is usually from 5 to 12, *Capreolus* and *Hydropotes* having the fewest. In *Giraffa* the number is upwards of 180. The nature of the horns and horn-cores has been already explained; in the majority of genera these appendages are present in both sexes, although much larger in the male (see p. 310).

The *Bovidæ*, or hollow-horned Ruminants (Cavicornia), form a most extensive family, with members widely distributed throughout the Old World, with the exception of the Australian region; but in America they are less numerous, and confined to the Arctic and northern temperate regions, no species being indigenous either to South or Central America. There is scarcely any natural and well-defined group in the whole class which presents greater difficulties of subdivision than this; consequently zoologists are as yet very little agreed as to the extent and boundaries of the genera into which it should be divided. For the present the genera provisionally adopted may be arranged under a number of sections or groups, which some writers regard as subfamilies. The series may be commenced with the Antelopes, the greater number of which are now characteristic of the Ethiopian region.

Alcelaphine Section.—Includes large African Antelopes, of which the type genus ranges into Syria; generally characterised by their great height at the withers as compared with the rump. Skull with large frontal sinuses, extending into the horn-cores, and the horns lyre-shaped or recurved, and more or less approximated at the base. No large pits at apertures of supraorbital foramina in frontals; upper molars hypsodont and narrow. Horns in both sexes. General colour mostly uniform.

Alcelaphus.[1]—If *Damalis* be included, this genus is represented by some nine or ten living species. Head more or less long and narrow, with the muffle moderately broad and naked. Nostrils approximated, edged with stiff hairs. Horns compressed and ringed at the base, more or less lyrate, and bent back at the tips. Hoofs small. Tail of moderate length, and heavy. Two mammæ.

[1] Blainville, *Bull. Soc. Philom.* 1816, p. 75.

In the typical forms, such as the Bubaline Antelope (*A. bubalinus*), the Harte-beest (*A. caama*, Fig. 137), and the Tora Antelope (*A. tora*, Fig. 138), the horns, which present the peculiar curvature shown in the figures, are situated on a crest at the vertex of the skull, and the facial portion of the cranium is greatly elongated. The Harte-beest, which is found throughout Central and Southern Africa, stands nearly 5 feet high at the withers, and is a somewhat ungainly looking animal, with short hair, which is grayish-brown above and nearly white beneath. In the Pliocene of the Siwalik Hills in Northern India there occur remains of an *Alcelaphus* (*A. palæindicus*)

FIG. 137.—The Harte-beest (*Alcelaphus caama*).

in which the skull had the long facial portion characteristic of the typical group, while the horns approximate to those of the Bontebok. The Blessbok (*A. albifrons*) and Bontebok (*A. pygargus*), belonging to the genus *Damalis* of many authors, have the facial portion of the skull shorter, the horns situated more in advance of the plane of the occiput, and inclining regularly backwards. Of the Blessbok Mr. C. J. Anderson observes that "it is of a beautiful violet colour, and is found in company with black Wildebeests and Springboks in countless thousands on the vast green plains of short crisp, sour grass occupying a central position in South Africa. Cattle and horses refuse to pasture on the grassy products of these plains, which afford sustenance to myriads of this Antelope, whose skin emits a most delicious and powerful perfume of flowers and sweet-smelling herbs." Since the time this was written these Antelopes have been greatly reduced in number. *A.* (*Damalis*) *hunteri*, from East Africa, appears to be allied to *A. senegalensis*, but in the more elongated facial portion of the skull approximates to the Harte-beest, and thus confirms the view that *Damalis* should not form a distinct genus.

FIG. 138.—Head of *Alcelaphus tora*. From Sclater, *Proc. Zool. Soc.* 1873, p. 762.

Connochætes.[1]—Head short and massive, with the muffle very broad and bristly. Nostrils widely separated, hairy within. Horns on the vertex of the skull, immediately over the occiput, approximated at base, cylindrical, bent outwards, and recurving upwards at the tip. Extremities of premaxillæ much expanded laterally, and firmly ankylosed. Vertebræ: C 7, D 14, L 6, S 4, C 16. Hoofs very narrow. Tail very long, covered throughout with long hairs. Four mammæ. Two species, *C. taurina* and *C. gnu* (Fig. 139),

[1] Lichtenstein, *Berlin Ges. Naturforsch. Freunde Magazin*, vol. vi. pp. 152, 165 (1814).

both from South Africa. The former, or Brindled Gnu, is distinguished by the absence of long hair on the face, the black (instead of white) tail, and the presence of dark vertical streaks on the shoulders; it is never found to the south of the Orange River.

The White-tailed Gnu stands about 4 feet 6 inches at the withers. These animals were formerly found in large herds, and are remarkable not only on account of their peculiar form, but also for their grotesque actions when alarmed. Some interesting observations have recently been published upon the mode of

FIG. 139.—The White-tailed Gnu (*Connochætes gnu*).

development of the horns of the Gnu,[1] from which it appears that in very young individuals the horns are straight and divergent, situated some distance below the vertex of the head, and separated by a wide hairy interval. These young horns form the straight tips of those of the adult, the basal downwardly curved portion being subsequently developed. In the fully adult animal the base of the horns forms a helmet-like mass on the forehead which completely obliterates the hairy frontal space of the young.

Cephalophine Section.—Small or medium-sized African and Indian Antelopes, with simple horns present only in the males, a more or less elongated suborbital gland, a lachrymal depression in the skull, and square-crowned upper molars (Fig. 140). Lateral hoofs well developed.

[1] F. E. Blaauw, *Proc. Zool. Soc.* 1889, p. 2.

Cephalophus.[1]—One pair of horns, arising far back on the frontals, conical, short, angulated at the base, and erect or recurved. Sub-orbital gland opening in the form of a slit, or as a row of pores. Auditory bulla divided by a distinct septum. Muffle large and moist. Tail very short. Head tufted. Upper molars of larger species with an accessory internal column. Dorsal vertebræ fourteen in number. Some sixteen species, confined to southern and tropical Africa.

The Duikerboks, as the members of this genus are called, are among the most graceful of the African Antelopes, the smallest species not being larger than a rabbit. The West African *C. sylvicultor* and *C. longiceps* are the largest species.

Tetraceros.[2]—Two pairs of conical horns, of which the anterior are much the smaller. Suborbital gland elongated, and lachrymal fossa very large. Upper molars (Fig. 140) without accessory internal column. One existing Indian species (*T. quadricornis*).

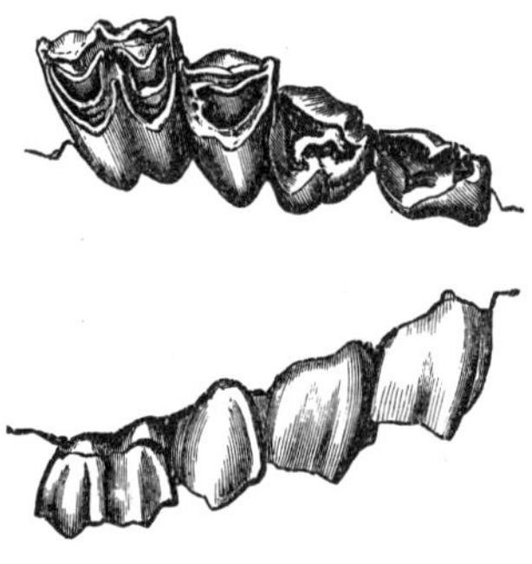

FIG. 140.—Palatal and outer aspects of the three right upper premolars and first molar of the Four-horned Antelope (*Tetraceros quadricornis*). From the *Palæontologia Indica*.

The Four-horned Antelope is found throughout the peninsula of India in jungle. The general colour is brown, lighter beneath and on the inside of the limbs. Remains of this species are found fossil in the cave-deposits of Madras, and a small Ruminant from the Pliocene of the Siwalik Hills has been provisionally referred to this genus.

Cervicaprine Section. — Small or large Antelopes now confined to the Ethiopian region, with horns present only in the males, lachrymal vacuity generally large, more or less distinct pits at the apertures of the supraorbital foramina in the frontals, and narrow upper molars in which there is no accessory internal column.

Neotragus.[3]—Distinguished from the next genus by having the crown of the head tufted, muzzle hairy, premaxillæ long and reaching the lachrymals, nasals very short, mesethmoid much ossified, third lobe of last lower molar either absent or very small, and the hinder lobe of the corresponding upper molar much reduced.

Three species, Salt's Antelope (*N. saltianus*), from Abyssinia, and also *N. kirki* and *N. damarensis*; the two latter having a small third lobe to the last molar. Writing of the first-named species,

[1] Hamilton-Smith, in *Griffith's Animal Kingdom*, vol. iv. p. 258 (1827). Taken to include *Grimmia*, *Terphone*, etc., of Gray.

[2] Leach, *Trans. Linn. Soc.* vol. xiv. p. 524 (1823).

[3] Hamilton-Smith, in *Griffith's Animal Kingdom*, vol. iv. p. 269 (1827).

Mr. W. T. Blanford[1] observes that "the *Beni-Israel*, or *Om-dig-dig*, one of the smallest Antelopes known, abounds on the shores of the Red Sea and throughout the tropical and subtropical regions of Abyssinia. It is occasionally, but rarely, found at higher elevations; I heard of instances of its being shot both at Serafie and Dildi, but it is not often seen above about 6000 feet. It inhabits bushes, keeping much to heavy jungle on the banks of water-courses, and is usually single, or in pairs, either a male and female or a female and young being found together; less often the female is accompanied by two young ones, which remain with her until full grown."

Nanotragus.[2]—Horns small, parallel with frontals, and rising immediately above postorbital process of frontals, in front of the fronto-parietal suture. Lachrymal fossa very large, suddenly descending in front of the orbit, and extending on to the maxilla; lachrymal vacuity small. Auditory bulla large and smooth, without internal septum. Nasals of moderate length. Crown of the head smooth; naked part of muffle small; aperture of suborbital gland small. Lateral hoofs small or absent. Nine species.[3]

The typical species is the Royal Antelope (*N. pygmæus*) of Guinea, the smallest existing representative of the Pecora. This species, together with *N. moschatus* and *N. tragulus* have no lateral hoofs, or tufts on the knees. In the *Scopophorine* group, comprising *N. scoparia*, *N. montanus*, and *N. hastatus*, both these appendages are present; while in the *Oreotragine* group (*N. melanotis* and *N. oreotragus*) the former are present and the latter absent.

Pelea.[4]—Horns rather small, compressed, upright, scarcely diverging, and placed immediately over the orbits. No suborbital gland, nor lachrymal fossa; premaxillæ not reaching nasals. Tail short and bushy. Colour uniform. One species—the Rehbok (*P. capreola*), South Africa, is nearly of the size of a Fallow Deer, although more resembling a Chamois in build and habits. The colour is of a uniform light gray. This animal inhabits bare rocky districts, and thus differs widely from the Water-buck and its allies.

Cobus.[5]—Large Antelopes, with the horns large, elongate, sublyrate, and ringed at the base, and with rudimentary suborbital glands. Skull with a deep frontal hollow, no lachrymal depression,

[1] *Geology and Zoology of Abyssinia*, p. 268.

[2] Sundevall, *Kongl. Vetensk. Akad. Handl.* for 1844, p. 191. Taken to include *Calotragus*, *Scopophorus*, *Nesotragus*, *Pediotragus*, and *Oreotragus* of Gray.

[3] See V. Brooke, *Proc. Zool. Soc.* 1872, pp. 642 and 875.

[4] Gray, *Cat. Ungulate Mamm. Brit. Mus.* p. 90 (1852).

[5] Andrew Smith, *Illustrations of Zoology of South Africa*, No. 12 (1840), "Kobus." Is taken to include *Adenota* and *Onotragus* of Gray.

large lachrymal vacuity, and the premaxillæ reaching the very long nasals. Tail long, with a ridge of hair above, and slightly tufted at the end. Colour uniform. Six species, African.

The Antelopes of this genus are water-loving animals, the Water-buck (*C. ellipsiprymnus*) and the Singsing (*C. defassus*) being well-known examples. Both these species are much alike, standing as much as 4 feet 6 inches at the withers. The Water-buck of South and Eastern Africa is characterised by the coarseness of its long hair; while in the Singsing of West and Central Africa the hair is remarkably fine and soft. Fossil Antelopes from the Pliocene of India are referred to *Cobus*. *Helicophora*, from the Lower Pliocene of Attica, is regarded as allied to *Cobus*, but it has no distinct supraorbital pits.

Cervicapra.[1]—An allied South African genus in which the tail is short and bushy and the premaxillæ do not reach the nasals. Three species.

The Reitbok (*C. arundineum*) is of a grizzly ochre colour; it stands nearly 3 feet in height, and has horns about 1 foot in length. The Nagor (*C. redunca*) is about 6 inches shorter, with horns of half the length, and fulvous brown above and white below; the West African *C. bohor* being rather larger.

Antilopine Section.—A large group of moderate-sized or small Antelopes, most abundant in the deserts bordering the Palæarctic, Oriental, and Ethiopian regions. Horns generally compressed and lyrate, or recurved, or cylindrical and spiral, ringed at base, sometimes present in both sexes. Skull with large pits at apertures of supraorbital foramina of frontals, and generally a distinct lachrymal fossa. Molars of upper jaw narrow, without inner accessory column, and resembling those of the Sheep and Goats. Tail moderate, compressed, hairy above.

Antilope.[2]—Horns, present only in the male, long, cylindrical, subspiral, and diverging. Suborbital gland large, with a somewhat linear opening; lachrymal depression of skull very large, and a small lachrymal fissure. Glands in the feet; lateral hoofs present. One species, India.

The well-known Black-buck (*A. cervicapra*) is found on open plains all over India, except in lower Bengal and Malabar. Old males are deep blackish-brown in colour on the back and sides and the outer surfaces of the limbs, the under parts and inner surfaces of the limbs white, and the back of the head, nape, and neck yellowish. Young males and females are fawn-coloured above. Very large herds are seen in the plains about Dehli and Mattra, which are said in some instances to reach to thousands. Horn-cores are found in the Pleistocene deposits of the valley of the

[1] De Blainville, *Bull. Soc. Philom.* 1816, p. 75. Syn. *Eleotragus*.

[2] Pallas, *Spicilegia Zoologica*, vol. i. p. 3 (1767).

Jumna which cannot be distinguished from those of the existing species.

Æpyceros.[1]—Horns compressed, lyrate, and wide-spreading; present only in male. No suborbital gland, or lachrymal depression in the skull. No lateral hoofs. Two species; one from South and the other from West Africa.

The Palla (*Æ. melampus*) is a large Antelope standing over 3 feet high at the withers, and readily distinguished by its dark red colour, gradually shading to white below. It is usually found on or near hills in herds of from twenty to thirty. *Æ. petersi* is from the Congo.

Saiga.[2]—Nose very large, convex, and inflated. Supraorbital gland present. Lachrymal fossa of skull small, and fissure absent; narial aperture very large; nasals extremely short; supraorbital pits rather small. Horns yellow, lyrate, of moderate length; present only in male. Vertebræ: C 7, D 13, L 6, S 4, C 10. One species, Eastern Europe and Western Asia.

The Saiga (*S. tartarica*) is a clumsily built and somewhat sheep-like Antelope inhabiting the steppes; it occurs fossil in the Pleistocene of France and England.

Pantholops.[3]—Allied in the characters of the head and skull to *Saiga*, but the nose less convex, the nostrils of the male more swollen, and the horns of that sex black, very long, compressed, and lyrate; those of female very short. One species, Central Asia.

The Chiru (*P. hodgsoni*) inhabits the highlands of Western Tibet and Turkestan. In the former area it generally goes in small herds of from three to six, and in the summer may be found grazing in early morning on the level spaces frequently found in the river valleys at elevations of about 15,000 feet. It is excessively shy and difficult to approach. The large size of the narial aperture in the skull of Chiru is suggestive of a connection with respiration at a high altitude, but this appears to be negatived by the occurrence of the same feature in the Saiga.

Gazella.[4]—Delicately built and sandy-coloured Antelopes, with lyrate or recurved horns, which may be absent in the female, and are always smaller and simpler in that sex than in the male. Skull with moderate lachrymal fossa, and a distinct lachrymal fissure. Vertebræ: C 7, D 13, L 6, S 4, C 14. Suborbital gland frequently small, and covered with hair. Face with a white streak running from the outer side of the base of each horn nearly down to the upper end of each nostril, cutting off a dark triangular central

[1] Sundevall, *Kongl. Vetensk. Akad. Handl.* for 1845, p. 271.

[2] Gray, *List Mamm. Brit. Mus.* p. 160 (1843).

[3] Hodgson, *Proc. Zool. Soc.* 1834, p. 81.

[4] De Blainville, *Bull. Soc. Philom.* 1816, p. 75. Is taken to include *Procapra* and *Tragops*.

patch, and bordered externally by a diffused dark line (see Fig. 121, p. 310). The Gazelles, of which there are some twenty-four existing species, are typically Palæarctic desert forms, the Springbok (*G. euchore*) being an outlying South African species. *G. picticaudata* and *G. gutturosa* are respectively found in Western Tibet and Mongolia, the former at great elevations. The majority of the Gazelles do not exceed 30 inches in height, although *G. mohr* is 36. Sir Victor Brooke classifies[1] the Gazelles as follows :—

A. No stripe on back; three lower premolars.
  a. White of rump not encroaching on the fawn of the haunches.
    I. Female with horns.
      1. Horns lyrate or sublyrate — *G. dorcas*, *G. isabella*, *G. rufifrons*, *G. lævipes*, *G. tilonura*, *G. naso.*
      2. Horns non-lyrate—*G. cuvieri*, *G. leptoceros*, *G. spekei*, *G. arabica*, *G. bennetti*, *G. fuscifrons*, *G. muscatensis.*
    II. Female without horns.
      G. subgutturosa, *G. gutturosa*, *G. picticaudata.*
  b. White of rump projecting forwards in an angle into the fawn colour of the haunches. Horns in both sexes.
      G. dama, *G. mohr*, *G. soemmerringi*, *G. granti* (Fig. 121), *G. thomsoni.*

B. A white stripe down the back, two lower premolars. Horns in both sexes.—*G. euchore.*

The East African *G. walleri* is an aberrant species, in which the females are hornless, which has been made the type of the genus *Lithocranius.* It is characterised by the extreme density of the horns and skull, the slenderness of the mandible, and the small size of the cheek-teeth, the upper molars being relatively broader and lower than usual. The cranium is remarkable for the shortness of its facial portion, the large size and production backwards of the supraoccipital, and for the circumstance that the long basicranial axis is nearly parallel with the plane of the palate.

Fossil species of *Gazella* are found in the Pliocene and Pleistocene deposits of Europe and India. *G. deperdita* (*brevicornis*), of the Lower Pliocene of France and Greece, appears to be a generalised species in which the lower molars frequently have accessory columns, traces of which are found in some of the existing forms.

Hippotragine Section.—Includes very large African Antelopes, with long horns, present in both sexes, which are placed over or behind the orbit, and are either recurved, straight, or subspiral. Skull with no distinct pits at apertures of supraorbital foramina in frontals, no lachrymal fossa, and only a small lachrymal fissure. No suborbital gland. Tail long, cylindrical, and tufted at the end.

[1] *Proc. Zool. Soc.* 1873, p. 537. Three species subsequently described are here added to the list.

Upper molars extremely hypsodont, very broad, and with large accessory columns, thus closely resembling those of the Oxen. Some authorities divide this section into two. In the Pliocene it occurs in India and Europe.

Hippotragus.[1]—Horns stout, rising vertically from a crest over the orbit at an obtuse angle to the plane of the nasals, then recurved; lachrymal fissure in some instances almost obliterated. Neck with an erect recurved mane. Tail very distinctly tufted. Four species, tropical Africa and south to the Cape.

The Sable Antelope (*H. niger*) is one of the best-known examples of this genus, occurring in South and East Africa. It stands upwards of 4½ feet in height at the withers, and, except for some white streaks on the face and the whole of the under surface of the body, is of a black colour. The Blaubok (*H. leucophæus*) is distinguished by the glaucous hue of the hair. The other species are the Equine Antelope (*H. equinus*) and Baker's Antelope (*H. bakeri*) from the Sudan, both closely allied, but the latter distinguished by its pale fulvous colour, pencilled ears, and black stripes on the shoulder.

Skulls of fossil Antelopes from the Pliocene of India have been referred to *Hippotragus* (*H. sivalensis*), and Sir V. Brooke suggests that the European Pliocene *Antilope recticornis* is not generically separable.

Oryx.[2]—Horns long, slender, nearly straight or somewhat recurved, rising behind the orbit, and inclining backwards in the plane of the nasals; lachrymal fossa distinct. Nape maned; tail long, and more haired than in *Hippotragus*. Four species, ranging over all the African deserts to Arabia and Syria.

The Gemsbok (*O. gazella*, Fig. 141), is a South African species characterised by its straight horns, the presence of a tuft of hair on the throat, as well as by the large patches and stripes of black on the head, back, limbs, and flanks. It stands nearly 4 feet in height at the shoulder, and the horns are 2 feet 9 inches in length. The colour of the upper part of the body is a rusty gray, and of the under part white, while these are separated from each other by a well-defined black band on either side. These bands unite on the breast, and are continued as a single black band until reaching the lower jaw, where they again divide and form two transverse bands on the head, terminating at the base of the horns. The head otherwise is white, as also are the limbs, with the exception of the thighs, which are black. The Gemsbok generally goes in pairs, or in small herds of three or four. The Beisa (*O. beisa*) of Abyssinia is distinguished by the absence of the tuft of hair on the throat. Writing of this

[1] Sundevall, *Kongl. Vetensk. Akad. Handl.* for 1844, p. 196.

[2] De Blainville, *Bull. Soc. Philom.* 1816, p. 75.

species in his *Geology and Zoology of Abyssinia*, Mr. W. T. Blanford observes that "the appearance of a herd of Oryx is very imposing. They are some of the most elegant and symmetrical of animals, the motions being those of a wild Horse rather than of an Antelope. Their favourite pace appears to be either a steady quick walk or a trot; they rarely break into a gallop unless greatly alarmed. When frightened they dash off, sometimes snorting and putting

FIG. 141.—The Gemsbok (*Oryx gazella*).

their heads down as if charging, raising their long tails, and looking very formidable. They are wary animals, though far less so than some other Antelopes. It is said that they frequently attack when wounded, and their long straight horns are most deadly weapons." The Arabian Beatrix Antelope (*O. beatrix*) is a much smaller animal, with the black markings confined to the head, fore limbs, and flanks. Finally, the Leucoryx (*O. leucoryx*) of North Africa, while agreeing in size with the Beatrix, differs by its curved horns and uniform coloration.

The extinct *Palæoryx*, of the Lower Pliocene of Europe and the Isle of Samos, appears to have been an ancestral form of *Oryx*, said to show some signs of affinity with *Hippotragus*.

Addax.[1]—Horns with the same inclination as in *Oryx*, but with a slight spiral twist. No mane on nape, but a slight one on the throat. Hoofs rounded. One species (*A. nasomaculatus*), from North Africa and Arabia, the colour of which is nearly white.

Tragelaphine Section.—Includes large, so-called Bovine, Antelopes now mainly characteristic of the Ethiopian region, but with one Oriental genus. Horns usually present in the male only (if developed in the female smaller), with a more or less distinct ridge in front, and usually twisted spirally, the front ridge twisting outwards from the base of the horn. Skull without lachrymal fossa, but with a large or small lachrymal fissure; usually large pits at the apertures of the supraorbital foramina on the frontals; premaxillæ reaching nasals. Muffle large and moist; nostrils approximated. Molars hypsodont or brachydont. Vertical white stripes frequently present on the body.

a. Hind limbs much shorter than the fore. Horns behind the orbit, short, conical, faintly angulated. Nose bovine. Body without vertical stripes. Molars (Fig. 123, p. 311) *hypsodont, with a large accessory column in those of the upper jaw. One Oriental genus.*

Boselaphus.[2]—The one genus of this subsection is represented only by the well-known Nilghai (*B. tragocamelus*) of India. The male stands over 4 feet in height at the shoulder, with horns about 8 inches in length; the hornless female being about one third smaller. Both sexes have a short erect mane, and the male has also a tuft of hair upon the throat. When adult the sexes are very different in colour, the male being of a dark iron gray or slate colour, approaching black on the head and legs, while the female and young are of a bright light brown or fawn colour. In both male and female at all ages the lips, chin, and under parts, as well as two transverse stripes on the inner sides of the ears and rings on the fetlocks, are white, and the mane and tip of the tail black. The Nilghai is one of the few Antelopes occurring in India, where it is found from near the foot of the Himalaya to the south of Mysore, though rare to the north of the Ganges and also in the extreme south. It is most abundant in Central India, and does not occur in Assam or the countries to the east of the Bay of Bengal. It frequents forests and low jungles, though often found in tolerably open plains, associating in small herds. One, or very often two, young produced at a birth. Fossil remains of species of this genus occur in the Pleistocene and Pliocene deposits of India.

b. Fore and hind limbs equal. Horns long, and spirally twisted. Nose cervine, and aperture of suborbital gland very small.

[1] Rafinesque, *Anal. Nat.* 1815, p. 56.

[2] De Blainville, *Bull. Soc. Philom.* 1816, p. 75. Syn. *Portax*, Hamilton-Smith.

Body generally striped. Molars brachydont, those of the upper jaw in existing forms with a small inner accessory column. Three existing Ethiopian genera.

Tragelaphus.[1]—Female hornless. Horns of males (Fig. 142) over

FIG. 142.—Head of *Tragelaphus gratus*. From Sclater, *Proc. Zool. Soc.* 1883, p. 36.

orbit, with one or two spiral turns, obscurely ridged, the posterior ridge being more developed than the anterior. Skull with small supraorbital pits, very small lachrymal fissure, and no deep intercornual depression in the frontals. Neck maned or smooth. Hoofs short or long. Coloration usually brilliant, differing markedly in the two sexes, and the white bands on the body, when present, numerous and distinct. Seven species.

[1] De Blainville, *Bull. Soc. Philom.* 1816, p. 75. Includes *Euryceros*, Gray.

The Harnessed Antelopes are among the handsomest of the whole group. The small Guib (*T. scriptus*) is not larger than a Goat, but *T. angasi* is 3 feet 4 inches in height at the shoulder. In *T. scriptus*, *T. angasi*, and *T. euryceros*, the two sexes differ in colour, the body is marked by white stripes descending from a white dorsal streak, and the hoofs are short; the third species differing from the others by the absence of a mane on the neck, back, and belly. *T. gratus* agrees with this group in coloration (the mane being

FIG. 143.—The Kudu (*Strepsiceros kudu*). From Sclater, *List of Animals in Zoological Society's Gardens*, 1883, p. 136.

absent), but differs in the extreme elongation of its hoofs. The Nakong, *T. spekei*, while having the long hoofs of *T. gratus*, has a perfectly plain body coloration, with a mane on the neck. The two species with elongated hoofs inhabit swampy districts, for which this peculiar structure is admirably adapted; and the Nakong, when frightened, will rush into the water and leave only its nostrils and the tips of the horns above the surface. The small Bushbuck (*T. sylvaticus*) of South Africa has no stripes, and short hoofs.

Strepsiceros.[1]—Females hornless. Horns (Fig. 143) more twisted than in *Tragelaphus*, forming an open spiral, with the anterior ridge

[1] Gray, *List. Mamm. Brit. Mus.* p. 155 (1843).

very strongly developed, and rising at an obtuse angle to the plane of the nasals. Skull with large supraorbital pits, large lachrymal fissure, and deep intercornual depression. Hoofs short. Body with white vertical stripes descending from a longitudinal dorsal streak. Two existing species.

The Kudu (*S. kudu*, Fig. 143) extends from South Africa to Abyssinia, and is only inferior in size to the Eland. The horns are about 4 feet in length, and form a very open spiral, and there is a fringe of long hair down the front of the neck. The Lesser Kudu (*S. imberbis*), of Somali-land is a much smaller form, without the fringe of hair on the neck, and with a much smaller axis formed by the spiral of the horns.

An imperfect skull from the Pliocene of Northern India has been referred to *Strepsiceros*.

Oreas.[1]—Females horned. Horns twisted on their own axis, with very strong ridges, inclining upwards and outwards in the plane of the nasals. General characters of skull as in preceding genus. Stripes on body, if present, very faintly marked. One existing species.

The Eland (*O. canna*) is the largest of all the Antelopes, the males standing nearly 6 feet at the withers. One variety from South Africa is of a uniform pale fawn colour, while the Central African form is of a bright tan colour, marked by a number of thin pale vertical stripes descending from a dark dorsal ridge—these markings fading more or less in the adults. The males have a large dewlap, a tuft of brown hair on the forehead, and a small mane on the neck. The straight black horns of the male are usually about 18 inches long. Elands were formerly extremely abundant in Southern and Eastern Africa, but their destruction has been so relentless that they have totally disappeared from extensive areas, and are daily becoming scarcer.

Portions of upper jaws from the Pliocene deposits of India appear to indicate the former existence in that area of large Antelopes closely allied to the Eland, but distinguished from the living species by the greater size of the inner accessory column in the upper molars.

Allied Extinct Types.—Large Antelopes with spirally twisted horns appear to have been common over Southern Europe in Pliocene times, but their exact affinity is in many cases difficult to determine. Of these, *Palæoreas*, which occurs in the Lower Pliocene of Europe and Algeria, appears to present affinities both to *Oreas* and *Strepsiceros*, and may have been the ancestral type from which these two genera are derived; the upper molars have well-developed accessory columns.

The so-called *Antilope torticornis*, of the French Pliocene,

[1] Desmarest, *Mammalogie*, p. 471 (1822).

resembles *Tragelaphus* in the greater development of the posterior as compared with the anterior ridge of the horn-cores, and has accordingly been referred to that genus. *Protragelaphus*, of the Lower Pliocene of Attica, differs from all the other types in the absence of the anterior ridge on the horn-cores and of the supraorbital pits, while it has a distinct lachrymal fossa.

In this place it will be convenient to notice certain fossil forms which do not accord with any of the existing sections of the family, and for the reception of which the *Palæotragine* section has been formed. In these types the horn-cores are laterally compressed like those of the modern Goats, but the upper molars resemble those of the brachydont Antelopes. The earliest of these genera, and the first representative of the Antelopes yet known, is *Protragoceros*, of the Middle Miocene of France, first described as *Antilope clavata*; *Palæotragoceros* and *Tragoceros*, of the Lower Pliocene, are distinguished by their larger horns and wider molars.

A remarkable large Antelope from the Lower Pliocene of the Isle of Samos, in the Turkish Archipelago, proposed to be described as *Criotherium*, appears to be unlike any other form. The horns, which are placed on the extreme vertex of the skull, are very short, tightly twisted, and project in front of the forehead. The upper molars have short and broad crowns, with no accessory column on the inner side.

Rupicaprine Section.—The Caprine Antelopes, as the typical members of this section may be termed, appear to connect the true Antelopes with the Goats. They are mostly small or medium-sized forms, inhabiting portions of the Palæarctic and Oriental regions, with one outlying North American genus. The typical forms present the following features. Horns present, and of nearly equal size in both sexes, rising behind the orbits, short, ringed at the base, conical or somewhat compressed, and recurved. Suborbital gland generally present, in some cases small. Build clumsy; hoofs large; tail short, tapering, hairy above. Skull with lachrymal fossa, but no fissure. Molars as in the Caprine section.

Rupicapra.[1]—Horns short and cylindrical, rising perpendicularly from the forehead for some distance, then bending sharply backwards and downwards, forming hooks with pointed tips. Premaxillæ not reaching the nasals. One species, Palæarctic.

The Gemse, or Alpine Chamois (*R. tragus*), inhabits the high mountains of Europe from the Pyrenees to the Caucasus. It stands about 2 feet in height at the withers. The body is covered in winter with long hair of a chestnut-brown colour, that of the head being paler, with a dark brown streak on each side. At other seasons the colour is somewhat lighter, in spring approaching to gray. Underneath the external covering the body is further

[1] De Blainville, *Bull. Soc. Philom.* 1816, p. 75.

protected from cold by a coat of short thick wool of a grayish colour. The tail is black; the ears are pointed and erect; the hoofs have the outer edges higher than the soles, and are thus admirably adapted for laying hold of the slightest projection or roughness on the face of the rocky precipices it frequents. The Chamois is gregarious, living in herds of fifteen or twenty, and feeding generally in the morning or evening. The old males, however, live alone, except in the rutting season, which occurs in October, when they join the herds, driving off the young males, and engaging in contests with

FIG. 144.—*Nemorhædus crispus.* From Sclater, *List of Animals in Zoological Society's Gardens*, 1883, p. 151.

each other that often end fatally. The period of gestation is twenty weeks, when the female, beneath the shelter of a projecting rock, produces one and sometimes two young. In summer the Chamois ascends to the limits of perpetual snow, being only outstripped in the loftiness of its haunts by the Ibex; and during that season it shows its intolerance of heat by choosing such browsing grounds as have a northern exposure.

Nemorhædus.[1]—Horns rounded, gradually recurving, without distinct hook at the end. Suborbital gland small or wanting; ears large; skull with a large lachrymal depression, and the premaxillæ not quite reaching the nasals. Some nine species, ranging from the Eastern Himalayas to North China and Japan, and southwards

[1] Hamilton-Smith, in *Griffith's Animal Kingdom*, vol. v. p. 352 (1827).

to Formosa, the Malay Peninsula, and Sumatra. The smallest species is the Himalayan Goral (*N. goral*). Of the larger forms we may mention the Himalayan Serow (*N. bubalinus*) the Cambing-Utan (*N. sumatrensis*) of Sumatra, and the Japanese *N. crispus* (Fig. 144). Of the Serow, Colonel Kinloch remarks that "it is a large and powerful beast. The body is covered with very coarse hair, which assumes the form of a bristly mane on the head and shoulders, and gives the beast a ferocious appearance, which does not belie its disposition. The colour is a dull black on the back, bright red on the sides, and white underneath, the legs also being dirty white. The ears are very large, the muzzle is coarse. The Serow has an awkward gait, but in spite of this can go over the worst ground; and it has perhaps no superior in going down steep hills. It is a solitary animal, and nowhere numerous."

Haploceros.[1]—The Rocky-Mountain Goat (*Haploceros montanus*), inhabiting the northern parts of California, appears to be very closely allied to *Nemorhœdus.* The horns are somewhat compressed at the base; there is no suborbital gland; and the ears are small. The hair, which is whitish in colour, is very long, and especially abundant in the region of the throat, shoulders, flanks, and tail. The animal is about the size of a large Sheep.

Budorcas.[2]—The Takin (*B. taxicolor*) of the Mishmi Hills in Assam, and an allied species from Eastern Tibet, are larger forms apparently related to *Nemorhœdus,* but with a much greater development of the horns. The horns of what is considered to be the male[3] arise from the vertex of the skull, and are nearly in contact in the middle line; they first bend outwards and downwards, and then suddenly upwards and backwards. Those regarded by Mr. Hume as referable to the female are directed at first outwards, and then gradually curve upwards and backwards, without any downward flexure or angulation. The horns of the male may be 2 feet in length, with a basal diameter of 13 inches. The muzzle is hairy, with a small naked muffle. There appear to be considerable seasonal and sexual variations in colour; the body being in some cases of a yellow dun, while in others it is a dusky, reddish-brown, with much black intermingled. The heads of large males are blackish.

Scarcely anything is known of the habits of the Takin, which never appears to have been seen alive by Europeans.

Caprine Section.—Both sexes with horns, but those of the female small. Horns usually compressed, triangular, with transverse ridges, and either curving backwards or spiral. Muzzle hairy, without naked muffle. Suborbital gland small or absent; lachrymal

[1] Hamilton-Smith, in *Griffith's Animal Kingdom*, vol. v. p. 354 (1827). Amended from "Aplocerus."

[2] Hodgson, *Journ. As. Soc. Bengal*, vol. xix. p. 65 (1850).

[3] See A. O. Hume, *Proc. Zool. Soc.* 1887, pp. 483-486.

fossa of skull present or absent. Tail short and flattened. Foot-glands frequently present. Molars very hypsodont; those of the upper jaw being narrow, without an accessory internal column. Mainly Palæarctic, but with some outlying forms.

This section includes the Goats and Sheep, which are so closely connected that it is difficult to give well-marked generic characters that will hold good for all the species. They seem to be one of

FIG. 145.—The Alpine Ibex (*Capra ibex*).

the latest developments of the *Bovidæ*, since they are unknown before the Pliocene period; and are essentially mountain forms.

Capra.[1]—Horns flattened from side to side, and either curving backwards (Fig. 145) or spirally twisted. No suborbital gland, and no lachrymal fossa in the skull. Foot-glands, if present, only in the fore feet. Chin more or less bearded. Males with a strong odour. Vertebræ: C 7, D 13, L 6, S 4, C 9-13. Some dozen species, ranging over all the higher mountains of Southern Europe, from Spain to the Caucasus; also found in Abyssinia, Persia, Sind, and Baluchistan, thence through the higher Himalaya, and so on to Tibet and Northern China. One outlying species occurs in the Nilgherries of Southern India.

[1] Linn. *Syst. Nat.* 12th ed. vol. i. p. 94 (1766).

The European Ibex or Steinbok (Fig. 145), which may be taken as a typical Goat, stands about $2\frac{1}{2}$ feet in height at the shoulder. In summer the hair is short and smooth, and of an ashy-gray colour, but a long coat is developed in winter. The horns of the male rise in a bold backward sweep from the forehead, and are characterised by the strong transverse ridges on the broad and flat anterior surface. They are said to be not more than some 2 feet in length, but these dimensions are greatly exceeded by the horns of the Himalayan Ibex. The Alpine Ibex lives at a greater height than the Chamois, spending the day just at the limit of perpetual snow, and descending at night to graze at lower levels. Both this and the Himalayan species generally live in small herds of from five to fifteen or more; they are wary animals, although not so much so as many of the wild Sheep. The following list, mainly taken from two papers by Mr. Sclater,[1] gives the distribution of the various species of Goats, with some remarks on their peculiarities:—

(1) *C. ibex*, confined to the Alps of Switzerland, Savoy, and the Tyrol, and now nearly extinct, except where artificially preserved. (2) *C. sibirica*, closely allied to the preceding, but with larger horns, occurs in the Altai Mountains, and throughout the Himalaya from Kashmir to Nipal, and northward towards Turkestan. (3) *C. sinaitica*, of the mountains of Upper Egypt, the Sinaitic Peninsula, and Palestine, is allied to the two preceding species, but has the horns somewhat more compressed, with a difference in the ridges on the front. (4) *C. caucasica*, a very distinct species, confined to the Caucasus, where it inhabits the western part of the Great Caucasus; with thick horns curving backwards and outwards in one plane, with the exception of their tips, which incline inwards.[2] (5) *C. pallasi* is an allied species from the Eastern Caucasus, distinguished, among other features, by the curvature of the horns, which lie flatter and twist more outward from the forehead, with a greater terminal inward bend. (6) *C. pyrenaica*, of the Pyrenees, and the higher ranges of Central Spain, Andalusia, and Portugal, is another nearly related species. (7) *C. ægagrus*, formerly abundant over the Grecian Archipelago, but now restricted in Europe to Crete and some of the Cyclades, is found throughout the mountains of Asia Minor and Persia, and thence to Baluchistan and Sind. The horns are thinner and sharper in front than in the Ibexes, and this species is generally regarded as the ancestral stock of the various breeds of domestic Goats. (8) *C. dorcas*, a Goat from the island of Jura, near Eubœa, has been described under this name, and is apparently nearly allied

[1] *Proc. Zool. Soc.* 1886, p. 314; and 1887, p. 552.

[2] Specimens referred by Dinnik to *C. caucasica* have been made the types of another species—*C. severtzovi.*

to *C. ægagrus*. (9) *C. walie*, an apparently well-characterised species from the highest ranges of Abyssinia. (10) *C. falconeri*; the Markhoor differs from all the preceding species by the spiral twisting of its horns, which attain enormous dimensions. It occurs in the Pir-Panjal range south of Kashmir, and thence into Afghanistan and the Suleiman range, and northwards to Astor, Gilgit, and Scardo (Baltistan). The specimens from the Suleiman range have the spiral of the horns very close, somewhat as in the Eland; while in those from Astor, Gilgit, and Scardo it is very open, as in the Kudu. The Pir-Panjal race occupies a somewhat intermediate position in this respect. (11) *C. jemlaica*, the Thar, inhabits suitable regions along the whole range of the Himalaya from Kashmir to Bhutan. Together with the next species, it differs from the more typical Goats in its short, thick, and much compressed horns, the anterior border of which is keeled, and the moist naked muffle. There are no glands in the fore feet. It was generically separated by Gray as *Hemitragus*. (12) *C. hylocrius*, the so-called Ibex of the Nilgherries, Anamallays, and other adjoining ranges of Southern India, is an outlying species, apparently allied to the preceding, but with somewhat different horns, in which the external angle in front is much rounded off.

Of fossil Goats we have but little knowledge. Remains of *C. pyrenaica* are found in cave-deposits at Gibraltar; and it is not improbable that the genus is represented in the Upper Pliocene of France. Several species occur in the Pliocene of India, *C. sivalensis* being apparently closely allied to *C. jemlaica*, while another has horns resembling those of *C. falconeri*, and it is possible that a third may be more nearly related to the Ibexes.

Ovis.[1]—Horns curving backwards and downwards in a bold sweep, with the tips everted, generally with more or less prominent transverse ridges, and brownish in colour. Suborbital gland and lachrymal fossa usually present, but generally small. Foot-glands in all the feet. Chin not bearded;[2] males without a strong odour. Vertebræ: C 7, D 13, L 6, S 4, C 10-14. Some twelve species, mainly Palæarctic, but extending into the adjacent portions of the Oriental region, and with one outlying species in North America.

The more typical Sheep are closely connected with the Goats by the Himalayan Bharal (*O. nahura*) and the Aoudad (*O. tragelaphus*) of Northern Africa, both these species having no suborbital gland and no lachrymal fossa, while their comparatively smooth and olive-coloured horns show a decided approximation to those of the Goats. Both present, however, the ovine character of glands in all the feet. In the typical Sheep the basioccipital of the skull is wider in front than behind, with the anterior pair of tubercles

[1] Linn. *Syst. Nat.* 12th ed. vol. i. p. 97 (1766).

[2] There may be a beard on the throat, as in *O. cycloceros*.

widely separated and much larger than the posterior pair. The Bharal, however, resembles the Goats in having an oblong basioccipital, with the posterior tubercles larger and more prominent than the anterior ones, both being situated in the same anteroposterior line. These transitions towards the caprine type are, however, not sufficient to support the view that the Bhàral should form the type of a distinct genus (*Pseudois*), more especially since some of the typical Sheep, like *O. canadensis*, have the lachrymal fossa of the skull very much reduced in size.

The distinction of the various permanent modifications under which wild Sheep occur is a matter of considerable difficulty. Trivial characters, such as size, slight variations in colour, and especially the form and curvature of the horns, are relied upon by different zoologists who have given attention to the subject in the discrimination of species, but no complete accord has yet been established. The most generally recognised forms are enumerated below.

The geographical distribution of wild Sheep is interesting. The immense mountain ranges of Central Asia, the Pamir and Thian-Shan of Turkestan, may be looked upon as the centre of their habitat. Here, at an elevation of 16,000 feet above the sea-level, is the home of the magnificent *Ovis poli*, named after the celebrated Venetian traveller Marco Polo, who met with it in his adventurous travels through this region in the thirteenth century. It is remarkable for the great size of the horns of the old rams and the wide open sweep of their curve, so that the points stand boldly out on each side, far away from the animal's head, instead of curling round nearly in the same plane, as in most of the other species. A Sheep from the same region, in which the horns retain their more normal development, has received the name of *O. karelini*, but, according to Mr. W. T. Blanford,[1] is not distinct specifically from *O. poli*. Eastward and northward is found the Argali (*O. argali*), with a wide and not very well determined range; it formerly occurred in the Altai, but is now found in Northern Mongolia. Still farther north, in the Stanovoi Mountains and Kamschatka, is *O. nivicola*, and away on the other side of Behring's Strait, in the Rocky Mountains and adjacent highlands of western North America, is the "Bighorn" or Mountain Sheep (*O. canadensis*), the only member of the genus found in that continent, and indeed—except the Bison, Musk-Ox, Mountain Goat (*Haplocèros*), and the Prong-buck (*Antilocapra*)—the only hollow-horned Ruminant, being like the rest obviously a straggler from the cradle of its race. The two last-named species are nearly allied, and are characterised by the slight development of the ridges on their horns and the very shallow lachrymal fossa. Turning southward from the point from which we started, and still a little to the east, in Nipal and Western Tibet,

[1] *Proc. Zool. Soc.* 1884, p. 326.

is the Himalayan Argali (*O. hodgsoni*), having massive and strongly curved horns, with bold ridges, like those of the true Argali. Indeed, were it not for their isolated areas there would appear to be no grounds for distinguishing these two closely allied forms, and it is not improbable that they are really identical. *O. brookei* appears to have been founded on a hybrid between *O. hodgsoni* and *O. vignei*. In the same districts, and also in Southern Ladak, there occurs the Bharal (*O. nahura*), with smaller, smoother, and more spreading horns. Passing in a south-westerly direction we find a series of smaller forms, *O. vignei* of Ladak, *O. cycloceros* of Northern

FIG. 146.—The Moufflon (*Ovis musimon*). From a living animal in the London Zoological Gardens.

India, Persia, and Baluchistan, *O. gmelini* of Asia Minor and Persia, *O. ophion*, confined to the elevated pine-clad Troodos Mountains of the island of Cyprus, and said at the time of the British occupation in 1878 to have been reduced to a flock of about twenty-five individuals, and *O. musimon*, the Moufflon of Corsica and Sardinia (see Fig. 146), believed to have been formerly also a native of Spain. In the three latter species the females are hornless. Lastly, we have the somewhat aberrant, Goat-like Aoudad (*O. tragelaphus*), of the great mountain ranges of North Africa, in which, as already mentioned, the skull and horns resemble those of the Bharal, although the tail is longer, and there is a thick fringe of long hair on the throat, chest, and fore legs.

We thus find that Sheep are essentially inhabitants of high mountainous parts of the world, for dwelling among which their wonderful powers of climbing and leaping give them special advantages. No species frequent by choice either level deserts, open plains, dense forests, or swamps. By far the greater number of species are inhabitants of the continent of Asia, one extending into North America, one into Southern Europe, and one into North Africa. No wild Sheep exist in any other part of the world, unless the so-called Musk-Ox of the Arctic regions, the nearest existing ally to the true Sheep, may be considered as one. Geologically speaking, Sheep appear to be very modern animals, or perhaps it would be safer to say that no remains that can be with certainty referred to the genus have been met with in the hitherto explored true Tertiary beds, which have yielded such abundant modifications of Antelopes and Deer. They are generally considered not to be indigenous in the British Isles, but to have been introduced by man from the East in prehistoric times. A fossil Sheep (*Ovis savigni*), apparently allied to the Argali, has, however, been described from the so-called Forest-bed of the Norfolk coast.

The Sheep was a domestic animal in Asia and Europe before the dawn of history, though quite unknown as such in the New World until after the Spanish conquest. It has now been introduced by man into almost all parts of the world where settled agricultural operations are carried on, but flourishes especially in the temperate regions of both hemispheres. Whether our well-known and useful animal is derived from any one of the existing wild species, or from the crossing of several, or from some now extinct species, is quite a matter of conjecture. The variations of external characters seen in the different domestic breeds are very great. They are chiefly manifested in the form and number of the horns, which may be increased from the normal two to four or even eight, or may be altogether absent in the female alone, or in both sexes; in the form and length of the ears, which often hang pendent by the side of the head; in the peculiar elevation or arching of the nasal bones in some Eastern races; in the length of the tail, and the development of great masses of fat at each side of its root, or in the tail itself; and in the colour and quality of the fleece.

Ovibos.[1]—This genus is generally considered to be a connecting link between the Caprine and Bovine sections, but should rather be regarded as an aberrant type of the former. Horns of adult male rounded, smooth, and closely approximated at their bases, where they are depressed and rugose; curving downwards, and then upwards and forwards. Muzzle caprine; no suborbital gland, no lachrymal fossa or fissure in skull; orbits tubular; a large narial aperture and very short nasals; premaxillæ not reaching nasals.

[1] De Blainville, *Bull. Soc. Philom.* 1816, p. 76.

Tail short, and molar teeth caprine. One existing and two fossil species, Palæarctic and Nearctic.

The animal commonly known as the Musk-Ox (*Ovibos moschatus*), though approaching in size the smaller varieties of Oxen, is in structure and habits closely allied to the Sheep, its affinities being well expressed by the generic name *Ovibos* bestowed upon it by De Blainville. The specific name, as also the common English appellatives "Musk-Ox," "Musk-Buffalo," or "Musk-Sheep," applied to it by various authors, refer to the musky odour which the animal exhales. This does not appear to be due to the secretion of a special gland, as in the case of the Musk-Deer; but it must be

FIG. 147.—The Musk-Ox (*Ovibos moschatus*).

observed that, except as regards the osteology, very little is known of the anatomy of this species. It about equals in size the small Welsh and Scotch cattle. The head is large and broad. The horns in the old males have extremely broad bases, meeting in the median line, and covering the brow and whole crown of the head. They are directed at first downwards by the side of the face and then turn upwards and forwards, ending in the same plane as the eye. Their basal halves are of a dull white colour, oval in section and coarsely fibrous; their middle part smooth, shining, and round; their tips black. In the females and young males the horns are smaller, and their bases are separated from each other by a space in the middle of the forehead. The ears are small, erect, and pointed, and nearly concealed in the hair. The space between the nostrils and the upper lip is covered with short close hair, as in Sheep and Goats, without any trace of the bare muffle of the Oxen. The greater part of the animal is covered with long brown hair, thick, matted, and curly on the shoulders, so as to give the appearance of a hump, but elsewhere straight and hanging down,—that of the sides, back, and haunches reaching as far as the middle of the legs and entirely concealing the very short tail. There is also a thick woolly underfur, shed in the summer. The hair on the lower jaw, throat, and chest is long and straight, and hangs down like a beard or dewlap, though there is no loose fold of skin in this situation as in Oxen. The limbs are stout and short, terminating in unsymmetrical hoofs, the external one being rounded, the internal pointed, and the sole partially covered with hair.

The Musk-Ox is at the present day confined to the most northern parts of North America, where it ranges over the rocky barren grounds between the 60th parallel and the shores of the Arctic Sea. Its southern range is gradually contracting, and it appears that it is no longer met with west of the Mackenzie River, though formerly abundant as far as Eschscholtz Bay. Northwards and eastwards it extends through the Parry Islands and Grinnell Land to North Greenland, reaching on the west coast as far south as Melville Bay; and it was also met with in abundance by the German polar expedition of 1869-70 at Sabine Island on the east coast. No trace of it has been found in Spitzbergen or Franz Joseph Land. As proved by the discovery of fossil remains, it ranged during the Pleistocene period over northern Siberia and the plains of Germany and France, its bones occurring very generally in river deposits along with those of the Reindeer, Mammoth, and Woolly Rhinoceros. It has also been found in Pleistocene gravels in several parts of England, as Maidenhead, Bromley, Freshfield near Bath, Barnwood near Gloucester, and also in the lower brick-earth of the Thames valley at Crayford, Kent.

It is gregarious in habit, assembling in herds of twenty or thirty head, or, according to Hearne, sometimes eighty or a hundred, in which there are seldom more than two or three full-grown males. The Musk-Ox runs with considerable speed, notwithstanding the shortness of its legs. Major H. W. Feilden, naturalist to the Arctic expedition of 1875, says: "No person watching this animal in a state of nature could fail to see how essentially ovine are its actions. When alarmed they gather together like a flock of sheep herded by a collie dog, and the way in which they pack closely together and follow blindly the vacillating leadership of the old ram is unquestionably sheep-like. When thoroughly frightened they take to the hills, ascending precipitous slopes and scaling rocks with great agility." They feed chiefly on grass, but also on moss, lichens, and tender shoots of the willow and pine. The female brings forth a single young one in the end of May or beginning of June after a gestation of nine months. According to Sir J. Richardson, "when this animal is fat its flesh is well tasted, and resembles that of the Caribou, but has a coarser grain. The flesh of the bulls is highly flavoured, and both bulls and cows when lean smell strongly of musk, their flesh at the same time being very dark and tough, and certainly far inferior to that of any other ruminating animal existing in North America." The carcase of a Musk-Ox weighs, exclusive of fat, above 3 cwt. On this subject, Major Feilden[1] says: "The cause of the disagreeable odour which frequently taints the flesh of these animals has received no elucidation from my observations. It does not appear to be confined to either sex, or to any particular season of the year; for a young unweaned animal, killed at its mother's side and transferred within an hour to the stew-pans, was as rank and objectionable as any. The flesh of some of these animals of which I have partaken was dark, tender, and as well flavoured as that of four-year old Southdown mutton."

Remains of two fossil species of this genus (*O. bombifrons* and *O. cavifrons*) have been described from Pleistocene beds in the United States, the one from Kentucky and the other from the Arkansas River. Both (if indeed they be valid species) appear closely allied to the living form.

Bovine Section.—Horns present and of nearly equal size in both sexes; in form rounded or angulated, placed on or near the vertex of the skull, extending more or less outwards, and curving upwards near the extremities; external surface comparatively smooth and never marked by prominent transverse ridges or knobs. Muzzle broad, with large naked muffle; nostrils lateral; no suborbital gland. Skull without any trace of lachrymal fossa or fissure. Tail long and cylindrical; generally tufted at the extremity, rarely hairy throughout. Males usually with a dew-lap on the throat. No foot-glands. Molar teeth extremely hypsodont; those of the upper jaw with a nearly square cross-section, and a large accessory inner column.

The section is abundantly represented in the Palæarctic, Oriental, and Ethiopian regions, with one Nearctic species and an outlying and aberrant species in Celebes.

Bos.[2]—The whole of the species of Oxen were included by Linnæus in the single genus *Bos*, and although the species have been distributed by modern zoologists in several genera—such as *Anoa*, *Bubalus*, *Bison*, *Poëphagus*, *Bibos*, and *Bos*—the characters by which they are separated are so slight that it seems, on the whole, preferable to retain the old genus in its original wide sense. Using then the term *Bos* in this sense, it will include all the representatives of the section—about a dozen in number—and may be divided into several groups.

[1] *Zoologist*, September 1877.

[2] Linn. *Syst. Nat.* 12th ed. vol. i. p. 98 (1766).

The first group includes the Buffaloes (genus *Bubalus*), chiefly characterised by their more or less flattened and angulated horns, which incline upwards and backwards, with an inward curve towards their tips, and are placed below the plane of the occiput, or vertex of the skull. The premaxillæ reach to the nasals, and the vomer is peculiar in being so much ossified as to join the posterior border of the palate. The back has a distinct ridge in the region of the withers; and the forehead is frequently convex. Oriental and Ethiopian region, and Celebes.

The most generalised representative of this group is the small Anoa (*B. depressicornis*) of Celebes, the type of the genus *Anoa* or *Probubalus*, which has the same cranial structure as in the more typical Buffaloes, to the young of which (as was pointed out by the late Professor Garrod) it presents a striking resemblance. Its colour is black; and the short and prismatic horns are directed upwards from the forehead. In the Pliocene Siwaliks of India there occur the remains of larger Buffaloes (*B. occipitalis* and *B. acuticornis*) closely allied to the Anoa, but with longer and more distinctly angulated horns. The still larger *B. platyceros* of the last-named deposits, in which the horns are wide-spreading and much flattened, appears to be in some respects intermediate between the preceding and following forms. The typical Indian Buffalo (*Bos buffelus*), which has been domesticated over South-East Asia, Egypt, and Southern Europe, is, in the wild state, a gigantic animal with enormous horns. These horns are longer, more slender, and more outwardly directed in the female than in the male; and in the former sex may have a length of more than 6 feet from base to tip. They are widely separated at their bases, the forehead is very convex, and the ears are not excessively large, and have no distinct fringe. These Buffaloes frequent swampy and moist districts in several parts of India, but it is in many instances difficult to decide whether they belong to really wild or to feral races. Very large skulls, specifically indistinguishable from those of the existing form, occur in the Pleistocene deposits of the Narbada valley in India; while an allied, if not specifically identical form, occurs in the Pliocene of the same country. There is some doubt whether *B. antiquus* of the Pleistocene of Algeria is most nearly related to the Indian or to the African species.

In Africa two species of Buffalo are recognised by Sir Victor Brooke,[1] namely the large *B. caffer*, occurring typically at the Cape, but said by this writer to range to Abyssinia, and the smaller *B. pumilus*, which seems to have a very wide distribution. The skulls of both these forms are shorter than in the Indian species, while the horns are also shorter, much more curved inwardly, and more approximated on the forehead. In the large typical form of

[1] *Proc. Zool. Soc.* 1873, p. 474.

B. caffer from South Africa the colour is black, the horns of the male are very thick, much reflected, and closely approximated on the forehead, where they form a helmet-like mass.[1] The large northern form described as *B. æquinoctialis* has the horns somewhat less thick, and thus approximates to the so-called *B. pumilus*.

The latter occurs typically in Western Africa, where it has also been described as *B. brachyceros*. In the typical form the horns are thinner and less reflected than in *B. caffer*, and in some specimens they are more widely separated on the forehead, and are marked at their bases by distinct rugæ. The colour is ruddy brown, inclining to rufous in one specimen. The skulls of Buffaloes from West Africa, probably referable to the form described as *B. centralis*, appear to connect *B. pumilus* with *B. caffer*, as shown by their larger size and the form of their horns; so that further observations are required to show whether the smaller form is really entitled to rank as a distinct species, or merely as a well-marked local race.

The second group comprises the Bisons, which are more nearly allied to the true Oxen, having similar rounded horns, but the skull being less massive, with a longer and more tapering frontal region, and a wider frontal diameter. The superior part of the forehead is transversely arched, the intercornual space elevated in the middle, the horns situated below the plane of the occiput, and the orbits more or less prominent. The premaxillæ do not extend upwards to reach the nasals. The Bisons (Fig. 148) have the body covered with short, crisp, woolly hair, while on the head and neck there is an abundance of much longer and darker hair, which forms a mane concealing the eyes, ears, and the bases of the horns. There is also a long beard beneath the chin; while a line of long hair extends from the head nearly to the tail, the latter being tufted at the extremity. The withers are much higher than the hind quarters, so that there is a kind of hump at the shoulders.

The group is represented by two species—the European and the American Bison. The former is the *Bos bonasus* of Linnæus, and is also identical with the *Bos bison* of Ray. The German name *Wisent* is the equivalent of the Greek *Bison*. The American species is the *Bos americanus* of Gmelin. Both species are closely allied, but the American Bison is slightly the smaller animal of the two, and is shorter and weaker in the hind quarters, with a smaller pelvis; its body is, however, more massive in front; and the hair on the head, neck, and fore quarters is longer and more luxuriant. A large bull American Bison, preserved in the Museum at Washington, stands 5 feet 8 inches in height at the withers. The European Bison appears to have been formerly

[1] Sir V. Brooke states that this species is distinguished from *B. pumilus* by the absence of a fringe to the ears, but specimens in the British Museum show that this is not the case.

abundant over a large portion of Europe in the Pleistocene period —the fossil race described as *B. priscus* not being specifically distinct; but at the present day it exists only in the primeval forests of Lithuania, Moldavia, Wallachia, and the Caucasus, where it is artificially preserved.

The American Bison formerly ranged over about one-third of the North American continent. Thus, to quote from Mr. Hornaday,[1] "starting almost at tide-water on the Atlantic coast, it extended across the Alleghany mountain system to the prairies along the Mississippi, and southward to the delta of that great system.

FIG. 148.—The American Bison (*Bos americanus*). After Hornaday.

Although the great plain country of the West was the natural home of the species, where it flourished most abundantly, it also wandered south across Texas to the burning plains of North-Eastern Mexico, westward across the Rocky Mountains into New Mexico, Utah, and Idaho, and northward across a vast treeless waste to the bleak and inhospitable shores of the Great Slave Lake itself." In consequence of the settlement of the country by Europeans the area inhabited by the Bison was gradually contracted, till about 1840 one mighty herd occupied the centre of its former range. The completion of the Union Pacific Railway in 1869 divided this great herd into a southern and a northern division, the former comprising a number of individuals estimated at nearly four millions, while the latter contained about a million and a half. Before 1880 the southern herd had, however, practically ceased to exist; while the same fate overtook the northern one in 1883. In 1889 some twenty stragglers in Texas represented the last of the southern herd; while there were a few others in Colorado, Wyoming, Montana,

[1] *The Extirpation of the American Bison*, 1889.

and Dakota. A herd of some two hundred wild individuals, derived from the northern herd, is preserved by the United States Government in the Yellowstone National Park; and it is believed that some five hundred of the race known as Wood-Bison exist in British territory; but with these exceptions this magnificent species is exterminated. The multitudes in which the American Bison formerly existed are almost incredible; the prairies being absolutely black with them as far as the eye could reach, and the numbers in the herds being, as we have said, reckoned by millions. Mr. Hornaday even considers that the whole of the game in South

FIG. 149.—The Yak (*Bos grunniens*), domestic variety.

Africa was never equal to the number of Bison on an equal area of the American prairies.

An extinct Bison from the Pleistocene of Texas, known as *Bos latifrons*, was probably the ancestor of the recent American species.

The Yak (*Bos grunniens*) appears to be allied both to the Bisons and the true Oxen, being distinguished from the former by the different position occupied by the long hair, which forms a fringe investing the shoulders, flanks, and thighs, and grows over the whole of the tail. In the skull the orbits are less tubular, the forehead flatter, and the premaxillæ less widely separated from the nasals. There is no distinct dewlap. Wild Yaks inhabit the higher regions of Chinese Tibet and the region of the Karakoram, as well as the more outlying parts of Ladak, such as the Changchemo valley. Owing, however, to incessant pursuit those now found within the territories of the Maharaja of Kashmir are stragglers from Chinese Tibet. The height of the Yak is somewhat lower than that of the larger domestic cattle. The colour of the wild race is black, tending to brown on the flanks; but many of the tame breeds which have been crossed with ordinary cattle have more or less white (Fig. 149), and it is the white tails of these half-breeds that are so esteemed in India as "chowries." Yaks are exceedingly intolerant of heat, and the wild ones always live at very great elevations. Tame Yaks are extensively used as beasts of burden in Tibet, where they are extremely valuable in crossing the high and desolate wastes of that region; they have, however, the great drawback that they refuse to eat corn, so that in districts where there is no grass it is frequently necessary to make forced marches with wearied beasts in order to prevent them (and thus the whole party) perishing from starvation.

The skull of an extinct species from the Pliocene of Northern India, described as *Bos sivalensis*, appears to indicate a species allied to the Yak.

With the Bibovine group we come to the consideration of three Oriental species which connect the preceding forms with the typical Oxen. The three species are the Gaur (*B. gaurus*) the Gayal (*B. frontalis*, Fig. 150) of India, and the Banteng (*B. sondaicus*) of Burma, Java, Bali, and Lambok. In this group, as in the true Oxen, there are thirteen pairs of ribs, against fourteen in the Bisons. All the three species are characterised by the great height of the spines of the anterior dorsal vertebræ, causing a prominent ridge down the back. The horns, which are of a greenish colour in the Gaur, are somewhat flattened, and after running outwards are directed upwards instead of backwards; they occupy the vertex of the skull. The frontals are more or less concave, the premaxillæ do not join the nasals, and the occipital aspect of the skull is characterised by the deep incisions made by the temporal fossæ. The lower part of the legs is white (Fig. 150), and the hoofs are comparatively small and pointed. The Gaur (*B. gaurus*) is the largest of the three species, and inhabits all the large forests of India from near Cape Comorin to the foot of the Himalaya; it is commonly known to sportsmen as the Indian Bison. It stands fully 6 feet in height at the withers, which are much elevated; and since the whole back is arched the line from the nose to the root of the tail forms an almost continuous curve. The most characteristic feature of the animal is, however, the large and convex intercornual frontal crest, which curves forward, and thus gives a concave profile to this part of the skull. As a rule the Gaur prefers hilly regions, although it is sometimes met with on the flat. It is very shy and readily frightened; and it has never been domesticated. The Gayal, or Mithan, of which a figure is given in woodcut 150, is at once distinguished from the Gaur by the straight line between the horns (which are black in colour), owing to the absence of the intercornual crest of the latter. The horns are also shorter, more rounded, and less curved. In the Indian Museum, Calcutta, there are, however, skulls which are to a great extent intermediate between those of typical Gaurs and those of typical Gayals, but these may belong to hybrids. The Gayal occurs in Assam, Chittagong, and adjacent districts, but it appears that these animals exist in a semi-domesticated condition, no wild race being known to Europeans, although it is probable that such may exist in the unexplored Mishmi Hills.

FIG. 150.—The Gayal (*Bos frontalis*). From Sclater, *List of Animals in Zoological Society's Gardens*, 1883.

The Banteng (*B. sondaicus*) is a smaller and lighter built animal than either of the preceding, with a longer and sharper head, and more rounded and slender horns. The dorsal ridge is, moreover, but slightly developed; while the bright dun colour of the body of the female readily distinguishes it from the darker hue of the Gaur and Gayal.

A fossil skull from the Pleistocene deposits of the Narbada valley, India, described as *Bos palæogaurus*, is believed to indicate a species nearly allied to the Gaur, if indeed it be specifically distinct.

The true Oxen, or Taurine group, are now represented solely by *Bos taurus* and *Bos indicus*. Both of these species are now known only by domesticated races, unless the herds of the former preserved at Chillingham and some other British parks are the survivors of an original wild race. The dorsal ridge of the Bibovine group is here wanting; the horns are rounded, with their extremities directed backwards, and are placed at the extreme vertex of the skull; while the long frontal region is nearly flat; the temporal fossæ scarcely intrude upon the occipital aspect of the skull; and the premaxillæ reach the nasals. The hoofs are large and rounded. It is known that wild Oxen were abundant in the forests of Europe at the time of Julius Cæsar, by whom they were described as the Urus, equal to the German Aurochs; and the large skulls found in turbary and Pleistocene deposits, and described under the name of *Bos primigenius*, can only be regarded as having belonged to the large original race of *B. taurus*, of which it has been thought the Chillingham cattle are smaller descendants.[1] The subfossil skulls described as *B. longifrons* and *B. frontosus* must also be looked upon as referable to smaller races of the same species. That the domestic cattle of Europe are descendants from the various races of the same original species there can be no doubt, but in the case of the humped cattle of India (*B. indicus*) it is quite probable that their origin may be, at least in part, different. The extinct *Bos namadicus*, of the Pleistocene deposits of India, was a species with the general characters of the Taurine group, but with an inclination to a flattening of the horns, and with an approximation to a Bibovine type of occiput, as well as with the separation of the premaxillæ from the nasals.

The earliest representatives of this group occur in the Pliocene of the Siwalik Hills in Northern India. One of these species (*B. planifrons*) appears to be allied to *B. namadicus*; but the other (*B. acutifrons*) was a gigantic species characterised by the sharp median angulation of the frontal region, and the pyriform section of the enormous horn-cores.

The extinct *B. elatus*, from the Upper Pliocene of France and Italy, is the representative of a generalised type, which may be known as the Leptobovine group. The males had rounded horn-cores widely separated at their bases, and placed low down on the forehead. The females (which have been described as *Leptobos*) were often or always hornless. The limbs were unusually slender. This group also occurs in the Pliocene of the Siwalik Hills.

[1] The late Mr. Alston, *Fauna of Scotland*, "Mammalia" (Glasgow, 1880), p. 25, considers that the Chillingham cattle are descendants of a race which had escaped from domestication.

Suborder PERISSODACTYLA

This is a perfectly well-defined group of Ungulate mammals, represented in the actual fauna of the world by only three distinct types or families—the Tapirs, the Rhinoceroses, and the Horses—poor in genera and species, and (except in the case of the two domesticated species of *Equus*, which have been largely multiplied and diffused by man's agency) not generally numerous in individuals, though widely scattered over the earth's surface. Palæontological

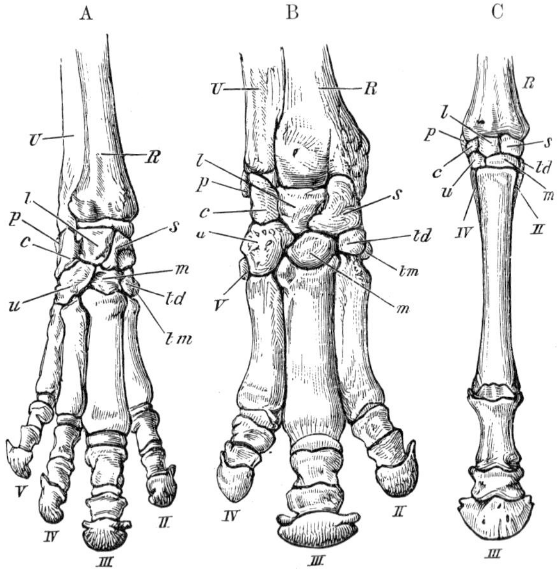

FIG. 151.—Bones of right fore foot of existing Perissodactyles. A, Tapir (*Tapirus indicus*), $\times\frac{1}{5}$; B, Rhinoceros (*Rhinoceros sumatrensis*), $\times\frac{1}{5}$; C, Horse (*Equus caballus*), $\times\frac{1}{8}$. *U*, ulna; *R*, radius; *c*, cuneiform; *l*, lunar; *s*, scaphoid; *u*, unciform; *m*, magnum; *td*, trapezoid; *tm* trapezium.—From Flower, *Osteology of Mammalia*.

records, however, show very clearly that these are but the surviving remnants of a very extensive and much-varied assemblage of animals, which flourished upon the earth through the Tertiary geological period, and which, if it could be reconstructed in its entirety, would not only show members filling up structurally the intervals between the existing apparently isolated forms, but would also show several marked lines of specialisation which have become extinct without leaving any direct successors.

The following are the principal characters distinguishing them from the Artiodactyla. Premolar and molar teeth in continuous series, with massive, quadrate, transversely ridged or complex crowns,—the posterior premolars often resembling the true molar in size and structure. Crown of the last lower molar commonly bilobed, and if a third lobe is present in this tooth it is wanting in the last lower milk-molar. Dorso-lumbar vertebræ never fewer than twenty-two, usually twenty-three in the existing species. Nasal bones expanded posteriorly. An alisphenoid canal. Femur with a third trochanter.[1] The middle or third digit on both fore and hind feet larger than any of the others, and symmetrical in itself, the free border of the ungual phalanx being evenly rounded (see Fig. 151). This may be the only functional toe, or the second and fourth may be subequally developed on each side of it. In the Tapirs and many extinct forms, the fifth toe also remains on the fore limb, but its presence does not interfere with the symmetrical arrangement of the remainder of the foot around the median line of the third or middle digit. Traces of a hallux have only been found in some extremely ancient and primitive forms. The astragalus has a pulley-like surface above for articulation with the tibia, but its distal surface is flattened and unites to a much greater extent with the navicular than with the cuboid, which bone is of comparatively less importance than in the Artiodactyla. The calcaneum does not articulate with the lower or distal extremity of the fibula. The stomach is always simple, the cæcum is large and capacious, the placenta diffused, and the mammæ are inguinal. The gall-bladder is invariably absent.

As regards the dentition, the whole of the premolar series may be preceded by milk-teeth; and it has been demonstrated in *Rhinoceros* that when there is no displacement of the first cheek-tooth that tooth is a persistent milk-molar; the same condition apparently holding good in *Palæotherium.* This feature indicates considerable dental specialisation, the milk-molars, according to the theory generally accepted by the leading English zoologists, being the acquired, and the premolars the original series. Another peculiar feature of the dentition of the Perissodactyla, very rarely met with among the Artiodactyla, is that the premolars tend to resemble the true molars; this feature occurring in all the existing genera, although not found in the earlier generalised types. The cheek-teeth of all the members of the suborder are primarily constructed on some modification of what is known as the lophodont plan. Thus the upper molars (Fig. 155, p. 375) have an outer antero-posterior wall from which proceed two transverse ridges, formed by the coalescence of the primitive inner and outer columns, towards the inner aspect of the crown; while in the lower molars there may be either two simple transverse ridges, or these ridges may be curved into crescents, coming into contact with one another at their extremities. Those forms having brachydont teeth show this plan of structure in its simplest modification; but in cases, as in the

[1] Wanting in the aberrant *Chalicotherium.*

Horse, where the teeth assume an extremely hypsodont form, the original plan is so obscured by infoldings of the enamel that it can only be traced with difficulty.

At the present day the Perissodactyla are sharply differentiated into Horses, Tapirs, and Rhinoceroses, but the knowledge already gained of the extinct representatives of the suborder shows such a close alliance between these groups that it is exceedingly difficult to make any satisfactory classification of the whole. This is of course exactly what might have been expected; and the same would doubtless be the case with all other groups if we knew as much of their past history as we do of that of the Perissodactyles.

The detailed account of the anatomy of the Horse given in the sequel will afford much information as to the general structure of the members of the suborder.

Family TAPIRIDÆ.

Both upper and lower cheek-teeth brachydont and simply bilophodont; hinder premolars as complex as the molars; last lower molar without third lobe; first upper cheek-tooth with a milk-predecessor.[1] Outer columns of upper molars conical. Four digits in the manus, and three in the pes.

Tapirus.[2]—Dentition $i\ \frac{3}{3}$, $c\ \frac{1}{1}$, $p\ \frac{4}{3}$, $m\ \frac{3}{3}$; total 42. Of the upper incisors, the first and second are nearly equal, with short, broad crowns; the third is large and conical, considerably larger than the canine, which is separated from it by an interval. Lower incisors diminishing in size from the first to the third; the canine, which is in contact with the third incisor, large and conical, working against (and behind) the canine-like third upper incisor. In both jaws there is a diastema between the canines and the commencement of the teeth of the cheek-series, which are all in contact. First upper premolar with a triangular crown, narrow in front owing to the absence of the anterior inner cusp. The other upper premolars and molars all formed on the same plan and of nearly the same size, with four roots and quadrate crowns, rather wider transversely than from before backwards, each having four cusps, connected by a pair of transverse ridges, anterior and posterior. The first lower premolar compressed in front; the others composed of a simple pair of transverse crests, with a small anterior and posterior cingular ridge.

Skull elevated and compressed. Orbit and temporal fossa widely continuous, there being no true postorbital process from the frontal bone. Anterior narial apertures very large, and extending high on the face between the orbits; nasal bones short, elevated,

[1] See W. N. Parker, *Proc. Zool. Soc.* 1882, p. 775.

[2] Cuvier, *Tableau Élément. de l'Hist. Nat.* p. 152 (1798); *ex* Brisson.

triangular, and pointed in front. Vertebræ: C 7, D 18, L 5, S 6, C about 12. Limbs short and stout. Fore feet with four toes, having distinct hoofs: the first is absent, the third the longest, the second and fourth nearly equal, the fifth the shortest and scarcely reaching the ground in the ordinary standing position. Hind feet with the typical Perissodactyle arrangement of three toes,—the middle one being the largest, the two others nearly equal. Nose and upper lip elongated into a flexible, mobile snout or short proboscis, near the end of which the nostrils are situated. Eyes rather small. Ears of moderate size, ovate, erect. Tail very short. Skin thick and but scantily covered with hair.

The existing species of Tapir may be grouped into two sections, the distinctive characters of which are only recognisable in the skeleton. (A) With a great anterior prolongation of the ossification of the nasal septum (mesethmoid), extending in the adult far beyond the nasal bones, and supported and embraced at the base by ascending plates from the maxillæ (genus *Elasmognathus*, Gill). Two species, both from Central America, *Tapirus bairdi* and *T. dowi*. The former is found in Mexico, Honduras, Nicaragua, Costa Rica, and Panama; the latter in Guatemala, Nicaragua, and Costa Rica. (B) With ossification of the septum not extending farther forward than the nasal bones (*Tapirus* proper). Three species, *T. indicus*, the largest of the genus, from the Malay Peninsula (as far north as Tavoy and Mergui), Sumatra, and Borneo, distinguished by its peculiar coloration, the head, neck, fore and hind limbs, being glossy black, and the intermediate part of the body white; *T. americanus* (*T. terrestris*, Linn.), the common Tapir of the forests and lowlands of Brazil and Paraguay (Fig. 152); and *T. roulini*, the Pinchaque Tapir of the high regions of the Andes. All the American species are of a nearly uniform dark brown or blackish colour when adult; but it is a curious circumstance that when young (and in this the Malay species conforms with the others) they are conspicuously marked with spots and longitudinal stripes of white or fawn colour on a darker ground.

The habits of all the kinds of Tapirs appear to be very similar. They are solitary, nocturnal, shy, and inoffensive, chiefly frequenting the depths of shady forests and the neighbourhood of water, to which they frequently resort for the purpose of bathing, and in which they often take refuge when pursued. They feed on various vegetable substances, as shoots of trees and bushes, buds, and leaves. They are hunted by the natives of the lands in which they live for the sake of their hides and flesh.

The singular fact of the existence of so closely allied animals as the Malayan and the American Tapirs in such distant regions of the earth, and in no intervening places, is accounted for by what is known of the geological history of the race; for the Tapirs must once have had a very wide distribution. There is no proof of their having lived in the Eocene epoch, but in deposits of Miocene and Pliocene date remains undistinguishable generically from the modern Tapirs, and described as *T. priscus*, *T. arvernensis*, etc., have been found in France, Germany, and in the Red Crag of Suffolk. Tapirs appear, however, to have become extinct in Europe before the Pleistocene period, since none of their bones or teeth have been found in any of the caverns or alluvial deposits in which those of Elephants, Rhinoceroses, and Hippopotamuses occur in abundance; but in other regions their distribution at this age was far wider than at present,

FIG. 152.—The American Tapir (*Tapirus americanus*).

as they are known to have extended eastward to China (*T. sinensis*, Owen) and westwards over the greater part of the southern United States of America, from South Carolina to California. Lund also distinguished two species or varieties from the caves of Brazil, one of which appears identical with *T. americanus*. Thus we have no difficulty in tracing the common origin in the Miocene Tapirs of Europe of the now widely separated American and Asiatic species. It is, moreover, interesting to observe how very slight an amount of variation has taken place in forms isolated during such an enormous period of time.

The anatomy of the soft parts of the Tapirs[1] conforms to the

[1] See J. Murie, *Journ. Anat. and Physiol.* vol. vi. p. 131, 1871; W. N. Parker, *Proc. Zool. Soc.* 1882, p. 768; and F. E. Beddard, *Proc. Zool. Soc.* 1889, p. 252.

general Perissodactyle type, as exemplified in the Rhinoceros and the Horse, although on the whole (as might have been expected) presenting a closer resemblance to the former. *T. americanus* differs from *T. indicus* by the absence, or at any rate the less development, of the intestinal valvulæ conniventes, the presence of a moderator band in the heart, the shape of the glans penis, and the more elongated cæcum, which is sacculated by four distinct longitudinal fibrous bands. The convolutions of the hemispheres of the brain of the Tapirs are simpler than in other Perissodactyles, thus tending to confirm the inferences which may be drawn from the skeleton and teeth as to the comparatively low or generalised organisation of these animals.

Palæotapirus.—This name has been applied to an imperfectly known form from the Upper Eocene Phosphorites of Central France, which is regarded by Dr. Filhol as referable to this family.

Family LOPHIODONTIDÆ.

Molars brachydont and bilophodont, those of the lower jaw with either straight or imperfectly crescentoid ridges; premolars smaller and usually simpler than the molars; last lower molar generally with a third lobe. Outer columns of upper molars conical or flattened. Digits usually as in the preceding family.

This family includes a number of more or less imperfectly known forms, all of which are extinct and apparently confined to the Eocene period, and ranging from the size of a Rabbit to that of a Rhinoceros. Although some of these appear to have died out without giving rise to more specialised forms, it is probable that this family contained the ancestral types from which most or all of the modern Perissodactyles have been derived. Only very brief mention can be made here of some of the leading genera. *Lophiodon*, of the Middle and Upper Eocene of Europe, with the dental formula, $i\,\frac{3}{3}$, $c\,\frac{1}{1}$, $p\,\frac{3}{3}$, $m\,\frac{3}{3}$, includes the largest representatives of the family, and is generally regarded as a stock which has died out without giving rise to later forms. The ridges of the lower molars are straight, and the last of these teeth has a third lobe; while the second transverse ridge of the last upper premolar is usually incomplete; the outer columns of the upper molars are flattened, as in the next genus. *Hyrachyus*, of the Upper Eocene of the United States, and probably also occurring in the French Eocenes, is an allied genus, with four premolars and no third lobe to the last lower molar; the fourth upper premolar having the two ridges uniting internally to form a crescent. This genus has been regarded as the ancestor of the Rhinocerotic *Hyracodon*. The genus *Hyracotherium* was established in 1839 by Owen for a small animal no larger than a Hare, the skull of which was found in the London Clay at Herne

Bay. A more nearly perfect specimen, apparently of the same species, was afterwards (in 1857) described under the name of *Pliolophus vulpiceps*, of which the skull is figured in the accompanying woodcut. Other forms referable to the same genus have been obtained from the Wasatch Eocene of the United States, and were described by Professor Marsh under the name of *Eohippus*. There were four premolars, the fourth being unlike the molars, and in the upper jaw having only one inner cusp. The upper molars are of the general type of those of *Lophiodon*, but have conical outer columns, and the anterior transverse ridge imperfect, while the ridges of the lower molars are crescentoid. *Systemodon* differs from *Hyracotherium*

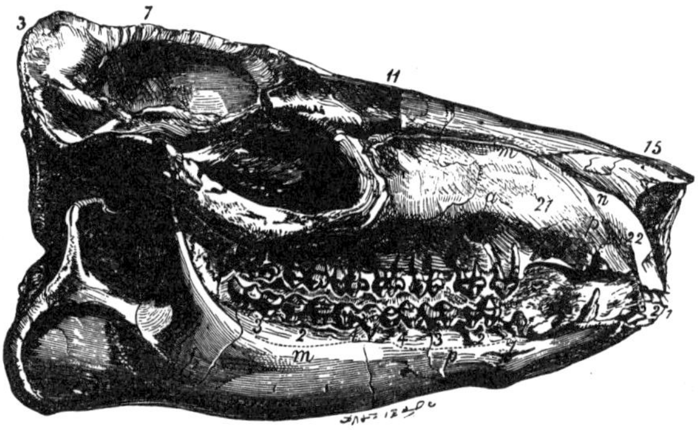

FIG. 153.—Right side of skull of *Hyracotherinum leporinum*, from the London Clay. ½ natural size. (After Owen.) 3, Occiput; 7, sagittal crest; 11, frontals; 15, nasals; 21, maxilla; 22, premaxilla; *d*, mandibular condyle; *a*, aperture of facial nerve; *p* 1-4, premolars; *m* 1-3, molars.

by the absence of a diastema between the first and second premolars; it occurs in the Wasatch Lower Eocene of the United States. In *Pachynolophus* (*Lophiotherium*, *Orotherium*, or *Orohippus*), which is common to the Middle and Upper Eocene of Europe and the Bridger Eocene of North America, the outer columns of the upper molars are flattened, and in some cases, at least, the last premolar resembles the molars, that of the upper jaw having two inner cusps.[1] This genus, indeed, so closely connects *Hyracotherium* with the genera *Epihippus* and *Anchilophus* as to show that the distinction between the *Lophiodontidæ* and *Palæotheriidæ* is really an arbitrary one. *Epihippus*, of the Upper Eocene of the United States, has both the third and fourth upper premolars as complex in the molars, and is distinguished from *Anchilophus* by the lower cusps and more imperfect transverse ridges of these teeth. The so-called *Orohippus agilis* belongs to this genus. *Isectolophus* is another American Eocene genus which may be provisionally placed in this family; it is regarded by Professors Scott and Osborn as connecting *Systemodon*

[1] The Swiss *P. siderolithicus* has only one cusp in the last upper premolar.

with the *Tapiridæ*; the fourth and probably the third upper premolar approximating in structure to the molars; the upper molars have conical outer columns. *Helaletes* is another closely allied form, with similar premolars, but with the outer columns of the upper molars flattened.

Family PALÆOTHERIIDÆ.

Molars (Fig. 155) brachydont, with the valleys between the ridges never filled with cement; upper premolars either simpler than

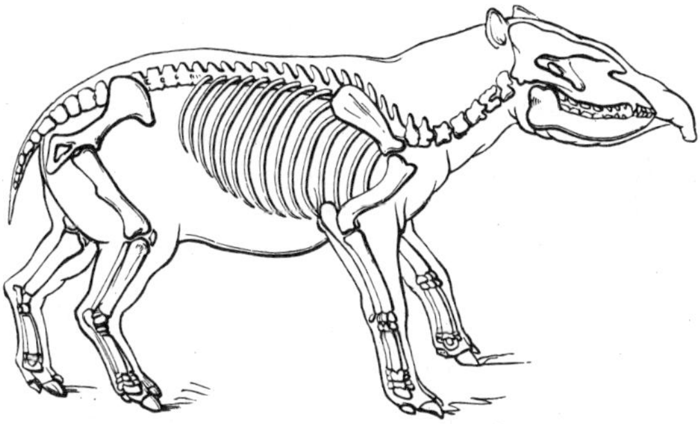

FIG. 154.—Restoration of *Palæotherium* (Upper Eocene). After Cuvier.

or as complex as the molars; lower molars with crescentoid ridges, and the last of the series with or without a third lobe. Outer columns of upper molars flattened. Orbit (at least usually) confluent with temporal fossa. Three digits on each foot. This family includes extinct genera ranging from the Middle and Upper Eocene to the Miocene, and passes so gradually into the following one that the maintenance of the two can only be supported on the ground of convenience. The typical genus, *Palæotherium*, was made known to science in the early part of the present century by Cuvier, who restored the skeleton (Fig. 154) with a short neck like that of the Tapirs, although it has been subsequently found that the neck was considerably longer. This genus (which may be taken to include *Paloplotherium*) ranges from

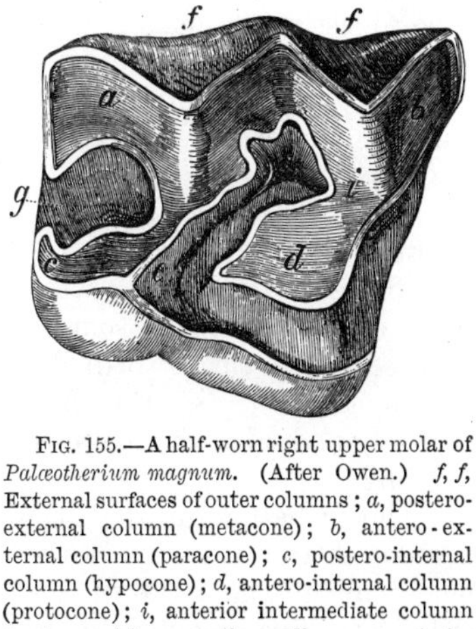

FIG. 155.—A half-worn right upper molar of *Palæotherium magnum*. (After Owen.) *f*, *f*, External surfaces of outer columns; *a*, postero-external column (metacone); *b*, antero-external column (paracone); *c*, postero-internal column (hypocone); *d*, antero-internal column (protocone); *i*, anterior intermediate column (protoconule); *e*, median valley; *g*, posterior valley.

the Middle to the Upper Eocene of Europe, and usually has the full typical dentition, although the first premolar may disappear. The last lower molar has a third lobe; and in the typical forms the last premolar is as complex as the molars, the diastema is short, and the canines are not large. In other forms, however, the hinder ridge of the fourth upper premolar may be aborted. The first upper cheek-tooth is generally a well-developed tooth, which may have a deciduous predecessor. *Anchilophus*, of the Upper Eocene of Europe, and *Anchitherium*, of the Miocene of Europe and North America, connect the preceding forms with the *Equidæ*. In the latter genus there is the full number of teeth, the last lower molar has almost completely lost the third lobe of *Anchilophus*, and the surfaces of the two outer lobes of the upper molars (Figs. 157, 158) lack the median vertical ridges of that genus. In the American species of *Anchitherium* (which have been described as *Mesohippus* and *Miohippus*) the lateral digits are larger than in the European Middle Miocene *Anchitherium aurelianense;* a mere splint represents the fifth metacarpal, and the meso- and ento-cuneiform of the tarsus do not unite as they do in the latter.

Family EQUIDÆ.

Molars hypsodont, with the outer columns of the upper ones flattened, the valleys completely filled with cement, and the enamel thrown into folds and plications; upper premolars as complex as molars, which they slightly exceed in size; ridges of lower molars crescentoid, and complicated by enamel-foldings; no distinct third lobe to last lower molar; summits of incisors with a central infolding of enamel. Orbit completely surrounded by bone. Digits three or one, but in the former case the median one is alone of functional importance; ulna and fibula incomplete; meso- and ento-cuneiform of tarsus united.

Such are the leading characters which serve to distinguish the existing Horses and their nearest fossil allies from the *Palæotheriidæ*. The Horse, as being the best known of the Perissodactyle Ungulates, is selected for a somewhat detailed description; but before proceeding to this it will be advisable to take a brief survey of the relations of the *Equidæ* to the extinct forms already noticed, and also of the modifications of the family at present existing.

The earliest form which can be certainly included in this line of descent is the American Lower Eocene genus *Phenacodus* (noticed below under the head of the suborder Condylarthra), in which there were five complete digits to the feet. From this form there is but a step to *Systemodon* and *Hyracotherium*, in which the functional digits of the manus were reduced to four, as in *Pachynolophus*

(Fig. 156, *a*), although one species retained a rudiment of the metacarpal of the pollex.

The transition from these animals of the Eocene period to the Horses of modern times has been accompanied by a gradual increase in size. The diminutive *Hyracotherium* of the Lower, and *Pachynolophus* of the Middle and Upper Eocene were succeeded in the Miocene period by the forms to which the name of *Anchitherium* has been given, of the size of sheep; these again in Pliocene times by *Hipparion* and *Protohippus*, as large as the modern donkeys; and it is mainly in the Pleistocene period that *Equidæ* occur which approach in size the existing Horse. Important structural modifications have also taken place, with corresponding changes in the

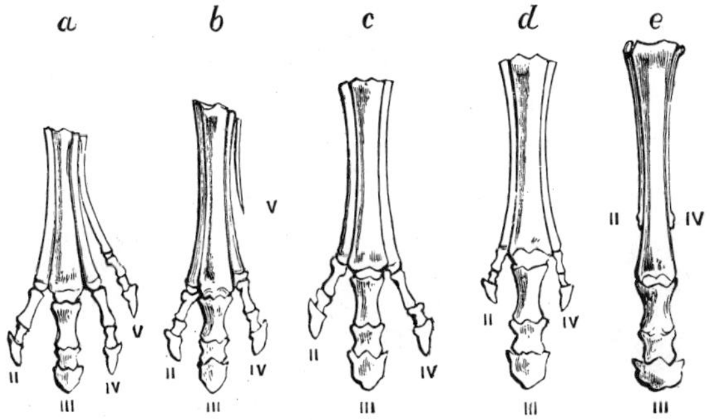

FIG. 156.—Successive stages of modification of the feet of extinct forms of Horse-like animals (chiefly from Marsh), showing gradual reduction of the outer and enlargement of the middle toe (III). *a*, *Pachynolophus* (Eocene); *b*, *Anchitherium* (Early Miocene); *c*, *Anchitherium* (Late Miocene); *d*, *Hipparion* (Pliocene); *e*, *Equus* (Pleistocene).

mode of life of the animal. Thus the neck has become elongated, the skull altered in form, the teeth greatly modified, and the limbs have undergone remarkable changes. The last two require to be described more in detail.

The teeth in the Eocene forms had, as mentioned above, the characteristic number of forty-four. This number has been retained throughout the series, at least theoretically; but one tooth on either side of each jaw, the anterior premolar, which in all the Eocene and Miocene species was well developed, persisting through the lifetime of the animal, is in all modern Horses rudimentary, functionless, and generally lost at an early period of life, evidently passing through a stage which must soon lead to its complete disappearance. The canines have also greatly diminished in size, and are rarely present in the female sex, so that practically a very large number of adult Horses of the present day have eight teeth less than the number possessed by their predecessors. The diastema or interval between the incisor and premolar teeth (of essential importance in the domesticated Horse to his master, as without it there would be no room for inserting the special instrument of subjugation to his commands, the bit) already existed in the earliest known forms, but has gradually increased in length. The incisors have undergone in comparatively recent times that curious change producing the structure more fully described hereafter, which distinguishes the Horse's incisors from those of all other known animals, with the exception of the extinct *Macrauchenia.* Lastly, the molars have undergone a remarkable series of modifications, much resembling in principle those that have taken place in several other groups of herbivorous animals. Distinctions in form which existed between the premolars, at least in the anterior part of the series, and the true molars have gradually disappeared, the teeth becoming all very uniform in the shape and structure of their grinding surface. The crowns of all these teeth

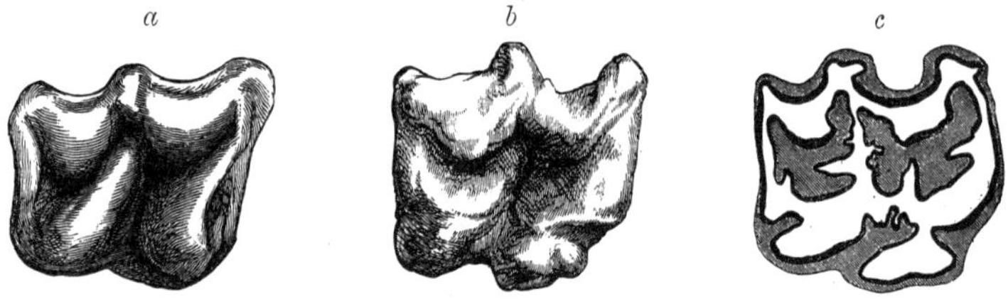

FIG. 157.—*a*, Grinding surface of unworn molar tooth of *Anchitherium; b*, corresponding surface of unworn molar of young Horse; *c*, the same tooth after it has been some time in use. The uncoloured portions are the dentine or ivory, the shaded parts the cement filling the cavities and surrounding the exterior. The black line separating these two structures is the enamel or hardest constituent of the tooth.

in the early forms were very short (see Fig. 158, *a*); there was a distinct constriction, or neck, between the crown and roots; and when the tooth was developing, as soon as the neck once rose fairly above the alveolar margin, the tooth remained permanently in this position. The term "brachydont" expresses this condition of teeth, the mode of growth of which may be illustrated by those of man. The free surface had two nearly transverse curved ridges, with valleys between (Fig. 157, *a*); but the valleys were shallow and had no deposit of cement filling them, the whole exposed surface of the unworn tooth being formed of enamel. When the ridges became worn down the dentine of the interior was exposed, forming islands surrounded by enamel. With the progress of time the crowns of the teeth gradually became longer, the valleys deeper, and the ridges not only more elevated but more curved and complex in arrangement. To give support to these high ridges and save them from breaking in use, the valleys or cavities between them became filled up to the top with cement, and as the crown wore down an admirable grinding surface consisting of patches and islands of the two softer substances, dentine and cement, separated by variously reduplicated and contorted lines of intensely hard enamel, resulted (Fig. 157, *c*). The crown continued lengthening until in the modern Horses it has assumed the form called "hypsodont" (Fig. 158, *b*). Instead of contracting into a neck, and forming roots, its sides continue parallel for a considerable depth in the socket, and as the surface wears away, the whole tooth slowly pushes up, and maintains the grinding edge constantly at the same level above the alveolus, much as in the perpetually growing Rodent's teeth. But in existing Horses there is still a limit to the growth of the molar. After a length is attained which in normal conditions supplies sufficient grinding surface for the lifetime of the animal, a neck and roots are formed, and the tooth is reduced to the condition of that of the brachydont ancestor. It is perfectly clear that this lengthening of the crown adds greatly to the power of the teeth as organs of mastication, and enables the animals in which it has taken place to find their sustenance among the comparatively dry and harsh herbage of the open plains, instead of being limited to the more succulent vegetable productions of the marshes and forests in which their predecessors probably dwelt.

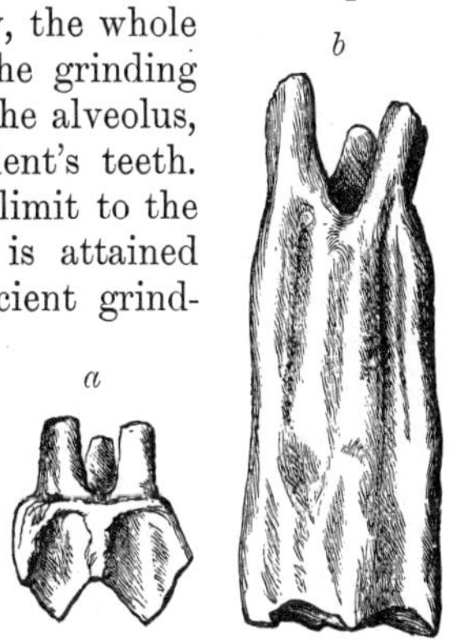

FIG. 158.—*a*, Outer view of second upper molar tooth of *Anchitherium* (brachydont form); *b*, corresponding tooth of Horse (hypsodont form).

The modifications of the limbs which took place *pari passu* with those of the teeth must have been associated with increased speed, especially over firm and unyielding ground. Short, stout legs, and broad feet, with numerous toes, spreading apart from each other when the weight of the creature is borne on them, are sufficiently well adapted for plodding deliberately over marshy and yielding surfaces, and the Tapirs and the Rhinoceroses, which in the structure of the limbs have altered but little from the primitive Eocene forms, still haunt the borders of streams and lakes and the shady depths of the forests, as was probably the habit of their ancient representatives, while the Horses are all inhabitants of the open plains, for life in which their whole organisation is in the most eminent degree adapted. The length and mobility of the neck, position of the eye and ear, and great development of the organ of smell, give them ample means of becoming aware of the approach of enemies, while the length of their limbs, the angles the different segments form with each other, and especially the combination of firmness, stability, and lightness in the reduction of all the toes to a single one, upon which the whole weight of the body and all the muscular power are concentrated, give them speed and endurance surpassing that of almost any other animal. When surprised, however, they are by no means helpless, both fore and hind feet becoming at need powerful weapons of defence.

If we were not so habituated to the sight of the Horse as hardly ever to consider its structure, we should greatly marvel at being told of a mammal so strangely constructed that it had but a single toe on each extremity, on the end of the nail of which it walked or galloped. Such a conformation is without a parallel in the vertebrate series, and is one of the most remarkable instances of specialisation, or deviation from the usual type, in accordance with particular conditions of life. It is clear, both from the structure of the foot itself, and also by an examination of the intermediate forms, that this toe corresponds to the middle or third digit of the complete typical or pentadactyle foot; and there is very strong evidence to show that by a gradual concentration of all the power of the limb upon this toe, and the concurrent dwindling away and final disappearance of all the others, the present condition of the Horse's foot has been produced.

Protohippus.[1]—In this Lower Pliocene North American genus (also described as *Merychippus*) the cheek-teeth resemble those of the generalised species of *Equus*, but have shorter crowns; while the milk-molars approximate to the permanent molars of *Anchitherium*. Each foot has three digits.

Hipparion.[2]—Upper cheek-teeth (Fig. 159), with the antero-

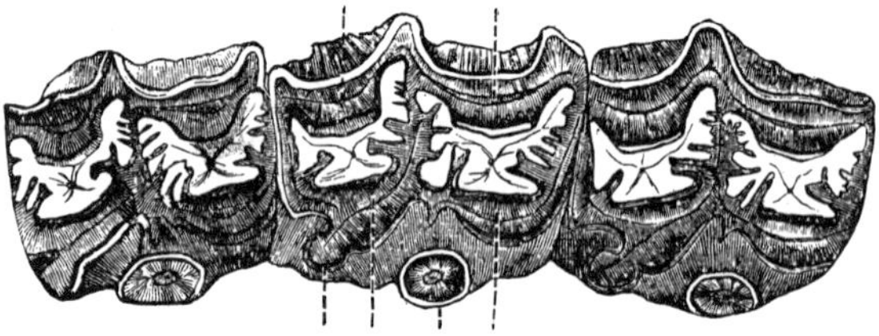

FIG. 159.—Three right upper cheek-teeth of *Hipparion*. *a*, Antero-external column; *b*, postero-external column; *c*, postero-internal column, or posterior pillar; *d*, antero-internal column, or anterior pillar; *f*, posterior intermediate column; *i*, anterior intermediate column. (From the *Palæontologia Indica*.)

internal column, or anterior pillar as it may be conveniently termed in this family, detached throughout the greater part of its height from the adjacent column. Either a single or three digits in each foot. First upper premolar large and persistent. This genus was very widely distributed in the Pliocene, occurring in Europe, Asia, and North America. In the typical European forms, and also in those

[1] Leidy, *Proc. Ac. Nat. Sci. Philad.* 1858, p. 26.

[2] Christol, *Ann. Sci. Indust. Mid. France*, vol. i. p. 180 (1832).

of North America, there were three digits in the feet (Fig. 156, *d*); but in the Indian *H. antilopinum* (separated by Cope as *Hippodactylus*) the lateral digits seem to have disappeared. There is some doubt whether or no *Hipparion* should occupy a place in the direct ancestry of the Horse, and Professor Cope suggests that while in America the intermediate place between *Anchitherium* and *Equus* was held by *Protohippus*, in Europe the same position was occupied by *Hipparion*—a view which involves the dual origin of the Horses of the New and Old Worlds.

Equus.[1]—Upper cheek-teeth with the anterior pillar (except in a very early stage of wear) joined by a narrow neck to the adjacent column (Fig. 157, *c*). Each foot with a single complete digit, but with remnants of the proximal portions of the second and fourth metapodials (Fig. 156, *e*); some extinct forms having claw-like rudiments of the terminal phalangeals of the lateral digits. First upper premolar very small or altogether absent in existing species, but in some fossil species larger and persistent; first lower premolar only occasionally developed in some fossil forms. Ears long. Tail long, with long hairs either at the end or throughout. A callosity on the inner side of the fore limb above the carpus.

Fossil Species.—In the Pleistocene Horses of South America described as *Hippidium*, as well as in the closely allied ones from North America for which the name *Pliohippus* has been proposed, the upper molars are shorter and more curved than in the existing species, while their anterior pillar is not longer antero-posteriorly than in *Hipparion*; the lateral claw-like hoofs persisting. Some of the European Pliocene species (like *E. stenonis*) agree with these species in the form of the grinding surface of the anterior pillar of the upper molars. In one of the species from the Lower Pliocene of India (*E. sivalensis*)—which was a contemporary of *Hipparion*—and in all the existing species, the grinding surface of the pillar in question is greatly elongated in the antero-posterior direction, as in Fig. 157, *c*.

Fossil remains of Horses are found abundantly in deposits of the most recent geological age in almost every part in America, from Eschscholtz Bay in the north to Patagonia in the south. In that continent, however, they became quite extinct, and no Horses, either wild or domesticated, existed there at the time of the Spanish conquest, which is the more remarkable as, when introduced from Europe, the Horses that ran wild proved by their rapid multiplication in the plains of South America and Texas that the climate, food, and other circumstances were highly favourable for their existence. The former great abundance of *Equidæ* in America, their complete extinction, and their perfect acclimatisation

[1] Linn. *Syst. Nat.* 12th ed. vol. i. p. 100 (1766).

when reintroduced by man, form curious but as yet unsolved problems in geographical distribution.

Existing Species.—The existing species of the genus are the following:—

The Horse, *Equus caballus*, is distinguished from the others by the long hairs of the tail being more abundant and growing quite from the base as well as the end and sides, and also by possessing a small bare callosity on the inner side of the hind leg, just below the "hock" or heel joint, in addition to the one on the inner side of the fore limb above the carpus, common to all the genus. The mane is also longer and more flowing, and the ears are shorter, the limbs longer, the hoofs broader, and the head smaller.

Though the existing Horses are not usually marked in any definite manner, or only irregularly dappled, or spotted with light surrounded by a darker ring, many examples are met with showing a dark median dorsal streak like that found in all the other members of the genus, and even with dark stripes on the shoulders and legs indicating "the probability of the descent of all the existing races from a single dun-coloured, more or less striped, primitive stock, to which our horses still occasionally revert."[1]

In Europe wild Horses were extremely abundant in the Neolithic or polished-stone period. Judging from the quantity of their remains found associated with those of the men of that time, the chase of these animals must have been among man's chief occupations, and they must have furnished him with one of his most important food supplies. The characters of the bones preserved, and certain rude but graphic representations carved on bones or reindeers' antlers, enable us to know that these Horses were rather small in size, and heavy in build, with large heads and rough shaggy manes and tails, much like, in fact, the present wild horses of the steppes of the south of Russia. They were domesticated by the inhabitants of Europe before the dawn of history, but it is doubtful whether the majority of the animals now existing on the Continent are derived directly from them, as it is more probable that they are descendants from Horses imported through Greece and Italy from Asia, derived from a still earlier domestication, followed by gradual improvement through long-continued attention bestowed on their breeding and training. Horses are now diffused by the agency of man throughout almost the whole of the inhabited parts of the globe, and the great modifications they have undergone in consequence of domestication and selective breeding are well exemplified by comparing such extreme forms as the Shetland pony, dwarfed by uncongenial climate, the thoroughbred racer, and the London dray-horse. In Australia,

[1] Darwin, *Variation of Animals and Plants under Domestication*, 1868, vol. i. chap. ii.

as in America, horses imported by the European settlers have escaped into the unreclaimed lands, and multiplied to a prodigious extent, roaming in vast herds over the plains where no hoofed animal ever trod before.

A wild Horse from Central Asia, named *E. prezevalskii*,[1] is described as having callosities on both limbs and broad hoofs like *E. caballus;* but the long hairs of the tail do not begin until about half way down its length. It also differs from *E. caballus* in having a short erect mane and no forelock; neither is there any dorsal stripe. The ears are of moderate size; the whole body is of a whitish-gray, paler beneath, and reddish on the head and upper parts of the limbs. If rightly described this form would appear to be intermediate between the true Horses and the Asses.

The second species is the domestic Ass (*E. asinus*), and the wild Asses of Africa (*E. asinus*, var. *africanus* and var. *somalicus*[2]). The domestic Ass, which is now nearly as widely diffused and useful to man as the Horse, was known in Egypt long before the latter, and is doubtless of African origin. The ears are long, the mane erect, the tail without long hairs at the base, and there are no callosities on the hind limbs. There is a dark dorsal stripe, and another across the shoulders; while the limbs are frequently banded. Of the wild forms the Nubian race (var. *africanus*) has distinct dorsal and shoulder stripes, but the rings on the limbs are often very indistinct; while in the Somali race the dorsal stripe is indistinct, and the shoulder stripe wanting, but the rings on the limbs are very boldly marked. Teeth and bones from a Pleistocene cavern deposit in Madras have been referred to *E. asinus*.

The Asiatic wild Asses, which roam in small herds in the open plains of Syria, of many parts of Persia, of the north-west of India, and the highlands of Tartary and Tibet, from the shores of the Caspian to the frontiers of China, differ from the last in being of a more rufous or isabelline colour, instead of pure gray, in wanting the dark streak across the shoulder, and having smaller ears. They have all a dark-coloured median dorsal stripe. Though it is considered probable by many zoologists that they form but a single species[3] (*E. hemionus*), they present such marked variations in size and form that they have commonly been divided into three—the Syrian Wild Ass (*E. hemippus*), the Onager (*E. onager*) from Persia, Baluchistan, the Punjab, Sind, and the desert of Kach, and the Kiang or Dzeggetai (*E. hemionus*) of the high table-lands of Tibet, where it is usually met with at an elevation of 15,000 feet and upwards above the sea-level. The last is considerably larger than

[1] See *Nature*, 21st August 1884, and *Zool. Garten.* vol. xxviii. p. 453.

[2] See Sclater, *Proc. Zool. Soc.* 1884, p. 542.

[3] See Blanford, *Zoology and Geology of Eastern Persia* (*Journeys of the Persian Boundary Commission*), p. 84.

either of the others, and differs from them in external appearance, having more the aspect of the horse. They are all remarkably swift, having been known to outstrip the fleetest Horse in speed.

Lastly, there are four striped species, all inhabitants of Africa. These constitute the genus *Hippotigris* of Hamilton-Smith, but they are not separable except by their coloration from the true Asses, and one of them, the Quagga (*E. quagga*), may be considered as intermediate. This animal was formerly met with in vast herds on the great plains of South Africa, between the Cape Colony and the Vaal River, but now, in common with most of the larger wild animals of that region, is becoming extremely scarce, owing to the

FIG. 160.—The Quagga (*Equus quagga*).

encroachments of European civilisation, if, indeed, it is not already extinct. In length of ears and character of tail it more resembles the Horse than it does the Ass, although it agrees with the latter in wanting the callosity on the inner side of the hind leg, just below the hock, characteristic of the Horse. The colour of the head, neck, and upper parts of the body is reddish-brown, irregularly banded and marked with dark brown stripes, stronger on the head and neck and gradually becoming fainter until lost behind the shoulder. There is a broad dark median dorsal stripe. The under surface of the body, the legs, and tail are nearly white, without stripes. The crest is very high, surmounted by a standing mane, banded alternately brown and white. Though never really domesticated, Quaggas have occasionally been trained to harness. The accompanying figure is reduced from a painting made from one of a pair which were driven in Hyde Park in the early part of the present century. The name is an imitation of the shrill barking neigh of the animal—"ouag-ga, ouag-ga," the last syllable very much prolonged. It must be remembered, however, in reading books of African travel that the same word is very commonly applied by hunters to Burchell's Zebra.

Of the Zebras proper, the one which was first known to Europeans, and was formerly considered the most common, is the True Zebra (*E. zebra*), sometimes called the Mountain Zebra. It inhabits the mountainous regions of the Cape Colony; but now, owing to the advances of civilised man into its somewhat restricted range, it has

FIG. 161.—True or Mountain Zebra (*Equus zebra*).

become very scarce, and is even, like the Quagga, threatened with extermination at no distant date. The second species, Burchell's Zebra (*E. burchelli*), still roams in large herds over the plains to the north of the Orange River, but in yearly diminishing numbers. Both species are subject to considerable individual variations in marking, but the following are the principal characters by which they can be distinguished.

E. zebra (Fig. 161) is the smaller of the two (about 4 feet high at the shoulders), and has longer ears, a tail more scantily clothed with hair, and a shorter mane. The general ground colour is white, and the stripes are black; the lower part of the face is bright brown. With the exception of the abdomen and the inside of the thighs, the whole of the surface is covered with stripes, the legs having narrow transverse bars reaching quite to the hoofs, and the base of the tail being also barred. The outsides of the ears have a white tip and a broad black mark occupying the greater part of the surface, but are white at the base. Perhaps the most constant and obvious distinction between this species and the next is the arrangement of the stripes on the hinder part of the back, where there are a number of short transverse bands passing from the median longitudinal dorsal stripe towards, and sometimes joining ith, the uppermost of the broad stripes which run obliquely across the haunch from the flanks towards the root of the tail. There is often a median longitudinal stripe under the chest.

FIG. 162.—Burchell's Zebra (*Equus burchelli*).

E. burchelli (Fig. 162) is a rather larger and more robust animal, with smaller ears, a longer mane, and fuller tail. The general ground colour of the body is pale yellowish-brown, the limbs nearly white, the stripes dark brown or black. In the typical form they do not extend on to the limbs or the tail; but there is a great variation in this respect, even in animals of the same herd, some being striped quite down to the hoofs (this form has been named *E. chapmani*). There is a strongly marked median longitudinal ventral black stripe, to which the lower ends of the transverse side stripes are usually united, but the dorsal stripe (also strongly marked) is completely isolated in its posterior half, and the uppermost of the broad haunch stripes runs nearly parallel to it. A much larger proportion of the ears is white than in the other species. In the middle of the wide intervals between the broad black stripes of the flanks and haunches fainter stripes are generally seen.

E. grevyi.—Under this name a Zebra has been described which was sent in 1882 to Paris from the Galla country, lying to the south of Abyssinia, the most northern locality in which Zebras have previously been met with. In many of its characters it resembles *E. zebra*, but the stripes are much finer and more numerous than in the typical examples of that species, and it has a strong, black, and isolated dorsal stripe. Even allowing for the great variations that are met with in the markings of animals of this group, the aberrant characters of this individual are quite sufficient to separate it specifically from the true Zebra of South Africa. Other similar specimens have been recently brought from the Somali country.

The flesh of the Zebras is relished by the natives as food, and their hides are very valuable for leather. Although the many attempts that have been made to break in and train these animals for riding or driving have sometimes been rewarded with partial success, they have never been domesticated in the true sense of the word.

There are thus at least seven modifications of the Horse type at present existing, sufficiently distinct to be reckoned as species by all zoologists, and easily recognised by their external characters. They are, however, all so closely allied that each will, at least in a state of domestication or captivity, breed with perfect freedom with any of the others. Cases of cross breeds are recorded between the Horse and the Quagga, the Horse and Burchell's Zebra, the Horse and the Hemionus or Asiatic wild Ass, the common Ass and the Zebra, the common Ass and Burchell's Zebra, the common Ass and the Hemionus, the Hemionus and the Zebra, and the Hemionus and Burchell's Zebra. The two species which are perhaps the farthest removed in general structure, the Horse and the Ass, produce, as is well known, hybrids or Mules, which in some qualities useful to man excel both their progenitors, and in some countries, and for certain kinds of work, are in greater requisition than either. Although occasional instances have been recorded of female Mules breeding with the males of one or other of the pure species, it is doubtful if any case has occurred of their breeding *inter se*, although the opportunities of doing so must have been great, as Mules have been reared in immense numbers for at least several thousands of years. We may therefore consider it settled that the different species of the group are now in that degree of physiological differentiation which enables them to produce offspring with each other, but does not permit of the progeny continuing the race, at all events unless reinforced by the aid of one of the pure forms.

The several members of the group show mental differences quite as striking as those exhibited by their external form, and more than perhaps might be expected from the similarity of their cerebral organisation. The patience of the Ass, the high spirit of the Horse, the obstinacy of the Mule, have long been proverbial. It is very remarkable that, out of so many species, two only should have shown any aptitude for domestication, and that these two should have been from time immemorial the universal and most useful companions and servants of man, while all the others remain in their native freedom to this day. It is, however, still a question whether this really arises from a different mental constitution causing a natural capacity for entering into relations with man, or whether it may not be owing to their having been brought gradually into this condition by long-continued and persevering efforts when the need of their services was keenly felt. It is quite possible that one reason why most of the attempts to add new species to the list of our domestic animals in modern times have ended in failure is that it does not answer to do so in cases in which existing species supply all the principal purposes to which the new ones might be put. It can hardly be expected that Zebras and Quaggas fresh from their native mountains and plains can be brought into competition as beasts of burden and draught with Horses and Asses, whose naturally useful qualities have been augmented by the training of thousands of generations of progenitors.

Not unfrequently instances occur of domestic Horses being produced with a small additional toe with complete hoof, usually on the inside of the principal toe, and, though far more rarely, three or more toes may be present. These malformations are often cited as instances of reversion to the condition of some of the earlier forms of equine animals previously mentioned. Such explanations, however plausible they appear at first sight, are nevertheless very doubtful. All the feet of polydactyle horses which we have examined bear little resemblance to those of *Hipparion* or *Anchitherium*, but look rather as if due to that tendency to reduplication of parts which occurs so frequently as a teratological condition, especially among domestic animals, and, whatever its origin, certainly cannot in many instances, as the cases of entire limbs superadded, or of six digits in man, be attributed to reversion.

Anatomy.—The anatomical structure of the Horse has been described in great detail in several works devoted to the subject, which will be mentioned in the bibliography, though these have generally been written from the point of view of the veterinarian rather than of the comparative anatomist. The limits of the present work will only admit of the most salient points being indicated, particularly those in which the Horse differs from the other Ungulata. Unless otherwise specified, it must be understood that all that is stated here, although mostly derived from observation upon the Horse, applies equally well to the other existing members of the group.

Skeleton.—The skull (Fig. 163) as a whole is greatly elongated, chiefly in consequence of the immense size of the face as compared with the hinder or true cranial portion. The basal line of the cranium from the lower border of the foramen magnum to the incisor border of the palate is very nearly straight. The orbit, of nearly circular form, though small in proportion to the size of the whole skull, is distinctly marked, being completely surrounded by a strong ring of bone with prominent edges. Behind it, and freely communicating with it beneath the osseous bridge (the postorbital

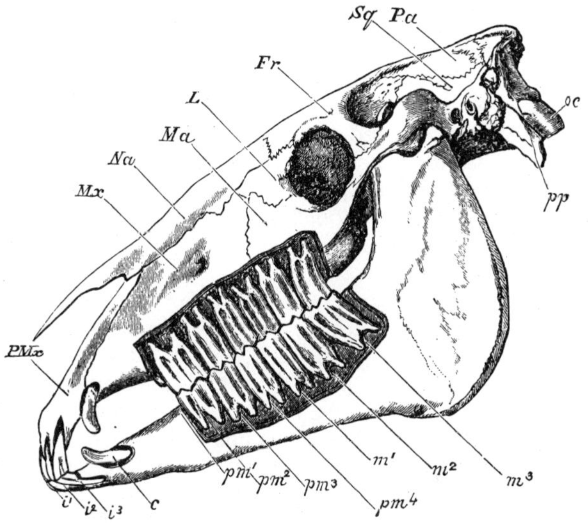

FIG. 163.—Side view of skull of Horse, with the bone removed so as to expose the whole of the teeth. *PMx*, Premaxilla; *Mx*, maxilla; *Na*, nasal; *Ma*, malar or jugal; *L*, lachrymal; *Fr*, frontal; *Sq*, squamosal; *Pa*, parietal; *oc*, occipital condyle; *pp*, paroccipital process; i^1, i^2, and i^3, the three incisors; *c*, the canine; pm^1, the situation of the rudimentary first premolar, which has been lost in the lower, but is present in the upper jaw; pm^2, pm^3, and pm^4, the three fully developed premolars; m^1, m^2, and m^3, the three true molars.

process of the frontal) forming the boundary between them, is the small temporal fossa occupying the whole of the side of the cranium proper, and in front is the great flattened expanse of the "cheek," formed chiefly by the maxilla, giving support to the long row of cheek-teeth, and having a prominent ridge running forward from below the orbit for the attachment of the masseter muscle. The lachrymal occupies a considerable space on the flat surface of the cheek in front of the orbit, and below it the jugal or malar does the same. The latter sends a horizontal or slightly ascending process backwards below the orbit to join the under surface of the zygomatic process of the squamosal, which is remarkably large, and, instead of ending as usual behind the orbit, runs forwards to join the greatly developed postorbital process of the frontal, and even forms part of the posterior and inferior boundary of the orbit, an arrangement not met with in other mammals. The closure of the orbit behind distinguishes the skull of the Horse from that of the Rhinoceros and Tapir, and also from all of the Perissodactyles of the Eocene period. In front of the cerebral cavity, the great tubular nasal cavities are provided with well-developed turbinal bones, and are roofed over by very large nasals, broad behind, and ending in front in a narrow decurved point. The opening of the anterior nares is prolonged backwards on each side of the face between the nasals and the elongated slender premaxillæ. The latter expand in front, and are curved downwards to form the semicircular alveolar border supporting the large incisor teeth. The palate is narrow in the interval between the incisor and cheek-teeth, in which are situated the large anterior palatine foramina. Between the cheek-teeth it is broader, and it ends posteriorly in a rounded excavated border opposite the hinder edge of the penultimate molar. It is mainly formed by the maxillæ, as the palatines are very narrow. The pterygoids are delicate slender slips of bone attached to the hinder border of the palatines, and supported externally by, and generally ankylosed to, the rough pterygoid plates of the alisphenoid, with no pterygoid fossa between. They slope very obliquely forwards, and end in curved, compressed, hamular processes. There is a distinct alisphenoid canal for the passage of the internal maxillary or main branch of the external carotid artery. The base of the cranium is long and narrow; the alisphenoid is very obliquely perforated by the foramen rotundum, but the foramen ovale is confluent with the large foramen lacerum medium behind. The glenoid surface for the articulation of the mandible is greatly extended transversely, concave from side to side, convex from before backwards in front, and hollow behind, and is bounded posteriorly at its inner part by a prominent postglenoid process. The squamosal enters considerably into the formation of the temporal fossa, and, besides sending the zygomatic process forwards, it sends down behind the meatus auditorius a post-tympanic process which aids to hold in place the otherwise loose tympano-periotic bone. Behind this the exoccipital gives off a very long paroccipital process. The periotic and tympanic are ankylosed together, but not with the squamosal. The former has a wide but shallow floccular fossa on its inner side, and sends backwards a considerable "pars mastoidea," which appears on the outer surface of the skull between the post-tympanic process of the squamosal and the exoccipital. The tympanic forms a tubular meatus auditorius externus directed outwards and slightly backwards. It is not dilated into a distinct bulla, but ends in front in a pointed styliform process; and completely embraces the truncated cylindrical tympanohyal, which is of great size, in correspondence with the large development of the whole anterior arch of the hyoid. This consists mainly of a long and compressed stylohyal, expanded at the upper end, where it sends off a triangular posterior process. The basihyal is remarkable for the long, median, pointed, compressed "glossohyal" process, which it sends forward from its anterior border into the base of the tongue. A similar but less developed process is found in the Rhinoceros. The mandible is largely developed, especially the region of the angle, which is expanded and flattened, giving great surface for the attachment of the masseter muscle. The condyle is greatly elevated above the alveolar border; its articular surface is very wide transversely, and narrow and convex from before backwards. The coronoid process is slender, straight, and inclined backwards. The horizontal ramus, long, straight, and compressed, gradually narrows towards the symphysis, where it expands laterally to form with the ankylosed opposite ramus the wide, semicircular, shallow alveolar border for the incisor teeth.

The vertebral column consists of seven cervical, eighteen dorsal, six lumbar, five sacral, and fifteen to eighteen caudal vertebræ. There may be nineteen rib-bearing vertebræ, in which case five only will be reckoned as belonging to the lumbar series. The odontoid process of the atlas is wide, flat, and hollowed above, as in the Ruminants. The bodies of the cervical vertebræ are elongated, strongly keeled, and markedly opisthocœlous, or concave behind and convex in front. Their neural laminæ are very broad, the spines almost obsolete, except in the seventh, and the transverse processes not largely developed. In the trunk vertebræ the opisthocœlous character of the centrum gradually diminishes. The spinous processes of the anterior thoracic region are high and compressed. To these is attached the powerful elastic ligament, *ligamentum nuchæ*, or "paxwax," which passing forwards in the middle line of the neck above the neural arches of the cervical vertebræ, to which it is also connected, is attached to the occiput and supports the weight of the head. The transverse processes of the lumbar vertebræ are long, flattened, and project horizontally outwards or slightly forwards from the arch. The metapophyses are moderately developed, and there are no anapophyses. The caudal vertebræ, except those quite at the base, are slender and cylindrical, without processes and without chevron-bones beneath. The ribs are eighteen or nineteen in number on each side, flattened, and united to the sternum by short, stout, tolerably well ossified sternal ribs. The sternum consists of six pieces; the anterior or presternum being extremely compressed, and projecting forwards like the prow of a boat. The segments which follow gradually widen, and the hinder part of the sternum is broad and flat.

As in all other Ungulates, there are no clavicles. The scapula is long and slender; the suprascapular border is rounded, and slowly and imperfectly ossified. The spine is very slightly developed; rather above the middle its edge is thickened and somewhat turned backwards, but it gradually subsides at the lower extremity without forming any acromial process. The coracoid process is a prominent rounded nodule. The humerus is stout and rather short, and has a double bicipital groove. The ulna is quite rudimentary, being only represented by little more than the olecranon. The shaft gradually tapers below, and is firmly ankylosed to the radius. The latter bone is of nearly equal width throughout. The three bones of the first row of the carpus (the scaphoid, lunar, and cuneiform) are subequal in size. The second row consists of a very broad and flat magnum, supporting the great third metacarpal, having to its radial side the trapezoid, and to its ulnar side the unciform, which are both small, and articulate distally with the rudimentary second and fourth metacarpals. The pisiform is large and prominent, flattened, and curved; articulating partly with the cuneiform and partly with the lower end of the radius. The large metacarpal is called in veterinary anatomy "cannon-bone"; the small lateral metacarpals, which gradually taper towards their lower extremities, and lie in close contact with the large one, are called "splint-bones." The single digit consists of a moderate-sized proximal (*os suffraginis*, or large pastern), a very short middle (*os coronæ*, or small pastern), and a wide, semi-lunar, ungual phalanx (*os pedis*, or coffin-bone). There is a pair of large nodular sesamoids behind the metacarpo-phalangeal articulation, and a single large transversely extended sesamoid behind the joint between the second and third phalanx, called the "navicular bone."[1]

The carpal joint, corresponding to the wrist of man, is commonly called the "knee" of the Horse, the joint between the metacarpal and the first phalanx the "fetlock," that between the first and second phalanges the "pastern," and that between the second and third phalanges the "coffin-joint."

In the hind limb the femur is marked, as in other Perissodactyles, by the presence of a "third trochanter," a flattened process, curving forwards, arising from the outer side of the bone, about one-third of the distance from the upper end. The fibula is reduced to a mere styliform rudiment of the upper end; its lower part being absent or completely fused with the tibia. The calcaneum has a long and compressed calcaneal process. The astragalus has a large flat articular surface in front for the navicular, and a very small one for the cuboid. The navicular and the external cuneiform bones are very broad and flat. The cuboid is small, and the internal and middle cuneiform bones are small and united together. The metapodials and phalanges resemble very closely those of the fore limb,

[1] This must not be confounded with the navicular of the tarsus.

but the principal metatarsal is more laterally compressed at its upper end than is the corresponding metacarpal. The joint between the femur and tibia, corresponding to the knee of man, is called the "stifle joint"; while that between the tibia and tarsus, corresponding to the ankle of man, is termed the "hock." The bones and joints of the foot have the same names as in the fore limb. The Horse is eminently "digitigrade," standing on the extremity of the single digit of each foot, which is kept habitually in a position approaching to vertical.

The muscles [1] of the limbs are modified from those of the ordinary mammalian type in accordance with the reduced condition of the bones and the simple requirements of flexion and extension of the joints, no such actions as pronation and supination, or opposition of digits, being possible or needed. The muscles, therefore, which perform these functions in other mammals are absent or rudimentary.

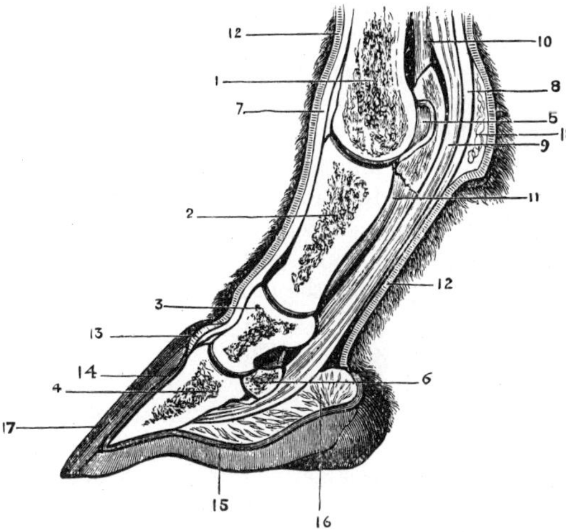

FIG. 164.—Section of foot of Horse. 1, Metacarpal bone; 2, first phalanx (*os suffraginis*); 3, second phalanx (*os coronæ*); 4, third or ungual phalanx (*os pedis*, or coffin-bone); 5, one of the upper sesamoid bones; 6, lower sesamoid or "navicular" bone; 7, tendon of anterior extensor of the phalanges; 8, tendon of superficial flexor (*fl. perforatus*); 9, tendon of deep flexor (*fl. perforans*); 10, suspensory ligament of fetlock; 11, inferior or short sesamoid ligament; 12, derma or skin of the foot, covered with hair, and continued into 13, the coronary cushion, 14, the podophyllous or laminar membrane, and 15, the keratogenous membrane of the sole; 16, plantar cushion; 17, hoof; 18, fatty cushion of fetlock.

Below the carpal and tarsal joints the fore and hind limbs correspond almost exactly in structure as well as function. On the anterior or extensor surface of the limb a powerful tendon (7 in Fig. 164), that of the anterior extensor of the phalanges (corresponding to the *extensor communis digitorum* of the arm and *extensor longus digitorum* of the foot of man) passes down over the metacarpal bone and phalanges, to be inserted mainly into the upper edge of the anterior surface of the last phalanx

[1] Want of space and of the necessary illustrations rendered it impossible to give an account of mammalian myology in the earlier chapters of this work.

or pedal bone. There is also a much smaller second extensor on the outer side of this in each limb, the lateral extensor of the phalanges. In the fore leg the tendon of this muscle (which corresponds with the *extensor minimi digiti* of man) receives a slip from that of the principal extensor, and is inserted into the first phalanx. In the hind leg (where it is the homologue apparently of the *peroneus brevis* of man) the tendon becomes blended with that of the large extensor.

A very strong ligamentous band behind the metapodium, arising from near the upper extremity of its posterior surface, divides into two at its lower end, and each division, being first connected with one of the paired upper sesamoid bones, passes by the side of the first phalanx to join the extensor tendon of the phalanges. This is called in veterinary anatomy the "suspensory ligament of the sesamoids," or of the "fetlock" (10 in Fig. 164); but its attachments and relations, as well as the occasional presence of muscular fibres in its substance, show that it is the homologue of the short flexor muscle of other mammals, curiously modified both in structure and function to suit the requirements of the Horse's foot. Behind or superficial to this are placed the two strong tendons of the long flexor muscles, the most superficial, or *flexor perforatus* (8), dividing to allow the other to pass through, and then inserted into the middle phalanx. The *flexor perforans* (9) is as usual inserted into the terminal phalanx. In the fore leg these muscles correspond with those similarly named in man. In the hind leg, the perforated tendon is a continuation of that of the plantaris, passing pulley-wise over the tuberosity of the calcaneum. The perforating tendon is derived from the muscle corresponding with the long flexor of man, and the smaller tendon of the oblique flexor (*tibialis posticus* of man) is united with it.

The hoof of the Horse corresponds to the nail or claw of other mammals, but is so constructed as to form a complete and very solid case to the expanded termination of the toe, giving a firm basis of support formed of a nonsensitive substance, which is continually renewed by the addition of material from within as its surface wears away by friction against the ground. The terminal phalanx of the toe is greatly enlarged and modified in form to support this hoof, and the size of the internal framework of the foot is further increased by a pair of lateral fibro-cartilaginous masses attached on each side to the hinder edges of the bone, and by a fibro-cellular and adipose plantar cushion in the median part. These structures are all enclosed in the keratogenous membrane or "subcorneous integument," a continuation of the ordinary derma of the limb, but extremely vascular, and having its superficial extent greatly increased by being developed into papillæ or laminæ. From this the horny material which constitutes the hoof is exuded. A thickened ring encircling the upper part, called coronary cushion (13), and the sole (15), are covered with numerous thickly set papillæ or villi, and take the greatest share in the formation of the hoof; the intermediate part constituting the front and side of the foot (14), corresponding with the wall of the hoof, is covered with parallel, fine longitudinal laminæ, fitting into corresponding depressions in the inner side of the horny hoof.

The horny hoof is divided into a wall or crust consisting of the front and sides, the flattened or concave sole, and the "frog," a triangular median prominence, notched posteriorly, with the apex turned forwards, situated in the hinder part of the sole. It is formed of pavement epithelial cells, mainly grouped in a concentric manner around the vascular papillæ of the keratogenous membrane, so that a section near the base of the hoof, cut transversely to the long axis of these papillæ, shows a number of small circular or oval orifices, with cells arranged concentrically round them. The nearer the surface of the hoof, or farther removed from the seat of growth, the more indistinct the structure becomes.

Small round or oval plates of horny epidermis called "chestnuts," growing like the hoof from enlarged papillæ of the skin, are found on the inner face of the fore limb, above the carpal joint, in all species of *Equidæ*, and in the Horse (*E. caballus*) alone similar formations occur near the upper extremity of the inner face of the metatarsus. Their use is unknown.

Behind the joint between the metapodium and the first phalanx is a prominence formed by the fatty cushion of the fetlock (18 in Fig. 164). On the middle of this is a small bare patch covered with thickened epidermis, the *ergot* or spur, generally concealed beneath the long hair which grows around it. This is the functionless vestige of the large callous pad found in this situation in the Tapir, and in fact in all mammals in which this part reaches the ground in walking.

Dentition.—The dentition of the Horse, when all the teeth are in place, is, as stated before, expressed by the formula $i\,\frac{3}{3}$, $c\,\frac{1}{1}$, $p\,\frac{4}{3}$, $m\,\frac{3}{3} = 42$. The incisors of each jaw are placed in close contact, forming a semicircle. The crowns are broad, somewhat awl-shaped, and of nearly equal size. They have all the great peculiarity, not found in the teeth of any other living mammal, of an involution of the external surface of the tooth (see Fig. 165) forming a deep fossa or pit, the bottom of which becomes partially filled up with cement. As the tooth wears, the surface, besides the external enamel layer as in an ordinary simple tooth, shows in addition a second inner ring of the same hard substance surrounding the pit, thus of course adding greatly to the efficiency of the tooth as an organ for biting tough, fibrous substances. This pit, generally filled in the living animal with particles of food, is conspicuous from its dark colour, and constitutes the "mark" by which the age of the horse is judged, as in consequence of its extending only to a certain depth, it becomes obliterated as the crown wears away, when the tooth assumes the character of an ordinary incisor, consisting only of a core of dentine surrounded by the external enamel layer. It is not quite so deep in the lower as in the upper teeth. The canines are either quite rudimentary or entirely absent in the female. In the male they are compressed, pointed, and smaller than the incisors, from which they are separated by a slight interval. The teeth of the cheek series are all in contact with each other, but separated from the canines by a considerable toothless space. The anterior premolars are quite rudimentary, often, especially in the lower jaw, not developed at all, and generally fall by the time the animal attains maturity, so that there are but six functional grinding teeth—three that have predecessors in the milk-dentition, and hence are considered as premolars, and three true molars, but otherwise, except the first and last of the series, not distinguishable in form or structure. These teeth in both upper and lower jaws are extremely long-crowned or hypsodont (Fig. 158), successive portions being pushed out as the surface wears away;—a process which continues until the animal becomes advanced in age. The enamelled surface is infolded in a complex manner (a modification of that found in other Perissodactyles, see Figs. 155, 167), the folds extending quite to the base of the crown, and the interstices being filled and the surface covered with a considerable mass of cement, which binds together and strengthens the whole tooth. As the teeth wear, the folded enamel, being harder than the other constituents—the dentine and cement—forms projecting ridges on the surface arranged in a definite pattern, which give it great efficiency as a grinding instrument (see Fig. 157, *b* and *c*). The free surfaces of the upper teeth are

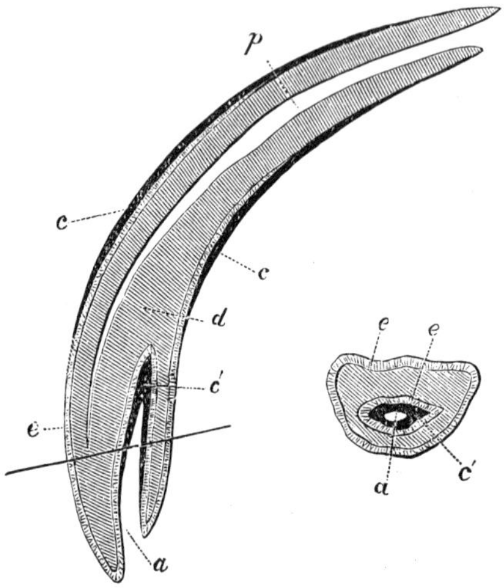

FIG. 165.—Longitudinal and transverse section of upper incisor of Horse. *p*, Pulp cavity; *d*, dentine or ivory; *e*, enamel; *c*, outer layer of cement; *c′*, inner layer of cement, lining *a*, the pit or cavity of the crown of the tooth.

quadrate, except the first and last, which are nearly triangular. The lower teeth are much narrower than the upper.

The milk dentition consists of $i\,\frac{3}{3}$, $c\,\frac{0}{0}$, $m\,\frac{3}{3} = 24$,—the canines and first or rudimentary premolars having apparently no predecessors. In form and structure they much resemble the permanent teeth, having the same characteristic enamel-foldings. Their eruption commences a few days after birth, and is complete before the end of the first year, the upper teeth usually appearing somewhat earlier than those of the lower jaw. The first teeth to appear are the first and second milk-molars (about five days), then the central incisor (from seven to ten days); this is followed by the second incisor (at one month), then by the third molar, and finally by the third incisor. Of the permanent teeth the first true molar appears a little after the end of the first year, followed by the second molar before the end of the second year. At about two and a half years the first premolar replaces its predecessor. Between two and a half and three years the first incisor appears. At three years the second and third premolars and the third true molar have appeared; at from three and a half to four years the second incisor; at four to four and a half years the canine; and, finally, at five years the third incisor, completing the permanent dentition. Up to this period the age of the horse is clearly shown by the state of the dentition, and for some time longer indications can be obtained from the wear of the incisor teeth, though this depends to a certain extent upon the hardness of the food or other accidental circumstances. As a general rule, the depression caused by the infolding of the surface of the incisor (the "mark") is obliterated in the first or central incisor at six years, in the second at seven years, and in the third at eight years. In the upper teeth, as the depressions are deeper, this obliteration does not take place until about two years later. After this period no certain indications can be obtained of the age of the horse from the teeth.

Digestive Organs.—The lips are flexible and prehensile. The membrane that lines them and the cheeks is quite smooth. The palate is long and narrow; its mucous surface has seventeen pairs of not very sharply defined oblique ridges, extending as far back as the last molar tooth, beyond which the velum palati extends for about 3 inches, having a soft corrugated surface, and ending posteriorly in an arched border without uvula. This embraces the base of the epiglottis, and shuts off all communication between the cavity of the mouth and the nasal passages, respiration being, under ordinary circumstances, carried on exclusively through the nostrils. Between the mucous membrane and the bone of the hard palate is a dense vascular and nervous plexus. The membrane lining the fauces is soft and corrugated. An elongated raised glandular mass, 3 inches long and 1 inch from above downwards, extending backwards from the root of the tongue along the side of the fauces, with openings on the surface leading into crypts with glandular walls, represents the tonsil. The tongue, corresponding to the general form of the mouth, is long and narrow. It consists of a compressed intermolar portion with a flat upper surface, broad behind and becoming narrower in front; and of a depressed anterior part rather shorter than the former, which is narrow behind but widens towards the evenly rounded apex. The dorsal surface generally is very soft and smooth. There are two large circumvallate papillæ near the base, rather irregular in form, about a quarter of an inch in diameter and half an inch apart. The conical papillæ are very small and close set, though longer and more filamentous on the intermolar portion. There are no fungiform papillæ on the dorsum, but a few not very conspicuous ones scattered along the sides of the organ.

Of the salivary glands the parotid is by far the largest; elongated in the vertical direction, and narrower in the middle than at either upper or lower extremity. Its upper extremity embraces the lower surface of the cartilaginous ear-conch; its lower end reaches the level of the inferior margin of the mandible, along the posterior margin of which it is placed. Its duct leaves the inferior anterior angle, at first descends a little, and runs forward under cover of the rounded inferior border of the mandibular ramus, then curves up along the anterior margin of the masseter muscle, becoming superficial, pierces the buccinator, and enters the mouth by a simple aperture opposite the middle of the crown of the third premolar tooth. It is not quite so thick as a goose-quill when distended, and nearly a foot in length.

The submaxillary gland is of very similar texture to the last, but much smaller; it is placed deeper, and lies with its main axis horizontal. It is elongated and slender, and flattened from within outwards. Its posterior end rests against the anterior surface of the transverse process of the atlas, from which it extends forwards and downwards, slightly curved, to beneath the ramus of the jaw. The duct which runs along its upper and internal border passes forwards in the usual course, lying in the inner side of the sublingual gland, to open on the outer surface of a distinct papilla, situated on the floor of the mouth, half an inch from the middle line, and midway between the lower incisor teeth and the attachment of the frænum linguæ. The sublingual is represented by a mass of glands lying just beneath the mucous membrane of the floor of the mouth on the side of the tongue, causing a distinct ridge, extending from the frænum backwards, and the numerous ducts opening separately along the summit of the ridge. The buccal glands are arranged in two rows parallel with the molar teeth. The upper ones are the largest, and are continuous anteriorly with the labial glands, the ducts of which open on the mucous membrane of the upper lip.

The stomach of the Horse is simple in its external form, with a largely developed right *cul de sac*, and is a good deal curved on itself, so that the cardiac and pyloric orifices are brought near together. The antrum pyloricum is small and not very distinctly marked off. The interior is divided by the character of the lining membrane into two very distinct portions, right and left. Over the latter the dense white smooth epithelial lining of the œsophagus is continued, terminating abruptly by a raised crenellated border. Over the right part (rather the larger portion) the mucous membrane has a grayish-red colour and a velvety appearance, and contains very numerous peptic glands, which are wanting in the cardiac portion. The œsophageal orifice is very small, and is guarded by a strong crescentic or rather horse-shoe-like band of muscular fibres, which is supposed to be the cause of the difficulty of vomiting in the Horse. The small intestine is of great length (80 to 90 feet), its mucous membrane being covered with numerous fine villi. The cæcum is of conical form, about 2 feet long and nearly a foot in diameter; its walls are sacculated, especially near the base, having four longitudinal fibrous bands; and its capacity is about twice that of the stomach. It lies with its base near the lower part of the abdomen, and its apex directed towards the thorax. The colon is about one-third the length of the small intestine, and very capacious in the greater part of its course. As usual, it may be divided into an ascending, transverse, and descending portion; but the middle or transverse portion is folded into a great loop, which descends as low as the pubis; so that the colon forms altogether four folds, generally parallel to the long axis of the body. The descending colon is much narrower than the rest, and not sacculated, and being considerably longer than the distance it has to traverse, is thrown into numerous folds.

The liver (Fig. 166) is tolerably symmetrical in its general arrangement, being divided nearly equally into segments by a well-marked umbilical fissure. Each segment is again divided by lateral fissures, which do not extend quite to the posterior border of the organ; of the central lobes thus cut off, the right is rather the larger, and has two fissures in its free border subdividing it into lobules. The extent of these varies, however, in different individuals, being not usually so marked as in the figure, which is from a fœtal specimen. The two lateral lobes are subtriangular in form. The Spigelian lobe is represented by a flat surface between the portal fissure and the posterior border, not distinctly marked off from the left lateral by a fissure of the ductus venosus, as this vessel is buried deep in the hepatic substance, but the caudate lobe is distinct and tongue-shaped, its free apex reaching nearly to the border of the right lateral lobe. In most works on the anatomy of the Horse this has been confounded with the Spigelian lobe of man. There is no gall-bladder (as in all other Perissodactyles), and the biliary duct enters the duodenum about 6 inches from the pylorus. The pancreas has two lobes or branches—a long one passing to the left and reaching the spleen, and a shorter right lobe. The principal duct enters the duodenum with the bile-duct, and there is often a second small duct which opens separately near to this.

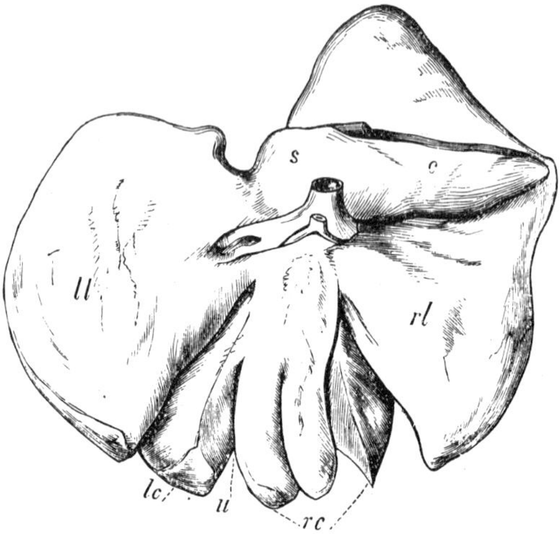

FIG. 166.—Under surface of the liver of the Horse. *u*, Umbilical fissure; *ll*, left lateral lobe; *lc*, left central lobe; *rc*, right central lobe; *rl*, right lateral lobe; *s*, Spigelian lobe; *c*, caudate lobe.

Circulatory and Respiratory Organs.—The heart has the form of a rather elongated and pointed cone. There is one anterior vena cava, formed by the union of the two jugular and two axillary veins. The aorta gives off a large branch (the anterior aorta) very near its origin, from which arise—first, the left axillary, and afterwards the right axillary and the two carotid arteries.

Under ordinary circumstances the Horse breathes entirely by the nasal passages, the communication between the larynx and the mouth being closed by the velum palati. The nostrils are placed laterally, near the termination of the muzzle, and are large and very dilatable, being bordered by cartilages upon which several muscles act. Immediately within the opening of the nostril, the respiratory canal sends off on its upper and outer side a diverticulum or blind pouch (called "false nostril") of a conical form, and curved, 2 to 3 inches in depth, lying in the notch formed between the nasal and premaxillary bones. It is lined by mucous membrane continuous with that of the nasal passage, but its use is not apparent. It is longer in the Ass than in the Horse. A similar structure is found in the Rhinoceros, and in a much more developed condition in the Tapir. Here may be mentioned the guttural pouches, large air sacs, diverticula from the Eustachian tubes, and lying behind the upper part of the pharynx. These are likewise found in other Perissodactyles, but their use is also still not clearly understood. The larynx has the lateral sacculi well developed, though entirely concealed within the alæ of the thyroid cartilage. The trachea divides into two bronchi, one for each lung.

Nervous System.—The brain differs little, except in details of arrangement of convolutions, from that of other Ungulates. The cerebral hemispheres are rather elongated and subcylindrical, the olfactory lobes are large and project freely in front of the hemispheres, and the greater part of the cerebellum is uncovered. The eye is provided with a nictitating membrane or third eyelid, at the base of which the ducts of the Harderian gland open.

Reproductive System.—The testes are situated in a distinct sessile or slightly pedunculated scrotum, into which they descend from the sixth to the tenth month after birth. The accessory generative glands are the two vesiculæ seminales, with the median third vesicle, or *uterus masculinus*, lying between them, the single bilobed prostate, and a pair of globular Cowper's glands. The penis is large, cylindrical, with a truncated, expanded, flattened termination. When in a state of repose it is retracted by a muscle arising from the sacrum, within the prepuce, a cutaneous fold attached below the symphysis pubis.

The uterus is bicornuate. The vagina is often partially divided by a membraneous septum or hymen. The mammæ (as in other members of the suborder), are two, inguinally placed. The surface of the chorion is covered evenly with minute villi, constituting a diffuse non-deciduate placenta. The period of gestation is eleven months.

Bibliography.—M. S. Arloing, "Organisation du pied chez le cheval," *Ann. Sci. Nat.* 1867, viii. pp. 55-81; H. Burmeister, *Los caballos fosiles de la Pampa Argentina*, Buenos Ayres, 1875; Chauveau and Arloing, *Traité d'anatomie comparée des animaux domestiques*, Paris, 1871, and English edition by G. Fleming, 1873; E. Cuyer and E. Alix, *Le Cheval*, 1886; A. Ecker, "Das Europäische Wildpferd und dessen Beziehungen zum domesticirten Pferd," *Globus*, Bd. xxxiv. Brunswick, 1878; Forsyth-Major, "Beiträge zur Geschichte der fossilen Pferde besonders Italiens," *Abh. Schw. Pal. Ges.* iv. pp. 1-16, pt. iv.; George, "Études zool. sur les Hémiones et quelques autres espèces chevalines," *Ann. Sci. Nat.* 1869, xii. p. 5; E. F. Gurlt, *Anatomische Abbildungen der Haussäugethiere*, 1824, and *Hand. der vergleich. Anat. der Haussäugethiere*, 2 vols. 1822; Huet, "Croisement des diverses espèces du genre cheval," *Nouv. Archives du Muséum*, 2d sér. tom. ii. p. 46, 1879; Leisering, *Atlas der Anatomie des Pferdes*, Leipsic, 1861; J. M'Fadyean, *The Anatomy of the Horse*, 1884; O. C. Marsh, "Notice of New Equine Mammals from the Tertiary Formation," *Am. Journ. of Science and Arts*, vol. vii. March 1874; Id. "Fossil Horses in America," *Amer. Naturalist*, vol. viii. May 1874; Id. "Polydactyle Horses," *Am. Journ. of Science and Arts*, vol. xvii. June 1879; Franz Müller, *Lehrbuch der Anatomie des Pferdes*, Vienna, 1853; R. Owen, "Equine Remains in Cavern of Bruniquel," *Phil. Trans.* vol. clix. (1870), p. 535; W. Percivall, *The Anatomy of the Horse*, 1832; G. Stubbs, *Anatomy of the Horse*, 1766. F. H. Huth's *Bibliographical Record of Hippology* (1887) contains a list of nearly four thousand works on Horses and Equitation, published in the various languages of the civilised world.

Family RHINOCEROTIDÆ.

Although the existing members of this family are readily distinguished from the other living representatives of the suborder by the simple crescentoid form assumed by the ridges of the lower cheek-teeth, yet it is exceedingly difficult to give a definition by which they can be distinguished from the *Lophiodontidæ*, from some members of which they are, indeed, probably derived. The outer columns of the upper molars (Fig. 167) are, however, so excessively flattened as to produce a continuous thick and nearly straight outer wall, which is often produced in advance of the anterior transverse ridge; both transverse ridges being but little curved, and intimately connected with the outer wall. The upper premolars are in most cases nearly or quite as complex as the molars, and the ridges of the lower cheek-teeth are crescentoid. The last lower molar has no third lobe. The height of the crowns of the cheek-teeth is variable. The skull is large, with the orbit confluent with the temporal fossa. There are either three or four digits in the manus, and three in the pes. One or more dermal horns are attached to the fronto-nasal region of the skull of existing forms, but these were wanting in some of the fossil species.

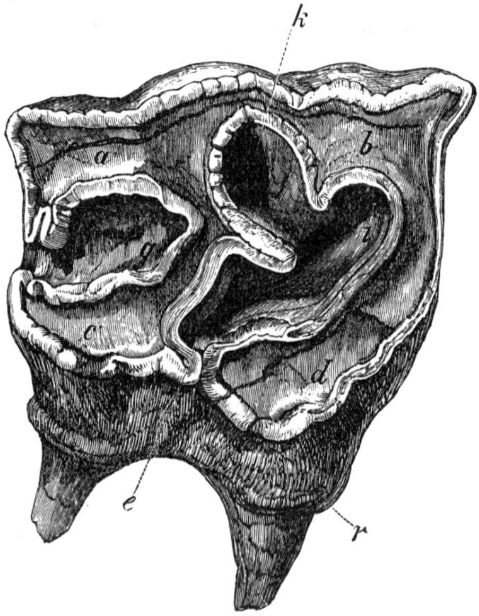

FIG. 167.—A partially worn second right upper molar of *Rhinoceros antiquitatis*. Letters as in Fig. 155 (p. 375), except *k*, which indicates a prolongation of the median valley. (After Owen.)

Rhinoceros.[1]—Incisors variable, reduced in number, often quite rudimentary, and early deciduous. Upper canines absent. Molar series, consisting of the full number of four premolars and three molars above and below, all in contact and closely resembling each other, except the first, which is much smaller than the rest and often deciduous; and the last, in which the hinder lobe is partly

[1] Linn. *Syst. Nat.* 12th ed. vol. i. p. 104 (1766).

aborted, so that the contour of the crown is triangular. Head large, skull elongated, elevated posteriorly into a transverse occipital crest. No postorbital processes. Nasal bones large and stout, co-ossified, and standing out freely above the premaxillæ, from which they are separated by a deep and wide fissure; the latter small, generally not meeting in the middle line in front, often quite rudimentary. Tympanics small, not forming a bulla. Brain cavity very small for the size of the skull. Vertebræ: C 7, D 19-20, L 3, S 4, C about 22. Limbs stout, and of moderate length. Three completely developed toes, with distinct broad rounded hoofs on each foot (Fig. 151, p. 368), some fossil forms having a fourth in the manus. Eyes small. Ears of moderate size, oval, erect, prominent, placed near the occiput. Skin very thick, in many species thrown into massive folds. Hairy covering scanty. When one horn is present it is situated over the conjoined nasal bones; when two, the hinder one is over the frontals. These horns, which are of a more or less conical form and usually recurved, often grow to a great length (three or even four feet), and are composed of a solid mass of hardened epidermic cells growing from a cluster of long dermal papillæ. The cells formed on each papilla constitute a distinct horny fibre, like a thick hair, and the whole are cemented together by an intermediate mass of cells which grow up from the interspaces between the papillæ. It results from this that the horn has the appearance of a mass of agglutinated hairs, which, in the newly growing part at the base, readily fray out on destruction of the softer intermediate substance; but the fibres differ from true hairs in growing from a free papilla of the derm, and not within a follicular involution of the same.

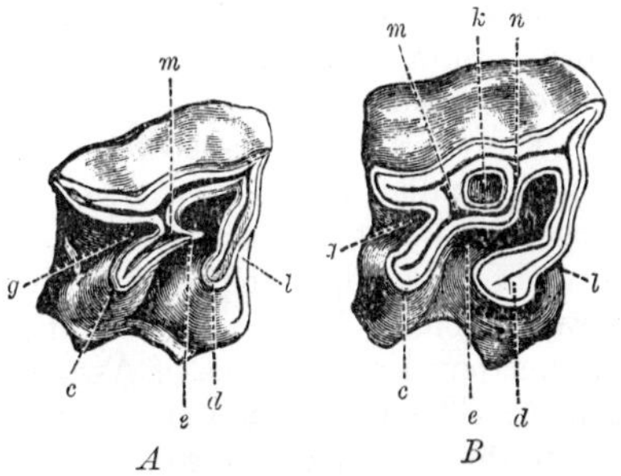

Fig. 168.—A partially worn second right upper molar of (*A*) *Rhinoceros sondaicus*, and (*B*) *R. unicornis*. *k*, Fossette cut off from median valley; *m*, crotchet; *n*, crista, or combing-plate; *e*, anterior valley; *l*, anterior intermediate column. Other letters as in Fig. 155, p. 375.

The large lower cutting teeth of the typical Rhinoceroses have been very generally regarded as incisors, but comparison with fossil allied types, in which three lower incisors and canines are present, leaves little doubt but that they are really canines. The upper molar teeth present some amount of specific variation; thus while one type (Fig. 168, *A*) has only a simple "crotchet" projecting from the posterior transverse ridge into the median valley, in others (Fig. 168, *B*) this crotchet joins a "crista," or "combing-plate," projecting from the outer wall to cut off a distinct fossette from the median valley. Occasionally, however (as in Fig. 167), the crotchet and combing-plate do not completely join, although the fossette is distinctly indicated. The first upper premolar may occasionally be preceded by a milk-tooth. The Rhinoceroses differ from the Horses and agree with the Tapirs in the direction of the cæcum.

The living species of *Rhinoceros* are all animals of large size, but of little intelligence, generally timid in disposition, though ferocious when attacked and brought to bay, using the nasal horns as weapons, by which they strike and toss their assailant. Their sight is dull, but their hearing and scent are remarkably acute. They feed on herbage, shrubs, and leaves of trees, and, like so many other large animals which inhabit hot countries, sleep the greater part of the day, being most active in the cool of the evening or even during the night. They are fond of bathing and wallowing in water or mud. None of the species have been domesticated. Animals of the group have existed in both the Old and New Worlds since the latter part of the Eocene period. In America they all became extinct before the end of the Pliocene period. In the Old World their distribution has become greatly restricted, and they are no longer found in Europe and North Asia, but only in Africa and portions of the Indian and Indo-Malayan regions.

Existing Species.—The existing (as well as many of the extinct) species of Rhinoceroses naturally divide into three groups, which are regarded by some zoologists as of generic value.

Rhinocerotic, or Typical Group.—The adults with a single large compressed incisor above on each side, and occasionally a small lateral one; below, a very small incisor and a very large, procumbent, pointed canine. Nasal bones pointed in front. A single nasal horn. Skin very thick, and raised into strong, definitely arranged ridges or folds.

There are two well-marked species of one-horned Rhinoceroses. (1) The Indian Rhinoceros, *R. unicornis* (Fig. 169) of Linnæus,[1] the largest and best known, from being the most frequently exhibited alive in England, is at present only met with in a wild state in the terai region of Nipal and Bhutan, and in the upper valley of the Brahmaputra or province of Assam, though it formerly had a wider range. The first Rhinoceros seen alive in Europe since the time when these animals, in common with nearly all the large remarkable beasts of both Africa and Asia, were exhibited in the Roman

[1] Many authors use Cuvier's name, *R. indicus*, in preference to this, on the ground that there are more than one species with one horn, forgetting that the name substituted is equally inconvenient, as more than one species live in India. The fact of a specific name being applicable to several members of a genus is no objection to its restriction to the first to which it was applied; otherwise changes in old and well-received names would constantly have to be made in consequence of new discoveries.

shows, was of this species. It was sent from India to Emmanuel, King of Portugal, in 1513; and from a sketch of it, taken in Lisbon, Albert Dürer composed his celebrated but rather fanciful engraving, which was reproduced in so many old books on natural history. Both in this and the following species the post-glenoid and post-tympanic processes of the squamosal bone of the skull unite below so as to completely surround the external auditory meatus. The molar teeth are hypsodont, and have a horizontal plane of wear; those of the upper jaw (Fig. 168, *b*) being characterised by the presence of a combing-plate joining the crotchet, and

FIG. 169.—Indian Rhinoceros (*Rhinoceros unicornis*). This figure, and also figures 170, 172, are reduced from drawings by J. Wolf, from animals living in the London Zoological Society's Gardens.

the absence of a distinct buttress at the antero-external angle. The stomach departs from the ordinary Perissodactyle type. The small intestine is beset over most of its surface with long and fine villi; and the Spigelian lobe of the liver is well developed. There is a gland behind the foot. Teeth from the Pleistocene of the Narbada valley in India apparently indicate the existence of the Indian Rhinoceros at that epoch. (2) The Javan Rhinoceros (*R. sondaicus*, Fig. 170) is a smaller form, readily distinguished by dental and internal characters, as well as by the different arrangement of the plications of the skin (as seen in the figures); the horn in the female appears to be very little developed, if not altogether absent. This species has a more extensive geographical range, being found in the Bengal Sunderbans near Calcutta, Burma, the Malay Peninsula, Java, Sumatra, and probably Borneo. The molar teeth have shorter crowns than in the preceding species, and wear into ridges; those of the upper jaw (Fig. 168, *a*) having no combing-plate, and a strongly-marked buttress at the antero-external angle (not distinctly shown in the figure). The visceral anatomy, according to Beddard,[1] does not differ materially from that of the next species. In respect to its dentition and anatomical characters this species is indeed more nearly allied to the Sumatran than to the Indian Rhinoceros; and thereby indicates that the division of the existing Rhinoceroses into separate genera is not advisable.

Fig. 170.—Javan Rhinoceros (*Rhinoceros sondaicus*).

Ceratorhine Group.—The adults with a moderate-sized compressed incisor above, and a laterally placed, pointed, procumbent canine below, which is sometimes lost in old animals. Nasal bones narrow and pointed anteriorly. A well-developed nasal, and a small frontal horn separated by an interval. The skin thrown into folds, but these not so strongly marked as in the former group. The smallest living member of the family, the Sumatran Rhinoceros, *R. sumatrensis*, Cuvier, now represents this group. Its geographical range is nearly the same as that of the Javan species, though not extending into Bengal; but it has been found in Assam, Chittagong, Burma, the Malay Peninsula, Sumatra, and Borneo. So far as can be determined during the life of the type specimen, it appears that the hairy form from Chittagong, described as *R. lasiotis*, is only a variety of this species.[2] The molar teeth of the Sumatran Rhinoceros are almost indistinguishable from those of the Javan species,

[1] *Trans. Zool. Soc.* vol. xii.; see also *Proc. Zool. Soc.* 1889, p. 9.

[2] See Beddard and Treves, *Proc. Zool. Soc.* 1889, p. 9.

and reference has already been made to the resemblance between the visceral anatomy of these species.[1] The form of the stomach is very similar to that of the Horse. The liver (Fig. 171) has a comparatively large caudate lobe, but is chiefly remarkable for the peculiar shape of the Spigelian lobe, which mainly consists of a thin strip of tissue, 8 inches long, $\frac{3}{4}$ inch wide, and $\frac{1}{4}$ inch deep. The small intestine, in place of the villi of *R. unicornis*, has throughout the greater part of its length a uniform series of thin and nearly or quite continuous transverse foldings, like the valvulæ conniventes of the human small intestine. There is no gland behind the foot.

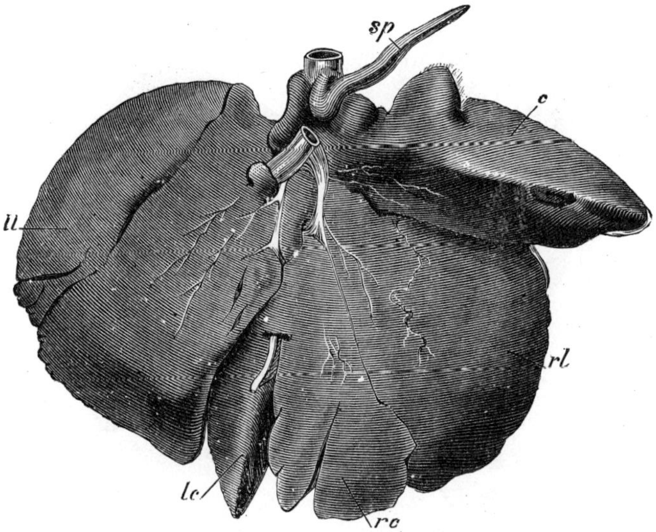

FIG. 171.—Posterior aspect of the liver of *Rhinoceros sumatrensis*. *rc*, Right central lobe; *rl*, right lateral lobe; *lc*, left central lobe; *ll*, left lateral lobe; *c*, caudate lobe; *sp*, Spigelian lobe. (From Garrod, *Proc. Zool. Soc.* 1873, p. 102.)

The post-glenoid and post-tympanic processes of the squamosal do not unite below the auditory meatus. The presence of a lateral nasal diverticulum, like that of the Horses and Tapirs, has been verified only in this species, although it doubtless occurs in the others.

Atelodine Group.—In the adults the incisors and canines quite rudimentary or entirely wanting. Nasal bones thick, rounded and truncated in front. Well-developed anterior and posterior horns in close contact. Skin without any definite permanent folds.

The two well-marked existing species are peculiar to the African continent.

[1] For the internal anatomy of *R. sumatrensis* see Garrod, *Proc. Zool. Soc.* 1873, p. 92; and Beddard and Treves, *loc. cit.*

The common Two-horned Rhinoceros, *R. bicornis*, is the smaller of the two, with a pointed prehensile upper lip, and a narrow compressed deep symphysis of the lower jaw. It ranges through the wooded and watered districts of Africa, from Abyssinia in the north to the Cape Colony, but its numbers are yearly diminishing, owing to the inroads of European civilisation, and especially of English sportsmen. It feeds exclusively upon leaves and branches of bushes and small trees, and chiefly frequents the sides of wood-clad rugged hills. Specimens in which the posterior horn has attained a length

FIG. 172.—Common African Rhinoceros (*Rhinoceros bicornis*).

as great as, or greater than, the anterior have been separated under the name of *R. keitloa*, but the characters of these appendages are too variable to found specific distinctions upon. The Common African Rhinoceros is far more rarely seen in menageries in Europe than either of the three Oriental species, but one has lived in the gardens of the London Zoological Society since 1868. The molar teeth of this species are of the general type of those of *R. sondaicus*, having no combing-plate to join the crotchet in those of the upper jaw. The conch of the ear is much rounded at its extremity, and edged by a fringe of short hairs; while the nostrils are somewhat rounded. The eye is placed immediately below the posterior horn.[1] Both in this and the following species the post-glenoid and post-tympanic processes of the squamosal do not unite below the

[1] These external points of distinction from *R. simus* are taken from a paper by Sclater in the *Proc. Zool. Soc.* 1886, p. 143.

auditory meatus. Nothing is known of the anatomy of the soft parts of either of them.

Burchell's or the Square-mouthed Rhinoceros (*R. simus*), sometimes called the White Rhinoceros, though the colour (dark slate) is not materially different from that of the last species, is the largest of the whole group, and differs from all the others in having a square truncated upper lip and a wide, shallow, spatulate symphysis to the lower jaw. In conformity with the structure of the mouth, this species lives entirely by browsing on grass, and is therefore more partial to open countries or districts where there are broad grassy valleys between the tracts of bush. It is only found in Africa south of the Zambesi, and of late years has become extremely scarce, owing to the persecutions of sportsmen; indeed, the time of its complete extinction cannot be far off. No specimen of this species has ever been brought alive to Europe. Mr. F. C. Selous[1] gives the following description of its habits from extensive personal observation:—

"The square-mouthed rhinoceros is a huge ungainly-looking beast, with a disproportionately large head, a large male standing 6 feet 6 inches at the shoulder. Like elephants and buffaloes they lie asleep during the heat of the day, and feed during the night and in the cool hours of early morning and evening. Their sight is very bad; but they are quick of hearing, and their scent is very keen; they are, too, often accompanied by rhinoceros birds, which, by running about their heads, flapping their wings, and screeching at the same time, frequently give them notice of the approach of danger. When disturbed they go off at a swift trot, which soon leaves all pursuit from a man on foot far behind; but if chased by a horseman they break into a gallop, which they can keep up for some distance. However, although they run very swiftly, when their size and heavy build is considered, they are no match for an average good horse. They are, as a rule, very easy to shoot on horseback, as, if one gallops a little in front of and on one side of them, they will hold their course, and come sailing past, offering a magnificent broadside shot, while under similar circumstances a prehensile-lipped rhinoceros will usually swerve away in such a manner as only to present his hind-quarters for a shot. When either walking or running, the square-mouthed rhinoceros holds its head very low, its nose nearly touching the ground. When a small calf accompanies its mother it always runs in front, and she appears to guide it by holding the point of her horn upon the little animal's rump; and it is perfectly wonderful to note how in all sudden changes of pace, from a trot to a gallop or *vice versâ*, the same position is always exactly maintained. During the autumn and winter months (*i.e.* from March to August) the square-mouthed

[1] *Proc. Zool. Soc.* 1881, p. 726.

rhinoceros is usually very fat; and its meat is then most excellent, being something like beef, but yet having a peculiar flavour of its own. The part in greatest favour among hunters is the hump, which, if cut off whole and roasted just as it is in the skin, in a hole dug in the ground, would, I think, be difficult to match either for juiciness or flavour."

The molar dentition is of the type obtaining in *R. unicornis*, so that in this respect *R. simus* has the same relation to *R. bicornis* as is presented by *R. unicornis* to *R. sondaicus*. The ear-conch of the Square-mouthed Rhinoceros is very large, elongated, and pointed at its extremity, which bears only a slight tuft of hair; it is much expanded in the middle, and the lower portion has its edges united to form a short tube. The nostrils have a long slit-like aperture; and the eye is situated behind the posterior horn.

Extinct Species.—Using the generic term *Rhinoceros* in its widest signification, a very large number of fossil forms may be referred to it, the earliest of which date from the Upper Eocene (Oligocene) Phosphorites of Central France. Only a few of the more important of these types can, however, be even mentioned in this place.

In the Pliocene Siwaliks of India *R. sivalensis* appears to have been the direct ancestor of *R. sondaicus*; while *R. palæindicus* was probably nearly related to *R. unicornis*, although the upper molars had not developed a combing-plate.

R. schleirmacheri, of the Lower Pliocene of Europe, falls into the Ceratorhine group, although differing from *R. sumatrensis* by the union of the post-glenoid and post-tympanic processes of the squamosal beneath the auditory meatus. The Middle Miocene *R. sansaniensis* was a closely allied if not identical form.

The Atelodine group was very widely spread in past epochs. Thus the huge *R. platyrhinus* of the Indian Pliocene, and the equally large *R. antiquitatis* of the Pleistocene of Europe, were specialised forms with a dentition resembling that of *R. simus*, to which they were probably allied. An upper molar of *R. antiquitatis*—the so-called Tichorine, or Woolly Rhinoceros—is shown in the woodcut on p. 402. Of this species nearly whole carcases, with the thick woolly external covering, have been discovered associated with those of the Mammoth, preserved in the frozen soil of the north of Siberia. In common with some other extinct species it had a solid median wall of bone supporting the nasals, from which it is inferred that the horns were of a size and weight surpassing that of the modern species. In the Lower Pliocene of Attica *R. pachygnathus* appears to have been closely allied to *R. bicornis*. Several species, such as *R. leptorhinus* (Fig. 173), *R. megarhinus*, and *R. etruscus*, occur in the European Pleistocene which do not present a marked relationship to any of the living forms. This group is also represented in the Pleistocene of Southern India by the small *R. deccanensis* and *R. karnuliensis*.

In the Upper Miocene, or Lower Pliocene, of North America numerous Rhinoceroses with incisor teeth occur which have no nasal horn, although in those forms of which the limbs are known the fore feet resembled those of existing species in having only three digits. These species have been generically separated as *Aphelops*, but so closely do they resemble existing Rhinoceroses that at one time Professor Cope proposed to refer the hornless female of *R. sondaicus* (described by Lesson as *R. inermis*) to the same genus. If these American types be included in *Rhinoceros* there seems no valid reason for separating the European Lower Pliocene and Miocene forms described as *Aceratherium*, at least some of which have

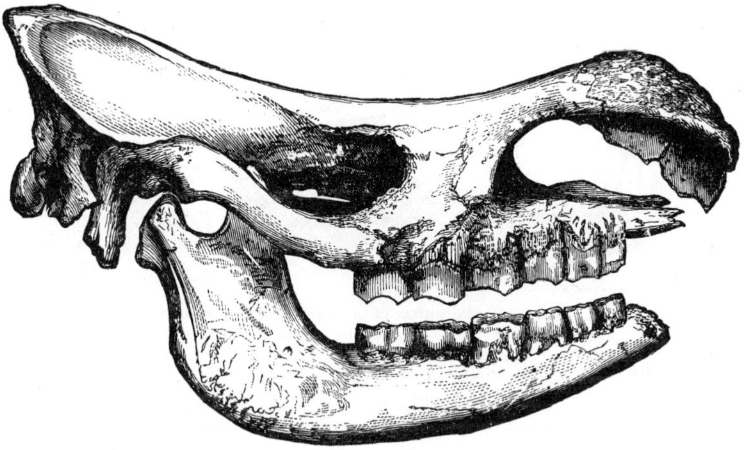

FIG. 173.—Skull of *Rhinoceros leptorhinus*, from the Pleistocene of Essex. About ⅛ natural size.

four digits in the manus. This group is represented in the Upper Eocene Phosphorites of France, and also by a very large species in the Pliocene of India. Lastly, *R. minutus*, of the Lower Miocene of France, and an allied North American species are distinguished by carrying a pair of very small horns placed transversely across the nasals, from which feature it has been proposed that they should be separated generically as *Diceratherium*.

Extinct Generic Types.—The Tertiary deposits of different parts of the world have yielded remains of many extinct forms more or less closely related to the Rhinoceroses, and some of which should certainly be included in the same family; although others perhaps form the types of one or more distinct families. One of the most remarkable of these extinct types is the huge *Elasmotherium*, from the Pleistocene of Siberia, in which the dentition was reduced to two premolars and three molars on either side of each jaw. The structure of the skeleton is essentially rhinocerotic, the skull having an ossified nasal septum, and a huge frontal prominence for the support of a very large horn. The teeth are extremely hypsodont, with the enamel plicated to a remarkable degree, and unlike those of *Rhinoceros*. The genus is evidently a very specialised one.

The other genera we have to notice are more generalised types. Of these the North American *Hyracodon*, with the full typical number of teeth, and without nasal horn, appears to connect the Rhinoceroses with the Lophiodont *Hyrachyus*. The genera *Amynodon* and *Metamynodon* (Fig. 174), from the American Tertiaries, are forms allied to the Rhinoceroses, with the full number of incisors and canines, and the hinder lobe of the last upper molar not aborted. The lower canines are either upright, or less proclivous than in the Rhinoceroses; in *Metamynodon* the premolars are reduced to $\frac{3}{2}$. Molar teeth from the Phosphorites of Central France, described

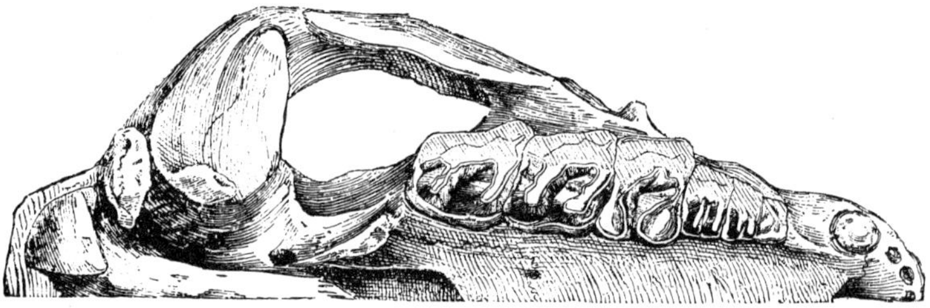

FIG. 174.—Right half of the palatal surface of the cranium of *Metamynodon planifrons*, from the Upper Miocene of North America. (After Scott and Osborn.)

under the name of *Cadurcotherium*, are constructed on the general plan of those of the Rhinoceroses, although distinguished by their extreme narrowness; this type of tooth being very similar to that found in *Homalodontotherium* from Tertiary deposits in Patagonia. The latter has the full number of teeth, without any diastema in the series. Until we have some knowledge of the skeleton of these remarkable forms nothing definite can be said as to their serial position.

Families LAMBDOTHERIIDÆ, CHALICOTHERIIDÆ, AND TITANOTHERIIDÆ.

These families contain a large number of more or less nearly related extinct types from Tertiary beds of both the Old and New Worlds, some of which present most remarkable deviations from the ordinary Ungulate structure. All are characterised by their brachydont molars, which depart widely from the normal lophodont type. The upper molars consist of four columns, of which the two external ones are expanded to form an outer wall; the posterior pair being connected in some cases by an oblique transverse ridge, while there may be traces of an anterior ridge. The premolars are simpler.

Lambdotheriidæ.—This family is confined to the Upper Eocene and Miocene of North America, where it is represented by *Lambdotherium*, *Palæosyops*, and *Limnosyops;* it presents the normal type of foot structure, and all the genera except the first have the full complement of teeth. There were four digits in the manus. The last lower molar has a third lobe. *Limnosyops* differs from *Palæosyops* in having two inner columns to the last upper molar.

Chalicotheriidæ.—The genus *Chalicotherium*, which is found in the Tertiaries of Europe, Asia, and North America, differs so remarkably in the structure of the feet from all other Ungulates that it has been proposed to regard it as the representative of a distinct order, Ancylopoda. The molars are, however, almost indistinguishable from those of the preceding and following families; while the cervical vertebræ and portions of the limbs are of a Perissodactyle type. On the other hand, the femur has lost its third trochanter; while the phalanges are strangely modified, the terminal ones forming long curved claws, while the others (Fig. 175) have strong ginglymoid distal articulations. These phalanges were, indeed, long regarded as referable to Edentates, being described in Europe as *Macrotherium*, and in the United States as *Morotherium* and *Moropus*. *Ancylotherium*, of the Grecian Pikermi beds, is founded upon phalanges which indicate an allied genus. The Indian species of *Chalicotherium* is distinguished by the loss of the incisors and the upper canine; while all the species want the first premolar.

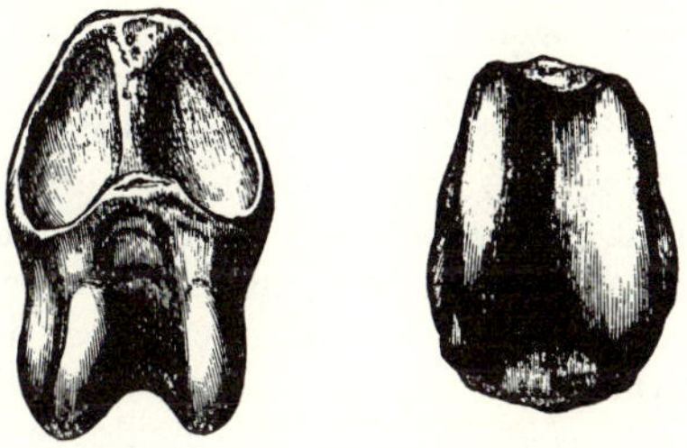

FIG. 175.—Anterior and distal aspects of a phalangeal bone of *Chalicotherium sivalense*. (From the *Palæontologia Indica*.)

Titanotheriidæ.—This exclusively North American family includes gigantic forms closely allied to the *Lambdotheriidæ*, but with the last upper premolar as complex as the molars, and frequently with large bony protuberances in the nasal region. The best known genus, *Titanotherium* (*Menodus*,[1] *Brontotherium*, *Symborodon*, *Allops*, etc.), may either have the full complement of teeth, or the incisors may be reduced to $\frac{2}{0}$. The canines and incisors are small, and there is no diastema when the full dental series is developed. The skull is very like that of the Rhinoceroses; but has a transverse pair of large bony prominences on the nasal region, varying considerably in shape and size in the different species, which in the living animal were probably covered with horny sheaths. The third

[1] This name is the earliest, but is preoccupied.

trochanter of the femur was aborted. These huge animals—inferior in size only to the Elephant—appear to have been abundant in the United States during the Miocene period.

Family MACRAUCHENIIDÆ.

This extinct South American family is best known by the genus *Macrauchenia*, as represented by *M. patachonica* and *M. boliviensis*, which are apparently from Pleistocene formations. They are very singular and specialised forms, quite out of the line of descent of any of the existing Perissodactyles, and the steps by which they are connected with the rest of the group have not yet been discovered. Of the larger species, *M. patachonica*, the skeleton is completely known. It had the full number of forty-four teeth, forming an almost uninterrupted series. The cervical vertebræ resemble those of the Camels in the position of the vertebrarterial canal, but the ends of the centra are flat, and not opisthocœlous as in the allied forms. In some of the limb characters it resembles the *Equidæ*, but in the articulation of the fibula with the calcaneum it agrees with the Artiodactyles. The structure of the feet is, however, distinctly Perissodactylate, there being three toes on each. The teeth approximate to a Rhinocerotine structure; and the incisors have an infolding of the enamel of their crowns, as in those of the Horses. The nares open on the top of the skull, and it is probable that the muzzle was produced into a short proboscis. Several other South American forms have been referred to this family, some of which have received distinct generic names, but further evidence is required before many of them can be accepted. Possibly *Homalodontotherium* should be placed here.

Family PROTEROTHERIIDÆ.

Proterotherium.—Here may be noticed certain very remarkable Perissodactyles from the South American Tertiaries, for which the name *Proterotherium* has been proposed. The cheek-teeth are so like those of *Anchitherium* that they have been described under that name. The upper jaw has one pair of canine-like incisors and no canines, while the lower jaw carries two pairs of incisors. In the skull the orbits were completely closed, as in the Horses. The feet were tridactyle, like those of *Hipparion*, but the tarsus was constructed on an Artiodactyle type.

SUBUNGULATA.

By far the greater number of the Subungulata are extinct, and of many of those whose former existence has been revealed, chiefly by the labours of the American palæontologists, our knowledge is at present necessarily imperfect, though daily extending. It will only be possible here to give details of some of the more interesting or best-known forms.

The characters by which the skeleton of the feet of the Subungulata are distinguished from those of the Ungulata Vera have been already mentioned on p. 275. In addition to these it may be observed that the feet frequently have five functional digits, and may be plantigrade; while the upper surface of the astragalus is generally flattened, instead of presenting the strongly-marked pulley-like ridges and groove so characteristic of the Ungulata Vera.

Suborder HYRACOIDEA.

Family HYRACIDÆ.

This division is constituted to receive a single family of mammals, the affinities of which have long constituted a puzzle to

FIG. 176.—*Hyrax capensis.*

zoologists. They were first placed among the Rodents, to which animals their small size and general appearance and habits give them much superficial resemblance. Cuvier's investigations into their anatomical structure, and especially their dental characters, led him to place them among the Ungulates, near the genus *Rhinoceros*, a position long accepted by many zoologists. Further knowledge of their organisation and mode of development caused Milne-Edwards, Huxley, and others to disassociate them from this connection, and, failing to find any agreement with any other known forms, to place them in an order entirely apart. Palæontology has thrown no light upon the affinities of this anomalous and isolated group, as no extinct animals possessing their distinctive characters have as yet been discovered.

The dentition, according to the usual interpretation, consists only of incisors and molars, the formula in all known species being $i\,\frac{1}{2}$, $c\,\frac{0}{0}$, $p\,\frac{4}{4}$, $m\,\frac{3}{3}$. The upper incisors have persistent pulps, and are curved longitudinally, forming a semicircle as in Rodents. They are, however, not flattened from before backwards as in that order, but prismatic, with an antero-external, an antero-internal, and a posterior surface, the first two only being covered with enamel; their apices are consequently not chisel-shaped, but sharp pointed. They are preceded by functional, rooted milk-teeth. The outer lower incisors, which should perhaps be regarded rather as canines, have long tapering roots, but not of persistent growth. They are straight, procumbent, with awl-shaped, trilobed crowns. Behind the incisors is a considerable diastema. The molars and premolars are all contiguous, and formed almost exactly

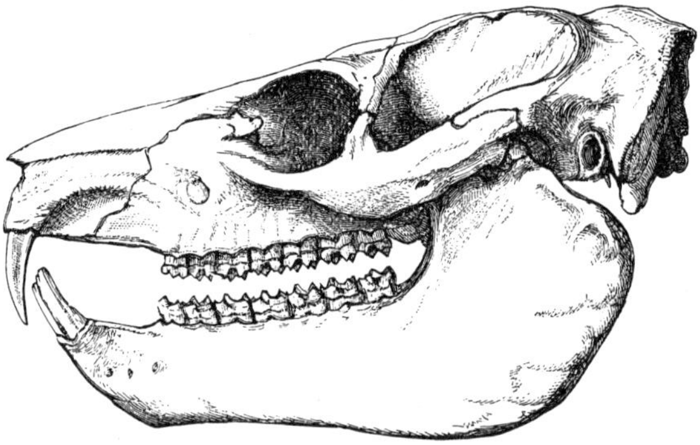

FIG. 177.—Skull and dentition of *Dendrohyrax dorsalis*. × $\frac{2}{3}$.

on the pattern of some of the Perissodactyle Ungulates. The hyoid arch is unlike that of any known mammal. The dorsal and lumbar vertebræ are very numerous, 28 to 30, of which 21 or 22 bear ribs. The tail is extremely short. There are no clavicles. In the fore foot the three middle toes are subequally developed, the fifth is present, but smaller, and the hallux is rudimentary, although, in one species at least, all its normal bones are present. The ungual phalanges of the four outer digits are small, somewhat conical, and flattened in form. The carpus has a distinct os centrale. There is a slight ridge on the femur in the place of a third trochanter. The fibula is complete, thickest at its upper end, where it generally ankyloses with the tibia. The articulation between the tibia and astragalus is more complex than in other mammals, the end of the malleolus entering into it. The hind foot is very like that of *Rhinoceros*, having three well-developed toes. There is no trace of a hallux, and the fifth metatarsal is represented only by a small nodule. The ungual phalanx of the inner (or second) digit is deeply cleft, and has a peculiar long curved claw, the others have short broad nails. The stomach is formed upon much the same principle as that of the Horse or Rhinoceros, but is more elongated transversely and divided by a constriction into two cavities—a large left *cul de sac*, lined by a very dense white epithelium, and a right pyloric cavity, with a very thick, soft, vascular lining. The intestinal canal (Fig. 178) is long, and has an arrangement perfectly unique among mammals, indeed among vertebrated animals, for, in addition to the ordinary short, but capacious and sacculated cæcum (*cm*) at the commencement of the colon, there is, lower down, an additional pair of large, conical, pointed, supplemental cæca (*c*). The liver is much subdivided, and there is no gall-bladder. The brain resembles that of the typical Ungulates far more than the Rodents. The testes are permanently abdominal. The ureters open into the fundus of the bladder, as in some Rodents. The female has six teats, of which four are inguinal and two axillary; and the placenta is zonary, as in the Elephant and Carnivora.

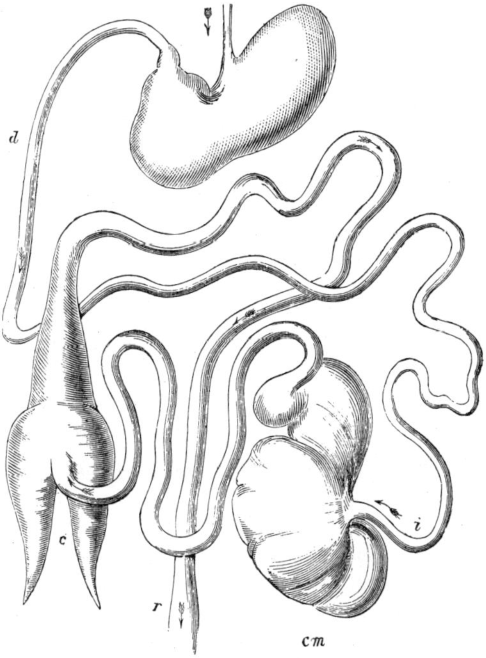

FIG. 178.—Diagrammatic view of the alimentary canal of *Hyrax capensis*, the intestines being somewhat abbreviated. *d*, Duodenum; *i*, ileum; *cm*, cæcum; *c*, supplemental colic cæca; *r*, rectum.

There are two distinct forms of Hyrax, differing both in structure and habits, which may be accorded generic rank.

Hyrax.[1]—Molar teeth having the same pattern as those of

[1] Hermann, *Tab. Affinit. Anim.* p. 115 (1783). It has recently been proposed to substitute the earlier name *Procavia* in lieu of *Hyrax*. The anatomy of Hyrax was first described by Pallas (*Spicilegia Zoologica*). Besides minor memoirs, two detailed accounts of its structure have appeared—one by Brandt, in *Mém. Acad. Nat. Scien. St. Pétersbourg*, 7ième sér. vol. xiv. No. 2, 1869; and

Rhinoceros. Interval between upper incisors less than the width of the teeth. Lower incisors slightly notched at the cutting edge. Vertebræ: C 7, D 22, L 8, S 6, C 6. Of this form the earliest known species, *H. capensis* (Fig. 176) is the type. There are several other species, as *H. habessinicus* and *syriacus*, from Eastern Africa and Syria. They inhabit mountainous and rocky regions, and live on the ground.

Dendrohyrax.[1]—Molar teeth having the same pattern as *Palæotherium* (except that the third lower molar has but two lobes). Interval between upper incisors exceeding the width of the teeth. Lower incisors with very distinctly trilobed crowns. Vertebræ: C 7, D 21, L 7, S 5, C 10. The members of this section frequent the trunks and large branches of trees, sleeping in holes. There are several species, not distinctly defined, from western and south Africa, as *D. arboreus* and *D. dorsalis.* The members of both groups appear to have a power like that possessed by the Lizards called Geckos of clinging to vertical surfaces of rocks and trees by the soles of their feet.

It should be added that some writers separate three of the African species usually included in *Hyrax* (viz. *H. bocagei*, *H. bakeri*, and *H. blainvillei*) under the designation of *Heterohyrax.*[2]

Suborder PROBOSCIDEA.

This name has been appropriated to a well-marked group of animals, presenting some very anomalous characters, allied in many respects to the typical Ungulata, but belonging neither to the Artiodactyle nor Perissodactyle type of that order. It has been thought that they possess some, though certainly not very close, affinities with the Rodentia, and also with the Sirenia. It is certain, however, that the two species of Elephant, which are the sole living representatives of the group, stand quite alone among existing mammals, differing widely from all others in many points of their structure. In some respects, as the skull, proboscis, and dentition, they are highly specialised; but in others, as in the presence of two anterior venæ cavæ and in the structure of the limbs, they retain a low or generalised condition. A considerable series of extinct forms, extending back through the Pliocene and Miocene epochs, show the same type under different modifications, and in still more

another by George, in *Annales des Sciences Naturelles*, 6ième sér. tom. i. 1874, in which references to all the previous literature will be found. The mechanism by which the sole of the foot is enabled to adhere to smooth surfaces is fully described by G. E. Dobson, *Proc. Zool. Soc.* 1876, p. 526.

[1] Gray, *Ann. Mag. Nat. Hist.* ser. 4, vol. i. p. 48 (1868).

[2] See a paper by J. V. Barboza du Bocage, in the *Jorn. Sci. Phys. Nat. Lisboa* (2), vol. i. p. 186 (1889), where a list of all the known species will be found.

generalised outlines; and certain forms from the Eocene of North America, if their affinities are rightly interpreted, appear to link the true Proboscidea to some unknown primitive type of Ungulata.

The following are the principal characters common to existing, and, by inference, to the extinct, Proboscidea. The nose extended into a long, muscular, very flexible and prehensile proboscis, at the end of which the nostrils are situated, and from which the name given to the group is derived. The teeth consisting of ever-growing incisors of very great size, but never exceeding one pair in each jaw, and often present in one jaw only; no canines; large and transversely ridged molars. No clavicles. Limbs strong, the upper segment, especially in the hind limb, the longer. Radius and ulna distinct, the latter articulating extensively with the carpus. Fibula and tibia distinct. Astragalus very flat on both surfaces. Manus and pes short, broad, and massive, each with five toes, though the outer pair may be more or less rudimentary, all encased in a common integument, though with distinct, broad, short hoofs. Third digit the largest. Two anterior venæ cavæ entering the right auricle. Stomach simple. A capacious cæcum. Testes permanently abdominal. Uterus bicornuate. Placenta nondeciduate and zonary. Mammæ two, pectoral.

With regard to the teeth, the incisors,[1] which project largely out of the mouth, and are commonly called "tusks," are of an elongated conical form, and generally curved. They are composed mainly of solid dentine, the fine elastic quality and large mass of which renders it invaluable as "ivory" for commerce and the arts. A peculiarity of the dentine of most Proboscidea is that it shows, in transverse fractures or sections, striæ proceeding in the arc of a circle from the centre to the circumference in opposite directions, and forming by their decussations curvilinear lozenges, as in the "engine-turning" of the case of a watch. The enamel-covering in existing species is confined to the extreme apex, and very soon wears off, but in some extinct species it forms persistent longitudinal bands of limited breadth. The tusks have small milk-predecessors, shed at an early age.

The molar teeth present a remarkable series of modifications, from the comparatively simple form in *Dinotherium*, with two or three strongly pronounced transverse ridges and a normal mode of succession, to the extremely complex structure and anomalous mode

[1] These teeth are by some writers classed as canines, as their roots are implanted in the maxillæ; but, as in Rodents, they are originally developed in the gum covering the premaxillæ, in which bones their primitive alveoli are sunk. As growth proceeds, however, firm support for such massive and weighty bodies can only be obtained by their roots gradually sinking through the premaxillæ into the great and specially modified alveolar processes of the maxillæ, but this does not vitiate their homology with the incisors of other mammals.

of replacement found in the true Elephants. The intermediate conditions occur in the various species of *Mastodon*. In this genus the enamel-covered transverse ridges of each tooth are generally more numerous than in *Dinotherium*, and often complicated by notches dividing their edge or by accessory columns attached to them, but in the unworn tooth they stand out freely on the surface of the crown, with deep valleys between (Fig. 179, I). In the Elephants the ridges are still further increased in number, and consequently narrower from before backwards, and are greatly extended

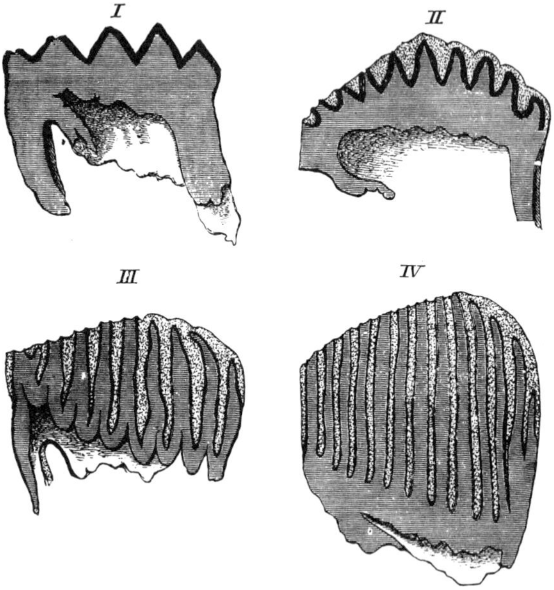

FIG. 179.—Longitudinal sections of the crown of a molar tooth of various Proboscideans, showing stages in the gradual modification from the simple to the complex form. I, *Mastodon americanus*; II, *Elephas insignis*; III, *Elephas africanus*; IV, *Elephas primigenius*. The dentine is indicated by transverse lines, the cement by a dotted surface, and the enamel is black.

in vertical height, so that, in order to give solidity to what would otherwise be a laminated or pectinated tooth, it becomes necessary to envelop and unite the whole in a large mass of cement, which completely fills up the valleys, and gives a general smooth appearance to the organ when unworn; but as the wear consequent upon the masticating process proceeds, the alternate layers of tissue of different hardness—cement, dentine, and enamel—which are disclosed upon the surface form a fine and very efficient triturating instrument. The modification of the tooth of a Mastodon into that of an Elephant is therefore precisely the same in principle as that of the molar of a Palæotherium into that of a Horse, or of the corresponding tooth of one of the primitive Artiodactyles into that of an Ox. The intermediate stages, moreover, even in the present state of our knowledge, are so numerous that it is not possible to draw a definite line between the two types of tooth structure (see Fig. 179, I, II, III, IV).

As regards the mode of succession, that of modern Elephants is, as before mentioned, very peculiar. During the complete lifetime of the animal there are but six molar teeth on either side of each jaw, with occasionally a rudimentary one in front, completing the typical number of seven. The last three represent the true molars of ordinary mammals; those in front appear to be milk-molars, which are never replaced by permanent successors, but the whole series gradually moves forwards in the jaw, and the teeth become worn away and their remnants cast out in front, while development of others proceeds behind. The individual teeth are so large, and the processes of growth and destruction by wear take place so slowly, that not more than one, or portions of two, teeth are ever in place and in use on either side of each jaw at one time, and the whole series of changes coincides with the usual duration of the animal's life. On the other hand, the Dinotherium, the opposite extreme of the Proboscidean series, has the whole of the molar teeth in place and use at one time, and the milk-molars are vertically displaced by premolars in the ordinary fashion. Among Mastodons transitional forms occur in the mode of succession as well as in structure, many species showing a vertical displacement of one or more of the milk-molars, and the same has been observed in one extinct species of Elephant (*E. planifrons*) as regards the posterior of these teeth.

All known Proboscideans are animals of comparatively large dimensions, and some are the most colossal of land mammals. The head is of great proportionate size; and, as the brain case increases but little in bulk during growth, while the exterior wall of the skull is required to be of great superficial extent to support the trunk and the huge and ponderous tusks, and to afford space for the attachment of muscles of sufficient size and strength to wield the skull thus heavily weighted, an extraordinary development of air-cells takes place in the cancellous tissue of nearly all the bones of the cranium (Fig. 180). These cells are not only formed in the walls of the cranium proper, but are also largely developed in the nasal bones and upper part of the premaxillæ and maxillæ, the bones forming the palate and the basicranial axis, and even extend into the interior of the ossified mesethmoid and vomer. Where two originally distinct bones come into contact, the cells pass freely from one to the other, and almost all the sutures become obliterated in old animals. The intercellular lamellæ in the great mass which surrounds the brain cavity superiorly and laterally mostly radiate from the inner to the outer table, but in the other bones their direction is more irregular. Like the similar but less developed air-cells in the skulls of many other mammals, they all communicate with the nasal passages, and they are entirely secondary to the original growth of the bones, their development having scarcely commenced in the new-born animal, and they gradually enlarge as the growth of the creature proceeds towards maturity. The nasal bones are very short, and the anterior narial aperture is situated high in the face. The zygomatic arch is slender and straight, the jugal bone being small, and forming only the middle part of the arch, the anterior part of which (unlike that of typical Ungulates) is

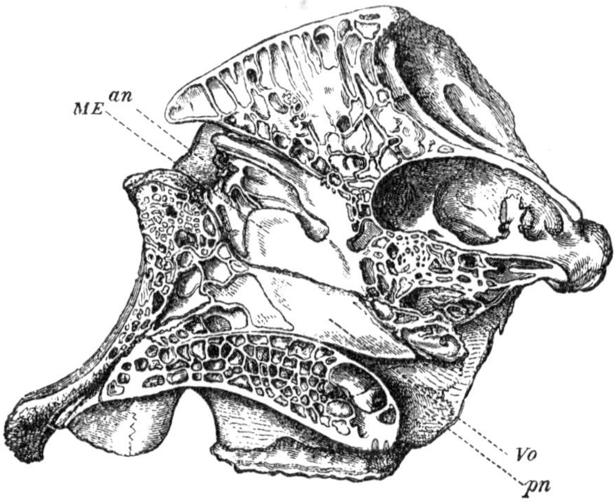

FIG. 180.—A vertical section of the skull of the African Elephant (*Elephas africanus*) taken to the left of the middle line, and including the vomer (*Vo*) and the mesethmoid (*ME*). *an*, Anterior, and *pn*, posterior narial aperture. $\frac{1}{12}$ natural size. (From Flower's *Osteology of the Mammalia.*)

formed only by the maxilla. The maxillo-turbinals are but rudimentary, the elongated proboscis supplying their place functionally in warming and clearing from dust the inspired air.

The neck is very short. The limbs are long and stout, and remarkable for the great length of the upper segment (especially the femur) as compared with the distal segment, the manus, and pes. It is owing to this and the vertical position of the femur that the knee-joint in the hind leg is placed much lower, and is more conspicuous externally than in most quadrupedal mammals; and this having been erroneously compared with the hock-joint or ankle of typical Ungulates, the popular fallacy that the joints of the Elephant's leg bend in a contrary direction to that of other mammals has arisen. There is no round ligament in the hip-joint, or third trochanter to the femur. The radius and ulna are distinct, though fixed in a crossed or prone position. The fibula also is quite distinct from the tibia. The feet are short and broad, the carpal and tarsal bones being very square, with flattened surfaces for articulation; the astragalus especially differs from that of typical Ungulates in its flatness, in the absence of a distinct pulley-like articular surface at either extremity, and in having no articular facet for the cuboid. The fibula articulates with the calcaneum, as in Artiodactyles. Of the five toes present on each extremity (see Fig. 98), the middle one is somewhat the largest, and the lateral ones smallest, and generally wanting (especially in the hind foot) the complete number of phalanges. The ungual phalanges are all small, irregular in form, and late in ossification. The whole are encased in a common integument, with a flat, subcircular, truncated sole, the only external indication of the toes being the broad oval nails or hoofs arranged in a semicircle around the front edge of the sole. The hind foot is smaller and narrower than the front. The liver is small and simple, and there is no gall-bladder. In form the brain resembles that of the Rodents and other lower orders of mammals, the cerebellum being entirely behind and uncovered by the cerebrum, but the hemispheres of the latter are richly convoluted.

The Proboscidea are exclusively vegetable feeders, living chiefly on leaves and young branches of forest trees and various kinds of herbage, which they gather and convey to their mouth by the very mobile proboscis, an organ which combines in a marvellous manner strength with dexterity of application, and is a necessary compensation for the shortness and inflexibility of the neck, as by it many of the functions of the lips of other animals are performed. By its means the Elephant is enabled to drink without bending the head or limbs; the end of the trunk being dipped into the stream or pool, a forcible inspiration fills the two capacious air-passages in its interior with water, which, on the tip of the trunk being turned upwards and inserted into the mouth, is ejected by a blowing action, and swallowed; or if the animal wishes to refresh and cool its skin, it can throw the water in a copious stream over any part of its surface. Elephants can also throw dust and sand over their bodies by the same means and for the same purpose, and wild animals have been frequently observed fanning themselves with leafy boughs held in the trunk. The species are at present limited in their geographical distribution to the Ethiopian and Oriental regions, but they formerly had a far more extensive range.

Family ELEPHANTIDÆ.

Cheek-teeth succeeding one another in an arc of a circle, and portions of only two, or at most three, of the hinder teeth in use at any one time. Premolars frequently lost, and in any case of no functional importance.

Elephas.[1]—Dentition: $i\,\frac{1}{0}$, $c\,\frac{0}{0}$, $dm\,\frac{3}{3}$, $m\,\frac{3}{3} = 26$. The incisors variable, but usually of very large size, especially in the male sex, directed somewhat outwards, and curved upwards, without enamel except on the apex before it is worn. The molars composed of numerous flattened enamel-covered plates or ridges of dentine, projecting from a common many-rooted base, surrounded and united together by cement, and extending straight across the crown, without (in most forms) any median division into inner and outer columns. The number of plates increases from the anterior to the posterior molar in regular succession, varying in the different species, but the third and fourth (or the last milk-molar and the first true molar), and these only, have the same number of ridges, which always exceeds five. Premolars nearly always wanting. Skull of adult very high and globular. Mandible ending in front in a short, deflected, and spout-like symphysis. Vertebræ: C 7, D 19-21, L 3-4, S 4, C 26-33.

The existing species of the genus differ so much that they have been referred by some writers to distinct genera; fossil forms show, however, such a transition from the one to the other that it is scarcely possible to regard them even as the representatives of distinct groups.

In the well-known Indian or Asiatic Elephant (*E. indicus*) the average number of plates of the six successive molar teeth is expressed by the "ridge-formula," 4, 8, 12, 12, 16, 24. The plates are compressed from before backwards, the anterior and posterior surfaces (as seen in the worn grinding face of the tooth, Fig. 181) being nearly parallel. Ears of moderate size. Upper margin of the end of the proboscis developed into a distinct finger-like process, much longer than the lower margin. Five nails on the fore feet, and four (occasionally five) on the hind feet.

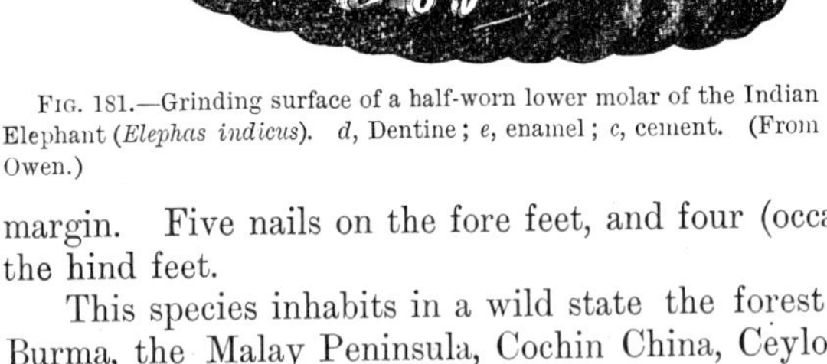

FIG. 181.—Grinding surface of a half-worn lower molar of the Indian Elephant (*Elephas indicus*). *d*, Dentine; *e*, enamel; *c*, cement. (From Owen.)

This species inhabits in a wild state the forest lands of India, Burma, the Malay Peninsula, Cochin China, Ceylon, and Sumatra. The elephants from the last-named islands, presenting some variations from those of the mainland, have been separated under the name of *E. sumatranus*, but the distinction has not been satisfactorily estab-

[1] Linn. *Syst. Nat.* 12th ed. vol. i. p. 48 (1766).

lished. The appearance of the Asiatic Elephant is familiar to all. Though rarely breeding in captivity, it has been domesticated from the most remote antiquity, and is still extensively used in the East as a beast of burden. In the wild state it is gregarious, associating in herds of ten, twenty, or more individuals, and though it may, under certain circumstances, become dangerous, it is generally inoffensive and even timid, fond of shade and solitude and the neighbourhood of water. The height of the male at the shoulder when full grown is usually from 8 to 10 feet, but occasionally as much as 11. The female is somewhat smaller.

In the African Elephant (*E. africanus*) the molars (Fig. 182) are of coarse construction, with fewer and larger plates and thicker enamel. Ridge-formula: 3, 6, 7, 7, 8, 10. The plates not flattened, but thicker in the middle than at the edges, so that their worn grinding surfaces are lozenge-shaped. Ears very large. The

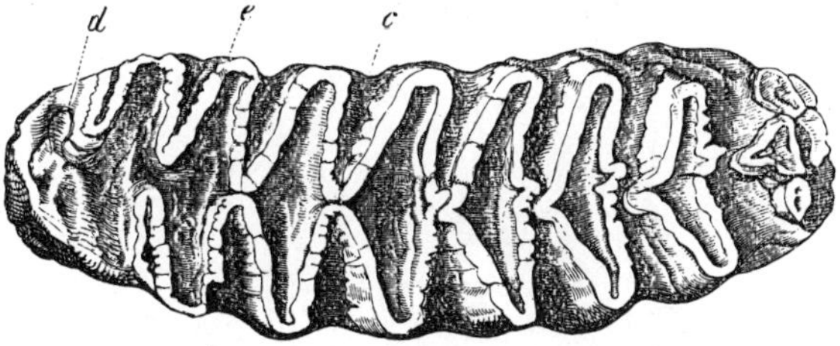

FIG. 182.—Grinding surface of a partially worn right upper molar of the African Elephant (*Elephas africanus*). Letters as in the preceding figure. The left side of the figure is the front of the tooth, and the lower side the outer border. (From Owen.)

upper and lower margins of the end of the trunk forming two nearly equal prehensile lips. But three hoofs on the hind foot. This species now inhabits the wooded districts of the whole of Africa south of the Sahara, except where it has been driven away by human settlements. Fossil remains of Pleistocene age, undistinguishable specifically, have been found in Algeria, Spain, and Sicily. It was trained for war and show by the ancient Carthaginians and Romans, and recent experience of the species in captivity in England shows that it is as intelligent as its Asiatic relative, if not more so, while surpassing it in courage, activity, and obstinacy. Nevertheless, in modern times, no people in Africa have been sufficiently civilised or enterprising to care to train it for domestic purposes. It is hunted chiefly for the sake of the ivory of its immense tusks, of which it yields the principal source of supply to the European market, and the desire to obtain which is rapidly leading to the extermination of the species. In size the male African elephant often surpasses that of Asia, but the female is usually smaller. The circumference of the fore foot is half the height at the shoulder, a circumstance which enables the hunters to judge from the footprints the exact size of the animals of which they are in pursuit. The African Elephant also differs from its Indian congener in having tusks in both sexes, whereas in the latter the male only is so armed. Moreover, the eye is relatively larger, the forehead more convex, and the colour somewhat darker. Whereas the Indian Elephant frequents the depths of forests and seldom leaves their shade during the daytime, the following account by Sir Samuel Baker indicates different habits in the African species. This traveller observes: "In Africa, the country

FIG. 183.—African Elephant (*Elephas africanus*). From a young specimen in the London Zoological Gardens.

being generally more open than in Ceylon, the Elephant remains throughout the day either beneath a solitary tree or exposed to the sun in the vast prairies, where the thick grass attains a height of from nine to twelve feet. The general food of the African Elephant consists of the foliage of trees, especially mimosas. Many of the mimosas are flat-headed, about thirty feet high, and the richer portion of the foliage confined to the crown. Thus the Elephant, not being able to reach to so great a height, must overturn the tree to obtain the coveted food. The destruction caused by a herd of Elephants in a mimosa forest is extraordinary, and I have seen trees uprooted of so large a size that I am convinced no single elephant could have overturned them. I have measured trees four feet six inches in circumference and about thirty feet high uprooted by elephants. The natives have assured me that the elephants mutually assist each other, and that several engage together in the work of overturning a large tree."

Extinct Species of Elephant.—Abundant remains of Elephants are found embedded in alluvial gravels, or secreted in the recesses of caves, into which they have been washed by streams and floods, or dragged as food by Hyænas and other carnivorous inhabitants of these subterranean dens. Such remains belonging to the Pleistocene and Pliocene periods have been found in many parts of Europe, including the British Isles, in North Africa, throughout the North American continent from Alaska to Mexico, and extensively distributed in Asia, where the deposits of the sub-Himalayan Siwalik

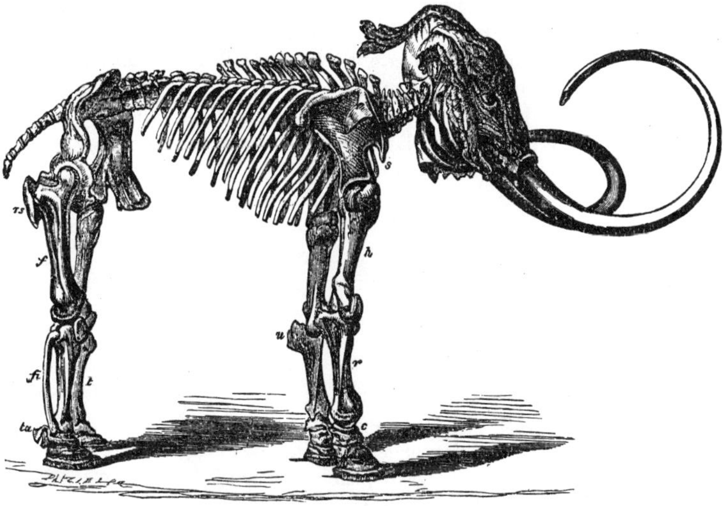

FIG. 184.—Restored skeleton of the Mammoth (*Elephas primigenius*). From Tilesius in *Mém. Acad. Imp. Sc. St. Pétersbourg*, vol. v. (1815). *s*, Scapula; *h*, humerus; *r*, radius; *u*, ulna; *c*, carpus; *rs*, ischium; *f*, femur; *t*, tibia; *fi*, fibula; *ta*, tarsus.

Hills, and equivalent deposits in the Punjab, Perim Island,[1] and Burma, belonging to the earliest Pliocene, are rich in the remains of Elephants of varied form. These species are chiefly known and characterised at present by the skulls and teeth; some of the latter resemble the existing Indian and some the African type, but the majority are between the two, and make the distinction between the two existing species as of generic importance quite impracticable. Others again approach so closely in the breadth and coarseness of the ridges and paucity of cement to *Mastodon* as to have been placed by some zoologists in that genus. These form the subgenus called *Stegodon* by Falconer, and may be regarded as a distinct group of the genus.

[1] In the Gulf of Cambay,—not the island of the same name in the Red Sea.

Among the best known extinct Elephants is *E. primigenius*, the Mammoth,[1] very closely resembling the existing Indian species, and one of the most recently extinct and extensively distributed of all the fossil forms. Probably no animal which has not survived to the historic period has left such abundant and well-preserved evidence of its former existence. The discovery of immense numbers, not only, as in the case of most extinct creatures, in the form of fragmentary bones and teeth, but often as more or less nearly entire carcases, or "mummies," as they may be called, with the flesh, skin, and hair *in situ*, in the frozen soil of the tundras of Northern Siberia, has for a long time given great interest to the species, and been the cause of many legendary stories among the natives of the lands in which they occur. Among these one of the most prevailing is that the Mammoth was, or still is, an animal which passes its life habitually in burrows below the surface of the ground, and immediately dies if by any chance it comes into the upper air.

Of the whole group the Mammoth is in many respects, as in the size and form of the tusks, and especially the characters of the molar teeth, the farthest removed from the primitive Mastodon-like type, while its nearest surviving relative, *E. indicus*, has retained the slightly more generalised characters of the Mammoth's contemporaries of more southern climes, *E. columbi* of America, and *E. armeniacus* of the Old World, if, indeed, it can be specifically distinguished from them.

The tusks or upper incisor teeth were doubtless present in both sexes, but probably of smaller size in the female. In the adult males they often attained the length of from 9 to 10 feet measured along the outer curve. Upon leaving the head they were directed at first downwards and outwards, then upwards and finally inwards at the tips, and generally with a tendency to a spiral form not seen in other species of Elephant. Different specimens, however, present great variations in curve, from nearly straight to an almost complete circle.

It is chiefly by the characters of the molar teeth that the various extinct modifications of the Elephant type are distinguished. Those of the Mammoth (Fig. 185) differ from the corresponding organs of allied species in the great breadth of the crown as compared with the length, the narrowness and close approximation of the ridges, the thinness of the enamel and its straightness, parallelism, and absence of "crimping," as seen on the worn surface, or in a horizontal section of the tooth. Dr. Falconer gave the prevailing

[1] The word Mammoth was introduced into the languages of Western Europe about two centuries ago from the Russian, and is thought by Pallas and Nordenskiöld to be of Tartar origin, but others, as Witzen, Strahlenburg, and Howorth, have endeavoured to prove that it is a corruption of the Arabic word *Behemoth*, or great beast.

"ridge-formula" as 4, 8, 12, 12, 16, 24. Dr. Leith Adams, working from more abundant materials, has shown, however, that the number of ridges of each tooth, especially those at the posterior end of the series, is subject to very great individual variation, ranging in each tooth of the series within the following limits: 3 to 4, 6 to 9, 9 to 12, 9 to 15, 14 to 16, 18 to 27, excluding the small plates called talons at each end of the tooth. Besides these variations in the number of ridges or plates of which each tooth is composed, the thickness of the enamel varies so much as to have given rise to a distinction between a "thick-plated" and a "thin-plated" variety—the latter being most prevalent among the specimens from the Arctic regions, and most distinctively characteristic of the species. From the specimens with thick enamel plates the transition to the other species or varieties mentioned above, including *E. indicus*, is almost imperceptible.

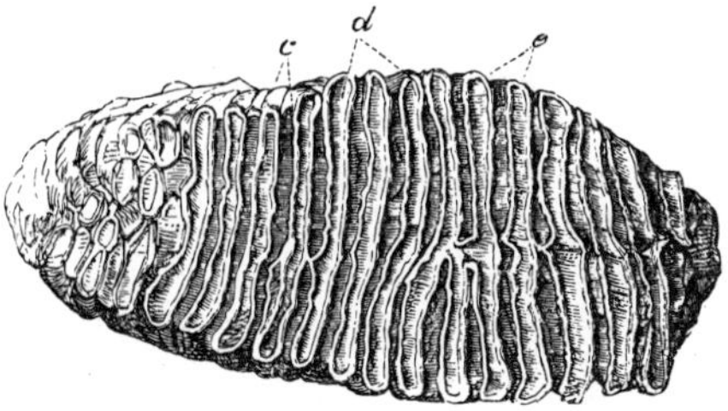

FIG. 185.—Grinding surface of upper molar of the Mammoth (*Elephas primigenius*). *c*, Cement; *d*, dentine; *e*, enamel. (From Owen.)

The bones of the skeleton generally more resemble those of the Indian Elephant than of any other known species, but the skull differs in the narrower summit, narrower temporal fossæ, and more prolonged incisive sheaths required to support the roots of the enormous tusks. Among the external characters by which the Mammoth was distinguished from either of the existing species of Elephant was the dense clothing, not only of long coarse outer hair, but also of close woolly under hair, of a reddish-brown colour, evidently in adaptation to the colder climate which it inhabited. This character, for a knowledge of which we are indebted to the well-preserved remains found in Northern Siberia, is also represented in the rude but graphic drawings of prehistoric age found in caverns in the south of France.[1] In size different individuals varied considerably, but the average height does not appear to have exceeded that of either of the existing species of Elephant.

The geographical range of the Mammoth was very extensive. There is scarcely a county in England in which some of its remains have not been found either in alluvial deposits of gravel or in caverns, and numbers of its teeth are from time to time dredged

[1] The best known of these is the etching upon a portion of tusk found in the cave of La Madelaine in the Dordogne, figured in Lartet and Christy's *Reliquiæ Aquitanicæ*, and in many other works bearing on the subject of the antiquity of man.

up from the bottom of the sea by the fishermen who ply their trade in the German Ocean, having been washed out of the water-worn cliffs of the eastern counties of England. In Scotland and Ireland its remains are less abundant, but they have been found in vast numbers at various localities throughout the greater part of Central Europe (as far south as Santander in Spain and Rome), Northern Asia, and the northern part of the American continent, though the exact distribution of the Mammoth in the New World is still a question of debate. It has not hitherto been met with in any part of Scandinavia or Finland.

In point of time, the Mammoth belongs exclusively to the Pleistocene epoch, and it was undoubtedly contemporaneous with man in France, and probably elsewhere. There is evidence to show that it existed in Britain before, during, and after the glacial period.

As before indicated, it is in the northern part of Siberia that its remains have been found in the greatest abundance, and in quite exceptional conditions of preservation. For a very long period there has been from that region a regular export of Mammoth ivory in a state fit for commercial purposes, both eastward to China and westward to Europe. In the middle of the tenth century an active trade was carried on at Khiva in fossil ivory, which was fashioned into combs, vases, and other objects, as related by Abu 'l Kásim, an Arab writer of that period. Middendorff reckoned that the number of tusks which have yearly come into the market during the last two centuries has been at least a hundred pairs, and Nordenskiöld, from personal observation, considers this calculation as probably rather too low than too high. They are found at all suitable places along the whole line of the shore between the mouth of the Obi and Behring Straits, and the farther north the more numerous do they become, the islands of New Siberia being now one of the most favourite collecting localities. The soil of Bear Island and of Liachoff Islands is said to consist only of sand and ice with such quantities of Mammoth bones as almost to compose its chief substance. The remains are not only found around the mouths of the great rivers, as would be the case if the carcases had been washed down from more southern localities in the interior of the continent, but are imbedded in the frozen soil in such circumstances as to indicate that the animals had lived not far from the localities in which they are now found, and they are exposed either by the melting of the ice in unusually warm summers or by the washing away of the sea cliffs or river banks by storms or floods. In this way the bodies of more or less nearly perfect animals, often standing in the erect position, with the soft parts and hairy covering entire, have been brought to light.

References to the principal recorded discoveries of this kind, and to the numerous speculations to which they have given rise, both among ignorant peasants and learned academicians, will be found in Nordenskiöld's *Voyage of the Vega* (English translation, vol. i. 1881, p. 398 *sq.*) and a series of papers in the *Geological Magazine* for 1880 and 1881, by H. H. Howorth, as well as in a separate work on the Mammoth by the same writer. For the geographical distribution and anatomical characters, see Falconer's *Palæontological Memoirs*, vol. ii. 1868; Boyd Dawkins, "*Elephas primigenius*, its Range in Space and Time," *Quart. Journ. Geol. Soc.* xxxv. p. 138 (1879); and Leith Adams, "Monograph of British Fossil Elephants," part ii., *Palæontographical Society* (1879).

E. antiquus, of the European Pleistocene, has a lower ridge-formula than in the Mammoth, the molars being narrower, and approximating to those of the African Elephant in structure. Small allied forms occur in the rock-fissures and caverns of Malta, and have been described as *E. mnaidriensis* and *E. melitensis*; some of the individuals of the latter not exceeding 3 feet in height. The European *E. meridionalis* is a southern form of somewhat earlier age, very common in the Upper Pliocene of Italy and France, and also in the so-called Forest-bed of the Norfolk coast. It attained very large dimensions, its height being estimated at upwards of 15 feet. The ridge-formula is lower than in *E. antiquus*, the molars are broad, with the worn enamel-discs generally expanded in the middle, and the enamel itself is crenulated.

Elephant remains are very abundant in the Pleistocene and Pliocene deposits of India, those from the latter beds being the oldest representatives of the genus. Of these the Pleistocene *E. namadicus* appears closely allied to *E. antiquus*, from which it is distinguished by a bold ridge across the forehead. Among the Pliocene forms *E. hysudricus* may be an ancestral type allied to the Indian Elephant; while *E. planifrons* is closely related to *E. meridionalis*, although retaining the ancestral feature of developing premolars.

The Stegodont group is peculiar to the eastern parts of the Old World, and, as already observed, connects the true Elephants intimately with the Mastodons. The molars (Fig. 179, II) are characterised by the lowness of the ridges, while the intervening valleys may have but little cement, and there may be a more or less distinct longitudinal groove in the crown dividing each ridge into an inner and an outer moiety. In species like *E. insignis* the ridge-formula is nearly the same as in *E. meridionalis*, but in *E. clifti* some of the molars carry only six ridges, and premolars were present, so that we thus have such a complete transition to the next genus that it is very difficult to know where to draw the line between the two.

Mastodon.[1]—Dentition: $i\ \frac{1}{1\text{ or }0}$, $c\ \frac{0}{0}$, $dm\ \frac{3}{3}$, $m\ \frac{3}{3}$. Upper incisors

[1] Cuvier, *Ann. du Muséum*, vol. viii. p. 270 (1806).

very large, as in *Elephas*, sometimes with longitudinal bands of enamel, more or less spirally disposed. Lower incisors variable; when present comparatively small and straight, sometimes persistent, sometimes early deciduous, and in some species never present. Grinding surface of molars with transverse ridges, the summits of which are divided more or less into conical or mammillary cusps, and often with secondary or additional cusps between and clustering against the principal ridges; enamel thick; cement very scanty, never filling up the interspaces between the ridges. The third, fourth, and fifth cheek-teeth (*i.e.* the last milk-molar, and the first and second molars) having the same number of ridges,[1] which never exceeds five.

In the upper jaw the incisors, though of large size, were apparently never so much curved as in some species of Elephant, and they often have longitudinal bands of enamel, more or less spirally disposed upon their surface, which are not met with in Elephants. Lower incisors were present throughout life in some species, which have the symphysis of the lower jaw greatly elongated to support them (as in *M. angustidens*, *M. pentelici*, and *M. longirostris*). In the common North American species (*M. americanus*) the mandibular symphysis is short, but it may have a small incisor on one side. In other species no inferior tusks have been found, at all events in adult life (see figure of *M. arvernensis*).

The molar teeth increase in size from before backwards, but as many as three of these teeth may be in place in each jaw at one time. There is in many species a true vertical succession, affecting either the third, or the third and second, or (in *M. productus*) the first, second, and third of the six molariform teeth. These three are therefore reckoned as milk-molars, and their successors as premolars, while the last three, which are never changed, correspond to the true molars of those animals in which the typical dentition is fully developed. The study of the mode of succession of the teeth in the different species of Mastodons is particularly interesting, as it exhibits so many stages of the process by which the very anomalous dentition of the modern Elephants may have been derived by gradual modification from the typical heterodont and diphyodont dentition of the ordinary mammal. It also shows that the anterior molars of Elephants do not correspond to the premolars of other Ungulates, but to the milk-molars, the early loss of which in consequence of the peculiar process of horizontal forward-

[1] This, and the larger number of ridges in the latter, are the only absolute distinctions which Falconer could find between *Mastodon* and *Elephas* (*Palæont. Memoirs*, ii. p. 9), and it is clear that they are somewhat arbitrary. The line between the two genera is drawn at this point more as a matter of convenience for descriptive purposes than as indicating any great natural break in the sequence of modifications of the same type.

moving succession does not require, or allow time for, their replacement by premolars. It must be noted, however, that, in the Mastodon in some respects the least specialised in tooth-structure, the *M. americanus* of North America, no vertical succession of the

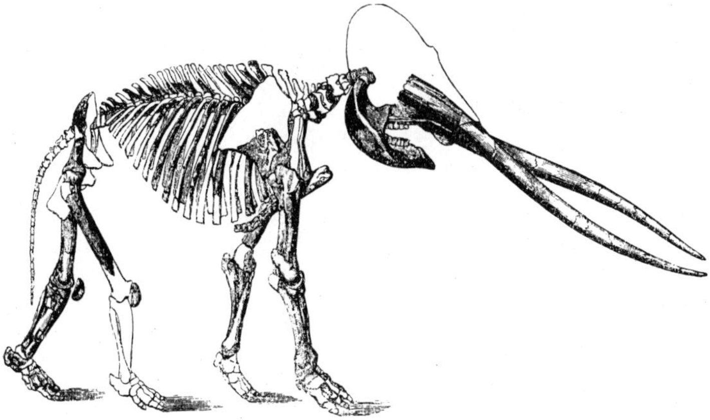

FIG. 186.—Restoration of the skeleton of *Mastodon arvernensis*, from the Pliocene of Europe. (After Sismonda.)

molars has yet been observed, although vast numbers of specimens have been examined.

The Mastodons have fewer ridges on their molar teeth than the Elephants; the ridges are also less elevated, wider apart, have a thicker enamel-covering, and scarcely any cement filling up the space between them. Sometimes (as in *M. americanus*) the ridges are simple transverse wedge-shaped elevations, with straight or concave edges. In other species the summits of the ridges are more or less subdivided into conical cusps, and may have accessory cusps clustering around them (as in *M. arvernensis*, see Fig. 187). When the apices of these are worn by mastication, their surfaces present circles of dentine, surrounded by a border of enamel, and as the attrition proceeds different patterns are produced by the union of the bases of the cusps, a trilobed or trefoil form being characteristic of some species (Fig. 188).

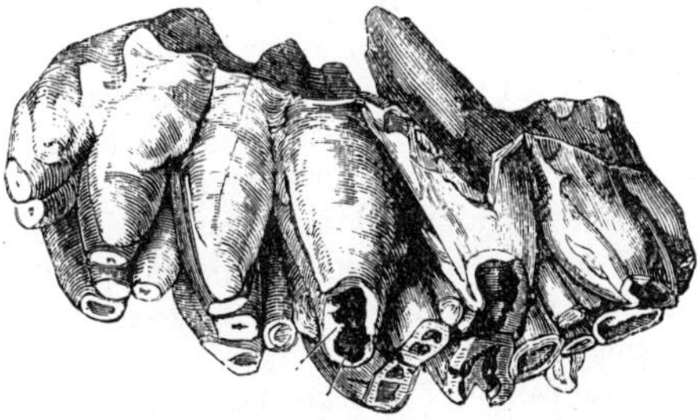

FIG. 187.—Oblique side and crown view of the last upper molar of *Mastodon arvernensis*. (From Owen.)

As already mentioned, certain of the molariform teeth of the middle of the series in Mastodons have the same number of principal ridges, those in front of them having fewer and those behind a greater number. These teeth were distinguished as "intermediate" molars by Dr. Falconer, and are three in number, namely the last milk-molar and the first and second true molars (or the third, fourth, and fifth of the whole series). The number of ridges on these intermediate molars is nearly always three or four, and the tooth in front has usually one fewer and that behind one more, so that the ridge-formula of most Mastodons can be reduced either to 1, 2, 3, 3, 3, 4, or 2, 3, 4, 4, 4, 5. The former characterises the section called *Trilophodon* (of which an intermediate molar is shown in Fig. 188), and the latter that called *Tetralophodon* by Dr. Falconer. These divisions are very useful, as under one or the other all the present known species of Mastodon can be ranged, but observations upon a larger number of individuals have shown that the number of ridges upon the teeth is not quite so constant as implied by the formulæ given above. Their exact enumeration is even difficult in many cases, as "talons" or small accessory ridges at the hinder end of the teeth occur in various stages of development, until they take on the character of true ridges. Transitional conditions have also been shown, at least in some of the teeth, between the trilophodont and the tetralophodont forms, and again between the latter and what has been called a "pentalophodont" type, which leads on towards the condition of dental structure characteristic of the true Elephants.

FIG. 188.—Grinding surface of the partially worn last left lower milk-molar of *Mastodon angustidens*, from the Upper Miocene of India. The lower side of the figure is the outer border of the tooth.

The range of the genus *Mastodon* in time was from the middle of the Miocene period to the end of the Pliocene in the Old World, when it became extinct; but in America several species—especially the one best known, owing to the abundance of its remains, which has been variously called *M. americanus*, *M. ohioticus*, and *M. giganteus*—survived to a late Pleistocene period.

The range in space will be best indicated by the following list of some of the better known species. (1) Trilophodont series—*M. angustidens*,[1] *borsoni*, *pentelici*, *turicensis*, from Europe; *M. falconeri*

[1] Also found beyond the extreme north-western frontier of India.

and *pandionis*, from India; *M. americanus, obscurus*, and *productus*, North America; and *M. cordillerum* and *humboldti*, South America. (2) Tetralophodont series—*M. arvernensis, M. longirostris*, from Europe; *M. latidens, sivalensis*, and *perimensis*, from India; *M. mirificus*, from North America. *Mastodon arvernensis* and *M. longirostris*, together with a trilophodont species, occur in the crag-deposits of Norfolk and Suffolk.

Family DINOTHERIIDÆ.

An extinct family distinguished from the *Elephantidæ* by the whole series of permanent cheek-teeth being in use at the same time.

Dinotherium.[1]—Dentition of adult: $i\ \frac{0?}{1}$, $c\ \frac{0}{0}$, $p\ \frac{2}{2}$, $m\ \frac{3}{3} = 22$; all present at the same time, there being no horizontal succession, but the premolars replacing milk-teeth in the ordinary manner. The presence or absence of upper incisors has not yet been clearly ascertained. Lower incisors large, conical, descending, and slightly curved backwards, implanted in a greatly thickened and deflected beak or prolongation of the symphysis. In section they do not show the decussating striæ characteristic of Mastodons and Elephants. Crowns of molars carrying strong transverse, crenulated ridges, with deep valleys between, much resembling the lower ones of the Tapirs. Ridge-formula of the permanent molar series: 2, 2, 3, 2, 2. The three ridges of the first true molar are constant in both upper and lower jaws, although it is quite an anomalous character among Proboscideans for this molar to have more ridges than those which come behind it. The last milk-molar has also three ridges, the

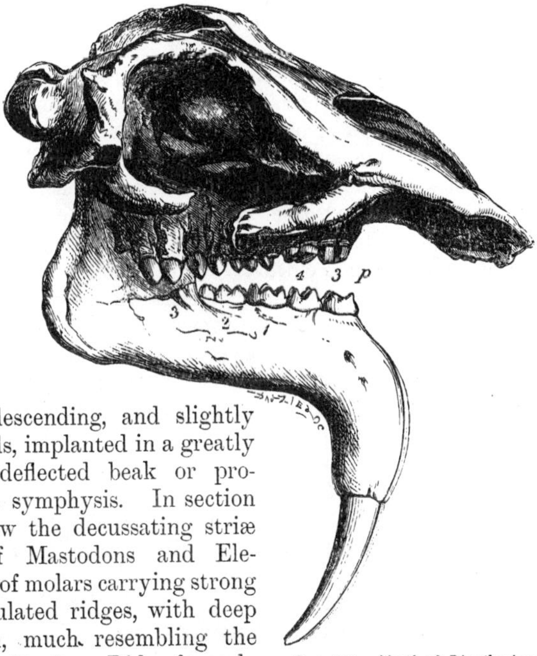

FIG. 189.—Skull of *Dinotherium giganteum*, from the Lower Pliocene of Eppelsheim, Hessen-Darmstadt. (After Kaup.) *p*, 3, 4, premolars; 1, 2, 3, molars.

[1] Kaup, *Isis*, vol. xxii. p. 401 (1829).

penultimate but two. The cranium is much depressed, with comparatively little development of air-cells. The remainder of the skeleton is imperfectly known, but apparently agrees in its general characters with that of the other Proboscideans.

Remains of *Dinotherium giganteum*, an animal of elephantine proportions, strikingly characterised by the pair of huge tusks descending nearly vertically from the front of the lower jaw, were first discovered at Eppelsheim, near Darmstadt, and described by Kaup. They have since been met with in various Lower Pliocene and higher Miocene formations in the south of Germany, France, Greece, and Asia Minor. Closely allied forms also occur in the Lower Pliocene and Upper Miocene of India, but none are known from America.

Suborder AMBLYPODA.

Uintatherium.[1]—Among the most remarkable of the comparatively recent discoveries in the higher Eocene formations of the western states of North America has been one of a group of animals of huge size, approaching that of the largest existing Elephants, presenting a combination of characters quite unlike those known among other recent or extinct creatures, and of which there were evidently many species living contemporaneously, but all of which became extinct before the close of the Eocene period. To form some idea of their appearance, we must imagine animals very elephantine in general proportions and in the structure of their limbs. The feet had five short toes. The tail, as in the Elephants, was long and slender, but the neck, though still short, was not so much abbreviated as in the Proboscideans, and there is no evidence that these animals possessed a trunk. The head differed greatly from that of the Elephants, being long and narrow, more like that of a Rhinoceros, and, as in that animal, was elevated behind into a great occipital crest, and it had developed upon its upper surface three pairs of conspicuous, laterally diverging protuberances—one pair in the parietal region, one on the maxillaries in front of the orbits, and one (much smaller) near the fore part of the elongated nasal bones. Whether these were merely covered by bosses of callous skin, as the rounded form and ruggedness of their extremities would indicate, or whether they formed the bases of attachment for horns of still greater extent, like those of the Rhinoceros or of the Cavicorn Ruminants, can only be a matter of conjecture. There were no upper incisors, but usually three on each side below, of comparatively small size, as was also the lower canine. A huge, compressed, curved, sharp-pointed canine tusk, very similar in form

[1] Leidy, *Proc. Ac. Nat. Sci. Philad.* 1872, p. 169.

and position to that of the Musk-Deer, descended from each side of the upper jaw. These were present in both sexes, but very much smaller in the female, as was also the flange-like process of the lower jaw by which they were guarded. Behind these, and at some distance from them, were on each side above and below six cheek-teeth, of comparatively small size, placed in continuous series, each with a pair of oblique ridges conjoined internally and diverging externally in a V-like manner, and provided with a stout basal cingulum. The normal dental formula was therefore $i\,\frac{0}{3},\ c\,\frac{1}{1},\ p\,\frac{3}{3},\ m\,\frac{3}{3} = 34$; and the dentition had thus already attained a remarkable degree of specialisation, although the brain was smaller and more rudimentary in characters than in almost any other

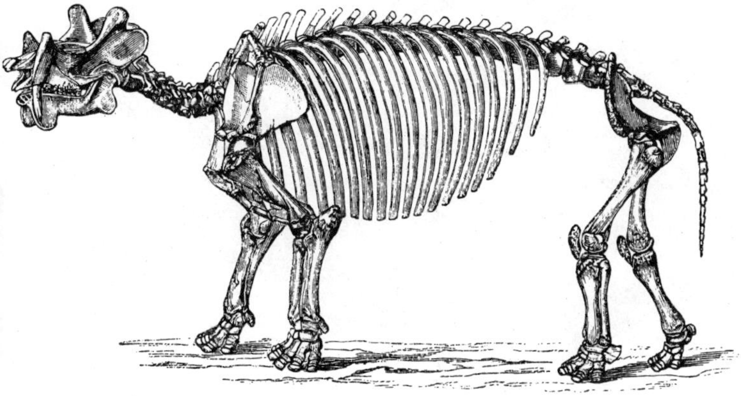

FIG. 190.—Skeleton of *Uintatherium mirabile.* $\frac{1}{30}$ natural size. (From Marsh, *Am. Journ. Sci.* vol. xii. pl. 2.)

known mammal. In its comparative length and the absence of a third trochanter the femur of these animals resembles that of the Proboscidea. The first discovered evidences of the existence of animals of this group were described by Leidy in 1872, under the name of *Uintatherium* (from the Uinta mountains, near which they were found). Subsequently the names *Dinoceras*, *Tinoceras*, *Loxolophodon*, etc., have been applied to various members of the group, but the characters by which they are distinguished do not seem of sufficient importance to allow of their separation from the type genus *Uintatherium*.[1]

Coryphodon.[2]—Another interesting form referred to this suborder is *Coryphodon*, which appears to connect the *Uintatheriidæ* with the most primitive Perissodactyla. It was first described by Owen in

[1] For detailed descriptions and figures of this group, see Marsh, "Monograph of the Dinocerata," *Rep. U.S. Geol. Surv.* vol. x. (1884).

[2] Owen, *Brit. Foss. Mamm. and Birds*, p. 299 (1846).

1846 from a fragment of a jaw from the London Clay. Other remains were afterwards discovered in France, and lately in great abundance, indicating many species from the size of a Tapir to that of a Rhinoceros, in the Lower and Middle Eocenes of New Mexico and Wyoming in the United States. *Coryphodon* had forty-four teeth; the canines of both jaws were large and sharp pointed, and the molars had strongly pronounced oblique ridges. The general proportions were those of a Bear, but the tail was of moderate length, and the feet short and wide. The femur had a third trochanter; and the cranium was devoid of protuberances.

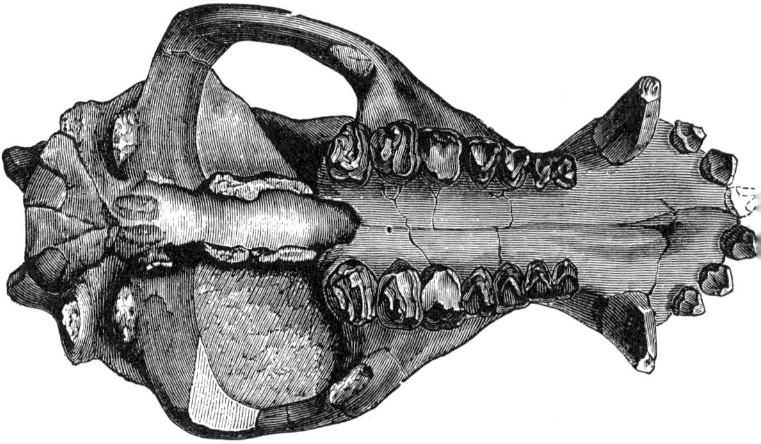

FIG. 191.—Palatal aspect of the cranium of *Coryphodon hamatus*, from the Wasatch Eocene of New Mexico. ⅜ natural size. (After Cope.)

The genus should be regarded as the type of a distinct family *Coryphodontidæ*.

Suborder CONDYLARTHRA.

The term Condylarthra has been proposed by Professor Cope for a number of generalised and mostly comparatively small Ungulates, which were probably allied both to the Perissodactyla and Artiodactyla, but present characters separating them from those divisions as commonly defined. In the structure of the carpus and tarsus these forms (which are chiefly known to us from the Eocene of the United States) come nearer to the Hyracoidea than to any other existing type. As a rule they have the full dental formula; the molars are brachydont, generally bunodont, and in many instances also tritubercular; while the premolars are always simpler than the molars.

The humerus is quite peculiar among Ungulates in having an

entepicondylar foramen; the femur has a third trochanter; and the form and relations of the astragalus are similar to those obtaining in the Carnivora. The feet are usually furnished with five functional digits, of which the ungual phalanges are pointed. In many respects the skeleton of these remarkably generalised Ungulates approximates so decidedly to a Carnivorous type as to have led palæontologists to conclude that the Ungulata and Carnivora are branches of an original common stock.

In this work space only permits of allusion to a few of the more important types of this group. *Periptychus*, which occurs in the lowest Eocene of New Mexico, is a bunodont type readily distinguished by the vertical flutings of the premolars, and the small size of the incisors and canines. It has been suggested that this genus is closely related to the stock of the bunodont Artiodactyla. Of greater interest is the genus *Phenacodus*, which is regarded as the lowest factor in the series from which the modern Horse has been evolved, where it holds the position immediately below *Hyracotherium* or *Systemodon* (see p. 374). One of the species was about the size of a Bull-dog, while another might be compared to a small Leopard. The structure of the cheek-teeth is such as might readily be modified into that obtaining in *Hyracotherium;* all the feet had five fully developed digits, and the tail was long. *Meniscotherium* and *Hyracodontotherium* are more specialised forms of somewhat later age, with a lophodont dentition; the latter genus being European.

Suborder TOXODONTIA.

In addition to the *Macraucheniidæ* and certain other forms noticed under the head of the Perissodactyla, the Tertiaries of South America have yielded some very remarkable forms of mammalian life, the nature and affinities of which have greatly puzzled all zoologists who have attempted to unravel them.

Nesodon and *Toxodon.*—Among these *Nesodon*, from Patagonia, has the full typical Eutherian number of teeth; the crowns of the incisors being short, and the molars having a complex rhinocerotic type of structure somewhat intermediate between *Homalodontotherium* (p. 412) and the following genus *Toxodon*. The typical species of *Nesodon* was about as large as a Sheep, but nothing more is known of it than the teeth and portions of the skull.

Toxodon is an animal about the size of a Hippopotamus; it was first discovered by Darwin, and many specimens have since been found in Pleistocene deposits near Buenos Ayres, and described by Owen, Gervais, and Burmeister. The teeth consist of large incisors, very small lower canines, and strongly curved molars, all with persistent roots, the formula being apparently $i\,\frac{2}{3}$, $c\,\frac{0}{1}$, $p\,\frac{4}{3}$, $m\,\frac{3}{3} = 38$.

The cranial characters exhibit a combination of those found in both Perissodactyles and Artiodactyles, but the form of the hinder part of the palate and the absence of an alisphenoid canal belong to the latter; and the tympanic, firmly fixed in between the squamosal and the exoccipital, ankylosed to both, and forming the floor of a long upward-directed meatus auditorius, is so exactly like that of the Suina that it is difficult to believe it does not indicate some real affinity to that group. These characters seem to outweigh in importance those by which some zoologists have linked *Toxodon* to the Perissodactyla, and the absence of the third trochanter and the articulation of the fibula with the calcaneum tell in the same direction. According to the recent observations of Ameghino the hind feet were certainly tridactylous, and the front feet probably so. The earlier allied genera *Protoxodon* and *Adinotherium* are definitely known to have tridactylous front and hind feet, which conform to the Perissodactylate type, the bones of the proximal and distal rows of the carpus interlocking. *Acrotherium*, which has similar feet, differs from all other Ungulates, and indeed from all Eutherians except some individuals of the existing carnivorous genus *Otocyon*, in having eight cheek-teeth, five of which have been reckoned as premolars.

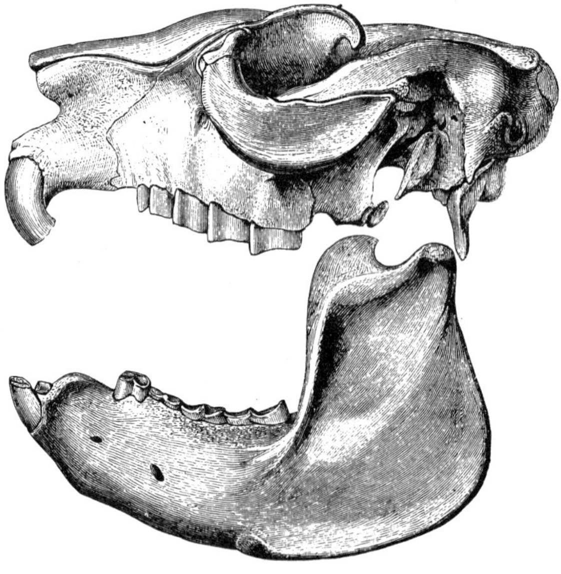

FIG. 192.—Cranium and Lower Jaw of *Typotherium cristatum*. $\frac{1}{3}$ natural size. From Gervais

Typotherium.—*Typotherium* (Fig. 192), also called *Mesotherium* from the same locality as *Toxodon*, was an animal rather larger tha a Capybara, and of much the same general appearance. Its skeleton is completely known, and shows a singular combination of characters, resembling *Toxodon* or a generalised Ungulate on the one hand, and the Rodents, especially the *Leporidæ*, on the other. In the presence of clavicles it differs from all known Ungulates, and in having two pairs of lower incisors from all Rodents. The teeth are $i\ \frac{1}{2}$, $c\ \frac{0}{0}$, $p\ \frac{2}{1}$, $m\ \frac{3}{3} = 24$.

From the Tertiaries of various parts of South America a number of forms more or less closely allied to *Toxodon* and *Typotherium* have been recently described, but as many of them are very imperfectly known, and there is much doubt as to their generic position, it will be unnecessary to refer to them further.

It will thus be seen that, although our knowledge of many of these forms is still very limited, we may trace among them a curious chain of affinities, which would seem to unite the Ungulates on the one hand with the Rodents on the other; but further materials are required before we can establish with certainty so important a relationship, one which, if true, would alter materially some of the prevailing views upon the classification of mammals.

Group TILLODONTIA.

Here may be noticed a remarkable group of animals, called by Marsh, Tillodontia, the remains of which are found abundantly in

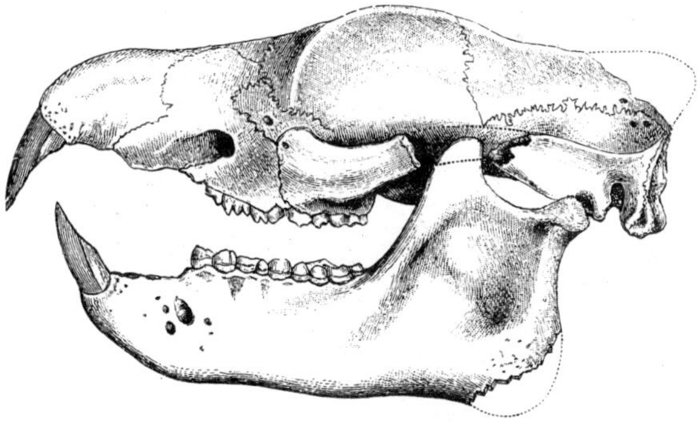

FIG. 193.—Skull of *Tillotherium fodiens*. $\frac{1}{3}$ natural size. From Marsh.

the Lower and Middle Eocene beds of North America. They seem to combine the characters of the Ungulata, Rodentia, and Carnivora. In the genus *Tillotherium* of Marsh (probably identical with the previously described *Anchippodus* of Leidy) the skull (Fig. 193) resembled that of the Bears, but the molar teeth were of the Ungulate type, while the large incisors were very similar to those of the Rodents. The dental formula is $i\ \frac{2}{2}$, $c\ \frac{1}{1}$, $p\ \frac{3}{4}$, $m\ \frac{3}{3}$. The first pair of incisors was very small; the upper molars were tritubercular, while the lower ones had crescentoid ridges as in *Palæotherium*. The skeleton resembled that of the Carnivores, but the scaphoid and lunar bones were distinct, and there was a third trochanter on the femur. The feet were plantigrade, and each had five digits, all with long pointed claws. In the allied genus *Stylinodon* all the teeth were rootless. Some forms were as large as a Tapir.

These, with other more or less closely allied animals, such as *Calamodon* and *Psittacotherium*, constituting a group called Tæniodonta, are included by Cope in his large order Bunotheria, to which also the existing Insectivora are referred. The dentition of some of these forms makes a remarkable approximation towards a Rodent type, while it has been suggested that there are also signs of remote Edentate affinities. The constantly increasing knowledge of these annectant forms adds to the difficulty so often referred to in this work of establishing anything like a definite classification of the heterodont mammals. An incisor tooth from the Swiss Eocene has recently been referred to *Calamodon*.

Bibliography of Ungulata.—In addition to the works and memoirs mentioned under the different sections of the order, the following may be referred to:—W. Kowalevsky, "Monographie des genus Anthracotherium," *Palæontographica* 1873; Id. "Sur l'Anchitherium aurelianense et sur l'histoire paléontologique des Chevaux," *Mém. de l'Acad. Imp. des Sciences de St. Pétersbourg*, 1873; Id. "On the Osteology of the Hyopotamidæ," *Philosophical Transactions*, 1873; L. Rütimeyer, "Versuch einer natürlichen Geschichte des Rindes," etc., *Neue Denks. der allgem. Schweiz. Gesellsch. für Naturwissenschaften*, 1867; Id. "Die Rinder der Tertiär-Epoche," *Abhand. der Schweiz. Paläont. Gesellsch.* 1877 and 1878; Id. "Beiträge zu einer Natürliche Geschichte der Hirsche," *ibid.* 1880-1881; C. J. Forsyth-Major, "Beiträge zur Geschichte der Fossilen Pferde," *ibid.* 1880; M. Schlosser, "Beiträge zur Kenntniss der Stammesgeschichte der Hufthiere und Versuch einer Systematik der Paar-und Unpaarhufer," *Morph. Jahrb.* 1886; E. D. Cope, "The Perissodactyla," *Amer. Natural.* 1887; M. Pavlow, "Études sur l'histoire paléontologique des Ongulés," *Bull. Soc. Imp. Naturalistes Moscow*, 1887-1890. W. B. Scott and H. F. Osborn, "The Mammalia of the Uinta Formation," *Trans. Amer. Phil. Soc.* vol. xvi. (1889).

CHAPTER X

THE ORDER RODENTIA

THE Rodentia, or Rodents, form a well-defined order, readily distinguished by their large scalpriform incisors and the absence of any trace of canines. The existing forms are mostly of comparatively small size, and are generally of terrestrial habits, although a few are arboreal or natatorial. The dentition is diphyodont; the mandible never has more than a single pair of incisors; the premolars are always below the full number, being very generally $\frac{1}{1}$, or altogether wanting. The feet are plantigrade or semi-plantigrade, generally with five digits, and usually unguiculate, although occasionally of a subungulate type. Clavicles are present as a rule, although they may be imperfect or rudimentary.

The upper incisors resemble the lower in growing uninterruptedly from persistent pulps, and, except in the suborder Duplicidentata, agree with them in number; the premolars and molars may be rooted or rootless, with tuberculated or laminated crowns, and are arranged in an unbroken series. The orbits communicate freely with the temporal fossæ; the condyle of the mandible is elongated in the antero-posterior direction, and, through the absence of a postglenoid process to the squamosal, admits of a backward and forward motion of the jaw. The intestine (except in the *Myoxidæ*) has a large cæcum; the testes are inguinal or abdominal; the uterus is two-horned, the cornua either opening separately into the vagina or uniting to form a corpus uteri; the placenta is discoidal and deciduate; and the smooth cerebral hemispheres do not extend backwards so as to cover any part of the cerebellum.

The Rodents include by far the greatest number of species, and have the widest distribution of any of the orders of terrestrial mammals, being in fact cosmopolitan, although more abundant in some parts than in others. The total number of known existing species exceeds 900. South America may be regarded as their headquarters at the present day; while in Australia and Madagascar they are represented only by a few genera. All the Rodents are exclusively herbivorous, and the whole of them gather their food by gnawing. They present considerable diversity of habits. Thus the Squirrels are arboreal, and some of them provided with a parachute, for taking flying leaps from tree to tree; the Hares are cursorial; the Jerboas agile jumpers; the Mole-Rats fossorial; while the Beavers and Water-Voles are aquatic. In spite, however, of this diversity of habits the Rodents present a remarkable similarity in general structure; so much so, indeed, that the characters employed for distinguishing the various families and genera are comparatively trivial, and of slight structural importance. The skull of the Rodents is characterised by the invariable presence of the zygomatic arch, of which the middle portion is formed by the jugal (Fig. 7, p. 37); and, as already mentioned, the orbit communicates freely with the temporal fossa. There is invariably a long diastema separating the incisors from the cheek-teeth; and, with the exception of the Duplicidentata, the glenoid cavity of the squamosal is elongated antero-posteriorly. Postorbital processes of the frontals exist only in the Squirrels, Marmots, and Hares; in all other genera they are rudimentary or altogether absent; the zygoma never sends upwards a corresponding process; the lachrymal foramen is always within the orbital margin; in many species the infraorbital foramen is very large (in some as large as the orbit), and transmits part of the great masseter muscle (Fig. 194, *m*), by means of which the jaws are worked. The zygomatic arch varies in its degree of development, and the position of the jugal therein is used as a distinguishing character for grouping the families; the nasals are, with few exceptions, large, and extend far forwards; the parietals are moderate, and there is generally a distinct interparietal. The palate is narrow from before backwards—this being especially pronounced in the Hares, where it is reduced to a mere bridge between the premolars; while in other cases, as in the Mole-Rats (*Bathyergina*), it is extremely narrow transversely, its width being less than that of one of the molar teeth. Auditory bullæ are always present, and generally large; in some genera, as in the Gerbilles and Jerboas, there are also supplemental

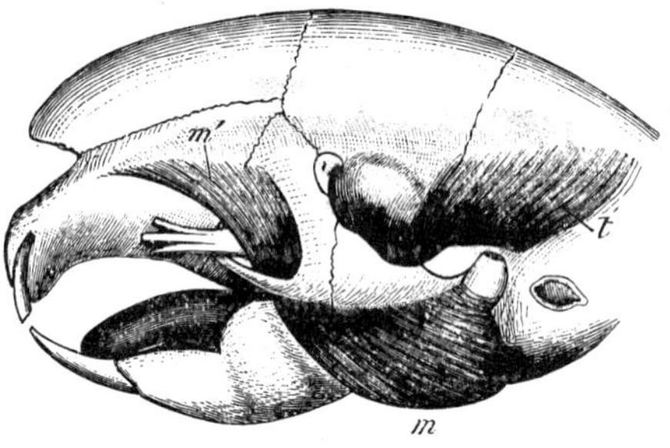

FIG. 194.—Skull of *Hystrix cristata* (juv.) *t*, Temporal muscle; *m*, masseter; *m'*, portion of masseter transmitted through the infraorbital foramen, the superior maxillary nerve passing outwards between it and the maxillary bone.

mastoid bullæ forming great hemispherical bony swellings at the back of the skull (see Fig. 7, *Per*); and in these genera, and in the true Hares, the meatus auditorius is tubular and directed upwards and backwards. The mandible is characterised by the abruptly narrowed and rounded symphysial part supporting the large incisors, as well as by the small size of the coronoid process and the great development of the angular portion.

The dental formula varies from $i\ \frac{2}{1}$, $c\ \frac{0}{0}$, $p\ \frac{3}{2}$, $m\ \frac{3}{3}$ (total 28) in the Duplicidentata to $i\ \frac{1}{1}$, $c\ \frac{0}{0}$, $p\ \frac{0}{0}$, $m\ \frac{2}{2}$ (total 12) in *Hydromys*, *Xeromys*, and one species of *Heterocephalus*; but in the great majority of forms it is very constant, $i\ \frac{1}{1}$, $c\ \frac{0}{0}$, $p\ \frac{(0-1)}{(0-1)}$, $m\ \frac{3}{3}$ being very typical. Only in the Duplicidentata is there a second pair of upper incisors, which are of very small size, and situated immediately behind the large normal pair. This group is also peculiar in that the enamel of the incisors is not confined to their anterior surfaces, but extends partially on to their sides. It is by reason of the thick layer of enamel on their anterior surface and its absence from the posterior surface that the incisors maintain their sharp chisel-like edge, which is so essentially characteristic of the order. Both the upper and the lower incisors are regularly curved—the curvature being somewhat greater in the upper ones—and since they grew continuously from persistent pulps, it is quite evident that should any accident, such as the loss of one of them, or displacement by fracture of the jaw, prevent the regulation of the length by attrition against one another, the unopposed tooth will gradually curve upon itself until a complete circle or more has been formed, the tooth, perhaps, passing during its growth through some part of the animal's head. The molars, as already mentioned, may be rooted or rootless, tuberculated or laminated; this diversity of structure occurring even in the same family. When there are more than three cheek-teeth those in front of the last three have succeeded milk-teeth, and must therefore be considered premolars. In some species, as in the Agoutis (*Dasyproctidæ*), the milk-teeth are long retained, while in the allied Cavies (*Caviidæ*) they are shed before birth.

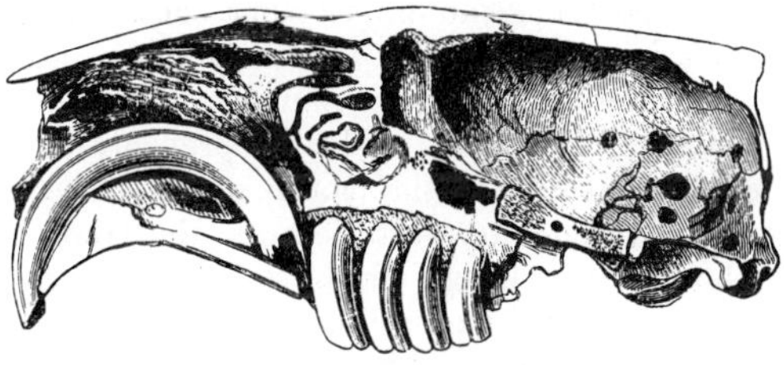

FIG. 195.—Vertical and longitudinal section through skull of the Beaver (*Castor fiber*) showing the cerebral cavity, the greatly developed turbinal lamellæ, the mode of implantation of the large incisor, and the curved rootless molars.

There are generally nineteen dorso-lumbar vertebræ (thirteen dorsal and six lumbar), their form varying in the different genera.

In the cursorial and leaping species the lumbar transverse processes are generally very long, and in the Hares there are large compressed hypapophyses. The caudal vertebræ exhibit great variety in structure, being in a rudimentary condition in the Guinea-Pig, while in the Jumping Hares and prehensile-tailed Porcupines they are of very large dimensions. The scapula is usually narrow, with a long acromion; the clavicles may be altogether absent or imperfect, as in the Porcupines, Cavies, and Hares, but in most species they are well developed. In all existing forms the humerus has no entepicondylar foramen, and the radius and ulna are distinct. In most species the manus has five digits, with phalanges normally developed; the pollex being rarely rudimentary or absent. The pelvis has well-developed ischia and pubes, meeting in a long, and usually bony, symphysis. The femur varies considerably in form, but generally has a well-defined third trochanter; in the Sciurine and Hystricine Rodents the tibia and fibula are distinct, but in the Rats and other Murines, and in the Hares, these bones are united, often high up; the pes is much more variable than the manus, the digits varying in number from five, as in the Squirrels and Rats, to four, as in the Hares, or even three, as in the Capybara, Viscacha, and Agouti; in the *Dipodidæ* the metatarsals are greatly elongated, and in some of the species, as in the Jerboas, they are ankylosed together.

The mouth is divided into two cavities communicating by a constricted orifice, an anterior one containing the large incisors, and a posterior one in which the molars are placed; the hairy integument of the face being continued inwards behind the incisors. This peculiar arrangement evidently prevents substances not intended for food getting into the mouth, as when the animal is engaged in gnawing through an obstacle. In the Hares and Pacas the inside of the cheeks is hairy, and in some species, as in the Pouched Rats and Hamsters, there are large internal cheek-pouches lined with the hairy integument, which open near the angles of the mouth and extend backwards behind the ears. In the New World Pouched Rats (*Geomyidæ*) the pouches open externally on the cheeks. The tongue presents little variability in length, being always short and compressed, with an obtuse apex never protruded beyond the incisors. In most species there are three circumvallate papillæ at the base; and the apical portion is generally covered with small filiform papillæ, some of which in the Porcupines (*Hystrix*) become greatly enlarged, forming toothed spines. The stomach varies in form from the simple oval sac of the Squirrel to the complex ruminant-like organ of the Lemming. In the Water-Vole (*Arvicola amphibius*) and the Agouti (*Dasyprocta aguti*) it is strongly constricted between the œsophagus and pylorus. In the common Dormouse the œsophagus immediately before entering the stomach is much dilated, forming a large egg-shaped sac with thickened glandular walls; and in some other species, as in *Lophiomys imhausi* and in the Beaver, glandular masses are attached to and open into the cardiac or pyloric pouches. The alimentary canal (Fig. 196) of all Rodents, with the exception of the Dormice (*Myoxidæ*), has a cæcum, which is often of great length and sacculated, as in the Hares, Water-Voles, and Porcupines. In some instances, as in the Hamster and Water-Vole, the long colon is spirally twisted upon itself near its commencement. The liver is typically divided in all, but the lobes are variously subdivided in the different species (in *Capromys* they are divided into minute lobules); and the gall-bladder, though present in most, is absent in a few. In most species the penis (which is generally provided with a bone) can be more or less completely retracted within the fold of integument surrounding the anus, where it lies curved backwards upon itself under cover of the integument. It may, however, be carried forward some distance in front of the anal orifice, from which in the breeding season, as in the Voles and Marmots, the prominent testicular mass separates it. The testes in the rutting season form projections in the groins, but (except in the Duplicidentata) do not completely leave the cavity of the abdomen. Prostatic glands and, except in the Duplicidentata, vesiculæ seminales are present in all. The uterus may be double, each division opening by a separate aperture into a common vagina, as in *Leporidæ*, *Sciuridæ*, and *Hydrochœrus*, or completely two-horned, as in most species. The mammæ vary in number and position from the single abdominal pair of the Guinea-Pig to the ten thoracico-abdominal pairs found in some of the Rats. In the *Octodontidæ* the mammæ are placed high up on the sides of the body.

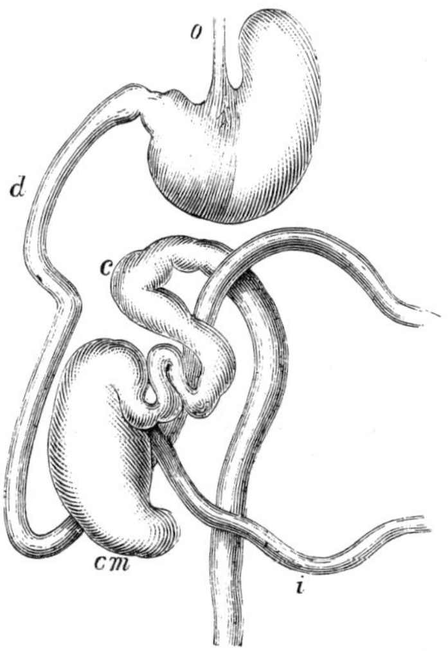

FIG. 196.—Alimentary canal of Rat (*Mus decumanus*), the greater part of the small intestine being omitted. *o*, Œsophagus; *d*, duodenum; *i*, ileum; *cm*, cæcum; *c*, colon.

The peculiar odour evolved by many Rodents is due to the secretions of special glands, which may open either into the prepuce, as in *Mus*, *Arvicola*, *Cricetus*, etc., or into the rectum, as in *Arctomys* and *Aulacodus*, or into the passage common to both, as in the Beaver, or again, into pouches opening near the anus, as in the Hare, Agouti, and Jerboa.

The integument is generally thin, and the panniculus carnosus (the sheet of muscle underlying the skin) rarely much developed. The fur varies exceedingly in character. Thus it may be very fine and soft, as in the Chinchillas and Hares, in others more or less replaced by spines on the upper surface, as in the Spiny-Rats and Porcupines; in several genera, as in *Xerus*, *Acanthomys*, *Platacanthomys*, *Echinothrix*, *Loncheres*, and *Echinomys*, the spines are flattened. In the muscular structures the chief peculiarities are noticeable in the comparatively small size of the temporal muscles, and in the great double masseters (Fig. 194), which are the principal agents in gnawing; the digastrics also are remarkable for their well-defined central tendon, and in many species their anterior bellies are united between the mandibular rami; the cleidomastoid generally arises from the basioccipital, and the pectoralis major is connected with the latissimus dorsi; in the Porcupines and Hares the tendons of the flexor digitorum longus and flexor hallucis longus are connected in the foot, while in the Rats and Squirrels they are separate, and the flexor digitorum longus is generally inserted into the metatarsal of the hallux.[1]

Rodents are tolerably well represented in a fossil condition from the period of the Upper Eocene, while if *Decticadapis*, of the Lower Eocene of Rheims, is rightly referred to it the order dates from the oldest Tertiary. All the fossil forms at present known are, however, essentially true Rodents, and afford no clue as to the relations of the order with other mammals. The remote affinities of the Rodents to the Proboscidea, as well as their more marked resemblances to *Typotherium*, have been already mentioned. Whether there is a real genetic affinity (as Professor Cope suggests) with the Tillodontia cannot be decided with the evidence at present available.

Suborder SIMPLICIDENTATA.

Only one pair of upper incisors, having their enamel confined to their front surfaces. Incisive foramina moderate and distinct; fibula not articulating with the calcaneum. Testes abdominal, and descending periodically only into a temporary sessile scrotum.

Section SCIUROMORPHA.

Zygomatic arch slender, chiefly formed by the jugal, which is not supported by a long maxillary process extending backwards beneath it; postorbital processes of frontal present or absent;

[1] See G. E. Dobson, *Journ. Anat. Phys.* vol. xvii.

infraorbital opening small (except in *Anomalurus*); mandible with the angular part arising from the inferior surface of the bony socket of the lower incisor; clavicles well developed; fibula distinct.

Family ANOMALURIDÆ.

Arboreal forms, having their limbs connected by a cutaneous expansion supported by a cartilaginous process arising from the olecranon; tail long and hairy, with large imbricated scales on its

FIG. 197.—*Anomalurus fulgens.* From Alston, *Proc. Zool. Soc.* 1875.

inferior surface near the root; sixteen pairs of ribs; no postorbital processes on the frontals; $p\ \frac{1}{1}$; molars not tuberculate, with transverse enamel-folds. Confined to the Ethiopian region.

Anomalurus,[1] with several species from West and Central Africa, alone represents the family. The peculiar caudal scales, which evidently assist the animal in climbing, and the position of the cartilaginous support of the parachute, are well shown in Fig. 197. All the species but two are from Western Africa; *A. orientalis* occurs near Zanzibar, and *A. pusillus* is from the equatorial regions of that

[1] Waterhouse, *Proc. Zool. Soc.* 1842, p. 124.

continent. According to Mr. O. Thomas,[1] the latter "little animal is most nearly allied to the West-African *A. beecrofti*, but differs from that species in its duller and less yellow upper side, in the entire absence of rufous on its neck and belly, and, as from all the other described species, in its diminutive size."

Family SCIURIDÆ.

Arboreal or terrestrial forms, with cylindrical hairy tails, without scales, and with twelve or thirteen pairs of ribs. Skull (Figs. 198, 199) with distinct postorbital processes; infra-orbital opening small; palate broad; $p\,\frac{2}{1}$; first upper premolar very small or deciduous; molars rooted, tubercular.

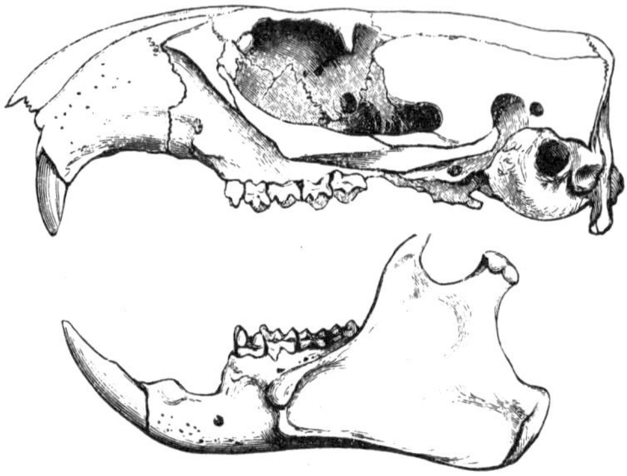

FIG. 198.—Lateral view of skull of American Marmot (*Arctomys monax*).

Subfamily **Sciurinæ.**—Incisors compressed; form slender; tail long and hairy. Cosmopolitan (excluding Australian region).

This subfamily includes the true Squirrels, of which seven existing genera are usually recognised.

Sciurus.[2]—Tail long and bushy; ears generally well developed, pointed, often tufted; feet adapted for climbing, the anterior having four digits and a rudimentary pollex, and the posterior with five digits, all of which have long, curved, and sharp claws. Mammæ, from four to six. Skull (Fig. 199) lightly built, with long postorbital processes. Penultimate upper premolar, when present, minute.

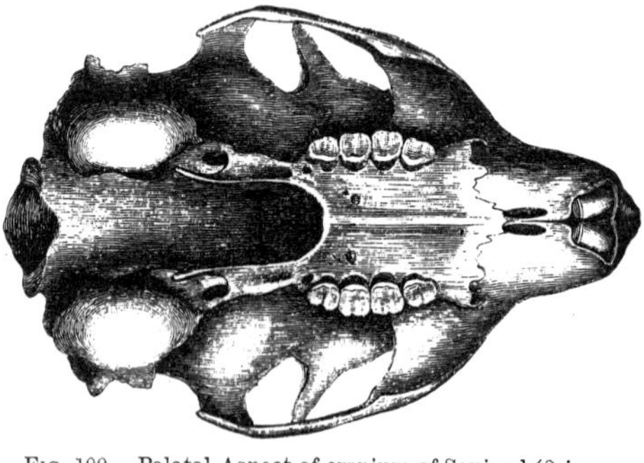

FIG. 199.—Palatal Aspect of cranium of Squirrel (*Sciurus bicolor*). Natural size.

[1] *Proc. Zool. Soc.* 1882, p. 8.

[2] Linn. *Syst. Nat.* 12th ed. vol. i. p. 86 (1766).

True Squirrels are found in most of the temperate and tropical regions of the world, exclusive of Madagascar and the Australian region. They are, however, most abundant in the Malayan part of the Oriental region, and attain their largest size and most brilliant coloration in the tropics. Their size is very variable, so that whereas *S. soricinus*, of Borneo, is no larger than a Mouse, *S. bicolor*, of the Malayan region, is nearly as large as a Cat. The common

FIG. 200.—Burmese Squirrel (*Sciurus pygerythrus*). After Anderson.

English Squirrel (*S. vulgaris*) is found over the whole of the Palæarctic region, reaching in one direction from Ireland to Japan, and in the other from the north of Italy to Lapland; its remains occur in the Norfolk "Forest-bed." In the Malayan region "nearly all the numerous species are brilliantly marked, and many are ornamented with variously coloured longitudinal stripes along their bodies. One of the commonest and best known of the striped species is the little Indian Palm-Squirrel (*S. palmarum*), which in large numbers runs about every Indian village. Another Oriental species (*S. caniceps*) presents almost the only known instance among mammals of the temporary assumption during the breeding season of a distinctly ornamental coat, corresponding to the breeding-plumage of birds. For the greater part of the year the animal is of a uniform gray colour; but about December its back becomes a brilliant orange-yellow, which lasts until about March, when it is again replaced by gray. The Squirrel shown in Fig. 200 is a native of Burma and Tenasserim, and is closely allied to *S. caniceps*, but goes through no seasonal change of colour.

"The number of species in the genus is about 75, of which 3 belong to the Palæarctic, 15 to the Ethiopian, about 40 to the Oriental, and 16 to the combined Nearctic and Neotropical regions" (Thomas).

Fossil species referred to *Sciurus* are found in the European Tertiaries down to the Phosphorites of Central France, while others occur in the White River Miocene of the United States.

Rhithrosciurus.[1]—A very striking Squirrel, confined to Borneo, and as yet only known from three or four examples, has been separated generically under this name. The general shape of its skull is very different from that of other Squirrels; but its most peculiar characteristic is the presence of from seven to ten minute parallel vertical grooves running down the front face of its incisors; no other Squirrel having really grooved incisors, and no other member of the whole order incisive grooves resembling these. Its premolars number $\frac{1}{1}$, and its molars are simpler and less ridged than in the other genera. This Squirrel (*R. macrotis*) is far larger than the English, with an enormously long bushy tail, long tufted ears, and black and white bands down its sides.

Xerus.[2]—Fur coarse and spiny. Claws long and comparatively straight. Ear-conchs minute or absent. Skull with the postorbital processes short and directed backwards, the bony palate prolonged considerably behind the tooth-row, and the external ridge on the front face of the anterior zygomatic root more developed, and continued much farther upwards than in *Sciurus*. Premolars $\frac{2}{1}$; molars as in *Sciurus*. Mammæ two. This genus contains four well-marked species, known as Spiny Squirrels, all natives of Africa. They are terrestrial in their habits, living in burrows which they dig for themselves. *X. getulus*, a striped species of North Africa, has much the size and appearance of the Indian Palm-Squirrel; all the others are a little larger than the English Squirrel.

Tamias.[3]—All the members of this genus are characterised by the possession of internal cheek-pouches, and by their style of coloration; being ornamented on the back with alternate light and dark bands. Their skulls are slenderer and lighter than those of the

[1] Gray, *Ann. Mag. Nat. Hist.* ser. 3, vol. xx. p. 272 (1867).
[2] Hemprich and Ehrenberg, *Symbol. Phys. Mamm.* vol. i. (1832).
[3] Illiger, *Prodromus Syst. Mamm.* p. 83 (1811).

true Squirrels, from which they differ in several unimportant details. There is only one functional premolar—the small anterior one usually found in *Sciurus* being either absent altogether or quite small and functionless. There are some four well-defined species, all found in North America, one (*T. asiaticus*) extending also through Siberia into Eastern Europe.[1] They are generally known as Ground-Squirrels, but in America, where they are among the commonest and best known of the indigenous Rodents, as "Chipmunks." The members of this genus seem to lead into the genus *Spermophilus*, so that the division of the *Sciuridæ* into two subfamilies, although convenient for classification, is rather artificial.

Remains of *Tamias*, probably belonging to existing species, occur in the Pleistocene deposits of Europe and Nebraska.

Pteromys[2] and *Sciuropterus*.[3]—The Flying Squirrels, although incapable of true flight, can yet float through the air for considerable distances by the aid of an extension of skin connecting their fore and hind limbs, and forming a sort of parachute. This parachute is merely a lateral extension of the ordinary skin of the body, which passes outwards between the limbs and terminates at the wrists and ankles. In addition to the lateral membrane there is a narrow and inconspicuous one passing from the cheek along the front of the shoulder to the front of the wrist, and another—at least in the larger species—stretching across behind the body from ankle to ankle and involving the base of the tail. The Flying Squirrels are divided into three genera. Of those with a normal dentition *Pteromys* contains the larger and *Sciuropterus* the smaller species. The two differ in certain details of dentition, as well as in the greater development in the former of the expanded membranes, especially of the "interfemoral" or posterior membrane, which in the latter is almost wholly absent. In *Pteromys* the tail is cylindrical and comparatively thin, while in *Sciuropterus* it is broad, flat, and laterally expanded, and evidently compensates for the absence of the interfemoral membrane by acting as a supplementary parachute. In appearance Flying Squirrels resemble the other forms, although they are even more beautifully coloured. Their habits, food, etc., are also very similar to those of the true Squirrels, except that they are more decidedly nocturnal, and are therefore less often seen by the traveller; their peculiar shrill cry is, however, well known to all who have camped out in the regions which they inhabit. Their mode of flight is precisely similar to that of the Flying Phalangers of Australia. Of each of the two genera there are about thirteen or fourteen species, all

[1] Some American zoologists have recently proposed to raise a large number of the forms usually regarded as local races to the rank of species.

[2] Cuvier, *Leçons d'Anatomie Comp.* (1800).

[3] Cuvier, *Ann. du Muséum*, vol. x. p. 126 (1825).

natives of the Oriental region, except that one of *Sciuropterus* is found in North America, and another in Siberia and Eastern Europe.

Eupetaurus.[1]—Externally as in *Pteromys*, except that the claws are less sharp. Skull with a more produced muzzle than in the latter, more distinct supraorbital notches, longer anterior palatal foramina, and a shorter bony palate. Cheek-teeth differing from those of all other *Sciuridæ* in their hypsodont character. One large species (*E. cinereus*), from Gilgit and adjacent districts on the extreme north-west of Kashmir territory. This fine Flying Squirrel is chiefly known by one entire specimen and some imperfect skins.

Extinct Genera.—The genera *Pseudosciurus* and *Sciuroides*, from the Upper Eocene of Europe, have the molar teeth more elongated than in *Sciurus*. *Gymnoptychus* with $p\ \frac{1}{1}$, from the North American Miocene, approximates in the structure of its molars to *Tamias*. *Meniscomys* ($p\ \frac{2}{1}$), from the latter deposits, together with *Sciurodon* of the French Phosphorites, are regarded as Squirrels showing signs of affinity with the *Haplodontidæ*.

Subfamily **Arctomyinæ.**—Incisors not compressed; typically the form stout, and the tail comparatively short. This subfamily comprises burrowing forms which may be collectively known as Marmots; as already mentioned, they are so intimately connected with the preceding subfamily that the division into two groups is purely a matter of convenience. They are confined to the Palæarctic and Nearctic regions.

Arctomys.[2]—External form stout and heavy, ears short, tail short and hairy, cheek-pouches rudimentary or absent. Fore feet with four well-developed digits, and a rudimentary pollex provided with a flat nail. Skull (Fig. 198) large and heavy, with the post-orbital process stout, and at right angles to the axis. Incisors broad and powerful. First upper premolar nearly as large as the second. Molar series nearly parallel, scarcely converging behind at all.

The various species of true Marmot, which exceed a dozen in number, are all much alike in general appearance, ranging in size from about 15 to 25 inches in length, with tails from 3 to 12 inches long.

The Alpine Marmot (Fig. 201) is peculiar to Europe, being found in the Alps, Pyrenees, and Carpathians; its remains occur in European Pleistocene deposits. *A. bobac* occurs in Eastern Europe and Siberia. Several species (*e.g. A. monax*, Fig. 198) are found in the Nearctic region, and many in Kashmir and Central Asia. The long-tailed Red Marmot (*A. caudatus*) is a fine Himalayan species, which may be seen on the mountain passes to the north of the valley of Kashmir, as soon as the snow begins to disappear,

[1] O. Thomas, *Journ. As. Soc. Bengal*, vol. lvii. p. 256 (1888).

[2] Schreber, *Säugethiere*, vol. iv. p. 721 (1792).

sitting at the entrance to its burrow, which is generally beneath a rhubarb plant.

The following account of the habits of the Alpine Marmot is given by Professor Blasius: "Marmots live high up in the snowy regions of the mountains, generally preferring exposed cliffs, whence they may have a clear view of any approaching danger, for which, while quietly basking in the sun or actively running about in search of food, a constant watch is kept. When one of them raises the cry of warning, the loud piercing whistle so well known to travellers

FIG. 201.—Alpine Marmot (*Arctomys marmotta*). After Brehm.

in the Alps, they all instantly take to flight and hide themselves in holes and crannies among the rocks, often not reappearing at the entrance of their hiding-places until several hours have elapsed, and then frequently standing motionless on the look-out for a still longer period. Their food consists of the roots and leaves of various Alpine plants, which, like squirrels, they lift to their mouths with their fore paws. For their winter quarters they make a large round burrow, with but one entrance, and ending in a sleeping-place thickly lined with hay. Here often from ten to fifteen Marmots pass the winter, all lying closely packed together fast asleep until the spring."

Cynomys.[1]—Size and form intermediate between *Arctomys* and *Spermophilus*. Ears and tail short. Cheek-pouches shallow. Fore

[1] Rafinesque, *Amer. Monthly Mag.* vol. ii. p. 45 (1817).

feet with five claws, that on the pollex as large as that on the fifth toe. Skull (Fig. 202) heavily built, with the postorbital processes directed outwards. Dentition (as shown in Fig. 202) remarkably heavy, the molar teeth differing from those of *Arctomys* and *Spermophilus* by having three instead of two transverse grooves on their crowns. First premolar nearly as large as the second. Molar series strongly convergent behind.

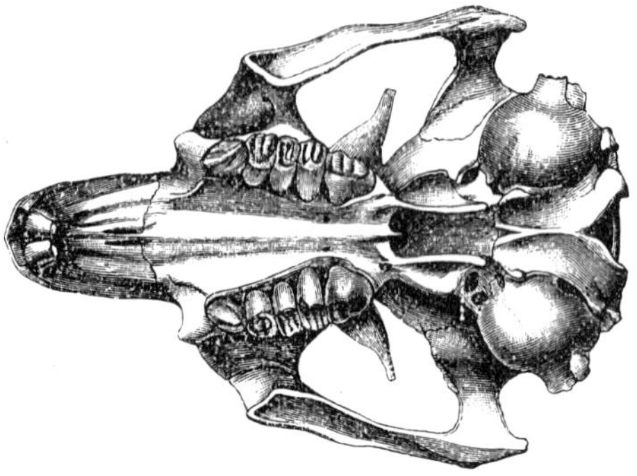

FIG. 202.—Palatal aspect of the cranium of the Prairie Marmot (*Cynomys ludovicianus*).

Two species of Prairie Marmots, or, as they are often called, "Prairie Dogs," are found in North America. They live together in large communities, inhabiting burrows excavated at short distances apart, and feeding on the buffalo-grass which covers the plains. The small burrowing owl (*Athene cunicularia*) and the rattlesnake are often found inhabiting their burrows; the former probably availing itself of the convenience of a ready-made habitation, the latter coming there to feed on the young Marmots.

Spermophilus.[1]—Size much smaller than in either of the preceding genera; form more slender and squirrel-like. Tail very variable, from 1 to 8 or 9 inches in length. Cheek-pouches always present. Fore feet with four well-developed toes and a rudimentary pollex, of which the claw may be either present or absent. Skull more lightly built than in the other preceding genera, with the postorbital processes slender and directed backwards. Molar series nearly parallel, as in *Arctomys*, but all these teeth much smaller and lighter; first premolar simply rounded, never more than about one-third of the size of the second.

The Pouched Marmots, or Sousliks, have nearly the same distribution as *Tamias*, and are represented by a considerable number of species. They present a far greater range of variation than is found among the true Marmots, some of them, such as the European species, being scarcely as large as a common squirrel, almost entirely without external ears, and with the tail reduced to a mere stump, barely an inch long, while others are more than three times this size, with large and often tufted ears, and long bushy squirrel-like tails. Professor Blasius gives the following details of the habits of the common European Souslik (*S. citillus*):

[1] F. Cuvier, *Mém. du Muséum*, vol. vi. p. 293 (1822).

"It lives in dry treeless plains, especially on a sandy or clayey soil, and is never found either in forests or on swampy ground. It forms burrows, often 6 or 8 feet deep, in which food is stored up and the winter sleep takes place. Each burrow has but one entrance, which is closed up when winter approaches,—a second hole, however, being previously formed from the sleeping-place to just below the surface of the ground. The second hole is opened the next year, and used as the ordinary entrance, so that the number of closed-up holes round a burrow gives an indication of the length of time that it has been occupied. Sousliks ordinarily feed on roots, seeds, berries, etc., but occasionally also on animal food, preying readily on eggs, small birds, and mice, the remains of these latter being often found in their burrows. They bring forth in the spring from four to eight young ones, which, if taken early, may be easily tamed. They are often eaten by the peasants, the inhabitants of the Russian steppes considering their flesh an especial delicacy."

Remains of *Spermophilus* are not uncommon in European Tertiary deposits, some belonging to living and others to extinct species.

Extinct Genera.—*Plesispermophilus*, from the Upper Eocene Phosphorites of Central France, appears to be closely allied to the Sousliks. *Plesiarctomys* (*Sciuravus* or *Paramys*), which is common to the Middle Tertiaries of Europe and North America, appears to be a generalised form, showing some resemblance both to *Arctomys* and *Sciurus*, but with tritubercular upper molars and no postorbital processes to the skull; in the latter respect agreeing with the next family. In the size of the preorbital vacuity the skull resembles the Hystricomorpha.

Family HAPLODONTIDÆ.

Distinguished from the *Sciuridæ* by the absence of postorbital processes to the frontals, the depressed skull, and the rootless cheek-teeth. Premolars $\frac{2}{1}$; the penultimate upper one small.

Haplodon.[1]—*H. rufus* and *H. major*, of North America, west of the Rocky Mountains, are the only representatives of the family; their habits are similar to those of *Cynomys*.

Family CASTORIDÆ.

Skull massive, without postorbital processes, the angle of the mandible rounded, and the cheek-teeth rootless, with re-entering enamel-folds. Premolars $\frac{1}{1}$. Habits natatorial.

Castor.[2]—The upper molars are subequal, each with one internal

[1] Richardson, *Zool. Journ.* vol. iv. p. 334 (1829). Amended.

[2] Linn. *Syst. Nat.* 12th ed. vol. i. p. 78 (1766).

and two external enamel-folds; the stomach has a large glandular mass situated to the right of the œsophageal orifice; the anal and urethro-genital orifices open within a common cloaca; the tail is broad, horizontally flattened, and naked; and the hind feet are webbed. One or two species, Palæarctic and Nearctic.

Zoologists are not yet of accord as to whether the European and American Beavers should be regarded as distinct species or as local races; the general concensus of opinion being in favour of the latter view.

The European Beaver (*C. fiber*) was at one time an inhabitant of the British Isles, having been found, according to Pennant, in certain Welsh rivers so late as the twelfth century, while subfossil remains of it occur in the peat-beds of many parts of the country. In Scandinavia Beavers are still found in the neighbourhood of Arendal. Isolated pairs are occasionally met with on the banks of the Rhone, Weser, and Elbe; and a considerable number are kept in a park belonging to the Emperor of Austria, on the banks of the Danube. They also occur sparingly in Russia and Poland, in the streams of the Ural Mountains, and in those which flow into the Caspian. They live in burrows on the banks of rivers, like the Water-Rat, and show little of the architectural instinct so conspicuous in the American form, but this may be owing to unfavourable external conditions rather than to want of the faculty; for there is a well-authenticated instance of a colony of Beavers, on a small stream near Magdeburg, whose habitations and dam were exactly similar to those found in America.

The American Beaver (*C. canadensis*) extends over that part of the American continent included between the Arctic circle and the tropic of Cancer; owing, however, to the gradual spread of population over part of this area, and still more to the enormous quantity of skins that, towards the end of last and the beginning of the present century, were exported to Europe, numbering about 200,000 annually, this species is in imminent danger of extirpation. It is distinguished from the European Beaver by the shorter and somewhat wider nasals.

Remains of extinct species of *Castor* occur in the Pliocene of Europe, and in the North American Miocene; the one from the last-mentioned deposits being of small size, and separated by some writers as *Eucastor*.

Extinct Genera.—A very large Beaver known as *Trogontherium* (*Diobroticus*), and distinguished by the nature of the enamel-folds of the molars, occurs in the Upper Pliocene and Pleistocene of Europe. *Chalicomys* (*Steneofiber*) is a considerably smaller form from the Miocene of Europe and the United States, distinguished from all existing Rodents by the presence of an entepicondylar foramen in the humerus. *Palæocastor*, of the North American Miocene, is allied.

Section Myomorpha.

Skull (Fig. 203), with slender zygomatic arch, in which the jugal seldom extends far forwards, being usually supported by the long zygomatic process of the maxilla; no postorbital process; infraorbital vacuity variable; angle of mandible, except in the *Bathyerginæ*, rising from the inferior surface of the incisive alveolus. Clavicles well developed, except in *Lophiomys*. Tibia and fibula united.

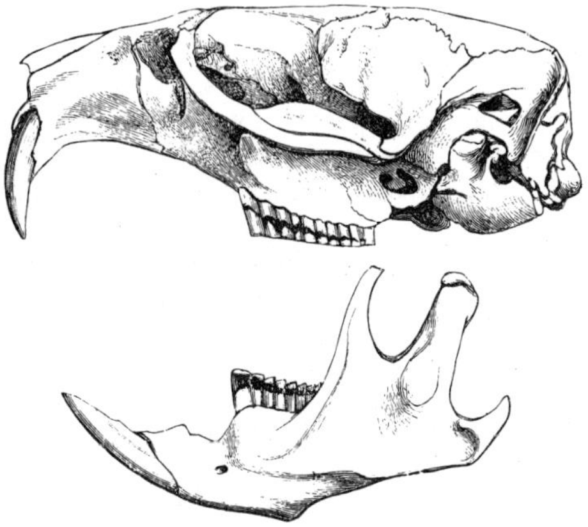

Fig. 203.—Side view of skull of *Fiber zibethicus*, natural size.

Family Myoxidæ.

Small arboreal forms, with long hairy tails, large eyes and ears, and short fore limbs. No cæcum in the intestine. Skull with narrow frontals, a high and narrow infraorbital vacuity of moderate size, and a long and slender coronoid process to the mandible. Premolars $\frac{1}{1}$; molars rooted, with transverse enamel-folds.

The Dormice form a natural family allied to the Squirrels in form and habits, and confined to the Palæarctic and Ethiopian regions. The absence of the cæcum distinguishes them from all other members of the order. They are usually divided into the following five genera, but some of these are of very doubtful value, and it might be preferable to retain *Muscardinus* and include all the others in *Myoxus*.[1]

Myoxus.[2]—Represented by the European *M. glis*, and characterised by the bushy distichous tail, simple stomach, and the large size and complex enamel-folds of the molars, which have flat crowns.

Eliomys.[3]—Tail tufted and distichous; stomach simple; and the molars small, with concave crowns and indistinct enamel-folds. Some seven species, Ethiopian and Palæarctic.

Graphiurus.[4]—Tail short, cylindrical, and tufted at the end;

[1] For a monograph of the *Myoxidæ*, see C. L. Reuvens, *Die Myoxidæ*, etc., 4to, Leyden, 1890.

[2] Schreber, *Säugethiere*, vol. iv. p. 824 (1792).

[3] Wagner, *Abh. baier. Akad.* vol. iii. p. 179 (1843).

[4] F. Cuvier, *Mammifères*, 60me livr. (1845).

molars very small, with the enamel-folds almost absent. Some three Ethiopian species.

Claviglis.[1]—Represented by one West African species, said to be distinguished from all other forms by the shorter tail, which is more distinctly pencilled. The right to generic distinction is, however, very problematical.

Muscardinus.[2]—Includes the Common Dormouse (*M. avellanarius*) of Europe, distinguished by the cylindrical bushy tail, and thickened glandular walls of the cardiac extremity of the œsophagus; the molars have flat crowns, with complex enamel-folds.

Fossil Dormice.—Using the generic term *Myoxus* in a more extended sense than the above, it has existed in Europe from the date of the Upper Eocene. A species nearly as large as a Guinea-Pig, with very complex molars, is common in the Pleistocene of Malta.

Family LOPHIOMYIDÆ.

The genus *Lophiomys*,[3] represented only by *L. imhausi* (Fig.

FIG. 204.—*Lophiomys imhausi.* From Milne-Edwards.

204) of North-East Africa, differs from the typical *Muridæ* in having the temporal fossæ roofed over by a thin plate of bone, rudimentary clavicles, and an opposable hallux. On these grounds it has been made the type of a family, but since all the features are Murine—the dentition being that of a typical Cricetine—it

[1] Jentink, *Notes Leyd. Mus.* vol. x. p. 41 (1888).
[2] Kaup, *Entwickl. Europ. Thierwelt*, p. 139 (1829).
[3] A. Milne-Edwards, *L'Institut*, vol. xxxv. p. 46 (1867).

appears doubtful whether that distinction is justifiable. The hair forms a crest along on the back, and is of a peculiar structure. The habits of this Rodent are arboreal.

Family MURIDÆ.

Skull (Fig. 203) with contracted frontals; a short and slender jugal, generally reduced to a splint between the zygomatic processes of the maxilla and squamosal; the lower root of the former process more or less flattened into a perpendicular plate; typically, the infraorbital vacuity tall, and wide above and narrow below. Lower incisors compressed; no premolars;[1] molars rooted, or rootless, tuberculate, or with angular enamel-folds. Pollex rudimental; tail generally nearly naked and scaly. Habits various, but mostly terrestrial.

This large and cosmopolitan family, which includes more than a third of the existing Rodents, is represented by about forty genera.

Subfamily **Hydromyinæ.**—Molars $\frac{2}{2}$ in number, rooted, and divided into transverse lobes. Represented by two Australasian genera.

Hydromys.[2]—External form modified for an aquatic life. Tip of muzzle extensively haired, so that the nostrils can be closed. Skull with the infraorbital vacuity crescentic, scarcely narrowed below, and its external wall without the perpendicular zygomatic plate characteristic of most of the family; incisive foramina very small.

Two species, with habits like those of the Water Voles, are known from Australia, Tasmania, and New Guinea. In the typical *H. chrysogaster* the colour of the back is black, with an admixture of golden-coloured hairs; the belly being of a dark golden hue.[3]

Xeromys.[4]—External form Murine. Tip of muzzle as in *Mus*, not as in *Hydromys*. Toes unwebbed. Tail scaly, very finely haired. Skull as in *Mus*, with the exception of the rounding of the supraorbital edges. Teeth as in *Hydromys*.

Represented by *X. myoides*, of Queensland; a species about twice the size of the Common Mouse. This genus serves to connect *Hydromys* with the other Murines, although it is difficult to say to which group it comes nearest.

Subfamily **Platacanthomyinæ.**—Molars rooted, with transverse

[1] *Sminthus* is referred to the *Dipodidæ*.

[2] Geoffroy, *Ann. du Muséum*, vol. vi. p. 81 (1805).

[3] For the anatomy of this animal see B. C. A. Windle, *Proc. Zool. Soc.* 1887, p. 53.

[4] O. Thomas, *Proc. Zool. Soc.* 1889, p. 247.

laminæ. Flattened spines mingled with the hair; tail thickly haired. Represented by one genus.

Platacanthomys.[1]—The one representative of this genus is *P. lasiurus*, found in the clefts of rocks and hollow trees in Southern India at elevations of about 3000 feet. This elegant little animal closely resembles a Dormouse; the tail and body having a length of 6 inches.

Subfamily **Gerbillinæ.**—Incisors narrow; molars with transverse laminæ (Fig. 205). Auditory bullæ very large in most cases. Hind limbs elongated. Tail usually long and hairy. Ranges over the Palæarctic, Oriental, and Ethiopian regions.

Gerbillus.[2]—Upper incisors grooved; first molar with three laminæ, second with two, and third with one. There are some sixty species, with a range coextensive with that of the family. The Gerbils, with their large and bright eyes and long tufted tails, are very graceful creatures, inhabiting sandy plains, where they form extensive burrows. Remains of existing species are found in cavern-deposits in Madras (Fig. 205).

FIG. 205.—The left ramus of the mandible of *Gerbillus indicus*, with an enlarged view of the molars, from a cavern deposit in Madras. (From the *Palæontologia Indica*.)

Pachyuromys.[3]—The African genus *Pachyuromys* is distinguished by the very large size of the auditory bulla, as well as by the short and fleshy tail, which is club-shaped. The incisors are narrow and faintly grooved.

Mystromys,[4] *Otomys*,[5] and *Dasymys.*[6]—These genera, also from South Africa, differ from *Gerbillus* in the form of the molars, and are represented by a few species.

Malacomys.[7]—The one known species of this genus is from the Gaboon, and is in some respect intermediate between the true Gerbils and the Rats. Thus the dentition and feet are those of the former, but the long scaly tail resembles that of the latter.

Subfamily **Phlæomyinæ.**[8]—This subfamily is represented only by *Phlœomys*[9] *cumingi*, of the Philippine Islands, in which the incisors are very broad, the molars are divided into transverse laminæ, and the claws are large. The muzzle is blunt; the ears are hairy

[1] Blyth, *Proc. As. Soc. Bengal*, vol. xxviii. p. 289 (1859).
[2] Desmarest, *Nouv. Dict. d'Hist. Nat.* vol. xxiv. p. 22 (1804).
[3] Lataste, *Le Nat.* vol. i. p. 314 (1880).
[4] Wagner, *Wiegmann's Archiv*, 1841, p. 132.
[5] F. Cuvier, *Dents des Mammifères*, p. 168 (1825).
[6] Peters, *Monatsber. Ak. Berlin*, 1875, p. 12.
[7] A. Milne-Edwards, *Bull. Soc. Philom.* sér. 6, vol. xi. p. 9 (1877).
[8] *Nesocia* was included by Alston in this subfamily.
[9] Waterhouse, *Proc. Zool. Soc.* 1839, p. 108.

externally; the tail is moderate, and thickly haired; and the auditory bullæ are very small. The first upper molar has three, and the others two laminæ.

Subfamily **Dendromyinæ.**—Incisors convex in front; molars $\frac{3}{3}$, rooted and tuberculated. Ears hairy; claws long. Confined to the Ethiopian region.

Dendromys.[1]—A small Rodent, with the habits of a Dormouse, characterised by its grooved incisors, slender form, and long scaly tail, which is sparsely haired. Two other Murines described as *Steatomys*[2] and *Lophuromys*[3] are referred to this subfamily. The first is of plump form, with a rather short and thickly haired tail, and grooved incisors. The latter resembles *Steatomys* in form, but has fine flattened bristles instead of fur, and plain incisors.

Subfamily **Cricetinæ.**—Molars $\frac{3}{3}$, tuberculate and rooted, with the tubercles of the upper ones arranged in two longitudinal rows (Fig. 206, *B*). This subfamily has an almost cosmopolitan distribution, and appears to include the most generalised members of the family, from which the more specialised *Murinæ* have been evolved.

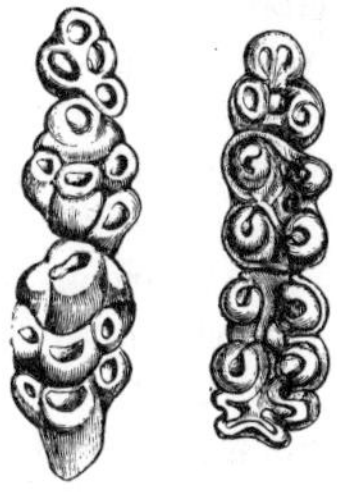

A B

Fig. 206.—Left upper molars of *Mus* (*A*) and *Cricetus* (*B*).

Cricetus.[4]—According to the arrangement proposed by Mr. O. Thomas[5] this genus is taken to include both the Hamsters of the Old World (*Cricetus* proper) and the white-footed or Vesper Mice (*Hesperomys*) of the New. Cheek-pouches are frequently present, and may be very large. The first molar (Fig. 206, *B*) generally has six tubercles. The tail may be very short.

This large and unwieldy genus may be divided into a number of groups or subgenera. The typical group includes the Hamsters of the Old World, characterised by the large size of their cheek-pouches, the walls of which are connected with muscles arising from the lumbar vertebræ. The tail is remarkable for its shortness. The best-known species is *C. frumentarius*, inhabiting Europe and Northern Asia. The American forms, which range over the whole of that continent, comprise a number of subgenera, of which the following are the most important. *Rhipidomys*, including Dormouse-like forms with long tails and a dentition like that of the typical group; *Oryzomys*, represented by Murine species; *Calomys*, with short tail and Hamster-like body; *Vesperimus*, with only five tubercles on the first molar; *Onychomys*, in which the tail is extremely

[1] Andrew Smith, *S. African Quart. Journ.* vol. ii. p. 158 (1834).
[2] Peters, *Reise n. Mossambique*, vol. i. p. 162 (1852).
[3] Peters, *Monatsber. Ak. Berlin*, 1874, p. 234.
[4] Cuvier, *Règne Animal*, vol. i. p. 198 (1817).
[5] *Proc. Zool. Soc.* 1888, p. 133.

short and Hamster-like, and the form is Arvicoline; *Scapteromys*, of Murine form with a long and hairy tail; *Phyllotis*, with a shorter tail; *Habrothrix*, an Arvicoline group, with a short and thinly haired tail; and *Oxymycterus*, distinguished from the preceding by having a nail instead of a claw on the pollex. With regard to the distribution of these forms Mr. Thomas[1] remarks that in South America as we proceed southwards there is a general tendency "to a disappearance of the tropical and northern Mouse- and Dormouse-like subgenera *Rhipidomys*, *Vesperimus*, and *Oryzomys*, with the appearance and increase of the Vole- and Hamster-like *Habrothrix* and *Calomys*—a change that is curiously paralleled in the Old World by the gradual supercession of *Mus* and *Myoxus* in favour of *Arvicola* and *Cricetus* as we go northwards from tropical to temperate and arctic regions." One species has spines in the fur.

Remains of *Cricetus* are abundant in the Pleistocene cavern-deposits of Brazil, where a number of the forms are referable to existing species; the genus is also represented in the Miocene of North America and Europe, the species from the former area having been described as *Eomys*, and those from the latter as *Cricetodon*.

Holochilus[2] (*Nectomys*).—The Rats of this genus are allied to the American forms of *Cricetus*, but have the third upper molars proportionately larger and the skull more stoutly built. This genus is confined to Brazil, and contains about six species, some of which are the largest indigenous Rats of America. Two species are aquatic in their habits, and have short webs between the toes of their hind feet.

Sigmodon[3] differs from *Cricetus* in the pattern of the molar teeth. It contains one species only, the Rice-Rat, *S. hispidus*, ranging from the United States to Ecuador.

Rhithrodon,[4] and *Ochetodon*.[5]—These are more or less like *Cricetus*, but with grooved upper incisors. The first is a South-American genus, and contains five Rat-like species, one from Venezuela, another from Peru, and the other three from Patagonia. The second consists of three North American mice, of about the size and proportions of the English Wood-Mouse (*Mus sylvaticus*).

Neotoma.[6]—A peculiar North American genus, in which the teeth simulate the prismatic appearance of those of the *Arvicolinæ*. There are four species known as Wood-Rats, all of about the size of *Mus decumanus*; one of them (*N. cinerea*) having a tail almost as bushy as a Squirrel's while the other three have ordinary scaly Rat-like tails.

[1] *Proc. Zool. Soc.* 1884, p. 451.

[2] Brandt, *Mém. Acad. Imp. St. Pétersbourg*, sér. 3, vol. iii. p. 428 (1835).

[3] Say and Ord, *Journ. Acad. Philad.* vol. iv. p. 352 (1825).

[4] Waterhouse, *Proc. Zool. Soc.* 1837, p. 29.

[5] Coues, *Proc. Acad. Philad.* 1874, p. 184.

[6] Say and Ord, *Journ. Acad. Philad* vol iv. p. 346 (1825).

Fossil remains of *Neotoma* from cavern-deposits in Pennsylvania are not improbably referable to the existing Florida Rat (*N. floridana*). *Paciculus*, from the Miocene of the United States, is regarded as an allied extinct genus with enamel-folds to the molars.

Hypogeomys.[1]—This and the following genera are confined to Madagascar, where they are the sole representatives of the Rodentia. *Hypogeomys* is a very peculiar form of large size, with long ears, feet, and tail. There is only one species, *H. antimena*, a fawn-coloured Rat about 9 inches long.

Nesomys.[2]—Contains two species of long-haired Rats, more or less rufous in colour, about the size of the Brown Rat.

Brachytarsomys.[3]—Represented only by *B. albicauda*, a pretty velvety-haired fawn-coloured Rat, with short feet and a long tail.

Hallomys.[4]—The only species (*H. audeberti*) is very like a *Nesomys*, but has much longer hind feet.

Eliurus.[5]—Represented by one small Dormouse-like species, characterised by its nearly naked and short ears, and long tail, of which the proximal third is scaly, and the remainder covered with long hair. The pollex is rudimental, but the hallux well developed.

Subfamily **Arvicolinæ.**—Molars usually imperfectly rooted or rootless, and composed of two longitudinal rows of triangular prisms placed alternately (Fig. 207). Tail moderate or short. Common to the Palæarctic and Nearctic regions.

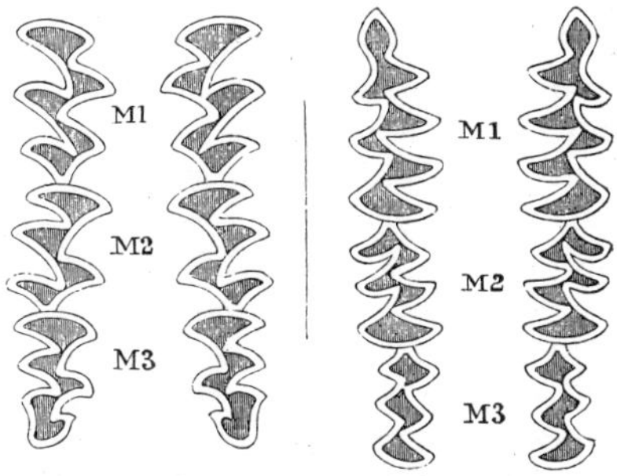

FIG. 207.—Upper (*A*) and lower (*B*) molars of the Water-Vole (*Arvicola amphibius*).

The Voles, as the members of this group are commonly termed, are so closely connected with the Cricetines that they may be regarded merely as a branch of that subfamily which has attained a peculiarly specialised type of molar dentition. The Voles are externally distinguished, as a rule, from true Rats and Mice by their more clumsy and heavy build and less graceful movements; by the small size of their eyes, the bluntness of the muzzle, the small ears, and the shorter limbs and tail.

[1] Grandidier, *Rev. and Mag. Zool.* 1869, p. 388.

[2] Peters, *Sitzber. Ges. Nat. Freunde*, 1870, p. 54 (1871).

[3] Günther, *Proc. Zool. Soc.* 1875, p. 79.

[4] Jentink, *Notes Leyd. Mus.* vol. i. p. 107, note 27 (1879).

[5] Milne-Edwards, *Ann. Sci. Nat.* sér. 6, vol. xx. art. 1, *bis*, p. 1 (1886).

Phenacomys.[1]—A North American genus distinguished by its rooted molars, and thus connecting the typical forms with Cricetines like *Neotoma.* Several species have been described by Dr. C. H. Merriam.

Arvicola.[2]—The type genus *Arvicola* has rootless molars, and naked soles to the feet. It includes over forty species inhabiting Europe, North America, and Asia, a few species entering into the northern limits of the Oriental region in India. Three species of the genus are found in the British Isles, of which the following account is given by Mr. O. Thomas :—

The common Water-Vole (*A. amphibius*) is as large as the Brown Rat. Its fur is long, soft, and thick, of a uniform grizzled brown all over, except when, as is not uncommon, it is wholly black. The tail is about half the length of its head and body, and the hind feet are unusually long and powerful, although not webbed, and have five rounded pads on their lower surfaces. Its molar teeth (see Fig. 207) present the following number of prismatic spaces :—in the upper jaw the first, or anterior, has 5, the second 4, and the third 4, of which the last is very irregular in shape, and is sometimes itself divided into two, making 5 in all; in the lower jaw the first has 7 spaces, of which the 3 anterior are generally not fully separated from one another, the second has 5, and the third 3. These numbers for the different teeth are taken as the characters of the subgenus *Paludicola* of Dr. Blasius, by whom this method of subdividing the genus was first introduced. The Water-Vole is one of the commonest English mammals, and is perhaps the most often actually seen of all, owing to its diurnal habits. It frequents rivers and streams, burrowing deeply into their banks, and in this way often causing considerable damage. Its food consists almost wholly of water-weeds, rushes, and other vegetable substances, but, like so many other Rodents, it will also occasionally eat animal food, in the shape of insects, mice, or young birds. The female during the warm season of the year has three or four litters, each of from two to seven young. The range of the Water-Vole extends over the whole of Europe and North Asia, from England to China, but it is not found in Ireland. The common Field-Vole, or short-tailed Field-Mouse (*A. agrestis*), representing the subgenus *Agricola*, is about the size of a House-Mouse, but with a short stumpy body, and a tail only about one third the length of the head and body combined. Its hind feet have six pads on their inferior surfaces. The colour is dull grizzled brown above, and grayish-white below. Its molar teeth have respectively 5, 5, and 6 prismatic spaces above, and 9, 5, and 3 below. The

[1] Merriam, *Fauna of North America*, No. 2, p. 28 (1889).

[2] Lacépède, *Mém. de l'Institut*, vol. iii. p. 495 (1801). Many writers employ the earlier name *Microtus* for the true Voles.

Field-Vole is one of the commonest of our smaller mammals, and frequents fields, woods, and gardens in enormous numbers, often doing very considerable damage in the latter, owing to its fondness for garden produce of all kinds. It is spread over the whole of Great Britain from the Hebrides southwards. Abroad its range extends from Finland to North Italy and from France and Spain to Russia. The Bank-Vole (*A. glareolus*) resembles in size and general appearance the common Field-Vole, but may be distinguished by its more or less rusty or rufous-coloured back, its larger ears, and the relatively longer tail, which attains to about half the length of the head and body. Its molar teeth present characters so different from those of all other Voles as to have caused it to be regarded as belonging to an entirely distinct genus, for which the name of *Evotomys* has been used. Their chief distinction lies in the fact that, unlike those of all other Voles, their pulp-cavities close up in adult life, and they form distinct roots, more resembling those of the ordinary Rats and Mice. The enamel-spaces of these teeth number respectively 5, 4, and 5 above, and 7, 3, and 3 below. The habits of this species are in every way similar to those of the Field-Vole. Its range in Great Britain extends northwards to Morayshire, beyond which it has not yet been observed. It is also found all along the north temperate zone from France to China, and is replaced in North America by a closely allied animal known as *A. gapperi*. It is probable, however, that both *A. gapperi* and *A. glareolus* are only southern climatic offshoots of a still more northern species, the *A. rutilus* of Northern Europe, Siberia, and Arctic America.

Fossil remains of *Arvicola* are common in European Pleistocene deposits, and they have also been obtained from the Upper Pliocene of the Norwich Crag.

Synaptomys.[1]—Represented by one North American species, having grooved upper incisors, skull and molars like those of *Myodes*, with the external characters of *Arvicola*.

Myodes.[2]—Distinguished from *Arvicola* by the more clumsy build, convex obtuse head, extremely short and Rabbit-like tail, short ears, small feet, the soles of which are furred, elongated claws, and thick fur, as well as by the breadth and massiveness of the skull, in which the zygomatic arch has a laminar expansion and the palate a peculiar contour; while the root of the lower incisor does not extend behind the last molar, the upper incisors are bevelled, and not grooved, and the molars have a characteristic pattern, which cannot be well explained without a figure.

The Lemmings, as the members of the genus are commonly called, are represented by the Norwegian Lemming (*M. lemmus*, Fig.

[1] Baird, *Mamm. North America*, pp. xliv. 558 (1857).

[2] Pallas, *Zoogr. Rosso-Asiat.* vol. i. p. 173 (1811).

208), and the North American *M. obensis*. Different individuals of the Norwegian Lemming vary considerably both in size and colour, but its usual length is about 5 inches, and its soft fur yellowish brown, marked with spots of dark brown and black. It has a short, rounded head, obtuse muzzle, small bead-like eyes, and short rounded ears, nearly concealed by the fur. The tail is very short. The feet are small, each with five claws, those of the fore feet strongest, and fitted for scratching and digging. The usual dwell-

FIG. 208.—The Lemming (*Myodes lemmus*).

ing place of the Lemmings is in the highlands or fells of the great central mountain chain of Norway and Sweden, from the southern branches of the Langfjeldene in Christiansand-stift to the North Cape and the Varangerfjord. South of the Arctic circle they are, under ordinary circumstances, exclusively confined to the plateaus covered with dwarf birch and juniper above the conifer region, though in Tromsö-amt and in Finmarken they occur in all suitable localities down to the level of the sea. The nest is formed under a tussock of grass or a stone, constructed of short dry straws, and usually lined with hair. The number of young in each nest is generally five, sometimes only three, but occasionally seven or eight, and at least two broods are produced annually. Their food is entirely vegetable, especially grass-roots and stalks, shoots of the dwarf birch, reindeer-lichens, and mosses, in search of which they form, in winter, long galleries through the turf or under the snow. They are restless, courageous, and pugnacious little animals. When suddenly disturbed, instead of trying to escape they will sit upright, with their back against a stone or other coign of vantage, hissing and showing fight in a very determined manner (Fig. 208).

The circumstance which has given more popular interest to the Lemming than to a host of other species of the same order of animals is that certain districts of the cultivated lands of Norway and Sweden, where in ordinary circumstances they are quite unknown, are occasionally and at very uncertain intervals, varying from five to twenty or more years, literally overrun by an army of these little creatures, which steadily and slowly advance, always in the same direction, and regardless of all obstacles, swimming across streams and even lakes of several miles in breadth, and committing considerable devastation on their line of march by the quantity of food they consume. In their turn they are pursued and harassed by crowds of beasts and birds of prey, as bears, wolves, foxes, dogs, wild cats, stoats, weasels, eagles, hawks, and owls, and never spared by man; even the domestic animals not usually predaceous, as cattle, goats, and reindeer, are said to join in the destruction, stamping them to the ground with their feet, and even eating their bodies. Numbers also die from diseases apparently produced from overcrowding. None ever return by the course by which they came, and the onward march of the survivors never ceases until they reach the sea, into which they plunge, and swimming onwards in the same direction as before perish in the waves. These extraordinary and sudden appearances of vast bodies of Lemmings, and their singular habit of persistently pursuing the same onward course of migration, have given rise to various speculations, from the ancient belief of the Norwegian peasants, shared in by Olaus Magnus, that they fall down from the clouds, to the almost equally untenable hypothesis, ingeniously maintained by the late Mr. W. D. Crotch, that they are acting in these migrations in obedience to an instinct inherited from vastly ancient times, and are still seeking the congenial home in a supposed submerged Atlantis, to which their ancestors of the Miocene period were wont to resort when driven from their ordinary dwelling-places by crowding or scarcity of food. The principal really ascertained facts regarding these migrations seem to be as follows. When any combination of circumstances has occasioned an increase in the numbers of the Lemmings in their ordinary dwelling-places, impelled by the restless or migratory instinct possessed in a less developed degree by so many of their congeners, a movement takes place at the edge of the elevated plateau, and a migration towards the lower-lying land begins. The whole body moves forward slowly, always advancing in the same general direction in which they originally started, but following more or less the course of the great valleys. They only travel by night; and, staying in congenial places for considerable periods, with unaccustomed abundance of provender, notwithstanding all the destructive influences to which they are exposed, they multiply excessively during their journey, having families still more numerous and more frequently than in their usual homes. The progress may last from one to three years, according to the route taken, and the distance to be traversed until the sea-coast is reached, which in a country so surrounded by water as the Scandinavian peninsula must be the ultimate goal of such a journey. This may be either the Atlantic or the Gulf of Bothnia, according as the migration has commenced from the west or the east side of the central elevated plateau. Those that finally perish in the sea, committing what appears to be a voluntary suicide, are only acting under the same blind impulse which has led them previously to cross smaller pieces of water with safety.

Cuniculus.[1]—Cranial and incisive characters those of *Myodes*, in the main, but the molars more of an Arvicoline type, the first upper one differing from that of all other members of the family in having seven prisms. Externally of the general shape of *Myodes*, but distinguished by the absence of external ears, the shortness and dense furring of the feet, the obsolete pollex with rudimentary nail, and the great length of the two middle claws of the manus. Represented by one species, the Banded Lemming (*C. torquatus*), of the Arctic region.

Remains of both *C. torquatus* and *Myodes lemmus* occur in British Pleistocene deposits.

Fiber.[2]—Closely allied to *Arvicola*, both externally and in cranial and dental characters, but with the tail nearly as long as the body (apart from the head), compressed, nearly naked, and reticulate. Feet incompletely webbed, and the whole body adapted for a thoroughly aquatic life.

The Musk-Rat or Musquash (*F. zibethicus*, Fig. 209) is the only representative of this genus, and the largest member of the subfamily, the head and body being about 12 inches in length. It is rather a heavily built animal, with a broad head, no distinct neck, and short limbs; the eyes are small, and the ears project very little beyond the fur. The fore limbs have four toes and a rudimentary thumb, all with claws; the hind limbs are larger, with five distinct toes, united by short webs at their bases. The tail is laterally compressed, nearly naked, and scaly. The hair much resembles that of a beaver, but is shorter; it consists of a thick soft under-

[1] Wagler, *Isis*, 1832, p. 1220.

[2] Cuvier, *Leçons d'Anatomie Compar.* tab. 1 (1800).

fur, interspered with longer stiff, glistening hairs, which overlie and conceal the former on the upper surface and sides of the body. The general colour is dark umber-brown, almost black on the back and gray below. The tail and naked parts of the feet are black. The musky odour from which it derives its name is due to the secretion of a large gland situated in the inguinal region, and present in both sexes.

The Musk-Rat is peculiar to America, being extensively distributed in suitable localities in the northern part of the continent, extending from the Atlantic to the Pacific, and from the Rio Grande

FIG. 209.—The Musk-Rat (*Fiber zibethicus.*)

to the barren grounds bordering the Arctic Seas. It is aquatic in its habits, living on the shores of lakes and rivers, swimming and diving with great facility, feeding on the roots, stems, and leaves of water-plants, or on fruits and vegetables which grow near the margin of the streams it inhabits. Musk-Rats are most active at night, spending the greater part of the day concealed in their burrows dug out of the bank, consisting of a chamber with numerous passages, all of which open under the surface of the water. For winter quarters they build more elaborate houses of conical or dome-like form, composed of sedges, grasses, and similar materials plastered together with mud. As their fur is an important article of commerce, large numbers are annually killed, being either trapped or speared at the mouths of their holes.

The skull of the Musk-Rat is shown in Fig. 203 (p. 459); its structure is essentially Arvicoline, but the squamosals are greatly expanded, with a corresponding reduction of the parietal and interparietal, and the interorbital constriction of the frontals attains its greatest development. Fossil remains of *Fiber* occur in the North American Pleistocene.

Neofiber.[1]—This genus, while agreeing with *Fiber* in the characters of the skull and teeth, differs by the cylindrical tail, and the normal form of the feet, in which the toes are not bent laterally at an angle with the sole. The single species *N. alleni*, commonly known as the Round-tailed Musk-Rat, is found in Florida, and is much less completely aquatic in its habits than *Fiber*. Its colour is brown above, and silvery-white mixed with rufous below, the sides of the body gradually shading from brown to rufous, the forehead and the tip of the nose are black, while the tail is rufous mingled with black.

Subfamily **Siphneinæ.** — Includes two genera of Mole-like Rodents with an *Arvicoline* dentition, but with the body thoroughly

FIG. 210.—*Siphneus armandi.* (From Milne-Edwards.)

adapted for a subterranean life, the limbs and tail being very short, and the external ears rudimentary. Both are Palæarctic.

Ellobius.[2]—The Russian *E. talpinus*, the typical representative of the genus, has short claws, and comes nearest to the *Arvicolinæ*. *E. fuscocapillus* is from Afghanistan.

Siphneus.[3]—This genus (Fig. 210) includes species inhabiting Northern and Central Asia, and is characterised by the great length of the claws of the manus. Remains of an existing species occur

[1] True, *Proc. U.S. Nat. Mus.* vol. vii. p. 170 (1884).

[2] Fischer, *Zoognosia*, vol. iii. p. 72 (1814).

[3] Brants, *Het. Geslact der Muizen*, p. 20 (1827).

in the Pleistocene of the Altai, while an extinct one has been described from the Pliocene of North China.

Subfamily **Deomyinæ.**—Represented only by the undermentioned genus, in which the bituberculate anterior and tricuspidate middle ridge of the first upper molar presents a condition intermediate between that obtaining in the *Cricetinæ* and that of the *Murinæ*.

Deomys.[1]—Externally as in *Mus.* Pollex with a narrow nail; hind feet elongate. Infraorbital vacuity of skull triangular, not narrowed below. Upper incisors with a pair of minute grooves. First upper molar with seven distinct tubercles, of which three are placed on the middle ridge, and two on each of the others. One species, *D. ferrugineus*, from the Lower Congo, an animal about the size of the Common Mouse.

Subfamily **Murinæ.**—Molars rooted and tuberculated; those of the upper jaw with three longitudinal rows of tubercles (Fig. 206, *A*).

This group includes the true Rats and Mice, and may be regarded as more specialised than the *Cricetinæ*. All the members of the group closely resemble one another, and are light and active, with large ears, bright eyes, and long and scaly tails. Their coloration, in conformity with the fossorial and nocturnal habits of most of the forms, is sombre, and their movements are remarkably agile and graceful.

Fig. 211.—The Australian Brown-footed Rat (*Mus fuscipes*). After Gould.

Mus.[2]—Incisors narrow, without grooves. Structure of molars as in Fig. 206, *A* (p. 463). Incisive foramina of skull long; coronoid process of mandible well developed.

[1] O. Thomas, *Proc. Zool. Soc.* 1888, p. 130.

[2] Linn. *Syst. Nat.* 12th ed. vol. i. p. 79 (1766).

Ears and eyes large; muzzle naked at the extremity. Fur soft, in some cases intermingled with spines. Pollex with a short nail in place of a claw. No cheek-pouches. Tail long, nearly naked, with rings of overlapping scales. Vertebræ: C 7, D 13, L 6, S 4, C 26-32.

This genus is the largest in the whole mammalian class, comprising not less than 130 species, ranging over the whole of the Old World, with the noteworthy exception of Madagascar. On the whole, the species are more numerous in tropical than in temperate regions, and very few occur in cold countries. Many of the species living in warm climates have flattened spines mingled with the fur; these spines being shed in winter, when a warmer covering is necessary, and replaced by hair. Five species occur in England, which are briefly noticed below; and it may be observed that none of the species are much larger than *M. decumanus* or smaller than *M. minutus*. As a rule the habits of the species are similar to those of the English forms, but a few are arboreal, while others again, like the one represented in Fig. 211, are aquatic. The earliest known representatives of the genus (excluding *Acanthomys gaudryi* of the Lower Pliocene Pikermi beds of Attica) occur in the Pleistocene of Europe.

The Brown or Norway Rat (*M. decumanus*) is a heavily built animal, growing to 8 or 9 inches in length, with a bluff rounded head, small ears (Fig. 212, *A*), and a comparatively short tail, which is always shorter than the head and body combined, and generally not longer than the body alone. The colour is a uniform grayish-brown above and white below, the ears, feet, and tail being flesh coloured. Black varieties, which are often mistaken for true Black Rats, are by no means rare, but the differences in size and proportions form a ready means of distinguishing the two. The Brown Rat is believed to be a native of Western China, where a race (*M. humiliatus*) has been discovered so like it as to be practically indistinguishable. Both this, and the next species agree in their

A

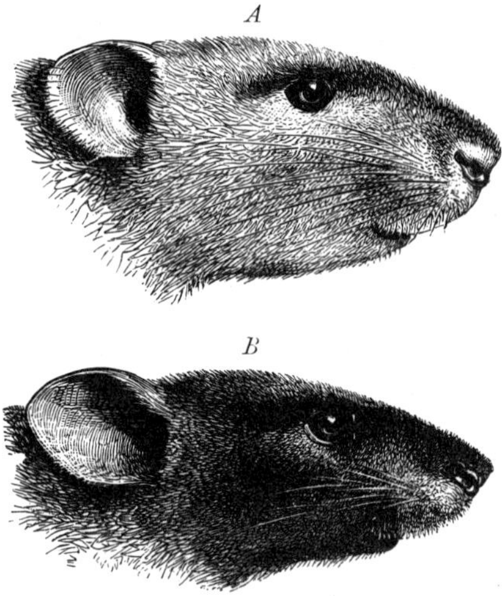

B

FIG. 212.—*A*, Head of Brown Rat (*M. decumanus*). *B*, Head of Black Rat (*Mus rattus*).

predaceous habits, omnivorous diet, and great fecundity. They bear four or five times in the year from four to ten blind and naked young, which are in their turn able to breed at an age of about six months; the time of gestation being about twenty days.

The Black Rat (*M. rattus*) is a smaller and more lightly built species, generally not more than 7 inches in length, with a slender head (Fig. 212, *B*), large ears, and a thin tail of about 8 or 9 inches in length. The colour is usually a glossy bluish-black, somewhat lighter below; but in the tropical variety described as *M. alexandrinus* the general colour is gray or rufous, and the belly white. The disposition of the Black Rat is milder than that of *M. decumanus*, and the white and pied rats kept as pets mostly belong to this species. In many localities where it was formerly abundant it has been entirely superseded by *M. decumanus*, but it is said that in some parts of Germany it has been lately reasserting itself.

M. musculus, the Common House-Mouse, is, like the Brown Rat, originally a native of Asia, whence it has spread to all the inhabited parts of the globe. Its habits and appearance are too well known to need any description.

M. sylvaticus, the Wood or Long-tailed Field-Mouse, is very common in many parts of England, often taking to barns and outhouses for shelter during the winter. It is of about the same size and proportions as *M. musculus*, but of a bright reddish-gray colour, with a pure white belly.

M. minutus, the Harvest-Mouse, is the smallest of the European Mice, seldom exceeding $2\frac{1}{2}$ or 3 inches in length. It is of a yellowish-red colour, with comparatively short ears and tail. It lives entirely away from human habitations, generally dwelling in grass or corn-fields, where it builds a globular nest of dried grass of the size of a cricket-ball, in which the young are nurtured.

Nesocia.[1]—General characters those of *Mus*, but the incisors and molars very much wider, and the tubercles of the latter more connected by transverse ridges, thus producing a laminated type of structure.

This genus has been placed by some writers in a distinct subfamily with *Phlœomys*, but Mr. O. Thomas regards it as so closely allied to *Mus* that even its generic separation may be open to question. It comprises several species, mostly spread over Southern Asia, ranging from Palestine to Formosa, and from Kashmir to Ceylon, but *N. scullyi* is found in Turkestan. The great Indian Bandicoot-Rat (*N. bandicota*) is the largest representative of the subfamily, often exceeding a foot in length. *N. bengalensis* is remarkable for possessing no less than eighteen mammæ. Fossil remains of *Nesocia* occur in the Pleistocene of Madras and in the

[1] Gray, *Ann. Mag. Nat. Hist.* vol. x. p. 264 (1842). Amended from *Nesokia*.

Pliocene of Northern India; those from the first-named deposits being referable to existing species.

Golunda.[1]—Like *Mus*, but with a distinct groove down the front of the upper incisors. There are only three species, one from Western India, one from West Africa, and the other from Eastern Africa.

Uromys.[2]—Differs from *Mus* in having the scales of the tail not overlapping, but set edge to edge, so as to form a sort of mosaic work. There are about six species of *Uromys*, spread over the northern part of the Australian region from the Aru Islands to Queensland.

Chiruromys.[3]—Externally like *Mus*, but with the terminal portion of the tail without scales above, quite naked, transversely wrinkled, and prehensile. Scales of remainder of tail more or less pentagonal, and arranged in oblique diagonal series. Supraorbital vacuity of skull without projecting plate in external wall. Incisive foramina short and narrow; auditory bulla small. Upper molars very complex, with the tubercles (of which there are eleven in the first tooth) low, and distinctly arranged in transverse rows. Known only by *C. forbesi*, from mountains in New Guinea, which must be regarded as a specialised form very similar in outward appearance to *Uromys cervinipes*.

Hapalotis.[4]—Hind limbs elongated. Incisive foramina very large. No coronoid process to the mandible. This genus is confined to Australia, where there are about fifteen species known. They are pretty little animals, with long ears and tail, and in many respects resemble the Jerboas, whose place they seem to take in the sandy Australian deserts. Remains of *H. albipes* occur in the Pleistocene of New South Wales.

Mastacomys.[5]—Like *Mus*, but with the molars remarkably broadened, and with only four mammæ. The single species of the genus is as yet only known from Tasmania, though it has been found fossil in New South Wales; it is somewhat similar in size and general appearance to the English Water-Vole, but has much longer and softer fur.

Acanthomys.[6]—Fur almost entirely composed of flattened spines. Teeth and skull as in *Mus*, but the coronoid process of mandible very small. There are six species of Spiny-Mice known, all of about the size of the Common Mouse. They are found in Syria,

[1] Gray, Charlesworth's *Mag. Nat. Hist.* vol. i. p. 586 (1837). Syn. *Pelomys*, Peters (1852).

[2] Peters, *Monatsber. Ak. Berlin*, 1867, p. 343.

[3] O. Thomas, *Proc. Zool. Soc.* 1888, p. 237.

[4] Lichtenstein, *Darst. neu. Säugethiere*, pt. iv. pl. 29 (1829).

[5] O. Thomas, *Ann. Mag. Nat. Hist.* ser. 5, vol. ix. p. 413 (1882).

[6] Geoffroy, *Ann. Sci. Nat.* sér. 2, vol. x. p. 126 (1840). *Acomys*.

Palestine, and Eastern Africa as far south as Mozambique. *A. dimidiatus* presents the appearance of a little Hedgehog when its spines are erected; it inhabits the stony deserts of Arabia Petræa and Palestine, and feeds on bulbs. A fossil Mouse (*A. gaudryi*) referred to this genus occurs in the Lower Pliocene of Attica.

Echinothrix.[1]—A very remarkable rat with an extremely elongated muzzle, all the bones of the face being much produced. The incisors are faintly grooved. The only species is *E. leucura*, an animal of about the size of the Brown Rat, with its fur thickly mixed with spines. It is found in Celebes.

Typhlomys.[2]—This genus is represented by a single species from China, which resembles a House-Mouse in size and general appearance, but has smaller ears, while the eyes are so reduced in size as to be totally concealed by the long eyelashes.

Cricetomys[3] and *Saccostomus.*[4]—These two African genera have been—from the presence of cheek-pouches—usually placed in the neighbourhood of *Cricetus*, but their molars are of the Murine type. *Cricetomys* is said to have grooved upper incisors, and is represented only by *C. gambianus.* There are two species of *Saccostomus.*

Pithechirus.—A small Rodent from Sumatra and Java described under this name is a true Mouse, having nothing to do with *Chiropodomys*, to which it has been compared.

Family SPALACIDÆ.

Mole-like forms, with very small or rudimentary eyes and ear-conchs, large claws, and short or rudimentary tail. Form cylindrical. Incisors large; premolars present or absent; molars rooted, with re-entering enamel-folds; palate narrow.

Subfamily **Spalacinæ.**—Angular part of the mandible arising from the lower edge of the socket of the lower incisor. No premolars.

Spalax.[5]—Represented by the great Mole-Rat (*S. typhlus*) of South-Eastern Europe, in which the eyes are completely covered by the skin.

Rhizomys.[6]—Eyes uncovered, although very minute; small naked ear-conchs; and a short partially hairy tail. Includes several species from Northern India, Tibet, China, Burma, Malaya, and Eastern Africa. A fossil species occurs in the Pliocene Siwaliks of Northern India.

[1] Gray, *Proc. Zool. Soc.* 1867, p. 599. Amended from *Echimys.*

[2] Milne-Edwards, *Bull. Soc. Philom.* sér. 6, vol. xi. p. 9 (1877).

[3] Waterhouse, *Proc. Zool. Soc.* 1840, p. 2.

[4] Peters, *Monatsber. Ak. Berlin*, 1846, p. 258.

[5] Güldenstädt, *Nov. Comment. Petrop.* vol. xiv. art. i. p. 409 (1770).

[6] Gray, *Proc. Zool. Soc.* 1830, p. 95.

Subfamily **Bathyerginæ.**—Angular part of the mandible arising from the side of the socket of the lower incisor. Premolars absent or present. Confined to the Ethiopian region.

Bathyergus.[1]—Upper incisors strongly grooved; $p\ \frac{1}{1}$, $m\ \frac{3}{3}$; no ear-conchs; very powerful claws. One species (*B. maritimus*), from South Africa, attaining a length of about 10 inches.

Georychus[2] and *Myoscalops.*[3]—Upper incisors without grooves. *Georychus*, with some half dozen species, generally has $p\ \frac{1}{1}$; *Myoscalops*, with one species, usually has $p\ \frac{3}{3}$, and the second toe of the foot is the longest. In *Georychus* the premolar may be wanting, and some examples of *Myoscalops* have only two teeth of this series.

Heterocephalus.[4]—Small and nearly naked forms, with small head, small eyes, no ear-conchs, moderately long tail, and powerful fore feet provided with a pair of large pads; $p\ \frac{0}{0}$, $m\ \frac{(2-3)}{(2-3)}$. Two species. These very remarkable little Rodents are regarded by Mr. O. Thomas as very closely allied to *Georychus*, but specialised, and, so to speak, somewhat degraded for a purely subterranean life, for which their hairless body is peculiarly adapted. They are found in Somali-land, where they burrow in the sandy soil.

Family GEOMYIDÆ.[5]

Terrestrial or fossorial forms, with large cheek-pouches opening on the cheeks outside the mouth. Squamosal much expanded, and the jugal extending forwards to the lachrymal. $P\ \frac{1}{1}$; molars rooted or rootless, with transverse laminæ. Nearctic and Neotropical regions.

Subfamily **Geomyinæ.**—Incisors broad; mastoid not appearing on the top of the skull; eyes small; ear-conch rudimentary; limbs short, subequal. Habits fossorial.

Geomys.[6]—Upper incisors deeply grooved. The common North American Pouched-Rat or "Pocket-Gopher" (*G. bursarius*) inhabits the plains of the Mississippi and lives in burrows. Several other species are recognised from the Southern United States, Mexico, and Central America. The genus is represented in the Pleistocene and Pliocene of the United States.

Thomomys.[7]—Upper incisors plain. Represented by two species,

[1] Illiger, *Prodromus Syst. Mamm.* p. 86 (1811).

[2] Illiger, *loc. cit.* p. 87.

[3] O. Thomas, *Proc. Zool. Soc.* 1890, p. 448 = *Heliophobius*; Peters, *Monatsber. Ak. Berlin*, 1846, p. 243.—Preoccupied.

[4] Rüppel, *Mus. Senkenb.* vol. i. Säugeth. p. 99 (1834).

[5] Including the *Saccomyidæ* of Coues.

[6] Rafinesque, *Amer. Monthly Mag.* vol. ii. p. 45 (1817).

[7] Wied, *Nova Acta Ac. Cæs. Leop.-Car.* vol. xix. pt. i. p. 383 (1839).

with numerous varieties found all over Canada and North America west of the Rocky Mountains. Remains referred to an existing species occur in the Pliocene of Oregon. *Entoptychus*, from the Miocene of the United States, is an allied genus, with broad incisors and rootless molars.

Subfamily **Heteromyinæ.**—Incisors narrow; mastoid appearing largely on the top of the skull; eyes and ears moderate or large; hind limbs and tail elongated. Habits terrestrial.

Dipodomys.[1]—This genus is characterised by the rootless molars. It is best known by *D. phillipsi*, the Kangaroo-Rat of the desert regions east of the Rocky Mountains, having habits like those of the Jerboas. The typical forms have four toes in the pes; but in others, which it has been proposed to separate as *Dipodops*, there are five: *D. ordi* and *D. agilis* belong to the latter group.

Perognathus[2] and *Heteromys.*[3]—In both these genera, which are represented by species of very small size, the molars are rooted; the latter being distinguished by the presence of flattened spines mingled with the fur, and having species ranging into South America. According to Dr. C. H. Merriam the forms described as *Cricetodipus* are not separable from *Perognathus;* while Dr. Coues considers that *Saccomys* was founded upon a species of *Heteromys*. *Pleurolichus*, from the Miocene of the United States, is regarded as an extinct genus allied to *Heteromys*.

Family DIPODIDÆ.

Terrestrial forms usually with four upper cheek-teeth, and typically with the following characters. Incisors compressed; molars with transverse enamel-folds; infraorbital vacuity of skull (Fig. 7, p. 37) large and rounded; jugal ascending in front to the lachrymal; and the mastoid part of the auditory bulla usually very large.

Subfamily **Sminthinæ.**—Molars rooted; $p\ \frac{1}{0}$, $m\ \frac{3}{3}$. Skull with the infraorbital vacuity widest below, and the incisive palatal foramina long. Limbs short. Palæarctic.

Sminthus.[4]—Represented by the Rat-like *S. vagans* from Northern Europe and Asia, in which the ears are rather long and pointed, the tail is covered with short hairs and nearly as long as the body, while the molars present a somewhat complicated pattern. This genus has generally been regarded as an aberrant member of the *Muridæ*, but was transferred in 1887 to the present family by Dr. H. Winge.

[1] Gray, *Ann. Mag. Nat. Hist.* vol. vii. p. 521 (1840).
[2] Wied, *Nova Acta Ac. Cæs. Leop.-Car.* vol. xix. pt. i. p. 369 (1839).
[3] Desmarest, *Mammalogie*, p. 313 (1820).
[4] Keyserling und Blasius, *Wirbelthiere Europ.* p. 38 (1840).

Subfamily **Zapodinæ**.—Molars rooted; $p\ \frac{1}{1}$, $m\ \frac{3}{3}$; cervical vertebræ free; hind limbs elongated; metatarsals separate; hind feet with five digits. Nearctic region.

Zapus.[1]—The American Jumping-Mouse (*Z. hudsonianus*) extends over almost the whole North-American continent from Labrador to Mexico.

Subfamily **Dipodinæ**.—Molars rooted; $p\ \frac{(0-1)}{(0-1)}$, $m\ \frac{3}{3}$; cervical vertebræ more or less ankylosed; hind limbs elongated; metatarsals united; hind feet with only three functional digits. Palæarctic and Ethiopian regions.

This subfamily includes the true Jerboas, and contains three genera: *Dipus*[2] with three toes, and *Alactaga*[3] and *Platycercomys*[4] with five, the outer two not reaching to the ground. The latter is distinguished by the absence of premolars, and comprises many species extending from Siberia to Nubia.

Remains of the existing *Alactaga decumana*[5] occur in the Pleistocene of Germany, and those of *Zapus hudsonianus* in the corresponding strata of the United States. *Platycercomys* has been recorded from the Pleistocene of Northern Asia.

Subfamily **Pedetinæ**.—Molars rootless; cervical vertebræ free; hind limbs elongated; metatarsals separate; hind feet with four digits. Vertebræ: C 7, D 12, L 7, S 3, C 30. Ethiopian region.

Pedetes,[6] the Cape Jumping-Hare (*P. caffer*), by far the largest species of the family, extends from Mozambique and Angola to the Cape of Good Hope.

Section Hystricomorpha.

Skull (Fig. 213) with a stout zygomatic arch; jugal not supported below by a continuation of the maxillary zygomatic process; infraorbital vacuity large; mandible with the angular part arising from the outer side of the bony socket of the lower incisor. Clavicles perfect or imperfect; fibula distinct. One premolar in each jaw.

Family Octodontidæ.

Clavicles complete. Skull with long incisive foramina extending into the maxillæ; and usually an inferior angle to the jugal. Molars with external and internal enamel-folds; $p\ \frac{1}{1}$, except in *Ctenodactylus*. Mammæ placed high up on the sides of the body. Confined to the Ethiopian and Neotropical regions, with the excep-

[1] Coues, *Bull. U.S. Geol. Surv. Terrs.* ser. 2, No. 5, p. 253 (1873). Syn. *Jaculus*, Wagler.
[2] Gmelin, *Syst. Nat.*, vol. i. p. 157 (1788).
[3] F. Cuvier, *Proc. Zool. Soc.* 1836, p. 141.
[4] Brandt, *Bull. Ac. St. Pétersbourg*, 1844, p. 209.
[5] = *A. jaculus*, Auct.
[6] Illiger, *Prodromus Syst. Mamm.* p. 81 (1811).

tion of one species of *Echinomys* which ranges into Central America. Habits mostly terrestrial, but occasionally fossorial or natatorial.

Subfamily **Ctenodactylinæ.**—Molars semi-rooted; jugal as in *Dipodidæ;* the two inner toes of the hind feet with a horny comb and rigid bristles. Ethiopian region.

Ctenodactylus.[1]— Represented only by *C. gundi* from North Africa, on the borders of the Sahara. Has no premolars; each foot has four digits; the hind limbs are rather longer than the fore; the ears small; and the tail reduced to a stump. This animal is about the size of the Water-Vole, and dwells on rocky ground, its habits

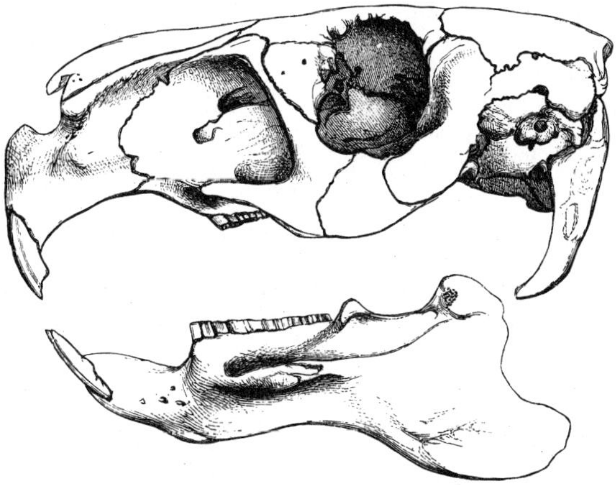

FIG. 213.—Skull of *Hydrochœrus capybara* (reduced).

being diurnal. The peculiar comb-like inner toes are employed for dressing the fur.

Pectinator.[2]—Closely allied to the preceding, but with a minute premolar in each jaw; and a moderately long and bushy tail. One species (*P. spekei*), from Somali-land.

Subfamily **Octodontinæ.**—Molars semi-rooted or rootless, with simple enamel-folds; fur soft. There are some six existing genera, including Rat-like species, all of which are South American, except *Petromys*, which is Ethiopian.

Octodon.[3]—Upper and lower molars alike; ears moderate; tail of medium length and tufted. Vertebræ: C 7, D 12, L 7, S 4, C 25. Typically represented by *C. cumingi* of Chili and Peru, with other species from Chili and Bolivia. They live in large communities.

[1] Gray, *Spicilegia Zoologica*, p. 10 (1830).

[2] Blyth, *Journ. As. Soc. Bengal*, vol. xxxiv. p. 294 (1855).

[3] Bennett, *Proc. Zool. Soc.* 1832, p. 46.

Habrocoma.[1]—Lower molars more complex than the upper; ears large; and fur extremely soft. Two Bolivian species.

Schizodon.[2]—One species, inhabiting elevated spots in the Southern Andes, and characterised by the enamel-folds of the upper molars meeting in the middle line. The external characters are much the same as in *Ctenomys*, but the ears are larger and the claws shorter.

Ctenomys.[3]—Incisors broad; molars rootless, with kidney-shaped crowns; last molar small and cylindrical; eyes and ears very small; claws larger than the toes. Some four species. Fossil remains are common in the Pleistocene of Buenos Ayres and the cavern-deposits of Brazil. Habits fossorial.

Spalacopus.[4]—Represented by two Chilian species, distinguished from the preceding genus by the rudimentary ears. These rodents store up magazines of food in their burrows.

Petromys.[5]—The South African *P. typicus* is closely allied to *Spalacopus*, but differs by its harsh fur, the shortness of the pollex. and the somewhat bushy tail. The teeth are semi-rooted, with single inner and outer enamel-folds, nearly meeting in the middle.

Subfamily **Echinomyinæ.**—Molars semi-rooted or rootless, with deep and curved enamel-folds; fur more or less harsh, frequently mixed with spines; tail generally long. One Ethiopian genus, and the remaining nine or so Neotropical. Many of the species are of large size, some being arboreal and others aquatic.

Myopotamus.[6]—Incisors very large; molars with two internal and two external enamel-folds in the upper, and three internal and one external in the lower jaw, last molar the largest; ears moderate; tail about two-thirds the length of the head and body, scaly, and sparsely haired; hind feet webbed; five digits. Vertebræ: C 7, D 13, L 6, S 4, C 25. The well-known Coypu (*M. coypu*), the only existing representative of this genus, is one of the largest living members of the order, and attains a length of about 2 feet. It is common in South America, living in burrows near water, and feeding on aquatic plants. Fossil remains of the genus occur in the caverns of Brazil, as well as in the Tertiaries of Argentina.

Capromys.[7]—This genus comprises arboreal forms from the West Indies allied to the Coypu, but, according to Dr. G. E. Dobson, showing signs of affinity with the *Hystricidæ.* The incisors are smaller than in the Coypu, and the upper molars have one internal

[1] Waterhouse, *Proc. Zool. Soc.* 1837, p. 30. Amended from *Abrocoma.*
[2] Waterhouse, *Proc. Zool. Soc.* 1841, p. 91.
[3] De Blainville, *Bull. Soc. Philom.* 1826, p. 62.
[4] Wagler, *ibid.* p. 1219.
[5] Andrew Smith, *S. African Quart. Journ.* vol. ii. p. 2 (1831).
[6] Geoffroy, *Ann. du Muséum*, vol. vi. p. 81 (1805).
[7] Desmarest, *Mém. Soc. d'Hist. Nat.* vol. i. p. 44 (1822).

and two external enamel-folds; the ears are comparatively small; the tail usually of considerable length, and the general form somewhat Rat-like. The typical *C. pilorides* is somewhat smaller than the Coypu, and is confined to Cuba; it is remarkable for the sub-division of the lobes of the liver into a number of lobules. *C. brachyurus* and *C. prehensilis* are also confined to Cuba. In Jamaica the genus is represented by *C. melanurus*, which is somewhat smaller than a Rabbit, and has no secondary lobulation of the liver.[1]

Aulacodus.[2]—Upper incisors with three deep grooves; molars as in *Capromys*. Fur very harsh; tail moderate, sparsely haired; manus with rudimentary pollex, and small fifth digit; pes with no hallux, and rudimental fifth digit. One species (*A. swinderianus*), from Western and Southern Africa, which attains a length of nearly 2 feet, and dwells in burrows.

Plagiodon.[3]—Allied to *Capromys*, but with the enamel-folds of the molars very complex, and forming a kind of zig-zag pattern in those of the upper jaw. Represented only by *P. ædium* of Hayti and Jamaica.

Loncheres[4] and *Echinomys.*[5]—These genera include small South American species, in most of which flattened lanceolate spikes are mingled with the fur. The majority of the species occur in Guiana and Brazil, but one species of *Echinomys* has been recorded from Central America. Fossil remains of both genera occur in the cavern-deposits of Brazil.

Mesomys.[6]—This genus resembles *Loncheres* externally, but the pollex has a short curved claw, and there are no spines in the fur.

Dactylomys.[7]—A Brazilian genus presenting the following distinctive features. Ears short; tail long and scaly; pollex minute; third and fourth digits of manus elongated, with short convex nails. Incisors flat; molars divided into two lobes, each of which has a single enamel-fold. Represented by two species, *D. typus* and *D. amblyonyx*, both of which seem to be rare and but little known. In the elongation of some of the digits *Dactylomys* recalls *Chiromys* among the Primates.

Cercomys.[8]—This South American genus is usually placed near *Carterodon*, from which it is readily distinguished by the pointed muzzle and the plain incisors.

[1] For description and anatomy of this species see Dobson, *Proc. Zool. Soc.* 1884, p. 233.

[2] Temminck, *Monographies des Mammifères*, vol. i. p. 245 (1827).

[3] Cuvier, *Ann. Sci. Nat.* sér. 2, vol. vi. p. 347 (1836). Amended.

[4] Illiger, *Prodromus Syst. Mamm.* p. 90 (1811).

[5] Desmarest, *Nouv. Dict. d'Hist. Nat.* vol. x. p. 45 (1817). Amended from *Echimys*.

[6] Wagner, *Wiegmann's Archiv*, 1845, pt. 2, p. 145.

[7] Geoffroy, *Ann. Sci. Nat.* sér. 2, vol. x. p. 126 (1838).

[8] F. Cuvier, *Mammifères*, 6me livr. (1829).

Carterodon.[1]—This genus, which was originally described upon the evidence of skulls from the Brazil caves, but subsequently found living, is readily distinguished by the broad and grooved incisors. The upper molars have one inner and two outer enamel-folds; those of the lower jaw being the reverse of this.

Fossil Forms.—Remains of the existing genus *Loncheres* occur in the Brazilian cave-deposits, which also yield the extinct *Dicolpomys.* A large number of fossil *Octodontidæ* from the Tertiaries of South America have been described under many generic names, but it will be sufficient to mention that *Phloramys* and *Pithanotomys* are considered to be allied to *Ctenomys;* while *Morenia*, *Orthomys*, and *Trilodon* show affinity to *Myopotamus*. *Pellegrinia*, from the Pleistocene of Sicily, may be allied both to *Ctenodactylus* and *Octodon.*

Family THERIDOMYIDÆ.

This extinct family, which is represented in the Tertiaries of Europe and the United States, comprises several genera of comparatively small Rodents, which are regarded by Dr. Schlosser as nearly related to the *Octodontidæ*, although connected by *Archæomys* with the *Chinchillidæ.* The dental formula is the same as in the *Octodontidæ.* In the typical genus *Theridomys*, from the Lower Miocene and Upper Eocene of Europe, the molars are rooted, and have three or four re-entering enamel-folds, which form isolated discs on the worn crowns. *Syllophodus*, from the Miocene of the United States, is closely allied. *Protechinomys* and *Trechomys* are genera from the Phosphorites of Central France with rooted molars; while in *Archæomys* of the same deposits the molars are rootless, with the enamel-folds dividing their crowns into laminæ, as in the Chinchillas.

Family HYSTRICIDÆ.

Build stout. Limbs subequal. A number of long and stout spines in the integument. Facial portion of skull short and broad, and the jugal without an inferior angle. Molars with external and internal enamel-folds; completely or partly rooted.

Subfamily **Synetherinæ.**—Molars rooted; clavicles complete; upper lip not cleft; soles tuberculated; pollex absent; four mammæ; tail generally prehensile; spines mixed with long hairs. This group is confined to America, all the forms except one being arboreal, and their habits less strictly nocturnal than in the next subfamily. There are three genera.

Erethizon.[2]—Represented by the common Canadian Porcupine

[1] Waterhouse, *Nat. Hist. of Mamm.* vol. ii. p. 351 (1848).

[2] F. Cuvier, *Dents des Mammifères*, p. 256 (1825).

(*E. dorsatus*), a stout heavily-built animal, with long hairs almost or quite hiding the spines; four anterior and five posterior toes; and a short stumpy tail. It is a native of the greater part of Canada and the United States where there is any remnant of the original forest left. Remains of *Erethizon* occur in cavern-deposits in Pennsylvania.

Synetheres.[1]—This genus contains some eight or ten species, known as Tree Porcupines (Fig. 214), found throughout the tropical

FIG. 214.—The Tree Porcupine (*Synetheres prehensilis*).

parts of South America, and one of them extending northwards into Mexico. They are of a lighter build than the Ground Porcupines, are covered with short, close, many-coloured spines, often mixed with hairs, and their tails are always prehensile. Their hind feet have only four toes, owing to the suppression of the hallux; but they have a peculiar fleshy pad on the inner side of the foot, between which and the toes boughs and other objects can be firmly grasped as with a hand. Vertebræ: C 7, D 17, L 5, S 3, C 36. An extinct species of this genus has been described from the cavern-deposits of Brazil.

[1] F. Cuvier, *Mém. du Muséum*, vol. ix. p. 413 (1822). "Sinéthère."

Chætomys.[1]—Distinguished by the shape of its skull and the greater complexity of its teeth. It contains only one species (*C. subspinosus*), a native of the hottest parts of Brazil.

Subfamily **Hystricinæ.**—Molars semi-rooted; clavicles incomplete; soles smooth; a rudimentary pollex; six mammæ; tail not prehensile. Now confined to the Old World, where they occur in Southern Europe, Africa, India, and the Malay Archipelago as far eastwards as Borneo. Habits terrestrial and nocturnal. Three genera.

Hystrix.[2]—This genus is readily characterised by the inflated skull, in which the nasal chamber is often considerably larger than

FIG. 215.—The Common Porcupine (*Hystrix cristata*).

the brain-case, and by the short tail, tipped with numerous slender stalked open quills, which make a loud rattling noise when the animal moves. Vertebræ: C 7, D 15, L 4, S 4, C 12. The best-known member is the Common Porcupine (*H. cristata*, Fig. 215), which occurs throughout Southern Europe and North and West Africa, but is replaced in South Africa by *H. africæ-australis*, and in India by the Hairy-nosed Porcupine (*H. leucura*).

The following account of the habits of the last-named species is from Dr. Jerdon: "*Hystrix leucura* is found over a great part of India, from the lower ranges of the Himalayas to the extreme south, but does not occur in lower Bengal, where it is replaced by *H. bengalensis.* It forms extensive burrows, often in societies, in the sides of hills, banks of rivers and nullas, and very often in the

[1] Gray, *Proc. Zool. Soc.* 1843, p. 21.
[2] Linn. *Syst. Nat.* 12th ed. vol. i. p. 76 (1766).

dams of tanks, and in old mud walls, etc. In some parts of the country they are very destructive to various crops, potatoes, carrots, and other vegetables. They never issue forth till after dark, but now and then one will be found returning to his lair in daylight. Dogs take up the scent of the Porcupine very keenly, and on the Nilghiris I have killed many by the aid of dogs, tracking them to their dens. They charge backwards at their foes, erecting their spines at the same time, and dogs generally get seriously injured by their strong spines, which are sometimes driven deeply into the assailant. The Porcupine is not bad eating,—the meat, which is white, tasting something between pork and veal."

Besides these three large crested species of *Hystrix*, there are four or five smaller species without nuchal crests occurring in North-East India and in the Malay region, from Nipal to Borneo.

Fossil species of *Hystrix* occur in the Pleistocene and Pliocene of India, and in Europe from the Upper Pliocene to the Middle Miocene, being perhaps also represented in the French Phosphorites. Remains from the Pliocene and Miocene of the United States have been referred to this genus, and if rightly determined are of especial interest from a distributional point of view.

Atherura.[1]—The Brush-tailed Porcupines are much smaller animals than the last, characterised by their long tails tipped with bundles of peculiar flattened spines. Of the three species two are found in the Malay region and one in West Africa. A fossil species occurs in the cavern-deposits of Madras.

Trichys.[2]—This genus contains but one Bornean species (*T. guentheri*), externally very like an *Atherura*, but differing from the members of that genus in many important cranial characters.

Family CHINCHILLIDÆ.

Terrestrial forms, with elongated hind limbs, bushy tails, very soft fur, and complete clavicles. Jugal without an inferior angle, and extending forwards to the lachrymal; palate contracted in front and deeply emarginate behind; incisors short, and the molars divided by continuous enamel-folds into transverse laminæ. Neotropical region. This family includes only three existing species, divided into as many genera.

Chinchilla.[3]—In this genus the fore feet have five and the hind four digits, the tail is long and bushy, and the auditory bullæ are enormous, appearing on the top of the skull. The one species (*C. lanigera*) is restricted to the alpine zones of the Andes from the north of Peru to the south of Chili. It is a Squirrel-like Rodent,

[1] Cuvier, *Règne-Animal*, 2d ed. vol. i. p. 215 (1829). "Atherure."

[2] Günther, *Proc. Zool. Soc.* 1876, p. 739.

[3] Bennett, *Gardens, etc. Zool. Soc.* pt. i. p. i. (1829).

about 10 inches in length, the tail somewhat exceeding 5 inches, and the ears very large. Its fur is greatly valued on account of its extreme softness and delicate gray colour.

Lagidium[1] and *Lagostomus*.[2]—*Lagidium* has four digits in both fore and hind feet, and *Lagostomus* three only in the hind feet, and the auditory bullæ are much smaller than in the preceding genus. *Lagidium* has the same distribution as *Chinchilla*; while *Lagostomus*, as represented by the Viscacha (*L. trichodactylus*), is found in the Pampas from the Uruguay River to the Rio Negro. The Viscachas live in burrows, generally in large numbers, and are nocturnal in their habits. Remains referable to the existing species, as well as others which appear to belong to extinct forms, occur in the Pleistocene deposits of South America.

Extinct Genera. — Several Rodents from the South American Tertiaries more or less closely allied to *Lagostomus* have been described by Dr. Ameghino under the names of *Prolagostomus*, *Pliolagostomus*, etc. The huge *Megamys* (*Potamarchus*), from the infra-Pampean deposits of Parana and Patagonia, is referred to this family, and has dimensions approximating to those of an Ox. Other fossil genera have received the names of *Epiblema* and *Tetrastylus*.

Family CASTOROIDIDÆ.

Castoroides.[3]—The large Beaver-like Rodent with the dimensions of a Bear from the Pleistocene of the United States described under this name is regarded by Dr. Coues as the type of a family. Its dentition is nearest to that of *Chinchilla* and *Hydrochœrus*, but some of the cranial characters are like those of the *Castoridæ*. The genera *Amblyrhiza* and *Loxomylus*, from the Pleistocene of the Antilles, appear to be allied types.

Family DASYPROCTIDÆ.

Terrestrial forms with subequal limbs, hoof-like claws, short or obsolete tail, and rudimentary clavicles. Mandibular masseteric ridge obsolete; palate broad; incisors long; molars semi-rooted, with external and internal enamel-folds. Neotropical region.

Dasyprocta.[4]—Includes several slender-limbed species, with three hind toes, commonly called Agoutis, inhabiting Central and South America, one (*D. cristata*) extending into the West-Indian Islands. Numerous fossil remains of this genus occur in the cavern-deposits of Brazil.

[1] Meyer, *Nova Acta Ac. Cæs. Leop.-Car.* vol. xvi. p. 576 (1833).
[2] Brooks, *Trans. Linn. Soc.* vol. xvi. p. 102 (1828).
[3] Foster, *Second Rep. Geol. of Ohio*, p. 81 (1838).
[4] Illiger, *Prodromus Syst. Mamm.* p. 93 (1811).

Cœlogenys.[1]—This genus is readily characterised by the presence of five hind toes, and the extraordinary development of its zygomatic arches, which are enormously expanded vertically, forming great convex bony capsules on the sides of the face, enclosing on each side a large cavity lined with mucous membrane, and communicating by a small opening with the mouth. The Paca C. *paca*) is about 2 feet long, and, like the species of *Dasyprocta*, lives generally in the forests or along the banks of rivers. This species appears to date from the epoch of the Pleistocene deposits of the Brazilian caves. A smaller species from Ecuador, living at elevations of from 6000 to 10,000 feet, has been described as *C. taczanowskii.*

Family DINOMYIDÆ.

Distinguished from the *Dasyproctidæ* by the cleft upper lip, rather long and bushy tail, the presence of four digits in both fore and hind feet, and the complete clavicles. The manubrium is broad; the optic foramina are confluent; the incisors broad; and the molars rootless, with enamel-folds dividing them into transverse laminæ.

Dinomys.[2]—The sole representative of this family is the Rodent known as *D. branicki*, of which hitherto only a single specimen has been obtained. This was captured in Peru, where it was found at daybreak walking about a courtyard; the inhabitants of the district were previously unacquainted with the species, from which its extreme rarity may be inferred. Externally it resembles much the Paca, having similar S-like nostrils; but in the laminated molars, and many features of the skeleton, it differs from all the other Rodents with hoof-like nails. It is regarded by its describer, the late Professor Peters, as a connecting link between the *Octodontidæ*, *Chinchillidæ*, *Dasyproctidæ*, and *Caviidæ*.

Family CAVIIDÆ.

Terrestrial or natatorial forms, with short incisors, strong mandibular masseteric ridges, long and curved paroccipitals, and palate contracted in front. Fore feet with four digits, hind feet with three. Clavicles imperfect. Molars divided by enamel-folds into transverse laminæ; milk-teeth shed before birth. Other characters as in *Dasyproctidæ*. Neotropical region.

Cavia.[3]—Limbs and ears short, subequal; tail none. Vertebræ: C 7, D 13, L 6, S 4, C 7. This genus includes several species widely

[1] F. Cuvier, *Ann. du Muséum*, vol. x. p. 203 (1807).
[2] Peters, *Monatsber. Ak. Berlin*, 1873, p. 551.
[3] Pallas, *Misc. Zool.* p. 30 (1766); *ex* Klein.

distributed throughout South America, extending even to the Straits of Magellan. The Restless Cavy (*C. porcellus*), which is found throughout Uruguay and Brazil, has been very generally regarded as the ancestral form of the domesticated Guinea-Pig. It is about 10 inches long, and weighs a little over a pound; its fur is long and of a nearly uniform grayish-brown colour. This species is rarely found in dry sandy localities, preferring marshes covered with aquatic plants, among which it lies concealed, feeding in the early morning and after sunset in the evening; but when the soil is dry it forms burrows. It lives in societies of from six to eighteen individuals, breeding but once a year, with one, or at most only two, young at a birth. The Guinea-Pig (probably a misnomer of Guiana-Pig) is larger than *C. porcellus*, and is regarded by Dr. Nehring as descended from another species, *C. cutleri*. It is white in colour, with irregular patches of reddish-brown and black. The Bolivian Cavy (*C. boliviensis*), found throughout the higher regions of Bolivia, usually at an elevation of 10,000 or 12,000 feet, is exceedingly shy, and lives in burrows, which in some districts are so numerous as to have completely undermined the soil. The Rock-Cavy (*C. rupestris*), distinguished by its short, blunt nails, is found in rocky situations throughout Brazil, and is much sought after for its flesh. The Southern Cavy (*C. australis*), common along the coast of Patagonia, forms deep burrows, with several outlets, in sandy declivities. Remains of existing species of *Cavia* are found in the cavern-deposits of Lagoa Santa, Brazil.

Dolichotis.[1]—Characterised by the great length of the ears and the short tail. The palate is so much contracted in front that the premolars of opposite sides touch by their antero-internal edges. Vertebræ: C 7, D 12, L 8, S 3, C 10.

The Patagonian Cavy (*D. patachonica*)—the only living representative of the genus—is rather larger than a Hare, which it somewhat resembles in external appearance. It inhabits the dry sterile districts of Patagonia and La Plata, disappearing wherever the country becomes more humid. This animal burrows in the earth, although in districts where the Viscacha is found it is said to avail itself of the works of the latter. Unlike other cavies, its eyes are protected from the glare of the sun by prominent eyelashes. The body is covered with a long dense fur of a rusty colour. Two young are produced at a birth. Three species of *Dolichotis* have been described from the Brazilian cave-deposits, one of which is probably not really separable from the existing form.

Hydrochœrus.[2]—A large aquatic form with all the feet fully webbed; the skull (Fig. 213, p. 481) large, with enormous par occipital processes; and the molars very complex, the third upper

[1] Desmarest, *Mammalogie*, p. 360 (1822).

[2] Erxleben, *Syst. Règ. Animal*, p. 191 (1777); *ex* Brisson.

one having some twelve transverse laminæ. Upper incisors grooved. Vertebræ: C 7, D 14, L 6, S 3, C 8.

The Capybara (*H. capybara*) is the largest existing Rodent, and the only living representative of the genus. It is a bulky and stoutly built animal, and attains a length of about 4 feet. The body is covered with long and coarse hair, reddish-brown above and brownish-yellow beneath. Capybaras are found over the whole of the eastern part of South America, and to the westward range into Bolivia and Peru. They frequent the borders of rivers and lakes, concealing themselves among reeds and other water plants. Remains of *Hydrochœrus* are found in the cavern-deposits of Brazil, which are probably referable to the existing species; one extinct species from the Pleistocene of Buenos Ayres is estimated to have attained a length of 5 feet, while *H. magnus* of the same deposits was of still larger dimensions. The genus is also represented in the Pleistocene of South Carolina and the infra-Pampean beds of Parana.

Extinct Genera.—A number of South American fossil Rodents have been referred to extinct genera of *Caviidæ*. Thus *Plexochœrus*, from the Tertiary of Argentina, differs from *Hydrochœrus* in having only nine laminæ in the last upper molar; *Cardiomys*, *Cardiatherium*, etc., from the infra-Pampeans are also stated to be allied to *Hydrochœrus*, while *Contracavia*, of the same deposits, is related to *Cavia*, but of larger size. *Microcavia*, again, from the Pleistocene of Argentina, is regarded as connecting *Cavia* with *Dolichotis*. The Tertiary European genera *Issiodoromys* and *Nesocerodon* are apparently referable to the present family.

Suborder DUPLICIDENTATA.

Two pairs of incisors in the upper jaw (the second very small, and placed directly behind the large first pair), the enamel of which extends round to their posterior surfaces. At birth there are three pairs of these incisors, but the outer one on each side is soon lost. Incisive foramina large and usually confluent; bony palate very narrow from before backwards; no true alisphenoid canal; fibula ankylosed to the tibia, and articulating with the calcaneum. Testes permanently external. This suborder includes the Picas, Hares, and Rabbits, all of which are strictly terrestrial.

Family LAGOMYIDÆ.

Complete clavicles, subequal limbs, no external tail, and short ears. Skull depressed, frontals contracted and without postorbital processes; $p\,\frac{1}{1}$ or $\frac{2}{2}$; molars rootless, with transverse enamel-folds. Palæarctic and Nearctic.

Lagomys.[1]—Represented by about a dozen species of small

[1] Cuvier, *Tabl. Élément. de l'Hist. Nat.* p. 132 (1798).

Guinea-Pig-like animals, inhabiting chiefly the mountainous parts of Northern Asia (from 11,000 to 14,000 feet), one species only being known from South-East Europe, and another from the Rocky Mountains.

The Picas, or Tailless Hares, live in holes among the rocks of their native mountains, and are agile and shy little creatures. The genus is well represented through the upper and middle Tertiaries. It has been proposed to separate those fossil forms with $p\ \frac{2}{1}$ as *Myolagus*, and those with $p\ \frac{1}{1}$ as *Titanomys*, but this seems scarcely advisable.

Family LEPORIDÆ.

Imperfect clavicles, elongated hind limbs, short recurved tail, and long ears. Skull (Fig. 216) compressed, frontals with large wing-shaped post-orbital processes $p\ \frac{3}{2}$; molars as in the *Lagomyidæ*. Cosmopolitan (except Australasia). Vertebræ: C 7, D 12, L 7, S 4, C 13-15.

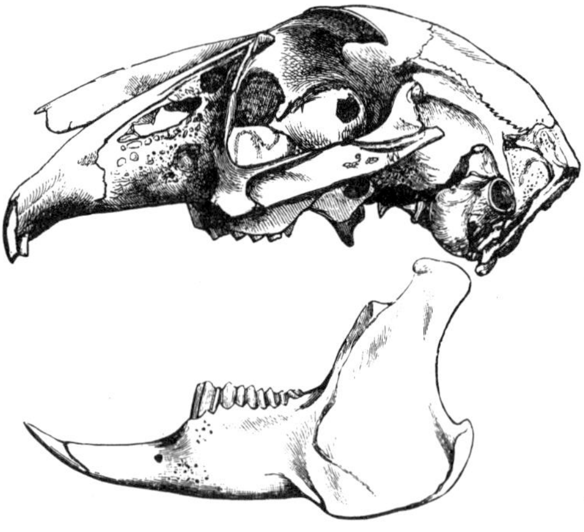

FIG. 216.—Skull of Hare (*Lepus timidus*).

Lepus.[1] — The single genus *Lepus* includes about twenty species, all of which resemble one another in general external characters. In all the fore limbs have five and the hind only four digits, and the soles of the feet are densely clothed with hairs similar to those covering the legs; the inner surface of the cheeks is also hairy. Although the family has such a wide distribution, the greater number of the species are restricted to the Palæarctic and Nearctic regions, only a single species (*L. brasiliensis*) extending into South America, where it has existed since the date of the Pleistocene deposits of the Brazilian caves.

The Common Hare (*L. timidus*[2]) may be taken as a typical example of the genus, and is characterised by the great length of

[1] Linn. *Syst. Nat.* 12th ed. vol. i. p. 77 (1766).

[2] From the absence of the Common Hare in Scandinavia it is considered probable that the name *L. timidus* was really applied to the Mountain Hare, and some writers accordingly use the name *L. europæus* for the former.

the ears and hind limbs. It is found in all parts of Europe except the north of Russia, the Scandinavian peninsula, and Ireland. Its fur is usually of a tawny gray colour above and white beneath, with the upper surface of the short tail and the tips of the ears black. The colour of the fur differs, however, considerably in different latitudes and at different seasons of the year; showing a tendency to become white during winter in northern countries, while assuming a reddish-yellow hue in the more genial climate of southern Europe. The Hare is a nocturnal animal, remaining during the day on its "form," as the slight depression is called which it makes in the open field, usually among grass.

FIG. 217.—The Common Hare (*Lepus timidus*).

The Mountain Hare (*L. variabilis*) is found throughout the northern part of the Palæarctic region, ranging from Ireland in the west to Japan in the east, and also occurring in several of the more southerly mountain ranges, such as the Pyrenees, the Alps, and the Caucasus. It is smaller than the common species, with a smaller and more rounded head, and shorter ears, tail, and hind limbs. In cold climates the colour of the whole animal changes in the winter to a pure white

FIG. 218.—The Mountain Hare (*Lepus variabilis*).

(as in Fig. 218), with the exception of the tips of the ears, which remain black. In Ireland no winter change of colour takes place.

The Rabbit (*L. cuniculus*), speaking of the wild race only, is distinguished from the Hare externally by its smaller size, shorter ears and feet, the absence or reduction of the black patch at the tip of the ears so characteristic of the Hare, and by its grayer colour. The skull is smaller and lighter, with a slenderer muzzle and a longer and narrower palate. Besides these characters, however, the Rabbit is sharply separated from the Hare by the fact that it brings forth its young naked, blind, and helpless; to compensate

FIG. 219.—The Rabbit (*Lepus cuniculus*).

for this, it digs a deep burrow in the earth in which they are born and reared, while the young of the Hare are born fully clothed with fur, and able to take care of themselves in the "form" in which they are born. The weight of the Rabbit is from $2\frac{1}{2}$ to 3 lbs., although individuals perfectly wild have been recorded up to more than 5 lbs. Its general habits are too well known to need a detailed description here. It breeds from four to eight times a year, bringing forth each time from three to eight young. Its period of gestation is about thirty days, and it begins to breed when six months old. It attains to an age of about seven or eight years.

The geographical distribution of the Rabbit presents many most interesting peculiarities. It is believed to be originally a native of the western half of the Mediterranean basin only, and still abounds in Spain, Sardinia, Southern Italy, Sicily, Greece, Tunis, and Algeria; and many of the Islands adjoining these countries are quite overrun with it. Thence it has spread, partly by man's agency, northwards throughout temperate Western Europe, increasing rapidly wherever it gains a footing; and this extension is still going on, as is shown by the case of Scotland, in which sixty years ago Rabbits were little known, while they are now found in all suitable localities up to the extreme north. It has also gained admittance into Ireland, and now abounds there as much as in England. Out of Europe the same extension of range has been going on. In New Zealand and Australia Rabbits, introduced either for profit or sport, have increased to such an extent as to form one of the most serious pests that the farmers have to contend against, as the climate and soil seem to suit them perfectly, and their natural enemies are too few and too lowly organised to keep their numbers within reasonable bounds. In other cases Rabbits introduced into islands have become or remained more or less distinct from their parent stock; thus the Rabbits both of the Falkland Islands and of Jamaica still show traces of their descent from domesticated varieties, and have never reverted to the ordinary brownish-gray type. And again, as was pointed out by Mr. Darwin,[1] the Rabbits in the island of Porto Santo, near Maderia, whose ancestors were introduced from Spain in 1418 or 1419, have formed quite a distinct diminutive race, barely half the bulk or weight of English Rabbits, and differing in certain slight details of colour and habits.

Bibliography of Rodentia.—G. R. Waterhouse, "Observations of the Rodentia," *Mag. Nat. Hist.* iii. (1839); Id. *Ann. Nat. Hist.* viii. and x. (1839-42); Id. "On the Geographical Distribution of the Rodentia," *Proc. Zool. Soc.* 1839, pp. 162-174; Id. *Natural History of the Mammalia*, vol. ii. "Rodentia" (1848); Gervais, *Dic. Univ. d'Hist. Nat.* xi. p. 202 (1848); Brandt, "Untersuchungen über die craniologischen Entwickelungsstufen und Classification der Nager der Jetzwelt," *Mém. de l'Acad. Impér. de St. Pétersbourg* (1855): Lilljeborg, *Systematisk Œfversight af de Gnagnde Däggdjuren*, Upsala, 1866; Alston, "On the Classification of the Order *Glires*," *Proc. Zool. Soc.* 1876, pp. 61-98; Trouessart, "Catal. de Rongeurs, Vivants et Fossiles," *Bullet. Soc. d'Études Scient. d'Angers*, 1880-81; Coues and Allen, "Monographs of North American Rodentia," *United States Geol. Surv. of Territories*, vol. xi. (1877); Winge, "Rodentia pa Lagos Santa, Brazil," *Mus. Lund.* vol. iii. (1887); various papers by Peters in *Monatsber. Ak. Berlin*, and by Alston, Anderson, Blanford, Dobson, Milne-Edwards, Thomas, and others, in *Proc. Zool. Soc.*, *Journ. Asiat. Soc. Beng.*, *Ann. Mag. Nat. Hist.*, etc.

[1] *Variations of Animals and Plants*, 2d ed. vol. i. p. 119.

CHAPTER XI

THE ORDER CARNIVORA

THOUGH the existing Carnivora as at present restricted[1] form a very natural and well-defined order among the Mammalia, it is difficult to find any important common diagnostic characters by which they can be absolutely separated; so that, as in the case of so many other natural groups, it is by the possession of a combination of various characters that they must be distinguished. Thus they are all unguiculate, and never have less than four well-developed toes on each foot, with nails more or less pointed, rarely rudimentary or absent. The pollex and hallux are never opposable to the other digits. They are regularly diphyodont and heterodont, and their teeth are always rooted.[2] Their dentition consists of small pointed incisors, usually three in number, on either side of each jaw, of which the first is always the smallest and the third the largest, the difference being most marked in the upper jaw; strong conical, pointed, recurved canines; cheek-teeth variable, but generally, especially in the anterior part of the series, more or less compressed, pointed, and trenchant; if the crowns are flat and tuberculated they are never complex or divided into lobes by deep inflexions of enamel. The condyle of the lower jaw is a transversely placed half-cylinder working in a deep glenoid fossa of corresponding form. The brain varies much in relative size and form, but the hemispheres are never destitute of well-marked convolutions (Fig. 23, p. 71). The stomach (Fig. 234) is always simple and pyriform. The cæcum is either absent or short and simple (Fig. 235), and the colon is not sacculated, or greatly wider than the small intestine. Vesiculæ seminales are never present. Cowper's glands are present

[1] The Feræ of Linnæus included all the then known species of the modern orders Carnivora, Insectivora, and Marsupialia.

[2] The tusks of the Walrus, altogether so aberrant in its dentition, are partial exceptions to this statement, but in old individuals the pulp-cavity fills up, and they cease to grow.

in some, absent in other groups. The uterus is bicornuate. The mammæ are abdominal, and very variable in number. The placenta is deciduate, and almost always zonary. The clavicle is often entirely absent, and when present is never complete. The humerus often has an entepicondylar foramen. The radius and ulna are distinct. The scaphoid and lunar bones are united into one, and there is never a distinct os centrale in the adult. The fibula is always a distinct slender bone.

Several of these characters are, however, not applicable to all the members of the extinct group of Carnivores for which the name Creodonta has been proposed, as will be noticed in the sequel.

The large majority of the species composing this order subsist chiefly upon some variety of animal food, though many are omnivorous, and some few chiefly, though not entirely, vegetable eaters. The more typical forms live altogether on recently-killed warm-blooded animals, and their whole organisation is thoroughly adapted to a predaceous mode of life. In conformity with this manner of obtaining their subsistence they are generally bold and savage in disposition, though some species are capable of being domesticated, and when placed under favourable circumstances for the development of such qualities exhibit a very high degree of intelligence and fidelity. The existing representatives of the order are naturally divided into two suborders, the members of the one being the more typical, and mainly terrestrial in their mode of life; while those of the other are aberrant, having the whole of their organisation specially modified for living habitually in water. These are called respectively the True, or Fissiped, and the Pinniped Carnivora.

Suborder CARNIVORA VERA.

Generally adapted for terrestrial progression and mode of life, though some may be partially aquatic in their habits. The fore limbs never have the first digit, or the hind limbs the first and fifth digits, longer than the others. Incisors $\frac{3}{3}$ on each side, with very rare exceptions. Cerebral hemispheres more or less elongated; always with three or four gyri on the outer surface forming arches above each other, the lowest surrounding the Sylvian fissure. The molar series of teeth have not the uniform characters of those of the Pinnipedia. There is always one tooth in each jaw which is specially modified, and to which the name of "sectorial" or "carnassial" tooth has been applied. The teeth in front of this are more or less sharp pointed and compressed; while those behind it are broad and tuberculated. The characters of the carnassial teeth deserve special attention, as, though fundamentally the same throughout the suborder, they are greatly modified in different genera. The upper carnassial is the most posterior of the teeth which have predecessors, and is therefore reckoned as the last premolar (*p* 4 of the typical dentition). It consists essentially of a more or less compressed blade supported on two roots and an inner tubercle supported by a distinct root (see Fig. 220). The blade when fully developed has three cusps or lobes (1, 2, and 3), but the anterior is always small, and often absent. The middle lobe is conical, high, and pointed; the posterior lobe has a compressed straight knife-like edge. The inner tubercle (4) varies very much

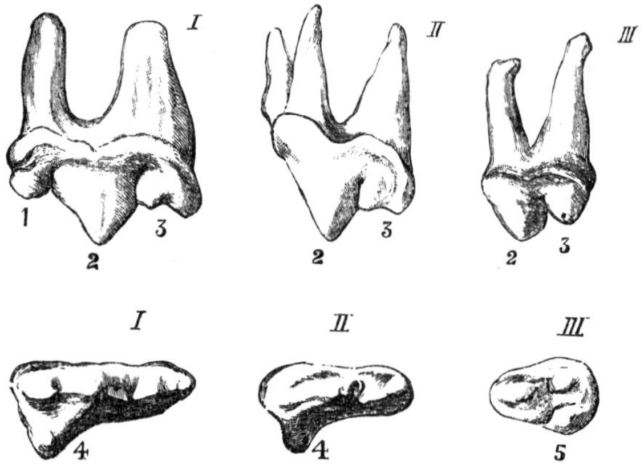

FIG. 220.—Left upper carnassial teeth of Carnivora. I, *Felis*; II, *Canis*; III, *Ursus*. 1, Anterior, 2, middle (paracone), and 3, posterior (metacone) cusp of blade; 4, inner tubercle (protocone) supported on distinct root; 5, inner cusp posterior in position, and without distinct root, characteristic of the *Ursidæ*.

in extent, but is generally placed near the anterior end of the blade, though sometimes it is median in position. In the *Ursidæ* alone both the inner tubercle and root are wanting, and there is often a small internal and posterior cusp (5) without root. In this aberrant family also the carnassial is relatively to the other teeth much smaller than in the rest of the Carnivora. The lower carnassial (see Fig. 221) is the most anterior of the teeth without predecessors in the milk-series; it is therefore reckoned the first true molar (*m* 1). It has two roots supporting a crown, consisting when fully developed of a compressed bilobed blade (1 and 2), a heel, or talon (4), and an inner cusp (3). The lobes of the blade, of which the hinder (2) is the larger, are separated by a notch, generally prolonged into a linear fissure. In the most specialised Carnivora, as the *Felidæ* (I), the blade alone is developed, both talon and inner cusp being absent or rudimentary. In others, as

Meles (V) and *Ursus* (VI), the heel is greatly developed, broad, and tuberculated. The blade in these cases is generally placed obliquely, its flat or convex (outer) side looking forwards, so that the two lobes are almost side by side, instead of anterior and posterior. The inner cusp (3) is generally conical, pointed, and placed to the inner side of the hinder lobe of the blade. The special characters of these teeth are more disguised in the Sea Otter (*Latax*) than in any other form, but even in it they can be traced.

The homology of the various parts of the Carnivorous carnassial

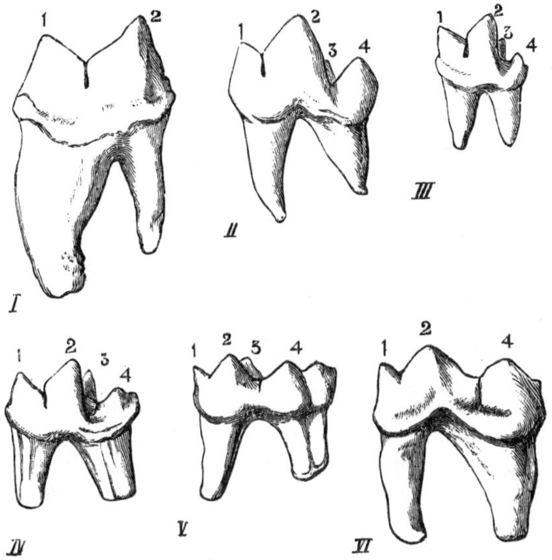

FIG. 221.—Left lower carnassial teeth of Carnivora. I, *Felis*; II, *Canis*; III, *Herpestes*; IV, *Lutra*; V, *Meles*; VI, *Ursus*. 1, Anterior lobe (paraconid) of blade; 2, posterior (protoconid) lobe of blade; 3, inner cusp (metaconid); 4, talon (hypoconid). It will be seen that the relative size of the two roots varies according to the development of the portion of the crown they have respectively to support.

with the primitive tritubercular type (p. 30) is indicated in the figures. It may be observed, however, that the anterior lobe of the three-lobed upper carnassial is an element added on to the more primitive two-lobed type. When the talon of the lower carnassial, as in *Canis*, consists of a large outer and small inner cusp, the latter (not seen in the figure) is the entoconid.

The toes are nearly always armed with large, strong, curved, and tolerably sharp claws, ensheathing the ungual phalanges, and held more firmly in their places by broad laminæ of bone reflected over their attached ends from the bases of the phalanges. In some forms, most notably the *Felidæ*, these claws are retractile; that is to say, the ungual phalanx, with the claw attached, folds back in the fore foot into a sheath by the outer or ulnar side of the middle phalanx of the digit, being retained in this position when the animal is at rest by a strong elastic ligament. In the hind foot the ungual phalanx is retracted on to the top, and not the side of the middle phalanx. By the action of the deep flexor muscles, the ungual phalanges are straightened out, the claws protruded from their sheath, and the soft "velvety" paw becomes suddenly converted into a most formidable weapon of offence. The habitual retraction of the claws preserves their points from wear in ordinary progression.

The skeleton of the Lion represented in Fig. 15 (p. 45) illustrates the digitigrade mode of progression of the *Felidæ*, as well as the essential characters of the bony framework of a typical Carnivore.

The Fissipedal Carnivora were divided by Cuvier into two groups, according to the position of the feet in walking,—the Plantigrada, or those that place the whole of the soles to the ground, and the Digitigrada, or those that walk only on the toes; and the difference between these groups was considered of equal importance to that which separated the Pinnigrada or Seals from both of them. The distinction is, however, quite an artificial one, since every intermediate condition exists between the extreme typical plantigrade gait of the Bears and the truly digitigrade walk of the Cats and Dogs; in fact, the greater number of the Carnivora belong to neither one form nor the other, but may be called "subplantigrade"; often when at rest applying the whole of the sole to the ground, but keeping the heel raised to a greater or less extent when walking.

An amended classification of the existing forms is into three distinct sections, of which the Cats, the Dogs, and the Bears may be respectively taken as representatives, and which are hence called Æluroidea, Cynoidea, and Arctoidea. This division is founded mainly on characters exhibited by the base of the skull, but is corroborated by the structure of other parts.[1] The presence or absence of a bridge of bone, covering the external carotid artery in a part of its course by the side of the alisphenoid bone, and enclosing the "alisphenoid canal" (see Fig. 8, p. 38), a character to which the late Mr. H. N. Turner first drew attention, might seem unimportant at first sight, but it is curiously constant in certain groups, which we have other reasons, derived often from a combination of less

[1] See Flower, "On the Value of the Characters of the Base of the Cranium in the Classification of the Order *Carnivora*," *Proc. Zool. Soc.* 1869, p. 4; Mivart, "On the Classification and Distribution of the *Æluroidea*," *ibid.* 1882, pp. 135 and 459; see also *The Cat, an Introduction to the Study of Backboned Animals, especially Mammals*, by the same author, 1881.

easily definable characters, to regard as natural. It is therefore generally mentioned in the following family definitions.

It must, however, be stated that while the arrangement is a convenient one as regards the existing Carnivores, it will not hold good when the fossil forms are included. Thus there is ample evidence to show that the Dogs and Bears were formerly so intimately connected that in a palæontological classification the *Canidæ* cannot be satisfactorily separated from the *Ursidæ;* while in another direction the *Canidæ* were closely allied to the ancestral *Viverridæ.* The most important objection against this classification is, however, the apparent intimate connection exhibited by fossil forms between the *Viverridæ* and the *Mustelidæ*, which, so far as the present evidence goes, tends to show that the latter are derived from the former. If this be eventually fully proved, it would seem to indicate that the Arctoidea are not a natural group; and that the resemblances between the *Ursidæ* and *Mustelidæ* have been independently acquired, in the course of the descent of the one family from a Canoid, and of the other from a Viverroid stock.

Section ÆLUROIDEA.

The Æluroidea or Cat-like Carnivores include the *Felidæ*, *Viverridæ*, *Proteleidæ*, and *Hyænidæ*. The existing representatives of this section present the following common features. Auditory bulla (Fig. 222) much dilated, rounded smooth, thin-walled, and (except in the *Hyænidæ*) divided into two chambers by a septum. Bony auditory meatus short. Paroccipital process applied to, and spread over the hinder part of the bulla (Fig. 222). Mastoid process never very salient, and often obsolete. Carotid canal (Fig. 8, p. 38, *car*) small, sometimes very inconspicuous. Condyloid and glenoid foramina concealed or wanting. Cæcum small, rarely absent. Os penis generally small and irregular (large in *Cryptoprocta*). Cowper's glands present; prostate distinctly lobed. Some details of the anatomy of the soft parts will be found under the head of *Genetta.*

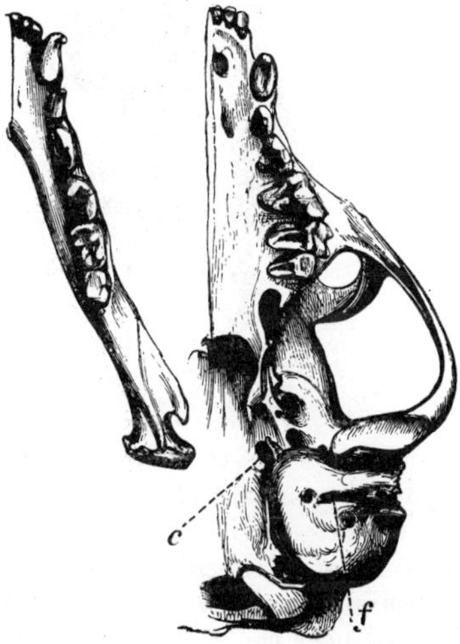

FIG. 222.—Left side of the palatal aspect of the cranium and mandible of the Suricate (*Suricata tetradactyla*). *c*, Carotid foramen; *f*, fissure in floor of auditory meatus. From Mivart, *Proc. Zool. Soc.* 1882, p. 184.

Family FELIDÆ.

In all the forms, both recent and fossil, which can be included in this family the canines are strongly developed, there are never more than one upper and two lower molars, and the three lower incisors are placed in the same horizontal line. With one exception, the humerus has an entepicondylar foramen.

The following characters are common to all the existing members. True molars reduced to one above and below, that of the upper jaw very small and transversely extended. Only two inferior premolars. Upper carnassial with three lobes to the blade; lower without talon or inner cusp. Auditory bulla not externally constricted. No alisphenoid canal. Carotid canal very minute. Digits 5-4. Dorsal vertebræ 13.

Felis.[1]—The whole structure of the animals of this genus exhibits the Carnivorous type in its fullest perfection. Dentition: $i\ \frac{3}{3}$, $c\ \frac{1}{1}$, $p\ \frac{3}{2}$, $m\ \frac{1}{1}$; total 30. A distinctly cusped inner tubercle to the upper carnassial. Claws completely retractile. The upper anterior premolar (*p.* 2), always small, and may be absent without any other modification in the dental or other structures. Such a variation should not therefore be considered as of generic importance. Incisors very small. Canines large, strong, slightly recurved, with trenchant edges and sharp points, and placed wide apart (Fig. 223). Premolars compressed and sharp pointed. The most posterior in the upper jaw (the carnassial), a very large tooth, consisting of a sub-compressed blade, divided into three unequal lobes supported by two roots, with a very small inner tubercle placed near the front end of the tooth and supported by a distinct root (Fig. 220). The upper true molar a very small tubercular tooth placed more or less transversely at the inner side of the hinder end of the last. In the lower jaw the true molar (carnassial) reduced to the blade alone, which is very large, trenchant, and

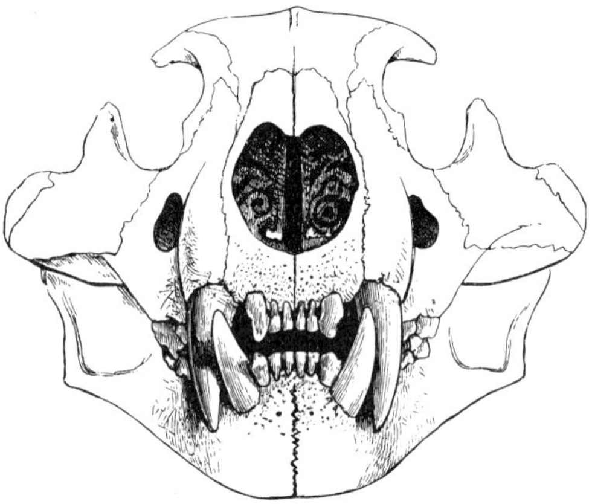

FIG. 223.—Front view of skull of Lion (*Felis leo*).

[1] Linn, *Syst. Nat.* 12th ed. vol. i. p. 60 (1766).

much compressed, divided into two subequal lobes. Occasionally it has a rudimentary talon, but never an inner cusp. The skull is generally short and rounded, though proportionally more elongated in the larger forms. The facial portion is especially short and broad, and the zygomatic arches are very wide and strong. The auditory bullæ are large, rounded, and smooth. Vertebræ: C 7, D 13, L 7, S 3, C 13-29. Clavicles better developed than in other Carnivora, but not articulating with either the scapulæ or sternum. Limbs digitigrade. Anterior feet with five toes, the third and fourth nearly equal and longest, the second slightly and the fifth considerably shorter; the pollex still shorter, not reaching as far as the metacarpo-phalangeal articulation of the second. Hind feet with only four toes. The third and fourth the longest, the second and fifth somewhat shorter and nearly equal; the hallux represented only by the rudimentary metatarsal bone. The claws all very large, strongly curved, compressed, very sharp, and exhibiting the retractile condition in the highest degree. The tail varies greatly in length, being in some a mere stump, in others nearly as long as the body. Ears of moderate size, more or less triangular and pointed. Eyes rather large. Iris very mobile, and with a pupillary aperture which contracts under the influence of light in some species to a narrow vertical slit, in others to an oval, and in some to a circular aperture. Tongue thickly covered with sharp-pointed, recurved horny papillæ. Cæcum small and simple.

As in structure so in habits, the Cats may be considered the most specialised of all the Carnivora. All the known members of the genus feed, in the natural state, almost exclusively on warm-blooded animals which they have themselves killed. One Indian species (*F. viverrina*) preys on fish and even (it is said) on freshwater molluscs. Unlike the Dogs, they never associate in packs, and rarely hunt their prey in open ground, but from some place of concealment wait until the unsuspecting victim comes within reach, or with noiseless and stealthy tread, crouching close to the ground for concealment, approach near enough to make the fatal spring. In this manner they frequently attack and kill animals considerably exceeding their own size. They are mostly nocturnal, and the greater number, especially the smaller species, more or less arboreal. None are aquatic, and all take to the water with reluctance, though some may habitually haunt the banks of rivers or pools, because they more easily obtain their prey in such situations.

The numerous species of the genus are very widely diffused over the greater part of the habitable world, though most abundant in the warm latitudes of both hemispheres. No species are, however, found in the Australian region, or in Madagascar. Although the Old-World and New-World Cats (except perhaps the Northern Lynx) are all specifically distinct, no common structural character has been pointed out by which the former can be separated from the latter. On the contrary, most of the minor groups into which the genus has been divided have representatives in both hemispheres.

Notwithstanding the considerable diversity in external appearance and size between different members of this extensive genus, the structural differences are but slight, and so variously combined in different species that the numerous attempts hitherto made to subdivide it are all unsatisfactory and artificial. The principal differences are to be found in the form of the cranium, especially of the nasal and adjoining bones, the completeness of the bony orbit posteriorly, the development of the first upper premolar and of the inner tubercle of the upper carnassial, the length of the tail, the form of the pupil, and the condition and coloration of the fur, especially the presence or absence of tufts or pencils of hair on the external ears. Writing in 1881 Professor Mivart [1] gave the number of existing species of *Felis* as 48, but by Mr. Blanford's reduction of the number of Indian species [2] the list may now be diminished to some 41. The following account is chiefly devoted to some of the more important and better known species.

A. *Old World Species.*—The Lion (*F. leo*, Fig. 224) has been well known to man from the earliest historic times. Its geographical habitat made it familiar to all the races among whom human civilisation took its origin, and its strongly marked physical and moral characteristics have rendered it proverbial, perhaps to an exaggerated degree, and have in all ages afforded favourite types for poetry, art, and heraldry. The literature of the ancient Hebrews abounds in allusions to the Lion; and the almost incredible numbers that are stated to have been provided for exhibition and destruction in the Roman amphitheatres (as many as six hundred on a single occasion by Pompey, for example) show how abundant these animals must have been within accessible distance of the capital of the world.

The geographical range of the Lion was once far more extensive than at present, even within the historic period covering the whole of Africa, the south of Asia, including Syria, Arabia, Asia Minor, Persia, and the greater part of Northern and Central India, and also the south-eastern portion of Europe, as shown by the well-known story told by Herodotus of the attacks by Lions on the Camels which carried the baggage of the army of Xerxes on its march through the country of the Pæonians in Macedonia. The very circumstantial account of that historian shows that the animal in his time ranged through the country south of the Balkans, through Roumania to the west of the River Carasu, and through Thessaly as far

[1] *The Cat*, pp. 392-426 (1881).

[2] *Fauna of British India*, "Mammalia," pp. 56-90 (1888).

south as the Gulf of Lepanto and the Isthmus of Corinth, having as its western boundary the River Potamo and the Pindus mountains. The whole of the evidence relating to the existence of Lions in Europe, and to their retreat from that continent shortly before the commencement of the Christian era, has been collected in the article on "*Felis spelæa*" in Boyd Dawkins and Sandford's *British Pleistocene Mammalia* (1868). Fossil remains attest a still wider range, as it is shown in the same work that there is absolutely no osteologi-

FIG. 224.—Lion and Lioness, after a drawing by Wolf in Elliot's Monograph of the *Felidæ*.

cal or dental character by which the well-known Cave Lion (*F. spelæa*), so abundantly found in cave-deposits of the Pleistocene age in Western Europe, can be distinguished from the existing *F. leo*.

At the present day the Lion is found in localities suitable to its habits, and where not exterminated (as it probably was in Europe) by the encroachments of man, throughout Africa from Algeria to the Cape Colony, and in Mesopotamia, Persia, and some parts of the north-west of India. According to Blanford,[1] Lions are still very numerous in the reedy swamps bordering the Tigris and Euphrates, and also occur on the west flanks of the Zagros mountains and the oak-clad ranges near Shiraz, to which they are

[1] *Zoology and Geology of Eastern Persia* (1876).

attracted by the immense herds of swine which feed on the acorns. The Lion nowhere exists in the table-land of Persia, nor is it found in Baluchistan. In India, where it is verging on extinction, it appears now to be confined to parts of Kattywar and Rajputana, though within the present century its range extended through the north-west part of India, from Bahawalpur and Sind to at least the Jumna (about Delhi), southward as far as Khandesh, and in Central India through the Saugor and Narbada territories, Bundelkund, and as far east as Palamau. It was extirpated in Harriana about 1824. One was killed at Rhyli, in the Dumaoh district, Saugor and Narbada territories, so late as in the cold season of 1847-48; and one was shot in 1810 near Kot-Deji, Sind.[1]

The great variations in external characters which different Lions present, especially in the colour and the amount of mane, has given rise to the idea that there are several species, or at all events distinct varieties peculiar to different localities. It was at one time supposed, on the authority of Captain Walter Smee,[2] that the Lion of Gujerat differed essentially from that of Africa in the absence of a mane, but subsequent evidence has not supported this view, which was probably founded upon young specimens having been mistaken for adults. Lions from that district as well as from Babylonia, which have lived in the gardens of the London Zoological Society, have had as fully developed manes as any other of the species. Mr. F. C. Selous[3] has shown that in South Africa the so-called Black-maned Lion and others with yellow scanty manes are found, not only in the same locality, but even among individuals of the same parentage.

The Lion belongs to a well-defined group, containing the largest members of the genus, and differing from the others in the well-marked character that the anterior cornu of the hyoid arch is but little ossified, so that this arch is connected with the cranium by a long ligament, instead of by a continuous chain of bones, and by the less important one that the pupil of the eye, when contracted, is a circular hole, instead of a vertical slit as in the cat. The Lion agrees with the Tiger and the Leopard in these respects, but differs from them in its uniform style of colouring, and from all the other *Felidæ* in the arrangement of its hairy covering; thus the hair of the top of the head, chin, and neck, as far back as the shoulder, is not only very much longer, but also differently disposed from the hair elsewhere, being erect or directed forwards, and so constituting the characteristic ornament called the mane. There is also a tuft of elongated hairs at the end of the tail, one upon each elbow, and in most lions a copious fringe along the middle line of the under

[1] See Blanford, *Fauna of British India*, "Mammalia," p. 57 (1888).

[2] *Transactions of the Zoological Society*, vol. i. p. 165 (1835).

[3] *A Hunter's Wanderings in Africa*, 1881, p. 258.

surface of the body, wanting, however, in some examples.[1] It must, however, be observed that these characters are peculiar to the adults of the male sex only, and that young lions show indications of the darker stripes and mottlings so characteristic of the greater number of the members of the genus.

The usual colour of the adult is yellowish-brown, but it may vary from a deep red or chestnut brown to an almost silver gray. The mane, as well as the long hair of the other parts of the body, sometimes scarcely differs from the general colour, but it is usually darker and not unfrequently nearly black. The mane begins to grow when the animal is about three years old, and is fully developed at five or six.

In size the Lion is only equalled or exceeded by the Tiger among the existing *Felidæ*; though both species present great variations, the largest specimens of the latter appear to surpass the largest Lions. A full-sized South African Lion, according to Selous, measures slightly less than 10 feet from nose to tip of tail, following the curves of the body. Harris gives 10 feet 6 inches, of which the tail occupies 3 feet. The Lioness is about a foot less. The tongue, like that of the other species of the genus, is long and flat, and remarkable for the development of the papillæ of the anterior part of the dorsal surface, which (except near the edge) are modified so as to resemble long, compressed, recurved, horny spines or claws; these, near the middle line, attaining the length of one-fifth of an inch. They give the part of the tongue on which they occur the appearance and feel of a coarse rasp, and serve the purpose of such an instrument in cleaning the flesh from the bones of the animals on which the Lions feed.

The habits of the Lion in a state of nature are fairly well known from the united observations of numerous travellers and sportsmen who have explored those districts of the African continent in which it is still common. It lives chiefly in sandy plains and rocky places interspersed with dense thorn-thickets, or frequents the low bushes and tall rank grass and reeds that grow along the sides of streams and near the springs where it lies in wait for the larger herbivorous animals on which it feeds. Although it is occasionally seen abroad during the day, especially in wild and desolate regions, where it is subject to but little molestation, the night is, as in the case of so many other predaceous animals, the period of its greatest activity. It is then that its characteristic roar is chiefly heard, as thus graphically described by Gordon Cumming:—

[1] Mr. Selous, whose opportunities for obtaining evidence upon this subject were very large, says that in the region of South Africa, between the Zambesi and the Limpopo rivers, he never saw a lion with any long hair under the body, and that the manes of the wild lions of that district are far inferior in development to those commonly seen in menageries in Europe.

"One of the most striking things connected with the Lion is his voice, which is extremely grand and peculiarly striking. It consists at times of a low, deep moaning, repeated five or six times, ending in faintly audible sighs; at other times he startles the forest with loud, deep-toned, solemn roars, repeated in quick succession, each increasing in loudness to the third or fourth, when his voice dies away in five or six low muffled sounds very much resembling distant thunder. At times, and not unfrequently, a troop may be heard roaring in concert, one assuming the lead, and two, three, or four more regularly taking up their parts, like persons singing a catch. Like our Scottish stags at the rutting season, they roar loudest in cold frosty nights; but on no occasions are their voices to be heard in such perfection, or so intensely powerful, as when two or three troops of strange Lions approach a fountain to drink at the same time. When this occurs, every member of each troop sounds a bold roar of defiance at the opposite parties; and when one roars, all roar together, and each seems to vie with his comrades in the intensity and power of his voice. The power and grandeur of these nocturnal concerts are inconceivably striking and pleasing to the hunter's ear."

"The usual pace of a Lion," C. J. Andersson[1] says, "is a walk, and, though apparently rather slow, yet, from the great length of his body, he is able to get over a good deal of ground in a short time. Occasionally he trots, when his speed is not inconsiderable. His gallop—or rather succession of bounds—is, for a short distance, very fast—nearly or quite equal to that of a horse. Indeed, unless the steed has somewhat the start when the beast charges, it will be puzzled to escape. Many instances are on record of horsemen who have incautiously approached too near to the Lion, prior to firing, who have been pulled down by him before they could get out of harm's way. Happily, however, the beast soon tires of the exertion of galloping, and unless his first rush succeeds he, for the most part, soon halts and beats a retreat." "The Lion, as with other members of the feline family," the same writer tells us, "seldom attacks his prey openly, unless compelled by extreme hunger. For the most part he steals upon it in the manner of a cat, or ambushes himself near to the water or a pathway frequented by game. At such times he lies crouched upon his belly in a thicket until the animal approaches sufficiently near, when, with one prodigious bound, he pounces upon it. In most cases he is successful, but should his intended victim escape, as at times happens, from his having miscalculated the distance, he may make a second or even a third bound, which, however, usually prove fruitless, or he returns disconcerted to his hiding-place, there to wait for another opportunity." His food consists of all the larger herbivorous animals of the country in which

[1] *The Lion and the Elephant*, 1873, p. 19.

he resides—buffaloes, antelopes, zebras, giraffes, or even young elephants or rhinoceroses, though the adults of these latter he dare not attack. In cultivated districts the cattle, sheep, and even human inhabitants are never safe from his nocturnal ravages. He appears, however, as a general rule, only to kill when hungry or attacked, and not for the mere pleasure of killing, as with some other carnivorous animals. Moreover, he by no means limits himself to animals of his own killing, but, according to Selous, often prefers eating game that has been killed by man, even when not very fresh, to taking the trouble to catch an animal himself. All books of African travel and sport abound with stories, many of which are apparently well authenticated, of the lion's prodigious strength, as exemplified by his being able to drag off a whole ox in his mouth to a long distance, even leaping fences and dykes with it.

The Lion appears to be monogamous, a single male and female continuing attached to each other irrespectively of the pairing season. At all events the Lion remains with the Lioness while the cubs are young and helpless, and assists in providing her and them with food, and in educating them in the art of providing for themselves. The number of cubs at a birth is from two to four, usually three. They are said to remain with their parents till they are about three years old. The following account by an eye-witness gives a good idea of Lion family life[1] :—

"I once had the pleasure of, unobserved myself, watching a lion family feeding. I was encamped on the Black Umfolosi in Zululand, and towards evening, walking out, about half a mile from camp, I saw a herd of zebra galloping across me, and when they were nearly 200 yards off, I saw a yellow body flash towards the leader, and saw him fall beneath the lion's weight. There was a tall tree about 60 yards from the place, and anxious to see what went on, I stalked up to it, while the lion was still too much occupied to look about him, and climbed up. He had by this time quite killed the beautifully striped animal, but instead of proceeding to eat it, he got up and roared vigorously, until there was an answer, and in a few minutes a lioness, accompanied by four whelps, came trotting up from the same direction as the zebra, which no doubt she had been to drive towards her husband. They formed a fine picture as they all stood round the carcase, the whelps tearing it and biting it, but unable to get through the tough skin. Then the lion lay down, and the lioness driving her offspring before her did the same four or five yards off, upon which he got up, and, commencing to eat, had soon finished a hind leg, retiring a few yards on one side as soon as he had done so. The lioness came up next and tore the carcase to shreds, bolting huge

[1] Hon. W. H. Drummond, *The Large Game and Natural History of South and South-East Africa*, 1875, p. 278.

mouthfuls, but not objecting to the whelps eating as much as they could find. There was a good deal of snarling and quarrelling among these young lions, and occasionally a stand-up fight for a minute, but their mother did not take any notice of them, except to give them a smart blow with her paw if they got in her way. . . . There was now little left of the zebra but a few bones, which hundreds of vultures were circling round waiting to pick, while almost an equal number hopped awkwardly about on the ground within 50 or 60 yards of it, and the whole lion family walked quietly away, the lioness leading, and the lion, often turning his head to see that they were not followed, bringing up the rear."

Though not strictly gregarious, Lions appear to be sociable towards their own species, and often are found in small troops, sometimes consisting of a pair of old Lions, with their nearly full-grown cubs, but occasionally of adults of the same sex; and there seems to be good evidence that several Lions will associate together for the purpose of hunting upon a preconcerted plan. As might be supposed, their natural ferocity and powerful armature are sometimes turned upon one another; combats, often mortal, occur among male Lions under the influence of jealousy; and Andersson relates an instance of a quarrel between a hungry Lion and Lioness over the carcase of an Antelope which they had just killed, and which did not seem sufficient for the appetite of both, ending in the Lion not only killing, but even devouring his mate. Old Lions, whose teeth have become injured with constant wear, often become "man-eaters," finding their easiest means of obtaining a subsistence in lurking in the neighbourhood of villages, and dashing into the tents at night and carrying off one of the sleeping inmates. Lions differ from most of the smaller *Felidæ* in never climbing trees; indeed, as mentioned before, they are rarely found in forests.

With regard to the character of the Lion, those who have had opportunities of observing it in its native haunts differ greatly. The exaggerated accounts of early writers as to its courage, nobility, and magnanimity have led to a reaction, which causes some modern authors to speak of it in language quite the reverse, and to accuse it of positive cowardice and all kinds of meanness. Livingstone goes so far as to say, "Nothing that I ever learned of the lion could lead me to attribute to it either the ferocious or noble character ascribed to it elsewhere," and he adds that its roar is not distinguishable from that of the ostrich. Of course these different estimates depend to a great extent upon the particular standard of the writer, and also upon the circumstance that Lions, like other animals, undoubtedly show considerable individual differences in character, and behave differently under varying circumstances. They are certainly not so reckless as to be entirely devoid of the instinct of self-preservation, and if one, perhaps satiated with a good meal the night before, unexpectedly disturbed in the day-time, will occasionally retreat when confronted, even by an unarmed man, that is scarcely a reason for assigning cowardice as one of the characteristics of the species. The latest authority, Selous, while never denying the daring courage of the Lion when hungry or provoked, and vindicating the awe-inspiring character of the roar of several Lions in unison, when heard at close quarters, as the grandest sound in nature, says with regard to its outward aspect:—

"It has always appeared to me that the word 'majestic' is singularly inapplicable to the lion in its wild state, as when seen by daylight he always has a stealthy furtive look that entirely does away with the idea of majesty. To look majestic a lion should hold his head high. This he seldom does. When walking he holds it low, lower than the line of his back, and it is only when he first becomes aware of the presence of man that he sometimes raises his head and takes a look at the intruder, usually lowering it immediately, and trotting away with a growl. When at bay, standing with open mouth and glaring eyes, holding his head low between his shoulders, and keeping up a continuous low growling, twitching his tail the while from side to side, no animal can look more unpleasant than a lion; but there is then nothing majestic or noble in his appearance."

Notwithstanding this evidently truthful description of the animal when seen under what may be called unfavourable circumstances, no one with an eye for beauty can contemplate the form of a fine specimen of a Lion, at all events in a state of repose, even though in the confinement of a menagerie, without being impressed with the feeling that it is a grand and noble-looking animal.

The Tiger (*F. tigris*) is so closely related to the Lion that it is chiefly by external characters that the two species are distinguished. There are, however, slight distinctions in the proportionate size of the lower teeth, the general form of the cranium, and the relative length of the nasal bones and ascending processes of the maxillaries by which the skull of the Lion and Tiger can be easily discriminated by the practised observer.

Although examples of both species present considerable variations in size, and reliance cannot always be placed upon alleged dimensions, especially when taken from skins stripped from the body, it seems well ascertained that the length of the largest-sized Bengal Tiger may exceed that of any Lion. According to Mr. W. T. Blanford,[1] adult males measure from $5\frac{1}{2}$ to $6\frac{1}{2}$ feet from the nose to the root of the tail; the tail itself measuring some 3 feet in length. Measured along the curves of the head and back to the tip of the tail, males usually give a length of from 9 to 10 feet,

[1] *Fauna of British India*, "Mammalia," p. 59 (1888).

but some specimens reach to 12 feet. The female is somewhat smaller, and has a lighter and narrower head. The Tiger has no mane, but in old males the hair of the cheeks is rather long and spreading. The ground colour of the upper and outer parts of the head, body, limbs, and tail is a bright rufous fawn, and these parts are beautifully marked with transverse stripes of a dark, almost black colour. The markings vary much in different individuals, and even on the two sides of the same individual. The under parts of the body, the inside of the limbs, the cheeks, and a large spot over each eye are nearly white. The Tigers which inhabit

FIG. 225.—The Tiger (*Felis tigris*).

hotter regions, as Bengal and the south Asiatic islands, have shorter and smoother hair, and are more richly coloured and distinctly striped than those of Northern China and Siberia, in which the fur is longer, softer, and lighter coloured.

The Tiger is exclusively Asiatic, but has a very wide range in that continent, having been found in almost all suitable localities south of a line drawn from the river Euphrates, passing along the southern shores of the Caspian and Sea of Aral by Lake Baikal to the Sea of Okhotsk. Its most northern range is the territory of the Amur, its most southern the islands of Sumatra, Java, and Bali. Westward it reaches to Turkish Georgia and eastward to the island of Saghalin. It is absent, however, from the great elevated plateau of Central Asia, nor does it inhabit Ceylon,

Borneo, or the other islands of the Indo-Malayan Archipelago, except those above mentioned. Its absence from Ceylon leads Mr. Blanford to conclude that the Tiger has only recently migrated into Southern India.

The principal food of the Tiger in India is cattle, deer, wild hog, and pea-fowl, and occasionally human beings. The regular "man-eater" is generally an old Tiger whose vigour is passed, and whose teeth are worn and defective; it takes up its abode in the neighbourhood of a village, the population of which it finds an easier prey than the larger or wilder animals named above. Though chiefly affecting grassy plains or swamps, it is also found in forests, and seems to be fond of haunting the neighbourhood of old ruins. As a rule, Tigers do not climb trees; but when pressed by fear, as during an inundation, they have been known to do so. They take to the water readily and are good swimmers. The Tigers of the Sundarbans (Ganges delta) continually swim from one island to the other to change their hunting-grounds for deer. The following extract on the habits of the Tiger is taken from Sir J. Fayrer's *Royal Tiger of Bengal* (1875):—

"The tigress gives birth to from two to five, even six cubs; but three is a frequent number. She is a most affectionate and attached mother, and generally guards and trains her young with the most watchful solicitude. They remain with her until nearly full grown, or about the second year, when they are able to kill for themselves and begin life on their own account. Whilst they remain with her she is peculiarly vicious and aggressive, defending them with the greatest courage and energy, and when robbed of them is terrible in her rage; but she has been known to desert them when pressed, and even to eat them when starved. As soon as they begin to require other food than her milk, she kills for them, teaching them to do so for themselves by practising on small animals, such as deer and young calves or pigs. At these times she is wanton and extravagant in her cruelty, killing apparently for the gratification of her ferocious and bloodthirsty nature, and perhaps to excite and instruct the young ones, and it is not until they are thoroughly capable of killing their own food that she separates from them. The young tigers are far more destructive than the old. They will kill three or four cows at a time, whilst the older and more experienced rarely kill more than one, and this at intervals of from three or four days to a week. For this purpose the tiger will leave its retreat in the dense jungle, proceed to the neighbourhood of a village or gowrie, where cattle feed, and during the night will steal on and strike down a bullock, drag it into a secluded place, and then remain near the 'marrie,' or 'kill,' for several days, until it has eaten it, when it will proceed in search of a further supply, and, having found good hunting ground in the vicinity of a village or gowrie, continue its ravages, destroying one or two cows or buffaloes a week. It is very fond of the ordinary domestic cattle, which in the plains of India are generally weak, half-starved, under-sized creatures. One of these is easily struck down and carried or dragged off. The smaller buffaloes are also easily disposed of; but the buffalo bulls, and especially the wild ones, are formidable antagonists, and have often been known to beat the Tiger off, and even to wound him seriously."

In many districts of India the number of Tigers has been very considerably diminished of late years. In some other countries they appear, however, to be on the increase; thus according to one of the administration reports of Java laid before the Dutch Chambers, portions of that island are being depopulated through Tigers. In 1882 the population of a village in the south-west of the Bantam province was removed and transferred to an island off the coast in consequence of the trouble caused to the people by Tigers. These animals have now become an intolerable pest in parts of the same province. The total population is about 600,000, and, in 1887, sixty-one were killed by Tigers, and in consequence of the dread existing among the people, it has been proposed to deport the inhabitants of the villages most threatened to other parts of the country where Tigers are not so common, and where they can pursue their agricultural occupations with a greater degree of security. At present they fear to go anywhere near the borders of the forest. The people seem disinclined, or they lack the means and courage, to attack and destroy their enemy, although considerable rewards are offered by Government for the destruction of beasts of prey. In 1888 the reward for killing a Royal Tiger was raised to two hundred florins. It appears also that the immunity of the Tiger is in part due to superstition, for it is considered wrong to kill one unless he attacks first or otherwise does injury

The Leopard (*F. pardus*, Fig. 226), although belonging to the same restricted group as the two preceding species, is distinguished from both by its inferior size, and its coloration. The animal now commonly known as the Leopard was called Pard (πάρδος and πάρδαλις) or Panther (πάνθηρ) by the ancients. Leopard (*leo-pardus*) is a later term, originally applied, it is believed, to the Cheeta or Hunting Leopard, upon the supposition that it was a creature intermediate between the Lion and the true Pard. If so it has been completely transferred to the more common species, and though in this sense a perfectly unnecessary and unmeaning term, has gradually superseded those by which this was originally known. Pard, so commonly used by Elizabethan authors, is now nearly obsolete in the English language, and Panther has either become synonymous with Leopard, or is used vaguely for any similar large feline animal, even the Puma of America.

Owing to their extensive geographical range, and the great variations, both in size, form, and coloration to which Leopards are subject, zoologists have scarcely decided whether all the forms popularly referred to this animal should be regarded as specifically alike, or whether they should constitute several distinct species, but the prevailing opinion is in favour of the former view. The

FIG. 226.—The Leopard (*Felis pardus*).

attempts to separate a larger and more robust variety, under the name of Panther, from a smaller and more graceful form, to which the term Leopard might properly be restricted, have failed, owing to the existence of intermediate conditions which cannot be assigned definitely to either one or the other form.[1] The most marked anatomical difference yet noted in different varieties of leopard is in the length of the tail as compared with that of the body, even the number of the caudal vertebræ showing variation, though within what limits, and whether correlated with other characters, has not yet been clearly ascertained. The fur of those specimens which inhabit the most northern confines of its range of distribution, as

[1] See W. T. Blanford, *Fauna of British India*, "Mammalia," p. 69 (1888).

North China, is longer and softer, and the markings are consequently less distinct than on those from more congenial climates, and the well-marked variation thus produced has given rise to the idea of specific distinction.

The size of different individuals, as before said, varies greatly, the head and body usually measuring from 3½ to 4½ feet in length, and the tail from 2½ to 3 feet, but specimens have been met with which fall short of or exceed these limits. The ground colour of the fur varies from a pale fawn to a rufous buff, graduating into a pure white on the under parts and inside of the limbs. This is spotted over with dark brown or black; the spots on the back and sides being arranged in rosettes or broken rings, which vary greatly in size and distinctness in different individuals, but are without the central spot seen in those of the Jaguar. The spots on the under parts and limbs are simple and blacker than those on the other parts of the body. The bases of the ears behind are black, the tips buff. The upper side of the tail is buff, spotted with broken rings like the back, its under surface white with simple spots. The hair of the cubs is longer than that of the adults, its ground colour less bright, and its spots less distinct. Perfectly black Leopards, which, however, in certain lights show the characteristic markings on the fur, are not uncommon. These appear to be examples of melanism, occurring as individual variations, sometimes in one cub out of a litter of which the rest are normally coloured, and therefore not indicating a distinct race, much less a species. These are met with chiefly in Southern Asia. We are not aware of any recorded case from Africa, though there seems no reason why they should not occur.

In habits the Leopard resembles the other large Cat-like animals, yielding to none in the ferocity and bloodthirstiness of its disposition. It is exceedingly quick and active in its movements, but seizes its prey by waiting in ambush or stealthily approaching to within springing distance, when it suddenly rushes upon it and tears it to the ground with its powerful claws and teeth. It preys upon almost any animal it can overcome, such as antelopes, deer, sheep, goats, monkeys, peafowls, and is said to have a special liking for dogs. It not unfrequently attacks human beings in India, chiefly children and old women, but instances have been known of a Leopard becoming a regular "man-eater." When favourable opportunities occur, it often kills many more victims than it can devour at once, apparently to gratify its propensity for killing, or only for the sake of their fresh blood. It generally inhabits woody districts, and can climb high trees with facility if necessary for its safety when hunted, but usually lives on or near the ground, among rocks, bushes, and roots and low branches of large trees.

The present geographical range of the Leopard is very extensive, as it is met with in various suitable localities, where not too much interfered with by human cultivation, throughout the greater part of Africa from Algeria to the Cape Colony, and through the whole of the South of Asia from Palestine to China, including all India south of the Himalaya, and the islands of Ceylon, Java, Sumatra, and Borneo. Fossil bones and teeth, indistinguishable from those of existing Leopards, have been found in cave-deposits of Pleistocene age in Spain, France, Germany, and England. The evidence of the former existence of the Leopard in England is described at length by Boyd Dawkins and Sanford in their *British Pleistocene Mammalia.*[1]

The Ounce, or Snow Leopard (*F. uncia*), inhabits the highlands of Central Asia, from the lofty mountains of Tibet to the southern parts of Siberia, at altitudes of from 9000 to 18,000 feet above the sea. It is about the size of the common Leopard, but lighter in colour, with longer fur, less distinct spots, and a long thick tail. Its skull differs in shape from that of all the other *Felidæ;* the facial portion being very broad, the nasal bones especially being wide and depressed, and the zygomatic arches very strong and deep. The Clouded Tiger (*F. nebulosa*[2]) is a beautifully marked species, with elongated head and body, long tail, and rather short limbs. The canine teeth are proportionally longer than in any existing member of the genus. It is thoroughly arboreal, and is found in the forests of South-East Asia and the islands of Sumatra, Java, Borneo, and Formosa. *F. serval*, the Serval, from South Africa, is yellow with black spots, and has a short tail and large ears. Numerous smaller species called Tiger Cats and Wild Cats, of which the Oriental *F. marmorata* (Fig. 227) is a good example, are found throughout the warmer parts of Asia and Africa. The Wild Cat of Europe, *F. catus*, still inhabits the mountainous and wooded parts of Great Britain.

FIG. 227.—The Marbled Cat (*Felis marmorata*). From Blanford, *Mammalia of British India*, p. 74, after Elliot.

The Caffre Cat (*F. caffra*[3]), of Africa and Southern Asia, was the species held in veneration by the ancient Egyptians, and immense

[1] *Monographs of the Palæontographical Society*, 1872.

[2] Syn. *F. macrocelis*. [3] Syn. *F. maniculata* and *caligata*.

numbers of its mummified remains have recently been found in Egypt, whence they have been imported in large quantities to this country for manure. This species is generally regarded as the main ancestral stock from which the European Domestic Cat has been derived; one of the arguments in support of this opinion being that the whole of the sole of the hind foot of *F. caffra* is black, and that the same feature obtains in the darker varieties of the Domestic Cat; while in *F. catus* there are only spots of black upon this portion of the limb. Remains of the Caffre Cat occur in the Pleistocene cave-deposits at Gibraltar. The Indian *F. rubiginosa* is the smallest species of Cat.

The Caracal or Persian Lynx (*F. caracal*) is an animal about the size of a fox, of slender build, with a moderately long tail, reaching down to the heels. It is of a uniform vinous or bright fulvous brown colour above, and is paler, sometimes almost white, beneath. It is quite or almost entirely unspotted. The tail has a black tip, and the ears are black externally, long, upright, pointed, and surmounted by a pencil of fine black hairs. It inhabits Central and North-West India, Persia, Arabia, Syria, and the greater part of Africa.

The true Lynxes comprise various species or varieties found in the northern and temperate regions of both the Old and New World, all larger than the true Wild Cats, with long limbs, short stumpy tail, ears tufted at the tip, and pupil of the eye linear when contracted. Their fur is generally long and soft, varying, however, according to season and locality, and always longish upon the cheeks. Their colour is always light brown or gray, and generally more or less spotted with a darker shade. The naked pads of the feet are more or less covered by the hair that grows between them. The skull and skeleton do not differ markedly from those of the other cats, but the small anterior upper premolar tooth found in many other species is usually wanting; and the lower carnassial has a rudimental talon. Their habits are exactly those of the other Wild Cats, and they are exceeded by none in the untameable savageness of their disposition. They capture their prey in the same manner, either lying in wait, or noiselessly stealing within reach, and then making a sudden rush or spring upon it. Their food consists of any mammals or birds which they can overpower. In inhabited countries they commit extensive ravages upon sheep, lambs, and poultry. Lynxes generally frequent rocky places and forests, being active climbers, and passing much of their time among the branches of the trees. Their skins are of considerable commercial value.

Zoologists are by no means agreed at present as to the specific distinctions, if any really exist, between the various modifications of this group. As many as eight species are sometimes recognised, four belonging to the Old and four to the New World. The former are *F. lynx*, of Scandinavia, Russia, Northern Asia, and till lately the forest regions of Central Europe; though not an inhabitant of Britain during the historic period, its remains have been found in cave-deposits of Pleistocene age; *F. cervária*, Siberia; *F. pardina*, Turkey, Greece, Sicily, Sardinia, and Spain; and *F. isabellina*, Tibet. The American varieties are *F. canadensis*, the most northern species, and *F. rufa*, the American Wild Cat or Bay Lynx, extensively distributed from the Atlantic to the Pacific throughout nearly the whole latitude of the United States, but replaced in Texas and

FIG. 228.—European Lynx (*Felis lynx*). From a drawing by Wolf in Elliot's *Monograph of the Felidæ*.

southern California by *F. maculata*, and in northern Oregon and Washington territory by *F. fasciata*.

In both cases, as might be supposed, specimens obtained from the more southern climates are shorter in their fur, more brightly coloured, and more distinctly spotted than those from colder regions. When only a few individuals of each most markedly different form are examined the distinctions are sufficiently evident. The occurrence, however, of transitional or intermediate forms makes it extremely difficult to draw the line between the different varieties or species, or to assign definite characters by which they can be separated. Wherefore it is best at present to accept the so-called species as only provisional, and wait until more abundant materials, with fuller knowledge of the localities from which they are derived, and of the variations due to age, sex, season, and climate, have been more carefully studied. We shall then probably come to the conclusion that all or nearly all the existing forms of northern Lynxes, whether American or Eurasian, belong to what may fairly be called a species, which is becoming by degrees differentiated into several more or less strongly marked local varieties. Mr. W. T. Blanford has indeed shown that the Tibetan Lynx (*F. isabellina*) is inseparable from *F. lynx*; the specimens from Gilgit being intermediate in colour between the typical forms of the two races. On the other hand, from the evidence of cranial characters, Professor Mivart is disposed to regard *F. pardina* as a valid species.

FIG. 229.—The Puma (*Felis concolor*).

B. *New World Species.*—The Puma or Couguar (*F. concolor*, Fig. 229), commonly called "Panther" in the United States, is about the size of a Leopard, but of an uniform brown colour. It usually measures from nose to root of tail about 40 inches, the tail being rather more than half that length. The head is rather small compared with that of other Cats and has no mane. The ears are large and rounded. The tail is cylindrical, with some bushy elongation of the hairs near the end, but not forming a distinct tuft as in the Lion. The general colour of all the upper parts and sides of the adult is a tawny yellowish-brown, sometimes having a gray or silvery shade, but in some individuals dark or inclining to red. The lower parts of the body, inner surface of the limbs, the throat, chin, and upper lip are dirty white; the outside of the ears, particularly at their base, and a patch on each side of the muzzle black; the end of the tail dusky. The young are, when born, spotted with dusky brown and the tail ringed; these markings gradually fading, and quite disappearing before the animal becomes full-grown.

The Puma has an exceedingly wide range of geographical distribution, extending over a hundred degrees of latitude, from Canada in the north to Patagonia in the south, and was formerly pretty generally diffused in suitable localities from the Atlantic to the Pacific Ocean, but the advances of civilisation have in recent years considerably curtailed the extent of the districts which it inhabits. In Central America it is still common in the dense forests which clothe the mountain ranges as high as 8000 or 9000 feet above the sea-level, where the hideous sound of its howling is said to be almost continuously heard at night during the breeding season. Though an expert climber, it is by no means confined to wooded districts, being frequently found in scrub and reeds along the banks of rivers, and even in the open pampas and prairies. Its habits much resemble those of the rest of the group to which it belongs; and, like the Leopard, when it happens to come within reach of an abundant and easy prey, as the sheep or calves of an outlying farming station, it kills far more than it can eat, either for the sake of the blood only or to gratify its propensity for destruction. It rarely attacks man, and, when pursued, escapes if possible by ascending lofty trees. Several instances have occurred of Pumas becoming tame in captivity. Edmund Kean, the celebrated actor, had one which followed him about like a dog. When caressed they express their pleasure by purring like a domestic cat.

F. onca, the Jaguar, is a larger and more powerful animal than the last, and more resembles the Leopard in its colours. It also is found in both North and South America, but with less extensive range, reaching northwards only as far as Texas, and southwards nearly to Patagonia. It climbs as well as the Puma, and preys to a great extent upon monkeys. Several allied smaller elegantly spotted forms inhabiting the intratropical regions of America are commonly included under the name of Ocelot or Tiger Cat, though zoologists are still undecided whether under this designation several distinct species have not been confused, or whether all the Ocelots are to be referred to a single species (*F. pardalis*) showing great individual or racial variation. Their fur has always a tawny yellow or reddish-gray ground colour, and is marked with black spots, aggregated in streaks and blotches, or in elongated rings enclosing an area which is rather darker than the general ground colour. They range through the wooded parts of tropical America, from Arkansas in the north as far south as Paraguay, and in their habits resemble the other smaller members of the Cat tribe, being ready climbers and exceedingly bloodthirsty.

F. yaguarundi, rather larger than the Domestic Cat, with an elongated head and body, and of a uniform brownish-gray colour, ranges from Matamoras to Paraguay. *F. eyra* is a small Cat, very Musteline in form, having an elongated head, body, and tail, and short limbs, and is also of a uniform light reddish-brown colour. It is a native of South America and Mexico. *F. pajeros* is the Pampas Cat. The American Lynxes have been already noticed with those of the Old World.

C. *Fossil Species.*—It has been already incidentally mentioned

FIG. 230.—The Ocelot (*Felis pardalis*).

that several of the existing species of *Felis*, such as the Lion, Leopard and Caffre Cat, are met with in a fossil condition in the European Pleistocene deposits, and it may be added that the Pardine Lynx has left its remains in the cavern-deposits of Gibraltar. The caves of Brazil have yielded remains of the Jaguar and Ocelot; while the Puma is found in the Pleistocene of the United States. Existing species now inhabiting India are met with in cavern-deposits in Madras. In the Pliocene Siwaliks of Northern India the huge extinct *F. cristata* shows characters connecting it both with the Tiger and the Jaguar; and the same deposits also contains the remains of a small species of the size of *F. bengalensis*. In Europe numerous species occur in the Upper and Lower Pliocene, some of which were as large as a Leopard. *F. atrox* and *F. augusta*, of the Pliocene of the United States, were of the dimensions of the Lion.

Cynælurus.[1]—The Cheeta or Hunting Leopard (*C. jubatus*) is distinguished from the other *Felidæ* by the inner tubercle of the upper carnassial, though supported by a distinct root, having no salient cusp upon it; by the tubercular molar being more in a line with the other teeth; and by the claws being smaller, less curved, and less completely retractile, owing to the feebler development of the elastic ligaments. The skull is short and high, with the frontal region broad and elevated in consequence of the large development of the frontal air-sinuses. The head is small and round, the body light, the limbs and tail long. Its colour is pale yellowish-brown with small black spots. The Cheeta is less savage and more easily tamed than most of the Cats. In Asia it has been trained for the chase of the Antelope. It has rather an extensive geographical range from the Cape of Good Hope, throughout Africa and the south-western parts of Asia, as far as Southern India.

Extinct Genera.—A number of forms are gradually becoming known, especially through the researches of American palæontologists, which, though evidently animals of the same general type, and therefore to be placed in or near the family *Felidæ*, depart so much in various details of structure that they must be referred to different genera. As one of the points in which *Felis* manifests its specialisation is the reduction of the number of the molar series of teeth, with concomitant shortening of the jaws, it might be supposed that in the earlier and perhaps ancestral forms these teeth would be more numerous and approach more nearly to the primitive or typical number of the heterodont mammals, viz. seven on each side. This is actually the case. Similarly we find that many of these forms exhibit a less specialised structure of the teeth themselves, as is shown by the absence of the anterior lobe of the upper carnassial, and the retention of the hind talon in the corresponding lower tooth. Again, some of them have an alisphenoid canal in the skull; while the femur may have a third trochanter, and the claws be very imperfectly retractile.

An extremely generalised form is the small *Proælurus*, from the Upper Eocene and Lower Miocene, with $p\ \frac{4}{4}$, $m\ \frac{1}{2}$, an alisphenoid canal, and a third trochanter to the femur. *Dinictis*, of the North American Miocene, is a larger allied form, with $p\ \frac{3}{3}$, $m\ \frac{1}{2}$; the upper carnassial having no anterior lobe, and the ungual phalanges being devoid of bony sheaths. The characters of the base of the skull, and the form and relations of the astragalus, differ very considerably from *Felis*. *Pseudælurus*, from the French Miocene, is another very generalised Feline, in which there may be either three or four

[1] Wagler, *Syst. Amphib.* etc. p. 30 (1830).

premolars, and the lower carnassial may retain its inner cusp. *Ælurictis*, of the French Phosphorites, with $p\ \frac{3}{(3-4)}$, $m\ \frac{1}{(1-2)}$, together with several American Miocene genera, such as *Nimravus* ($p\ \frac{3}{2}$, $m\ \frac{1}{2}$), *Archælurus* ($p\ \frac{3}{(3-4)}$, $m\ \frac{1}{2}$), *Pogonodon* ($p\ \frac{3}{3}$, $m\ \frac{1}{1}$), and *Hoplophoneus* ($p\ \frac{(2-3)}{2}$, $m\ \frac{1}{1}$), approach more closely to the modern Cats, although many or all of them retain the alisphenoid canal, and have not yet developed the anterior lobe to the upper carnassial, or lost the talon to the lower one. *Hoplophoneus* has a descending flange to the mandible; and its scapholunar bone has a line indicating its dual origin; while the femur still retains the third trochanter, of which all traces are lost in the modern Cats.

On the other hand, some of the extinct *Felidæ* show a most remarkable tendency towards a specialisation not occurring in any of the surviving members of the family, viz. an enormous development of the upper canines, with which is usually associated an expansion downwards and flattening of the anterior part of the ramus of the lower jaw, on the outer side of which the canine lies, when the mouth is closed. In *Machærodus nœogeus*, the Sabre-toothed Tiger, from the caves of Brazil and also from Pleistocene deposits near Buenos Ayres, an animal about the size of a Tiger, these teeth are 7 inches in length, greatly compressed, and finely serrated on the trenchant anterior edges. Similar serrations are seen on a much fainter scale in the unworn teeth of modern Tigers. Many modifications of this commonly-called "machærodont" type have been met with both in the Old and New World. In *M. cultridens*, of the Upper Pliocene of Italy and France, the upper canine is long and narrow, with smooth cutting edges; the smaller form described as *M. meganthereon* being apparently the female of this species. *M. crenatidens*, of the same deposits, is distinguished by the shorter and broader upper canine, in which both edges are strongly serrated; the same feature occurring in the closely allied or identical *M. latidens* of the English cavern-deposits. The Italian Pliocene form described as *M. nestianus* has serrations only on the hinder edge of the upper canine, and the third lower premolar is separated by a long interval from the fourth. *M. necator*, of the Pleistocene of South America, is remarkable as being the only member of the family in which the humerus has no entepicondylar foramen. A very remarkable form, *Eusmilus*, from the Upper Eocene Phosphorites of Central France, differs from all other known Felines in having only two pairs of incisors in the lower jaw, and a small canine separated by a very long diastema from the cheek-teeth, which consist only of one premolar and one sectorial true molar. The lower jaw is enormously expanded towards the symphysis to protect the large upper canines. This animal then, although of Eocene age, appears to form the culminating development of the sabre-toothed or machærodont dentition, the most specially carnivorous type of structure known.

Other species of *Machærodus* are found in the Pliocene deposits of Europe and Asia. The accompanying woodcut exhibits the last two upper teeth of the Indian *M. sivalensis*, from which it will be seen that the inner tubercle of the carnassial is much reduced in size, while the molar is very minute.

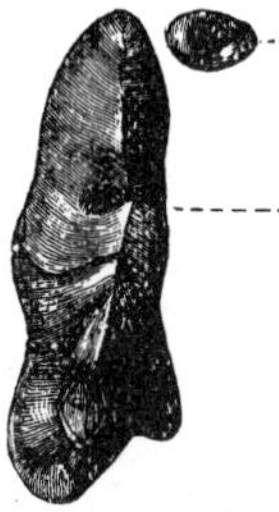

FIG. 231.—Oral surface of the left upper carnassial and molar of *Machærodus sivalensis*.

Family VIVERRIDÆ.

Premolars $\frac{3}{3}$ or $\frac{4}{4}$. Molars $\frac{1}{1}$ or $\frac{2}{2}$. Upper carnassial usually without an anterior lobe, and the lower one with a well-developed talon; second lower incisor (as in all the following families) raised above the level of the first and third. Auditory bulla externally constricted, and divided by a septum. An alisphenoid canal (with very rare exceptions). Carotid canal distinct as a groove on the side of the bulla. Humerus usually with an entepicondylar foramen. Digits usually 5-5, but sometimes the pollex or hallux or both may be wanting. Dorsal vertebræ 13 or 14. Limited in distribution to the Old World.

The subfamily **Cryptoproctinæ** contains the single genus *Cryptoprocta*.[1] Dentition: $i\ \frac{3}{3}$, $c\ \frac{1}{1}$, $p\ \frac{4}{4}$, $m\ \frac{1}{1}$; total 36. The teeth generally closely resemble those of the *Felidæ*. The first premolar of both jaws is very minute and early deciduous. The upper carnassial has a very small inner tubercle, quite at the anterior part of the tooth. The true molar is very small and placed transversely. The lower carnassial has a large trenchant bilobed blade, and a very minute talon, but no inner cusp. Skull generally like that of *Felis*, but proportionately longer and narrower. Orbit widely open behind. Vertebræ: C 7, D 13, L 7, S 3, C 29. Body elongated. Limbs moderate in size. Feet subplantigrade; five well-developed toes on each, with sharp, compressed, retractile claws. Ears moderate. Tail long and cylindrical.

The only known species, *C. ferox*, the "Foussa" of the Malagasy, is peculiar to Madagascar, being the largest carnivorous animal in the island. It is about twice the size of the common Cat (5 feet from nose to end of tail), with short close fur of nearly uniform pale brown. Little is as yet known of its habits, except that it is nocturnal, frequently attacks and carries off goats, and especially kids, and shows great ferocity when wounded, on which account it is much dreaded by the natives.

The remaining numerous specific and generic modifications found

[1] Bennett, *Trans. Zool. Soc.* vol. i. p. 137 (1833).

in the existing animals belonging to this family seem to arrange themselves mainly into two tolerably distinct groups, distinguishable by the characters of the auditory bulla and neighbouring parts of the base of the skull, and by the structure of the feet. The one form has the genus *Viverra* or Civet Cats for its most typical representative, and the other *Herpestes* or the Ichneumons.

Subfamily **Viverrinæ.**—Auditory bulla oval, or rather conical, broad and truncated and not everted behind, narrow in front and more or less compressed at the sides. The outer or anterior chamber very small and flat. The meatus with scarcely any inferior lip, its orifice being close to the tympanic ring. Paroccipital process triangular, its apex projecting slightly beyond the bulla. Claws strongly curved and more or less retractile. Perineal scent-glands generally present.

This subfamily includes both Ethiopian and Oriental forms, but the former are the more numerous.

The typical section, which includes five genera, has the following characters. Dentition: $i\ \frac{3}{3}$, $c\ \frac{1}{1}$, $p\ \frac{4}{4}$, $m\ \frac{2}{2}$ ($\frac{1}{2}$ in *Prionodon*); total 40. Skull elongated; facial portion small and compressed. Orbits well-defined but incomplete behind. Teeth always sectorial, never very small. Vertebræ: C 7, D 13, L 7 (or D 14, L 6), S 3, C 22-30. Body elongated and compressed. Head pointed in front; ears rather small. Extremities short. Feet small and rounded. Toes short, five on each foot. First toe both on fore and hind feet much shorter than the others. Palms and soles covered with hair, except the pads of the feet and toes, and in some species a narrow central line on the under side of the sole, extending backwards nearly to the heel. Tail moderate or long; usually marked with dark and light rings. A pair of large glandular follicles situated on the perineum (in both sexes), and secreting in most species an oily substance of a peculiarly penetrating odour.

The numerous species of this section form a large series, the two extremes of which differ considerably, but the several genera into which they may be divided blend so into one another that it is difficult to differentiate them sharply.

All the animals of this section are, for their size, extremely active, fierce, and rapacious. They feed chiefly on small mammals and birds.

Viverra.[1]—This includes the largest species. The teeth (Fig. 232) are stouter and less compressed than in the other genera; the second upper molar being especially larger. The auditory bulla smaller and more pointed in front. Body shorter and stouter; limbs longer; tail shorter, tapering. Under side of tarsus completely covered with hair. Claws longer and less retractile. Fur rather long and loose, and in the middle line of the neck and back usually

[1] Linn. *Syst. Nat.* 12th ed. vol. i. p. 63 (1766).

elongated so as to form a sort of crest or mane; neck with a black gorget. Pupil circular when contracted. Perineal glands greatly developed. These characters apply especially to *V. civetta*, the African Civet, or "Civet-Cat" as it is commonly called, an animal rather larger than a common Fox, and an inhabitant of intra-tropical Africa. *V. zibetha*, the Indian Civet, of about equal size, inhabits Bengal, China, the Malay Peninsula, and adjoining islands. *V. tangalunga*, from Java, Sumatra, Borneo, and the Philippines,

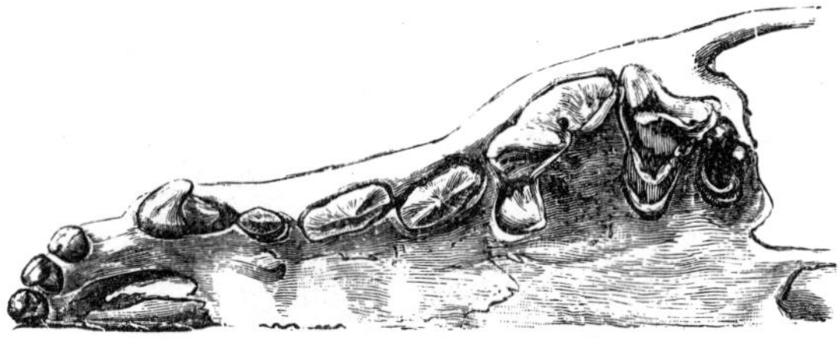

FIG. 232.—The left upper dentition of the Indian Civet (*Viverra zibetha*). From the *Palæontologia Indica*.

and *V. megaspila*, from Burma, are smaller but nearly allied animals; the latter being more distinctly spotted than either of the others. From these species and the next the civet of commerce, once so much admired as a perfume in England, and still largely used in the East, is obtained. The animals are kept in cages, and the odoriferous secretion collected from the interior of the perineal follicles with a spoon or spatula.

The Rasse or Lesser Indian Civet (*V. malaccensis*) may be regarded as the representative of a distinct group of *Viverra*, although often referred to a separate genus (*Viverricula*). The size of this animal is smaller than in the typical group, the build is slighter, the muzzle finer, the claws sharper and more curved, and there is no erectile mane along the back. Generally there is an alisphenoid canal in the skull; and the anterior chamber of the auditory bulla is much more inflated than the hinder one, so that the apparent length of the whole bulla is increased. This species is found over the greater part of India, and extends to the Malay Peninsula and Southern China.

Large species of *Viverra* occur in the Pleistocene and Pliocene of India, and also in the Pliocene of France, which approximate in some characters of the dentition to the extinct genus *Ictitherium*, mentioned at the end of the family. Species of this genus have also been described from the Miocene and Upper Eocene of Europe. The Lower Miocene *V. antiqua* has an alisphenoid canal, and all the other cranial characters of the typical forms.

Fossa.[1]—The Fossa of Madagascar comes so close to the Rasse

[1] Gray, *Proc. Zool. Soc.* 1864, p. 518.

that its right to generic distinction seems doubtful. There is, however, no scent-pouch. The limbs are slender; and there are two small bare spots on the sole of the hind foot, above the plantar pads. There is no dark line along the back; the throat gorget of *Viverra* is absent; and in the tail the spots only tend to form rings, which are not complete. The skull has an alisphenoid canal, and a large bulla as in the typical group of *Viverra*.

Genetta.[1]—The Genettes are smaller animals, with more elongated and slender bodies, and shorter limbs than the Civets. Skull elongated and narrow. Auditory bulla large, elongated, rounded

FIG. 233.—The Common Genet (*Genetta vulgaris*).

at both ends. Teeth compressed and sharp pointed. The inner side of the third upper premolar has a tubercle not present in the previous genus, and the talon of the lower carnassial is larger. Pupil contracting to a linear aperture. Tail long, slender. Fur short and soft, spotted or cloudy. Under side of the tarso-metatarsus with a narrow longitudinal bald streak. No pouch for storing the secretion of the scent-gland. *G. vulgaris*, the common Genet (Fig. 233), is found in France south of the river Loire, Spain, South-Western Asia, and Africa from Barbary to the Cape. *G. felina*, *senegalensis*, *tigrina*, and *pardalis* are other named species, all African in habitat.

A few details (taken from Professor Mivart's memoirs on the

[1] Cuvier, *Règne-Animal*, vol. i. p. 156 (1817).

Æluroidea) of the anatomy of the soft parts of the Genet may be given as illustration of these parts in the Carnivora generally, and of this family and genus in particular. The salivary glands are shown in Fig. 19 (p. 56), and these conform to the general type prevalent in the Æluroidea. Thus there is a distinct zygomatic gland; the parotid with its (Steno's) duct is well developed; and there is a small submaxillary gland. The stomach (Fig. 234), while conforming to the simple type characteristic of the Carnivora, is much larger than in the Cat; it is characterised by the presence of some strongly marked internal folds near the pyloric extremity, which stop suddenly at a point where the stomach makes an abrupt constriction and flexure. Beyond this point there are three other longitudinal folds; and the pyloric valve is small. The allied genera present modifications from this form of stomach. The cæcum (Fig. 235) is short, thick, and pointed. The liver (Fig. 236) much resembles that of the Cat, but differs in that the left lateral lobe is undivided, although having a small groove on its posterior or abdominal aspect, while the cystic fissure is less deep, and situated more to the right. The caudate lobe is relatively longer, has a deep concavity, and runs uninterruptedly into the Spigelian; the latter being relatively somewhat larger than in the Cat, with a deep groove dividing the proximal third from the distal two-thirds. In *Viverra* the right lateral and right central lobes are nearly equal in size. The variations in the form of the liver

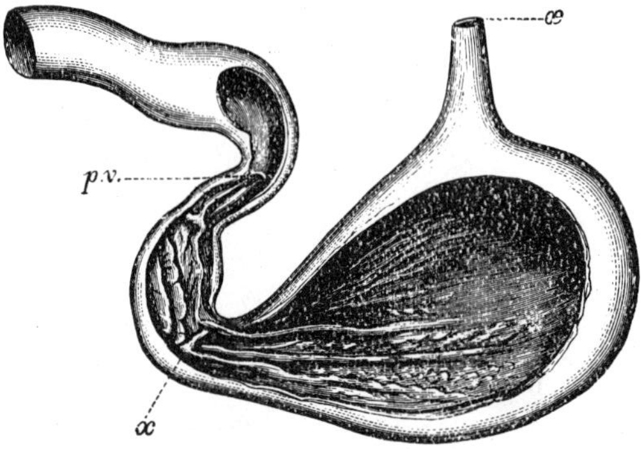

FIG. 234.—Stomach of Genet cut open. *œ*, Œsophagus *pv*, pyloric valve; *x*, sudden bend where the internal folds are interrupted. (From Mivart, *Proc. Zool. Soc.* 1882, p. 505.)

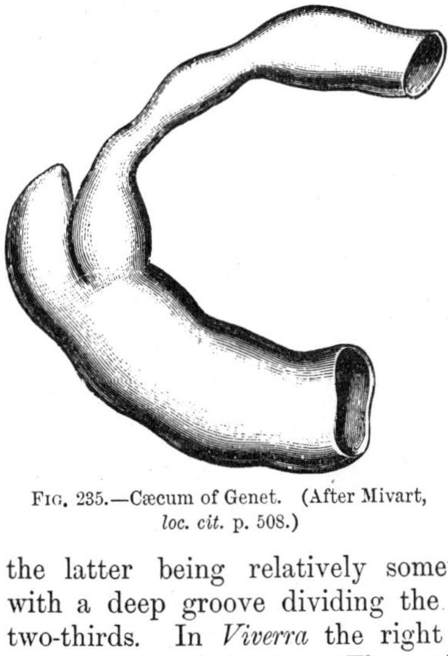

FIG. 235.—Cæcum of Genet. (After Mivart, *loc. cit.* p. 508.)

of the allied genera are detailed in Professor Mivart's memoir. The brain of the Genet is shown in Fig. 23 (p. 71); the small depression *d* placed on the superior lateral gyrus appears to be the sole representative of the distinct crucial sulcus which dis-

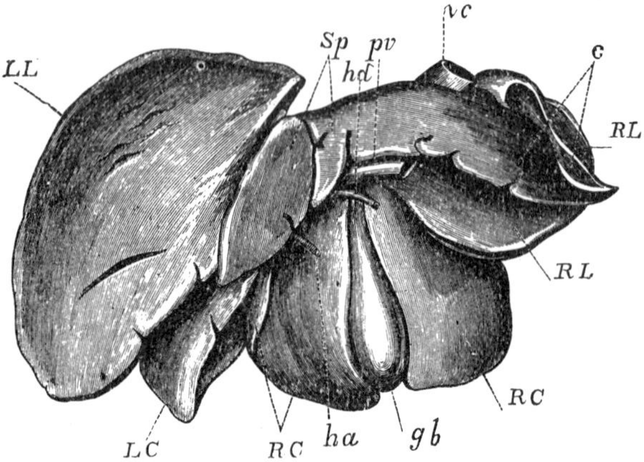

FIG. 236.—Abdominal aspect of the liver of the Genet. *c*, Caudal lobe; *gb*, gall-bladder; *ha*, hepatic artery; *hd*, hepatic duct; *LC*, left central lobe; *LL*, left lateral lobe; *pv*, portal vein; *RC*, right central lobe; *RL*, right lateral lobe; *Sp*, Spigelian lobe; *vc*, vena cava. (From Mivart, *Proc. Zool. Soc.* 1882, p. 510.)

tinguishes the brains of the *Felidæ* from those of all other members of the Æluroidea.

Prionodon.[1]—This and the following genus comprise the beautiful Linsangs (Fig. 238), which are distinguished from the preceding genera by the loss of the second upper molar, which is, however, very small in some of the Genets. In the present genus the ground colour is whitish or yellowish with brown or black markings, which may either form broad continuous patches across the hinder part of the body, or may be broken up into spots. The tail is very long, the limbs comparatively short, and the fur very short and close. The pollex and hallux are well developed; the claws are almost completely retractile; and the tarsus and metatarsus are completely haired. The pupil is round. The cæcum (Fig. 237) is remarkably small. This genus is exclusively Oriental, and comprises *P. gracilis* from Borneo, Java, and (?) Sumatra, *P. pardi-*

FIG. 237.—Cæcum of *Prionodon.* (From Mivart, *Proc. Zool. Soc.* 1882, p. 508.)

[1] Horsfield, *Zool. Research. Java* (1824).—*Prionodontidæ.*

color from Nipal, and *P. maculosus* from Tenasserim; the head and body of the latter measuring from 18 to 20 inches in length. Speaking of *P. pardicolor*, Mr. Hodgson observes that it is "equally at home on trees and on the ground; it dwells and breeds in the hollows of decayed trees. It is not gregarious at all, and preys chiefly upon small birds, which it is wont to pounce upon from the

FIG. 238.—The African Linsang (*Poiana poënsis*). From Mivart, *Proc. Zool. Soc.* 1882, p. 160.

cover of the grass. The times of breeding are said to be February and August, and the litter to consist of two young, there being two litters each year."

Poiana.[1]—This African genus, represented solely by one species, *P. poënsis* (Fig. 238), from Fernando Po, is very closely allied to the preceding, but the spots are smaller, and show no tendency to run into transverse bands or stripes, except in the region of the head and shoulder; while the sole of the foot has a narrow bald band running up towards the tarsus, as in *Genetta*. The length

[1] Gray, *Proc. Zool. Soc.* 1864, p. 520.

of the head and body is 38 inches, and that of the tail about 40 inches. It is probable that this animal should really be regarded as a slightly aberrant species of the genus *Prionodon.*

The five following genera differ in several important respects from all the preceding, and collectively constitute the *Paradoxurine* section of Professor Mivart. With the exception of one African form, they are mainly Oriental. In this section the auditory bulla is frequently in two portions, the posterior moiety in one case being unossified, and it is always much narrowed in front (Fig. 239). The palate (as in the figure) may be much produced behind the molars; and the teeth are often but slightly sectorial, and may be very small. The long tail is in most cases not ringed.

Paradoxurus.[1]—Dentition: $i\ \frac{3}{3}$, $c\ \frac{1}{1}$, $p\ \frac{4}{4}$, $m\ \frac{2}{2}$; total 40. The blunt and rounded form of the cusps of the hinder premolar and the molar teeth distinguishes this genus from most of the members of the family. Vertebræ: C 7, D 13, L 7, S 3, C 29-36. Head pointed in front. Ears small, rounded. Body long. Limbs moderate. Palms and soles almost entirely naked, and joining the foot-pads without the intervention of any hairy space. Claws completely retractile. Pupil vertical. Tail long, non-prehensile; in the Indian species without rings. The Paradoxures or Palm-Civets are less strictly carnivorous than the other members of the family. They are mostly about the size of the common Cat, or rather larger, and are partly arboreal in their habits. The species are rather numerous, and present considerable variations in the details of the form and size of their molar teeth; in only a few does the bony palate extend behind the molars. They are restricted geographically to Southern Asia and the Indo-Malayan archipelago. The best known species[2] are *P. niger*, *P. hermaphroditus*, *P. jerdoni*, *P. aureus*, *P. grayi* from India and Burma, *P. philippinensis* of the Philippines, *P. larvatus* of Southern China and Formosa, *P. leucomystax* of the Malay Peninsula, Sumatra, and Borneo, and *P. musschenbroeki* of Celebes. The name *Paradoxurus* was applied from the mistaken notion that the tail was prehensile. Mr. Blanford[3] gives the following account of the habits of *P. niger:* "The common Palm-Civet, Tree-Cat, or Toddy-Cat, is a familiar animal in most parts of India, though, being thoroughly nocturnal in its habits, it is but rarely seen in the daytime. It is arboreal, passing the day generally in trees, either coiled up in the branches, or in a hole in the trunk, and in places where cocoa-nut palms are common it frequently selects one of them for a residence. Mango groves are also a favourite resort. It not unfrequently takes up its abode in the thatched roofs of houses; Jerdon found a large colony

[1] F. Cuvier, *Hist. Nat. des Mammifères*, No. 186 (1821).

[2] See W. T. Blanford, *Proc. Zool. Soc.* 1885, p. 780.

[3] *Fauna of British India*, "Mammalia," p. 108 (1888).

established in the rafters of his own house in Tellicheri. It even occurs in large towns; I have known of one being caught in the middle of Calcutta."

Arctogale.[1]—This genus—represented only by *A. trivirgata* of Java, and *A. leucotis* of Burma, Tenasserim, Sumatra, Java, etc.—is chiefly distinguished from *Paradoxurus* by the extremely small size of the cheek-teeth (Fig. 239), which are often not in contact with one another; the upper carnassial being almost triangular in shape. Palate frequently convex longitudinally between the carnassials, and greatly produced behind the last molar, with a very narrow bony aperture of the posterior nares. The soles of the feet are still more naked than in *Paradoxurus*; and the pollex and hallux are more divergent. In *A. leucotis* the length of the head and body is 26·5 inches, and the tail 27 inches. In many specimens the three dorsal stripes are much less distinctly marked than in others, and tend to break up into spots; while the general coloration is considerably lighter.

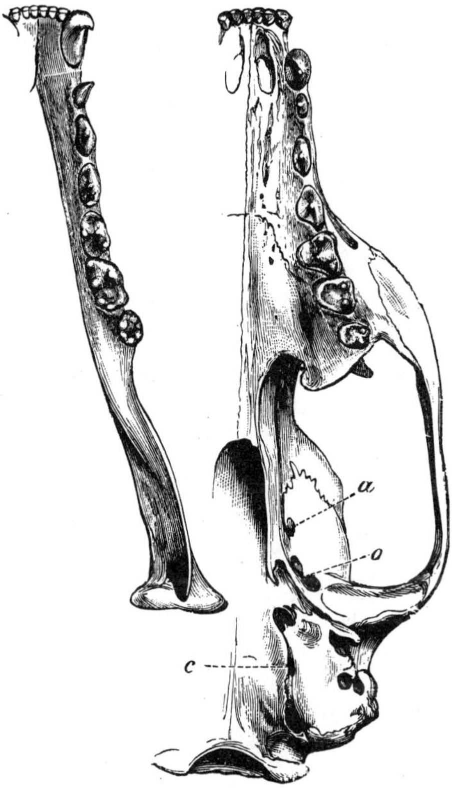

FIG. 239.—Palatal aspect of the left side of the cranium and mandible of *Arctogale leucotis*. *a*, Anterior opening of alisphenoid canal; *o*, foramen ovale; *c*, carotid canal }. (From Mivart, *Proc. Zool. Soc.* 1882, p. 165.)

Hemigale,[2] another modification of the Paradoxure type, contains one species, *H. hardwickei*, from Borneo and Malacca, an elegant-looking animal, smaller and more slender than the Paradoxures, of light gray colour, with transverse broad dark bands across the back and loins; the proximal portion of the tail being ringed. The tarsus is hairy. The general cranial

[1] Gray, *Proc. Zool. Soc.* 1864, p. 542, *ex* Petero.

[2] Jourdan, *Comptes Rendus*, vol. v. p. 442 (1837). Amended.

characters are those of *Paradoxurus*, but the auditory bulla is ankylosed into a single piece.

Arctictis.[1]—Dentition: $i\ \frac{3}{3}$, $c\ \frac{1}{1}$, $p\ \frac{4}{4}$, $m\ \frac{2}{2}$; total 40. The posterior upper molar and the first lower premolar very often absent. Cheek-teeth generally small and rounded, with a distinct interval between them, but formed generally on the same pattern as *Paradoxurus*. Vertebræ: C 7, D 14, L 5, S 3, C 34. Body elongated. Head broad behind, with a small pointed face. Whiskers long and numerous. Ears small, rounded, but clothed with a pencil of long hairs. Eyes small. Limbs short. Soles and palms broad, entirely naked. Tail very long and prehensile; thickly covered with long hair. Fur long and harsh. Cæcum extremely small. But one species is known, *A. binturong*, the Binturong, an inhabitant of Southern Asia from Nipal through the Malay Peninsula to the islands of Sumatra and Java. Although structurally agreeing closely with the Paradoxures, its tufted ears, long, coarse, and dark hair, and prehensile tail give it a very different external appearance. It may be regarded as a very aberrant Paradoxure, connected, so far as dental characters are concerned, with *Paradoxurus* by means of *Arctogale*. The bony palate also extends considerably behind the last molar, as in the latter. The Binturong is slow and cautious in its movements, chiefly if not entirely arboreal, and appears to feed on vegetable as well as animal substances.

Nandinia[2] contains one species, *N. binotata*, a somewhat aberrant Paradoxure, from West Africa. It is rather smaller than the true Paradoxures, with smaller and more pointed molar teeth, and no cæcum. The wall of the hinder chamber of the auditory bulla remains through life unossified.

The dentition appears to be of a more decidedly carnivorous type than in the other members of the section.

Cynogale.[3]—This remarkable genus is regarded by Professor Mivart as representing a third section of the *Viverrinæ;* it contains one species, *C. bennetti* (described by S. Müller under the name of *Potamophilus barbatus*), from Borneo, Sumatra, and the Malay Peninsula. This is a curious Otter-like modification of the Viverrine type, having semi-aquatic habits, both swimming in the water and climbing trees, living upon fish, crustacea, small mammals, birds, and fruit. The number and general arrangement of its teeth are as in *Paradoxurus*, but the premolars are peculiarly elongated, compressed, pointed and recurved, somewhat as in the Seals, though the molars are tuberculated. The head is elongated,

[1] Temminck, *Prospectus de Monographies des Mammifères*, March 1824; *Monographies*, vol. i. p. xxi. (1827).

[2] Gray, *List of Mamm. Brit. Mus.* p. 54 (1843).

[3] Gray, *Proc. Zool. Soc.* 1836, p. 88.

the muzzle broad and depressed. Whiskers very long and abundant. Ears small and rounded. Toes short and slightly webbed at the base. Tail short, cylindrical, covered with short hair. Fur very dense and soft, of a dark brown colour, mixed with black and gray. Humerus without entepicondylar foramen.

Subfamily **Herpestinæ.**—Auditory bulla very prominent, and somewhat pear-shaped, the posterior chamber being large, rounded, and generally with its greatest prominence to the outer side. The anterior chamber considerably dilated, and produced into a short inferior wall to the auditory meatus, in which is a depression or vacuity just below the centre of the opening of the meatus. Sometimes this vacuity is continued into the meatus, forming a narrow fissure. The paroccipital process does not project beyond the bulla, but is spread out and lost (in adult animals) on its posterior surface. Toes straight; claws lengthened, exserted, non-retractile. No perineal glands. The dentition is always of a markedly sectorial type; and the orbit may be surrounded by bone. Very generally the anus opens into a sac-like depression. The majority of the genera are Ethiopian; the type genus alone extending into the Oriental and Palæarctic regions.

Herpestes.[1]—Dentition: $i\ \frac{3}{3}$, $c\ \frac{1}{1}$, $p\ \frac{4}{4}$, sometimes $\frac{3}{3}$, $m\ \frac{2}{2}$; total 40 or 36. Teeth of molar series generally with strongly developed, sharply-pointed cusps. Skull elongated, constricted behind the orbits. Face short and compressed. Frontal region broad and arched. Postorbital processes of frontal and jugal bones well developed, generally meeting so as to complete the circle of the orbit behind. Vertebræ: C 7, D 13, L 7, S 3, C 21-26. Head pointed in front. Ears short and rounded. Body very long and slender. Extremities short. Five toes on each foot, the first, especially that on the hind foot, very short. Toes free, or but slightly palmated. Palms generally naked. Distal portion of soles naked, under surface of tarsus and metatarsus usually clothed with hair, but considerable specific variation in this respect. Tail long or moderate, generally thick at the base, and sometimes covered with more or less elongated hair. The longer hairs covering the body and tail almost always annulated. This genus contains a very large number of animals commonly called Ichneumons, or in India Mungooses, varying in size from that of a large Cat down to a Weasel. They are widely distributed over the African continent and the southern parts of Asia, especially India and the Indo-Malayan archipelago, one species occurring also in Spain. They are mostly terrestrial in their habits, feeding on small mammals and birds, reptiles, especially snakes, eggs of birds and reptiles, and also insects. Some species are partially domesticated, being used to keep houses clear of rats, mice, and

[1] Illiger, *Prodromus Syst. Mamm.* p. 135 (1811).

snakes. *H. ichneumon* was a sacred animal to the ancient Egyptians. They vary considerably in appearance, some, as *H. galera* and *H. urva* (Fig. 240), are larger and heavier, with stouter body, longer limbs, and stronger teeth. The common Indian Mungoose (*H. mungo*) is considerably smaller than the Egyptian form; its fur is of a pale gray colour, the hairs being largely white ringed, while the cheeks and throat are more or less reddish. Like the Egyptian species, it is frequently domesticated, and put to a similar use. It is especially serviceable in India as a serpent-killer, destroying not only the eggs and young of these creatures, but attacking without hesitation and killing the most venomous

FIG. 240.—The Crab-eating Mungoose (*Herpestes urva*). From Blanford, *Mammalia of British India*, p. 130.

adult snakes. The fact that it invariably survives those encounters has led to the belief that it either enjoys immunity from the effects of snake-poison, or that after being bitten it has recourse, as the natives maintain, to the root of a plant as an antidote. Neither of these suppositions has stood the test of scientific examination, for it has been found that when actually bitten it falls a victim to the poison as rapidly as other mammals, while there is no trustworthy evidence of its seeking a vegetable antidote. The truth seems to be that the Mungoose, by its exceeding agility and quickness of eye, avoids the fangs of the snake while fixing its own teeth in the back of the reptile's neck. One large species, believed to be from Africa, recently described as *H. grandis*, is remarkable for the extreme complexity of the cusps on the molars, and also for the absence of an entepicondylar foramen to the humerus; the latter feature also occurring in the allied *H. albicaudatus*. The Oriental *H. urva* (Fig. 246) is stated to be somewhat aquatic in habits, and to feed on frogs and crabs.

Remains of the small *H. nipalensis* occur in the cavern-deposits of Madras. Viverroids from the Miocene and Upper Eocene of Europe, which agree with *Herpestes* in the presence of an inner tubercle to the third upper premolar and of a hinder cusp to the fourth lower premolar, have been referred to the existing genus. The species which have been separated generically under the three following names are very closely allied to *Herpestes*.

Helogale,[1] premolars $\frac{3}{3}$, without diastema between first and second; soles of feet completely naked. Contains two small South-African species, *H. parvula* and *H. undulata*.

Bdeogale[2] contains also two small Ichneumon-like animals, *B. crassicauda* and *puisa*, differing from *Herpestes* proper in having only four toes on each foot, both pollex and hallux being absent. The orbit is nearly complete, the tail of moderate length and rather bushy.

Cynictis.[3]—Pollex present, but hallux absent. Skull shorter and broader than in *Herpestes*, rather contracted behind the orbits, which are large and complete behind. Face short. Anterior chamber of the auditory bulla very large. Front claws elongated. *C. penicillata*, from South Africa. The cæcum (Fig. 241) of this genus is longer than in any other member of the family.

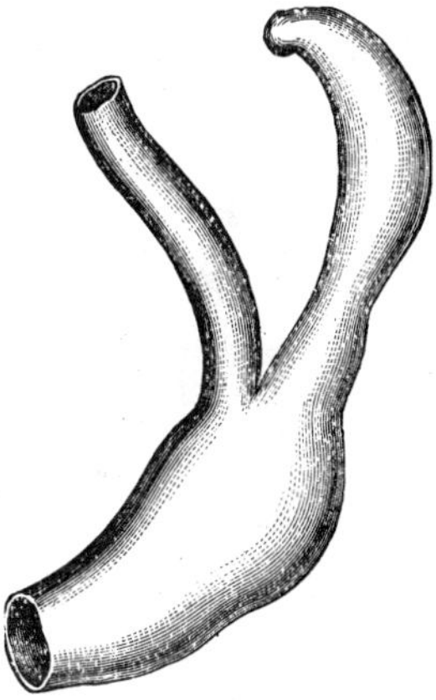

FIG. 241.—Cæcum of *Cynictis penicillata*. (From Mivart, *Proc. Zool. Soc.* 1882, p. 508.)

All the foregoing Herpestines have the nose short, with its under surface flat, bald, and with a median longitudinal groove. The remaining forms have the nose more or less produced, with its under side convex, and a space between the nostrils and the upper lip covered with close adpressed hairs, and without any median groove.

Rhinogale.[4]—Toes 5-5. Claws of fore feet short, compressed, acute. Under surface of tarsus hairy. Palate flat. Founded on a single specimen from East Africa, *R. melleri*.

Crossarchus.[5]—Dentition: $i\ \frac{3}{3}$, $c\ \frac{1}{1}$, $p\ \frac{3}{3}$, $m\ \frac{2}{2}$; total 36. Snout elongated. Toes 5-5. Claws on fore feet long and curved. Hallux very short. Under surface of tarsus naked. Tail shorter than the body, tapering. Palate flat. Fur harsh. Species: *C.*

[1] Gray, *Proc. Zool. Soc.* 1861, p. 308.

[2] Peters, *Mith. Ges. Nat. Freunde Berlin*, 19th November 1850.

[3] Ogilby, *Proc. Zool. Soc.* 1833, p. 48.

[4] Gray, *Proc. Zool. Soc.* 1864, p. 573.

[5] F. Cuvier, *Hist. Nat. des Mammifères*, No. 199 (1825).

obscurus, the Kusimanse, a small burrowing animal from West Africa, of uniform dark brown colour; *C. fasciatus*; *C. zebra*; and *C. gambianus.*

Suricata.[1]—A more distinct genus than any of the above. The dental formula as in the last, but the teeth of the cheek-series remarkably short in the antero-posterior direction, corresponding with the shortness of the skull generally (Fig. 222). Orbits complete behind. Vertebræ: C 7, D 15, L 6, S 3, C 20. Though the head is short and broad, the nose is pointed and rather produced and movable. Ears very short. Body shorter and limbs longer than in *Herpestes.* Toes 4-4, the pollex and hallux being absent. Claws on fore feet very long and narrow, arched, pointed, and subequal. Hind feet with much shorter claws, soles hairy. Tail rather shorter than the body. One species only is known, the Suricate, *S. tetradactyla*, a small gray-brown animal, with dark transverse stripes on the hinder part of the back, from South Africa. The cæcum is short.

Galidictis,[2] *Galidea*,[3] and *Hemigalidea*[4] are names of three slight generic modifications of the Viverrine type, allied to the *Herpestinæ*, but placed by Mivart in a distinct subfamily, *Galidictiinæ.* They are all characterised by the absence of the alisphenoid canal in the skull, as well as of the entepicondylar foramen to the humerus; and are inhabitants of Madagascar. The best known, *Galidea elegans*, is a lively Squirrel-like little animal with soft fur and a long bushy tail, which climbs and jumps with agility. It is of a chestnut-brown colour, the tail being annulated with darker brown. The cæcum (Fig. 242) is remarkable for its comparative length and pointed termination. *Hemigalidea* is distinguished by the absence of rings on the tail. *Galidictis vittata* and *striata* chiefly differ from the Ichneumons in their coloration, being gray with parallel longitudinal stripes of dark brown.

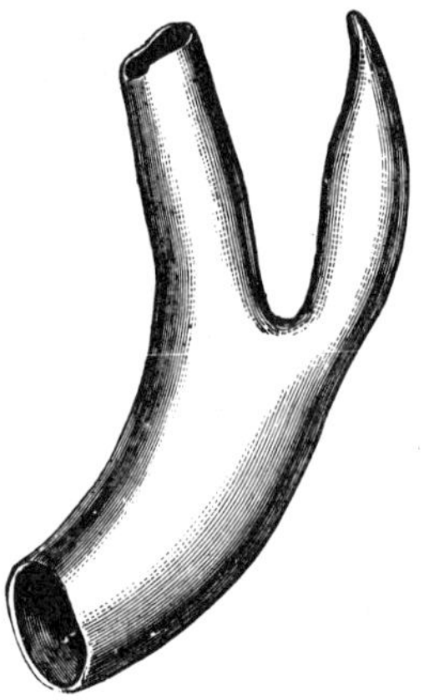

FIG. 242.—Cæcum of *Galidea elegans.* (From Mivart, *Proc. Zool. Soc.* 1882, p. 508.)

Eupleres[5] is another form, also from Madagascar, which has been placed in a subfamily apart. It differs remarkably from all the other *Viverridæ* in the weak development of the jaws and the

[1] Desmarest, "Tabl. Méth. Mamm." in *Nouv. Dict. d'Hist. Nat.* vol. xxiv. (1804).

[2] Geoffroy, *Comptes Rendus*, 1837, p. 578.

[3] Geoffroy, *Mag. de Zool.* 1839, pp. 27, 37.

[4] Doyère, *Ann. Sci. Nat.* vol. iv. p. 281 (1835).

[5] Jourdan, *Comptes Rendus*, 1837, p. 422. Amended.

small size of the teeth (Fig. 243), in consequence of which it was, when first discovered, placed in the order Insectivora. Dentition: $i\,\frac{3}{3}$, $c\,\frac{1}{1}$, $p\,\frac{4}{4}$, $m\,\frac{2}{2}$; total 40. Vertebræ: C 7, D 13, L 7, S 3, C 20. No alisphenoid canal; an entepicondylar foramen to the humerus. But one species is known, *E. goudoti*.

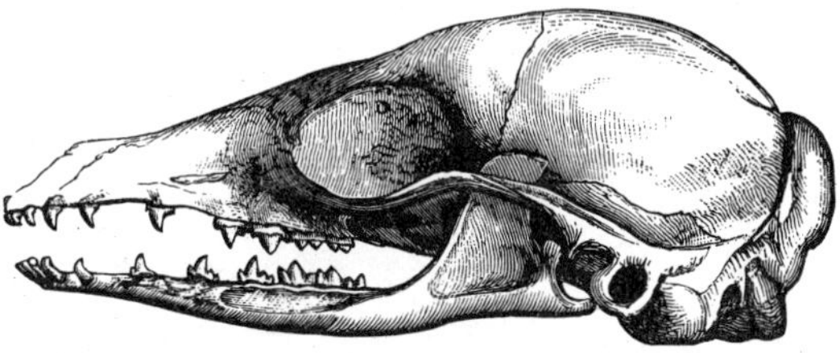

FIG. 243.—Skull of *Eupleres goudoti*. $\frac{4}{5}$ natural size. Mus. Roy. Coll. Surgeons.

Extinct Genera.—The Tertiaries of the Old World have yielded several genera allied to the existing Viverroids, some of which show decided signs of affinity with other families. Of these the Lower Miocene *Amphictis* appears to be nearly related to *Viverra*, but is distinguished by the form of the second lower molar, which is longer and has two distinct roots. *Palæoprionodon*, of the French Phosphorites, has a dentition very like that of *Prionodon*, the molars being reduced to $\frac{1}{2}$; the skull has an alisphenoid canal and the general basal characters of the *Viverridæ*, but resembles the *Mustelidæ* in the presence of a glenoid foramen and in the position of the condylar foramen. In *Stenoplesictis*, of the same deposits, the dental formula is $i\,\frac{3}{3}$, $c\,\frac{1}{1}$, $p\,\frac{4}{4}$, $m\,\frac{2}{2}$; and although the skull has a complete septum in the bulla, yet some of the cranial and dental features approximate so decidedly towards those of the extinct *Mustelidæ*, as to lead some authorities to refer the genus to that family. The most probable explanation of this resemblance is that the Musteloids have originated from generalised Viverroids allied to *Stenoplesictis*. The Lower Pliocene *Ictitherium* differs from all other Viverroids in the presence of three distinct lobes to the upper carnassial, and thereby connects the other members of the family so closely with the *Hyænidæ* that it is practically impossible to draw up a definition which will distinguish the two families.

The North American Eocene genera *Miacis* and *Didymictis* are generally regarded as representing a separate family—*Miacidæ*—with affinities both to the *Viverridæ* and *Canidæ*.

Family PROTELEIDÆ.

Skull with no alisphenoid canal; and the auditory bulla divided into two distinct chambers. Dorsal vertebræ 15. Molars $\frac{1}{1}$. Premolar and molar teeth very small and simple in character.

Proteles.[1]—This genus contains but a single species, *P. cristatus*,

[1] Geoffroy, *Mém. du Muséum*, vol. xi. p. 354 (1824).

the Aard-Wolf or Earth-Wolf of the Dutch colonists of the Cape, an animal nearly allied to the Hyænas, but remarkably modified in its dentition, the molar teeth being very small, placed far apart, and almost rudimentary in character (Fig. 244). The canines are long and rather slender. The dental formula is $i\,\frac{3}{3}$, $c\,\frac{1}{1}$, p and $m\,\frac{4}{3 \text{ or } 4}$; total 30 or 32. Vertebræ: C 7, D 15, L 5, S 2, C 24. The fore feet with five toes; the pollex though short, with a distinct claw. The hind feet with four subequal toes. Claws all strong, blunt, subcompressed, and non-retractile. The general external appearance is very like that of a small Striped Hyæna, but the muzzle is more pointed and the ears larger. It has a copious mane of long hair, capable of being erected when the animal is excited, along the middle line of the neck and back. It is a native of South Africa, and is a burrowing nocturnal animal, feeding on decomposing animal substances, larvæ, and termites. Observations upon specimens in captivity indicate that it has neither inclination nor power to attack or feed upon living vertebrated animals.

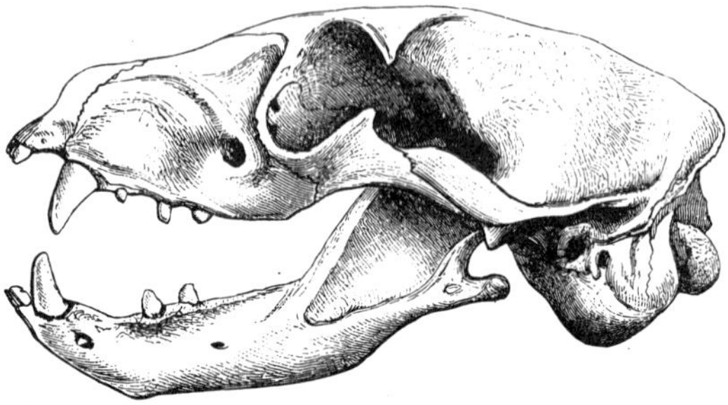

FIG. 244.—Skull and Dentition of the Aard-Wolf (*Proteles cristatus*). ½ natural size.

Some writers regard *Proteles* as representing a subfamily of the *Hyænidæ*.[1]

Family HYÆNIDÆ.

Skull with no alisphenoid canal; and the auditory bulla not divided by a septum into two chambers. Dorsal vertebræ 15. Molars usually $\frac{1}{1}$, but in some fossil forms $\frac{1}{2}$, or $\frac{2}{2}$, the second lower molar being very small; upper carnassial with three distinct lobes; lower carnassial with a large blade and small talon. No entepicondylar foramen to the humerus. This family is confined to the Old World, where it is now represented by a single genus, which, although evidently nearly related to the *Viverridæ*, is sufficiently distinct to be regarded as not referable to that family. The extinct *Ictitherium*, however, as already mentioned, connects the more generalised members of the *Hyænidæ* very closely with the *Viverridæ*.

Hyæna.[2]—Dentition in existing forms usually $i\,\frac{3}{3}$, $c\,\frac{1}{1}$, $p\,\frac{4}{3}$, m

[1] For Anatomy of *Proteles* see Flower, *Proc. Zool. Soc.* 1869, p. 474.

[2] Zimmermann, *Specimen Zoologiæ Geographicæ*, p. 365 (1777).

$\frac{1}{1}$; total 34. Teeth, especially canines and premolars, very large, strong, and conical. Upper carnassial (Fig. 245) with a very large, distinctly trilobed blade and a moderately developed inner tubercle placed at the anterior extremity of the blade. Molar very small, and placed transversely close to the hinder edge of the last, as in the *Felidæ*. Lower carnassial consisting of little more than the bilobed blade. Zygomatic arches of cranium very wide and strong. Sagittal crest high, giving attachment to very powerful biting muscles. Orbits incomplete behind. Vertebræ: C 7, D 15, L 5, S 4, C 19. Limbs rather long, especially the anterior pair, digitigrade, four subequal toes on each, with stout non-retractile claws. Pollex and hallux only represented by rudimentary metacarpal and metatarsal bones. Tail rather short. A large post-anal median glandular pouch, into which the largely developed anal scent glands pour their secretion.

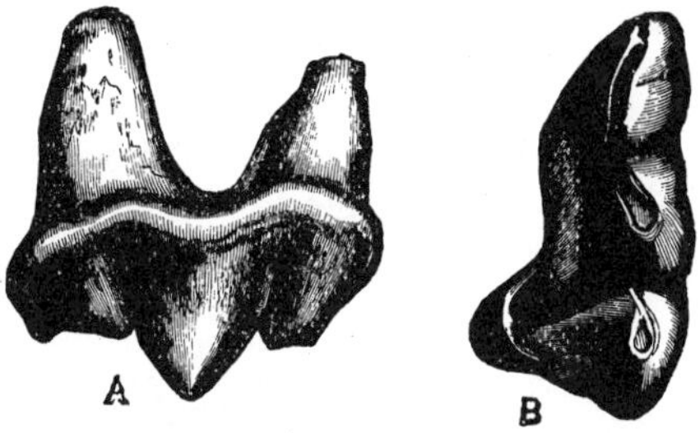

FIG. 245.—Outer (*A*) and palatal (*B*) aspects of the right upper carnassial tooth of the Striped Hyæna (*Hyæna striata*). From the *Quart. Journ. Geol. Soc.*

The three existing species of Hyæna are divisible into two sections, to which some zoologists assign generic rank, but fossil forms show such a transition between these two types as to render any such division impracticable.

The typical or *Euhyænine* group presents the following distinctive features. Upper molar moderately developed and three-rooted. An inner cusp and hind talon more or less developed on the lower molar. Ears large, pointed. Hair long, forming a mane on the back and shoulders. *H. striata*, the Striped Hyæna (Fig. 246) of Northern Africa and Southern Asia. *H. brunnea*, of South Africa, in some respects intermediate between this and the next group.

The Striped Hyæna is dirty gray in colour, with narrow transverse tawny or blackish stripes on the body and legs; the length of the head and body is $3\frac{1}{2}$ feet, and that of the tail, with its hair, $1\frac{1}{2}$ feet. It occurs throughout peninsular India, where it is most common in open hilly districts, and in North Africa. Mr. Blanford[1] gives the following account of its habits: "It is a nocturnal animal, and although an occasional individual may be met with returning to its den in the early morning, its rambles are usually commenced after sunset and ended before sunrise. During the night it roams far and wide, and no tracks of wild animals are more common in the countries where it is found than its unmistakable footprints, very like a dog's

[1] *Fauna of British India*, "Mammalia," p. 133 (1888).

in shape, but with the marks of the hind feet conspicuously smaller than those of the fore feet. Unlike the Spotted Hyæna, the Striped species appears to be solitary in its habits, and it is rare to meet with more than two together. The principal food of the Hyæna consists of the carcases of animals that have died of disease or been killed by beasts of prey, and very often it carries off portions of the body to its den. I once shot one that was carrying away the hind leg of a Nilghai. The powerful jaws and large teeth are admirably adapted for crushing bones, which are consumed by

FIG. 246.—The Striped Hyæna (*Hyæna striata*).

Hyænas, after the flesh has been picked off by vultures and jackals. Occasionally sheep or goats, and more often dogs, are carried off by Hyænas, and the latter at all events are often taken alive to the animal's den." The Striped Hyæna is essentially a cowardly animal, and one that is much more silent than *H. crocuta*. Remains of *H. striata* are found in the cavern-deposits of the south of France, and also in the Upper Pliocene of the Val d'Arno in Tuscany, and in the English Red Crag.

The *Crocutine* group presents the following characters. Upper molar extremely small, two- or one-rooted, often deciduous. Lower molar without trace of inner cusp, and with an extremely small talon. Ears moderate, rounded. Hair not elongated to form a mane. *H. crocuta*, the Spotted Hyæna (Fig. 247), from Africa south of the Sahara. In dental characters as well as in its visceral anatomy, especially as regards the reproductive organs of the female,[1] this species may be considered as by far the more

[1] The anatomical peculiarities of *Hyæna crocuta* have been fully elucidated in

specialised form. The Spotted Hyæna is a larger and bolder animal than the Striped species, hunting in packs, and uttering very frequently its unearthly cry. The coloration consists of dark brown spots on a yellowish ground. It was formerly very common at the Cape. Remains of a large race of this species are exceedingly common in the cavern-deposits of Europe, where they were first described under the name of *Hyæna spelæa;* teeth have also been met with in the Norfolk Forest-bed, and in cavern-deposits in Madras—the latter locality being exceedingly interesting from a distributional point of view.

In addition to the remains of existing species, to which refer-

FIG. 247.—The Spotted Hyæna (*Hyæna crocuta*).

ence has been already made, there were numerous extinct forms of *Hyæna* in the upper Tertiaries of Europe, from the horizon of the Lower Pliocene Pikermi beds of Greece upwards. In the Crocutine group *H. colvini* of the Pliocene of India (Fig. 248), and *H. robusta* of that of Italy, appear to have been ancestral forms allied to *H. crocuta;* the former being distinguished by the loss of the first upper premolar. *H. eximia*, of the Pikermi beds, is a more generalised form, in which the first lower premolar (lost in existing forms) is retained. In the typical group, *H. arvernensis* and *H. perrieri*, of the Upper Pliocene of the Continent, approximate to *H. brunnea;* although *H. perrieri* makes a farther step towards the Crocutine group by the loss of the inner cusp in the lower carnassial. The extinct *Hyænic-*

a series of papers by Morrison Watson in the *Proceedings of the Zoological Society* for 1877, 1878, 1879, and 1881, in which references to previous authors on the subject will be found.

tine group, as represented by the Indian *H. sivalensis* and the Grecian *H. græca*, connects *H. striata* with *Palhyæna.* Both are characterised by the presence of a small second lower molar behind the carnassial; while *H. græca* also has four lower premolars. Still more generalised is the *Lychyænine* group, comprising *H. macrostoma* of India and *H. chæretis* of the Pikermi beds; in these forms the muzzle was longer, and the premolars much more compressed than in the existing species, thus making a very decided approach to the *Viverridæ.* There were four lower premolars; the lower carnassial had an inner cusp, and it is probable that there was a second lower molar; while the first upper molar was placed partially behind the

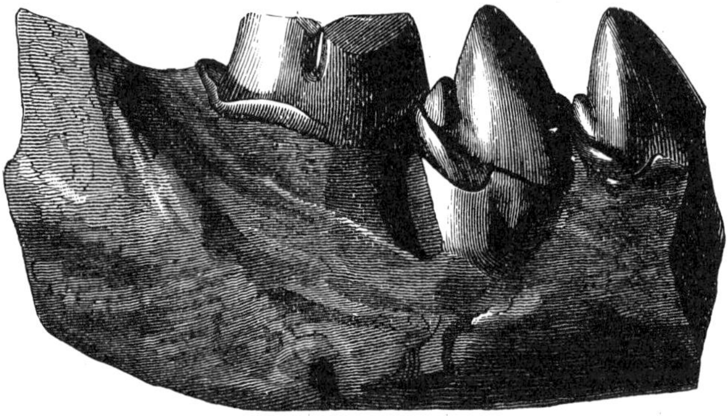

FIG. 248.—Outer view of part of the right ramus of the mandible of *Hyæna colvini*, showing the third and fourth premolars and the carnassial. (From the *Palæontologia Indica.*)

carnassial. The Lower Pliocene *Palhyæna hipparionum*, in which the dental formula is $i\ \frac{3}{3}$, $c\ \frac{1}{1}$, $p\ \frac{4}{4}$, $m\ \frac{2}{2}$, is a smaller form with long jaws and compressed premolars which approaches so closely to the Viverroid genus *Ictitherium* as to show pretty clearly how the Hyænas have been gradually modified from that stock.

Section CYNOIDEA.

Family CANIDÆ.

This section contains the single family of the *Canidæ*, or Dog-like animals, which appear to hold an intermediate position between the other two sections, retaining also many of the more generalised characters of the ancient members of the order. The structure of the auditory bulla and adjacent parts of the bones of the skull is intermediate between that of the Æluroid and Arctoid forms. In the number and arrangement of the teeth they more nearly approach the primitive heterodont type than any other existing Carnivora.

A cæcum is always present, sometimes short and simple, but when long it is folded upon itself in a characteristic manner.

The characters of the base of the cranium are shown in Fig. 8

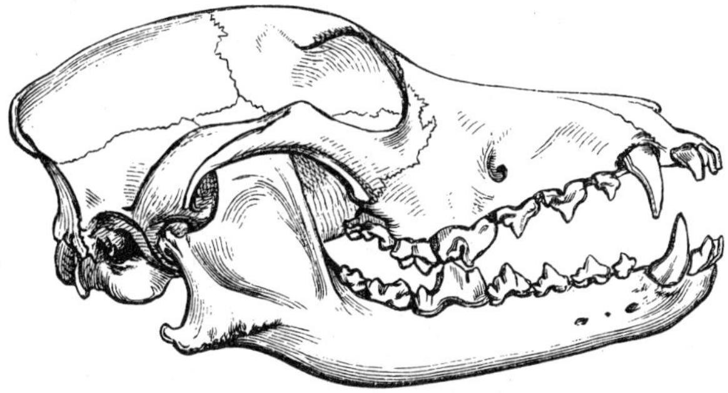

FIG. 249.—Right lateral aspect of the skull of the Dog (*Canis familiaris*).

(p. 38), where it will be seen that the auditory bulla is inflated, although it has only a rudimental internal septum; the paroccipital process, although in contact with the bulla, is prominent, and there is a large glenoid foramen. In all the existing forms the humerus has lost the entepicondylar foramen; the crowns of the upper molars are triangular in shape (Fig. 251), and the blade of the upper carnassial consists of two lobes.

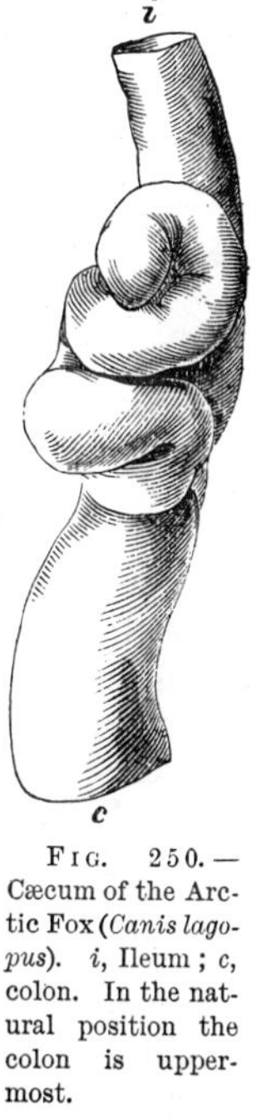

FIG. 250.—Cæcum of the Arctic Fox (*Canis lagopus*). *i*, Ileum; *c*, colon. In the natural position the colon is uppermost.

In the alimentary canal the cæcum (Fig. 250) is extremely characteristic. It is a simple appendage of nearly uniform width (about equal to that of the ileum) attached to the side of the canal, just beyond the ileo-cæcal valve, and with a rounded termination. In a Dog of average size it is 5 or 6 inches long if uncoiled, but it is normally folded by its mesenteric attachments backwards and forwards several times on itself by the side of the ileum, after the manner shown in the figure.

The existing Dogs form a very compact group, with numerous species closely resembling each other in essential characters, though differing considerably externally. The most marked differences are slight variations in the number of the true molar teeth, which exceed the usual number in the Cape Long-eared Fox (*Otocyon*), and fall short of it in some other less aberrant forms to which the names of *Icticyon* and *Cyon* have been given, and a diminution in the number of toes in the Cape Hunting Dog (*Lycaon*), which has 4-4, instead of 5-4 as in the remainder of the family.

After taking these away, there remain a great number of animals called Dogs, Wolves, Jackals, and Foxes, varying from one another only in the characters of the tail, ears, fur, form of the pupil, and some trifling peculiarities of skull and teeth, upon which some authors have divided them into many genera. These divisions are, however, extremely difficult, if not impossible, to define, on account of the numerous gradual transitions from one form to the other.

Canis.[1]—It appears on the whole convenient to retain all the species, with the exception of *Otocyon*, *Icticyon*, and *Lycaon*, in the old genus *Canis*, the most prominent characters of which are the following. Teeth, usually $i\ \frac{3}{3}$, $c\ \frac{1}{1}$, $p\ \frac{4}{4}$, $m\ \frac{2}{3}$; total 42. The absence of the last upper molar ($\overline{m\ 3}$), alone distinguishes this from the generalised dentition of heterodonts, and this tooth is occasionally present in one species (*C. cancrivorus*). In certain Asiatic species (*C. primævus* and its allies), which on this account have been separated to form the genus *Cyon* of Hodgson, the last lower molar ($\underline{m\ 3}$) appears to be constantly absent. The milk-dentition is $di\ \frac{3}{3}$, $dc\ \frac{1}{1}$, $dm\ \frac{3}{3}$; total 28,—the first permanent premolar having no predecessor. The teeth of both permanent and milk or temporary series are figured on p. 26, Fig. 3, from the outer aspect, while the woodcut 251 shows the palatal aspect of the hinder upper teeth. The upper carnassial ($\overline{p\ 4}$) consists of a stout blade, of which the anterior lobe is almost obsolete, the middle lobe large, conical, and pointed backwards, and the posterior lobe in the form of a compressed ridge; the inner tubercle is very small, and placed quite at the fore part of the tooth. The first molar is more than half the antero-posterior length of the carnassial, and considerably wider than it is long; its crown consists of two prominent conical cusps, of which the anterior is the larger, and a low broad inward prolongation, supporting two more or less distinct cusps and a raised inner border. The second molar resembles the first in general form, but is considerably smaller. The lower carnassial ($\underline{m\ 1}$) is a very large tooth, with a strong compressed bilobed blade, the hinder lobe being considerably the larger and more pointed, a small but distinct inner cusp placed at the hinder margin of the posterior lobe of the blade, and a broad, low, tuberculated talon,

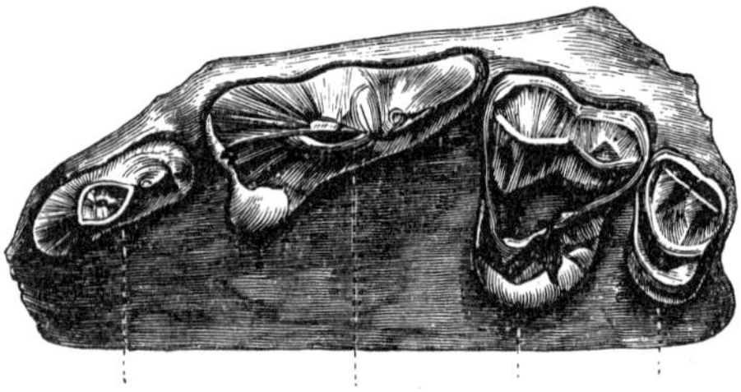

FIG. 251.—The last four left upper teeth of an extinct Wolf (*Canis cautleyi*). From the *Palæontologia Indica.*

[1] Linn. *Syst. Nat.* 12th ed. vol. i. p. 56 (1766).

or heel, occupying about one-third of the whole length of the tooth. The second molar is less than half the length of the first, with a pair of cusps placed side by side anteriorly, and a less distinct posterior pair. The third is an extremely small and simple tooth, with a subcircular tuberculated crown and single root.

The cranium (Fig. 249) is more or less elongated, the facial portion tapering forwards and compressed. The jaws are elongated, and the zygomata moderately strong. The postorbital processes of the frontal short, leaving the orbit widely open posteriorly. Vertebræ: C 7, D 13, L 7, S 3, C 17-22. Clavicles present, but very rudimentary. Limbs of moderate proportions, digitigrade. Feet short; five toes on the fore foot, the pollex much shorter than the others, and not reaching to the ground. Four toes on the hind foot, the hallux being represented by a rudiment of the metatarsal.[1] All the toes are provided with exserted, non-retractile, slightly curved, and blunt claws, which, being exposed, become worn at the tips. Tail moderate, or rather long, generally somewhat bushy. The pupil of the eye, when contracted, is in some species round, in others elliptical and vertical.

This extensive genus may be considered as truly cosmopolitan. One or more species occur in every part of the American continent from Greenland to Patagonia and the Falkland Isles; and similarly, in the Old World, Europe, Africa, and Asia, with most of the large islands adjacent, and even Australia, have their wild Dogs, though in the last case they may belong to a feral race, introduced originally by man. They are generally sociable animals, hunting their prey in packs. Many species burrow in the ground; none habitually climb trees. Though mostly carnivorous, feeding chiefly on animals they have chased and killed themselves, many, especially among the smaller species, eat garbage, carrion, insects, and also fruit, berries, and other vegetable substances. The species are very numerous, and, as in most other large genera, very ill-defined, few zoologists agreeing as to which of the many slightly different modifications should be considered as local varieties and which true species. Perhaps the best cranial character by which the different members of the genus can be distinguished is that pointed out by Burmeister, viz. that in the animals generally called Dogs, Wolves, and Jackals the postorbital process of the frontal bone is regularly smooth and convex above, with its extremity bent downwards, whereas in Foxes this process is hollowed above, with its outer margin (particularly of the anterior border) somewhat raised. This modification coincides in the main with that upon which Professor

[1] In Domestic Dogs a hallux is frequently developed, though often in a rudimentary condition, the phalanges and claw being suspended loosely in the skin, without direct connection with the other bones of the foot; it is called by dog-fanciers the "dew claw."

Huxley[1] has based his division of the group into two parallel series, the Thooids or Lupine forms and Alopecoids or Vulpine forms, which he characterises by the presence of frontal air-sinuses in the former, which not only affect the external contour but to a still greater degree the shape of the anterior part of the cranial cavity, and the absence of such sinuses in the latter. The pupil of the eye when contracted is round in most members of the first group, and vertically elliptical in the others, but more observations are required before this character can be absolutely relied upon. The form and length of the tail is often used for the purposes of classification, but its characters do not coincide with those of the cranium, since many of the South American *Canidæ* have the long bushy tails of Foxes and the skulls of Wolves. Taking into account various combinations of these and other minor characters, the species may be arranged in the following groups, which some authors have considered as of generic importance.

A. *Thooid or Lupine Series.*—The typical group, or *Canis* proper, contains the largest members of the genus, the true Wolves of the northern parts of both Old and New Worlds (*C. lupus*, etc.), the Jackals of Southern Asia and Africa (*C. aureus*, *mesomelas*, etc.), and the various breeds of the domestic Dog (*C. familiaris*). The true Wolves are (excluding some varieties of the domestic Dog) the largest members of the genus, and have a wide geographical range, extending over nearly the whole of Europe and Asia, and North America from Greenland to Mexico, but they are not found in South America or Africa, being replaced in both of these continents by various species of Jackals and Foxes. As might be expected from this extensive range, and the varied character of the climatic conditions of the countries they inhabit, they present great diversities of size, length and thickness of fur, and coloration, although resembling each other in all important structural characters. These differences have given rise to a supposed multiplicity of species, expressed by the names of *C. lupus*, *C. lycaon* (Central Europe), *C. laniger* and *C. niger* (Tibet), *C. pallipes* (India), *C. occidentalis*, *C. nubilis*, *C. mexicanus*, etc., of North America, but it is very doubtful whether some of these ought to be distinguished as other than local varieties. Mr. W. T. Blanford, in his recent work on the mammals of India, regards the two forms from Tibet mentioned above as inseparable from *C. lupus*. In North America there is a very distinct smaller species, called the Coyote or Prairie Wolf (*C. latrans*); and perhaps the Japanese Wolf (*C. hodophylax*) may also be distinct, although, except for its smaller size and shorter legs, it is scarcely distinguishable from the common species. Though generally distributed throughout the Indian peninsula, the Indian

[1] *Proc. Zool. Soc. Lond.*, 1880, p. 238. See also Mivart, *Dogs, Jackals, Wolves, and Foxes; a Monograph of the Canidæ* (1890).

Wolf (*C. pallipes*), which is rather smaller and slighter than *C. lupus*, is not found in Ceylon, nor in Burma and Siam. The ordinary colour of the Common Wolf is a yellowish or fulvous gray, but specimens have been met with almost pure white and others entirely black. In northern countries the fur is longer and thicker, and the animal generally larger and more powerful than in the southern portion of its range; this being especially the case with the Tibetan races. The habits of the Wolf are similar everywhere, and it is still, and has been from time immemorial, especially known to man in all the countries it inhabits as the devastator of his flocks of sheep. They do not catch their prey by lying in ambush, or stealing up close to it and making a sudden spring as the Cat tribe do, but by fairly running it down in open chase, which their speed and remarkable endurance enable them to do; and usually, except during summer, when the young families of cubs are being separately provided for by their parents, they assemble in troops or packs, and by their combined and persevering efforts are able to overpower and kill even such great animals as the American Bison. It is singular that such closely allied species as the Domestic Dog and the Arctic Fox are among the favourite prey of Wolves, and, as is well known, children and even full-grown people are not unfrequently the objects of their attack when pressed by hunger. Notwithstanding the proverbial ferocity of the Wolf in a wild state, many instances are recorded of animals taken when quite young becoming perfectly tame and attached to the person who has brought them up, when they exhibit many of the ways of a Dog. They can, however, rarely be trusted by strangers.

The history of the Wolf in the British Isles and its gradual extirpation has been thoroughly investigated by Mr. J. E. Harting in his work on *Extinct British Animals*, from which the following account is abridged: To judge by the osteological remains which the researches of geologists have brought to light, there was perhaps scarcely a county in England or Wales in which, at one time or another, wolves did not abound, while in Scotland and Ireland they must have been still more numerous. The fossil remains which have been discovered in Britain are not larger than, nor in any way to be distinguished from, those of European wolves of the present day. Wolf-hunting was a favourite pursuit of the ancient Britons as well as of the Anglo-Saxons. In Athelstan's reign these animals abounded to such an extent in Yorkshire that a retreat was built by one Acehorn, at Flixton, near Filey, wherein travellers might seek refuge if attacked by them. As is well known, great efforts were made by King Edgar to reduce the number of wolves in the country, but, notwithstanding the annual tribute of 300 skins paid to him during several years by the king of Wales, he was not altogether so successful as has been commonly imagined.

In the reign of Henry III the number of wolves in some parts of the country was sufficient to induce the king to make grants of land to various individuals upon the express condition of their taking measures to destroy these animals wherever they could be found. In Edward II's time the king's forest of the Peak, in Derbyshire, is especially mentioned as infested with wolves, and it was not until the reign of Henry VII (1485-1509) that wolves appear to have become finally extinct in England. This, however, is rather a matter of inference from the cessation of all mention of them in local records than from any definite evidence of their extirpation. Their last retreat was probably in the desolate wolds of Yorkshire. In Scotland, as might be supposed from the nature of the country, the wolf maintained its hold for a much longer period. There is a well-known story of the last of the race being killed by Sir Ewen Cameron of Lochiel in 1680, but there is evidence of wolves having survived in Sutherlandshire and other parts into the following century (perhaps as late as 1743), though the date of their final extinction cannot be accurately fixed. In Ireland, in Cromwell's time, wolves were particularly troublesome, and said to be increasing in numbers, so that special measures were taken for their destruction, such as the offering of large rewards for their heads, and the prohibition (in 1652) of the exportation of "wolf-dogs," the large dogs used for hunting the wolves. The active measures taken then and later reduced their numbers greatly, so that towards the end of the century they became scarce, but, as in the case of the sister island, the date of their final disappearance cannot now be ascertained. It has been placed, upon the evidence of somewhat doubtful traditions, as late as 1766.

Remains of *C. lupus* are common in the European Pleistocene; while the Indian Pliocene *C. cautleyi*, of which the upper teeth are shown in Fig. 251, was probably the ancestor of *C. pallipes. C. neschersensis*, of the Upper Pliocene of France, was a smaller extinct Wolf. A lower jaw from the French Pleistocene, described under the name of *Lycorus*, has only three premolars, but evidently belongs to the Wolf.

The Jackals are smaller than the Wolves, with the bushy tail about one-third the length of the head and body, and the carnassials relatively shorter as compared with the tubercular molars. The Common Jackal (*C. aureus*, Fig. 252) has a very wide distribution, ranging from South-Eastern Europe through South-Western Asia to India and Burma, and also occurring in Northern Africa; being replaced in the Ethiopian region by closely allied species. Remains indistinguishable from *C. aureus* occur in the Pliocene Siwaliks of Northern India. Jackals hunt at night in packs, uttering the piercing cries so well known to all who have resided in countries where these animals are found.

The origin of the Domestic Dog, with its numerous breeds, has been the subject of much controversy. Some naturalists believe it to be a distinct species, descended from one that no longer exists in a wild state; others have sought to find its progenitors in some one of the wild or feral races, either of true Dogs, Wolves, or Jackals; while others again believe that it is derived from the mingling of two or more wild species or races. It was probably the earliest animal domesticated by man, and few if any other species have undergone such an extraordinary amount of variation in size, form, and proportion of limbs, ears, and tail—

FIG. 252.—The Jackal (*Canis aureus*).

variations which have been perpetuated and increased by careful selective breeding. The Dingo or Australian Dog is met with wild, and also as the domestic companion of the aboriginal people. Dogs were also in the possession of the natives of New Zealand and other islands of the Pacific, where no placental mammals exist naturally, on their discovery by Europeans in the last century.

The second group includes the wild Dogs of the south-east of Asia, described as *Cyon*, and distinguished by slight modifications as *C. rutilans*, *C. dukhunensis*, and *C. javanicus*, and differing from the above in wanting the small last lower tubercular molar. This difference reduces the number of the teeth to the same as in *Viverra*, and is precisely paralleled by some of the species of the extinct genus *Cynodictis* mentioned below. The muzzle is shorter than in other species, and the facial profile is slightly convex instead of concave. The mammæ are also 12 or 14 instead of the normal 10; while there is long hair between the foot-pads. Wild Dogs inhabit not only the whole of the Oriental region, but extend into Central Asia as far north as the Altai and Amurland (*C. alpinus*). *C. dukhunensis* ranges from the forest regions of peninsular India to Gilgit and Western Tibet, where it must inhabit open country. In their general form, and more especially the shortness of the legs, these animals come nearer to the Jackals than to the Wolves. They hunt their prey in packs. Remains of species of this group occur in the cavern-deposits of the Continent, and have been described under the name of *C. europæus*.

A group for which the name *Lycalopex* has been proposed comprises certain South American *Canidæ*, distinguished from *Canis* proper by their longer tails and Fox-like aspect:—*C. cancrivorus*, *C. brasiliensis*, *C. melampus*, *C. vetulus*, *C. fulvicaudus*, *C. azaræ*, *C. magellanicus*, *C. griseus*. The last three have been further separated (under the name of *Pseudalopex*) on account of slight differences in the relative size of the molar teeth, and of their pupil being elliptical when contracted. *Nyctereutes* (one species, *C. procyonides*, from Japan and North-East Asia) has no claims to generic distinction but such as are founded upon its long loose fur, short ears, and short bushy tail, which give it some superficial resemblance to a Raccoon.

B. *Alopecoid or Vulpine Series.*—The *Vulpine* group (*Vulpes*) includes the true Foxes, of which there are numerous varieties and species, spread over North America, Eurasia, and Africa, which have been described under the names of *C. vulpes* (*Vulpes alopex*), the common Fox of Europe; *C. niloticus*, *adustus*, and *variegatus*, Africa; *C. flavescens*, *montanus*, *bengalensis*, *japonicus*, *corsac*, Asia; *C. fulvus*, *macrurus*, *velox*, North America. Mr. Blanford[1] concludes, however, that the Asiatic *C. flavescens* and *C. montanus*, and very probably the North American Cross-Fox (*C. fulvus*) are merely local races of *C. vulpes*, distinguished by certain peculiarities of coloration. The English Fox measures about 2 feet in length exclusive of the tail, which is about a foot long. Its fur is of a reddish-brown colour above, and more or less white beneath; the back of the ears and the fore part of the limbs are black, and the tip of its bushy tail is white. Its long, sharp muzzle, erect pointed ears, and sharp eye, give it the well-known appearance of sagacity and cunning. The Fox is a solitary animal, inhabiting a burrow, which it either excavates for itself, or obtains by ejecting the badger or the rabbit. So averse, indeed, is the Fox to dig for itself, that when foiled in its attempts to dispossess the badger, it has been known to take up its quarters with the latter, and it can be induced to make its home in artificial burrows constructed of stone

[1] *Fauna of British India*, "Mammalia," pp. 153, 154 (1888).

and earth for the purpose of facilitating the operation of digging out the cubs. The Fox also occurs in woods, and even in the open country without burrows, lying in its "cover" by day and stealing forth at night in search of its prey. Remains of the Common Fox occur not unfrequently in the Pleistocene deposits of Europe. The Indian *C. bengalensis* is a very much smaller and well-marked species.

The tail of the above forms is clothed with soft fur and long hair, uniformly mixed; from them Baird distinguishes, under the name of *Urocyon*, other species which have a concealed erect mane of stiff hairs along the upper line of the tail. These have also a shorter muzzle and a wide space between the temporal crests; they are *C. virginianus* and *C. littoralis*, both from North America. The Arctic Fox (*C. lagopus*, genus *Leucocyon*, Gray) has the tail very full and bushy and the soles of the feet densely furred below. Its colour changes according to season from bluish-gray to pure white.

Certain small elegant African Foxes (*C. zerda, famelicus*, and *chama*), with very large ears and corresponding large auditory bullæ, have been separated under the name of *Fennecus*, and are commonly known as Fennecs.

The earliest undoubted occurrence of the genus *Canis* seems to be in the Upper Miocene of Switzerland, where it is represented by the Fox-like *C. œningensis*. In the Upper Pliocene of France *C. megamastoides* is said to be allied to the Foxes and Jackals, but with some signs of affinity to the extinct *Cynodictis*. In the Pliocene Siwaliks of India there occurs *C. curvipalatus*, of the size of a small Fox, which appears to have certain resemblances to *Otocyon*.

Lycaon.[1]—This genus resembles in most of its characters the Dogs of the Lupine series, but the teeth are rather more massive and rounded, the skull is shorter and broader, and there are but four toes on each limb, as in *Hyæna*. The one species, *L. pictus*, the Cape Hunting Dog (Fig. 253) from South and East Africa, is very distinct externally from all the other *Canidæ*. It is nearly as large as a Mastiff, with large, broadly ovate erect ears, and singularly coloured, being not only variable in different individuals, but unsymmetrically marked with large spots of white, yellow, and black. It presents some curious superficial resemblances to *Hyæna crocuta*, perhaps a case of mimetic analogy. It hunts its prey in large packs. A lower jaw from a cave-deposit in Glamorganshire, which agrees with that of the existing form in the presence of an anterior cusp to the last lower premolar, has been made the type of a distinct species (*L. anglicus*).

Icticyon.[2]—The Bush-Dog (*I. venaticus*), from Guiana and Brazil, is a species about the size of a Fox, with close hair, and short legs

[1] Brookes, *Griffith's Animal Kingdom*, vol. v. p. 151 (1827).
[2] Lund, *K. Danks. Vid. Selsk. Afhand.* vol. xi. p. 62 (1845).

and tail, distinguished from all other Dogs by the reduction of the molar teeth to $\frac{1}{2}$, and their comparatively small size. The lower carnassial is also characterised by the loss of the inner cusp of the blade, and the secant form of its hind talon; both these features indicating a specialised type. Remains of the Bush-Dog are found in the Pleistocene cavern-deposits of Brazil, and were originally described under the name of *Speothos*.

Otocyon.[1]—Dentition: $i\ \frac{3}{3},\ c\ \frac{1}{1},\ p\ \frac{4}{4},\ m\ \frac{3\text{ or }4}{4}$; total 46 or 48. The molar teeth are thus in excess of any other living heterodont

FIG. 253.—The Cape Hunting Dog (*Lycaon pictus*).

mammal. They have the same general characters as in *Canis*, with very pointed cusps. The lower carnassial shows little of its typical characters, having five cusps on the surface; these can, however, be identified as the inner cusp, the two greatly reduced and obliquely placed lobes of the blade, and two cusps on the talon. The skull generally resembles that of the smaller Foxes, particularly the Fennecs. The auditory bullæ are very large. The hinder edge of the mandible has a very peculiar form, owing to the great development of an expanded, compressed, and somewhat inverted subangular process. Vertebræ: C 7, D 13, L 7, S 3, C 22. Ears very large. Limbs rather long. Toes 5-4. One species, *O. megalotis*, from South Africa, rather smaller than a common Fox.

Professor Huxley looks upon this as the least differentiated or most primitive existing form of the family, regarding the presence

[1] Lichtenstein, *Wiegmann's Archiv.* 1838, vol. i. p. 290.

of the four molar teeth as a survival of a condition of the dentition exhibited by the common ancestors of the existing *Canidæ* and the existing carnivorous Marsupials. There is, however, at present no palæontological proof of this, as none of the numerous fossil forms of *Canidæ* yet discovered have more than the normal number of molars.

Extinct Genera.—A large number of fossil Carnivora have been described from various Tertiary deposits which are more or less closely allied to the existing *Canidæ*, although, as already mentioned, connecting the latter so closely on the one hand with the *Viverridæ* and on the other hand with the *Ursidæ*, that it is almost, if not quite impossible to say where one family begins and the other ends. A few only of the more important of these annectant types will be mentioned here. *Temnocyon*, of the Miocene of the United States, is a true Dog, which agrees with *Icticyon* in having a secant hind talon to the lower carnassial, but preserves a generalised character in having an entepicondylar foramen to the humerus. An extremely interesting form is *Cynodictis*, of the Middle Tertiaries of Europe and the United States, which (as now restricted by Dr. Schlosser) includes a number of species mostly not larger than Foxes. The dental formula is generally the same as in *Canis*, but (as in that genus) the last lower molar may be absent. The teeth are very like those of the *Viverridæ*, the lower carnassial never being greatly elongated antero-posteriorly, and its inner cusp being situated immediately on the inner side of the hinder lobe of the blade, instead of somewhat behind it, as is the case in most Dogs. In the skull the auditory bulla is inflated, but is said to have no distinct septum; while the humerus invariably has an entepicondylar foramen. It is suggested that *Cynodictis* is not far removed from the ancestral type of many of the Viverroids and Canoids, and may itself have been derived from the undermentioned genus *Amphicyon*. M. Boule considers, indeed, that from the resemblance of the Pliocene *Canis megamastoides* (p. 553) to *Cynodictis* we ought to regard the Foxes and Jackals as the descendants of *Cynodictis*, while the Wolves have been derived directly from *Amphicyon*. The last named genus, which includes some species as large as a Bear, is found in the Upper Eocene and Lower Miocene of Europe, and is represented in the Miocene of the United States by the allied *Daphænus*. It is characterised by the presence of three upper molars—thus bringing up the dental formula to the full Eutherian number; by the five digits on all the feet, which were plantigrade; and by the presence of a third trochanter to the femur and an entepicondylar foramen to the humerus. The teeth are essentially those of a dog, and the base of the skull is also dog-like, although it is highly probable that the auditory bulla had no trace of a septum. According, however, to Dr. Filhol[1] the minute foramina

[1] *Arch. Mus. Lyon.* vol. iii. art. 1, p. 85 (1881).

described by Professor Cope[1] in the postparietal and mastoid which occur in *Ursus*, but are said to be absent in *Canis*, are present in *Amphicyon*. So far, however, as we can see, the presence or absence of those foramina cannot be regarded as diagnostic of *Ursus* and *Canis*, although they are generally more strongly developed in the former. *Amphicyon* may, indeed, be considered as a very generalised Dog, with affinities to the Bears in the structure of its limbs. *Dinocyon* is a still larger form, from the Middle Miocene of France, which, so far as its teeth are concerned, connects *Amphicyon* with the Ursoid genus *Hyænarctus* so closely as to render it absolutely impossible to indicate any characters of family importance by which they can be distinguished. The upper carnassial of *Dinocyon* is unknown. For other genera, see p. 562.

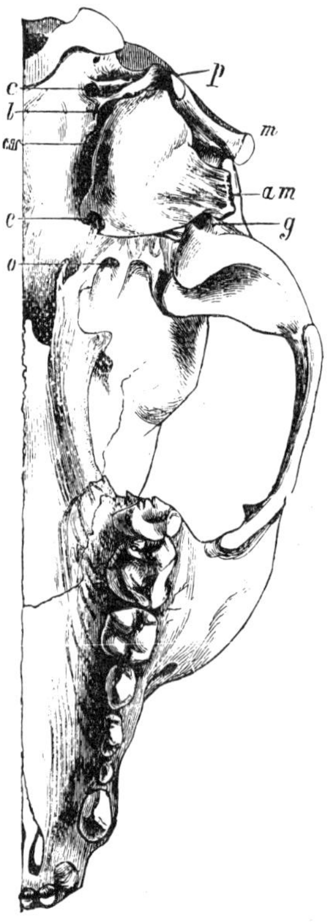

FIG. 254.—Right half of the palatal aspect of the cranium of the Raccoon (*Procyon lotor*). Letters as in Fig. 8, p. 38. (From the *Proc. Zool. Soc.* 1869, p. 10.)

Section ARCTOIDEA.

This section includes a considerable number of forms which agree in the essential characteristics of the structures of the base of the cranium and reproductive organs, and in the absence of a cæcum to the intestinal canal. They have no Cowper's glands, but there is a rudimentary prostate and a large cylindrical penial bone; while all the members of the group have five completely developed toes on each foot. Considerable diversity is found in the characters of the base of the skull in the various forms, but the following features are common to all. The cavity of the auditory bulla is simple, and has no trace of a dividing septum; the inferior lip of the auditory meatus (*am*, Fig. 254) is considerably prolonged; the paroccipital process (*p*) of the exoccipital is more or less triangular, directed backwards, outwards, and downwards, and standing quite apart from the bulla; the mastoid process (*m*) of the periotic is always widely separated from the paroccipital, and generally very prominent; the carotid foramen (*car*) is large, and placed on the inner margin

[1] *Proc. Amer. Phil. Soc.* vol. xviii. p. 452 (1880).

of the bulla, usually near the middle, but occasionally more posteriorly; the condyloid foramen is distinct and exposed, and never sunk into a common opening with the foramen lacerum posticum; and the glenoid foramen is always present, and usually conspicuous. The alisphenoid canal is absent except in *Ursus*, *Melursus*, and *Ælurus*.

It has been already observed (p. 501) that the evidence of fossil forms, so far as it goes, is not in favour of the Arctoidea being a natural group; so that its retention must be regarded as a somewhat provisional measure, largely based on its convenience. The group may be divided into the three families, *Ursidæ*, *Procyonidæ*, and *Mustelidæ*.[1]

Family URSIDÆ.

In existing forms the true molars $\frac{2}{3}$, with broad, flat tuberculated crowns. Typically the three anterior premolars of both jaws rudimentary and often deciduous. Fourth upper premolar (carnassial) with no third or inner root. An alisphenoid canal (except in *Æluropus*). Skull with the auditory bulla depressed, and scarcely at all inflated. Feet plantigrade. No entepicondylar foramen to the humerus. Kidneys conglomerate. Geographical distribution extensive.

Ursus.[2]—Dentition: $i\ \frac{3}{3}$, $c\ \frac{1}{1}$, $p\ \frac{4}{4}$, $m\ \frac{2}{3}$; total 42. The three anterior premolars above and below one-rooted, rudimentary, and frequently wanting. Usually the first (placed close to the canine) is present, and after a considerable interval the third, which is situated close to the other teeth of the molar series. The second is very rarely present in the adult state. The fourth (upper carnassial) differs essentially from the corresponding tooth of other Carnivores in wanting the inner tubercle supported by a distinct root. Its sectorial characters are very slightly marked, and it is much smaller than the first molar. The crowns of both the true molars are longer than broad, with flattened, tuberculated, grinding surfaces. The second has a large backward prolongation or heel. The lower carnassial has a small and indistinct blade and greatly developed tubercular heel. The second molar is of about the same length, but with a broader and more flattened tubercular crown. The third is smaller. The milk-teeth are comparatively small, and shed at an early age. Skull more or less elongated. Orbits small and incomplete behind. Palate prolonged considerably behind the last molar tooth. Vertebræ: C 7, D 14, L 6, S 5, C 8-10. Body heavy. Feet broad, completely plantigrade; the five toes on each foot all well developed, and armed with long compressed and

[1] For full details of the Arctoidea see Mivart, *Proc. Zool. Soc.* 1885, p. 340.

[2] Linn. *Syst. Nat.* 12th ed. vol. i. p. 69 (1766).

moderately curved non-retractile claws. Palms and soles naked. Tail very short. Ears moderate, erect, rounded, hairy. Fur generally long, soft, and shaggy.

The Bears are all animals of considerable bulk, and include among them the largest members of the order. Though the species are not numerous, they are widely spread over the earth's surface (but absent from the Ethiopian and Australian regions, and only represented by one species in the Neotropical region), and differ much among themselves in their food and manner of life. They are mostly omnivorous or vegetable feeders, and even the Polar

FIG. 255.—Head of the Brown Bear (*Ursus arctos*). From Sclater, *Proc. Zool. Soc.* 1867, p. 817.

Bear, usually purely carnivorous or piscivorous, devours grass with avidity in summer. The various species may be arranged in the following groups:—

Thalassarctine Group.—Head comparatively small, molar teeth small and narrow. Soles more covered with hair than in the others. This group is represented only by the well-known Polar or White Bear (*U. maritimus*) of the Arctic regions, which is one of the few mammals which are completely white at all seasons of the year.

The typical, or *Ursine*, group includes a number of species, of which the Common Brown Bear (*U. arctos*) is the best known example. This species is an exceedingly variable one, and has a very wide range in the Palæarctic region; the Syrian form described as *U. syriacus*, as well as the Hairy-eared Bear (*U. piscator*, Fig. 255) of North-Eastern Asia, and the Snow-Bear (*U. isabellinus*) of

Kashmir and Nipal, not being specifically separable. The Brown Bear hibernates in cold regions, and in the Himalaya keeps to comparatively high regions, emerging from its winter lair in March, April, or May, according to the season and elevation, to feed on the numerous bulbous plants which abound in the regions it inhabits. Both the Syrian and Himalayan varieties are generally of lighter colour and smaller size than the typical European form. Bears were at one time found in the British Isles, from which, however, they have been long since exterminated. They are still found in the Pyrenees, and are comparatively abundant in parts of Norway, Hungary, and Russia. In the Kashmir Himalaya they were very abundant in some districts a few years ago, one of the present writers having in 1874 seen no less than seven examples at one time from the top of a mountain ridge; of late years their numbers have, however, been greatly diminished. The Brown Bear, although with strong powers of smelling, is very slow of sight and hearing, and in the Himalaya it is easy to approach so near that they may be shot with a smooth-bore gun. The Grizzly Bear (*U. horribilis*) of North America is so closely allied to the Brown Bear that some writers think it should only rank as a very well-marked local variety. The Black Bears of the Himalaya (*U. torquatus*), Japan (*U. japonicus*), and North America (*U. americanus*) belong to this group. The Himalayan species ranges from Persia to Assam, and thence to China and Formosa. In the greater part of this area it is essentially a forest animal, and may be found in autumn in the forests of the Kashmir valley feeding upon chestnuts and other fruits. It is also exceedingly fond of maize, mulberries, and walnuts; and a few years ago it was no very uncommon sight to see three or even five of these bears up a single mulberry or walnut tree in Kashmir. The Spectacled Bear (*U. ornatus*) of the Peruvian Andes is another member of this group.

The *Helarctine* group is represented only by the Malay Bear or Sun Bear (*U. malayanus*), in which the head is short and broad; the molar teeth are comparatively broad (but the length still exceeding the breadth), the tongue is very long and extensile, and the fur short and smooth. This small species inhabits the Malay Peninsula, Sumatra, Java, Borneo, Tenasserim, Arakan, Chittagong, and the Garo hills of India; it inhabits forest districts, and is an expert climber.

The earliest known occurrence of the genus is in the Lower Pliocene of the Indian Siwalik Hills; where it is represented by *U. theobaldi*, which was probably the ancestor of the existing *Melursus.* The genus is represented in the Upper Pliocene of Europe by the small *U. etruscus;* and in the Pleistocene by the existing *U. arctos*, as well as by the great extinct Cave-Bear (*U. spelæus*), distinguished by the complexity of the crowns of the molars and the total loss of the three anterior premolars in the adult condition. Remains of Bears are also found in cavern-deposits in the north of Africa. The small *U. namadicus*, from the Pleistocene of the Narbada valley, India, may have been allied to *U. malayanus*.

Melursus.[1]—This differs from the true Bears in the first upper incisor being absent or shed at a very early age, in the very small size of the other teeth, in the very large extensile lips, the deep concavity of the palate, and other minor characters. The one species, *M. labiatus*, the well-known Sloth-Bear of India, feeds chiefly on black ants, termites, beetles, fruit, honey, etc. This species inhabits peninsular India, from near the Himalaya to Cape Comorin and Ceylon, and its remains are found in the cavern-deposits of Madras. The black hair is very long and coarse; there is a light horse-shoe-shaped mark on the chest (as in *Ursus torquatus*), and the extremity of the muzzle is of an ashy gray.

FIG. 256.—*Æluropus melanoleucus*. (From Milne-Edwards.)

Æluropus.[2]—Dentition: $i\,\frac{3}{3}$, $c\,\frac{1}{1}$, $p\,\frac{4}{3}$, $m\,\frac{2}{3}$; total 40. Premolars large, increasing in size from first to last, and two-rooted except the first. First upper molar with quadrate crown, broader than long; second larger than the first. Cranium with zygomatic arches and sagittal crest immensely developed, and ascending ramus of mandible very high, giving greater spaces for attachments of temporal muscle than in any other existing member of the order. Facial portion

[1] Meyer, *Uebersicht d. neu. Zool. Entdeckungen*, etc. p. 155 (1793).

[2] A. Milne-Edwards, *Nouv. Arch. du Muséum*, vol. vii. *Bull.* p. 88 (1871). Amended from "Ailuropus."

short. Bony palate not extending behind the last molar tooth. No alisphenoid canal. Feet bear-like, but soles more hairy, and perhaps less completely plantigrade. Fur long and thick. Tail very short. One extremely rare species, *A. melanoleucus* (Fig. 256), discovered by Père David in 1869, in the most inaccessible mountains of Moupin in Eastern Tibet. Said to feed principally on roots, bamboos, and other vegetables. It is of the size of a small Brown Bear, of a white colour, with ears, spots round the eyes, shoulders and limbs black. In the large size and complex crowns of the upper premolars this genus differs very markedly from the true Bears. The fourth upper premolar (carnassial) makes no approach to the markedly sectorial type presented by the corresponding tooth of *Hyænarctus*, its structure being, on the whole, more like that of *Ælurus*.

Extinct Genera.—The genus *Arctotherium* includes some very large Bear-like animals from the Pleistocene of South America and California, in which the dentition departs less widely from a normal carnivorous type than in the true Bears. Thus the upper carnassial (Fig. 257) is relatively larger than in *Ursus*; while the crowns of the upper molars are broader and shorter. The humerus is said to have an entepicondylar foramen. *Hyænarctus*, of the Miocene and Pliocene of Europe and Southern Asia, has the crowns of the upper molars either square or triangular; the upper carnassial having three distinct lobes to the blade, while the lower carnassial is practically indistinguishable from that of the Dog-like *Dinocyon* (p. 556). The proximal extremity of the ulna differs from that of *Ursus* in having a long olecranon, and thereby resembles the corresponding bone of the Dogs. Indeed all the characters at present available tend to show a complete passage from the Tertiary Dog-like animals, through *Dinocyon*, *Hyænarctus*,

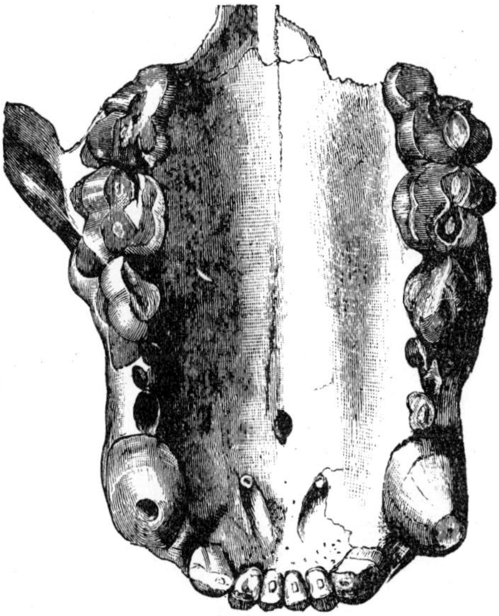

Fig. 257.—Palate of *Arctotherium bonariense*, Pleistocene, South America—¼ natural size. (From the *Palæontologia Indica*.)

and *Arctotherium*, to the true Bears. Most of the species of *Hyænarctus* were of very large dimensions, but smaller forms occur in the Miocene. *Cephalogale*, of the Continental Tertiaries, is a genus represented by several species of medium size showing evident signs of affinity with *Hyænarctus*. The upper molars have subtriangular crowns, while the carnassial is short, and has two comparatively low lobes. Here also may be mentioned several other genera, apparently more or less closely allied to the present group, some of which are regarded by Dr. Schlosser as showing marked signs of affinity to the *Procyonidæ*. Among these are *Simocyon* from the Pliocene of Europe, with $p\ \frac{2}{2-4}$, $m\ \frac{2}{2}$; and *Enhydrocyon* of the North American Miocene, with $p\ \frac{3}{3}$, $m\ \frac{2}{2}$, a secant talon to the lower carnassial, and a very short skull. The Miocene *Ælurodon* comprises several large North American forms, having a trilobed upper carnassial like that of *Hyænarctus*, and a dental formula similar to that of the latter and *Canis Prohyæna* is founded upon a much-worn jaw of *Ælurodon*. *Hyænocyon*, of the Miocene of the United States, with $p\ \frac{3}{3}$, $m\ \frac{1}{2}$, appears to be an allied form, also having a trilobed upper carnassial.

Family PROCYONIDÆ.

True molars $\frac{2}{2}$, tuberculated or multicuspid; upper carnassial short and broad. Alisphenoid canal absent, except in *Ælurus*. Feet plantigrade. Tail generally annulated. In some cases an entepicondylar foramen to the humerus. Typically American, but with the outlying Oriental genus *Ælurus*.

Ælurus.[1]—Dentition: $i\ \frac{3}{3}$, $c\ \frac{1}{1}$, $p\ \frac{3}{4}$, $m\ \frac{2}{2}$; total 38. First lower premolar very minute and deciduous. Molars (Fig. 259) remarkable for their great transverse breadth and the numerous cusps of their crowns. Vertebræ: C 7, D 14, L 6, S 3, C 18. Skull (Fig. 259) high and compressed, very convex, with the facial portion short, the palate convex antero-posteriorly, and the ascending ramus of mandible extremely high. Head round. Face short and broad. Ears large, erect, pointed. Limbs stout, with large sharp semi-retractile claws. Tail nearly as long as body, cylindrical, annulated, and clothed with long hairs. Fur long and thick. One existing species, *Æ. fulgens*, the Panda (Fig. 258), an animal rather larger than a Cat, found in the South-East Himalaya, at heights of from 7,000 to 12,000 feet above the sea, among rocks and trees, and chiefly feeding on fruits and other vegetable substances. Its fur is of a remarkably rich reddish-brown colour, darker below.

The genus *Ælurus* has been made the type of a distinct family,

[1] F. Cuvier, *Hist. Nat. des Mammifères* (1825). Amended from "Ailurus." For anatomy, see Flower, *Proc. Zool. Soc.*, 1870, p. 752.

but its relationship to the Raccoons is regarded by Mr. W. T. Blanford [1] as sufficiently close to admit of its being included in the same family. According to this zoologist the Panda often sleeps coiled up like a Cat, with the bushy tail over its head, but at other times resting on its legs with the head tucked under the chest and between the fore legs, after a manner said to be common with the Raccoons. Although by no means strictly nocturnal, these animals sleep much during the day, and roam out in search of food in the morning and evening. The young are born in a very helpless

Fig. 258.—The Panda (*Ælurus fulgens*). The dark nasal stripe shown in this figure is generally absent. (From Sclater, *Proc. Zool. Soc.* 1869, p. 408.)

condition, and remain for a long period concealed in the holes of trees or rocks.

Fossil remains of a species of *Ælurus* (*Æ. anglicus*) have been obtained from the English Pliocene Crag deposits which indicate an animal of about one and half times the size of *Æ. fulgens.* The first evidence of this fossil species was afforded by part of the mandible with the last molar in place, and the subsequent discovery of an entire first upper molar renders full confirmation of the generic determination. This distribution of *Ælurus* is very important, as showing how its area may have once approximated to that of the ancestors of the American representatives of the family. It is probable that the genus existed in India during the Siwalik period.

[1] *Fauna of British India*, "Mammalia," p. 189 (1888).

The whole of the undermentioned genera are American, and are characterised by the absence of an alisphenoid canal in the skull.

Procyon.[1]—Dentition: $i\ \frac{3}{3}$, $c\ \frac{1}{1}$, $p\ \frac{4}{4}$, $m\ \frac{2}{2}$; total 40. The molar teeth broad and tuberculated (Fig. 259). The upper carnassial with three cusps along the outer margin, and a very broad bicuspid inner tubercle, giving an almost quadrate form to the crown. First molar with a large tuberculated crown, rather broader than long;

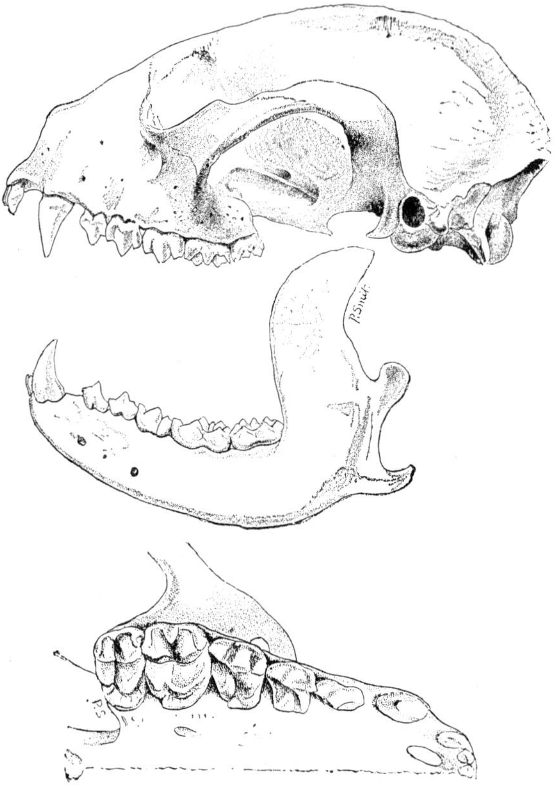

FIG. 259.—Lateral view of skull and right half of palate of *Ælurus fulgens*. (From Blanford, *Mammalia of British India*, p. 190.)

second considerably smaller, with transversely oblong crown. Lower carnassial with an extremely small and ill-defined blade, placed transversely in front, and a large inner cusp and hind talon. Second molar as long as the first, but narrower behind, with five obtuse cusps. Vertebræ: C 7, D 14, L 6, S 3, C 16-20. Body stout. Head broad behind, but with a pointed muzzle. Limbs

[1] Storr, *Prodromus Meth. Mamm.* p. 35 (1780).

plantigrade, but in walking the entire sole is not applied to the ground as it is when the animal is standing. Toes, especially of the fore foot, very free, and capable of being spread wide apart. Claws compressed, curved, pointed, and non-retractile. Tail moderately long, cylindrical, thickly covered with hair, annulated, non-prehensile. Fur long, thick, and soft. The well-known Raccoon [1] (*Procyon lotor*, Fig. 260) of North America is the type of this genus. It is a clumsy thickly-built animal about the size of a Badger, with a coat of long coarse grayish-brown hairs, short ears, and a bushy black and white ringed tail. Its range extends over the whole of

FIG. 260.—The Raccoon (*Procyon lotor*).

the United States, and stretches on the west northwards to Alaska and southwards into Central America, where it attains its maximum size. The following notes on the habits of the Raccoon are taken from Dr. C. H. Merriam's *Mammals of the Adirondack Region* :—

"Raccoons are omnivorous beasts, and feed upon mice, small birds, birds' eggs, turtles and their eggs, frogs, fish, crayfish, molluscs, insects, nuts, fruits, maize, and sometimes poultry. Excepting the bats and flying squirrels, they are the most strictly nocturnal of all our mammals, and yet I have several times seen them abroad on cloudy days. They haunt the banks of ponds and streams, and find much of their food in these places, such as

[1] A corruption of the North American Indian "arrathkune" or "arathcone." The French *raton* or *raton laveur*, German *Waschbär*, and other European names are derived from a curious habit the Raccoon has of dipping or washing its food in water before eating it.

crayfish, mussels, and fish, although they are unable to dive and pursue the latter under water, like the otter and mink. They are good swimmers, and do not hesitate to cross rivers that lie in their path. . . . The Raccoon hibernates during the severest part of the winter, retiring to its nest rather early, and appearing again in February or March, according to the earliness or lateness of the season. It makes its home high up in the hollow of some large tree, preferring a dead limb to the trunk itself. It does little in the way of constructing a nest, and from four to six young are commonly born at a time, generally early in April in this region. The young remain with the mother about a year."

The South-American *P. cancrivorus*, the Crab-eating Raccoon, is very similar to *P. lotor*, but differs by its much shorter fur, larger size, proportionally more powerful teeth, and other minor characters. It extends over the whole of South America, as far south as the Rio Negro, and is very common in all suitable localities. Its habits are similar to those of the North-American species. Fossil remains of *Procyon* have been described from the Pleistocene deposits of the United States.

Bassaris.[1]—A form closely allied to *Procyon*, but of more slender and elegant proportions, with a sharper nose, longer tail, and more digitigrade feet, and with teeth otherwise like, but smaller, and more sharply denticulated. It was formerly, but erroneously, placed among the *Viverridæ*. Two species:—*B. astuta*, from the southern parts of the United States and Mexico, and *B. sumichrasti*, from Central America.

Bassaricyon.[2]—This name has been given to a distinct modification of the Procyonine type of which at present only two examples are known, one from Costa Rica and the other from Ecuador, which, appearing to be different species, have been named *B. gabbi* and *B. alleni*. They much resemble the Kinkajou (*Cercoleptes*) in external appearance, but the skull and teeth are more like those of *Procyon* and *Nasua*.

Nasua.[3]—Dentition as in *Procyon*, but the upper canines are larger and more strongly compressed, and the molars smaller. The facial portion of the skull is more elongated and narrow. Vertebræ: C 7, D 14, L 6, S 3, C 22-23. Body elongated and rather compressed. Nose prolonged into a somewhat upturned, obliquely truncated, mobile snout. Tail long, non-prehensile, tapering, annulated. These animals, commonly called Coatis or Coati-Mundis, live in small troops of eight to twenty, are chiefly arboreal, and feed on fruits, young birds, eggs, insects, etc. Recent researches have reduced the number of supposed species to two, *N. narica* of Mexico

[1] Lichtenstein, *Isis*, 1831, p. 512.

[2] Allen, *Proc. Ac. Nat. Sci. Philad.* 1876, p. 20.

[3] Storr, *Prodromus Meth. Mamm.* p. 35 (1780).

and Central America, and *N. rufa* of South America from Surinam to Paraguay. Remains of this genus, mostly referable to the existing species, occur in the cavern-deposits of Brazil.

Cercoleptes.[1]—Dentition: $i\ \frac{3}{3}$, $c\ \frac{1}{1}$, $p\ \frac{3}{3}$, $m\ \frac{2}{2}$; total 36. Molars with low flat crowns, very obscurely tuberculated. Skull short and rounded, with flat upper surface. Vertebræ: C 7, D 14, L 6, S 3, C 26-29. Clavicles present, but in a very rudimentary condition. Head broad and round. Ears short. Body long and musteline. Limbs short. Tail long, tapering, and prehensile. Fur short and soft. Tongue long and very extensile. But one species of this somewhat aberrant genus is known, *C. caudivolvulus*, the Kinkajou, found in the forests of the warmer parts of South and Central America. It is about the size of a Cat, of a uniform, pale, yellowish-brown colour, nocturnal and arboreal in its habits, feeding on fruit, honey, eggs, and small birds and mammals, and is of a tolerably gentle disposition and easily tamed.

Family MUSTELIDÆ.

True molars $\frac{1}{2}$ (or $\frac{1}{1}$ in *Mellivora*[2]). No alisphenoid canal. In the upper molar the inner tubercular portion is always longer in the antero-posterior direction than the secant external portion; the degree of inflation of the auditory bulla is but slight; and the palate is generally much produced behind the last molars, as is the case with the members of the preceding family. The postglenoid process of the cranium is generally considerably curved over the glenoid fossa, so as to hold very tightly the condyle of the mandible. The humerus may or may not have an entepicondylar foramen. Except in the Otters, the kidneys resemble those of the *Procyonidæ* in being of simple structure.

This family is a large and widely distributed one, especially in the northern temperate regions of the earth. The different genera, which are very difficult to arrange in any natural order, are rather artificially divided, chiefly according to the characters of their feet and claws, into the Otter-like (Lutrine), Badger-like (Meline), and Weasel-like (Musteline) forms.

Subfamily **Lutrinæ.**—Feet short, rounded (except the hind feet of *Latax*). Toes webbed. Claws small, curved, blunt. Head broad and much depressed. Upper molar large and quadrate, with its inner tubercular portion much expanded antero-posteriorly (Fig. 261). Kidneys conglomerate. Habits aquatic.

Lutra.[3]—Dentition: $i\ \frac{3}{3}$, $c\ \frac{1}{1}$, $p\ \frac{4}{3}$, $m\ \frac{1}{2}$; total 36. Upper

[1] Illiger, *Prodromus Syst. Mamm. et Avium*, p. 127 (1811).

[2] Also in two other species noticed below. One extinct Otter has two upper molars.

[3] Erxleben, *Syst. Regn. Animal*, p. 445 (1777).

carnassial with a trenchant tricuspid blade, and a very large inner lobe, hollowed on the free surface, with a raised sharp edge, and extending along two-thirds or more of the length of the blade. True molar large, with a quadricuspidate crown, broader than long. First upper premolar very small, and in some cases absent (Fig. 261). Skull broad and depressed, contracted immediately behind the orbits. Facial portion very short; brain case large. Vertebræ: C 7, D 14-15, L 6-5, S 3, C 20-26. Body very long. Ears short and rounded. Limbs short. Feet more or less completely webbed; claws usually well developed on all the toes, although they may be rudimentary or absent. Tail long, thick at the base and tapering, rather depressed. Fur short and close. The humerus may or may not have an entepicondylar foramen. In conformity with the shape of the skull, the posterior part of the brain is expanded laterally.

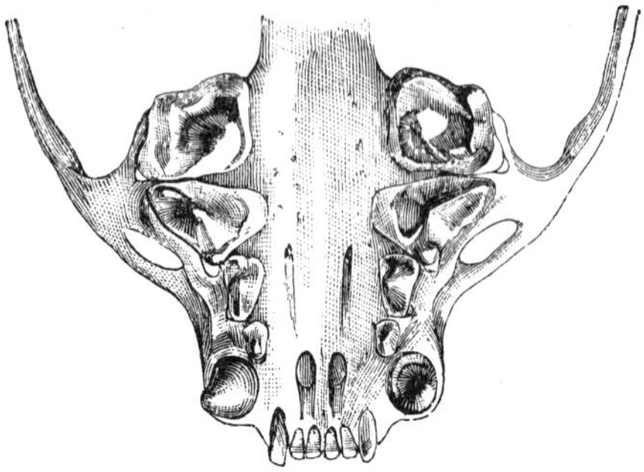

FIG. 261.—Palate of *Lutra cinerea*. (From the *Palæontologia Indica*.)

The Common British Otter (*L. vulgaris*), as the type of the genus, may be described somewhat fully. It has an elongated, low body, short limbs, short broad feet, with five toes on each, connected together by webs, and all with short, moderately strong, compressed, curved, pointed claws. Head rather small, broad, and flat; muzzle very broad; whiskers thick and strong; eyes small and black; ears short and rounded. Tail a little more than half the length of the body and head together, very broad and strong at the base, and gradually tapering to the end, somewhat flattened horizontally. The fur is of very fine quality, consisting of a short soft under fur of a whitish-gray colour, brown at the tips, interspersed with longer, stiffer, and thicker hairs, very shining, grayish at the base, bright rich brown at the points, especially on the upper parts and outer surface of the legs; the throat, cheeks, under parts and inner surface of the legs brownish-gray throughout. Individual Otters vary much in size; but the total length from the nose to the end of the tail averages about 3½ feet, of which the tail occupies 1 foot 3 or 4 inches. The weight of a full-sized male is from 18 to 24 lbs., that of a female about 4 lbs. less.

As the Otter lives almost exclusively on fish, it is rarely met with far from water, and usually frequents the shores of brooks rivers, lakes, and, in some localities, the sea itself. It is a most expert swimmer and diver, easily overtaking and seizing fish in the water, but when it has captured its prey it brings it to shore to devour it. When lying upon the bank it holds the fish between its fore-paws, commences at the head, and then eats gradually towards the tail, which it is said always to leave. The female produces three to five young ones at a time, in the month of March or April, and brings them up in a nest formed of grass or other herbage, usually placed in a hollow place in the bank of a river, or under the shelter of the roots of some overhanging tree. The Common Otter is found in localities suitable to its habits throughout Great Britain and Ireland, though far less abundantly than formerly, for, being very destructive to fish, and thus coming into keen competition with those who pursue the occupation of fishing either for sport or for gain, it is rarely allowed to live in peace when once its haunts are discovered. Otter-hunting with packs of hounds of a special breed, and trained for the purpose, was formerly a common pastime in the country. When hunted down and brought to bay by the dogs, the Otter is finally despatched by long spears carried for the purpose by the huntsmen.

The Common Otter ranges throughout the greater part of Europe and Asia, the Indian *L. nair* not being distinct. A closely allied but larger species, *L. canadensis*, is extensively distributed throughout North America, where it is systematically pursued by professional trappers for the value of its fur. The Common Otter is regularly trained by the natives of some parts of Bengal to assist them in fishing, by driving the fish into the nets. In China Otters are taught to catch fish, being let into the water for the purpose attached to a long cord.

Otters are widely distributed over the earth, and, as they are much alike in size and coloration, their specific distinctions are by no means well defined.[1] Besides those mentioned above, the following may be noticed. In the Oriental region there are *L. ellioti*[2] of India, *L. sumatrana* of the Malay countries, and *L. cinerea* ranging over the greater part of the region. The latter species (often known as *L. leptonyx*) is of small size, with a short head, and rudimentary claws, which may be absent; it was at one time regarded as generically distinct, under the name of *Aonyx*. The upper true molar (Fig. 261) is characterised by the great development of its inner tubercular portion, and the first upper premolar is absent. In the Ethiopian region there are two species, *L. capensis* and *L. maculicollis*. Of the Neotropical forms it will suffice to mention the small *L. felina* and the large *L. brasiliensis*. The latter is by far the largest of the existing forms, and is characterised by the presence of a prominent flange-like ridge along each lateral

[1] See Thomas, *Proc. Zool. Soc.* 1889, p. 190.

[2] The synonymy of this species is not settled, and the adoption of the name given here only preliminary.

margin of the tail, on which account it was referred by Dr. Gray to a distinct genus, with the name of *Pteronura sambachi.* It should be observed that all Otters have a very distinct inner cusp to the blade of the lower carnassial, but that the relative size of this cusp varies in the different species.

Extinct Otters.—Several species of fossil Otters have been described. Thus in the Indian Siwaliks we have *L. palæindica*, which is closely allied to *L. sumatrana*, and a larger form described as *L. bathygnathus.* The Pliocene of Hessen-Darmstadt yields *L. hessica;* while *L. dubia*, of the Middle Miocene of France, is a species characterised by the small size of the inner cusp of the lower carnassial—a character in which it resembles those Tertiary forms described as *Trochictis*, which are believed to connect *Lutra* with the *Mustelinæ.* Two very large Otters, respectively from the Indian Siwaliks and the Italian Miocene, named *L. sivalensis* and *L. campanii*, may be regarded either as representing a very distinct *Enhydriodont* group of *Lutra* or as referable to a separate genus *Enhydriodon.* They are characterised by certain peculiarities in the structure of the teeth, and the second upper premolar may be absent in the Indian form. Lastly, the genus *Potamotherium* contains a small Otter (*P. valetoni*) from the Lower Miocene of the Continent, which differs from all other known *Mustelidæ* in having a minute second upper true molar. This species is evidently a very generalised form approximating to the *Viverridæ* in its dental formula, and also in the characters of the teeth themselves. The brain, as recently described by Dr. Filhol, differs from that of *Lutra* and other Mustelines in the great relative width of the anterior extremity of the hemispheres and olfactory lobes, and also in the disposition of the sulci, in both of which respects it more nearly resembles the *Viverridæ.*

Latax.[1]—Dentition: $i\ \frac{3}{2}$, $c\ \frac{1}{1}$, $p\ \frac{3}{3}$, $m\ \frac{1}{2}$; total 32. Differs from all other existing Carnivora in having but two incisors on each side of the lower jaw, the one corresponding to the first (very small in the true Otters) being constantly absent. Though the molar teeth generally resemble those of *Lutra* in their proportions, they differ very much in the exceeding roundness and massiveness of their crowns and bluntness of their cusps. Feet webbed. Fore feet small, with five subequal toes, furnished with short compressed claws; palms naked. Hind feet very large, depressed, and fin-like. The phalanges flattened as in the Seals. The fifth toe the longest and stoutest, the rest gradually diminishing in size to the first, all with moderate claws. Tail moderate, cylindrical, and obtuse; about one-fourth the length of the head and body.

[1] Gloger, *Nova Acta Ac. Cæs. Leop.-Car.* vol. xiii. pt. 2, p. 511 (1827): Syn. *Enhydra;* Fleming, *Philosophy of Zoology*, vol. ii. p. 187 (1822). Preoccupied by *Enhydris*, Merrem, *Tent. Syst. Amphib.* p. 140 (1820).

The Sea-Otter (*L. lutris*, Fig. 262) is the sole representative of this genus. The entire length of the animal from nose to end of tail is about 4 feet, so that the body is considerably larger and more massive than that of the English Otter. The skin is peculiarly loose, and stretches when removed from the animal so as to give the idea of a still larger creature than it really is. The pellage is remarkable for the preponderance of the beautifully soft woolly under fur, the longer stiffer hairs being very scanty. The general colour is a deep liver brown, everywhere silvered or frosted with the hoary tips of the longer hairs. These are, however, removed when the skin is dressed for commercial purposes.

Fig. 262.—The Sea-Otter (*Latax lutris*). From Wolf, *Proc. Zool. Soc.* 1865, pl. vii.

Sea-Otters are only found upon the rocky shores of certain parts of the North Pacific Ocean, especially the Aleutian Islands and Alaska, extending as far south on the American coast as Oregon; but, owing to the unremitting persecution to which they are subjected for the sake of their skins, which rank among the most valuable known to the furrier, their numbers are greatly diminishing, and, unless some restriction can be placed upon their destruction, such as that which protects the Fur-Seals of the Pribyloff Islands, the species is threatened with extermination, or, at all events, excessive scarcity. When this occurs, the occupation of five thousand of the half-civilised natives of Alaska, who are dependent upon Sea-Otter hunting as a means for obtaining their living, will be gone. The principal hunting grounds at present are the little rocky islets and reefs around the island of Saanach and the Chernobours, where they are captured by spearing, clubbing, or nets, and recently by the more destructive rifle bullet. They do not feed on fish, like the true Otters, but on clams, mussels, sea-urchins, and crabs, for the mastication of which the blunt cusps of their teeth are admirably suited. The female brings forth but a single young one at a time, apparently at any season of the year. They are excessively shy and wary, and all attempts to rear the young ones in captivity have hitherto failed.

Subfamily **Melinæ.**—Feet elongated. Toes straight. Claws non-retractile, slightly curved, subcompressed, blunt; those of the fore foot especially large. Upper molar variable. Kidneys simple. Habits mostly terrestrial and fossorial.

Mephitis.[1]—Dentition: $i\ \frac{3}{3}$, $c\ \frac{1}{1}$, $p\ \frac{3}{3}$, $m\ \frac{1}{2}$; total 34. Upper molar larger than the carnassial, subquadrate, rather broader than long. Lower carnassial with talon less than half the length of the whole tooth. Bony palate terminating posteriorly opposite the hinder border of the last molar tooth. Facial portion of skull short and somewhat truncated in front. Vertebræ: C 7, D 16, L 6, S 2, C 21. Head small. Body elongated. Limbs moderate, subplantigrade. Ears short and rounded. Tail long, abundantly clothed with very long fine hair. Anal glands largely developed. The secretion of these glands, which can be discharged at the will of the animal, has an intolerably offensive odour, which circumstance has rendered the Skunks, as they are commonly called, proverbial. They are strictly nocturnal animals, terrestrial and burrowing, feeding chiefly on small mammals, birds, reptiles, insects, worms, roots, and berries. All the known species have a prevalent black colour, varied by white strips or spots on the upper part (Fig. 263). They generally carry the body much arched, and the tail erect, the long loose hair of which waves like a plume over the back. There are three species, all inhabitants of the American continent, over which they have an extensive range.

The Common Skunk (*M. mephitica*, Fig. 263) is an animal of about the size of a small Cat, ranging from Hudson's Bay to Guatemala. The following account of its habits is given by Dr. C. H. Merriam in his *Mammals of the Adirondack Region*:—

"The skunk preys upon mice, salamanders, frogs, and the eggs of birds that nest on or within reach from the ground. At times he eats carrion, and if he chances to stumble upon a hen's nest the eggs are liable to suffer; and once in a while he acquires the evil habit of robbing the hen-roost, but as a rule skunks are not addicted to this vice. Of all our native mammals perhaps no one is so universally abused and has so many unpleasant things said about it as the innocent subject of the present biography; and yet no other species is so valuable to the farmer. Pre-eminently an insect-eater,

[1] Cuvier, "Tabl. de Classif." in *Leçons d'Anat. Compar.* vol. i. (1800).

he destroys more beetles, grasshoppers, and the like than all our other mammals together, and in addition to these he devours vast numbers of mice. He does not evince that dread of man that is so manifest in the great majority of our mammals, and when met during any of his circumambulations rarely thinks of running away. He is slow in movement and deliberate in action, and does not often hurry himself in whatever he does. His ordinary gait is a measured walk, but when pressed for time he breaks into a low shuffling gallop. It is hard to intimidate a skunk, but when once really frightened he manages to get over the ground at a very fair pace.

FIG. 263.—The Common Skunk (*Mephitis mephitica*).

Skunks remain active throughout the greater part of the year in this region, and hibernate only during the severest portion of the winter. They differ from most of our hibernating mammals in that the inactive period is apparently dependent solely on the temperature, while the mere amount of snow has no influence whatever upon their movements. Skunks, particularly when young, make very pretty pets, being attractive in appearance, gentle in disposition, interesting in manners, and cleanly in habits—rare qualities indeed! They are playful, sometimes mischievous, and manifest considerable affection for those who have the care of them. Their flesh is white, tender, and sweet, and is delicious eating. Skunks have large families, from six to ten young being commonly raised each season; and as a rule they all live in the same hole until the following spring."

The two ducts leading from the anal glands open at the tips of two small conical papillæ placed in such a position that the animal can protrude them externally, and can thus guide the direction of the jet of nauseous fluid, which can be propelled by the powerful muscles surrounding the glands to a distance of from 8 to 12 feet.

The Long-tailed Skunk (*M. macrura*), from Central and Southern Mexico, has two lateral stripes, and a longer and more bushy tail than the common species. *M. putorius*, of the Southern United States and thence southwards to Yucatan and Guatemala, is of a much smaller size, with four interrupted white lateral stripes, and a skull differing considerably in form from that of the type species. It is regarded by some writers as representing a distinct genus, *Spilogale;* and has been recently divided by Dr. C. H. Merriam into several nominal species.

Conepatus.[1]—The Skunk of tropical America (*C. mapacito*), ranging from Texas to Chili and Patagonia, differs considerably from the true Skunks, although in colour it is almost precisely similar to the common species, with which it also agrees in the variation of the relative development of the black and white. Its build is heavier than that of *Mephitis;* the snout and head are more Pig-like; and the nostrils open downwards and forwards instead of laterally on the sides of the muzzle. The skull also has many special characters, and the teeth are different in shape and, as a rule, in number also, the first minute premolar of *Mephitis* being almost invariably absent, so that the dental formula is $i\ \frac{3}{3}$, $c\ \frac{1}{1}$, $p\ \frac{2}{3}$, $m\ \frac{1}{2}$; total 32.

Remains of *Conepatus*, which have been referred to three species, are found in the cavern-deposits of Brazil.

Arctonyx.[2]—Dentition: $i\ \frac{3}{3}$, $c\ \frac{1}{1}$, $p\ \frac{4}{4}$, $m\ \frac{1}{2}$; total 38. Incisor line curved, the outer teeth being placed posteriorly to the others. Lower incisors proclivous. First premolars often rudimentary or absent. Upper molar much larger than the carnassial, longer in the antero-posterior direction than broad; lower carnassial with a very large, low, tuberculated talon. Cranium elongated and depressed; face long, narrow, and concave above. Bony palate extending as far backwards as the level of the glenoid fossa; palatal bones dilated; suborbital foramina very large. Vertebræ: C 7, D 16, L 4, S 4, C 20. Snout long, naked, mobile, and truncated, with large terminal nostrils, much like that of a Pig. Eyes small. Ears very small and rounded. Body compressed rather than depressed. Limbs of moderate length and digitigrade in walking.

[1] Gray, *Ann. Mag. Nat. Hist.* ser. 2, vol. i. p. 581 (1837).

[2] F. Cuvier, *Hist. Nat. des Mammifères* (1825).

Tail moderate, tapering. A full soft under fur, with longer, bristly hairs interspersed. The best-known species is *A. collaris*, the Sand-Badger, or *Bhálu-soor*[1] (*i.e.* Bear-pig) of the natives, found in the mountains of the north-east of India and Assam. It is rather larger than the English Badger, higher on its legs, and very Pig-like in general aspect, of a light gray colour, with flesh-coloured snout and feet; and is nocturnal and omnivorous in habits. The imperfectly known *A. taxoides* from Assam and Arakan, and perhaps China, is a much smaller species. A third form probably exists in Eastern Tibet. Professor Mivart remarks that the brain-case of *Arctonyx* is narrower than in any other Arctoid; while the palate is relatively longer than in any other Carnivore except *Procyon*; and the metatarsus is relatively shorter than in any other member of the order.

Mydaus.[2]—Dentition as in the last genus, but the cusps of the teeth more acutely pointed. Cranium elongated, face narrow and produced. Suborbital foramen small, and the palate, as in all the succeeding genera of this group, produced backwards about midway between the last molar tooth and the glenoid fossa. Vertebræ: C 7, D 14-15, L 6-5, S 3, C 12. Head pointed in front; snout produced, mobile, obliquely truncated, the nostrils being inferior. Limbs rather short and stout. Tail extremely short, but clothed with rather long bushy hair. Anal glands largely developed, and emitting an odour like that of the American Skunks. One species, *M. meliceps*, the Teledu, a small burrowing Badger, found in the mountains of Java at an elevation of 7000 or more feet above sea-level.

Meles.[3]—Dentition: $i\ \frac{3}{3}$, $c\ \frac{1}{1}$, $p\ \frac{4}{4}$, $m\ \frac{1}{2}$; total 38. The first premolar in both jaws extremely minute and often deciduous. Upper molar very much larger than the carnassial, subquadrate, as broad as long. Lower carnassial with a broad, low, tuberculated talon, more than half the length of the whole tooth. The postglenoid processes of the skull are so strongly developed, and the glenoid fossa is so deep, that the condyle of the lower jaw is firmly held in its place even after all the surrounding soft parts are removed. Vertebræ: C 7, D 15, L 5, S 3, C 18. Muzzle pointed. Ears very short. Body stout, broad. Limbs short, strong, subplantigrade. Tail short. The best-known species is the common Badger (*M. taxus*) of Europe and Northern Asia, still found in many parts of England, where it lives in woods, is nocturnal, burrowing, and very omnivorous, feeding on mice, reptiles, insects, fruit, acorns, and roots. Other nearly allied species, *M. leucurus* and *M. chinensis*, are found in continental Asia, *M. canescens* in Persia, and *M. anakuma* in Japan.

The appearance of the common Badger is too well known to

[1] Possibly the name should be Bálu-soor (Sand-pig).

[2] F. Cuvier, *Hist. Nat. des Mammifères* (1825).

[3] Storr, *Prodromus Meth. Mamm.* p. 34 (1780).

need description, but it may be mentioned that a full-grown individual stands about a foot in height at the shoulder, and measures from $2\frac{1}{2}$ to 3 feet in length. The young are born in a naked and blind condition, usually in litters of three or four. It appears that the usual period of gestation is about eleven and a half months, but instances are recorded where the period has been protracted to upwards of fifteen months.

Fossil remains of the common Badger are found in the Pleistocene deposits of Europe, while extinct species have been described from the Lower Pliocene beds of Maragha, in Persia.

Taxidea.[1]—Dental formula as in *Meles*, except that the rudimentary anterior premolar appears to be always wanting in the upper jaw. The upper carnassial much larger in proportion to the other teeth. Upper molar about the same size as the carnassial, triangular, with the apex turned backwards. Talon of lower carnassial less than half the length of the tooth. Skull very wide in the occipital region; the lambdoidal crest very greatly developed, and the sagittal but slightly, contrary to what obtains in *Meles*. Vertebræ: C 7, D 15, L 5, S 3, C 16. Body very stoutly built and depressed. Tail short. The animals of this genus are peculiar to North America, where they represent the Badgers of the Old World, resembling them much in appearance and habits. *T. americana* is the common American Badger of the United States; *T. berlandieri*, the Mexican Badger, is perhaps only a local variety.

Mellivora.[2]—Dentition: $i\ \frac{3}{3}$, $c\ \frac{1}{1}$, $p\ \frac{3}{3}$, $m\ \frac{1}{1}$; total 32. Upper carnassial large, with its inner tubercle quite at the anterior end of the blade, as in the following genera; molar much smaller and transversely extended, having a very small outer and a larger rounded inner lobe. Talon of lower carnassial very small, scarcely one-fourth of the whole length of the tooth, and with but one cusp; lower tubercular molar absent. Vertebræ: C 7, D 14, L 4, S 4, C 15. Body stout, depressed. Limbs short, strong. Head depressed, nose rather pointed. External ears rudimentary. Tail short. The animals of this genus are commonly called Ratels. *M. indica* from India, and *M. ratel* (Fig. 264) from South and West Africa, have nearly the same general appearance and size, being rather larger than a common Badger. Their coloration is peculiar, all the upper surface of the body, head, and tail being ashy gray, while the lower parts, separated by a distinct longitudinal boundary line, are black. The two species may be distinguished by the circumstance that the African one has a distinct white line round the body at the junction of the gray of the upper side with the black of the lower, while in the Indian form this line is absent; the teeth also of the former are, on the whole, larger, rounder, and heavier than those of

[1] Waterhouse, *Proc. Zool. Soc.* 1838, p. 154.

[2] Storr, *Prodromus Meth. Mamm.* p. 34 (1780).

the latter. In spite of these differences the two are, however, so nearly allied that they might almost be considered as local races of a single widely spread species.

The following account of the Indian species is extracted from Dr. Jerdon's *Mammals of India:* "The Indian badger is found throughout the whole of India, from the extreme south to the foot of the Himalayas, chiefly in hilly districts, where it has greater facilities for constructing the holes and dens in which it lives; but also in the north of India in alluvial plains, where the banks of

FIG. 264.—The African Ratel (*Mellivora ratel*).

large rivers afford equally suitable localities wherein to make its lair. It is stated to live usually in pairs, and to eat rats, birds, frogs, white ants, and various insects, and in the north of India it is accused of digging out dead bodies, and is popularly known as the grave-digger. It doubtless also, like its Cape congener, occasionally partakes of honey. It is often very destructive to poultry, and I have known of several having been trapped and killed whilst committing such depredations in Central India and in the northern Circars. In confinement the Indian badger is quiet and will partake of vegetable food, fruits, rice, etc."

A fossil species of *Mellivora*, apparently closely allied to the existing forms, occurs in the Pliocene Siwaliks of India. The same deposits have also yielded remains of an extinct genus described as *Mellivorodon*.

Helictis.[1]—Dentition: $i\ \frac{3}{3},\ c\ \frac{1}{1},\ p\ \frac{4}{4},\ m\ \frac{1}{2}$; total 38. Upper carnassial with a large bicuspid inner tubercle; upper molar smaller, wider transversely than in the antero-posterior direction. Lower carnassial with talon about one-third the length of the tooth. Skull elongated, rather narrow and depressed. Facial portion especially narrow. Infra-orbital foramen very large. Head rather small and produced in front, with an elongated, obliquely truncated, naked snout. Ears small. Body elongated. Limbs short. Tail short or moderate, bushy. Several species are described (*H. orientalis*, *personata* [Fig. 265], *moschata*, *subaurantiaca*), all from Eastern Asia; they are all small animals compared with the other members of the subfamily, climbing trees with agility and living much on fruit and berries as well as on small mammals and birds. The two first named species occur in British India, *H. orientalis* also ranging into Java; the Chinese *H. subaurantiaca* is brilliantly coloured in the region of the throat.[2]

FIG. 265.—*Helictis personata*. (From Blanford, *Mammalia of British India*, p. 175.)

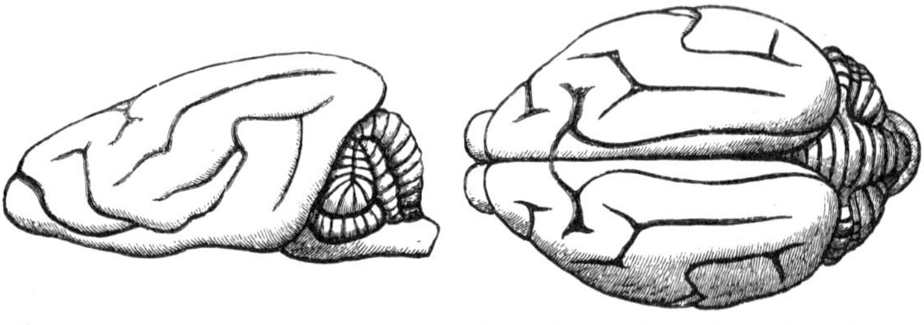

FIG. 266.—Left lateral and superior aspect of the brain of *Helictis sabaurantiaca*. (From Garrod, *Proc. Zool. Soc.* 1879, p. 307.)

The brain of *Helictis*, represented in the accompanying figure, shows the general type of cerebral structure characteristic of the *Mustelidæ*. The brain of this genus differs, however, from that of every other Carnivore in that the hippocampal gyrus rises to the surface on either side of the great longitudinal fissure, in

[1] Gray, *Proc. Zool. Soc.* 1831, p. 94.

[2] Garrod, *ibid.* 1879, pl. xxix.

consequence of which there is no crucial fissure, and the so-called "Ursine lozenge," so characteristic of the Arctoidea, is incomplete behind. The superior gyrus, as in *Ictonyx* and *Mustela*, ceases at the superior posterior angle of the hemisphere.

Ictonyx.[1]—Dentition: $i\ \frac{3}{3}$, $c\ \frac{1}{1}$, $p\ \frac{3}{3}$, $m\ \frac{1}{2}$; total 34. In general characters the teeth much resemble those of the Polecats (*Mustela*), being more delicately cut and sharply cusped than in most of the foregoing. Upper molar smaller than the carnassial, narrow from before backwards. Lower carnassial with a small narrow talon and distinct inner cusp. General form of body Musteline. Limbs short. Fore feet large and broad, with five stout, nearly straight, blunt, and non-retractile claws, of which the first and fifth are considerably shorter than the others. Tail moderate, with longer hairs towards the end, giving it a bushy appearance. Hairs generally long and loose. The best-known species of this genus, *I. zorilla*, the Cape Polecat, was placed by Cuvier in the genus *Mustela*, and by Lichtenstein in *Mephitis;* and in many characters it forms a transition between these genera. It is about the size of an English Polecat, but conspicuous by its coloration, having broad, longitudinal bands of dark brown, alternating with white. Its odour is said to be as offensive as that of the American Skunks. From the Cape of Good Hope it ranges as far north as Senegal. Another species, *I. frenata*, from Sennaar and Egypt, has been described.

Subfamily **Mustelinæ.**—Toes short, partially webbed; claws short, compressed, acute, curved, often semiretractile. Upper molar of moderate size, wide transversely. Kidneys simple. Terrestrial and arboreal in habits.

Galictis.[2]—Dentition: $i\ \frac{3}{3}$, $c\ \frac{1}{1}$, $p\ \frac{3}{3}$, $m\ \frac{1}{2}$; total 34. Molars small but stout. Upper carnassial with the inner tubercle near the middle of the inner border of the tooth. Lower carnassial with talon small, and inner cusp small or absent. Body long. Limbs short; claws non-retractile. Palms and soles naked. Head broad and depressed. Tail of moderate length. The best-known species are *G. vittata*, the Grison (genus *Grisonia*, Gray), and *G. barbara*, the Tayra (genus *Galera*, Gray), both South American; *G. allamandi* is an intermediate form.

Remains of *Galictis* occur in the Pleistocene cavern-deposits of Brazil, and also in the Pleistocene of North America.

Mustela.[3]—Dentition: $i\ \frac{3}{3}$, $c\ \frac{1}{1}$, $p\ \frac{3-4}{3-4}$, $m\ \frac{1}{2}$; total 34 or 38. Upper carnassial with inner tubercle close to the anterior edge of the tooth. Molar nearly as large as carnassial. Lower carnassial with small or no inner cusp. Vertebræ: C 7, D 14, L 6, S 3, C 18-23. Body long and slender. Limbs short, digitigrade. Feet

[1] Kaup, *Thierreich*, vol. i. p. 352 (1835). [2] Bell, *Proc. Zool. Soc.* 1837, p. 45.
[3] Linn. *Syst. Nat.* 12th ed. vol. i. p. 66 (1766).

rounded; toes short, with compressed, acute, semiretractile claws. Tail moderate or long, more or less bushy.

The genus *Mustela*, as restricted by Cuvier (*Règne Animal*, 1817), contains a very natural assemblage of animals commonly called Martens, Sables, Polecats, Stoats, Ermines, and Weasels, all closely allied in structure and habits. A structural division, however, occurs between the two first-named and all the others, especially shown in the presence of an additional small premolar tooth on each side of the jaw; and, availing himself of this and some other minor characters, Cuvier divided the genus into two subgenera, for the first of which he retained the name of *Mustela*, and to the second assigned that of *Putorius*. Three years later Nilsson (*Skand. Fauna*, 1820) definitely constituted the two groups into genera, applying to the first the name of *Martes*, by which the animals composing it had been generally designated by the Latin-writing zoologists of the preceding century, and keeping *Mustela* for the more typical Weasels and their immediate allies. Later zoologists have been divided between the nomenclature of Cuvier, which has the priority, and that of Nilsson, which on other grounds is preferable. Those who adopt the latter affirm that Cuvier's names, being only used by him in a subgeneric sense, and not binominally, need not be applied generically, but this is contrary to the practice usually followed in such cases; and therefore, if the original genus be divided, the name *Mustela* should be retained for the Martens, and *Putorius* for the Polecats and Weasels. Here, however, the genus will be employed in its wider sense, and divided into two groups.

The typical group of the Martens[1] presents the following distinctive features. Body long, slender, and very flexible, though less so than in the true Weasels. Head somewhat triangular; muzzle pointed, the nose extending a little beyond the lips; eyes large and prominent; ears conspicuous, broad, somewhat triangular, rounded at the ends, furred outside and in. Limbs short; feet rounded; toes short, five on each foot, all with short, compressed, curved, sharp-pointed claws; soles densely furred between the naked pads. Tail moderately long, more or less bushy. Outer fur long, strong, and glossy; a very abundant soft under fur. Skull elongated and depressed. Facial portion moderate and rather compressed. Zygomata arched and wide, but slender. Postorbital processes small. Auditory bullæ large, but not very globose. Mandible with a strong triangular vertical coronoid process and a well-developed angular process. Premolars $\frac{4}{4}$. Upper incisors in a straight transverse line, rather long and

[1] By all old authors of authority, as Ray, Pennant, Shaw, and Fleming, the word is written "Martin," but this form of spelling is now generally reserved by way of distinction for the bird. The term "Marten-Cat," often used, is a misnomer.

compressed; first and second subequal, third considerably larger. Lower incisors very small, especially the first, and crowded together, the second placed rather behind the others. Canines long and sharp-pointed. Upper premolars: first very small, with simple crown and one root; second and third nearly equal in size and two-rooted, with simple compressed sharp-pointed crowns, with very slightly developed accessory cusps; fourth (the carnassial) with blade consisting chiefly of the central and posterior lobes, the anterior being rudimentary, inner tubercle small and confined to the anterior part of the tooth. True molar tubercular, about twice as wide transversely as in the antero-posterior direction, having an outer, more elevated, but smaller portion, bearing three blunt tubercles; to the inner side of this the crown is contracted, and its surface deeply hollowed; it then expands again into a broad low lobe, with the central part elevated, and a raised, even, semicircular, slightly crenated internal border. Lower premolars: first very small, simple, and one-rooted; second, third, and fourth increasing slightly in size, with high compressed pointed crowns and posterior accessory cusps, best marked in the third. First molar (carnassial) with well-marked bilobed blade, talon scarcely more than one-third of the length of the tooth, and a very small inner cusp. Second molar small, single-rooted, with a low, flattened, subcircular or oval tubercular crown.

In geographical distribution the Martens are limited to the northern hemisphere, ranging throughout the greater part of the temperate regions of both Old and New Worlds, as far north as conditions of existence suited to their habits are met with, and southwards in America to 35° N. lat., while in Asia one species is met with as far in this direction as the island of Java.

The various species appear to be very similar in their habits. They live in woods and rocky places, and are thoroughly arboreal, spending most of their time in trees, although descending to the ground in quest of prey. They climb with great facility, and are agile and graceful in their movements. Some species are said occasionally to resort to berries and other fruit for food, but as a rule they are strictly carnivorous, feeding chiefly on birds and their eggs, small mammals, as squirrels, hares, rabbits, and moles, but chiefly mice of various kinds, of which they destroy great numbers, and occasionally snakes, lizards, and frogs. In proportion to their size they are among the most bloodthirsty of animals, though less so than the true Weasels. The female usually makes her nest of moss, dried leaves, and grass in the hollow of a tree, but sometimes in a hole among rocks or ruined buildings, and produces several young at a birth, usually from four to six. Though wild and untameable to a great degree if captured when fully grown, when taken young they are very docile, and have frequently been made pets of, not having the strong unpleasant odour of the smaller *Mustelidæ*. The common European Marten appears to have been partially domesticated by the Greeks and Romans, and to have been used to keep houses clear from rats and mice before cats were introduced.[1] In the same way, according to Hodgson, the Yellow-bellied Weasel (*M. cathia*) "is exceedingly prized by the Nipalese for its service in ridding houses of rats. It is easily tamed; and such is the dread of it common to all Murine animals that not one will approach a house where it is domiciled." It is, however, to the great value attached to the pelts of these animals that their importance to man is chiefly due. Though all yield fur of serviceable quality, the commercial value varies immensely, not only according to the particular species from which it is obtained, but according to individual variation, depending upon age, sex, season, and other trifling circumstances. The skins from northern regions are more full and of a finer colour and gloss than those from more temperate climates, as are those of animals killed in winter compared with the same individuals in the summer season. The caprices of fashion have, moreover, set wholly factitious values upon slight shades of colour, recognised and named by experienced furriers, but not indicating any specific or other distinctions of which zoologists have any cognisance. Enormous numbers of animals are annually caught, chiefly in traps, to supply the demand of the fur trade, Siberia and North America being the principal localities from which they are obtained.

With the exception of the Pekan (*M. pennanti*) all the Martens are so much alike in size, general colouring, and cranial and dental characters that the discrimination of the species, and assignment of the proper geographical distribution to each, has been a subject which has sorely perplexed the ingenuity and patience of zoologists. The following description by Dr. Elliott Coues of the external characters of the American Pine Marten (*M. americana*) will apply almost equally well to most of the others: "It is almost impossible to describe the colour of the Pine Marten, except in general terms, without going into the details of the endless diversities occasioned by age, sex, season, or other incidents. The animal is 'brown,' of various shades from orange or tawny to quite blackish; the tail and feet are ordinarily the darkest, the head lightest, often quite whitish; the ears are usually rimmed with whitish; on the throat there is usually a large tawny-yellowish or orange-brown patch, from the chin to the fore legs, sometimes entire, sometimes broken into a number of smaller, irregular blotches, sometimes wanting, sometimes prolonged on the whole under surface, when the animal is

[1] See Rolleston, "On the Domestic Cats, *Felis domesticus* and *Mustela foina*, of Ancient and Modern Times," *Journal of Anatomy and Physiology*, vol. ii. p. 47, 1868.

bicolor like a Stoat in summer. The general 'brown' has a grayish cast, as far as the under fur is concerned, and is overlaid with rich lustrous blackish-brown in places where the long bristly hairs prevail. The claws are whitish; the naked nose pad and whiskers are black. The tail occasionally shows interspersed white hairs, or a white tip."

The species generally recognised as distinct are the following, the first five belonging to the Old and the last two to the New World :—

M. foina, the Beech Marten, Stone Marten, or White-breasted Marten.—Distinguished from the following by the greater breadth of the skull, and some minute but constant dental characters, by

FIG. 267.—The Pine Marten (*Mustela martes*).

the dull grayish-brown colour of the fur of the upper parts, and the pure white of the throat and breast. It inhabits the greater part of the continent of Europe, but is more southern than the next in its distribution, not being found in Sweden or Norway, nor, according to the investigations of Mr. Alston, in the British Isles, although included in their fauna by all earlier writers; to the eastward it ranges into Afghanistan and the Himalaya.

M. martes, the Pine Marten (Fig. 267).—Outer fur rich dark brown; under fur reddish-gray, with clear yellow tips; breast spot usually yellow, varying from bright orange to pale cream-colour or yellowish-white. Length of head and body 16 to 18 inches; of tail (including the hair) 9 to 12 inches. This species is extensively distributed throughout northern Europe and Asia, and was formerly common in most parts of Great Britain and Ireland. Though commonly called "Pine Marten," it does not appear to have any special preference for coniferous trees, except that, inasmuch as they constitute the greater proportion of the forests of the countries which it inhabits, it is more often met with in them than in any other. With regard to its recent occurrence in the British Isles, Mr. Alston writes in the *Proc. Zool. Soc.* 1879 :—

"Although greatly reduced in numbers by persecution, it still maintains its ground in the wilder districts of Scotland, the north of England, Wales, and Ireland; and occasionally specimens are killed in counties where the species was thought to have been long extinct. In Scotland it is still found, though comparatively rarely, in the Lews and in most of the Highland mainland counties, being perhaps most abundant in Sutherland and Ross-shire, especially in the deer forests. In the Lowlands a Marten is now a very great rarity; but a fine example was killed in Ayrshire in the winter of 1875-76. In the north of England Mr. W. A. Durnford says the species is still plentiful in the wilder parts of Cumberland, Westmoreland, and Lancashire, and in Lincolnshire. several have been recorded, the latest killed in 1865, by Mr. Cordeaux. In Norfolk one was shot last year; and I have myself examined a fine example which was shot in Hertfordshire, within 20 miles of London, in December 1872. In Dorsetshire the last is said to have been killed in 1804; but a specimen occurred in Hampshire about forty years ago, and another in Surrey in 1847. In Ireland the following counties were enumerated by Thompson as habitats of this species: Donegal, Londonderry, Antrim, Down, Armagh, Fermanagh, Longford, Galway, Tipperary, Cork, and Kerry. The *Cat-crann* is probably now a rarer animal in Ireland than it was when Thompson wrote; but it still exists in various districts, especially in County Kerry, whence the society has received several living examples; and Professor A. Leith Adams states that it has been seen of late years even in county Dublin."

M. zibellina, the Sable (German, *Zobel* and *Zebel*; Swedish, *sabel*; Russian, *sobel*, a word probably of Turanian origin).—Closely resembling the last, if indeed differing from it except in the quality of the fur, which is the most highly valued of that of all the group. Found chiefly in Eastern Siberia.

M. flavigula, the Indian Marten.—Inhabits the southern slopes of the Himalaya, the Nilgiri Hills, the interior of Ceylon, the Malay Peninsula, and Java. The coloration of this species is very striking, the upper parts being blackish-brown, and the throat and breast yellow or orange, in the bright coloured variety. It differs from the other species in having the soles of the feet more or less naked.

M. melampus.—Japan.

M. americana, the North-American Sable or Marten.—A species so closely allied to the European Pine Marten and Asiatic Sable that it is very difficult to assign constant distinguishing characters between them. The importance of the fur of this animal as an article of commerce may be judged of from the fact that 15,000 skins were sold in one year by the Hudson's Bay Company as long ago as 1743, and the more recent annual imports into Great Britain have exceeded 100,000. It is ordinarily caught in wooden traps of very simple construction, being little enclosures of stakes or brush in which the bait is placed upon a trigger, with a short upright stick supporting a log of wood, which falls upon its victim on the slightest disturbance. A line of such traps, several to a mile, often extends many miles. The bait is any kind of meat, a mouse, squirrel, piece of fish, or bird's head. It is principally trapped during the colder months, from October to April, when the fur is in good condition, as it is nearly valueless during the shedding in summer. Dr. Coues tells us that, notwithstanding the persistent and uninterrupted destruction to which the American Sable is subjected, it does not appear to diminish materially in numbers in unsettled parts of the country. It holds its own partly in consequence of its shyness, which keeps it away from the abodes of men, and partly because it is so prolific, bringing forth six to eight young at a litter. Its home is sometimes a den under ground or beneath rocks, but oftener the hollow of a tree, and it is said frequently to take forcible possession of a squirrel's nest, driving off or devouring the rightful proprietor.

M. pennanti, the Pekan or Pennant's Marten, also called Fisher Marten, though there appears to be nothing in its habits to justify the appellation.—This is the largest species of the group, the head and body measuring from 24 to 30 inches, and the tail 14 to 18 inches. It is also more robust in form than the others, its general aspect being more that of a Fox than a Weasel; in fact, its usual name among the American hunters is "Black Fox." Its general colour is blackish, lighter by mixture of brown or gray on the head and upper fore part of the body, with no light patch on the throat, and unlike the other Martens generally darker below than above. It was generally distributed in wooded districts throughout the greater part of North America, as far north as Great Slave Lake, 63° N. lat., and Alaska, and extending south to the parallel of 35°; but at the present time it is almost exterminated in the settled parts of the United States east of the Mississippi.

Fossil remains of a Marten from the Pliocene Siwaliks of India indicate a species which cannot be distinguished from those now inhabiting the same region; while remains of *M. martes* occur in European cavern-deposits, and in the fens of Cambridgeshire.

With the *Putoriine* group (genus *Putorius*) we come to those smaller forms distinguished by having only three premolars in each jaw, by the absence of an inner cusp to the blade of the lower carnassial, as well as by certain external characters. This group contains a few species known as Minks, differing from the rest by slight structural modifications, and especially by their semiaquatic habits. They are distinguished from the Polecats, Stoats, and Weasels, which constitute the remainder of the group, by the facial part of the skull being narrower and more approaching in form that of the Martens, by the premolar teeth (especially the anterior one in the upper jaw) being larger, by the toes being partially webbed, and by the absence of hair in the intervals between the naked pads of the soles of the feet. The two best-known species, so much alike in size, form, colour, and habits that although they are widely separated geographically some zoologists question their specific distinction, are *M. lutreola*, the *Nörz* or *Sumpfotter* (Marsh-Otter) of Eastern Europe, and *M. vison*, the Mink of North America. The former inhabits Finland, Poland, and the greater part of Russia, though not found east of the Ural Mountains. Formerly it extended westward into Central Germany, but it is now very rare, if not extinct, in that country. The latter is found in places which suit its habits throughout the whole of North America. Another form, *M. sibirica*, from Eastern Asia, of which much less is known, appears to connect the true Minks with the Polecats.

For the following description, chiefly taken from the American form (though almost equally applicable to that of Europe), we are mainly indebted to Dr. Coues's *Fur-bearing Animals of North America*. In size it much resembles the English Polecat,—the length of the head and body being usually from 15 to 18 inches, that of the tail to the end of the hair about 9 inches. The female is considerably smaller than the male. The tail is bushy, but tapering at the end. The ears are small, low, rounded, and scarcely project beyond the adjacent fur. The pellage consists of a dense, soft, matted under fur, mixed with long, stiff, lustrous hairs on all parts of the body and tail. The gloss is greatest on the upper parts; on the tail the bristly hairs predominate. Northern specimens have the finest and most glistening pellage; in those from southern regions there is less difference between the under and over fur, and the whole pellage is coarser and harsher. In colour different specimens present a considerable range of variation, but the animal is ordinarily of a rich dark brown, scarcely or not paler below than on the general upper parts; but the back is usually the darkest, and the tail is nearly black. The under jaw, from the chin about as far back as the angle of the mouth, is generally white. In the European Mink the upper lip is also white, but as this occasionally occurs in American specimens it fails as an absolutely distinguishing character. Besides the white on the chin, there are often other irregular white patches on the under parts of the body. In very rare instances the tail is tipped with white. The fur, like that of most of the animals of the group to which it belongs, is an important article of commerce.

The principal characteristic of the Mink in comparison with its congeners is its amphibious mode of life. It is to the water what the other Weasels are to the land, or Martens to the trees, being as essentially aquatic in its habits as the Otter, Beaver, or Musk-Rat, and spending perhaps more of its time in the water than it does on land. It swims with most of the body submerged, and dives with perfect ease, remaining long without coming to the surface to breathe. It makes its nest in burrows in the banks of streams, breeding once a year about the month of April, and producing five or six young at a birth. Its food consists of frogs, fish, freshwater

FIG. 268.—The Common Polecat (*Mustela putorius*).

molluscs and crustaceans, as well as mice, rats, musk-rats, rabbits, and small birds. In common with the other animals of the genus, it has a very peculiar and disagreeable effluvium, which, according to Coues, is more powerful, penetrating, and lasting than that of any animal of the country except the Skunk. It also possesses the courage, ferocity, and tenacity of life of its allies. When taken young, however, it can be readily tamed, and lately Minks have been extensively bred in captivity in America, both for the sake of their fur and for the purpose of using them in like manner as Ferrets in England, to clear buildings of rats.

The Polecats include four species confined to the northern hemisphere, the best known of which is the Common Polecat (*M. putorius*, Fig. 268). The Ferret is a domesticated variety of this species, generally of a yellowish-white colour; whereas the Wild

Polecat is dark brown above and black beneath, the face being variegated with dark brown and white markings.

The skull is rough, strongly ridged, and of a far more powerful type than that of the Stoats, Weasels, or Martens; being in the female much smaller and lighter than in the male. The fur, which is long, coarse, and of comparatively small value, changes its colour very little, if at all, at the different seasons of the year.

The distribution and habits of this species have been described by Blasius, the following being an abstract of his account. The Polecat ranges over the greater part of Europe, reaching northwards into Southern Sweden, and in Russia to the region of the White Sea. It does not occur in the extreme South, but is common everywhere throughout Central Europe. In the Alps it ranges far above the tree-line during the summer, but retreats in winter to lower ground. In fine weather it lives either in the open air, in holes, fox-earths, rabbit-warrens, under rocks, or in wood-stacks, while in winter it seeks the protection of deserted buildings. During the day it sleeps in its hiding-place, sallying forth at night to plunder dovecots and hen-houses. It climbs but little, and shows far less activity than the Marten. It feeds ordinarily on small mammals, such as rabbits, hamsters, rats, and mice, on such birds as it can catch, especially poultry and pigeons, and also on snakes, lizards, frogs, fish, and eggs. Its prey is devoured only in its lair, but, even though it can carry away but a single victim, it commonly kills everything that comes in its way, often destroying all the inhabitants of a hen-house in order to gratify its passion for slaughter. The pairing time is towards the end of the winter, and the young, from three to eight in number, are born in April or May, after a period of gestation of about two months. The young, if taken early, may be easily trained, like Ferrets, for rabbit-catching. The Polecat is very tenacious of life, and will bear many severe wounds before succumbing; it is also said to receive with impunity the bite of the adder. Its fetid smell has become proverbial.

Four other species of Polecats are known, viz.—The Siberian Polecat (*M. eversmanni*) of Western and Northern Asia is nearly allied to the European species, but the head and back are almost white, and the skull is stouter and more constricted behind the orbits. The Tibetan *M. larvata* is distinguished from the last by the presence of a process connecting the pterygoid with the auditory bulla, and by a difference in the shape of the upper molar. The American Polecat (*M. nigripes*), inhabiting the central plateau of the United States, and extending southwards into Texas, is another closely allied species, although some zoologists have made it the type of the genus *Cynomyonax*. Finally, the Mottled Polecat (*M. sarmatica*) is a species sparsely distributed in Eastern Europe and parts of Western Asia, but common in Southern Afghanistan.

Its skull, although smaller, resembles that of the common species; but the coloration is very different, all the upper parts being mottled with large irregular reddish spots on a white ground, and the under side, limbs, and tail deep shining black. The tail is long.

The Common Polecat occurs in a fossil condition in the cave-deposits of Europe.

The remaining members of the genus comprise the true Weasels and Stoats, which are of almost cosmopolitan distribution. In the Common Weasel (*M. vulgaris*, Fig. 269) the upper parts, outside of limbs and tail, are a uniform reddish-brown, the under parts pure

Fig. 269.—The Common Weasel (*Mustela vulgaris*).

white. In very cold regions, both in Europe and America, it turns completely white in winter, but less regularly and at a lower temperature than the Stoat, from which it is easily distinguished by its smaller size, and by its wanting the black end of the tail. The length of the head and body of the male is usually about 8 inches, that of the tail 2½ inches; the female is smaller.

This species is pretty generally distributed throughout Europe, Northern and Central Asia, British North America, and the northern portions of the United States. It possesses in a full degree all the active, courageous, and bloodthirsty disposition of the rest of the genus, but its diminutive size prevents it attacking and destroying any but the smaller mammals and birds. Mice, rats, voles, moles, and frogs constitute its principal food. It is generally found on or near the surface of the ground, but it can not only pursue its prey through very small holes and crevices of rocks and under dense tangled herbage, but follow it up the stems and branches of trees, or even into the water, swimming with perfect ease. It constructs a nest of dried leaves and herbage, placed in a hole in the ground or a bank or hollow tree, in which it brings up its litter of four to six (usually five) young ones. The mother will defend her young with the utmost desperation against any assailant, having been often known to sacrifice her own life rather than desert them.

The Stoat or Ermine (*M. erminea*) has nearly the same distribution as the Weasel, but in Asia it is said to extend into parts of the Kashmir Himalaya. Its size, as already mentioned, considerably exceeds that of the Weasel; and its most distinctive feature is the black tip at the end of the tail, which remains when the rest of the pellage turns white. The white winter skins from the northern regions of its habitat, where the fur is thick and close, form the well-known and valuable ermine of commerce. Remains of the Stoat are found in the Pleistocene cavern-deposits of Europe. The other species of Weasels are very numerous and widely distributed.

Extinct Mustelines.—A number of European Miocene Carnivores may be referred to the genus *Mustela* in its wider sense, and serve to confirm the propriety of this use of the term. Thus *M. sectoria* is a species of somewhat larger size than the Stoat, with $p\ \frac{4}{4}$, while in *M. angustifrons* the number of premolars is $\frac{3}{4}$, and in *M. mustelina* $\frac{4}{3}$; the latter species agreeing very closely in size with the Stoat. The extinct *Plesictis*, in which there are $p\ \frac{4}{4}$ and the lower carnassial has a large inner cusp, is distinguished from *Mustela* by the circumstance that the temporal ridges of the skull never unite to form a sagittal crest. Moreover, the inner tubercular portion of the upper molar (as in some of the Miocene species of *Mustela*) is shorter in an antero-posterior direction than the secant outer moiety; and the auditory bulla is more inflated than in *Mustela*, although it has no septum. Both these features indicate a decided approximation to the Viverroid genus *Stenoplesiotis* (p. 539); and since there are no well-marked characters of family value by which these two genera can be distinguished the available evidence points to a transition from the Viverroid to the Musteloid type. *Mustela larteti*, of the Middle Miocene of France, should perhaps be referred to *Ictonyx*.

Pœcilogale.[1]—This genus has been made for the reception of the South African *Mustela albinucha*, in which the coloration is similar to that of *Ictonyx*, but the number of cheek-teeth is usually reduced to $p\ \frac{2}{2}$, $m\ \frac{1}{1}$, although there may be a second lower molar. The auditory bulla is quite flat.

Lyncodon.[2]—This name has been proposed for a small Musteline

[1] O. Thomas, *Ann. Mag. Nat. Hist.* ser. 5, vol. xi. p. 370 (1883).

[2] Gervais, *Dict. Univ. d'Hist. Nat.* t. iv. p. 685 (1849).

from Patagonia, with $p\ \frac{2}{2}$, $m\ \frac{1}{1}$, which Mr. O. Thomas suggests may be nothing more than an aberrant southern form of *Mustela* (*Putorius*) *brasiliensis*. The auditory bulla is more inflated than in the typical Weasels. This animal is somewhat larger than the Stoat.

Gulo.[1]—Dentition: $i\ \frac{3}{3}$, $c\ \frac{1}{1}$, $p\ \frac{4}{4}$, $m\ \frac{1}{2}$; total 38. Crowns of the teeth very stout. Upper molar very much smaller than the carnassial. Lower carnassial large, with very small talon and no inner cusp. Third upper incisor unusually large, almost like a canine. The dentition, though really but a modification of that of the Weasels, presents a great general resemblance to that of the Hyæna. Palate prolonged somewhat behind the last molar. Humerus with an entepicondylar foramen. Vertebræ: C 7, D 15, L 5, S 3, C 15. Body

FIG. 270.—The Wolverene (*Gulo luscus*).

and limbs stoutly made. Feet large and powerful, subplantigrade, with large, compressed, much curved, and sharp-pointed claws. Soles of the feet (except the pads of the toes) covered with thick bristly hairs. Ears very small, nearly concealed by the fur. Eyes small. Tail short, thick, and bushy. Fur full, long, and rather coarse. The one species, the Wolverene or Glutton (*G. luscus*, Fig. 270), an inhabitant of the forest regions of Northern Europe, Asia, and America, much resembles a small Bear in appearance. It is a very powerful animal for its size, climbs trees, and lives on grouse, squirrels, hares, foxes, beavers, reindeer, and is said to attack even horses and cows. The Wolverene has a curious habit of stealing and secreting articles of which it can make no possible use, as is exemplified in the following instance related by Dr. Coues:

[1] Storr, *Prodromus Meth. Mamm.* p. 34 (1780).

"A hunter and his family, having left their lodge unguarded during their absence, on their return found it completely gutted—the walls were there, but nothing else. Blankets, guns, kettles, axes, cans, knives, and all the other paraphernalia of a trapper's tent had vanished, and the tracks left by the beast showed who had been the thief. The family set to work, and, by carefully following up all his paths, recovered, with some trifling exceptions, the whole of the lost property." The pairing season occurs in March, and the female, secure in her burrow, produces her young, four or five at a birth, in June or July. In defence of these she is exceedingly bold, and the Indians, according to Coues, "have been heard to say that they would sooner encounter a she-bear with her cubs than a carcajou (the Indian name of the glutton) under the same circumstances."

Fossil remains of the Wolverene are found in cavern and other Pleistocene deposits in various parts of Europe.

Suborder PINNIPEDIA.

The Eared-Seals, Walruses, and Seals differ from the rest of the Carnivora mainly in the structure of their limbs, which are modified for aquatic progression,—the two proximal segments being very short and partially enveloped in the general integument of the body; while the third segment, especially in the hinder extremities, is elongated, expanded, and webbed. There are always five well-developed digits on each limb. In the hind limb the two marginal digits (first and fifth) are stouter and generally longer than the others. The teeth also differ from those of the more typical Carnivora. The incisors are always fewer than $\frac{3}{3}$. The cheek series consists generally of four premolars and one molar of very uniform characters, with never more than two roots, and with conical, more or less compressed, pointed crowns, which may have accessory cusps, placed before or behind the principal one, but are never broad and tuberculated; and there is no differentiated carnassial tooth. The milk-teeth are very small and simple, and are shed or absorbed at a very early age, usually either before or within a few days after birth. The brain is relatively large; the cerebral hemispheres being broad in proportion to their length, with numerous and complex convolutions. There is a very short cæcum. The kidneys are divided into numerous distinct lobules. There are no Cowper's glands. The mammæ are either two or four, and abdominal in position. No clavicles. Tail always very short. Eyes very large and exposed, with flat cornea.

The animals of this group are all aquatic in their mode of life, spending the greater part of their time in the water, swimming and diving with great facility, feeding mainly on fish, crustaceans, and other marine animals, and progressing on land with difficulty.

They always come on shore, however, for the purpose of bringing forth their young. They are generally marine, but they occasionally ascend large rivers, and some inhabit inland seas and lakes, as the Caspian and Baikal. Though not numerous in species, they are widely distributed over the world, but occur most abundantly on the coasts of lands situated in cold and temperate zones. The suborder is divisible into three well-marked families: the *Otariidæ*, Fur-Seals or Sea-Bears, which form a transition from the Fissiped Carnivora to the Seals; the *Trichechidæ*, containing the Walrus; and the *Phocidæ* or typical Seals.

The resemblances between the skull and other parts of the body of the Fur-Seals and the Ursoid true Carnivora is suggestive of some genetic relationship between the two groups, and Professor Mivart[1] expresses the opinion that the one group is the direct descendant of the other. The same writer goes on to suggest that if the Eared-Seals have been derived from Bear-like Carnivores this need not necessarily hold good with the true Seals, which may have had another, and possibly Lutrine, origin. The presence of an alisphenoid canal in *Ursus* and the *Otariidæ*, and its absence in *Lutra* and the *Phocidæ*, together with other osteological features, are cited in support of this view; but although these resemblances and differences are certainly remarkable, yet much more evidence is required before the hypothesis can be accepted as even a probable one. It must, moreover, be borne in mind that the true Bears are a very modern group; and there is a possibility that the Pinnipeds may prove to have been independently derived from the extinct Carnivora noticed below under the name of Creodonta.

Family OTARIIDÆ.

When on land the hind feet are turned forwards under the body, and aid in supporting and moving the trunk as in ordinary mammals. A small external ear. Testes suspended in a distinct external scrotum. Skull with postorbital processes, and an alisphenoid canal. Angle of mandible inflected. Palms and soles of feet naked.

Otaria.[2]—Dentition: $i\ \frac{3}{2}$, $c\ \frac{1}{1}$, $p\ \frac{4}{4}$, $m\ \frac{1\text{ or }2}{1}$; total 34 or 36. First and second upper incisors small, with the summits of the crowns divided by a deep transverse groove into an anterior and a posterior cusp of nearly equal height; the third large and canine-like. Canines large, conical, pointed, recurved. Molars and premolars usually $\frac{5}{5}$, of which the second, third, and fourth are preceded by milk-teeth shed a few days after birth; sometimes (as in Fig. 271) a sixth upper molar (occasionally developed on one

[1] *Proc. Zool. Soc.* 1885, p. 497.

[2] Péron, *Voyage aux Terres Australes*, vol. ii. p. 37 note (1816).

side and not the other); all with similar characters, generally uniradicular; crown moderate, compressed, pointed, with a single principal cusp, and sometimes a cingulum, and more or less developed anterior and posterior accessory cusps. Vertebræ: C 7, D 15, L 5, S 4, C 9-14. Head rounded. Eyes large. Pinna of ear small, narrow, and pointed. Neck long, Skin of all the feet extended far beyond the nails and ends of the digits, with a deeply-lobed margin. The nails small and often quite rudimentary, especially those of the first and fifth toes of both feet, the best-developed and most constant being the three middle claws of the hind foot, which are elongated, compressed, and curved.

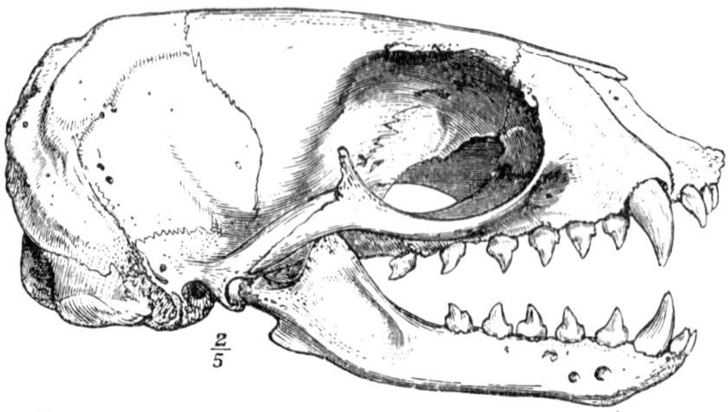

FIG. 271.—Skull of *Otaria forsteri*. (From Gray, *Proc. Zool. Soc.* 1872, p. 660.)

The Eared-Seals, commonly called Sea-Bears or Sea-Lions, are widely distributed, especially in the temperate regions of both hemispheres, though absent from the coasts of the North Atlantic. As might be inferred from their power of walking on all fours, they spend more of their time on shore, and range inland to greater distances, than the true Seals, especially at the breeding time, though they are obliged always to return to the water to seek their food. They are gregarious and polygamous, and the males are usually much larger than the females, a circumstance which has given rise to some of the confusion existing in the specific determination of the various members of the genus. Some of the species possess, in addition to the stiff, close, hairy covering common to all the group, an exceedingly fine, dense, woolly under fur. The skins of these, when dressed and deprived of the longer harsh outer hairs, constitute the "sealskin" of commerce, so much valued for wearing apparel, which is not the product of any of the true Seals. The best-known species are *O. stelleri*, the Northern Sea Lion, the largest of the genus, from the North Pacific, about 10 feet in length; *O. jubata*, the Patagonian or Southern Sea Lion (Fig. 272), from the Falkland Islands and Patagonia; *O. californiana*, from California, frequently exhibited alive in menageries in Europe; *O. ursina*, the common Sea-Bear or Fur-Seal of the North Pacific, the skins of which are imported in immense numbers from the Prybiloff Islands; *O. pusilla*, from the Cape of Good Hope; *O. forsteri* and others, from the coasts of Australia and various islands scattered over the southern hemisphere. These have been grouped by some zoologists into many genera, founded upon very trivial modifications of teeth and skull. In a recent memoir Mr. Beddard[1] concludes that if the genus be split up at all, it should be divided into *Otaria*, containing only *O. jubata* (with its numerous synonyms), and *Arctocephalus*, comprising all the other species. The latter group is distinguished by the more narrow and pointed nose, the longer ears, the palate not excavated nor truncated posteriorly, the presence of

FIG. 272.—The Patagonian Sea-Lion (*Otaria jubata*). From Sclater, *Proc. Zool. Soc.* 1866, p. 80.

a hook-like process to the pterygoids, and by the posterior border of the nasals extending behind the zygoma.

The following account of *O. ursina* in the Prybiloff Islands is taken, with slight verbal alteration, from Nordenskiöld's *Voyage of the Vega*: "The Sea-Bears are found year after year during summer at certain parts of the coast, known as 'rookeries,' where, collected in hundreds of thousands, they pass several months without the least food. The males or 'bulls' come first to the place, most of them in the month of May or in the beginning of June. The most violent conflicts, often with a deadly issue for one of the parties, now arise regarding the space of about a hundred square feet which each bull-seal considers necessary for his home. The

[1] "On the structure of Hooker's Sea-Lion (*Arctocephalus hookeri*)," *Trans. Zool. Soc.* vol. xii. p. 369 (1890).

strongest and most successful in fight retain the best places near the shore; the weaker have to crawl farther up on land, where the chances of getting a sufficient number of spouses are not particularly great. The fighting goes on with many feigned attacks and parades. At first the contest concerns only the proprietorship of the soil. The attacked, therefore, never follows his opponent beyond the area he has once taken up, but haughtily lays itself down, when the enemy has retired, in order to collect strength for a new combat. The animal in such a case grunts with satisfaction, throws himself on his back, scratches himself with his fore feet, attends to his toilet, or cools himself by slowly fanning with one of his hind feet; but he is always on the alert and ready for a new fight, until he is tired out and meets his match and is driven farther up from the beach. In the middle of June the females come up from the sea. At the water's edge they are received in a very gallant way by some strong bulls that have succeeded in securing for themselves places next the shore, and now are bent by fair means or foul on annexing the females for their harem. But scarcely is the female that has come up out of the water established with male No. 1 than he rushes towards a new female on the surface of the water. Male No. 2 now stretches out his neck and without ceremony lays hold of the female of No. 1, to be afterwards exposed to a similar trick by No. 3. In such cases the females are quite passive, never fall out with each other, and bear with patience the severe wounds they often get when they are pulled about by the combatants, now in one direction, now in another. All the females are finally distributed in this way after furious combats among the males, those of the latter who are nearest the beach getting from 12 to 15 consorts to their share. Soon after landing the females bring forth their young, which are treated with great indifference, and are protected by their adopted father only within the limits of the harem. Next comes the pairing season, and when it has passed there is an end to the arrangement and distribution into families at first so strictly maintained. The males, rendered lean by three months' absolute fasting, by degrees leave the rookery, which is left in possession of the Walruses and the young Sea Bears, including a number of young males that have not ventured to the place before. In the middle of September, when the young have learned to swim, the place is quite abandoned, with the exception of single animals that have for some reason remained behind."

Family TRICHECHIDÆ.

In many characters the single genus representing this family is intermediate between the *Otariidæ* and *Phocidæ*, but it has a completely aberrant dentition. It has no external ears, as in the

Phocidæ; but when on land the hind feet are turned forwards and used in progression, though less completely than in the *Otariidæ*. The upper canines are developed into immense tusks, which descend a long distance below the lower jaw. All the other teeth (Fig. 273), including the lower canines, are much alike, small, simple, and one-rooted, the molars with flat crowns. The skull is without postorbital process, but has an alisphenoid canal.

Trichechus.[1]—Dentition of young: $i\ \frac{2}{2}$, $c\ \frac{1}{1}$, p and $m\ \frac{5}{4}$. Many of these teeth are, however, lost early or remain through life in a rudimentary state concealed by the gums. The teeth which are usually developed functionally are $i\ \frac{1}{0}$, $c\ \frac{1}{1}$, $p\ \frac{3}{3}$, $m\ \frac{0}{0}$; total 18. Vertebræ: C 7, D 14, L 6, S 4, C 12. Head round. Eyes rather small. Muzzle short and broad, with on each side a group of long,

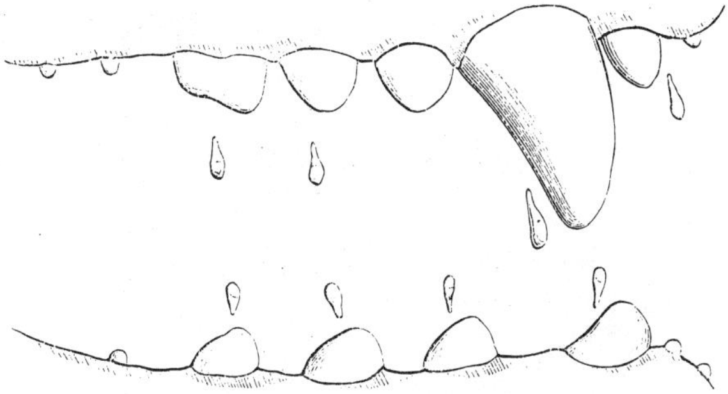

FIG. 273.—Diagram of the dentition of the Walrus (*Trichechus rosmarus*). The denticles placed apart from the others are milk-teeth, and disappear soon after birth. The small teeth in connection with the jaws frequently persist throughout life.

very stiff, bristly whiskers. The remainder of the hair-covering very short and adpressed. Tail very rudimentary. Fore feet with subequal toes, carrying five minute flattened nails. Hind feet with subequal toes, the fifth slightly the largest, having cutaneous lobes projecting beyond the ends as in *Otaria;* first and fifth with minute flattened nails; second, third, and fourth with large, elongated, subcompressed pointed nails.

Trichechus is the almost universally accepted generic name by which the Walrus or Morse[2] is known to zoologists, but some confusion has been introduced into literature by the revival of the nearly obsolete terms *Rosmarus* by some authors and *Odobænus* by others. *T. rosmarus* is the name of the species met with in the Arctic seas; that of the North Pacific, if distinct, is *T. obesus*. The preceding and following descriptions will apply equally to both.

[1] Linn. *Syst. Nat.* 12th ed. vol. i. p. 49 (1766).

[2] The former word is a modification of the Scandinavian *vallross* or *hvalros* ("whale-horse"), the latter an adaptation of the Russian name for the animal.

A full-grown male Walrus measures from 10 to 11 feet from the nose to the end of the very short tail, and is a heavy, bulky animal, especially thick about the shoulders. The soles of both fore and hind feet are bare, rough, and warty. The surface of the skin generally is covered with short, adpressed hair of a light, yellowish-brown colour, which, on the under parts of the body and base of the flippers, passes into dark reddish-brown or chestnut. In old animals the hair becomes more scanty, sometimes almost entirely disappearing, and the skin shows ample evidence of the rough life and pugnacious habits of the animal in the innumerable scars with which it is usually covered. It is everywhere more or less wrinkled,

FIG. 274.—The Walrus (*Trichechus rosmarus*).

but especially over the shoulders, where it is thrown into deep and heavy folds.

The tusks are formidable weapons of defence, but their principal use seems to be scraping and digging among the sand and shingle for the molluscs and crustaceans on which the Walrus feeds. They are said also to aid in climbing up the slippery rocks and ledges of ice on which so much of the animal's life is passed. Although this function of the tusks is affirmed by numerous authors, some of whom appear to have had opportunities of actual observation, it is explicitly denied by Malmgren.

Walruses are more or less gregarious in their habits, being met with generally in companies or herds of various sizes. They are only found near the coast or on large masses of floating ice, and rarely far out in the open sea; and, though often moving from one part of their feeding ground to another, they have no regular seasonal migrations. Their young are born between the months of

April and June, usually but one at a time, never more than two. Their strong affection for their young, and their sympathy for each other in times of danger, have been particularly noticed by all who have had the opportunity of observing them in their native haunts. When one of their number is wounded, the whole herd usually join in a concerted and intelligent defence. Although harmless and inoffensive when not molested, they exhibit considerable fierceness when attacked, using their great tusks with tremendous effect either on human enemies who come into too close quarters or on Polar Bears, the only other adversaries they can meet with in their own natural territory. Their voice is a loud roaring, and can be heard at a great distance; it is described by Dr. Kane as "something between the mooing of a cow and the deepest baying of a mastiff, very round and full, with its bark or detached notes repeated rather quickly seven or nine times in succession."

The principal food of the Walrus consists of bivalved molluscs, especially *Mya truncata* and *Saxicava rugosa*, two species very abundant in the Arctic regions, which it digs up from the mud and sand in which they lie buried at the bottom of the sea by means of its tusks. It crushes and removes the shells by the aid of its grinding teeth and tongue, swallowing only the soft part of the animal. It also feeds on other molluscs, sand-worms, star-fishes, and shrimps. Portions of various kinds of algæ or sea-weeds have been found in its stomach, but whether swallowed intentionally or not is still doubtful.

The commercial products of the Walrus are its oil, hide (used to manufacture harness and sole-leather and twisted into tiller ropes), and tusks. The ivory of the latter is, however, inferior in quality to that of the Elephant. Its flesh forms an important article of food to the Eskimo and Tchuktchis. Of the coast tribes of the last-named people the Walrus forms the chief means of support. "The flesh supplies them with food, the ivory tusks are made into implements used in the chase and for other domestic purposes, as well as affording a valuable article of barter, and the skin furnishes the material for covering their summer habitations, harness for their dog-teams, and lines for their fishing gear" (Scammon).

Geographically the Walrus is confined to the northern circumpolar regions of the globe, extending apparently as far north as explorers have penetrated, but its southern range has been much restricted of late in consequence of the persecutions of man. On the Atlantic coast of America it was met with in the sixteenth century as low as the southern coast of Nova Scotia, and in the last century it was common in the Gulf of St. Lawrence and on the shores of Labrador. It still inhabits the coast round Hudson's Bay, Davis Straits, and Greenland, where, however, its numbers are daily decreasing. It is not found on the Arctic coast of America between the 97th and

158th meridians. In Europe occasional stragglers have reached the British Isles, and it was formerly abundant on the coasts of Finmark. It is rare in Iceland, but Spitzbergen, Nova Zembla, and the western part of the north coast of Siberia are still constant places of resort, in all of which a regular war of extermination is carried on. The North Pacific, including both sides of Behring's Strait, northern Kamschatka, Alaska, and the Pribyloff Islands, are also the haunts of numerous Walruses, which are isolated from those of the North Atlantic by the long stretches of coast, both of Siberia and North America, where they do not occur. The Pacific Walrus appears to be as large as, if not larger than, that of the Atlantic; its tusks are longer and more slender, and curved inwards; the whiskers are smaller, and the muzzle (of the skull) relatively deeper and broader. These and certain other minor differences have induced some naturalists to consider it specifically distinct under the name of *Trichechus obesus*. Its habits appear to be quite similar to those of the Atlantic form. Though formerly found in immense herds, it is rapidly becoming scarce, as the methods of destruction used by the American whalers, who have systematically entered upon its pursuit, are far more certain and deadly than those of the native Tchuktchis, to whom, as mentioned before, the Walrus long afforded the principal means of subsistence.

Fossil remains of Walruses and closely allied animals have been found in the United States, and in England, Belgium, and France, in deposits of Pliocene age.

Family PHOCIDÆ.

The true Seals are the most completely adapted for aquatic life of all the Pinnipeds. When on land the hind limbs are extended behind them and take no part in progression, which is effected by a series of jumping movements produced by the muscles of the trunk, in some species aided by the fore limbs only. The palms and soles of the feet are hairy. There is no pinna to the ear, and no scrotum, the testes being abdominal. The upper incisors have simple, pointed crowns, and vary in number in the different groups. All the forms have well-developed canines and $\frac{5}{5}$ teeth of the cheek-series. In those species of which the milk-dentition is known, there are three milk molars (Fig. 275), which precede the second, third, and fourth permanent molars; the dentition is therefore $p\ \frac{4}{4}$, $m\ \frac{1}{1}$, the first premolar having as usual no milk-predecessor. The skull has no postorbital process and no alisphenoid canal; and the angle of the mandible is not inflected. The fur is stiff and adpressed, without woolly under fur.

Subfamily **Phocinæ.**—Incisors $\frac{3}{2}$. All the feet with five well-developed claws. The toes on the hind feet subequal, the first and fifth not greatly exceeding the others in length, and with the interdigital membrane not extending beyond the toes.

Halichœrus.[1]—Dentition: $i\,\frac{3}{2}$, $c\,\frac{1}{1}$, $p\,\frac{4}{4}$, $m\,\frac{1}{1}$; total 34. Crowns of molars large, simple, conical, recurved, slightly compressed,

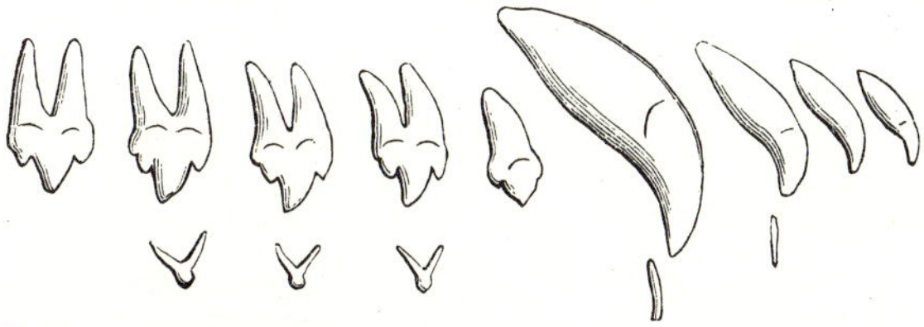

FIG. 275.—Upper permanent and deciduous dentition of the Greenland Seal (*Phoca grœnlandica*). The first and second deciduous incisors are already absorbed.

with sharp anterior and posterior edges, but without accessory cusps, except sometimes in the two hinder ones of the lower jaw. With the exception of the last one or two in the upper jaw and the last in the lower jaw they are all uniradicular. Vertebræ: C 7, D 15, L 5, S 4, C 14.

One species, *H. grypus*, the Gray Seal of the coasts of Scandinavia and the British Isles (see page 604.)

Phoca.[2]—Dental formula as the last. Teeth smaller and more pointed. Molars (Figs. 275 and 276) with two roots (except the first in each jaw); and their crowns with accessory cusps. Vertebræ: C 7, D 15, L 5, S 4, C 12-15. Head round and short. Fore feet short, with five very strong, subcompressed, slightly curved, rather sharp claws, subequal in length. On the hind feet the claws much narrower and less curved. The species of this genus are widely distributed throughout the northern hemisphere, and include *P. barbata*, the Bearded Seal; *P. grœnlandica*, the Greenland Seal; *P. vitulina*, the Common Seal (Fig. 277); and *P. hispida*, the Ringed Seal of the North Atlantic; *P. caspica*, from the Caspian and Aral Seas; and *P. sibirica*, from Lake Baikal.

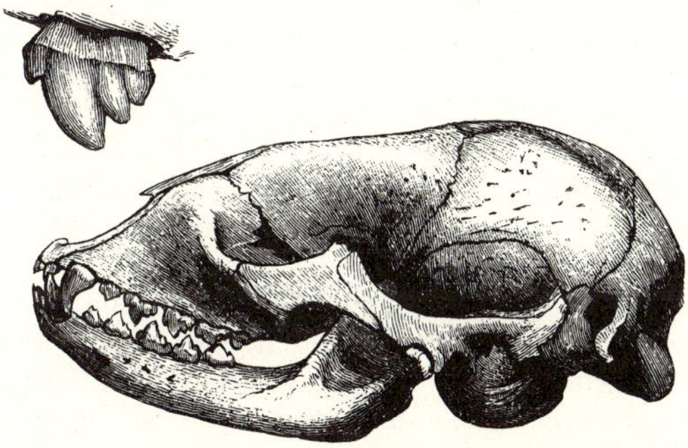

FIG. 276.—Skull of Common Seal, showing form of teeth.

[1] Nilsson, *Faun. Scandinav.* vol. i. p. 377 (1820).

[2] Linn. *Syst. Nat.* 12th ed. vol. i. p. 55 (1766).

Although the members of this subfamily swim and dive with the greatest ease, often remaining as much as a quarter of an hour or more below the surface, and are dependent for their sustenance entirely on living prey captured in the water, yet they frequently resort to sandy beaches, rocks, or ice-floes, either to sleep or to bask in the sun, and especially for the purpose of bringing forth their young. The latter appears to be the universal habit, and, strange as it may seem, the young seals—of some species at least—take to the water at first very reluctantly, and have actually to be taught to swim by their parents. The number of young produced is usually one annually, though occasionally two. They are at first covered with a coat of very thick, soft, nearly white fur, and until it falls off they do not usually enter the water. This occurs in the Greenland and Gray Seal when from two to three weeks old, but in the Common Seal apparently much earlier. One of this species born in the London Zoological Gardens had shed its infantile woolly coat and was swimming and diving about in its pond within three hours after its birth. The movements of the true Seals upon the ground or ice are very different from those of the Eared-Seals. Thus the hinder limbs (by which mainly they propel themselves through the water) are on land always perfectly passive, stretched backwards, with the soles of the feet applied to each other, and often raised to avoid contact with the ground. Sometimes the fore limbs are equally passive, being placed close to the sides of the body, and motion is then effected by a shuffling or wriggling action produced by the muscles of the trunk. When, however, there is any necessity for a more rapid mode of progression the animals use the fore paws, either alternately or simultaneously, pressing the palmar surface on the ground and lifting and dragging the body forwards in a succession of short jumps. In this way they manage to move so fast that a man has to step out beyond a walk to keep up with them; but such rapid action costs considerable effort, and they very soon become heated and exhausted. These various modes of progression appear to be common to all species so far as has been observed.

Most kinds of Seals are gregarious and congregate, especially at the breeding season, in immense herds. Such is the habit of the Greenland Seal (*Phoca grœnlandica*), which resorts in the spring to the ice-floes of the North Sea, around Jan Mayen Island, where about 200,000 are killed annually by the crews of the Scotch, Dutch, and Norwegian sealing vessels. Others, like the Common Seal of the British islands (*P. vitulina*), though having a wide geographical range, are never met with in such large numbers or far away from land. This species is stationary all the year round, but some have a regular season of migration, moving south in winter and north in summer. They are usually harmless, timid, inoffensive animals, though, being polygamous, the old males often fight desperately with each other, their skins being frequently found covered with wounds and scars. They are greatly attached to their young, and remarkably docile and easily trained when in captivity; indeed, although there would seem little in the structure or habits of the Seal to fit it by nature to be a companion of man, yet there is perhaps no wild animal which attaches itself so readily to the person who takes care of and feeds it. Seals appear to have much curiosity, and it is a very old and apparently well-

FIG. 277.—The Common Seal (*Phoca vitulina*).

attested observation that they are strongly attracted by musical sounds. Their sense of smell is very acute, and their voice varies from a harsh bark or grunt to a plaintive bleat. Seals feed chiefly on fish, of which they consume enormous quantities; some, however, subsist largely on crustaceans, especially species of *Gammarus*, which swarm in the northern seas, also on molluscs, echinoderms, and even occasionally sea-birds, which they seize when swimming or floating on the water.

Although the true Seals do not possess the beautiful under fur ("seal-skin" of the furriers) which makes the skin of the Sea-Bears so precious, yet their hides are still sufficiently valuable as articles of commerce, together with the oil yielded by their fat, to subject them to a devastating persecution, by which their numbers are being continually diminished.

Two species of seals only are met with regularly on the British coasts, the Common Seal and the Gray Seal. The former (Fig. 277) is a constant resident in all suitable localities round the Scottish, Irish, and English coasts, from which it has not been driven away by the molestations of man. Although, naturally, the most secluded and out-of-the-way spots are selected as their habitual dwelling-places, there are few localities where they may not be occasionally met with. Within the writers' knowledge one was seen not many years ago lying on the shingly beach at so populous a place as Brighton, and another was caught in the river Welland, near Stamford, 30 miles from the sea. They frequent bays, inlets, and estuaries, and are often seen on sandbanks or mudflats left dry at low tide, and, unlike some of their congeners, are not found on the ice-floes of the open sea, nor, though gregarious, are very large numbers ever seen in one spot. The young are produced at the end of May or beginning of June. They feed chiefly on fish, and the destruction they occasion among salmon is well known to Scottish fishermen. The Common Seal is widely distributed, being found not only on the European and American coasts bordering the Atlantic Ocean, but also in the North Pacific. It is from 4 to 5 feet in length, and variable in colour, though usually yellowish-gray, with irregular spots of dark brown or black above and yellowish-white beneath. The Gray Seal (*Halichœrus grypus*) is of considerably larger size, the males attaining when fully adult a length of 8 feet from nose to end of hind feet. It is of a yellowish-gray colour, lighter beneath, and with dark gray spots or blotches, but, like most other Seals, is liable to great variations of colour according to age. This species appears to be restricted to the North Atlantic, having been rarely seen on the American coasts, but not farther south than Nova Scotia; it is chiefly met with on the coasts of Ireland, England, Scotland, Norway, and Sweden, including the Baltic and Gulf of Bothnia, and Iceland, though it does not appear to range farther north. It is apparently not migratory, and its favourite breeding places are rocky islands; the young being born in the end of September or beginning of October.

Subfamily **Monachinæ**.—Incisors $\frac{2}{2}$. Cheek-teeth two-rooted, except the first. On the hind feet the first and fifth toes greatly exceeding the others in length, with nails rudimentary or absent.

Monachus.[1]—Dentition: $i\ \frac{2}{2}$, $c\ \frac{1}{1}$, $p\ \frac{4}{4}$, $m\ \frac{1}{1}$; total 32. Crowns of molars strong, conical, compressed, hollowed on the inner side, with a strongly marked lobed cingulum, especially on the inner side, and slightly developed accessory cusps before and behind. The first and last upper and the first lower molar considerably smaller than the others. Vertebræ: C 7, D 15, L 5, S 2, C 11. All the

[1] Fleming, *Philosophy of Zoology*, vol. ii. p. 187 (1822).

nails of both fore and hind feet very small and rudimentary. One species, *M. albiventer*, the Monk-Seal of the Mediterranean and adjacent parts of the Atlantic.

The other genera[1] of this section have the same dental formula, but are distinguished by the characters of the cheek-teeth and the feet. They are all inhabitants of the shores of the southern hemisphere.

Ogmorhinus.[2]—All the teeth of the cheek-series with three distinct pointed cusps, deeply separated from each other; of these the middle or principal cusp is largest and slightly recurved; the other two (anterior and posterior) are nearly equal in size, and have their apices directed towards the middle one. Skull much elongated. One species, *O. leptonyx*, the Sea-Leopard, widely distributed in the Antarctic and southern temperate seas.

Lobodon.[3]—Cheek-teeth with much-compressed elongated crowns and a principal recurved cusp, rounded and somewhat bulbous at the apex, and one anterior, and one, two, or three posterior, very distinct accessory cusps. One species, *L. carcinophaga.*

Pœcilophoca.[4]—Cheek-teeth small, with simple, subcompressed, conical crowns, having a broad cingulum, but no distinct accessory cusps. One species, *P. weddelli.*

Ommatophoca.[5]—All the teeth very small; those of the cheek-series with pointed recurved crowns, and small posterior and still less developed anterior accessory cusps. Orbits very large. Nails quite rudimentary on front, and absent on hind feet. The skull bears a considerable resemblance to that of the members of the next subfamily, towards which it may form a transition. There is one species, *O. rossi*, of which very little is known.

Subfamily **Cystophorinæ.**—Incisors $\frac{2}{1}$. Teeth of cheek-series generally one-rooted. Nose of males with an appendage capable of being inflated. First and fifth toes of hind feet greatly exceeding the others in length, with prolonged cutaneous lobes, and rudimentary or no nails.

Cystophora.[6]—Dentition: $i\ \frac{2}{1},\ c\ \frac{1}{1},\ p\ \frac{4}{4},\ m\ \frac{1}{1}$; total 30. The last molar has generally two distinct roots. Beneath the skin over the face of the adult male, and connected with the nostrils, is a sac which, when inflated, forms a kind of hood covering the

[1] For details of these and the other genera see Mivart, *Proc. Zool. Soc.* 1885, p. 486, *et seq.*

[2] Peters, *Monatsb. K. P. Akad. Wissensch. zu Berlin*, p. 393 (1875), substituted for *Stenorhynchus*, F. Cuvier; preoccupied for a genus of Crustacea.

[3] Gray, *Zoology of Erebus and Terror*, vol. i. p. 5 (1844).

[4] New name, Syn. *Leptonyx*, Gray, *Charlesworth's Mag. Nat. Hist.* vol. i. p. 582 (1837); preoccupied by Swainson, 1821.

[5] Gray, *Zoology of Erebus and Terror*, vol. i. p. 7 (1844).

[6] Nilsson, *Faun. Scandinav.* vol. i. p. 382 (1820).

upper part of the head. Nails present, though small, on the hind feet. One species, *C. cristata*, the Hooded or Bladder-Nose Seal of the Polar Seas.

Macrorhinus.[1]—Dentition as the last, but cheek-teeth of simpler character, and all one-rooted. All the teeth, except the canines, very small relatively to the size of the animal. Hind feet without nails. Vertebræ: C 7, D 15, L 5, S 3, C 11. Nose of adult male produced into a short tubular proboscis, ordinarily flaccid, but capable of dilatation and elongation under excitement. One species, *M. leoninus*, the Elephant Seal, or Sea-Elephant of the whalers, the largest of the whole family, attaining the length of nearly 20 feet. Formerly abundant in the Antarctic Seas, and also found on the coast of California.

Extinct Seals.—Remains of animals of this group have been found in late Miocene and Pliocene strata in Europe and America, the most abundant and best-preserved being those of the Pliocene Antwerp Crag, the subject of an illustrated monograph by Van Beneden. Nothing has, however, yet been discovered which throws any light upon the origin of the group, since all the extinct forms at present known come within the definition of the existing families; and, though annectant forms between these occur, there are as yet no transitions to a more generalised type of mammal. Indeed, all those of which the characters are best known belong to the completely developed Phocine or Trichechine, and not to the Otariine, type. The typical genus *Phoca* occurs in the Antwerp Crag, while remains of Seals provisionally referred to this genus are found in the Pliocene of the Crimea and the Miocene of Malta and Virginia. Of the other Antwerp forms *Callophoca* is said to be allied to *Phoca grœnlandica*, *Platyphoca* to *Phoca barbata*, *Phocanella* to *Phoca foetida*, *Gryphoca* to *Halichœrus*, *Palæophoca* and *Monatherium* to *Monachus*, and *Mesotaria* to *Cystophora*; while *Prophoca* does not appear to come very close to any existing form. It should be observed that it is extremely doubtful whether all these fossil Seals are really entitled to generic distinction.

Bibliography of Pinnipedia.—J. A. Allen, *History of North American Pinnipeds*, 1880; St. George Mivart, "Notes on the Pinnipedia," *Proc. Zool. Soc.* 1885, p. 484; P. J. Van Beneden, *Ossements fossiles d'Anvers*, in the *Mém. Acad. Roy. d. Belgique.*

Suborder CREODONTA.

The discovery of fossil remains in Eocene and early Miocene formations both in Europe and North America shows that numerous species of terrestrial carnivorous animals existed upon the earth during those periods which cannot be referred to either of the

[1] F. Cuvier, *Mém. du Muséum*, vol. xi. p. 200 (1824), "Macrorhine."

sections into which the order has now become broken up. By some zoologists these have been supposed to be Marsupials, or at least to show transitional characters between the Metatherian and Eutherian subclasses. By others they are looked upon as belonging altogether to the latter group, and as the common ancestors of existing Carnivores and Insectivores, or perhaps rather as descendants or relatives of such common ancestors, retaining more of the generalised characters than any of the existing species. They shade off almost insensibly into numerous other forms less distinctly carnivorous, to the whole of which, including the modern Insectivora, Cope (to whom we are indebted for much of our knowledge of the American extinct species) gives the name of BUNOTHERIA, those more specially related to the existing Carnivora forming the suborder

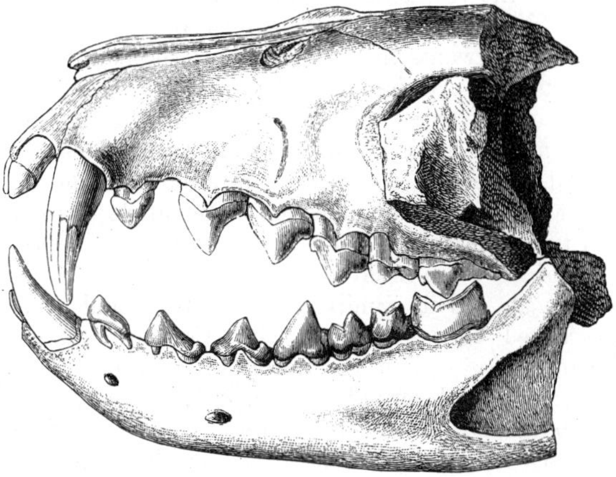

FIG. 278.—Anterior portion of the skull of *Hyænodon leptorhynchus*. (After Filhol.)

Creodonta. These are instances, however, in which the application of the principles of classification adopted in the case of existing species, of which the entire structure is known, and which have become divided into isolated groups by the extinction of intermediate forms, is almost impossible. If the generally accepted view of evolution is true, and the extreme modifications pass insensibly into each other by minute gradations (a view the palæontological proof of which becomes strengthened by every fresh discovery), there must be many of these extinct forms which cannot be assigned to definitely characterised groups. There are, however, some which stand out prominently from the others as formed on distinct types, having no exact representatives at present living on the earth.

The more typical Creodonts appear, however, to be so closely related to the true Carnivora through the extinct *Miacidæ* (p. 539), that it is on the whole advisable to regard them as representing a distinct suborder of Carnivora. In the strong development of the canines (Fig. 278) they are distinguished from the modern Insectivora; and they also differ from the latter and resemble the true Carnivores in the form of the incisors, the second one in the lower jaw (when three are present) being thrust up above the level of the other two in the manner obtaining in most of the modern Carnivora. Some of the most generalised forms included in the present group approximate so closely to the Condylarthrous Ungulates as to indicate that both groups have probably had a common origin.

The Creodonta as a whole are characterised by the small size of the brain, the absence of a single differentiated carnassial tooth, and the triangular form or secant character of their upper molars. In the carpus the scaphoid and lunar were usually distinct; the femur has a third trochanter; the upper or tibial surface of the astragalus usually wants the groove found in modern Carnivores; and the feet were plantigrade. The curious resemblance of the molars of many of these forms to those of the Marsupials may indicate a genetic relationship between the two groups; but, on the other hand, the presence of a full set of milk-teeth and the absence of palatal vacuities, or of an inflection of the angle of the mandible, sharply distinguishes them from that order. Space permits of a notice only of the more interesting forms.

Hyænodontidæ.—This family is taken to include some of the more specialised types, such as the European and American *Hyænodon* and *Oxhyæna* and the European *Pterodon.* In *Hyænodon* (Fig. 278) the dental formula is $i\ \frac{3}{3}$, $c\ \frac{1}{1}$, $p\ \frac{4}{4}$, $m\ \frac{2}{3}$; the fourth premolar above and the first true molar below being formed upon the "carnassial" plan, but the teeth behind these, instead of being tuberculated as in all existing Carnivora, repeat the characters of the carnassial, and also increase in size, especially in the lower jaw, from before backwards. The last lower molar differs from the two preceding teeth, and is very like the carnassial of *Felis.* The scaphoid and lunar of the carpus were fused together. Some species, as *H. leptorhynchus*, were as large as a Wolf, while others did not exceed a Fox in size. *Pterodon* is readily distinguished by having $m\ \frac{3}{3}$, by the larger size of the inner tubercles of the upper molars, and the similarity in the form of the three lower molars. In some species there were only two upper incisors, and the first lower premolar may be wanting. *Oxhyæna* is a specialised form with $i\ \frac{2-3}{0}$, $c\ \frac{1}{1}$, $p\ \frac{4}{4}$, $m\ \frac{2}{2}$, and a very long mandibular symphysis.

Proviverridæ.—The European and American genus *Proviverra* (*Cynohyænodon* or *Stypolophus*) may be regarded as representing a second family. The dental formula in this genus is the typical $i\ \frac{3}{3}$, $c\ \frac{1}{1}$, $p\ \frac{4}{4}$, $m\ \frac{3}{3}$, the upper molars have a large inner tubercle, while the lower molars are differentiated into a blade and talon, the blade having a large inner cusp. The upper teeth closely resemble the molars of *Dasyurus*, while the lower molars are like the lower carnassial of *Cynodictis* and *Viverra;* and thus indicate how the Creodonts may have passed into the true Carnivores through the extinct *Miacidæ*.

Arctocyonidæ and Mesonychidæ.—The first of these families is represented by *Arctocyon primævus*, one of the oldest known Tertiary mammals, from the lowest Eocene beds of La Fère, department of Aisne, France, and also by other species from corresponding beds at Rheims. The dental formula is $i\,\frac{3}{3}$, $c\,\frac{1}{1}$, $p\,\frac{?}{3-4}$, $m\,\frac{3}{3}$. The upper molars (Fig. 279) are tritubercular, with an incipient postero-internal column (hypocone); the lower are quadritubercular; and the premolars simple. The typical species was of large size, but the two of which the teeth are figured were considerably smaller. In the American *Mesonyx* the dental formula was the typical one, the jaws were comparatively short, the mandibular symphysis was elongated, the cheek-teeth were of simple structure, and resembled the premolars of many of the true Carnivora, and the astragalus had a grooved tibial surface and distinct distal facets for the cuboid and navicular, resembling in the latter respect the corresponding bone of a Perissodactyle Ungulate. The terminal phalanges had deeply fissured extremities, and are said to be more like those of Rodents than true Carnivores. *Mesonyx ossifragus* was larger than a Grizzly Bear. *Amblyctonus*, of the same deposits, differs by the smooth tibial face of the astragalus and the development of an anterior cusp to the lower molars.

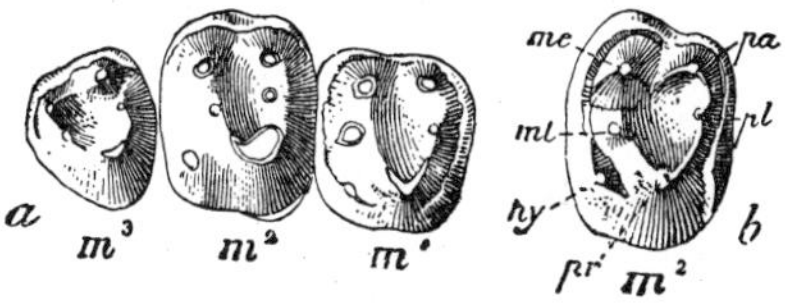

FIG. 279.—The three right upper molars of *Arctocyon dueli* (*a*), and the second of *A. gervaisi* (*b*); from the Lowest Eocene of Rheims. *pr*, protocone; *pa*, paracone; *me*, metacone; *hy*, hypocone; *ml*, metaconule; *pl*, paraconule. (From Osborn.)

CHAPTER XII

THE ORDER INSECTIVORA

THE Insectivora comprise a number of comparatively small mammals, generally of terrestrial, although rarely of arboreal or aquatic habits, and presenting the following common features. They are unguiculate, and have plantigrade or subplantigrade, and generally pentadactylate feet, in which the pollex and hallux are not opposable to the other digits. They are diphyodont and heterodont, and the teeth are rooted. The molars are studded with sharp cusps, the crowns of the upper molars being either quadrangular or triangular; there are never less than two incisors in either side of the mandible; and in many cases the incisors, canines, and anterior premolars are not clearly differentiated from one another (Fig. 280); the canines being usually weak. Clavicles are present, except in *Potamogale*. The body is clothed with fur or protected by an armature of spines; the testes are inguinal or placed near the kidneys, and are not received into a scrotum; the penis is pendent or suspended from the wall of the abdomen; the uterus is two-horned and with or without a distinct corpus uteri; the placenta is discoidal and deciduate; and the smooth cerebral hemispheres do not extend backwards over the cerebellum (Fig. 281). The projection of the muzzle far beyond the extremity of the lower jaw is a very general feature. The humerus generally has an entepicondylar

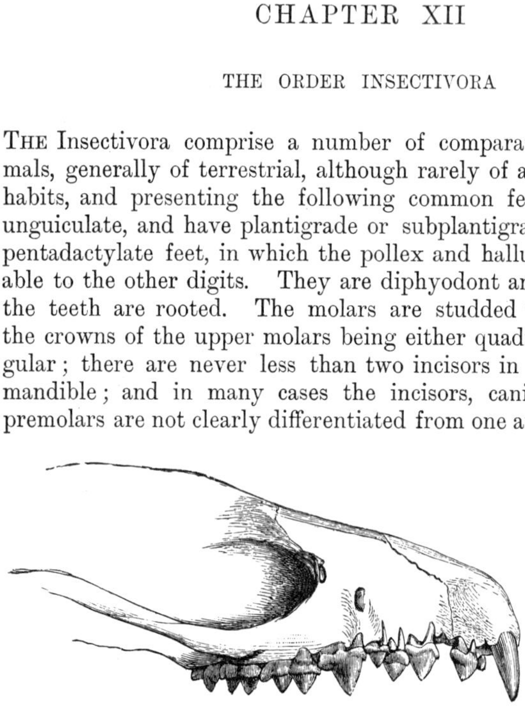

FIG. 280.—Right lateral aspect of the anterior portion of the cranium of *Erinaceus collaris*. Enlarged. (From Dobson, *Proc. Zool. Soc.* 1881, p. 403.)

foramen. Certain forms, such as *Talpa* and *Galeopithecus*, are unique among mammals in having ossified intercentra in the dorso-lumbar region of the vertebral column.

Representatives of this order are found throughout the temperate and tropical parts of both hemispheres (except South America and Australia), and exhibit much variety both in organisation and in habits. With the exception of the *Tupaiidæ*, all are nocturnal; the greater number are cursorial, but some (*Talpa*, *Chrysochloris*, *Oryzorictes*) are fossorial; some (*Potamogale*, *Nectogale*, *Myogale*) are natatorial, and a few (*Tupaiidæ*) arboreal; while the species of the aberrant genus *Galeopithecus* glide through the air like the Flying Squirrels. To the great majority the term insectivorous is strictly applicable, *Galeopithecus* alone being phytophagous; while *Potamogale* is said to feed on fish, and the different species of Moles live chiefly on worms. The general organisation of the Insectivora indicates a very low type, and were it not for the specialised character of their placentation and the tendency to lose the differentiated characters of the anterior teeth they might be regarded as closely allied to the ancestral type of many of the heterodont mammals. The strongly marked distinction of the canines from the incisors and anterior premolars in the Mesozoic and most of the Tertiary mammals (excepting some of the Ungulates) points, however, very decidedly to the conclusion that the want of definition between these teeth in many of the modern Insectivora is an acquired feature. Fossil forms apparently indicate a relationship on the one hand with the Creodont Carnivora, and on the other with the Lemuroid Primates; indeed it is in some instances impossible to say whether extinct genera are really Insectivores or Lemuroids.

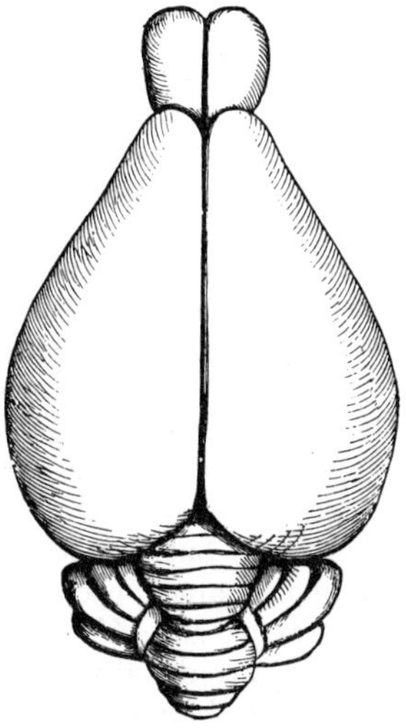

FIG. 281.—Upper surface of the brain of *Tupaia ferruginea*. (From Garrod, *Proc. Zool. Soc.* 1879, p. 304.)

In most Insectivora the cranial cavity is of small relative size, and in none is the brain-case elevated to any considerable extent above the facial line. The facial part of the skull is generally much produced, and the premaxillary and nasal bones are well developed. The zygomatic arch is usually slender or deficient, the latter being the case in most of the species; and postorbital processes of the frontals are found only in the *Galeopithecidæ*, *Tupaiidæ*, and *Macroscelididæ*. The number of dorsal vertebræ varies from 13 in *Talpa* to 19 in *Centetes;* that of the lumbar from 3 in *Chryso-*

chloris to 6 in *Talpa* and *Sorex*; and of the caudal from the rudimentary series of 8 in *Centetes* to the 40 or more of *Microgale*. Not less variable are the characters of the vertebræ themselves; the spinous processes often being very long in one and short in another species of the same genus. In the *Soricidæ* and *Myogale* the neural arches of the cervical vertebræ are very slender. In the *Soricidæ* and *Gymnura* the four anterior vertebræ develop large single hypapophyses. In *Galeopithecus* the centrum of each vertebra supports posteriorly a pair of intercentral ossifications; while in *Erinaceus*, *Myogale*, and *Talpa* small oval ossicles are found on the inferior surfaces of the lumbar interspaces. In *Erinaceus*, owing to the thickness of the neural cord in the cervical region and its abrupt termination, the diameter of the neural canal in the cervical and first two dorsal vertebræ greatly exceeds that of any of the succeeding vertebræ. The sternum is variable, but generally narrow, bilobate in front, and divided into segments. The pectoral girdle presents some remarkable adaptive modifications, most fully expressed in *Talpa*, having relation to the use of the fore limbs in burrowing; but in the Golden Moles (*Chrysochloris*) the forearm and manus alone become specially modified for this purpose. In *Galeopithecus* and *Macroscelides* the bones of the forearm (radius and ulna) are distally united. The manus has generally five digits, but in *Rhynchocyon* and in one species of *Oryzorictes* the pollex is wanting, while in the true Moles it is extremely modified. The femur has, in most species, a prominent ridge below the greater trochanter representing a third trochanter. In *Galeopithecus*, *Tupaia*, *Centetes*, *Hemicentetes*, *Ericulus*, and *Solenodon* the tibia and fibula are distinct, but in all the other genera more or less united together. The pes usually possesses five digits (rarely four by reduction of the hallux); and in some forms, as in the leaping species (*Macroscelides*, *Rhynchocyon*), the tarsal bones are greatly elongated. The form of the pelvis, and especially of the symphysis pubis, varies within certain limits; and these differences have been proposed by Leche as a basis for the classification of the families. Thus in the *Galeopithecidæ*, *Tupaiidæ*, and *Macroscelididæ* there is a long symphysis; in the *Erinaceidæ*, *Centetidæ*, and *Potamogalidæ* the symphysis is short; and in the *Soricidæ*, *Talpidæ*, and *Chrysochloridæ* there is none.

Space does not admit of attempting a sketch of the modifications of the muscular system, which will be found fully described in Dr. Dobson's *Monograph*, referred to in the bibliography. As to the nervous system, it has been already mentioned that the brain throughout the order presents a low type of organisation; in none of the members do the cerebral hemispheres present any trace of convolutions, nor do they extend backwards so as to cover the cerebellum, while the olfactory lobes are large and project in front, and the corpus callosum is short and thin. In the Hedgehogs (*Erinaceus*) the spinal column ends abruptly opposite the third or fourth dorsal vertebra in a slender filament, and the dorsal and lumbar nerves, given off in front of this point, are carried backwards in two compressed bundles occupying the suddenly narrowed spinal canal as far as the sacrum.

Owing to the similarity in the character of the food, the truly insectivorous species, forming more than nine-tenths of the order, present little variety in the structure of their digestive organs. Except in *Galeopithecus* the stomach is a simple, thin-walled sac; but in some, as in *Centetes* and allied genera, the pyloric and œsophageal openings are very close together. The intestinal canal has much the same calibre throughout, and varies from three (in the Shrews) to twelve times (in the Hedgehogs) the length of the head and body. In the arboreal genera, *Galeopithecus* and *Tupaia*, as well as in the *Macroscelididæ*, all of which probably feed in part on vegetable substances, most of the species possess a cæcum. The liver is deeply divided into lobes, the right and left lateral being cut off by deep fissures; and both the caudate and Spigelian lobes being generally well developed. The gall-bladder, which is usually large and globular, is placed on the middle of the posterior surface of the right central lobe.

In most of the members of the order (*Soricidæ*, *Centetidæ*, *Chrysochloridæ*) the penis is capable of being more or less completely retracted within the fold of integument surrounding the anus; in some (*Galeopithecidæ*, *Talpidæ*) it is pendent in front of the anus; while in others (*Macroscelididæ*, *Erinaceidæ*, *Solenodontidæ*) it is carried forwards and suspended from the abdominal wall. In the subfamily *Centetinæ* and *Chrysochloris* the testes lie immediately behind the kidneys, but in others more or less within the pelvis. During the rutting season they become greatly enlarged, forming protrusions in the inguinal region. Except in *Rhynchocyon* the uterine cornua are long and open into a short corpus uteri, which in many species (*Soricidæ*, *Talpidæ*, *Centetidæ*, *Chrysochloridæ*) is not separated from the vagina by a distinct os uteri. With the exception of *Galeopithecus* all Insectivora appear to be multiparous, the number of young at a birth varying from two to eight in *Erinaceus*, and from twelve to twenty in *Centetes*. The position of the mammary glands and the number of the teats vary greatly. Thus in *Galeopithecus* there are two pairs of axillary teats, and in *Solenodon* a single post-inguinal pair; but in most species they range from the thorax to the abdomen, varying from two pairs in *Gymnura* to twelve in *Centetes*. In *Chrysochloris* the thoracic and inguinal teats are lodged in deep cup-shaped depressions.

Odoriferous glands exist in many species. In most Shrews these glands occur on the sides of the body at a short distance behind the axilla, and their exudation is probably protective, since few carnivorous animals will eat the dead bodies of these creatures. In both species of *Gymnura* and in *Potamogale* large pouches are situated on either side of the rectum and discharge their secretions by ducts, opening in the first-named genus in front of, and in the latter within the margin of the anus. In *Centetes* the ducts of similarly situated racemose glands open by pores at the bottom of deep pits placed at either side of the anus.

The integument is thin, but in many species is lined by a muscular coat, which is probably more developed in the Hedgehogs (*Erinaceidæ*) than in any other mammal. In this family and the *Centetidæ* most of the species are protected by spines implanted in the panniculus carnosus muscle, and more or less replacing the fur of the upper surface of the body.

The order is usually divided into two suborders, but the very aberrant genus which constitutes the first might well be raised to ordinal rank. It has little in common with the true Insectivora, but as it certainly belongs to no other of the recognised mammalian orders it is retained among them chiefly to avoid the inconvenience of increasing the number of ordinal divisions for the sake of a single isolated form.

Suborder DERMOPTERA.

Upper and lower incisors compressed, multicuspidate, the lower deeply pectinated; fore and hind limbs connected by a broad integumentary expansion forming a parachute.

Family GALEOPITHECIDÆ.

In addition to the characters given under the head of the sub-order it may be mentioned that the orbit is nearly surrounded by bone, the zygomatic arches are well developed, the tympanic forms a bulla, the ulna is distally united with the radius, the tibia and fibula are distinct, the pubic symphysis is long, the penis is pendent, the testes are received into inguinal pouches, the mammæ are axillary, the uterus is two-horned, and there is a large cæcum.

Galeopithecus[1]—Dentition: $i\,\frac{2}{3}$, $c\,\frac{1}{1}$, $p\,\frac{2}{2}$, $m\,\frac{3}{3}$; total 34. Second upper incisor and canine with two roots. Two species—*G. volans* and *G. philippinensis*. The former, which is distinguished from the latter by the form of the upper incisors, has a total length of nearly 2 feet. The long and slender limbs are connected by a broad integumentary expansion extending outwards from the sides of the neck and body, and forming also a web between the fingers and toes as far as the base of the claws (Fig. 282); the hind limbs are

[1] Pallas, *Acta Acad. Sci. Imp. Petropolis*, vol. iv. pt. 1, p. 208 (1780).

further connected by a similar expansion passing outwards along the back of the feet to the base of the claws, and, inwardly, involving the long tail to the tip, forming a true interfemoral membrane, as in the Bats.

The two species of Flying Lemurs, as the representatives of this genus are commonly but erroneously called, live in the forests of the

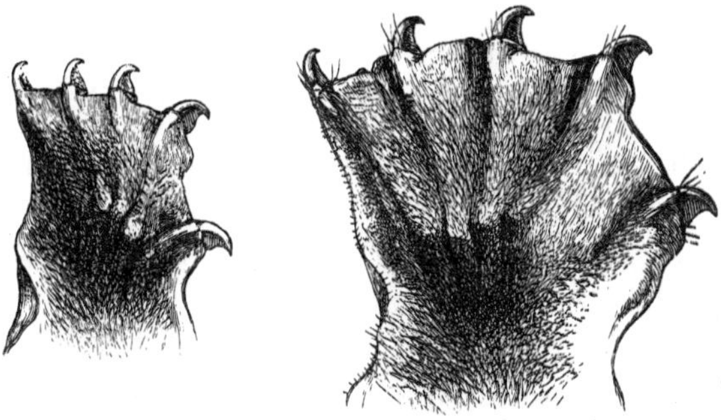

FIG. 282.—Feet of *Galeopithecus philippinensis*.

Malay Peninsula, Sumatra, Borneo, and the Philippine Islands, where they feed chiefly on the leaves and fruits of trees. Their habits are nocturnal, and during the daytime they cling to the trunks or limbs of trees, head downwards, in a state of repose. With the approach of night their season of activity commences, when they may be seen gliding from tree to tree supported on their cutaneous parachute, and they have been observed to traverse in this way a space of 70 yards with a descent of only about one in five.

Galeopithecus was referred by some of the older zoologists and anatomists to the Bats, and by others to the Lemurs, but Professor Peters's view, that it belongs to neither of these orders, and should be considered an aberrant Insectivore, has been very generally accepted, although, as mentioned above, the association is by no means a close one. Besides differing from the Bats in the form of the anterior limbs and of the double-rooted outer incisor and canine, it also contrasts strongly with them in the presence of a large sacculated cæcum, and in the great length of the colon, which is so remarkably short in all the Chiroptera. From the Lemurs, on the other hand, the form of the brain, the characters of the teeth, the structure of the skull, and the deciduate discoidal placenta completely separate it. In a recent elaborate memoir on the myology and affinities of *Galeopithecus* Dr. Leche [1] considers that we have in this genus an indication of the mode in which the Insectivora were modified into the Chiroptera, although it is completely off the direct

[1] *Ueber die Säugethiergattung Galeopithecus. Sv. Ak. Handl.* vol. xxi. pt. xi. (1886).

line of descent. The deeply pectinated crowns of the lower incisors of *Galeopithecus* are quite unique in the class, and the only approach to the double-rooted canine, except in *Erinaceus* and *Talpa*, is found among the Marsupials in *Perameles*, where the root of the canine is grooved.

Suborder INSECTIVORA VERA.

Upper and lower incisors conical, unicuspidate or with basal cusps only, the lower not pectinated; limbs free, formed for terrestrial progression.

The following table gives a key to the distinctive characters of the existing families:—

I. Upper molars broad, multicuspidate, with more or less well-defined W-shaped crowns.
  A. Symphysis pubis long; generally a cæcum; cerebral cavity comparatively large.
    a. Orbit encircled by bone; metatarsus moderate; arboreal. *Tupaiidæ.*
    b. Orbit not encircled by bone; metatarsus greatly elongated; terrestrial. *Macroscelididæ.*
  B. Symphysis pubis short or none; no cæcum; cerebral cavity small; skull without postorbital processes.
    a. First and second upper molars with a central fifth cusp.
      a′. Tympanic annular, not forming a bulla. *Erinaceidæ.*
    b. No central fifth cusp to upper molars.
      a′. Tympanic annular, not forming a bulla; no zygomatic arch. *Soricidæ.*
      b′. Tympanic forming a bulla; zygomatic arch developed. *Talpidæ.*
II. Upper molars narrow, with V-shaped crowns.
      a.′ Tympanic annular, not forming a bulla; zygomatic arch imperfect.
        a″. No clavicles. *Potamogalidæ.*
        b″. Clavicles well developed.
          a‴. Skull constricted between the orbits; penis suspended. *Solenodontidæ.*
          b‴. Skull not constricted; penis pendent, retractile. *Centetidæ.*
      b′. Tympanic forming a bulla; zygomatic arch well developed. *Chrysochloridæ.*

The second section, in which the molars are of the primitive tritubercular type, should probably be regarded as containing the most generalised representatives of the order; and it is noteworthy that the whole of them are confined to Africa, Madagascar, and the West Indies, whereas most of the first section are widely distributed over the Palæarctic and Oriental regions. None of the existing families of the second section are known in a fossil condition, although it is suggested that the extinct *Leptictidæ* includes allied types.

Family TUPAIIDÆ.

Skull with comparatively large brain-case, orbit surrounded by bone, well-developed zygomatic arch, perforated jugal, and a tympanic

FIG. 283.—The Pentailed Tree-Shrew (*Ptilocercus lowi*). From Gray, *Proc. Zool. Soc.* 1848. $\frac{1}{2}$ natural size.

bulla. Upper molar broad, with cusps arranged in a W. Pubic symphysis long; radius and ulna, and tibia and fibula separate; metatarsus only slightly longer than tarsus. Usually a short cæcum. Habits arboreal and diurnal. Confined to the Oriental region.

Tupaia.[1]—Dentition: $i\ \frac{2}{3}$, $c\ \frac{1}{1}$, $p\ \frac{3}{3}$, $m\ \frac{3}{3}$; total 38. Feet naked beneath, the sole furnished with projecting pads; claws moderate, curved, and sharp; head pointed; ears rounded; tail bushy, distichous, with short hair below. The Tree-Shrews, of which there are some nine species, are found in India, Burma, the Malay Peninsula, the Nicobars, Sumatra, Java, and Borneo. The species closely resemble one another, differing chiefly in size and in the

[1] Raffles, *Trans. Linn. Soc.* vol. xiii. p. 256 (1822).

colour and length of the fur. Their general appearance is very Squirrel-like. Their food consists of insects and fruit, which they usually seek in the trees, but also occasionally on the ground. When feeding they often sit on their haunches, holding the food, after the manner of Squirrels, between their fore-paws.

Ptilocercus.[1]—Represented only by the Pentailed Tree-Shrew (*P. lowi*, Fig. 283) of Borneo, in which the tail is of extraordinary length, with the proximal two-thirds naked, and the remaining third furnished with a bilateral fringe of long hairs, from which the genus takes its name.

Extinct Genera.—An Insectivore from the Middle Miocene of France, described as *Lantanotherium*, is said to be nearly allied to *Tupaia*. The genus *Parasorex*, from strata of similar age, has the dental formula $i\ \frac{3}{3}$, $c\ \frac{1}{1}$, $p\ \frac{4}{4}$, $m\ \frac{3}{3}$, and is regarded as connecting the present with the following family.

Family MACROSCELIDIDÆ.

Skull with comparatively large brain-case, strong zygomatic arch, a tympanic bulla, orbit surrounded by bone, imperforate jugal, and usually no postorbital process. Molars broad, with four cusps arranged in a W. Pubic symphysis long; proximal end of tibia and fibula united; radius and ulna united or separate; metatarsus much longer than tarsus. A large cæcum. Habits terrestrial, saltatorial, and nocturnal. The family is confined to Africa.

Macroscelides.[2]—Dentition: $i\ \frac{3}{3}$, $c\ \frac{1}{1}$, $p\ \frac{4}{4}$, $m\ \frac{2}{2-3}$; total 40 or 42. Distal extremity of radius and ulna united. Five digits in manus, and five or four in pes. This genus, which is taken to include *Petrodromus*, comprises ten species widely distributed throughout the African continent. All are closely related, resembling one another in general form, and even in the colour of the fur. They fall into two groups, distinguished by the presence or absence of a small third lower molar.[3] *M. tetradactylus* (Fig. 284), the type of the genus *Petrodromus*, differs from all the other species in the absence of the hallux, and of the third lower molar. These animals are commonly known as Jumping Shrews, and, like the following genus, have the muzzle much produced.

Rhynchocyon.[4]—Dentition: $i\ \frac{1}{3}$, $c\ \frac{1}{1}$, $p\ \frac{4}{4}$, $m\ \frac{2}{2}$; total 36. Upper incisor frequently shed in the adult. Radius and ulna distinct;

[1] Gray, *Proc. Zool. Soc.* 1848, p. 23.

[2] Andrew Smith, *S. African Quart. Journ.* vol. ii. No. 1, p. 64 (1833).

[3] The above correct formula of the dentition of this family has been recently worked out by O. Thomas, *Proc. Zool. Soc.* 1890, pp. 445, 446.

[4] Peters, *Bericht k. preuss. Ak. Wiss.* 1847, p. 36.

hind limbs relatively shorter, and proboscis longer than in the type genus; four digits in each foot. Four closely allied species have been described from East Africa. The head and body of the type

FIG. 284.—*Macroscelides tetradactylus.* ×½. (From Peters, *Reise nach Mossambique.*)

species measures about 8 inches in length; and the long tail is covered with a ringed skin, sparsely haired. Its habits are fossorial.

Family ERINACEIDÆ.

Skull with a small brain-case; no postorbital process; slender and occasionally imperfect zygomatic arch, and an annular tympanic, which does not form a bulla. Upper molars with four principal cusps and a small central median cusp. Acromion of scapula bifid; pubic symphysis short; radius and ulna free, but tibia and fibula united proximally. No cæcum; penis carried forward and suspended from the wall of the abdomen. Habits terrestrial. Found in the Palæarctic, Ethiopian, and Oriental regions.

Subfamily **Gymnurinæ**.—Palate completely ossified; pelvis very narrow; fur without spines.

Gymnura.[1]—Dentition: $i\ \frac{3}{3}$, $c\ \frac{1}{1}$, $p\ \frac{4}{4}$, $m\ \frac{3}{3}$; total 44. This genus, if *Hylomys* is rightly included, is represented by the two

[1] Horsfield and Vigors, *Zool. Journ.* vol. iii. p. 246 (1828).

species, *G. rafflesi* and *G. suilla*, from the Malay Penisula and Indian Archipelago. The former has the appearance of a large Rat with a long tail and head and projecting mobile snout; the latter, which is much smaller, with a short tail and small third upper premolar, has long been known under the name of *Hylomys suillus*, and classed with the *Tupaiidæ*. Both species present a very generalised type of dentition, in this respect occupying an almost central position in the order. *G. suilla* is represented in Mount Kina-Balu, Borneo, by a variety characterised by the presence of a dark dorsal streak. Many zoologists prefer to retain *Hylomys* as a distinct genus.

Subfamily **Erinaceinæ**.—Palate imperfectly ossified; pelvis wide; fur with spines.

Erinaceus.[1]—Dentition: $i\,\frac{3}{2}$, $c\,\frac{1}{1}$, $p\,\frac{3}{2}$, $m\,\frac{3}{3}$; total 36. The first pair of upper incisors (Fig. 285) are considerably larger than the others, and are widely separated from one another in the middle line; the canine is very similar to the third incisor; and, except in *E. europæus* (Fig. 285), each of these teeth is inserted by two distinct roots (Fig. 280, p. 610). The first lower incisor is large and proclivous. The number of vertebræ is C 7, D 15, L 6, S 3, C 11.

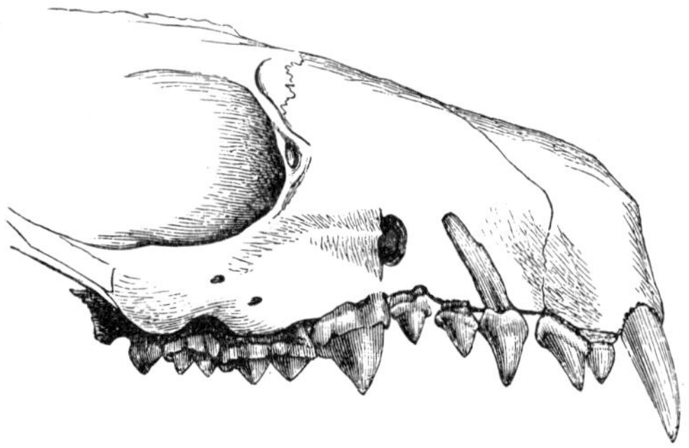

FIG. 285.—Right lateral aspect of the anterior portion of the skull of the Hedgehog (*Erinaceus europæus*). Enlarged. (From Dobson, *Proc. Zool. Soc.* 1881, p. 403.)

The Hedgehogs comprise nearly twenty species, distributed throughout Europe, Africa, and the greater part of Asia, but not found in Madagascar, Ceylon, Burma, Siam, the Malay Peninsula, or Australia. All the species resemble one another in the armature of spines investing the upper surface and sides of the body; and all possess the power of rolling themselves up into the form of a ball, protected on all sides by the strong spines; the dorsal integument being brought downwards and inwards over the head and tail, so as to include the limbs also, by the action of special muscles. The common Hedghog (*E. europæus*) is the most aberrant species, differing from all the rest in the peculiarly shaped and single-rooted third upper incisor and canine (Fig. 285), and in its very coarse, harsh fur. The dentition of the long-eared North Indian form, *E. collaris* (Fig. 280), may be considered characteristic of all the other species, the only important differences being found

[1] Linn. *Syst. Nat.* 12th ed. vol. i. p. 75 (1766).

in the variable size and position of the second upper premolar, which is very small, external, and deciduous in the Indian *E. micropus* and *pictus*. The former species, limited to South India, is further distinguished by the absence of the jugal bone. Of the African species, *E. diadematus*, with long frontal spines, is probably the commonest; while *E. albiventris* has been made the type of a separate genus on account of the total absence of the hallux.

The well-known European species feeds on insects, worms, slugs, mice, rats, lizards, snakes, etc., as well as on eggs, fruit, and roots. It hibernates during the winter. The young are usually produced in July or August in litters of not more than four, but there may be a second litter in October; and the period of gestation is believed not to exceed a month. The Indian, and probably also the African species, do not hibernate.

The existing *E. europæus* dates from the Pleistocene period, and extinct species of the genus are found in the Upper and Middle Miocene of the Continent.

Extinct Genera.—The French Lower Miocene genus, *Palæoerinaceus*, appears to be allied to *Erinaceus*, but is distinguished by the wider and completely ossified palate. In the Upper Eocene of Central France there are two genera, which appear to be most nearly allied to *Gymnura*, although connected by *Palæoerinaceus* with *Erinaceus*. Of these *Necrogymnurus*,[1] with which *Cayluxotherium* is apparently identical, has teeth like *Gymnura*, but an imperfectly ossified palate like *Erinaceus*; and the skull is remarkable for the peculiar rugose structure of the parietal and temporal regions. *Comphotherium* is distinguished by the presence of a cingulum to the lower molars, like that found in *Gymnura*.

Family SORICIDÆ.

Skull (Fig. 286) long and narrow, with no zygomatic arch or postorbital process, and the tympanic ring-like and not forming a bulla. Upper molars with the cusps arranged in a distinct W. No pubic symphysis. The tibia and fibula united. No cæcum. Habits usually terrestrial, rarely aquatic. Distribution extensive.

The Shrews are Rat-like or Mouse-like insectivores, with the body covered with hair, and the muzzle long and pointed. Their dentition (Fig. 286) is peculiar and characteristic. Thus the first upper incisor is large and hook-like, with a more or less developed basal cusp on the posterior border. Between this and the last premolar there are a variable number of small teeth, representing the other incisors, the canine, and the anterior premolars; although, owing to the early obliteration of the maxillo-premaxillary suture,

[1] Originally given incorrectly as *Neurogymnurus*.

their homology is exceedingly difficult to determine. Three molars are invariably present, of which the third is much the smallest. In the mandible there are always six teeth, but in one species of *Myosorex* there may be a seventh. The first lower incisor is usually directed horizontally forwards; the second incisor (formerly reckoned as the canine) is the smallest tooth of the series, the fourth premolar being slightly larger.

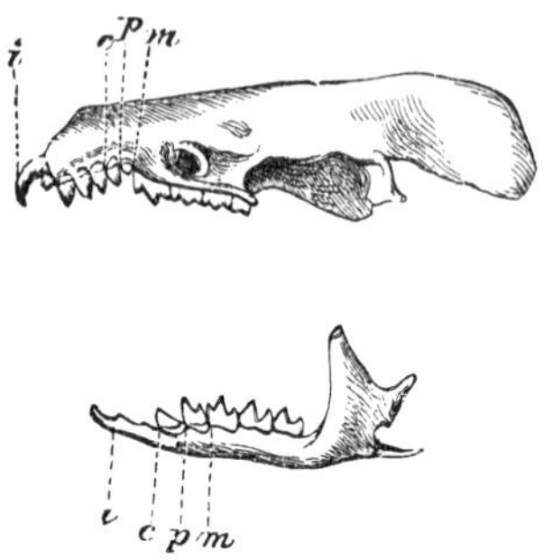

FIG. 286.—Left lateral view of the cranium and mandible of *Sorex verœpacis.* In the cranium—*i*, first incisor; *c*, fourth incisor; *p*, canine; *m*, fourth premolar: in the mandible—*i*, first incisor; *c*, second incisor; *p*, fourth premolar; *m*, first molar. (From Alston, *Proc. Zool. Soc.* 1877.)

This family, which includes considerably more than half the representatives of the order, has a distribution coextensive with the latter. Many classifications of this difficult group have been attempted, but according to the latest proposal of Dr. Dobson,[1] the genera may be divided into two subfamilies, distinguished by the apparently trivial character of the colour of the teeth.

Subfamily **Soricinæ.**—Summits of the teeth coloured red.

Sorex.[2]—Dentition: $i\ \frac{4}{2}$, $c\ \frac{1}{0}$, $p\ \frac{2}{1}$, $m\ \frac{3}{3}$; total 32. Openings of male and female generative organs separated from the anal orifice; penis cylindrical or tapering; ear well developed; tail long, covered with equal or subequal hairs.

It has been shown by Brandt that the position of the premaxillo-maxillary sutures in the type of the genus is between the fourth and fifth tooth, so that it appears that we must regard this genus as differing from all other Eutherian mammals in having four upper incisors. Dr. Dobson, in his paper quoted, classes the tooth here reckoned as the upper canine with the premolar series in all the Shrews. Habits terrestrial. Species numerous, inhabiting the Palæarctic and Nearctic regions.

Of the two species found in the British Isles the Common Shrew (*S. vulgaris*, Fig. 287) is by far the most common in England, and is about the size of the House Mouse, to which it approximates in general form. The body is clothed with close long fur, very soft and dense, and varying in colour from light reddish to dark brown above, rarely speckled or banded with white. The under surface of both the body and the tail is grayish. The basal four-fifths of all the hairs above and beneath are dark bluish-gray; the hairs of the tail are less densely set and coarser. On each side of the body, at a point about one-third of the distance between the elbow and the knee, may be found, especially in the rutting season,

[1] *Proc. Zool. Soc.* 1890, p. 49. [2] Linn. *Syst. Nat.* 12th ed. vol. i. p. 73 (1766).

a gland covered by two rows of coarse hairs. This secretes a peculiar fluid, on which the odour of the animal depends; this odour being evidently protective, and rendering the creature secure against the attacks of many predaceous animals.

The geographical range of the Common Shrew is exceedingly wide, extending eastwards through Europe and Asia (north of the Himalayas) to North America.

The Lesser Shrew (*S. pygmæus*[1]) is far less common in England and Scotland, although more abundant in Ireland, where *S. vulgaris* is unknown. It is distinguished from the latter not only by its inferior dimensions, but also by the circumstance that the third upper incisor is not longer than the fourth, and by the considerably shorter length of the forearm and manus. This species extends through Europe and Asia as far as the island of Saghalin. Both

FIG. 287.—The Common Shrew (*Sorex vulgaris*).

this and the preceding species generally live in wooded districts, making their nests under the roots of trees, or in slight hollows. The great mortality noticeable among the Shrews in the early part of the autumn is probably due to insufficiency of food. The breeding season extends from the latter part of April to the beginning of August. The young, which are blind, naked, and toothless at birth, are very quickly developed. The number in a litter is usually from five to seven, but may be as many as ten.

The Alpine Shrew (*S. alpinus*), which is restricted to the Alpine region of Central Europe, is slightly larger than the common species, from which it is distinguished by the longer tail, the length of which exceeds that of the head and body, by the fur being dark on both surfaces of the body, and also by the larger size of the upper canine.

In North America *S. bendirei* is by far the largest species of the genus; and, as in many other species of the same country, the fourth upper incisor is relatively small. In *S. hoyi* (separated by

[1] Syn. *S. minutus.*

some writers as *Microsorex*), of the same country, this tooth is rudimentary.

Other North American Shrews, which are regarded by some zoologists as generically distinct under the name of *Neosorex*, are aquatic, and thus take the place of the Old World genus *Crossopus*. These are *S. palustris* of the Rocky Mountains and *S. hydrodromus* of Unalaska Island, both of which resemble *Crossopus* in having the feet provided with swimming fringes, but agree with the other species of *Sorex* in their dentition and the character of the tail. The former species is about the size of *Crossopus fodiens*, while the latter is scarcely larger than *S. pygmæus*.

Soriculus.[1]—Dentition : $i\ \frac{4}{2}$, $c\ \frac{1}{0}$, $p\ \frac{1-2}{1}$, $m\ \frac{3}{3}$; total 30, or rarely 32. Opening of male or female generative organs forming with the anal orifice a shallow cloaca. Ear and tail as in *Sorex*. First upper incisor with an internal cusp. Habits terrestrial.

This genus is the only representative in the Oriental region of the *Soricinæ*, which are otherwise confined to the Palæarctic and Nearctic regions. The Indian and Burmese species comprise *S. nigrescens*, *S. caudatus*, and *S. macrurus*.

Notiosorex.[2]—Dentition : $i\ \frac{3}{2}$, $c\ \frac{1}{0}$, $p\ \frac{1}{1}$, $m\ \frac{3}{3}$; total 28. Tail moderate; first upper incisor without an inner cusp; other characters as in *Soriculus*. Habits terrestrial.

This American genus is represented by *S. crawfordi* and *S. evotis*, which are found in Central America and Mexico, and are thus some of the most southerly representatives of the Shrews in that continent. Their external appearance is very similar to that of the Old World genus *Crocidura*.

Blarina.[3]—Dentition : $i\ \frac{4-3}{2}$, $c\ \frac{1}{0}$, $p\ \frac{2}{1}$, $m\ \frac{3}{3}$; total 32 or 30. Ear truncated above; tail short; otherwise as in *Soriculus*. This group of so-called Earless or Short-tailed Shrews is mainly North American, the common forms being *B. dekayi* and *B. brevicauda*. The species vary considerably in size; and *B. mexicana* and *micrura* extend the range of the genus into Mexico and Guatemala. The following account of the habits of *B. brevicauda* is taken from Dr. Merriam's *Mammals of the Adirondack Region*: "The rigours of our northern winters seem to have no effect in diminishing its activity, for it scampers about on the snow during the severest weather, and I have known it to be out when the thermometer indicated a temperature of − 20° Fahr. It makes long journeys over the snow, burrowing down whenever it comes to an elevation that denotes the presence of a log or stump, and I am inclined to believe that at this season it must feed largely upon

[1] Blyth, *Journ. As. Soc. Bengal*, vol. xxiv. p. 36 (1855). [2] Coues, *Bull. U.S. Geol. Surv. Terrs.* vol. iii. p. 646 (1877). [3] Gray, *Proc. Zool. Soc.* 1837, p. 124.

the chrysalides and larvæ of insects that are always to be found in such places." Dr. Merriam has made the interesting discovery that the common short-tailed North American Shrew supplements its insectivorous fare by feeding on beech-nuts, which will account for the generally very worn state of the teeth in this species.

Crossopus.[1]—Dentition: $i\ \frac{3}{2}$, $c\ \frac{1}{0}$, $p\ \frac{2}{1}$, $m\ \frac{3}{3}$; total 30. Opening of male or female generative organs enclosed within the same ring as the anal orifice; penis broad, with lateral processes. Ears small, not truncated. Tail long, with an inferior fringe of elongated hair; feet also fringed. Habits aquatic. The Palæarctic Water-Shrew (*C. fodiens*) is considerably larger than the Common Shrew, from which it is readily distinguished externally by its shorter and much broader muzzle, comparatively smaller eyes, and larger feet adapted for swimming,—the sides of the feet and toes being provided with comb-like fringes of stiff hairs. The tail is longer than the body, and possesses a well-developed swimming fringe of moderately long, regularly arranged hairs, which extend along the middle of the flat under surface from the end of its basal third to its extremity. The fur of the body is long and very dense, varying much in colour in different individuals, and this has given rise to descriptions of many nominal species; the prevailing shades are dark brown, almost black, above, and more or less bright ashy tinged with yellowish beneath; sometimes in the same litter there are individuals with the under surface more or less dark coloured. In the number as well as in the shape of the teeth the Water-Shrew differs from the Common Shrew: there is a premolar less on each side above; the bases of the teeth are much more prolonged posteriorly; and their cusps are much less stained brown, so that in old individuals with worn teeth they often appear altogether white. This species resembles the otter in its aquatic habits, swimming and diving with great agility. It frequents rivers and lakes, making its burrows in the overhanging banks, from which when disturbed it escapes into the water. Its food consists of insects and their larvæ, small crustaceans, and probably the fry of small fishes. It is generally distributed throughout England, is less common in Scotland, and as yet it has not been recorded in Ireland; specimens have been obtained from many parts of Europe, and also from Asia as far eastward as the Altai Mountains.

Subfamily **Crocidurinæ.**—Teeth completely white.

Myosorex.[2]—Dentition: $i\ \frac{3}{2}$, $c\ \frac{1}{0}$, $p\ \frac{2}{1-2}$, $m\ \frac{3}{3}$; total 30 or 32. Penis cylindroid and tapering; male or female generative organs opening close to anal orifice, but not forming a cloaca. Ears well developed; tail long, clothed with equal or subequal hairs. Habits terrestrial.

This genus is typically represented by *M. varius*, a very small

[1] Wagler, *Isis*, 1832, p. 275.
[2] Gray, *Proc. Zool. Soc.* 1837, p. 124.

Shrew from the Cape, which is quite unique among the whole family in having a rudimental seventh pair of lower teeth.

Crocidura.[1]—Dentition: $i\ \frac{3}{2}$, $c\ \frac{1}{0}$, $p\ \frac{2-1}{1}$, $m\ \frac{3}{3}$; total 28 or 30. Male or female generative organs forming a short cloaca with the anal orifice. Tail long, with a mixture of long and short hairs. Other characters as in *Myosorex.* Habits terrestrial.

This Old World genus includes over seventy nominal species, which have been divided into four subgenera, *C. aranea* and *C. suaveolens* of Continental Europe, and *C. cœrulea* of India, being well-known forms. The species are very variable and difficult to discriminate. *C. aranea* has a very wide distribution, ranging from Central and Southern Europe to North Africa and Central Asia. The name Musk-Rat is popularly applied in India to *C. cœrulea*, which frequents houses at night, hunting round rooms for cockroaches and other insects, and occasionally uttering a sharp shrill cry. The strong musky odour of this animal arises from large glands situated beneath the skin of the side of the body, a short distance behind the fore limbs. This odour is so powerful and penetrating that it is popularly believed in India that if the animal runs over a corked bottle of wine or beer it will infect the fluid within. Jerdon says that certainly many bottles are met with quite undrinkable from the peculiar musky odour of their contents, but, rejecting the possibility of its passing through the glass, he attributes it to the corks having been infected previously to bottling, stating in corroboration of this view that he has never found the odour in liquors bottled in England.

Diplomesodon.[2]—Dentition: $i\ \frac{2}{2}$, $c\ \frac{1}{0}$, $p\ \frac{1}{1}$, $m\ \frac{3}{3}$; total 26. Tail moderate; soles of the feet hairy. Other characters as in *Crocidura.* Habits terrestrial.

This genus is represented only by *D. pulchellus* of the Kirghiz steppes, which is allied to the following form, although retaining the normal Shrew-like external contour.

Anurosorex.[3]—Dentition: $i\ \frac{2}{2}$, $c\ \frac{1}{0}$, $p\ \frac{1}{1}$, $m\ \frac{3}{3}$; total 26. Ear very short; tail rudimental or short; soles of feet naked. Other characters as in *Diplomesodon.*

The two species of this genus are Mole-like terrestrial forms, of which the typical *A. squamipes* occurs in Tibet, while *A. assamensis* is found in Assam. The latter species has the longer tail. The habits of both are probably fossorial.

Chimarrogale.[4]—Dentition: $i\ \frac{3}{2}$, $c\ \frac{1}{0}$, $p\ \frac{1}{1}$, $m\ \frac{3}{3}$; total 28. Penis broad, with lateral processes; male or female generative organs opening within the same integumentary ring as the anal orifice.

[1] Wagler, *Isis*, 1832, p. 275. [2] Brandt, in *Lehmann's Reise.-Zool. Anh.* p. 299 (1852). [3] Milne-Edwards, *Comptes Rendus*, vol. lxx. p. 341 (1870).
[4] Anderson, *Journ. As. Soc. Bengal*, vol. xlvi. p. 262 (1877).

Tail long, with an inferior fringe of elongated hairs; ears small; plantar callosities simple; toes free. Habits aquatic.

This genus includes *C. himalayica* of the Himalaya and *C. platycephalus* of Japan. Both have the feet fringed, and, together with the next genus, may be regarded as the eastern analogues of *Crossopus* among the red-toothed series; their structural resemblances to the latter, if Dr. Dobson's classification is a natural one, being probably due to adaptation for a similar mode of life.

Nectogale.[1]—Dentition: $i\ \frac{3}{2}$, $c\ \frac{1}{0}$, $p\ \frac{1}{1}$, $m\ \frac{3}{3}$; total 28. External

FIG. 288.—*Nectogale elegans*. (From Milne-Edwards, *Mammif. Tibet.*)

ears not forming a conch, valvular. Plantar callosities forming adhesive pads; toes webbed. Other characters as in *Chimarrogale*. Habits aquatic.

The sole representative of this genus is the Tibetan Water-Shrew (*N. elegans*, Fig. 288), which differs from all other members of the family by the webbed toes and the presence of the disc-like adhesive pads on the under surface of the feet, which are believed to enable the creature to hold on to smooth rocks or stones in the beds of the streams it inhabits. This species is probably more completely aquatic in its habits than the allied *Chimarrogale*.

Fossil Soricidæ.—Remains of existing species of *Sorex* or *Crossopus* occur in the Norfolk Forest bed, while an extinct species has been found in the Pleistocene of Sardinia. *Crocidura* occurs in the cavern-deposits of Madras. Shrews from the Miocene and Upper

[1] Milne-Edwards, *Comptes Rendus*, vol. lxx. p. 341 (1870).

Eocene of Europe have been referred to *Sorex* and the genus *Amphisorex*, which is a synonym of *Crossopus*.

Family TALPIDÆ.

Allied to the *Soricidæ*, but distinguished by the presence of a zygomatic arch and auditory bulla in the skull, and by the form of the teeth. The eyes are very small, and in some species covered with skin; the ears are short and concealed by the fur; the fore limbs are generally more or less modified for digging; there is no symphysis pubis; the intestine has no cæcum; the tibia and fibula are united; and the unicuspidate first upper and lower incisors are not extended horizontally forwards.

This family is connected with the *Soricidæ* by *Urotrichus* and *Uropsilus*. All the members are limited to the temperate regions of Europe, Asia, and North America; and the majority of them are of fossorial habits, although a few are aquatic or cursorial. The family has been divided into two subfamilies by Professor Mivart, and since this arrangement has been very generally adopted it will be followed here. From the presence of intermediate forms like *Scaptonyx* Dr. Dobson, in the second part of his *Monograph of the Insectivora*, has proposed a different arrangement, which, with the omission of some forms which are of not more than subgeneric value, is as follows:—

MYOGALÆ—*Myogale*.
CONDYLURÆ—*Condylura*.
SCALOPES { *Scapanus*. *Scalops*.
TALPÆ—*Talpa*.
UROTRICHI { *Scaptonyx*. *Urotrichus*.
UROPSILI—*Uropsilus*.

Subfamily **Myogalinæ**.—Clavicles and humerus moderately elongated; manus without falciform bone.

Myogale.[1]—Dentition: $i\ \frac{3}{3}$, $c\ \frac{1}{1}$, $p\ \frac{4}{4}$, $m\ \frac{3}{3}$; total 44. Feet webbed. Habits aquatic. This genus is represented by the two species *M. moschata* (Fig. 289) and *M. pyrenaica*, of which the former is by far the largest member of the family, its total length being about 16 inches. Its long proboscis-like snout projects far beyond the margin of the upper lip; the toes are webbed as far as the bases of the claws; and the long scaly tail is laterally flattened, so as to form a powerful instrument of propulsion when swimming. This species inhabits the banks of streams and lakes in South-East Russia, where its food consists of various aquatic insects. *M. pyrenaica*,

[1] Cuvier, "Tabl. de Classif." in *Leçons d'Anat. Compar.* vol. i. (1800).

living in a similar manner in the region of the Pyrenees, is very much smaller, has a round tail, and a proportionally longer snout. Fossil remains of *M. moschata* occur in the Norfolk Forest bed, and were originally described under the name of *Palæospalax*. The genus is also represented in the Middle and Lower Miocene of the Continent.

Urotrichus.[1]—Dentition: $i\,\frac{2}{1}$, $c\,\frac{1}{1}$, $p\,\frac{4}{3}$ or $\frac{3}{4}$, $m\,\frac{3}{3}$; total 36. Feet not webbed; manus broad. Habits fossorial. The Mole-Shrews,

FIG. 289.—The Desman (*Myogale moschata*). $\frac{1}{3}$ natural size.

as these animals are called, are represented by *U. talpoides* of the mountains of Japan and *U. gibbsi* of North America. These two species are small and closely allied animals; the American form (which it has been proposed to separate subgenerically as *Neurotrichus*) having $p\,\frac{3}{4}$.

Uropsilus.[2]—Dentition: $i\,\frac{2}{1}$, $c\,\frac{1}{1}$, $p\,\frac{3}{3}$, $m\,\frac{3}{3}$; total 34. Manus narrow; tail naked and scaly. Habits cursorial. The single species, *U. soricipes*, from the borders of Tibet, is a slate-coloured animal with the external form of a Shrew but the skull of a Mole.

[1] Temminck, *Fauna Japonica*, vol. i. p. 22 (1842).

[2] Milne-Edwards, *Arch. du Muséum*, vol. vii. Bull. p. 92 (1872).

Subfamily **Talpinæ**.—Clavicle and humerus very short and broad; manus with a large falciform bone.

A. First upper incisor much larger than the second (New World Moles).

Scalops.[1]—Dentition: $i\ \frac{3}{2}$, $c\ \frac{1}{0}$, $p\ \frac{3}{3}$, $m\ \frac{3}{3}$; total 36. Extremity of muzzle simple; hind feet webbed; tail short and nearly naked. Represented by three species in the United States.

Scapanus.[2]—Dentition: $i\ \frac{3}{3}$, $c\ \frac{1}{1}$, $p\ \frac{4}{4}$, $m\ \frac{3}{3}$; total 44. Extremity of muzzle simple. The two North American species of this genus resemble *Scalops* in general characters, but have a dentition like *Condylura*. The habits are like those of the latter, and the right to generic distinction is doubtful.

Condylura.[3]—Dentition: $i\ \frac{3}{3}$, $c\ \frac{1}{1}$, $p\ \frac{4}{4}$, $m\ \frac{3}{3}$; total 44. Extremity of muzzle surrounded by filiform appendages. The Star-nosed Mole (*C. cristata*) derives its name from the star-like ring of appendages at the extremity of the muzzle, with the nostrils in the centre. The general contour is Mole-like, but the tail is nearly as long as the body, and the manus is somewhat less powerful, with its terminal phalanges not cleft. The length of the head and body is about 5 inches. This species is common in parts of North America, and forms tunnels in the ground like the Common Mole.

B. First upper incisor scarcely larger than the second (Old World Moles).

Scaptonyx.[4]—Dentition: $i\ \frac{3}{2}$, $c\ \frac{1}{1}$, $p\ \frac{4}{4}$, $m\ \frac{3}{3}$; total 42. Manus moderately broad, as in *Urotrichus*. Represented only by *S. fusicaudatus* of Eastern Tibet, which may be regarded as connecting *Talpa* with *Urotrichus*, having the head of the former and the limbs of the latter.

Talpa.[5]—Dentition (usually): $i\ \frac{3}{3}$, $c\ \frac{1}{1}$, $p\ \frac{4}{4}$, $m\ \frac{3}{3}$; total 44. Manus extremely broad.

This genus includes the true Moles, of which the common English Mole[6] (*T. europœa*) is the type. This animal is about 6 inches in total length, of which rather more than one inch is occupied by the tail. The body is elongated and cylindrical, and, owing to the very anterior position of the fore limbs, the head appears to rest between the shoulders; the muzzle is long and obtusely pointed, terminated by the nostrils, which are close together; the minute eye is almost hidden by the fur; the ear is without a conch, and opens on a level with the surrounding integument. The fore limbs are rather short and very muscular, terminating in broad, naked, shovel-shaped feet, with the palms normally directed out-

[1] Cuvier, "Tabl. de Classif." in *Leçon d'Anat. Comp.* vol. i. (1800). [2] Pomel, *Arch. Sci. Phys. Nat.* vol. ix. p. 247 (1848). [3] Illiger, *Prodromus Syst. Mamm. et Avium*, p. 125 (1811). [4] Milne-Edwards, *N. Arch. du Muséum*, vol. vii. Bull. p. 92 (1872). [5] Linn. *Syst. Nat.* 12th ed. p. 73 (1766). [6] The following account is taken almost entirely from Dr. Dobson.

wards, and each with five subequal digits armed with strong flattened claws. The hind feet are long and narrow, and the toes are provided with slender claws. The body is densely covered with soft, erect, velvety fur, the hairs being uniform in length and thickness, except on the muzzle and short tail. The colour of the fur is generally black, with a more or less grayish tinge, or brownish-black, but various paler shades up to pure white have been observed.

The food of the Mole consists chiefly of the earth-worm, in pursuit of which it forms its well-known underground excavations. Its habits were many years ago studied and described by M. Henri le Court. Like many other mammals, the Mole has a lair to which it may retire for security. This consists of a central nest formed under a hillock, placed in some protected situation, as under a bank, or between the roots of trees. The nest, which is lined with dried grass or leaves, communicates with the main run by four passages, of which only one joins it directly, leading downwards for a short distance and then ascending again. The other three are directed upwards and communicate at regular intervals with a circular gallery constructed in the upper part of the hillock, which in turn communicates by five passages leading downwards and outwards with another much larger gallery placed lower down on a level with the central nest, from which passages proceed outwards in different directions, one only communicating directly with the main run, while the others, curving round, either soon join or end blindly. The main run is somewhat wider than the animal's body; its walls are smooth, and formed of closely compressed earth, the depth varying according to the nature of the soil, but ordinarily from 4 to 6 inches. Along this tunnel the animal passes backwards and forwards several times daily, and here traps are laid by mole-catchers for its capture. From the main run numerous passages are formed on each side, along which the animal hunts its prey, throwing out the soil in the form of mole-hills. The Mole is one of the most voracious of mammals, and, if deprived of food, is said to die in from ten to twelve hours. Almost any kind of flesh is eagerly devoured by captive Moles, which have been seen by various observers, as if maddened by hunger, to attack animals nearly as large as themselves, such as birds, lizards, frogs, and even snakes; toads, however, they will not touch, and no form of vegetable food attracts their notice. If two Moles be confined together without food, the weaker is invariably devoured by the stronger. Moles take readily to the water, in which respect they resemble their representatives on the North American continent. Bruce, writing in 1793, remarks that he saw a Mole paddling towards a small island in the Loch of Clunie, 180 yards from land, on which he noticed mole-hills.

The sexes come together about the second week in March, and

the young—generally from four to six in number—which are brought forth in about six weeks, quickly attain their full size.

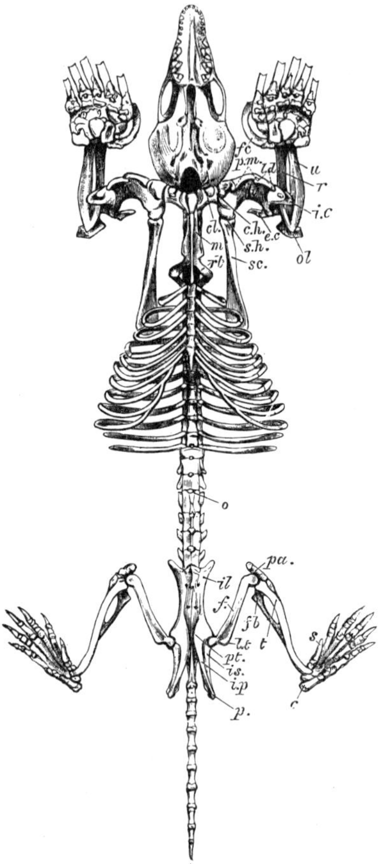

FIG. 290.—Skeleton of Mole $\times \frac{2}{3}$ (lower jaw removed to show base of skull). *c*, Calcaneum; *c.h.*, clavicular articulation of the humerus; *cl.*, clavicle; *e.c*, external condyle of humerus; *f.*, femur; *fb*, fibula; *fc*, falciform bone (radial sesamoid); *h*, humerus; *i.c*, internal condyle of humerus; *il*, left ilium; *i.p*, ramus of the ilium and pubis; *is.*, ischium; *l.d*, ridge of insertion of latissimus dorsi muscle; *l.t*, lesser trochanter; *m*, manubrium sterni; *o*, fourth intercentral ossicle; *ol*, olecranon; *p.*, pubis widely separated from that of the opposite side; *pa.*, patella; *p.m.*, ridge for insertion of pectoralis major muscle; *pt.*, pectineal eminence; *r*, radius; *rb*, first rib; *s*, plantar sesamoid ossicle corresponding to the radial sesamoid (os falciforme) in the manus; *sc.*, scapula; *s.h.*, scapular articulation of the humerus; *t*, tibia; *u*, ulna.

The Mole exhibits in the whole of its organisation a perfect adaptation to its peculiar mode of life. In the structure of the skeleton (Fig. 290) very striking departures from the typical mammalian form are noticeable. Thus the presternum is so much produced anteriorly as to extend forward as far as a vertical line from the second cervical vertebra, carrying with it the very short and almost quadrate clavicle, which is articulated with its anterior extremity and distally with the humerus; being also connected ligamentously with the scapula. The fore limbs are thus brought opposite the sides of the neck, and from this position a threefold advantage is derived: in the first place, as this is the narrowest part of the body, they add but little to the general width, which, if increased, would lessen the power of movement in a confined space; secondly, this position allows of a longer fore limb than would otherwise be possible, and so increases its power; and, thirdly, although the entire limb is relatively very short, its anterior position enables the animal, when burrowing, to thrust the claws so far forward as to be in a line with the end of the muzzle, the importance of which is evident. Posteriorly, the hind limbs are similarly removed out of the way by approximation of the hip-joints to the centre line of the body. This is effected by inward curvature of the innominate bones at the acetabula to such an extent that they almost meet in the centre, while the pubic bones are widely separated behind. The shortness of the fore limb is caused by the great reduction in the length of the humerus, which has lost all resemblance to its normal shape. In addition to the usual articulation with the glenoid cavity of the scapula, the humerus also has a separate articulation with the extremity of the clavicle. The bones of the manus are enormously expanded laterally; this expansion being increased by the large sickle-like bone on the radial side of the carpus, which is considered by some anatomists to represent the prepollex. The skull is long and tapering, with very slender zygomatic arches and elongated nasals, which are ankylosed together, and in advance of which the mesethmoid is more or less ossified. The vertebræ are usually C 7, D 13, L 6, S 6, C 10-12; all having very strong surfaces for mutual articulation. The upper incisors are chisel-like, and the canine has two roots; the first three upper premolars are simple and conical, but the fourth is much larger, and canine-like. In the mandible the incisors are small and somewhat proclivous, while the canine can only be distinguished from them by its position; the first lower premolar is larger than the others.

The Common Mole has an exceedingly wide distribution, ranging over the greater part of the Palæarctic region, where it is met with in places so widely sundered as England and Japan. It occurs in both the Himalaya and Altai mountains. In Ireland it is unknown, and in Scotland it extends as far north as Caithness. Eight species of the genus are recognised, which may be grouped, from the characters of their dentition, as follows, viz.: $i\ \frac{3}{3}$, $c\ \frac{1}{0}$, $p\ \frac{4}{4}$, $m\ \frac{3}{3}$, *T. wogura*; $i\ \frac{3}{3}$, $c\ \frac{1}{1}$, $p\ \frac{4}{4}$, $m\ \frac{3}{3}$, *T. europæa*, *cæca*, *longirostris*, *micrura*; $i\ \frac{3}{3}$, $c\ \frac{1}{1}$, $p\ \frac{3}{4}$, $m\ \frac{3}{3}$, *T. leucura*, *leptura*; $i\ \frac{3}{3}$, $c\ \frac{1}{1}$, $p\ \frac{3}{3}$, $m\ \frac{3}{3}$, *T. moschata*.

Except in *T. europæa*, the eyes are covered by a membrane. In *T. micrura* the short tail is concealed by the fur. *T. cæca* is found south of the Alps; the remaining species are Asiatic, and two only —*T. micrura* and *T. leucura*—occur south of the Himalaya. *T. moschata*, of Tibet, is regarded by some zoologists as generically distinct under the name of *Scaptochirus*.

Remains of *T. europæa* occur in the Norfolk Forest bed, while extinct species are found in the European Tertiaries as far down as the Lower Miocene, although it has been proposed to separate some of these forms generically. *Protalpa*, of the Upper Eocene Phosphorites of Central France, is very closely allied, but the structure of the humerus is somewhat less specialised.

Extinct Genera.—A number of extinct Insectivora from the European Tertiaries more or less closely allied to the Moles have been described, but since our knowledge of most of them is extremely imperfect their precise affinities are in many instances problematical. Of these the Lower Miocene *Tetracus* is said to have affinity both with *Myogale* and *Erinaceus*; while the forms described as *Mysarachne* and *Echinogale*, are considered to connect the present with the two preceding families. *Plesiosorex* is another Lower Miocene type known only by the mandible, in which there are ten teeth; it is generally referred to the *Myogalinæ*. The minute *Amphidozotherium*, of the French Phosphorites, is considered to be allied to *Urotrichus*.

Family ADAPISORICIDÆ

This extinct family is represented by the genera *Adapisorex* and *Adapisoriculus*, of the lowest Eocene of Rheims, which are regarded as allied to the *Soricidæ*, but somewhat more specialised. In the type genus the formula of the lower teeth is i 2, c 1, p 4, m 3; the incisors and canine being proclivous, and the molars (of which the last is small and without a third lobe) quadritubercular. *Adapisoriculus* is a smaller form with differently shaped molars.

Here also may be mentioned the genera *Orthaspidotherium* and *Pleuraspidotherium*, from the above-mentioned deposits, which are probably members of the present order. They appear to have been animals somewhat smaller than a Hedgehog, with quadritubercular upper molars (Fig. 291), and the hinder premolars more complex than those of the *Erinaceidæ*. In the first-named genus the dental formula is $i\ \frac{3}{3}$, $c\ \frac{1}{1}$, $p\ \frac{4}{4}$, $m\ \frac{3}{3}$; the third and fourth upper premolars having one outer column. *Pleuraspidotherium* has apparently only three premolars, of which the third and fourth (Fig. 291) have two outer columns. The humerus in both has no entepicondylar foramen, the femur has a third trochanter, and the astragalus is vertically perforated.

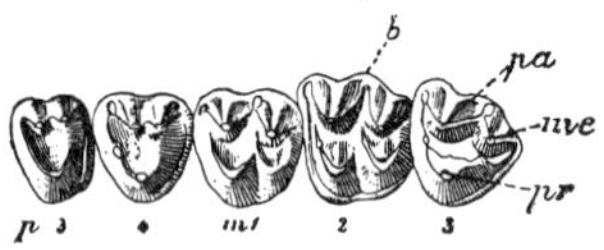

FIG. 291.—The last left upper cheek-teeth of *Pleuraspidotherium aumonieri*; from the Lowest Eocene of Rheims. *pr*, protocone; *me*, metacone; *pa*, paracone; *b*, cingulum-cusp. (From Osborn.)

Family POTAMOGALIDÆ.

Skull with a small brain-case, no zygomatic arch or postorbital process, and the tympanic annulate and not forming a bulla. Upper molars with the cusps arranged in a broad V, and somewhat intermediate in structure between those of the preceding and succeeding families. No clavicle; pubic symphysis ligamentous;

tibia and fibula typically united distally. No cæcum. Confined to the Ethiopian region.

Potamogale.[1]—Dentition: $i\ \frac{3}{3}$, $c\ \frac{1}{1}$, $p\ \frac{3}{3}$, $m\ \frac{3}{3}$; total 40. Represented only by *P. velox* of Western Equatorial Africa. This animal (Fig. 292) inhabits the banks of streams, and is thoroughly adapted for an aquatic life; it is nearly 2 feet in length, the tail measuring about half. The long cylindrical body is continued uninterruptedly into the thick laterally compressed tail, the legs are very short, and the toes are not webbed, progression through the water evidently depending wholly on the action of the powerful tail, while the limbs are folded inwards and backwards. The muzzle is broad and

FIG. 292.—*Potamogale velox.* × $\frac{1}{4}$. (From Allman, *Trans. Zool. Soc.* vol. vi. pl. i.)

flat, and the nostrils are protected by valves. The fur is dark brown above, the extremities of the hairs on the back being of a metallic violet hue by reflected light, beneath whitish. This curious animal was discovered by M. du Chaillu.

Geogale.[2]—Dentition: $i\ \frac{2}{2}$, $c\ \frac{1}{1}$, $p\ \frac{3}{2}$, $m\ \frac{3}{3}$; total 34. This genus is known solely by *G. aurita*, a small Mouse-like species from Madagascar, agreeing closely with *Potamogale* in the general form of the skull and teeth. The tibia and fibula are distinct, but it is not known whether a clavicle exists; and the material at present available is insufficient to definitely fix the natural position of the genus.

Family SOLENODONTIDÆ.

Skull with a small brain-case constricted between the orbits, no

[1] Du Chaillu, *Proc. Boston Soc. Hist. Nat.* vol. vii. p. 363 (1860).

[2] Milne-Edwards, *Ann. Sci. Nat.* vol. xv. p. 5 (1872).

zygomatic arch or postorbital process, and the tympanic annulated and not forming a bulla. Upper molars tritubercular, the cusps being arranged in a V. Pubic symphysis short; tibia and fibula distinct. Vertebræ: C 7, D 15, L 4, S 5, C 23. No cæcum. The penis is carried forwards and suspended from the abdomen; the testes are received into perineal pouches; the mammary glands are post-inguinal; the uterine cornua end in cæcal sacs.

Solenodon.[1]—Dentition: $i\,\frac{3}{3}$, $c\,\frac{1}{1}$, $p\,\frac{3}{3}$, $m\,\frac{3}{3}$; total 40. This genus, with *S. paradoxus* and *S. cubanus* (Fig. 293), from Hayti and Cuba

FIG. 293.—*Solenodon cubanus.* × $\frac{1}{5}$. (From Peters, *Abh. Akad. Berlin.*)

respectively, alone represents the family. These species, which differ chiefly in the colour and quality of the fur, have a remarkably long cylindrical snout, a long naked tail, feet formed for running, and the body clothed with long, coarse fur.

The position of the mammæ quite behind on the buttocks is unique among Insectivora. The first upper incisor is much enlarged, and this and the other incisors, canines, and premolars, closely resemble those of *Myogale;* the second lower incisor is, as in *Potamogale*, much larger than the anterior one, and is deeply hollowed out internally. While thus apparently showing relationship with the *Talpidæ*, the form of the crowns of the molar teeth connects them with the next family.

[1] Brandt, *Mém. Ac. Imp. St. Pétersbourg*, 1833, vol. ii. p. 459.

Family CENTETIDÆ.

Skull (Fig. 294) with a small cylindrical brain-case not constricted between the orbits, no zygomatic arch or postorbital process, and the tympanic annulate and not forming a bulla. Upper molars tritubercular. Pubic symphysis short; and the tibia and fibula either united or free. No cæcum. The penis is pendent and retractile within the fold of the integument surrounding the anus; the testes are abdominal; the mammæ are thoracic and ventral; and the uterine cornua are terminated by the Fallopian tubes. All the species are limited to Madagascar.

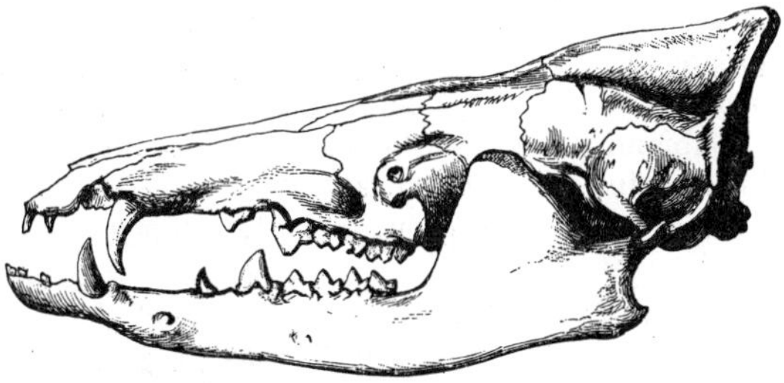

FIG. 294.—Left lateral view of the skull of the Tenrec (*Centetes ecaudatus*). Reduced.

Subfamily **Centetinæ.**—Tibia and fibula distinct; testes near kidneys; fur with spines.

Centetes.[1]—Dentition: $i\ \frac{2}{3}$, $c\ \frac{1}{1}$, $p\ \frac{3}{3}$, $m\ \frac{3}{3}$; total 38. Vertebræ: C 7, D 19, L 5, S 3, C 8. The single species is the well-known Tenrec (*C. ecaudatus*), characterised by the absence of a tail; it reaches a total length of from 12 to 16 inches, and is the largest known Insectivore. The adult males have long canines, the extremities of the lower pair being received into pits in front of the upper ones (Fig. 294). It is probably the most prolific of all mammals, since as many as twenty-one young are said to have been brought forth at a birth. The young have strong white spines arranged in longitudinal lines along the back, but these are lost in the adult animal, which is provided only with a nuchal crest of long rigid hairs. In rare instances a fourth upper molar may be developed.

Hemicentetes.[2]—Dentition: $i\ \frac{3}{3}$, $c\ \frac{1}{1}$, $p\ \frac{3}{3}$, $m\ \frac{3}{3}$; total 40. This genus is represented by the two species *H. semispinosus* (of which the skull is shown in Fig. 295) and *H. nigriceps*. It differs from *Centetes* by the presence of the third upper incisor, the much smaller canines, and by the form of the skull. Both species are very much smaller than *C. ecaudatus*, and the dorsal spines are retained in the adult state. Vertebræ: C 7, D 16, L 5, S 3, C 9.

[1] Illiger, *Prodromus Syst. Mamm. et Avium*, p. 124 (1811). [2] Mivart, *Proc. Zool. Soc.* 1871, p. 72.

Ericulus.[1]—Dentition: $i\,\frac{2}{2}$, $c\,\frac{1}{1}$, $p\,\frac{3}{3}$, $m\,\frac{3}{3}$; total 36. Vertebræ: C 7, D 17, L 6, S 4, C 9. The single species, *E. setosus*, is a Hedgehog-like animal, having the whole upper surface and the short tail densely covered with close-set spines. The facial bones are much shorter than in any of the preceding genera, and the first upper incisor is elongated, as in *Erinaceus*. Judging from the slight development of the cutaneous muscles compared with those of the true Hedgehogs, it is probable that complete involution of the body does not take place.

Subfamily **Oryzorictinæ.**—Tibia and fibula united; testes near urethra; fur without spines.

Microgale.[2]—Dentition: $i\,\frac{3}{3}$, $c\,\frac{1}{1}$, $p\,\frac{3}{3}$, $m\,\frac{3}{3}$; total 40. This genus

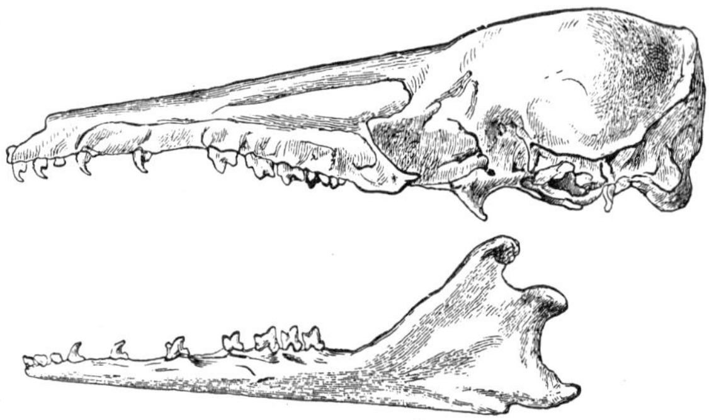

FIG. 295.—Skull of *Hemicentetes semispinosus*. × 2. (From Mivart, *Proc. Zool. Soc.* 1871.)

includes *M. longicaudata* and *M. cowani*, both of which are small Mouse-like species, the former with a tail double the length of the head and body, and having 43 caudal vertebræ; teeth like those of *Centetes ecaudatus*, but, owing to the comparatively much shorter muzzle, not separated by wide spaces, and the last premolar and molar with internal basal processes.

Oryzorictes.[3]—Represented by two species, *O. hova* and *O. tetradactylus*, the latter distinguished by the presence of only four digits in the manus, the three inner having long laterally compressed fossorial claws. The general form of the head and body of the two species known is like that of a Mole. These animals burrow in the rice-fields and do much damage to the crops.

Family CHRYSOCHLORIDÆ.

Skull conical, not constricted between the orbits, without post-orbital process, but with well-developed zygomatic arch and tympanic

[1] I. Geoffroy, *Ann. Sci. Nat.* sér. 2, vol. viii. p. 60 (1837). [2] Thomas, *Journ. Linn. Soc.—Zool.* vol. xvi. p. 319 (1882). [3] Grandidier, *Rev. and Mag. Zool.* 1870, p. 50.

bulla. Upper molars tritubercular, with the crowns very tall. No pubic symphysis; the tibia and fibula united. The eyes are covered by the hairy integument; the ears short and concealed by the fur; the internal generative organs are as in *Centetinæ;* the mammæ are thoracic and inguinal and placed in cup-shaped depressions. Habits fossorial. Confined to the southern part of the Ethiopian region, not extending to Madagascar.

This family is closely allied to the *Centetidæ*, occupying the same relative position with respect to that family that the *Talpidæ* does to the *Soricidæ*. Compared with the *Talpidæ*, we find the following differences in the structural adaptation to a fossorial life; the manubrium sterni is not anteriorly elongated, neither are the

FIG. 296.—The Golden Mole (*Chrysochloris obtusirostris*).

clavicles shortened; but this is compensated for by a deep hollowing out of the antero-lateral walls of the thorax, the ribs in these parts and the sternum being convex inwards. The long clavicles have their distal extremities pushed forward, and the concavities on the sides and inferior surface of the thorax lodge the thick muscular arms.

Chrysochloris.[1]—Dentition: $i\ \frac{3}{3}$, $c\ \frac{1}{1}$, $p\ \frac{3}{3}$, $m\ \frac{3-2}{3-2}$; total 40 or 36 Vertebræ: C 7, D 19, L 3, S 5, C 8. This genus includes some seven or eight South African species, commonly known as Golden Moles (Fig. 296). Those species, in which the molars are reduced to $\frac{2}{2}$, with a basal talon to the lower ones, and without a prominence in the temporal fossa, have been placed in a separate genus, *Chalcochloris*, by Professor Mivart. Nearly all the species have the fur of the upper surface of a brilliant metallic lustre, varying from golden bronze to green and violet of different shades. The manus

[1] Lacépède, *Mém. de l'Institut*, vol. iii. p. 493 (1801—read 1799).

has four digits, of which the two outer are small, while the middle ones are large, with immensely powerful claws.

Extinct Types.—The only fossil forms which can be referred to the section of the Insectivora with tritubercular molars are the *Leptictidæ*, of the Eocene and Miocene of North America. This family includes the genera *Leptictis*, *Mesodectes*, and *Ictops*, all of which are regarded by Dr. Schlosser as true Insectivora, although they were placed by Professor Cope with the Creodont Carnivora.

Bibliography of Insectivora.—Peters, *Reise nach Mossambique—Säugeth.* 1852; Id. "Ueber die Classification der Insectivora," *Monatsb. Akad. Wissensch. Berlin*, 1865, and other papers; Mivart, "On the Osteology of Insectivora," *Journ. Anat. and Phys.* 1867, 1868, and *Proc. Zool. Soc.* 1871; Gill, "Synopsis of Insectivorous Mammals," *Bull. Geol. and Geog. Survey, U.S.A.* Washington, 1875 (includes a general bibliography of the order); Dobson, *Monograph of the Insectivora, Systematic and Anatomical*, London, 1882-90.

CHAPTER XIII

THE ORDER CHIROPTERA

MAMMALS, having their fore limbs specially modified for flight. The forearm consists of a rudimentary ulna, and a long curved radius. The carpus has six bones supporting a small pollex and four greatly elongated fingers, between which and the sides of the body and the hinder extremities a thin expansion of the integument (the wing-membrane or patagium) is extended. The knee is directed backwards, owing to the rotation of the hind limb outwards by the wing-membrane; a peculiar elongated cartilaginous process (the calcar), rarely rudimentary or absent, arising from the inner side of the ankle-joint, is directed inwards, and supports part of the posterior margin of an accessory membrane of flight, extending from the tail or posterior extremity of the body to the hinder limbs (the inter-femoral membrane). The penis is pendent; the testes are abdominal or inguinal; the mammary glands thoracic and generally postaxillary; the uterus is simple or with more or less long cornua; the placenta discoidal and deciduate; and the smooth cerebral hemispheres do not extend backwards over the cerebellum. The dental series includes incisors, canines, premolars, and molars and never exceeds $i\,\frac{2}{3}$, $c\,\frac{1}{1}$, $p\,\frac{3}{3}$, $m\,\frac{3}{3}$; total 38.

The animals comprised in this order are at once distinguished by the presence of true wings, and this peculiarity is accompanied by other modifications of bodily structure having special relation to flight. Thus, in contrast to most other mammals, in which the hind limbs greatly preponderate in size over the fore, in the present order the fore limbs immensely exceed the short and weak hinder extremities. The thorax, as giving origin to the great muscles which sustain flight, and containing the proportionately large lungs and heart, is remarkably capacious, and the ribs are flattened and close together; the shoulder-girdle is also greatly developed in comparison with the weak pelvic bones.

Linnæus included the Bats among the Primates, mainly on

account of the number of their upper incisors, supposed to be always four, the thoracic position of the mammæ, and the pendent condition of the penis. Many other zoologists, taking into consideration the placental characters and the form of the uterus, have followed him; but it is evident that the situation of the mammæ is related to the necessarily central position of the young during flight, the shortness of the uterine cornua, observable in so many species, to the generally uniparous gestation requiring less room, while the discoidal deciduate placenta is equally present in and characteristic of the Insectivora, many species of which also have the penis pendent. Thus, the reasons for maintaining the Bats in this high position being disposed of, we find in the low organisation of their brain a proof of their inferior status; while furthermore, although they differ widely from all other mammals in external form, it is evident that this is only the result of special adaptation to aerial locomotion; and, taking into account their whole bodily structure, we may accept the view of Professor Huxley that they should merely be regarded as exceedingly modified Insectivora.

So thoroughly, however, has this adaptation for flight been carried out that of all animals the Bats are the least terrestrial, not one of them being equally well fitted for progression on the earth. This is due to the hind as well as the fore limbs being pressed into the service of aerial locomotion. Thus the hind limb is so rotated outwards by the wing-membrane that, contrary to what obtains in all other vertebrates, the knee is directed backwards, and corresponds in position to its serial homologue the elbow. It necessarily follows from this arrangement that when a Bat is on the ground it rests on all fours, having the knees directed upwards; while, in order to bring it into a position for forward progression, the foot rotates forwards and inwards on the ankle. Walking under these circumstances is at best only a kind of shuffle, and that this is fully recognised by the animal is evidenced by its great anxiety to take wing, or, if this be impracticable, to ascend to some point where it can hitch itself up by the claws of the hind legs in its usual position when at rest.

The bones of the skeleton are characterised by their slenderness and the great size of the medullary canals in those of the extremities. The vertebral column is short, and the vertebræ differ very slightly in number and form throughout the species. The general number of the dorso-lumbar vertebræ is 17, of which 12 are dorsal; the cervicals are very broad, but short from before backwards, their breadth being due to the great transverse diameter of the spinal canal rendered necessary by the comparatively large size of the spinal cord, which, after giving off the nerves to the fore limbs and thorax, rapidly diminishes in size, and in the lumbo-sacral region is reduced to a fine thread. Except in the frugivorous *Pteropodidæ*, the vertebræ, from the third cervical backwards, are devoid of neural spines. From the first dorsal to the last lumbar vertebra the spinal column forms a single curve backwards, which is most pronounced in the lumbar region. The centra of the vertebræ are but slightly movable upon each other, and in old individuals appear to become partially ankylosed together. The caudal vertebræ are simple cylindrical bones without processes; their number and length being extremely variable even in closely allied species; and the anterior caudals are generally united to the

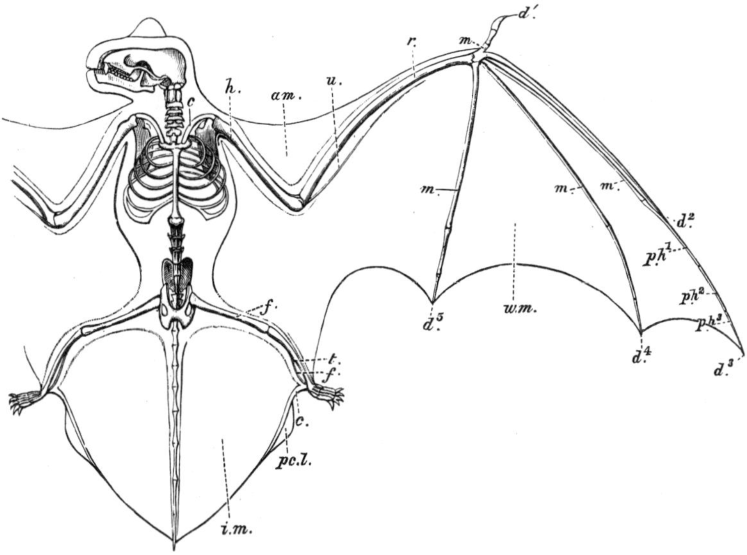

FIG. 297.—Skeleton and flying-membranes of the Noctule Bat (*Vesperugo noctula*). × $\frac{1}{3}$. *c*, Clavicle; *h*, humerus; *r*, radius; *u*, ulna (rudimentary); d^1, pollex; d^2, d^3, d^4, d^5, other digits of the manus supporting *wm*, the wing-membrane; *m*, *m*, metacarpal bones; ph^1, first phalanx; ph^2, second phalanx; ph^3, third phalanx; *am*, antebrachial membrane; *f*, femur; *t*, tibia; *fb*, fibula (rudimentary); *c*, calcar supporting *im*, the interfemoral membrane; *pcl*, post-calcaneal lobe.

ischial tuberosities. The relative development of the caudal vertebræ is, indeed, intimately correlated to the habits of the animals; the long tail in the insectivorous forms supporting and controlling the position of the large interfemoral membrane, which appears not only to aid their rapid motions when in pursuit of their prey by acting as a rudder, but also to assist in the capture and retention of the larger insects. In the frugivorous types, on the other hand, this is not required, and the tail is accordingly rudimentary or absent. In all Bats the presternum has a prominent keel for the attachment of the great pectoral muscles. In most species the ribs are much flattened, and in some they are partially ankylosed by their contiguous margins.

The skull is subject to considerable structural variations, even within the limits of a single family. Postorbital processes to the frontals are found only in the *Pteropodidæ* and some *Nycteridæ* and *Emballonuridæ*. *Pteropus leucopterus* and *Pteralopex* are peculiar in having the orbit completely surrounded by bone. A slender zygomatic arch is present, except in some of the *Phyllostomatidæ*.

The milk-teeth are peculiar in that they are utterly unlike those of the permanent series. They are slender, with sharp recurved cusps; and as a rule are shed at an early period (in the *Rhinolophidæ* before birth), but may coexist with some of the fully developed permanent teeth. The permanent teeth are subject to great variation of form, although they always have distinct roots. In the Insectivorous types they are acutely cusped, the cusps in those of the upper jaw being arranged in a more or less distinct W; but in the frugivorous forms, like the *Pteropodidæ* and some of the *Phyllostomatidæ*, the molars are longitudinally grooved or hollowed out.

The pectoral girdle maintains a very constant type. Thus the clavicle is very long, strong, and curved; and the scapula large, oval, triangular, with a long curved coracoid process. The humerus, though long, is scarcely two-thirds the length of the radius. The ulna is rudimentary, its proximal extremity, which articulates with but a small part of the humerus, being ankylosed to the radius; and immediately beyond the joint it is reduced to a slender splint-like bone, extending about as far as the middle of the radius. In all species a detached sesamoid bone exists in the tendon of the triceps muscle. The radius is very long, in some species actually equal to the length of the head and body. The proximal row of the carpus consists of a single bone formed by the united scaphoid, lunar, and cuneiform; which, with the extremity of the radius, forms the radio-carpal joint. In the distal row the trapezium, trapezoid, and magnum vary in size in the different families, the unciform appearing to be the most constant, and the pisiform being generally very small.

The manus is always furnished with five digits. The first, fourth, and fifth digits consist of a metacarpal and two phalanges; but in the second and third digits the number of phalanges is different in certain families. The pollex always terminates in a claw, which—like the proximal phalanx—is best developed in the frugivorous species. In most of the frugivorous *Pteropodidæ* the second digit is provided with a claw; but in all other Bats this and the remaining digits are unarmed. In the genus *Triænops* alone a very peculiar short bony process projects from the outer side of the proximal extremity of the terminal phalanx of the fourth digit. The relative development of the digits and their phalanges will be noticed under each family.

As might be expected from the small size of the posterior limbs, the pelvic girdle is relatively weak. The ilia are long and narrow. In the males of most species the pubic bones of opposite sides are very loosely united in front, while in females they are widely separated; and in the family *Rhinolophidæ* alone do these bones form a symphysis. The ileo-pectineal eminence develops a long pectineal process, which in the subfamily *Hipposiderinæ* is continued forwards to the anterior extremity of the ilium enclosing a preacetabular foramen unique among mammals. The acetabulum is small and directed outwards and slightly upwards; and with this is related the peculiar position of the hind limb already noticed as one of the chief characteristics of the order. The femur is slender and cylindrical, with a small head and very short neck, and scarcely differs in form throughout the order. The bones of the leg and foot are variable; in the subfamily *Molossinæ* alone is there a well-developed fibula, while in all other species this bone is either very slender, or cartilaginous and ligamentous in its upper third, or reduced to a small bony process above the heel, as in *Megaderma*, or altogether absent, as in *Nycteris*.

The foot consists of a very short tarsus, and of slender, laterally compressed toes, with much curved claws. The hallux is composed of a metacarpal, a proximal and an ungual phalanx, and is slightly shorter than the other four toes, each of which has an additional phalanx, except in the subfamily *Hipposiderinæ* and in the anomalous genera *Thyroptera* and *Myxopoda*, where all the toes have the same number of phalanges as the first digit, and are equal to it in length. In the genus *Chiromeles* the first digit is thumb-like and separated from the others, and in the typical *Molossinæ* the first and fifth digits are much thicker than the intermediate toes.

The most noticeable peculiarities in the myology of the order consist in the separated bands or slips into which the platysma is divided, and in the presence of the remarkable muscle termed occipito-pollicalis, which extends from the occipital bone to the base of the terminal phalanx of the pollex.

Although, as already mentioned, the brain presents a low type of organisation, yet probably no animals possess so delicate a sense of touch as the Chiroptera. It is undoubtedly this perceptive power which enabled the individuals deprived of sight, hearing, and smell, in Spallanzani's well-known experiments, to avoid the numerous threads hung across the rooms in which they were permitted to fly about. In the common Bats the tactile organs evidently exist, not only in the delicate vibrissæ which spring from the sides of the muzzle, but also in the highly sensitive and widely extended integumentary structures entering into the formation of the wing-membranes and ear-conchs; while in many other species, notably in the tropical Rhinolophine and Phyllostomatine Bats, peculiar foliaceous cutaneous expansions surrounding the nasal apertures or extending backwards behind them are added. These structures, collectively known as the "nose-leaf" (whence the term "leaf-nosed Bats"), have been shown by Dr. Dobson to be made up partly of the extended and thickened marginal integument of the nostrils, and partly of the highly differentiated glandular eminences occupying the sides of the muzzle, in which, in all the common Bats, the vibrissæ are implanted.

In all species of leaf-nosed Bats, and especially in the *Rhinolophidæ*, where the nasal appendages reach their highest development, the superior maxillary division of the fifth nerve is of remarkably large calibre. The nasal branch of this nerve, which is given off immediately beyond the infraorbital foramen, is by far the largest portion; the palpebral and labial branches consisting of a few slender nerve-fibres only. This branch passes forwards and upwards on the side of the maxilla, but soon spreads out into numerous filaments extending into the muscles and integument above, and into the base of the nose-leaf. The nerve supply of the nose-leaf is further augmented by the large nasal branch of the ophthalmic division of the fifth nerve. While the many foliations, elevations, and depressions which vary the form of the nose-leaf greatly increase the sensory surface supplied by the fifth nerve, and during rapid flight intensify the vibrations conveyed to it, the great number of sweat and oil glands which enter into its structure perform a function analogous to that of the glands of the auditory canal in relation to the membrana tympani in maintaining its surface in a highly sensitive condition. The nasal appendages of the Chiroptera may thus be regarded as performing the office of an organ of a very exalted sense of touch standing in the same relation to the nasal branches of the fifth nerve as the aural apparatus to the auditory nerve; for, as the latter organ collects and transmits the waves of sound, so the former receives impressions arising from vibrations communicated to the air by approaching objects.

In no order of mammals is the ear-conch so greatly developed or so variable in form. Thus in most of the insectivorous species the ears are longer than the head, while in some, as in the common Long-eared Bat (*Plecotus auritus*), their length nearly equals that of the head and body. The form of the conch is very characteristic of the various families; in most the tragus is remarkably large, in some extending nearly to the outer margin of the conch; and its function appears to be to cause undulations in the waves of sound, and so intensify and prolong them. It is worthy of notice that in the *Rhinolophidæ*, the only family of insectivorous Bats wanting the tragus, the auditory bullæ reach their greatest size, and the highly sensitive nasal appendages their highest development; and that in the typical group of the *Molossinæ* the ear-conch is divided by a prominent keel; and the antitragus is unusually large in those species in which the tragus is minute (see Fig. 298, *a*). In the frugivorous Bats the form of the ear-conch is very simple, and but slightly variable throughout the various types.

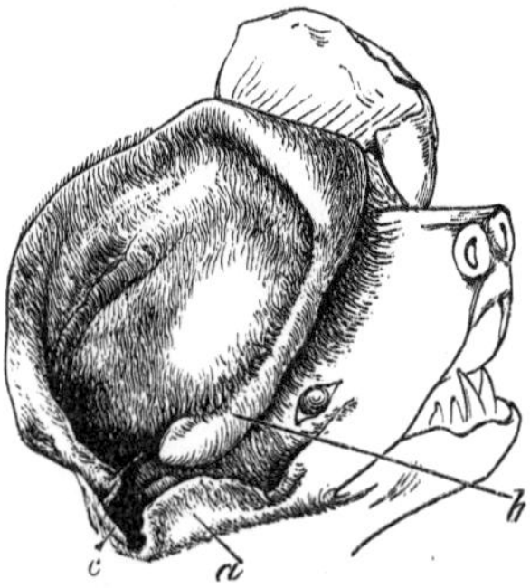

FIG. 298.—Head of *Molossus glaucinus*. (From Dobson, *Proc. Zool. Soc.* 1876.) *a*, Antitragus; *b*, keel of the ear-conch; *c*, notch behind antitragus.

In all Bats the ears are extremely mobile, each moving independently at the will of the animal. This has been observed even in the frugivorous *Pteropodidæ*, in which the peculiar vibratory movements first noticed in *Artibeus perspicillatus* may also be seen when the animals are alarmed.

The opening of the mouth is anterior in most species, but in many it is inferior, the extremity of the nose being more or less produced beyond the lower lip,—so much so indeed in the small South-American species *Rhynchonycteris naso* as to resemble that of the Shrews. The lips exhibit the greatest variety in form, which will be referred to under each family. The absence of a fringe of hairs is characteristic of all fruit-eating Bats, and probably always distinguishes them from the insectivorous species, which they may resemble in the form of their teeth and other respects.

The œsophagus is narrow in all species, and especially so in the sanguivorous Desmodont *Phyllostomatidæ*. The stomach presents two principal types of structure, which correspond respectively to the two great divisions of the order, the Megachiroptera and the Microchiroptera; in the former (with the exception of *Harpyia*) the pyloric extremity is more or less elongated and folded upon itself, in the latter it is simple, as in the Insectivora Vera; a third exceptional type is met with in the Desmodont *Phyllostomatidæ*, where the left or cardiac extremity is greatly elongated, forming a long narrow cæcum-like appendage. The intestine is comparatively short, varying from one and a half to four times the length of the head and body, being longest in the frugivorous and shortest in the insectivorous species. Only in *Rhinopoma microphyllum* and *Megaderma spasma* has a very small cæcum been found.

The liver is characterised by the great size of the left lateral lobe, which occasionally equals half the size of the whole organ; the right and left lateral fissures are usually very deep; in the Megachiroptera (*Harpyia* excepted) the Spigelian lobe is ill-defined or absent, and the caudate is generally very large; but in the Microchiroptera, on the other hand, the Spigelian lobe is very large, while the caudate is small, in most species forming a ridge only. The gall-bladder is generally well developed and attached to the right central lobe, except in the *Rhinolophidæ*, where it is connected with the left central.

In most species the hyoids are simple, consisting of a chain of slender, elongated, cylindrical bones connecting the small basi-hyoid with the cranium, while the pharynx is short, the larynx shallow with feebly developed vocal cords, and guarded by a short, acutely-pointed epiglottis, which in some genera (*Harpyia*, *Vampyrus*) is almost obsolete. In *Epomophorus*, however, we find a remarkable departure from the general type. Thus the pharynx is long and very capacious; the aperture of the larynx is far removed from the fauces, and, opposite to it, opens a canal,

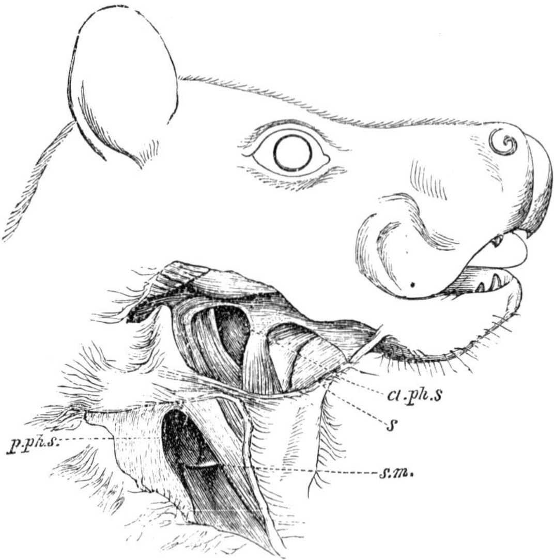

FIG. 299.—Head and neck of *Epomophorus franqueti* (adult male, natural size). The anterior (*a.ph.s*) and posterior (*p.ph.s*) pharyngeal sacs are opened from without, the dotted lines indicating the points where they communicate with the pharynx; *s*, thin membranous septum in middle line between the anterior pharyngeal sacs of opposite sides; *s.m.*, sterno-mastoid muscle separating the anterior from the posterior sac. (Dobson, *Proc. Zool. Soc.* 1881.)

leading from the narial chambers, and extending along the back of the pharynx; the laryngeal cavity is spacious and its walls are ossified; the hyoid bone is quite unconnected, except by muscle, with the cranium; the ceratohyals and epihyals are cartilaginous and greatly expanded, entering into the formation of the walls of the pharynx, and in the males of three species at least, supporting the orifices of a large pair of air-sacs communicating with the pharynx (Fig. 299).

In extent, peculiar modifications, and sensitiveness the cutaneous system reaches its highest development in this order. As a sensory organ its chief modifications in connection with the external ear and with the nasal and labial appendages have been described when referring to the nervous system. It remains therefore to consider its relative development as part of the organs of flight.

The extent and shape of the flying-membranes depend mainly on the form of the bones of the anterior extremities, and on the presence or absence of the tail. Certain modifications of these membranes, however, are met with which do not depend on the skeleton, but are related to the habits of the animals, and to the manner in which the wing is folded in repose.

These membranes consist of the "antebrachial membrane," extending from the point of the shoulder along the humerus and more or less of the forearm to the base of the pollex, the metacarpal bone of which is partially or wholly included in it; the "wing-membrane," which is spread out between the greatly elongated fingers, and extends along the sides of the body to the posterior extremities, generally reaching to the feet; and the "interfemoral membrane," the most variable of all, which is supported between the extremity of the body, the legs, and the calcar (Fig. 297).

The antebrachial and wing-membranes are most developed in those species fitted only for aerial locomotion, which when at rest hang with the body enveloped in the wings; but in the family *Emballonuridæ*, and especially in the subfamily *Molossinæ* (the species of which are the best fitted of all Bats for terrestrial progression), the antebrachial membrane is reduced to the smallest size, and is not developed along the forearm, leaving also the pollex quite free, and the wing-membrane is very narrow and folded in repose completely under the forearm. The relative development of the interfemoral membrane has been referred to above in describing the caudal vertebræ. Its small size in the frugivorous and sanguivorous species, in which its presence would be injurious as impeding their motions when searching for food as they hang suspended by their feet, is easily understood.

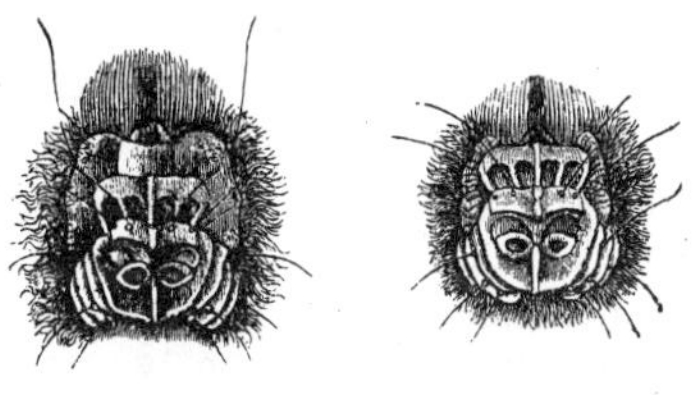

FIG. 300.—Frontal sac and nose-leaf in male and female of *Hipposiderus larvatus*. (Dobson, *Proc. Zool. Soc.* 1873.)

Odoriferous glands and pouches opening on the surface of the outer skin are developed in many species, but in most cases more so in males than in females, and thus constitute secondary sexual characters, which will be referred to when treating of the peculiarities of certain species.

All the fossil Chiroptera at present known are true Bats in every sense of the word, and therefore throw no light on the origin of the order. The earliest representatives of the order occur in beds of Upper Eocene (Lower Oligocene) age.

The order is divided by Dobson into the suborders Megachiroptera and Microchiroptera.

Suborder MEGACHIROPTERA.

Frugivorous Bats, generally of large size. Crowns of molars smooth, marked with a longitudinal groove (cuspidate in *Pteralopex*); bony palate continued behind the last molar, narrowing slowly backwards; three phalanges in the index finger, the third phalanx generally terminated by a claw; sides of the ear-conch forming a complete ring at the base; tail, when present, inferior to (not contained in) the interfemoral membrane; pyloric extremity of the stomach generally much elongated; the Spigelian lobe of the liver ill-defined or absent, and the caudate well developed.

Limited to the tropical and subtropical parts of the eastern hemisphere.

Mr. O. Thomas[1] considers that the ordinary type of molar dentition found in this suborder is a specialised adaptation from the cuspidate type of the Microchiroptera; the genus *Pteralopex* retaining the ancestral form of teeth.

Family PTEROPODIDÆ.

Since all the forms are included in this family its characters may be taken to be the same as those of the suborder.

Subfamily **Pteropodinæ.**—Tongue moderate; molars well developed.

Epomophorus.[2]—Dentition; $i\ \frac{2-1}{2}$, $c\ \frac{1}{1}$, $p\ \frac{2}{3}$, $m\ \frac{1}{2}$; total 28 or 26. Tail absent or very short, when present free from interfemoral membrane; second digit of manus clawed; premaxillæ united. This genus includes some seven species inhabiting Africa south of the Sahara. The head is large and long, and the lips are expansible, and frequently with peculiar folds. The ears have a white tuft of hair on the margin; and in the males of most species there are large glandular pouches in the skin of the side of the neck near the shoulder, from the mouth of which project long and coarse yellowish hairs, forming tufts on the shoulders, from which the genus takes its name. Another male secondary sexual character consists in the presence of a pair of large air-sacs extending outwards on each side from the pharynx beneath the integument of the neck, in the position shown in Fig. 299. These sacs are evidently capable of being greatly distended at the will of the animal, and their inflation probably occurs under the same

[1] *Proc. Zool. Soc.* 1888, p. 473.

[2] Bennett, *Trans. Zool. Soc.* vol. ii. p. 38 (1835).

circumstances that the wattles of male gallinaceous birds swell up, namely, when engaged in courting the females. Other remarkable conditions in which these Bats appear to differ from all other species occur in the form of the hyoid bones and larynx. These Bats appear to live principally on figs, the juicy contents of which their large lips and capacious mouths enable them to swallow without loss.

Pteropus.[1]—Dentition: $i\ \frac{2}{2}$, $c\ \frac{1}{1}$, $p\ \frac{3}{3}$, $m\ \frac{2}{3}$; total 34. This genus has more than forty species, and thus includes more than half the members of the family. All are of large size, and the absence of a tail, the long pointed muzzle (Fig. 301), and the woolly fur covering the neck render their recognition easy. They are commonly known as "Flying Foxes," or Fox-Bats; and one of the species (*P. edulis*) inhabiting Java measures 5 feet across the fully extended wings, and is thus the largest known species of the order. All the species closely resemble one another in dentition, and are mainly distinguished by the form of the ears and the quality of the fur. *P. scapulatus*, from North-East Australia, approaches the species of the second subfamily in the remarkable narrowness of its molars and premolars.

Fig. 301.—Head of Fox-Bat (*Pteropus personatus*). From Gray, *Proc. Zool. Soc.* 1866.

The range of this genus extends from Madagascar and the neighbouring islands through the Seychelles to India, Ceylon, Burma, the Malay Archipelago, Southern Japan, New Guinea, Australia, and Polynesia (except the Sandwich Islands, Ellice's Group, Gilbert's Group, Tokelau, and the Low Archipelago). Of the islands inhabited by it some are very small and remote from any continent, such as Savage Island in the South Pacific and Rodriguez in the Indian Ocean. Although two species inhabit the Comoro Islands, which are scarcely 200 miles from the African coast, not a single species is found in Africa; but in India, separated by thousands of miles of almost unbroken ocean, a species exceedingly closely allied to the common Madagascar Fox-Bat is abundant. The Malay Archipelago and Australia are their headquarters; and in some places they occur in countless multitudes. Mr. Macgillivray remarks of *P. conspicillatus*: "On the wooded slope of a hill on Fitzroy Island I one day fell in with this Bat in prodigious numbers, looking while flying in the bright sunshine (so unusual for a nocturnal animal) like a large flock of

[1] Geoffroy, *Ann. du Muséum*, vol. xv. p. 90 (1810).—*Ex.* Brisson.

rooks. On close approach a strong musky odour became apparent, and a loud incessant chattering was heard. Many of the branches were bending under their load of Bats, some in a state of inactivity,

FIG. 302.—Female and young of *Xantharpyia collaris*. (From Sclater, *Proc. Zool. Soc.* 1870, p. 127.)

suspended by their hind claws, others scrambling along among the boughs, and taking to wing when disturbed."

Xantharpyia.[1]—Dentition as in *Pteropus*, but a short tail present, and the fur on the back of the neck similar to that of the body.

[1] Gray, *List. Spec. Mamm. Brit. Mus.* pp. 37, 38 (1843): Syn. *Cynonycteris*.

This genus is represented by some nine species, which have a distribution very similar to that of *Pteropus*, except that they extend into Africa, and are not found in Australia and Polynesia. *X. ægyptiaca* inhabits the chambers of the Great Pyramid and other deserted buildings in Egypt, and is probably the species so generally figured in Egyptian frescoes. Fig. 302 exhibits an African species of this genus in the attitude assumed by the Fox-Bats when at rest.

Boneia.[1]—This genus, as represented by *B. bidens* of Borneo, differs from *Xantharpyia* in having only a single pair of upper incisors.

Cynopterus.[2]—Dentition: $i\,\frac{2}{2-1}$, $c\,\frac{1}{1}$, $p\,\frac{3}{3}$, $m\,\frac{2}{2}$; total 32 or 30. Muzzle short, grooved like *Pteropus* in front; tail and fur generally as in *Xantharpyia*, but the former sometimes wholly absent. This genus, with seven species, is almost limited to the Oriental region. *C. marginatus* is very common in India, and extremely destructive to ripe fruit of every description. Dr. Dobson states that "he gave to a specimen of this Bat obtained at Calcutta a ripe banana, which, with the skin removed, weighed exactly 2 ounces; the animal immediately, as if famished with hunger, fell upon the fruit, seizing it between the thumbs and the index fingers, and took large mouthfuls out of it, opening the mouth to the fullest extent with extreme voracity. In the space of three hours the whole fruit was consumed. Next morning the Bat was killed, and found to weigh one ounce, or half the weight of the food eaten in three hours. Indeed the animal when eating seemed to be a kind of living mill, the food passing from it almost as fast as devoured, and apparently unaltered, eating being, as it were, performed only for the pleasure of eating."

Harpyia.[3]—Dentition: $i\,\frac{1}{0}$, $c\,\frac{1}{1}$, $p\,\frac{2}{3}$, $m\,\frac{2}{2}$; total 24. Premaxillæ well developed and united in front; facial bones much elevated above the margin of the jaw, nostrils tubular (Fig. 303); body and limbs as in *Cynopterus*. Includes two species from the Austro-Malayan subregion, readily recognised by the peculiar tubular and projecting nostrils, as shown in the accompanying woodcut.

FIG. 303.—Head of *Harpyia major*. (From Dobson, *Proc. Zool. Soc.* 1877.)

Cephalotes.[4]—Dentition: $i\,\frac{1}{1}$, $c\,\frac{1}{1}$, $p\,\frac{2}{3}$, $m\,\frac{2}{3}$; total 28. Pre-

[1] Jentink, *Notes Leyd. Mus.* vol. i. p. 117 (1879).—Amended. [2] F. Cuvier, *Dents des Mammifères*, p. 39 (1825). [3] Illiger, *Prodromus Syst. Mamm. et Avium*, p. 118 (1811). [4] Geoffroy, *Ann. du Muséum*, vol. xvi. p. 99 (1810).

maxillæ separate in front; nostrils simple; muzzle short; index finger without a claw; tail short. Includes one species, having the same distribution as *Harpyia*. The wing-membrane arises from the middle line of the back, to which it is attached by a longitudinal very thin process of the integument; the wings are quite naked, but the back covered by them is clothed with hair.

Pteralopex.[1]—External characters as in *Pteropus;* ears short and hairy; wings arising from the middle line of the back. Muzzle very short; plane of orbit directed more upwards than in *Pteropus;* orbit surrounded by bone; sagittal crest strongly developed. Teeth cuspidate; upper incisors with broad posterior ledges; upper canine short and thick, with a stout secondary cusp in the middle of the posterior border, and two smaller postero-internal basal cusps; cheek-teeth short and broad, with their anterior and posterior basal ledges so developed and the main cusps so nearly conical as to obliterate the longitudinal grooving of *Pteropus*. Lower incisors very disproportionate, the outer pair being nearly twenty times the bulk of the inner; lower canine stout, with a simple posterior basal ledge. Represented by *P. atrata* of the Solomon Islands. As already mentioned, Mr. Thomas regards the dentition of this genus as the most generalised type found in the suborder.

Subfamily **Carponycteriinæ.**—Facial part of skull much produced; molars narrow, and scarcely raised above the gum; and the tongue exceedingly long, attenuated in the anterior third, and armed with long recurved papillæ near the tip.

Notopteris.[2]—Dentition: $i\ \frac{2}{1}$, $c\ \frac{1}{1}$, $p\ \frac{2}{3}$, $m\ \frac{2}{2}$; total 28. Index finger without a claw; wings arising from the middle line of the back; tail long; first upper premolar long, with two roots. The single representative of the genus, *N. macdonaldi*, inhabits the Fiji Islands, Aneiteum Island, and New Guinea. It is at once distinguished from all other Bats of this family by the length of its tail, which is nearly as long as the forearm.

Eonycteris.[3]—Dentition: $i\ \frac{2}{2}$, $c\ \frac{1}{1}$, $p\ \frac{3}{3}$, $m\ \frac{2}{3}$; total 34. First upper premolar small, with a single root. This genus is likewise represented by a single species (*E. spelæa*), from the Farm Caves, Moulmein, Burma, which has somewhat the appearance of *Xantharpyia;* but the absence of a claw to the index finger and the characteristic tongue and teeth at once distinguish it.

Carponycteris[4] and *Melonycteris*,[5] each with a single species, are closely allied; the index finger in both has a claw, and the number of the teeth is the same as in *Eonycteris*. *Carponycteris minima* is

[1] O. Thomas, *Ann. Mag. Nat. Hist.* ser. 6, vol. i. p. 155 (1888). [2] Gray, *Proc. Zool. Soc.* 1859, p. 36. [3] Dobson, *Journ. As. Soc. Bengal*, vol. xlii. p. 204 (1873). [4] New name: Syn. *Macroglossus*, F. Cuvier, *Dents des Mammifères*, p. 40 (1825). Preoccupied by *Macroglossum*, Scopoli, 1777. [5] Dobson, *Proc. Zool. Soc.* 1877, p. 119.

the smallest known species of the suborder, being much smaller than the common Noctule Bat of Europe, and its forearm scarcely longer than that of the Long-eared Bat. It is nearly as common in certain parts of India as *Cynopterus marginatus* (compared with which it is proportionally equally destructive to fruit), and extends eastward through the Malay Archipelago as far as New Ireland, where it is associated with *Melonycteris melanops*, distinguished from it by its larger size and the total absence of the tail.

Nesonycteris.[1]—Dentition: $i\ \frac{2}{1}$, $c\ \frac{1}{1}$, $p\ \frac{3}{3}$, $m\ \frac{2}{3}$; total 32. Allied to *Melonycteris*, but distinguished by the absence of the inner pair of lower incisors, and of a claw to the index finger. Tail wanting. Represented by *N. woodfordi*, of the Solomon Islands.

Callinycteris.[2]—Dentition: $i\ \frac{2}{2}$, $c\ \frac{1}{1}$, $p\ \frac{2}{2}$, $m\ \frac{3}{3}$; total 32. Allied to the preceding, but with a short tail; no claw to index. One species from Celebes.

Trygenycteris.[3]—Dentition: $i\ \frac{2}{2}$, $c\ \frac{1}{1}$, $p\ \frac{3}{3}$, $m\ \frac{2}{3}$; total 34. No external tail; a claw on index. One species from West Africa.

Suborder MICROCHIROPTERA.

Insectivorous (rarely frugivorous or sanguivorous) Bats, of comparatively small size. Crowns of molars acutely cusped, marked by transverse grooves; bony palate narrowing abruptly, not continued backwards laterally behind the last molar; one rudimentary phalanx (rarely two phalanges or none) in the index finger, which is never terminated by a claw; outer and inner sides of ear-conch commencing inferiorly from separate points of origin; tail, when present, contained in the interfemoral membrane, or appearing upon its upper surface; stomach simple (except in the Desmodont *Phyllostomatidæ*); Spigelian lobe of the liver very large, and the caudate generally small. Inhabit the tropical and temperate regions of both hemispheres. The members of this suborder may be divided into two sections.

Section VESPERTILIONINA.

Tail contained within the interfemoral membrane; the middle pair of upper incisors never large, and separated from each other by a more or less wide space. Middle finger with two osseous phalanges only (except in *Myxopoda aurita*, *Thyroptera tricolor*, and *Mystacops tuberculatus*). First phalanx of the middle finger extended (in repose) in a line with the metacarpal bone.

[1] O. Thomas, *Ann. Mag. Nat. Hist.* ser. 5, vol. xix. p. 417 (1887).

[2] Jentink, *Notes Leyd. Mus.* vol. xi. p. 209 (1889).

[3] New name: Syn. *Megaloglossus*; Pagenstecher, *J. B. Mus. Hamburg*, vol. ii. p. 125 (1885). Preoccupied by *Megaglossa*, Rond., 1865.

Family RHINOLOPHIDÆ.

In all the species of this family the nasal appendages are highly developed, and surround the sides of the nasal apertures, which are situated in a depression on the upper surface of the muzzle; the ears are large and generally separate, without trace of a tragus; the premaxillæ are rudimentary, suspended from the nasal cartilages, and supporting a pair of very small incisors; the molars have acute W-shaped cusps; the skull is large, and the nasal bones which support the large nasal cutaneous appendages are much expanded vertically and laterally; in the females a pair of teat-like appendages are found in front of the pubis; and the tail is long and produced to the posterior margin of the interfemoral membrane. This family is found in the temperate and tropical parts of the eastern hemisphere.

From whatever point of view the *Rhinolophidæ* may be considered, they are evidently the most highly organised of insectivorous Bats. In them the osseous and cutaneous systems reach the most elaborate development. Compared with those of the present family the bones of the extremities and the flying-membranes of other Bats appear coarsely formed, and even their teeth seem less perfectly fitted to crush the hard bodies of insects. The very complicated nasal appendages, which evidently act as delicate organs of special perception, here reach their highest development, and the differences in their form afford valuable characters in the discrimination of the species, which resemble one another very closely in dentition and in the colour of the fur.

Subfamily **Rhinolophinæ**.—First toe with two, other toes with three, phalanges each; ilio-pectineal spine not connected by bone with the antero-inferior surface of the ilium.

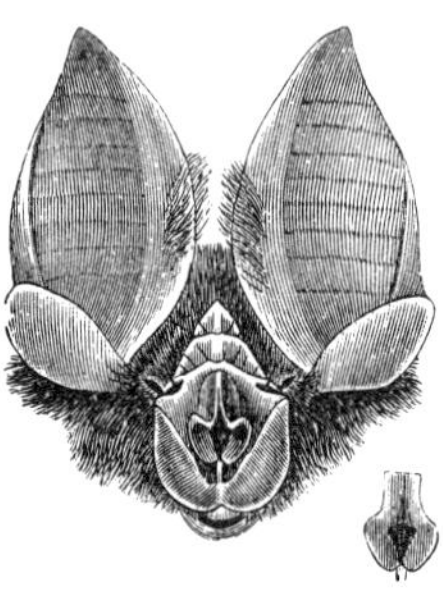

FIG. 304.—Head of Indian Horse-shoe Bat (*Rhinolophus mitratus*). (From Dobson, *Monogr. Asiat. Chiropt.*)

Rhinolophus.[1]—Dentition: $i\ \frac{1}{2}$, $c\ \frac{1}{1}$, $p\ \frac{2}{3}$, $m\ \frac{3}{3}$; total 32. Nose-leaf (Fig. 304) with a central process behind and between the nasal orifices, posterior extremity lanceolate, antitragus large. Includes more than twenty species. *R. luctus*, in which the forearm has a length of 3 inches, is the largest species, inhabiting elevated hill tracts in India and Malayana; *R. hipposiderus* of Europe, extending into South England and Ireland, forearm 1·5 inches, is one of the smallest; and *R. ferrum-equinum*, with the forearm 2·3 inches in length, represents the average size of the species, which are mainly distinguished from one another by the form of the nose-leaf. The

[1] Geoffroy, *Nouv. Dict. d'Hist. Nat.* vol. xix. p. 383 (1803).

last-named species extends from England to Japan, and southward to the Cape of Good Hope. The genus is represented in the Himalaya by the closely allied *R. tragatus*, distinguished by having three vertical grooves on the lower lip, in place of the single groove found in *R. ferrum-equinum*. *Rhinolophus* is represented in the Upper Eocene Phosphorites of Central France by *R. antiquus* and *R. dubius*; the former appears to have the same dental formula as in the existing species, but differs slightly in the structure of some of the lower molars, so that it is separated generically by some writers under the name of *Pseudorhinolophus*. The face is also longer than in existing forms, and there are certain differences in the structure of the skull. *Alastor*, from the same deposits, differs from *Rhinolophus* by the extreme shortness of the nasal region. *Palæonycteris*, from the Lower Miocene of France, is said to be allied to *Rhinolophus*, but the premolars are $\frac{3}{3}$, and the limb bones are stated to resemble those of *Molossus*.

Subfamily **Hipposiderinæ.**—Toes equal, of two phalanges each; ilio-pectineal spine united by a bony isthmus with a process derived from the antero-inferior surface of the ilium.

Hipposiderus.[1]—Dentition: $i\,\frac{1}{2}$, $c\,\frac{1}{1}$, $p\,\frac{2-1}{2}$, $m\,\frac{3}{3}$; total 30 or 28. Tail well developed. This genus, of which more than twenty species have been described, differs from *Rhinolophus* in the form of the nose-leaf, which is not lanceolate behind and is unprovided with a central process covering the nostrils. The largest species, *H. armiger*, appears to be the most northerly, having been taken at Amoy in China, and in the Himalaya at an elevation of 5,500 feet. Many of the species are provided with a peculiar frontal sac behind the nose-leaf, rudimentary in females (Fig. 305), which the animal can evert at pleasure; the sides of this sac secrete a waxy substance, and its extremity supports a pencil of straight hairs.

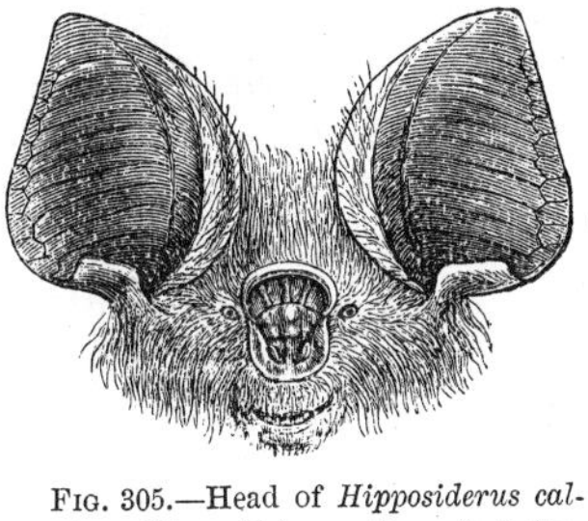

FIG. 305.—Head of *Hipposiderus calcaratus*. (From Dobson, *Proc. Zool. Soc.* 1877.)

Anthops.[2]—Like *Hipposiderus*, but with the tail rudimentary, consisting merely of three or four vertebræ hidden in the base of the interfemoral membrane. Nose-leaf very complicated, its upright transverse portion emarginate above, and the projections rounded

[1] Gray, *Proc. Zool. Soc.* 1834, p. 53. The Bats of this genus are usually described as *Phyllorhina*, but this use has been shown to be incorrect; see Blanford, *Proc. Zool. Soc.* 1887, p. 637.

[2] O. Thomas, *Ann. Mag. Nat. Hist.* ser. 6, vol. i. p. 156 (1888).

and hollowed behind, and their substance quite thin. Premolars $\frac{2}{2}$. Represented by *A. ornatus* of the Solomon Islands.

Mr. O. Thomas, the describer of this Bat, remarks that it is evidently more nearly allied to the preceding than to the succeeding genera, although it agrees with *Cœlops* in the rudimentary tail.

Rhinonycteris[1] and *Triænops*.[2]—These are two allied genera with well-developed tails; the former being represented by *R. aurantia* from Australia, and the latter by *T. persicus* from Persia and Eastern Africa. *Triænops* (Fig. 306) is characterised by the remarkable form of its nasal appendages and ears, and the presence of a peculiar osseous projection from the proximal extremity of the second phalanx of the fourth finger.

FIG. 306.—Head of *Triænops persicus*. × 2. (From Dobson, *Monogr. Asiat. Chiropt.*)

Cœlops.[3]—This genus is known only by a single species, *C. frithi*, from the Bengal Sunderbans, Java, and Siam (in the roof of the great pagoda at Laos); and is distinguished, not only by the form of its nose-leaf, but also by the great length of the metacarpal of the index finger, as well as by the shortness of the calcar and interfemoral membrane and the rudimental tail.

Family NYCTERIDÆ.

This small family, including only two genera of Bats of peculiar aspect, limited to the tropical and subtropical parts of the eastern hemisphere, is distinguished from the *Rhinolophidæ* by the presence of a distinct tragus to the ear, and by the premaxillæ being cartilaginous or small and separated from one another in front by a distinct space.

Megaderma.[4]—Dentition: $i\ \frac{0}{2}$, $c\ \frac{1}{1}$, $p\ \frac{2-1}{2}$, $m\ \frac{3}{3}$; total 28 or 26. This genus, which is represented by five species, is readily recognised by the absence of upper incisors, the cylindrical narrow muzzle surmounted by an erect naked cutaneous nose-leaf, the base of which conceals the nasal orifices, by the immense connate ears with large bifid tragi, and by the great extent of the interfemoral membrane, in the base of which the very short tail is concealed. *M. gigas* (Fig. 307), from Central Queensland (length of forearm 4·2

[1] Gray, *Proc. Zool. Soc.* 1847, p. 16. [2] Dobson, *Journ. As. Soc. Bengal*, vol. xl. p. 455 (1871). [3] Blyth, *Journ. As. Soc. Bengal*, vol. xvii. p. 251 (1848). [4] Geoffroy, *Ann. du Muséum*, vol. xv. p. 197 (1810).

inches), is not only the largest species of the genus but also of the suborder. *M. lyra*, common in India (forearm 2·7 inches), has been caught in the act of sucking the blood, while flying, from a small species of *Vesperugo*, which it afterwards devoured, so that it is probable that the Bats of this genus do not confine themselves to

FIG. 307.—*Megaderma gigas*. $\times \frac{1}{2}$. (From Dobson, *Proc. Zool. Soc.* 1880.)

insect prey alone, but also feed, when they can, upon the smaller species of Bats and other small mammals.

The Oriental *M. spasma* and *M. lyra* differ from the Ethiopian *M. cor* and *M. frons* in having two upper premolars instead of one, and also in the shape of the frontals and nasals.

Nycteris.[1]—Dentition: $i\ \frac{2}{3}$, $c\ \frac{1}{1}$, $p\ \frac{1}{2}$, $m\ \frac{3}{3}$; total 32. This genus, of which there are seven species, differs so much from *Megaderma* that it may be considered the type of a separate subfamily. As in that genus, the frontal bones are deeply hollowed out and expanded laterally, the muzzle presents a similar cylindrical form, and the lower jaw also projects, but the single elevated nose-leaf is absent, and instead of it the face is marked by a deep, longitudinal, sharp-edged groove extending from the nostrils (which are on the upper

[1] Geoffroy, *Nouv. Dict. d'Hist. Nat.* vol. xv. p. 501 (1803).

surface of the muzzle, near its extremity) to the low band connecting the bases of the large ears; the sides of this depression being margined as far back as the eyes by small horizontal cutaneous appendages. All the species resemble one another closely, and are mainly distinguished by the form of the tragus and the size and relative position of the second lower premolar. With the exception of *N. javanica*, they are all limited to the Ethiopian region.

Family VESPERTILIONIDÆ.

Nostrils opening by simple crescentic or circular apertures at the extremity of the muzzle, not surrounded by distinct foliaceous cutaneous appendages; premaxillæ small, lateral, and separated by a wide space in front; tragus distinct. In addition to these characters, it may be observed that the skull is of moderate size, the nasal and frontal bones not being much extended laterally or vertically, nor furrowed by deep depressions. The number of incisors varies from $\frac{2}{3}$ to $\frac{1}{3}$, rarely (in *Antrozous* only) $\frac{1}{2}$, premolars $\frac{3}{3}$, or $\frac{2}{2}$, or $\frac{1}{2}$, rarely (in *Vesperugo noctivagans* of North America) $\frac{2}{3}$; the upper incisors are small, separated by a wide space in the middle line, and placed in pairs or singly near the canine; the molars are well-developed, with acute W-shaped cusps.

This family, which may be regarded as occupying a central position in the suborder, includes the common simple-faced Bats of all countries, of which the well-known Pipistrelle and the Whiskered Bat (*Vespertilio mystacinus*) may be taken as familiar types, and its species number more than 150, or considerably more than one-third the total number of the known species of the entire order. The various genera may be conveniently grouped into the *Plecotine*, *Vespertilionine*, *Miniopterine*, and *Thyropterine* divisions.

In the *Plecotine* division, of which the common Long-eared Bat (*Plecotus auritus*) is the type, the crown of the head is but slightly raised above the face-line, the outermost upper incisor is close to the canine, and the nostrils are margined behind by grooves on the upper surface of the muzzle, or by rudimentary nose-leaves; the ears also are generally very large and united.

Plecotus.[1]—Dentition: $i\ \frac{2}{3}$, $c\ \frac{1}{1}$, $p\ \frac{2}{3}$, $m\ \frac{3}{3}$; total 36. Outer margin of ear-conch ending abruptly near the angle of the mouth, the inner margin with a more or less prominent rounded projection directed inwardly above the base; tragus very large, tapering upwards, with a lobe at the base of its outer margin, rounded, and placed half horizontally. This genus is represented by the European Long-eared Bat (*P. auritus*), and *P. macrotis*, restricted to North America. The latter is distinguished by the great size of

[1] Geoffroy, *Descript. de l'Egypte*, vol. ii. p. 112 (1812).

the glandular prominences of the sides of the muzzle, which meet in the centre above and behind the nostrils. *P. auritus* extends over the greater part of the Palæarctic region, occurring in Ireland in the west and the Himalaya in the east.

Synotus.[1]—Dentition: $i\ \frac{2}{3}$, $c\ \frac{1}{1}$, $p\ \frac{2}{2}$, $m\ \frac{3}{3}$; total 34. This genus is distinguished from the preceding by the loss of one lower premolar and by the outer margin of the ear being carried forwards above the mouth and in front of the eye; it includes the European Barbastelle Bat (*S. barbastellus*) and *S. darjilingensis* from the Himalaya.

Otonycteris.[2]—Dentition: $i\ \frac{1}{3}$, $c\ \frac{1}{1}$, $p\ \frac{1}{2}$, $m\ \frac{3}{3}$; total 30. The reduction in the number of upper incisors readily characterises this genus, which appears to connect the typical representatives of the section, through *Scotophilus*, with the Vespertilionine division. It is represented by a single species, *O. hemprichi*, from North Africa and the Himalaya.

Nyctophilus.[3]—Dentition: $i\ \frac{1}{3}$, $c\ \frac{1}{1}$, $p\ \frac{1}{2}$, $m\ \frac{3}{3}$; total 30. This and the following genera are distinguished from all the preceding by the presence of a rudimentary nose-leaf. The present genus contains *N. timoriensis* of the Australian region, and *N. microtis* of New Guinea.

Antrozous.[4]—Dentition: $i\ \frac{1}{2}$, $c\ \frac{1}{1}$, $p\ \frac{1}{2}$, $m\ \frac{3}{3}$; total 28. Readily distinguished from the other members of the whole family by having but two lower incisors, and from the other species of the section by the separate ears. The single species, *A. pallidus*, inhabits California.

The *Vespertilionine* division includes some nine-tenths of all the representatives of the family. They are distinguished from the preceding section by the simple nostrils, opening by crescentic or circular apertures at the extremity of the muzzle, the generally small size of the ears, and the absence of grooves on the forehead.

Vesperugo.[5]—Dentition: $i\ \frac{2-1}{3}$, $c\ \frac{1}{1}$, $p\ \frac{2-1}{2-3}$, $m\ \frac{3}{3}$; total 34, 30, or 36. This large genus comprises about one-third of the section, and is divided into groups or subgenera, according to the number of premolars and incisors; the latter varying from $\frac{2}{3}$ to $\frac{1}{3}$ in the subgenera *Scotozous* and *Rhogeëssa*, and the premolars from $\frac{2}{2}$ to $\frac{1}{2}$ (in the subgenus *Lasionycteris* $\frac{2}{3}$). The Bats of this genus are generally easily distinguished by their comparatively thickly formed bodies, flat broad heads and obtuse muzzles, short, broad, and triangular obtusely-pointed ears, obtuse and usually slightly incurved tragus, short legs, and by the presence in most species of a well-developed post-calcaral lobule. This lobule (which is supported by a cartil-

[1] Keyserling and Blasius, *Wirbelthiere Europ.* p. 55 (1840). [2] Peters, *Monatsber. Ak. Berlin*, 1859, p. 222. [3] Leach, *Trans. Linn. Soc.* vol. xiii. p. 78 (1822). [4] Allen, *Proc. Ac. Nat. Sci. Philad.* 1862, p. 247. [5] Keyserling and Blasius, *Wiegmann's Archiv*, 1839, p. 312.

aginous process derived from the calcar) may act as a kind of adhesive disc in securing the animal's grasp when climbing over smooth surfaces. *Vesperugo* probably contains the greatest number of individuals among the genera of Chiroptera, and, with the exception of *Vespertilio*, its species have also the widest geographical range, being almost cosmopolitan; and one of the species, the well-known Serotine (*V.* [*Vesperus*] *serotinus*) is remarkable as the only species of Bat known to inhabit both the Old and the New World; one (*V. borealis*) has been found close to the limits of the Arctic circle, and another (*V. magellanicus*) inhabits the cold and desolate shores of the Straits of Magellan, being doubtless the Bat referred to by Mr. Darwin in the *Naturalist's Voyage.* The Common Pipistrelle (*V. pipistrellus*), ranging over the greater part of the Palæarctic region, is the best known species.

Chalinolobus.[1]—This genus agrees with *Vesperugo* in the dental formula, but is readily distinguished by the presence of a well-defined lobe projecting near the angle of the mouth from the lower lip, and by the unicuspidate first upper incisor. The species fall into two subgenera—*Chalinolobus* proper, with $p\ \frac{2}{2}$, represented by *C. morio* from New Zealand, Tasmania, and Australia, and three other species from Australia; and *Glauconycteris*, with $p\ \frac{1}{2}$, limited to Southern and Equatorial Africa, and known by *C. argentatus* and two other species, the Bats of this subgenus being especially remarkable for their peculiarly thin membranes, traversed by very distinct reticulations and parallel lines.

Scotophilus.[2]—Dentition: $i\ \frac{1}{3},\ c\ \frac{1}{1},\ p\ \frac{1}{2},\ m\ \frac{3}{3}$; total 30. This genus comprises a comparatively small number of species nearly allied to *Vesperugo*, and some of which approach so closely to the aberrant types of the latter ranged under the subgenus *Scotozous*, as to render the definition of the present genus almost impossible.[3] The species are restricted to the tropical and subtropical regions of the eastern hemisphere, though widely distributed within these limits. The more typical species are distinguished especially by the single pair of unicuspidate upper incisors separated by a wide space and placed close to the canines, by the small transverse first lower premolar squeezed in between the canine and second premolar, and, generally, by their conical nearly naked muzzles and remarkably thick leathery membranes. *S. kuhli* is probably the commonest species of Bat in India, and appears often on the wing even before the sun has touched the horizon,

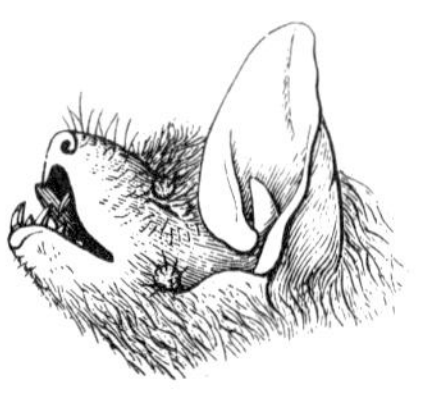

FIG. 308.—Head of *Scotophilus emarginatus.* (Dobson, *Monogr. Asiat. Chiropt.*)

[1] Peters, *Monatsber. Ak. Berlin*, 1866, p. 672.
[2] Leach, *Trans. Linn. Soc.* vol. xiii. p. 71 (1822).
[3] See O. Thomas, *Ann. Mus. Genova* (2), vol. ix. pp. 84-88 (1890).

especially when the white-ants are swarming, feeding eagerly upon them as they rise in the air. *S. gigas*, from Equatorial Africa, with the forearm measuring 3·4 inches, is by far the largest species. *S. albofuscus*, from the Gambia, which is readily distinguished from the other species by its white wings, is an aberrant form, in which the lower premolars are long and not crowded together, and thereby so closely resembles *Vesperugo* (*Scotozous*) *dormeri* as to render it almost impossible to distinguish *Scotophilus* and *Vesperugo*. The figured species is from India.

Nycticejus.[1]—This genus, with the same dental formula as *Scotophilus*, is distinguished by the first lower premolar not being squeezed in between the adjoining teeth, and by the comparatively much greater size of the last upper molar. It includes only the common North American *N. humeralis* (*crepuscularis*), a small Bat scarcely larger than the Pipistrelle. It seems, however, as pointed out by Mr. O. Thomas, that the discovery of *Scotophilus albofuscus* renders the generic distinctness of *Nycticejus* no longer tenable, and that the species should be known as *Scotophilus humeralis*.

Atalapha.[2]—Dentition: $i\ \frac{1}{3},\ c\ \frac{1}{1},\ p\ \frac{2-1}{2},\ m\ \frac{3}{3}$; total 32 or 30. The five species of this genus, which are confined to the New World, are generally characterised by the interfemoral membrane being more or less covered with hair (in the two commonest species, *A. noveboracensis* and *A. cinerea*, wholly covered), and by the peculiar form of the tragus, which is expanded above and abruptly curved inwards. These species have two upper premolars, of which the first is extremely small and quite internal to the tooth-row.

Harpyiocephalus.[3]—Dentition: $i\ \frac{2}{3},\ c,\ \frac{1}{1},\ p\ \frac{2}{2},\ m\ \frac{3}{3}$; total 34. This genus includes some eight species of small Bats distinguished by their prominent tube-like nostrils and hairy interfemoral membrane. *H. suillus*, from Java and neighbouring islands, is the best-known species, and another closely allied (*H. hilgendorfi*) has been described by Professor Peters from Japan. The remaining six species are known only from the Himalaya and Tibet. All appear to be restricted to the hill tracts of the countries in which they are found.

Vespertilio.[4]—Dentition: $i\ \frac{2}{3},\ c\ \frac{1}{1},\ p\ \frac{3}{3},\ m\ \frac{3}{3}$; total 38. Next to *Vesperugo*, this genus includes by far the largest number of species, amounting to over forty; it has, however, rather a wider geographical distribution in both hemispheres, one species at least being recorded from the Navigators' Islands. The species are easily recognised by the peculiar character of the upper incisors, the crowns of which diverge from each other; by the large number of premolars, of which the second upper one is always very small;

[1] Rafinesque, *Journ. de Physique*, vol. lxxxviii. p. 417 (1819). [2] Rafinesque, *Précis des Decouvértes et Trav. Somiol.* p. 12 (1814). [3] Gray, *Ann. Mag. Nat. Hist.* vol. x. p. 259 (1842) [4] Linn. *Syst. Nat.* 12th ed. vol. i. p. 46 (1766).

and by the oval elongated ear and narrow attenuated tragus. In the British Isles this genus is represented by four species, viz. Bechstein's Bat (*V. bechsteini*); the Reddish-Gray Bat (*V. nattereri*), of very local occurrence; Daubenton's Bat (*V. daubentoni*); and the Whiskered Bat (*V. mystacinus*).

Cerivoula.[1]—This genus, which has the same dental formula as *Vespertilio*, is distinguished by the parallel upper incisors, and the comparatively large size of the second upper premolar. Some ten species have been described from the Ethiopian and Oriental regions, of which *C. picta*, from India and the Indo-Malayan subregion, is the best-known, being well characterised by its brilliantly coloured orange fur and conspicuously marked membranes, which are variegated with orange and black. This genus includes the most delicately formed and most truly insectivorous, tropical, forest-haunting Bats, which appear to stand as regards the species of *Vespertilio* in a position similar to that occupied by *Chalinolobus* with respect to *Vesperugo*.

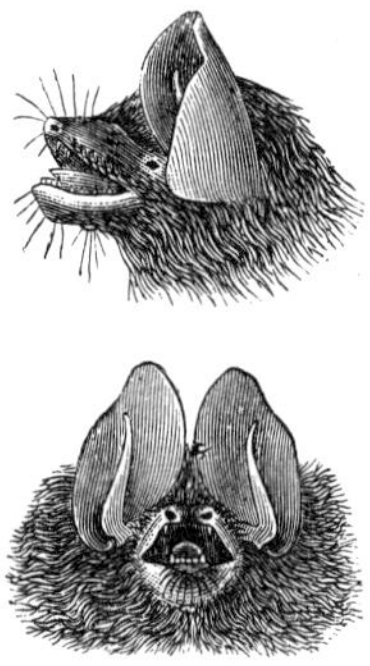

FIG. 309.—Side and front views of the head of *Cerivoula hardwickei*. (Dobson, *Monogr. Asiat. Chiropt.*)

The *Miniopterine* division includes only two genera, and is characterised by the great elevation of the crown of the head above the facial line, and by the upper incisors being separated from the canine and also in the middle line.

Natalus.[2]—This genus, while having the divisional characters mentioned above, agrees in the dental formula and its general external form with *Cerivoula*, from which it is distinguished by the short triangular tragus. It includes three species, restricted to South and Central America and the West Indies; the head of *N. micropus* being shown in Fig. 310.

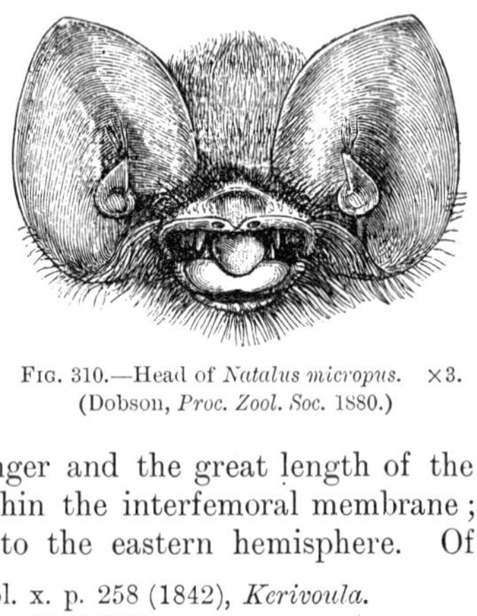

FIG. 310.—Head of *Natalus micropus*. ×3. (Dobson, *Proc. Zool. Soc.* 1880.)

Miniopterus.[3]—Dentition: $i\ \frac{2}{3}$, $c\ \frac{1}{1}$, $p\ \frac{2}{3}$, $m\ \frac{3}{3}$; total 36. In addition to the difference in the number of the teeth, this genus is distinguished by the shortness of the first phalanx of the middle finger and the great length of the tail, which is wholly contained within the interfemoral membrane; it includes four species, restricted to the eastern hemisphere. Of

[1] Gray, *Ann. Mag. Nat. Hist.* vol. x. p. 258 (1842), *Kerivoula.*

[2] Gray, *Mag. Zool. Bot.* vol. ii. p. 496 (1838).

[3] Bonaparte, *Fauna Italica*, fasc. xxi. (1837).

these the best-known, *M. schreibersi*, is very widely distributed, being found almost everywhere throughout the tropical and warmer temperate regions of the eastern hemisphere; specimens from Germany, Madagascar, Japan, and Australia differing in no appreciable respect from one another.

The last or *Thyropterine* division, which likewise comprises only two genera, is characterised by the presence of an additional osseous phalanx in the middle finger and an equal number of phalanges in the toes, and also by peculiar accessory clinging organs attached to the extremities.

Thyroptera.[1]—Dentition: $i\ \frac{2}{3}$, $c\ \frac{1}{1}$, $p\ \frac{3}{3}$, $m\ \frac{3}{3}$; total 38. In the single species *T. tricolor* of Brazil the clinging organs have the appearance of small, circular, pedunculated, hollow discs (Fig. 311), resembling in miniature the sucking cups of cuttle-fishes, and are

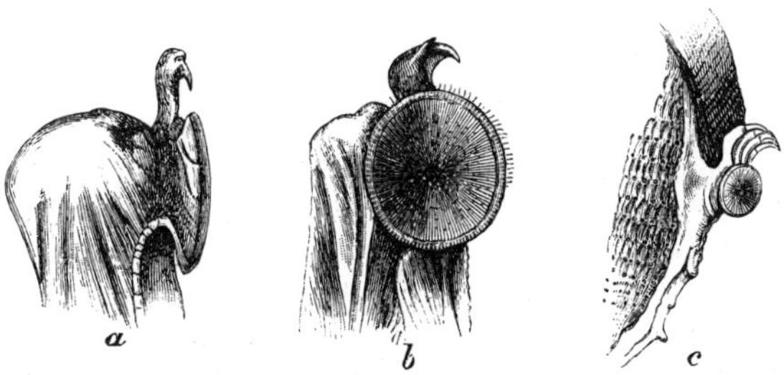

FIG. 311.—Suctorial discs in *Thyroptera tricolor*. *a*, Side and *b*, concave surface, of thumb-disc; *c*, foot with disc, and calcar with projections (all much enlarged). Dobson, *Proc. Zool. Soc.* 1876.

attached to the inferior surfaces of the thumbs and soles of the feet. With these the animal is enabled to maintain its hold when creeping over smooth vertical surfaces.

Myxopoda.[2]—The second genus is likewise represented only by a single species—*M. aurita* of Madagascar—and is distinguished from the preceding by the characters of the teeth and the form of the ears. The whole inferior surface of the pollex supports a large sessile horse-shoe-shaped adhesive pad, with the circular margin directed forwards and notched along its edge, and a smaller pad occupies part of the sole of the foot.

Fossil Vespertilionidæ.—It is not improbable that *Vesperugo* is represented in the Upper Eocene of the Paris basin by *V. parisiensis*, which appears to be allied to *V. serotina*, although it has been regarded by some writers as generically distinct, under the name of *Nyctitherium*. *Vesperugo* (*Nyctitherium*) also occurs in the Bridger Eocene of the United States; *Nyctilestes* from the same

[1] Spix, *Sim. and Vesp. Bresil*, p. 61 (1823).

[2] A. Milne-Edwards, *Bull. Soc. Philom.* sér. 7, vol. ii. p. 1 (1878).

deposits being an allied extinct genus. A number of European Miocene species have been referred to *Vespertilio*, but the term in these cases must be used in a somewhat wide sense. *Vespertiliavus*, of the Phosphorites of Central France, differs from *Vespertilio* in the proportions of its premolars.

Section EMBALLONURINA.

Tail perforating the interfemoral membrane and appearing on its upper surface, or produced considerably beyond the truncated membrane; the middle pair of upper incisors generally large and close together.

Family EMBALLONURIDÆ.

First phalanx of the middle finger folded (in repose) on the dorsal surface of the metacarpal bone (except in *Noctilio* and *Mystacops*). Nostrils opening by simple circular or valvular apertures at the extremity of the muzzle, not surrounded or margined by foliaceous cutaneous appendages; tragus distinct.

The *Emballonuridæ* are generally easily distinguished by the peculiar form of the muzzle, which is obliquely truncated, the nostrils projecting more or less in front beyond the lower lip; by the first phalanx of the middle finger being folded in repose forwards on the upper surface of the metacarpal bone; by the tail, which either perforates the interfemoral membrane or is produced far beyond it; and by the upper incisors, which are generally a single pair separated from the canine and also in the middle line. The family is cosmopolitan like the *Vespertilionidæ*, but rarely extends north or south of the thirtieth parallel of latitude.

Subfamily **Emballonurinæ.**—Tail slender, perforating the interfemoral membrane, and appearing upon its upper surface, or terminating in it; legs long, fibula very slender; upper incisors weak.

In the *Furipterine* division the tail terminates in the interfemoral membrane; the crown of the head is greatly elevated above the face-line; the thumb and first phalanx of the middle finger are very short; and the dentition is $i\ \frac{2}{3}$, $c\ \frac{1}{1}$, $p\ \frac{2}{3}$, $m\ \frac{3}{3}$; total 38.

Represented by two genera, *Furipterus*[1] and *Amorphochilus*,[2] each including one species of peculiar aspect; the latter distinguished from the former by the widely separated nostrils and the great extension backwards of the bony palate. Habitat South America.

In the typical or *Emballonurine* division part of the tail is included in the basal half of the interfemoral membrane, the remain-

[1] Bonaparte, *Faun. Ital.* vol. i. (1832-41): Syn. *Furia*, F. Cuvier, *Mém. du Muséum*, vol. xvi. p. 150 (1828). Preoccupied by Linn. 1766.

[2] Peters, *Monatsber. Ak. Berlin*, 1877, p. 185.

ing part passing through and appearing upon its upper surface; the crown of the head is slightly elevated; the pollex and first phalanx of the middle finger are moderately long; and the number of the premolars is always $\frac{2}{2}$.

Emballonura.[1]—Incisors $\frac{2}{3}$. Extremity of the muzzle more or less produced beyond the lower lip, forehead flat. Contains some five species, inhabiting islands from Madagascar through the Malay Archipelago to the Navigators' Islands.

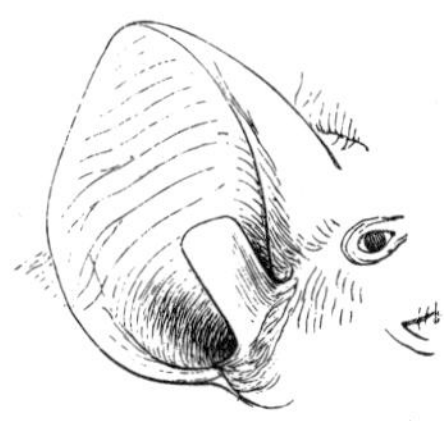

FIG. 312.—Ear of *Emballonura raffrayana*. ×2. (Dobson, *Proc. Zool. Soc.* 1878.)

Coleüra.[2]—Incisors $\frac{1}{3}$. Extremity of the muzzle broad, forehead concave. Has two species from East Africa and the Seychelles Islands.

Rhynchonycteris.[3]—This genus is distinguished from *Coleüra* by the much-produced extremity of the muzzle. The single species, *R. naso*, from Central and South America, is very common in the vicinity of streams throughout the tropical parts of these countries; it is usually found during the day resting on the vertical faces of rocks, or on the trunks of trees growing over the water, and, owing to the peculiar grayish colour of the fur covering the body and growing in small tufts from the antebrachial membrane, so as to counterfeit the weathered surfaces of rocks and the bark of trees, easily escapes notice. As the shades of evening approach it appears early on the wing, flying close to the surface of the water, and seizing the minute insects that hover over it.

Saccopteryx.[4]—Incisors $\frac{1}{3}$. Antebrachial membrane with a pouch opening on its upper surface. This genus contains six species from Central and South America. In the adult males a valvular longitudinal opening is found on the upper surface of the membrane, varying in position in different species. This opening leads into a small pouch (in some species large enough to hold a pea), the interior of which is lined with a glandular membrane secreting an unctuous substance of a reddish colour with a strong ammoniacal odour. The presence of this sac only in males indicates that it is a secondary sexual character analogous to the shoulder-pouches of *Epomophorus* and the frontal sacs of *Hipposiderus*. It is quite rudimentary in the females.

Taphozous.[5]—Incisors $\frac{1}{2}$; upper pair deciduous. This genus, represented by some ten species, inhabiting the tropical and subtropical parts of all the eastern hemisphere except Polynesia, forms the second group of this division, distinguished by the cartilaginous

[1] Temminck (Van der Hœven), *Tijdsch. Nat. Ges.* 1839, p. 22.

[2] Peters, *Monatsber. Ak. Berlin*, 1867, p. 479. [3] Peters, *loc. cit.* p. 477.

[4] Illiger, *Prodromus Syst. Mamm. et Avium*, p. 121 (1811).

[5] Geoffroy, *Descript. de l'Egypte*, vol. ii. p. 126 (1812).

premaxillaries, deciduous upper incisors, and the presence of only two lower incisors. Most of the species have a peculiar glandular sac (Fig. 313) placed between the angles of the lower jaw. This is a sexual character, for, while always more developed in males than in females, in some species, although distinct in the male, it is quite absent in the female. An open gular sac is wanting in both sexes in *T. melanopogon*, but about its usual position the openings of small pores may be seen, the secretion exuding from

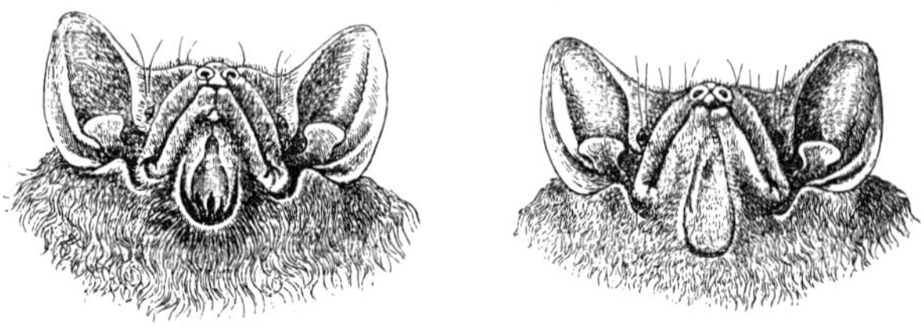

Fig. 313.—Heads of *Taphozous longimanus*, showing relative development of gular sacs in male and female. (Dobson, *Proc. Zool. Soc.* 1873.)

which probably causes the hairs to grow very long, forming the black beard found in many male specimens of this species.

In the *Diclidurine* division there is but a single genus, represented by two species.

Diclidurus.[1]—Dentition: $i\ \frac{1}{3}$, $c\ \frac{1}{1}$, $p\ \frac{2}{2}$, $m\ \frac{3}{3}$; total 32. Both species are from the Neotropical region, the typical *D. albus* ranging as far north as Central America. This Bat resembles the species of *Taphozous* in the form of the head and ears, but, besides other characters, differs from all other Bats in possessing a peculiar pouch, opening on the centre of the inferior surface of the interfemoral membrane; the extremity of the tail enters this, and perforates its fundus.

The *Noctilionine* division is likewise represented only by a single genus, with two species. This genus connects the present with the following family, possessing characters common to both, but also so many remarkable special peculiarities as almost to warrant the formation of a separate family for its reception.

Noctilio.[2]—Dentition: $i\ \frac{2}{1}$, $c\ \frac{1}{1}$, $p\ \frac{1}{2}$, $m\ \frac{3}{3}$; total 28. The two species *N. leporinus* and *N. albiventer* inhabit Central and South America. The typical *N. leporinus* is a Bat of very curious aspect, with strangely folded lips, erect cutaneous processes on the chin, and enormous feet and claws. The first upper incisors are close together, and so large as to conceal the small outer ones, while in the lower jaw there is one pair of small incisors. This apparent resemblance to a Rodent actually led Linnæus to remove this species from the Bats and place it in the Rodents. This Bat is remark-

[1] Wied, *Isis*, 1819, p. 1629. [2] Linn. *Syst. Nat.* 12th ed. vol. i. p. 88 (1766).

able for feeding on fish—a circumstance which has only recently been fully authenticated.

The remaining genus of this subfamily is regarded as representing another division, which may be known as the *Rhinopomatine* division.

Rhinopoma.[1]—This genus, represented by the single species *R. microphyllum*, might also be elevated to the rank of a family, for it is difficult to determine its exact affinities, a kind of cross relationship attaching it to the *Nycteridæ* on the one hand and to this family, in which it is here placed provisionally, on the other. This species, distinguished from all other Microchiroptera as well by the presence of two phalanges in the index finger as by its remarkably long and slender tail projecting far beyond the narrow interfemoral membrane, inhabits the subterranean tombs in Egypt and deserted buildings generally from North-East Africa to Burma.

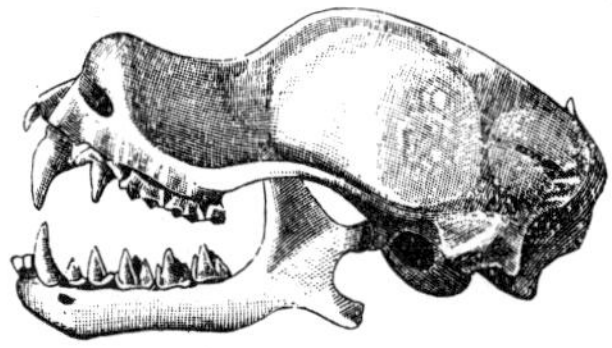

FIG. 314.—Skull of *Rhinopoma microphyllum*. ×2. (Dobson, *Monogr. Asiat. Chiropt.*)

Subfamily **Molossinæ.**—Tail thick, produced far beyond the posterior margin of the interfemoral membrane (except in *Mystacops*); legs short and strong, with well-developed fibula; upper incisors strong. This subfamily includes all the species of *Emballonuridæ* with short and strong legs and broad feet (whereof the first toe, and in most species the fifth also, is much thicker than the others, and furnished with long curved hairs), well-developed callosities at the base of the thumbs, and a single pair of large upper incisors occupying the centre of the space between the canines. In all the species the feet are free from the wing-membrane, which folds up perfectly under the forearm and legs; the interfemoral membrane is retractile, being movable backwards and forwards along the tail, and this power of varying its superficial extent must confer upon these Bats great dexterity in quickly changing the direction of their flight, as when obliged to double in pursuing their swift insect prey, which their extremely expansible lips evidently enable them to secure with ease. Like the preceding subfamily, the genera may be arranged in divisions, of which there are two.

The *Molossine* division is characterised by the production of the tail beyond the posterior margin of the interfemoral membrane; it includes three genera.

Chiromeles.[2]—Dentition: $i\ \frac{1}{1}$, $c\ \frac{1}{1}$, $p\ \frac{1}{2}$, $m\ \frac{3}{3}$; total 26. Hallux much larger than the other toes and separable from them, ears

[1] Geoffroy, *Descript. de l'Egypte*, vol. ii. p. 123 (1812).

[2] Horsfield, *Zool. Research Java* (1824).

separate. This genus is represented by a single species, *C. torquatus*, of large size (forearm 3·1 inches) and peculiar aspect, inhabiting the Indo-Malayan subregion. This Bat is nearly naked, a collar only of thinly spread hairs half surrounding the neck; and is further remarkable for its enormous throat-sac and curious nursing-pouches. The former consists of a great semicircular fold of skin forming a deep pouch round the neck beneath, and concealing the orifices of large subcutaneous pectoral glands, which discharge an oily fluid of insufferably offensive smell. The nursing-pouch is formed on each side by an extension of a fold of skin from the side of the body to the inferior surfaces of the humerus and femur. In the anterior part of this pouch the mammæ are placed.

Molossus.[1]—Dentition: $i\ \frac{1}{1-2},\ c\ \frac{1}{1},\ p\ \frac{1-2}{1},\ m\ \frac{3}{3}$; total 24 or 28. Upper incisors close together in the middle line. There are some ten species, restricted to the tropical and subtropical regions of the New World. The woodcut of the head of *M. glaucinus* (Fig. 315) exhibits the general physiognomy of the Bats of this genus. *M. obscurus*, a small species, is very common in tropical America. It inhabits the hollow trunks of palms and other trees, and also the roofs of houses. The males and females live apart (as, indeed, appears to be the case in most, if not in all, species of Bats). In the hollow trunk of a palm two colonies were discovered, one consisting of from 150 to 200 individuals, exclusively males, while the other was composed almost entirely of females.

FIG. 315.—Head of *Molossus glaucinus*. (Dobson, *Proc. Zool. Soc.* 1876.)

Nyctinomus.[2]—Dentition: $i\ \frac{1}{3-2},\ c\ \frac{1}{1},\ p\ \frac{2-1}{2},\ m\ \frac{3}{3}$; total 32 or 28. Upper incisors separated in the middle line. The genus contains about twenty-five species, inhabiting the tropical and subtropical parts of both hemispheres. The lips of the Bats of this genus are even more expansible than in *Molossus*, in many of the species (as in the woodcut of the head of *N. macrotis*, Fig. 316) showing vertical wrinkles. *N. tæniotis*, one of the largest species, alone extends into Europe, and has been taken as far north as Switzerland. *N. johorensis*, from the Malay Peninsula, is remarkable from the extraordinary form of its ears. *N. brasiliensis* is nearly as common as *Molossus obscurus* in tropical America, and extends farther north (California) and south than that species.

[1] Geoffroy, *Ann. du Muséum*, vol. vi. p. 154 (1805).
[2] Geoffroy, *Descript. de l'Egypte*, vol. ii. p. 114 (1812).

In the *Mystacopine* division the tail perforates the interfemoral membrane and appears upon the upper surface.

Mystacops.[1]—This genus includes only *M. tuberculatus* of New Zealand, where, together with *Chalinolobus tumorio*, it represents the whole indigenous mammalian fauna of the islands. There are three distinct phalanges in the middle finger; the greater part of the wing-membrane is exceedingly thin, but a narrow portion along the forearm, the sides of the body, and the legs is remarkably thick and leathery; beneath this thickened portion the wings are folded. With the wings thus encased, this species is the most quadrupedal of Bats. Other peculiarities of structure are found in the remarkable form of the claws of the thumbs and toes, which have each a small talon projecting from its concave surface near the base, also in the sole of the foot and inferior surface of the leg, as shown in Fig. 317. The plantar surface, including the toes, is covered with soft and very lax integument deeply wrinkled,

FIG. 316.—Head of *Nyctinomus macrotis*. (Dobson, *Proc. Zool. Soc.* 1876.)

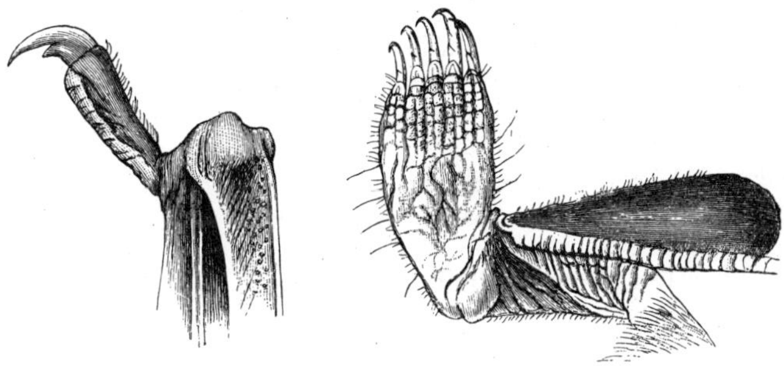

FIG. 317.—Pollex and leg and foot of *Mystacops tuberculatus*, enlarged. (Dobson, *Proc. Zool. Soc.* 1876.)

and each toe is marked by a central longitudinal groove with short grooves at right angles to it. The lax wrinkled integument is continued along the inferior flattened surface of the ankle and leg. These peculiarities appear to be related to climbing habits in the species.

Fossil Emballonuridæ.—In the cavern-deposits of Madras remains of the existing *Taphozous saccolæmus* are not uncommon; while in the corresponding beds of Brazil bones of a *Molossus*, probably referable to *M. temmincki*, now inhabiting the same region, are met with.

[1] New name: Syn. *Mystacina;* Gray, *Voyage of the "Sulphur,"* "Mamm.' p. 23 (1843). Preoccupied by *Mystacina*, Boie, 1822.

It has been suggested that remains from the Upper Eocene Phosphorites of Central France may indicate the existence of the genus *Taphozous* at that early epoch.

Family PHYLLOSTOMATIDÆ.

Middle finger with three well-developed bony phalanges; first phalanx of the middle finger short; nostrils in the front part of the cutaneous nasal appendages, or opening by simple apertures at the extremity of the muzzle; chin with warts or erect cutaneous ridges; premaxillæ well developed, united in front.

The members of this family are readily distinguished by the third phalanx in the middle finger, associated either with distinct cutaneous nasal appendages, or with well-developed first upper incisors, or with both. Unlike the *Rhinolophidæ*, their eyes are generally large; and the tragus is well developed, maintaining almost the same form throughout the species, however much the other parts of the body may vary. The fur is of a dull colour, and the face and back (in the *Stenodermatine* division especially) are often marked with white streaks, as in the *Pteropodidæ*, of which these Bats take the place in the western hemisphere. A few species, probably all those with the tail and interfemoral membrane well developed, feed principally on insects, while the greater number of the species of the *Vampirine* and *Glossophagine* divisions appear to live on a mixed diet of insects and fruits; and the *Desmodontine* division, of which two species only are known, are true bloodsuckers, and have their teeth and intestinal tract specially modified in accordance with their habits. The family is restricted to the tropical and subtropical parts of Central and South America.

Subfamily **Chilonycteriinæ**.—Nostrils opening by simple apertures at the extremity of the muzzle in front, not margined by a distinct nose-leaf; chin with expanded leaf-like appendages. It includes two genera.

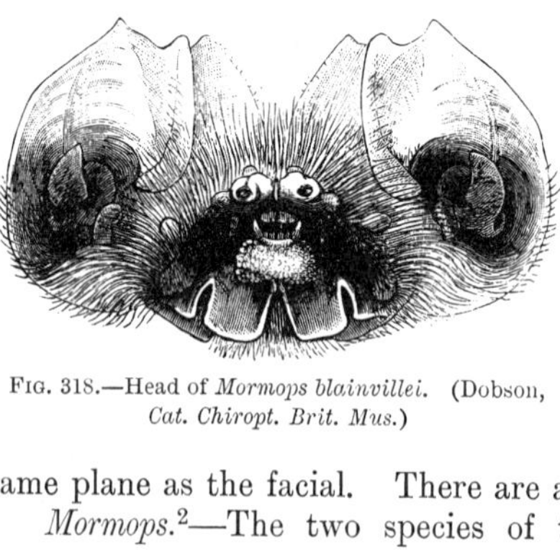

FIG. 318.—Head of *Mormops blainvillei*. (Dobson, *Cat. Chiropt. Brit. Mus.*)

Chilonycteris.[1]—Dentition: $i\ \frac{2}{2}$, $c\ \frac{1}{1}$, $p\ \frac{2}{3}$, $m\ \frac{3}{3}$; total 34. The crown of the head is moderately elevated above the facial line, and the basicranial axis is almost in the same plane as the facial. There are about half a dozen species.

Mormops.[2]—The two species of this genus are distinguished

[1] Gray, *Ann. Mag. Nat. Hist.* vol. iv. p. 4 (1839).

[2] Leach, *Trans. Linn. Soc.* vol. xiii. p. 76 (1820-22).—Amended.

from *Chilonycteris* by the great elevation of the crown of the head above the line of the face, as well as by the basicranial plane being nearly at right angles to the facial. Both species are noticeable for their peculiar physiognomy, as is shown in the accompanying woodcut (Fig. 318).

Subfamily **Phyllostomatinæ.**—Nostrils opening on the upper surface of the muzzle, the nasal apertures more or less surrounded or margined by well-developed cutaneous appendages, forming a distinct nose-leaf; chin with warts. The numerous genera, most of which can only be mentioned here by name, may be arranged under four divisions.

In the first or *Vampirine* division the muzzle is long and narrow in front; the distance between the eyes is generally less than, rarely equal to, that from the eye to the extremity of the muzzle; the nose-leaf is well developed, horse-shoe shaped in front, and lanceolate behind; interfemoral membrane well developed; tail generally distinct, rarely absent; inner margin of the lips not fringed. The dentition is: $i\ \frac{2}{2-1}$, $c\ \frac{1}{1}$, $p\ \frac{2}{2-3}$, $m\ \frac{3}{3}$; total 32. The cusps of the upper molars are usually well developed, and arranged in a W. Nearly all the species of this division appear to be insectivorous, so that the name applied to them must not be considered as having any relation to their habits. *Vampyrus spectrum*, a large Bat inhabiting Brazil, of forbidding aspect, which was long considered by naturalists to be sanguivorous in its habits, and named accordingly by Geoffroy, has been shown by the observations of modern travellers to be mainly frugivorous, and is considered by the inhabitants of the countries in which it is found to be perfectly harmless. It is the largest Bat in America, the length of the forearm being 4·2 inches. *Otopterus waterhousei* appears to prey occasionally on small species of Bats, like *Megaderma lyra* of the eastern hemisphere, which it resembles in many respects.

Lonchorhina,[1] *Otopterus,*[2] *and Dolichophyllum.*[3]—These three genera are characterised by the tail continuing to the hinder margin of the interfemoral membrane. *Lonchorhina* is represented by the single species *L. aurita*, in which the nose-leaf is much elongated, and the ear-conch and tragus are unusually large.

Vampyrus,[4] *etc.*—In all the remaining genera of this division the tail perforates the interfemoral membrane, so as to appear upon its upper surface. These genera are *Vampyrus, Lophostoma, Micronycteris,*[5]

[1] Tomes, *Proc. Zool. Soc.* 1863, p. 81. [2] New name: Syn. *Macrotus*; Gray, *Proc. Zool. Soc.* 1843, p. 21. Preoccupied by *Macrotis*, Dej. 1833. [3] New name: Syn. *Macrophyllum*; Gray, *Mag. Zool. Bot.* vol. ii. p. 489 (1838). Preoccupied by *Macrophylla*, Hope, 1837. [4] Leach, *Trans. Linn. Soc.* vol. xiii. pp. 74, 75 (1822). For the references to the other genera see Dobson, *Cat. Chiropt. Brit. Mus.* [5] Gray, *Proc. Zool. Soc.* 1866, p. 113. Syn. *Schizostoma*; Gervais, 1855. Preoccupied by Broun, 1835.

Trachyops, *Phylloderma*, *Phyllostoma*, *Anthorhina*,[1] *Mimon*, *Hemiderma*[2] and *Rhinophylla;* all, with the exception of the last, being distinguished from one another chiefly by the form of the skull and the presence or absence of the second lower premolar. *Trachyops*, *Phylloderma*, and the three last-named genera are each represented by a single species. *Phyllostoma hastatum*, in which the forearm has a length of 3·2 inches, and next in point of size to *Vampyrus spectrum*, is a well-known species in South America; *P. elongatum* (Fig. 319) differs in its smaller size and much larger nose-leaf. *Hemiderma brevicauda* is a small species, which forms a connecting link between this and the next division. *Rhinophylla pumilio*, the smallest known species of the family, is further distinguished by the narrowness of its molars, which do not form W-shaped cusps, and by the very small size of the last upper molar; characters connecting it with the *Stenodermatine* division.

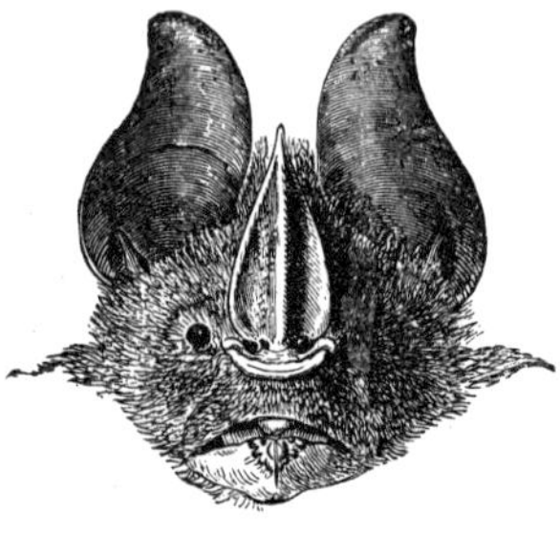

FIG. 319.—Head of *Phyllostoma elongatum*. (From Dobson, *Proc. Zool. Soc.* 1866.)

In the second or *Glossophagine* division of the subfamily the muzzle is long and narrow; the tongue remarkably long and extensible, much attenuated towards the tip, and beset with very long filiform recurved papillæ; lower lip with a wide groove above, and in front margined by small warts; nose-leaf small; tail short or absent. Dentition: $i\ \frac{1}{1}$, $c\ \frac{2}{2}$, $p\ \frac{2-3}{3-2}$, $m\ \frac{2}{3-2}$; teeth very narrow; molars with narrow W-shaped cusps, sometimes indistinct or absent; lower incisors very small or deciduous.

The ten species included in this division are arranged under seven genera,[3] distinguished principally by differences in the form and number of the teeth and the presence or absence of the zygomatic arch. The form and position of the upper incisors are extremely variable. In *Glossophaga* and *Phyllonycteris* the upper incisors form, as in the *Vampyrine* division, a continuous row between the canines; in *Monophylla* and *Leptonycteris*[4] they are separated into pairs by a narrow interval in front; while in *Lonchoglossa*, *Glossonycteris*, and *Chœronycteris* they are widely separated and placed in pairs near the canines. In the first four genera the lower incisors are present (at least up to a certain age), while in the last three

[1] New name: Syn. *Tylostoma;* Gervais, 1855. Preoccupied by Sharpe, 1849.
[2] Gervais, Castlenau's *Exped.-Zool.* p. 43 (1855): Syn. *Carollia*, Gray, 1838. Preoccupied by *Carolia*, Cantraine, 1837.
[3] The references to the genera of this and the following division will be found in Dobson's *Catalogue*.
[4] New name: Syn. *Ischnoglossa*, Saussure, 1860. Preoccupied by Kraatz, 1856.

they are deciduous even in youth. The zygomatic arch is wanting in *Phyllonycteris*, *Glossonycteris*, and *Chœronycteris*.

The typical species is *Glossophaga soricina*, which so closely resembles *Hemiderma brevicauda*, both in external form and dentition, that it has frequently been confounded with it. Its long fimbriated tongue, which it possesses in common with other species of the division, led Spix to describe it as a blood-sucker, believing that this organ was used to increase the flow of blood. This view is, however, without foundation, and from later observations it is evident that the peculiarly shaped tongue is used by the animal to lick out the pulpy contents of fruits having hard rinds. The food of the species of this division appears to consist of both fruit and insects, and the long tongue may also be used for extracting the latter from the deep corollæ of certain flowers. This type of tongue is shown in the woodcut of the head of *Chœronycteris* (Fig. 320); and it is paralleled among the Megachiroptera by the Carponycteriine *Pteropodidæ*.

FIG. 320.—Head of *Chœronycteris mexicana*, showing fimbriated tongue. (Dobson, *Cat. Chiropt. Brit. Mus.*)

The *Stenodermatine* division is characterised by the muzzle being very short and generally broad in front, the distance between the eyes nearly always exceeding (rarely equal to) that from the eye to the extremity of the muzzle; nose-leaf short, horse-shoe shaped in front, lanceolate behind (except in *Brachyphylla* and *Centurio*); interfemoral membrane always concave behind; tail none; inner margin of the lips fringed with conical papillæ. Dentition: $i\ \frac{2}{2-1}$, $p\ \frac{2}{2}$, $m\ \frac{3-2}{3-2}$; the number of the molars being either $\frac{3}{3}$, $\frac{2}{3}$, or $\frac{2}{2}$ in different species; premolars and molars very broad (except in *Sturnira*), the latter with concave or flat crowns margined externally by raised cutting-edges. Although the members of this division are usually distinguished from those of the Vampirine division by the peculiar shortness and breadth of the muzzle and the form of the molars, yet certain species of the latter closely resemble those of the former in external appearance, agreeing almost absolutely in the form of the nose-leaf, of the ears and tragus, and of the warts on the chin. These resemblances indicate that, while the form of the teeth and jaws has become modified to suit the nature of the food, the external characters, being but slightly affected by this cause, have remained much the same. The food of these Bats appears to be wholly or in great part fruit. The twenty species have been grouped into nine genera, distinguished by the form of the skull and teeth. *Artibeus*, with six species, includes the well-known frugivorous Bat, *A. perspicillatus*. Waterton believed that *A. planirostris*, a common Bat in British Guiana, usually found in the roofs of houses, and now known to be frugivorous, was the true blood-sucking Vampire. *Stenoderma achradophilum*, found in Jamaica and Cuba, associated with *Artibeus perspicillatus*, from which it is scarcely distinguishable externally except by its much smaller size, differs altogether in the absence of the horizontal plate of the palatal bones. *Sturnira lilium*, while agreeing with the above in the form of the nose-leaf and ears, differs from all the species of the family in its longitudinally-grooved molars, which resemble those of the *Pteropodidæ* more closely than those of any other Bats; and the presence of tufts of long differently coloured hairs over glands in the sides of the neck shows another common character still more remarkable, which can scarcely be considered the result of adaptive change. *Centurio senex* is the type of a genus distinguished from *Stenoderma* and other genera of this division by the absence of a distinct nose-leaf; its facial aspect, as shown in Fig. 321, is altogether bizarre.

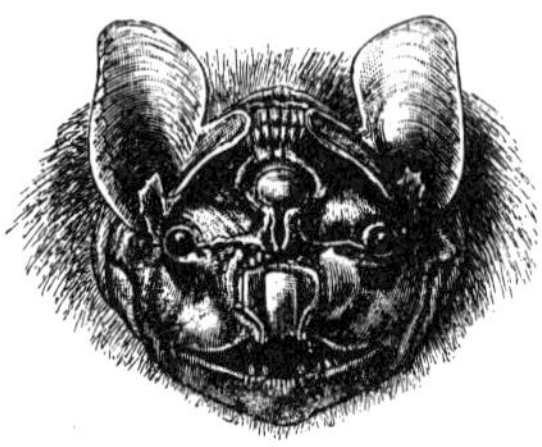

FIG. 321.—Head of *Centurio senex*. (Dobson, *Cat. Chiropt. Brit. Mus.*)

In the last or *Desmodont* division the muzzle is conical and short; there is a distinct nose-leaf; the interfemoral membrane is very short; and the tail is wanting. Dentition: $i\ \frac{1}{2}$, $c\ \frac{1}{1}$, $p\ \frac{2}{3}$, $m\ \frac{1-0}{1-0}$; total 24 or 20. Upper incisors very large, trenchant, occupying the whole space between the canines; premolars very narrow, with sharp-edged longitudinal crowns; molars rudimentary or wanting; stomach greatly elongated, intestiniform. There are only two genera, the single species of each of which are the true blood-sucking Vampires. They appear to be confined chiefly to the forest-clad parts, and their attacks on men and other warm-blooded animals were noticed by some of the earliest writers. Thus Peter Martyr (Anghiera), who wrote soon after the conquest of South America, says that in the Isthmus of Darien there were Bats which sucked the blood of men and cattle when asleep to such a degree as to kill them. Condamine, a writer of the eighteenth century, remarks that at Borja (Ecuador) and in other places they had entirely destroyed the cattle introduced by the missionaries. Sir Robert Schomburgk relates that at Wicki, on the river Berbice, no fowls could be kept on account of the ravages of these creatures, which attacked their combs, causing them to appear white from loss of blood. Although these Bats were known thus early to Europeans, the species to which they belonged were not determined until about sixty years ago, several of the large frugivorous species having been wrongly set down as blood-suckers and named accordingly; and it fell to the lot of Darwin to determine at least one of the blood-sucking species, the following being his account of the circumstances under which the discovery of the sanguivorous habits of *Desmodus rufus* was made: "The Vampire Bat is often the cause of much trouble by biting the horses on their withers. The injury is generally not so much owing to the loss of blood as to the inflammation which the pressure of the saddle afterwards produces. The whole circumstance has lately been doubted in England; I was therefore fortunate in being present when one was actually caught on a horse's back. We were bivouacking late one evening near Coquimbo, in Chili, when my servant, noticing that one of the horses was very restive, went to see what was the matter, and, fancying he could detect something, suddenly put his hand on the beast's withers and secured the Vampire."

These Bats present, in the extraordinary differentiation of the manducatory and digestive apparatus, a departure from the type of other members of the family unparalleled in any of the other orders of Mammalia, standing apart from all other mammals as being fitted only for a diet of blood, and capable of sustaining life upon that alone. Travellers describe the wounds inflicted by the large sharp-edged incisors as similar to those caused by a razor when shaving: a portion of the skin being shaved off and a large number of severed capillary vessels thus exposed, from which a constant flow of blood is maintained. From this source the blood is drawn through the exceedingly narrow gullet—too narrow for anything solid to pass—into the intestine-like stomach, whence it is probably gradually drawn off during the slow process of digestion, while the animal, sated with food, is hanging in a state of torpidity from the roof of a cave or the inner side of a hollow tree.

Desmodus.[1]—No true molar, and no calcar. The Common Vampire (*D. rufus*) is widely spread over the tropical and sub-tropical parts of Central and South America from Oaxaca to Southern Brazil and Chili. It is a comparatively small species, a little larger than the common Noctule, the head and body being about 3 inches in length, the forearm $2\frac{1}{2}$, with a remarkably long and strong thumb; it is destitute of a tail, and has a peculiar physiognomy, well represented in Fig. 322. The body is covered with rather short fur of a reddish-brown colour, but varying in shade; the extremities of the hairs being sometimes ashy. The teeth are peculiar and admirably adapted for the purposes for which they

FIG. 322.—Head of Vampire Bat (*Desmodus rufus*).

[1] Wied, *Beitr. Natgesch. Brasil*, vol. ii. p. 231 (1826).

are employed. The upper incisor is greatly enlarged, and of somewhat triangular shape (Fig. 323); the canine, although smaller than the incisor, is large and sharp; but the cheek-teeth are very small, with laterally compressed crowns rising but slightly above the level of the gum, their longitudinally disposed cutting-edges being continuous with the base of the canine and with each other. The lower incisors are small, bifid, and separated from the canine, with a space in front. The lower cheek-teeth are narrow, like those in the upper jaw, but the anterior tooth is slightly larger than the others, and separated by a small space from the canine. Behind the lower incisors the jaw is deeply hollowed out to receive the extremities of the large upper incisors. The exceedingly narrow œsophagus opens at right angles into the slender, intestine-like stomach, which, almost immediately terminates on the right, without a distinct pylorus, in the duodenum, but on the left forms a greatly elongated fundus, bent and folded upon itself, appearing at first sight like part of the intestines. This cardiac extremity of the stomach is, for a short distance to the left of the entrance of the œsophagus, still very narrow, but soon increases in size, till near its termination it attains a diameter quite three times that of the short pyloric portion. The length of this cardiac diverticulum of the stomach appears to vary from 2 to 6 inches, the size in each specimen probably depending on the amount of food obtained by the animal before it was captured.

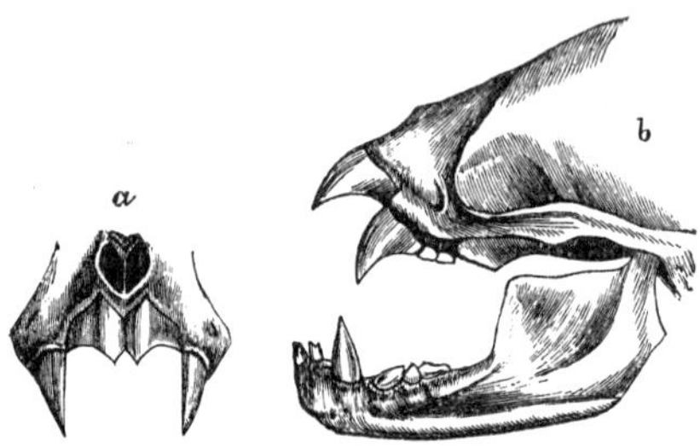

FIG. 323.—Dentition of *Desmodus rufus*. *a*, Front view of upper teeth; *b*, left lateral view of upper and lower teeth.

Diphylla.[1]—A small true molar in each jaw, and a rudimentary calcar. The single species *D. ecaudata* inhabits Brazil, and appears to be much less abundant than *Desmodus rufus*, from which, in addition to the characters already mentioned, it is distinguished by its slightly smaller size, the absence of a groove in the front of the lower lip, the non-development of the interfemoral membrane in the centre, and the peculiar form of the lower incisors, which are much expanded in the direction of the jaws and pectinated, forming a semicircular row touching each other, the outer pair being wider than the inner ones, and having six notches, the inner pair having only three notches.

Fossil Phyllostomatidæ.—Remains of *Vampyrus spectrum*, as well as of several species of *Phyllostoma* or closely allied types, are found

[1] Spix, *Sim. et Vesp. Brasil*, p. 68 (1823).

in the cavern-deposits of Brazil. The mandible of a large Bat from the Upper Eocene Phosphorites of Central France, described as *Necromantis*, has been referred to this family—a determination which, if confirmed, will be of great interest from a distributional point of view.

Bibliography of Chiroptera.—G. E. Dobson, *Catalogue of the Chiroptera in the Collection of the British Museum*, 1878, including descriptions of all the species of Bats then known; subsequent papers by the same author in *Rep. Brit. Assoc.*, *Proc. Zool. Soc.*, *Ann. Mag. Nat. Hist.*, and *Bull. Soc. Zool. de France;* by Peters in *Monatsb. Akad. Wiss. Berlin;* by O. Thomas in *Ann. Mag. Nat. Hist.*, *Proc. Zool. Soc.*, and *Ann. Mus. Genova;* and by J. Scully in *Ann. Mag. Nat. Hist.* and *Journ. As. Soc. Bengal;* H. A. Robin, *Recherches Anatomiques sur les Mammifères de l'Ordre des Chiroptères*, Paris, 1881; W. T. Blanford, "Notes on Indian Chiroptera," *Journ. As. Soc. Bengal*, vol. lviii. (1888). See also papers by Jentink, Bocage, and others.

CHAPTER XIV

THE ORDER PRIMATES

THIS order in the system of Linnæus includes Man, the Monkeys, the Lemurs, and the Bats. By common consent of all zoologists the last-named animals have been removed into a distinct order; but with regard to the association of the others there has been, and still is, much difference of opinion.

That all the Monkeys, from the highest Anthropoid Apes to the lowest Marmosets, form a natural and tolerably homogeneous group seems never to have been questioned; but whether the Lemurs on the one hand and Man on the other should be united with them in the same order are points of controversy. If, in accordance with the traditional views of zoologists, the former are still considered to be members of this order, they must form a sub-order apart from all the others, with which they have really very little in common except the opposable hallux of the hind foot, a character also met with in the Opossums, and which is therefore of very secondary importance.[1]

Using the term Primates in this wider sense it is not easy to give any precise definition of the order. The dentition is diphyodont and heterodont; the number of incisors being very generally $\frac{2}{2}$, and that of the molars, with the exception of the *Hapalidæ*, being $\frac{3}{3}$. The cheek-teeth are adapted for grinding, the molars being more complex than the premolars, and usually having four main tubercles, which may be either subconical or more or less compressed. The orbit is invariably surrounded by a ring of bone;

[1] For the arguments in favour of placing the Lemurs in a separate order see Milne-Edwards, "Observations sur quelques points de l'embryologie des Lemuriens et sur les affinités zoologiques de ces animaux," in the *Ann. des Sciences Nat.* October 1871; and P. Gervais, "Encephale des Lemures," in *Journ. de Zoologie*, tom. i. p. 7. For those for retaining them among the Primates, see Mivart, "On *Lepilemur* and *Chirogaleus*, and on the Zoological Rank of the Lemuroidea," in *Proc. Zool. Soc.* 1873, p. 484.

the clavicles are well developed; and the radius and ulna are never united. The scaphoid and lunar of the carpus, and commonly also the centrale, remain distinct from one another. There are usually five digits furnished with well-developed nails in both the manus and the pes; but the pollex may be rudimentary or wanting. The hallux, except in Man, is opposable to the other digits, and has a flat nail (absent in *Simia*); and the pollex, when present, is usually also more or less opposable. The terminal phalanges of the digits are flattened (except in the second digit of the pes of the Lemuroidea), and not cleft at their extremities. The fingers and toes generally do not taper towards their extremities, but (except in *Chiromys*) are dilated, flattened, and rounded at their tips. The humerus has no entepicondylar foramen, nor the femur a third trochanter. In the alimentary canal (Fig. 324) the stomach is generally simple, although sacculated in the subfamily *Semnopithecinæ* of the *Cercopithecidæ*; and there is always a cæcum, which is generally of large size. The placenta may be either non-deciduous, or discoidal and deciduous. There are always two mammæ in the pectoral region, except in *Chiromys*; and the testes descend into a scrotum.

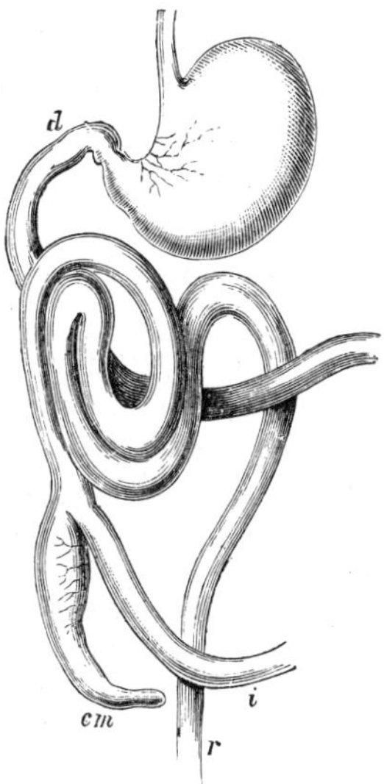

FIG. 324.—Alimentary canal of *Galago*, the greater part of the small intestine being omitted. *d*, duodenum; *i*, ileum; *cm*, cæcum; *r*, rectum.

The Lemuroidea are decidedly low in the scale of organisation, their placentation being of a lower type than that of the Insectivora; and all the Primates retain generalised features in their pentadactylate limbs and more or less bunodont cheek-teeth. In respect to cerebral characters and other features the higher representatives of the order have, however, acquired a specialisation clearly indicating their right to occupy the highest position in the animal kingdom. So far as the available material admits of forming an opinion, fossil forms appear to indicate an intimate connection between the Lemuroidea and Insectivora, so that in some cases it is almost impossible to determine whether an extinct type should be referred to the former or to the latter group. It is noteworthy that while in all existing Primates the upper molars are of a quadrituberculate type, in the extinct Lemuroid genus *Anaptomorphus* they are trituberculate.

Suborder LEMUROIDEA.

The Latin term *Lemur* was applied by Linnæus to the typical representatives of the present group of Primates, having been suggested by the nocturnal habits and strange ghost-like appearance of some of its members. As these animals had previously no vernacular appellation in English, this name has been generally adopted, and is now completely anglicised, making "Lemurs" in the plural. The French call them *Makis*, and the Germans *Halbaffen*, in allusion to their forming a transition from monkeys to ordinary quadrupeds. For the same reason they are called *Prosimiæ* by some systematic writers. When the name was bestowed by Linnæus only five species were known, of which one, *L. volans*, Linn. (*Galeopithecus volans* of modern writers), is now removed by common consent from the group. Notwithstanding the discovery of many new and curious forms, the Lemurs remain a very natural and circumscribed division of the animal kingdom, though no longer considered a single genus, but divided up into many genera and even families.

The existing species are not numerous, and do not diverge widely in their organisation or habits, being all of small or moderate size, all adapted to an arboreal life, climbing with ease, and, as they find their living, which consists of fruits, leaves, birds' eggs, small birds, reptiles, and insects, among the branches of the trees, they rarely have occasion to descend to the ground. None are aquatic, and none burrow in the earth. Many of the species, although by no means all, are nocturnal in their habits, spending the day in sleeping in holes, or rolled up in a ball, perched on a horizontal branch, or in the fork of a tree, and seeking their food by night. Their geographical distribution is very peculiar; by far the larger proportion of species, including all those to which the term "Lemur" is now especially restricted, being exclusively inhabitants of Madagascar, where they are so abundant and widely distributed that it is said by M. Grandidier, who has contributed more than any other traveller to enrich our knowledge of the structure and manners of these animals, that there is not a little wood in the whole island in which some of them cannot be found. From Madagascar as a centre a few species less typical in character extend through the African continent westward as far as Senegambia, and others are found in the Oriental region as far east as the Philippine Islands and Celebes.

The following are the essential characters by which the suborder as a whole is distinguished from the Anthropoidea. Skull with the orbit opening freely into the temporal fossa beneath the postorbital bar (except in *Tarsius*); and the lachrymal foramen situated externally to the margin of the orbit (Fig. 327). The pollex and hallux are always well developed, the latter being especially large; the second or index digit of the manus may be rudimentary; while in the pes the second digit invariably terminates in a long pointed claw. The cerebral hemispheres do not completely overlap the cerebellum, and are but slightly convoluted. The uterus is bicornuate. The placenta is non-deciduate, and either diffused or bell shaped—the whole of the chorion except the cephalic pole being covered with villi; and the allantois is of very great size. There may be abdominal mammæ. Except in *Chiromys*, the first pair of upper incisors are separated in the middle line. In marked contrast to the Anthropoidea, the middle or transverse portion of the colon is almost always folded or convoluted on itself. (See Fig. 324.)

In subdividing the group for the purpose of a more detailed description of the different animals of which it is composed it must first be noted that there are two very aberrant forms, each represented by a single species—the little *Tarsius* of the Indian archipelago, and the singular *Chiromys* or Aye-aye, which, though an inhabitant of Madagascar, the headquarters of the suborder, and living in the same forests and under the same external conditions as the most typical Lemurs, exhibits a most remarkable specialisation in the structure of its limbs and teeth, the latter being modified so as to resemble, at least superficially, those of the Rodents, in which order it was once placed. The differences between these two forms and the remaining Lemurs are so great that the whole suborder naturally divides itself into three families, the first of which may be again divided into four subfamilies.

Family LEMURIDÆ.

Upper incisors two on each side, small and separated by an interval in the middle line. Upper canine large, conical, compressed, and pointed. Premolars two or three, molars three on each side above and below, with numerous more or less pointed cusps. In the front of the lower jaw are on each side two or three closely approximated, long, slender teeth lying almost horizontally and projecting forwards. These are generally considered to represent the incisors and canine, but there is some doubt about their homologies, and they may be all considered as incisors, the canine being absent. The first lower premolar larger than those behind it, and shaped like a canine, of which it performs the function (Fig. 327). The orbit and temporal fossa widely continuous beneath the bar of bone (formed by the frontal and jugal) constituting the posterior boundary of the former cavity. The fibula well developed and distinct from the tibia. All the digits of both feet (except the second of the hind foot) with flat nails, and corresponding form of ungual phalanges.

Subfamily **Indrisinæ**.—The dentition of the adult consists of thirty teeth, usually expressed by the formula $i\ \frac{2}{1}$, $c\ \frac{1}{1}$, $p\ \frac{2}{2}$, $m\ \frac{3}{3}$; but, as indicated above, they may be $i\ \frac{2}{2}$, $c\ \frac{1}{0}$, $p\ \frac{2}{2}$, $m\ \frac{3}{3}$. In the milk-dentition there are twenty-two teeth, the true molars of course not being represented, but there are two additional teeth in the fore part of the lower jaw which have no successors in the permanent series. Hind limbs greatly developed, but the tarsus normal. Hallux of large size, and very opposable. The other toes united at their base by a fold of skin, which extends as far as the end of the first phalanx. Mammæ two, pectoral. Cæcum very large, and colon extremely long and spirally coiled.

The animals of this group are, as their organisation indicates, essentially arboreal, and feed exclusively on fruit, leaves, buds, and flowers. They are restricted geographically to the island of Madagascar. Among them are the largest members of the sub-order. A detailed and beautifully illustrated account of their characters, external and internal, and distribution and habits, is given in the *Histoire Naturelle de Madagascar*, by A. Grandidier and Alphonse Milne-Edwards (1875). The species are not numerous and are distributed into three genera.

Indris.[1]—Upper incisors subequal in size. Upper canine larger than the first premolar. Muzzle moderately long. Ears exserted. Carpus without an os centrale. Tail rudimentary. Vertebræ: C 7, D 12, L 9, S 4, C 9.

The only well-established species is the Indris (*I. brevicaudata*, Fig. 325), discovered by Sonnerat in 1780. It is the largest of the Lemurs, the length of the head and body being about 2 feet, and the tail 2 inches. It is very variable in colour, for although usually nearly black, marked with whitish spots principally in the lumbar region and forearm, individuals have been found quite white. It inhabits exclusively the forests of a part of the east coast of Madagascar, living in small troops of four or five in number, and resembling in most of its habits the animals of the next genus.

Propithecus.[2]—Second upper incisor much smaller than the first. Upper canine larger than the first premolar. Muzzle rather short. Ears short, concealed by the fur. An os centrale in the carpus. Tail long. Vertebræ: C 7, D 12, L 8, S 3, C 28.

The species are all subject to great variations in colour, which has led to much difficulty in discriminating them, and to much confusion of synonymy. Grandidier and Milne-Edwards recognise three as certainly distinct—*P. diadema*, *P. verreauxii*, and *P. coronatus* (Fig. 326). Some of these are to be found in almost

[1] Geoffroy, *Mag. Encyclop.* 2d ann. vol. i. p. 46 (1796), "Indri."

[2] Bennett, *Proc. Zool. Soc.* 1832, p. 20.

every part of the island of Madagascar, living in the woods in small bands of six or eight together, and feeding exclusively on buds, flowers, and berries. Their powerful hind limbs enable them to leap from tree to tree, often to a distance of 10 yards, without any apparent effort, and thus seeming to fly through the air. When obliged to descend to the ground to pass from one clump of trees

FIG. 325.—Indris (*Indris brevicaudata*). From Milne-Edwards and Grandidier, *Mammifères de Madagascar*, pl. 12.

to another they do not run on all fours, but stand erect, and throwing their arms above their heads progress by a series of short jumps, producing an effect which is described by travellers who have seen them thus in their native haunts as exceedingly ludicrous. They are not nocturnal, but most active in the morning and evening, remaining seated or coiled up among the branches during the heat of the day. They are naturally of a quiet and gentle disposition, and do not show much intelligence. All the species are also less vociferous than the true Lemurs, only when alarmed or angered making a noise which has been compared to the clucking of a fowl. Like the rest of the subfamily they never have more than a single young one at a time.

Avahis.[1]—Second upper incisor larger than the first. Upper

FIG. 326.—*Propithecus coronatus.* (From Milne-Edwards and Grandidier, *Mammifères de Madagascar*, pl. 7.)

canine scarcely larger than the first premolar. Muzzle very short. Ears very small and hidden in the fur, which is very soft and woolly. Carpus without an os centrale. Tail long. Vertebræ: C 7, D 11, L 9, S 3, C 23.

One species, *A. laniger*, the Woolly Lemur, or Avahis, considerably smaller than any of the last genus. It differs from them in

[1] Jourdan, *Mém. de l'Institut*, vol. ii. p. 231 (1834).

its habits, being quite nocturnal, and not associating in small troops, but being always met with either alone or in pairs. It is very slow in its movements, and rarely descends to the ground, but when it does it walks upright like the other *Indrisinæ*. It is found throughout the forests which clothe the mountains on the east coast of Madagascar, and also in a limited district on the north-west coast, the specimens from the latter locality being of smaller size and rather different in colour.

Subfamily **Lemurinæ**.—The dentition in the adult consists of thirty-six teeth, which, as usually enumerated, are $i\ \frac{2}{2}$, $c\ \frac{1}{1}$, $p\ \frac{3}{3}$, $m\ \frac{3}{3}$. In the fore part of the lower jaw are on each side three elongated, compressed, procumbent teeth, of which the outer, usually considered the homologue of the canine, is larger than the others. All the forms have long tails. Hind limbs not of the same disproportionate size as in the last group; and the cæcum much less developed. Tarsus but slightly elongated, the calcaneum being always less than one-fourth the length of the tibia. Toes of the hind feet free to the base. Habitat, Madagascar, and some of the adjacent Comoro Islands.

This group contains the typical Lemurs, or rather those to which the term is now chiefly restricted. Two somewhat aberrant members make it necessary to divide it into three genera.

Lemur.[1]—Upper incisors separated by an interval in the middle, and not in contact with each other or the canine, in front of which they are both placed. Muzzle elongated. Ears conspicuous and tufted. Mammæ two, pectoral. Vertebræ: C 7, D 12, L 7 (or D 13, L 6), S 3, C 27.

Animals much about the size of a common Cat, with Fox-like faces, soft thick fur, and long tails well clothed with hair. Not having the same disproportionate size of the limbs as the last group, they are much more quadrupedal in their actions, walking on the ground or running along the branches of trees on all four feet, but also jumping with marvellous agility. They are gregarious, living in small troops, are diurnal in their habits, but most active towards evening, when

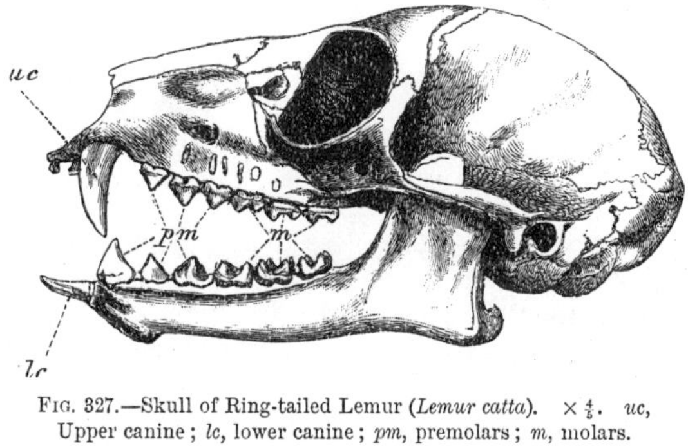

FIG. 327.—Skull of Ring-tailed Lemur (*Lemur catta*). × ½. *uc*, Upper canine; *lc*, lower canine; *pm*, premolars; *m*, molars.

[1] Linn. *Syst. Nat.* 12th ed. vol. i. p. 44 (1766).

they make the woods resound with their loud cries. They feed not only on fruits and buds, but also on eggs, young birds, and insects. When at rest or sleeping they generally coil their long, bushy tails around their bodies, apparently for the sake of the warmth it affords. They have either one or two young ones at a birth, which are at first nearly naked, and are

FIG. 328.—The Ring-tailed Lemur (*Lemur catta*).

carried about, hanging close to and almost concealed by the hair of the mother's belly. After a while they change their position and mount upon the mother's back, where they are carried about until they are able to climb and leap by themselves. Though no member of the *Indrisinæ* has as yet lived long enough in captivity to be brought alive to Europe, various species of *Lemurinæ* are commonly seen in menageries, and often breed in England. They present a great tendency to variation in their colouring, in consequence of which many nominal species have been made. The most distinct, and at the same time most beautiful, is the Ring-tailed Lemur (*L. catta*,

Fig. 328), of a delicate gray colour, and with a long tail marked with alternating rings of black and white. This is said by Mr. G. A. Shaw [1] to be an exception to all the other Lemurs in not being arboreal, but living chiefly among rocks and bushes. Pollen, however, says that it inhabits the forests of the south-west parts of Madagascar, living, like its congeners, in considerable troops, and not differing from them in its habits. He adds that it is extremely gentle, and active and graceful in its movements, and utters at intervals a little plaintive cry like that of a domestic cat. All the others have the tail of uniform colour. The largest species is *L. varius*, the Ruffed Lemur, sometimes black and white, and sometimes reddish-brown, the variation apparently not depending on sex or age, but on the individual. In *L. macaco* the male is black and the female red. *L. mongoz*, *L. collaris*, and *L. albifrons* are other well-known species.

Hapalemur.[2]—Upper incisors very small, subequal, separated widely in the middle line. Those of either side in contact with each other and with the canine, the posterior one being placed on the inside, and not in front of the latter. Muzzle very short and truncated. Mammæ four. There is apparently but one species, *H. griseus*, smaller than any of the true Lemurs, of a dark gray colour, with round face and short ears. It is quite nocturnal, and lives chiefly among bamboos, subsisting on the young shoots. A second species has been named *H. simus*, but it is doubtful if it is more than a variety.

Lepidolemur.[3]—Upper incisors absent or rudimentary. Muzzle more elongated than in the last. No distinct os centrale in the carpus. *L. mustelinus* is the best-known species. It has, at all events when adult, no upper incisors. It is rare, and like *Hapalemur* nocturnal in its habits. A second closely allied species, but with better developed premaxillæ, containing a pair of small styliform incisors, has been described by Peters [4] under the name of *Myxocebus caniceps*.

Subfamily **Galaginæ**.—Dentition as in *Lemurinæ*, from which the members of this subfamily are distinguished by the elongation of the tarsus, caused by a peculiar modification of the calcaneum and the navicular, the distal portion of the former and the whole of the latter having the form of almost cylindrical rods placed side by side, while the other bones retain nearly their normal form and proportion.

Chirogaleus.[5]—Last upper premolar very much smaller than the first molar, with only one external cusp. The animals included

[1] *Proc. Zool. Soc.* 1879, p. 132. [2] Gray, *Proc. Zool. Soc.* 1870, p. 829.

[3] I. Geoffroy, *Cat. Mus. Hist. Nat. Paris*, p. 75 (1851). Amended from *Lepilemur*. [4] *Monatsb. Ak. Berlin*, 1874, p. 690.

[5] Geoffroy, *Ann. du Muséum*, vol. xix. p. 171 (1812).

under this name appear to form a transition between the true Lemurs and the Galagos. The genus was originally established by Geoffroy St. Hilaire in 1812 for the reception of three species only known at that time by drawings made in Madagascar by the traveller Commerson. Subsequent discoveries have brought to light several others that may be referred to it, including one or two which are sometimes considered as forming a genus apart under the name of *Microcebus*. They are all small, some being less than a rat in size, long-tailed, and nocturnal in their habits. One of the largest, *C. furcifer*, is of a reddish-gray colour, and distinguished by a dark median stripe on its back which divides on the top of the head into two branches, one of which passes forwards above each eye. The most interesting peculiarity of these animals, a knowledge of which we owe to M. Grandidier, is that certain species (*C. samati*, *C. gliroides*, *C. milii*, etc.) during the dry season coil themselves up in holes of trees and pass into a state of torpidity like that of the hibernating animals in the winter of northern climates. Before this takes place an immense deposit of fat accumulates upon certain parts of the body, especially upon the basal portion of the tail, which has then dimensions corresponding to that of the well-known fat-tailed Sheep of the Cape, but which by the time they emerge from their torpor has acquired its normal proportions. The smallest species, to which many names have been given (*C. pusillus*, *rufus*, *smithi*, etc.), lives among the small branches on the tops of the highest trees, feeding on fruit and insects, and making nests which resemble those of birds.

Galago.[1]—Last upper premolar with two large external cusps, and nearly equalling the first molar in size. Calcaneum about one-third the length of the tibia, and the navicular much longer than the cuboid. Vertebræ: C 7, D 13, L 6, S 3, C 22-26. Tail long, and generally bushy. Ears large, rounded, naked, and capable of being folded at the will of the animal. Mammæ four, two pectoral and two inguinal.

The Galagos differ from all the Lemuroids previously mentioned, inasmuch as they are inhabitants, not of Madagascar, but of the African continent, being widely distributed in the wooded districts from Senegambia in the west to Abyssinia in the east, and as far south as Natal. They pass the day in sleep, but are very active at night, feeding on fruit, insects, and small birds. When they descend to the ground they sit upright, and move about by jumping with their hind legs, like jerboas and kangaroos. They are pretty little animals, varying in size from that of a small cat to less than a rat, with large eyes and ears, soft woolly fur, and long tails. There are several species, of which *G. crassicaudatus*, from Mozambique, is the largest. A similar species, or perhaps variety, from

[1] Geoffroy, *Mag. Encyclop.* 2d ann. vol. i. p. 49 (1796).

Angola is *G. montieri*. *G. garnetti, alleni, maholi, demidoffi*, and *senegalensis* are other recognised species. The last-mentioned was the first known to science, having been brought from Senegal by Adanson, and described in 1796 by Geoffroy, who adopted the name *Galago*, by which it was said to be called by the natives.

Subfamily **Lorisinæ**.—Dental formula as in *Lemurinæ*. Index finger very short, sometimes rudimentary and nailless. Fore and hind limbs nearly equal in length. Tarsus not specially elongated. Pollex and hallux diverging widely from the other digits, the hallux especially being habitually directed backwards. Tail short or quite rudimentary. Mammæ two, pectoral.

A small group of very peculiar animals, of essentially nocturnal habits, and remarkable for the slowness of their movements. They are completely arboreal, their limbs being formed only for climbing and clinging to branches, not for jumping or running. They have rounded heads, very large eyes, short ears, and thick, short, soft fur. They feed not only on vegetable substances, but, like many of the *Lemuridæ*, on insects, eggs, and also birds, which they steal upon while roosting at night. None of the species are found in Madagascar. One of the greatest anatomical peculiarities of these animals is the breaking up of the large arterial trunks of the limbs into numerous small parallel branches, constituting a *rete mirabile*, which is found also in the Sloths, with which the Loris are sometimes confounded on account of the slowness of their movements. The animals of this group are usually divided into four genera, though the characters by which they are separated are very trivial. There are more properly two natural divisions.

A. Characterised by the index finger being small, but having the complete number of phalanges, and by their Asiatic habitat.

These form the genus *Loris* of Geoffroy St. Hilaire (1796), *Stenops* of Illiger (1811), but they were in 1812 divided by Geoffroy into two genera, *Nycticebus* and *Loris*, a division which has been accepted by most modern zoologists.

Nycticebus.[1]—First upper incisor larger than the second, which is often early deciduous. Inner margins of the orbits separated from each other by a narrow flat space. Nasal and premaxillary bones projecting but very slightly in front of the maxillæ. Body and limbs stout. No external tail. Vertebræ: C 7, D 17, L 6, S 3, C 12. The species are *N. tardigradus*, the common Slow Lemur or Loris, of the Malay Countries, Sumatra, and Borneo; *N. javanicus*, of Java; and *N. cinereus* (Fig. 329) of Siam and Cochin China. The habits of all are much alike. They lead a solitary life in the recesses of large forests, chiefly in mountainous districts, where they sleep during the day in holes or fissures of large trees, rolled up into a ball, with the head between the hind legs. On the approach

[1] Geoffroy, *Ann. du Muséum*, vol. xix. pp. 162, 163 (1812).

of evening they awake; and during the night they ramble among the branches of trees, slowly and quietly, in search of their food, which consists of tender leaves and fruit, small birds, insects, and mice. When in quest of living prey they move noiselessly till quite close, and then suddenly seize it with one of their hands. The female produces but one young one at a time. *L. tardigradus* was placed by Linnæus at the head of the list of species of his genus *Lemur*, and its habits doubtless suggested the generic name which

FIG. 329.—The Gray Loris (*Nycticebus cinereus*). From A. Milne-Edwards, *N. Archives du Muséum*, vol. iii. pl. 3.

was transferred by Geoffroy to the less nocturnal and spectre-like Madagascar members of the group.[1]

Loris.[2]—Upper incisors very small and equal. Orbits very large, and only separated in the middle line above by a thin vertical plate of bone. Nasals and premaxillæ produced forwards considerably beyond the anterior limits of the maxillæ, and supporting a pointed nose. Body and limbs slender. No external tail. Vertebræ: C 7, D 14, L 9, S 3, C 6. This genus is represented only by the Slender Loris (*L. gracilis*) of Southern India and Ceylon (Fig. 330). This species is common in some of the forest regions of Southern India, and may be purchased in the bazaars at Madras, its eyes being regarded as a remedy by the natives for ophthalmic diseases. It is a strange-looking creature, about the size of a squirrel, of a yellowish-brown colour, with large, prominent eyes, pointed nose,

[1] For the anatomy of this genus, see J. L. C. Shroeder van der Kolk and W. Vrolik, "Recherches d'Anatomie comparée sur le genre *Stenops* d'Illiger," in *Bijdragen tot de Dierkunde*, Part I, Amsterdam, 1848-54.

[2] Geoffroy, *Mag. Encyclop.* 2d ann. vol. i. p. 48 (1796).

long thin body, long, angularly bent, slender limbs, and no tail. Its habits, according to Mr. W. T. Blanford,[1] are "very similar to those of *Nycticebus tardigradus*, except that the Slender Loris is rather quicker in its movements, though still slow in general. Like its ally, it is purely nocturnal and arboreal, living upon shoots and young leaves, insects, birds' eggs, birds, and lizards. It is said to be very fond of honey or syrup. It sleeps rolled up in a ball with its head between its legs, grasping its perch with its arms."

FIG. 330.—The Slender Loris (*Loris gracilis*). From Blanford, *Mammalia of British India*, p. 47.

B. Index fingers reduced to a mere tubercle without nail. Both the known species are from West Africa.

Perodicticus.[2]—A short tail, about a third of the length of the trunk. Two or three of the anterior dorsal vertebræ have very long slender spinous processes which in the living animal project beyond the general level of the skin, forming distinct conical prominences, covered only by an exceedingly thin and naked integument. The Potto, *P. potto*, is one of the oldest known members of the lemuroid group, having been described in 1705 by Bosman, who met with it in his voyage to Guinea. It was, however, lost sight of until 1825, when it was re-discovered in Sierra Leone, and fully described by Bennett in 1830 under the name of *Perodicticus geoffroyi*. Bennett's generic name has been retained, but the specific name bestowed by Gmelin, adopted from Bosman, has been restored. It is also found in the Gaboon. It is strictly nocturnal, and slower in its movements even than *Nycticebus tardigradus*, which otherwise it much resembles in its habits.

A second species, the Awantibo (*P. calabarensis*), rather smaller and more delicately made, with smaller hands and feet and rudimentary tail, constitutes the genus *Arctocebus* of Gray. It is found

[1] *Mammalia of British India*, p. 48 (1888).

[2] Bennett, *Proc. Zool. Soc.* 1839, p. 109.

at Old Calabar, and is very rare, only a few individuals having as yet been met with. Vertebrae: C 7, D 15, L 7, S 3, C 9.[1]

Family TARSIIDÆ.

Dentition: $i\ \frac{2}{1}$, $c\ \frac{1}{1}$, $p\ \frac{3}{3}$, $m\ \frac{3}{3}$; total 34. The first upper incisor large, and in contact with its fellow of the opposite side. Canine of moderate size. Molars with numerous pointed cusps. Lower canine semi-erect, its apex diverging from that of the single incisor. First lower premolar smaller than those behind it. Orbit to a large extent separated from the temporal fossa by a bony partition. Fibula slender, with its lower half confluent with the tibia. Second and third digits of the hind foot with compressed claws; all the other digits of both feet with flat nails. Calcaneum and navicular bone of the foot elongated as in the Chirogales and Galagos, but to a still greater extent. Colon short and not folded. Vertebræ: C 7, D 13, L 6, S 3, C 27.

Tarsius.[2]—The family contains the single genus *Tarsius*, of which but one species is known, *T. spectrum*, the Tarsier, a very singular little animal, rather smaller than an English squirrel, with very large eyes and ears, a long thin tail, tufted at the end, and immensely elongated tarsal portion of the foot, in allusion to which its generic name was given to it. It inhabits the forests of many of the islands of the Indo-Malayan archipelago, including Sumatra, Borneo, Celebes, and some of the Philippines, feeds chiefly on insects and lizards, sleeps during the day, but is tolerably active at night, moving chiefly by jumping from place to place, an action for which the structure of its hind legs, which present a curious resemblance to those of a frog, seems particularly well adapted. It is rare, not more than two being generally found together, and only brings forth one young at a time.[3]

Family CHIROMYIDÆ.

Dentition of adult: $i\ \frac{1}{1}$, $c\ \frac{0}{0}$, $p\ \frac{1}{0}$, $m\ \frac{3}{3}$; total 18. Incisors very large, compressed, curved, with persistent pulps and enamel only in front, as in Rodents. Teeth of cheek series with flat, very indistinctly tuberculated crowns. In the young the first set of teeth more resemble those of the normal lemurs, being $i\ \frac{2}{2}$, $c\ \frac{1}{0}$, $m\ \frac{2}{2}$, all very small. Orbit surrounded by a ring of bone posteriorly, beneath which it communicates freely with the temporal fossa. Fibula well

[1] For the anatomy of *P. potto*, see Van der Hoeven and Van Campen (*Ontleedkundige Onderzoek van den Potto van Bosman*, 1859) for *P. calabarensis*, Huxley, *Proc. Zool. Soc.* 1864, p. 314. [2] Storr, *Prodromus Meth. Mamm.* (1780).

[3] H. Burmeister, *Beiträge zur nähreren Kenntniss der gattung Tarsius*, 1846.

developed and distinct from the tibia. All the digits of both feet with pointed rather compressed claws, except the hallux, which has a flattened nail. Middle digit of the hand excessively attenuated. Vertebrae: C 7, D 12, L 6, S 3, C 27.

Chiromys.[1]—This family, like the last, is formed for the reception of a single genus, *Chiromys*,[2] containing one species, *C. madagascariensis*, the Aye-aye, an animal about the size of a cat, with a broad rounded head, short face, and large and naked ears. It has very large hands and long thin fingers with pointed claws, one of which (the middle or third) is remarkable for its extreme slenderness. The foot resembles that of the other lemurs in its large opposable hallux, with a flat nail, but all the other toes have pointed compressed claws, like that of the second toe in the *Lemurinæ* and the second and third in the *Tarsiidæ*. Tail long and bushy. General colour dark brown, the outer fur being long and rather loose, with a woolly undercoat. Mammæ two, inguinal in position. It is a native of Madagascar, where it was discovered by Sonnerat in 1780. The specimen brought to Paris by that traveller was the only one known until 1860. Since then many others have been obtained, and they may frequently be seen living in the gardens of the Zoological Society of London. Like so many of the Lemurs, the Aye-aye is completely nocturnal in its habits, living either alone or in pairs, chiefly in the bamboo forests. Observations upon captive specimens have led to the conclusion that it feeds principally on succulent juices, especially of the sugar-cane, which it obtains by tearing open the hard woody circumference of the stalk with its strong incisor teeth. It is said also to devour certain species of wood-boring caterpillars, which it obtains by first cutting down with its teeth upon their burrows, and then picking them out of their retreat with the claw of its attenuated middle finger. It

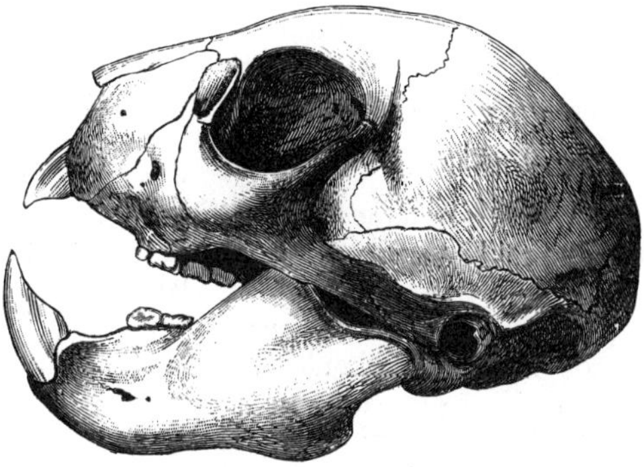

FIG. 331.—Skull of Aye-aye (*Chiromys madagascariensis*). × ⅘ Mus. Roy. Coll. Surgeons.

[1] Cuvier, "Table de Class." in *Leçons d'Anat. Comp.* vol. i. (1800).

[2] It was first named *Daubentonia* by Geoffroy; but this name was withdrawn by its author in favour of *Chiromys*, as it had been previously given to a genus in the vegetable kingdom. This would not, however, constitute preoccupation according to the modern rules of nomenclature.

constructs large ball-like nests of dried leaves, lodged in a fork of the branches of a tree with the opening on one side. The resemblance of its teeth to those so characteristic of the Rodentia caused it to be placed formerly in that order, and it was only when its anatomical characters were fully known that its true affinities with the Lemurs became apparent.[1]

Extinct LEMUROIDS.

The discoveries of the last few years have revealed the former existence, both in Europe and North America, of a number of extinct animals more or less closely allied to the living Lemurs, which are of especial interest as showing in some instances characters of a more generalised type than is the case with the living representatives of the suborder. It is, however, in some cases very difficult to determine whether these extinct forms should be referred to the Lemuroidea or Insectivora; and if those naturalists are right who regard these groups as survivors of a very generalised ancestral type of mammalian organisation, it is to be expected that as we recede in time we should find that the two groups show more and more marked signs of a natural connection. The earliest reference of one of these extinct Upper Eocene types to the Primates was made in 1862 by Professor L. Rütimeyer, of Basle, who described part of an upper jaw with three teeth from the so-called Bohnerz of Egerkingen, near Soleure in Switzerland, under the name of *Cœnopithecus lemuroides*, regarding the animal to which the specimen belonged as partaking of the characters both of the Lemurs and the American Monkeys. Most other palæontologists refused, however, to accept this determination, and it was not until many years later that the researches of Gaudry and Filhol showed not only that *Cœnopithecus* was indeed a true Lemuroid, but also that it was either identical with or closely allied to a form described by Cuvier in the early part of this century under the name of *Adapis* and regarded as referable to the Ungulata. Later researches have brought to light other Lemuroids in the Tertiaries of both the Old and the New World; and it is very noteworthy that all these types seem to have disappeared from both regions with the close of the upper portion of the Eocene period.

Among the more interesting of the forms which are generally regarded as true Lemuroids we may first mention a small species from the Quercy Phosphorites, of which the hinder cheek-teeth are shown in Fig. 332, *A*, which was originally described as *Necrolemur antiquus*, but appears to be generically identical with *Microchœrus*

[1] R. Owen, "On the Aye-aye," in *Trans. Zool. Soc.* 1862, vol. v. p. 33; W. Peters, "Ueber die Säugethier-Gattung *Chiromys*," in *Abhand. Königl. Akad. der Wissenschaften*, Berlin, 1865, p. 79.

erinaceus, of the upper Eocene of Hampshire, of which the corresponding teeth are shown in *B* of the same figure. In this genus, according to Dr. Schlosser, the dental formula is $i\ \frac{2}{1}$, $c\ \frac{1}{1}$, $p\ \frac{3}{3}$, $m\ \frac{3}{3}$, or the same as in the existing *Tarsius*; but it is not improbable that in some instances the first lower premolar may have been developed. The upper molars of *M. erinaceus* differ from those of *M. antiquus* by the simpler structure of their columns and the smaller size of the external cingulum, which lacks the median cusp found in the latter. The angle of the mandible is produced into a large hook-like flange which at once distinguishes the genus from all existing Lemurs; and the anterior lower premolar is not canine-like. *M. antiquus* is of very small size, but the larger *M. edwardsi* of the same deposits comes nearer in dimensions to *M. erinaceus*. The upper molars decrease in size from the first to the third, the first and second having a median cusp in the external cingulum, by which they are readily distinguished from the corresponding teeth of the under-mentioned genus *Hyopsodus*. The third upper molar differs from that of *Hyopsodus* by its small size and the abortion of its posterior columns. The skull approximates to that of the living genus *Galago*, exhibiting the same inflation of the auditory bulla. The upper molars are also not unlike one species of that genus, but the fourth upper premolar has but one outer cusp, as in *Chirogaleus*.

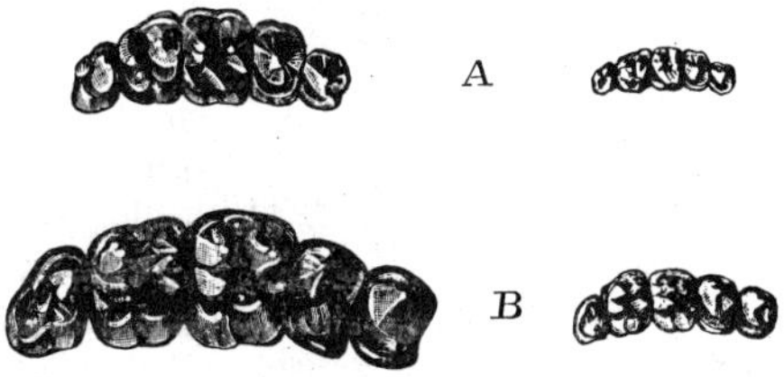

FIG. 332.—The last five right upper cheek-teeth of *Microchœrus antiquus* (*A*) and *Microchœrus erinaceus* (*B*). Twice natural size, and natural size.

The small *Anaptomorphus*, from the North American Eocene, has a skull of about the same size as that of the smallest species of *Microchœrus*, but the dental formula is $i\ \frac{2}{2}$, $c\ \frac{1}{1}$, $p\ \frac{2}{2}$, $m\ \frac{3}{3}$, and the upper molars are of the tritubercular type.

The well-known *Adapis* (*Aphelotherium* or *Palæolemur*), of the Upper Eocene of France and England, differs from all existing Lemuroids in possessing four premolars[1]; the dental formula being $i\ \frac{2}{2}$, $c\ \frac{1}{1}$, $p\ \frac{4}{4}$, $m\ \frac{3}{3}$. The fourth upper premolar has two outer cusps, and the upper molars (Fig. 333) resemble those of *Lepidolemur* and *Hapalemur*, while the lower canine is a well-developed tooth performing the usual function of biting against the canine of the upper jaw. The lower incisors have upright, spatulate crowns, as in the true Apes. The skull is said to approximate in contour to that of *Propithecus*. The typical *A. parisiensis* is of comparatively small size, but the species of which the upper cheek-teeth are shown in

[1] One specimen has been seen with only three lower premolars.

the woodcut is of much larger dimensions. The skull of *A. magna*, which measures upwards of 4 inches in length, resembles that of *A. parisiensis* in its general characters, but is modified much in the way that the skulls of larger animals differ from the smaller ones of the same natural group. Thus the brain-chamber and orbits are relatively smaller, the face larger, the muscular crests more developed, and the constriction between the cerebral and the facial portion of the skull more marked. These modifications remove the skull in its general characters still farther from the existing Lemurs—so much so that M. Filhol refers it and the other species of *Adapis* to a distinct zoological type, intermediate between the lemurs and the pachyderms, to which he gives the name of *Pachylemuriens*, but later researches do not support this view. As mentioned above, it has been suggested that *Cænopithecus lemuroides* is inseparable from *Adapis parisiensis*, but the postero-internal column of the upper molars is said to be larger. The genera *Tomitherium* and *Notharctus*, of the Eocene of the United States, appear to be allied to *Adapis*, but the second has a larger lower canine. The same deposits have also yielded more or less imperfect remains of other forms departing more widely from the existing Lemuroid type. Of these *Hyopsodus*, of the Wasatch and Bridger Eocene of the United States, has the dental formula $i\,\frac{2}{2}$, $c\,\frac{1}{1}$, $p\,\frac{4}{4}$, $m\,\frac{3}{3}$. The quadritubercular upper molars have well-developed accessory intermediate columns (protoconule and metaconule), and thus resemble those of *Microchœrus;* the external surfaces of the outer columns of their teeth being flattened, with vertical ridges and a distinct cingulum. The third upper molar has its postero-internal column (hypocone) partly aborted, but is otherwise as well developed as the preceding molars. *Microsyops*, of the North American Eocene, appears to have been an allied form in which there were probably only three premolars.

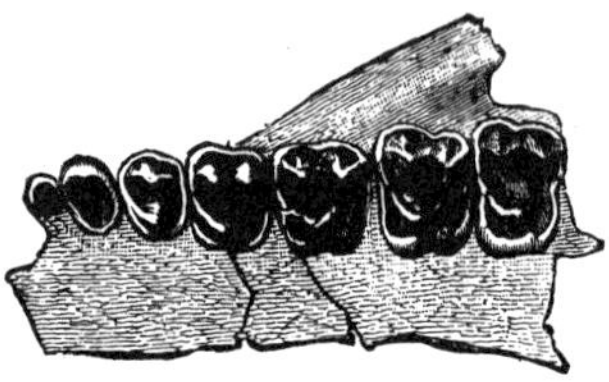

FIG. 333.—The left upper cheek-teeth of *Adapis magna*, from the Upper Eocene of Hampshire.

The genera *Protoadapis* and *Plesiadapis*, from the lowest Eocene of Rheims, may not improbably be regarded as primitive Lemuroids. The lower molars are quinquetubercular, and not unlike those of *Microsyops;* the dental formula of the lower jaw is i 2, c 1, p 3–4, m 3 in the first-named genus, but in the second the dentition is reduced to $i\,\frac{2}{1}$, $c\,\frac{1}{0}$, $p\,\frac{2}{2}$, $m\,\frac{3}{3}$. In *Plesiadapis* the lower

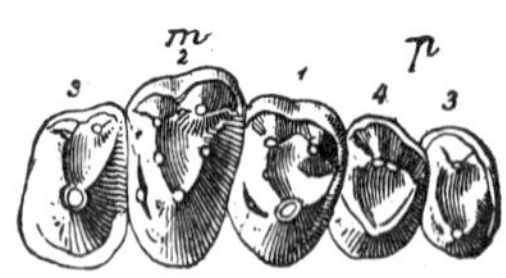

FIG. 334.—The right upper cheek-teeth of *Plesiadapis remensis;* from the Lowest Eocene of Rheims. × $\frac{3}{2}$. *p*, 3, 4, premolars; *m*, 1, 2, 3, molars. (From Osborn.)

and the first upper incisor are enlarged, the upper molars (Fig. 334) tritubercular, and the lower quadritubercular. *Indrodon*, of the lowest Eocene of the United States, resembles *Plesiadapis* in its tritubercular upper molars, and appears to have a nearly similar dental formula. *Mixodectes*, of the same deposits, was probably a more or less closely allied type. *Pelycodus* of the Wasatch Eocene of North America, in which the hallux was not opposable, and *Cryptopithecus* of the German Eocene, may be regarded as very generalised Lemuroids.

Bibliography.—Besides the works and memoirs on particular families and genera referred to above, see St. G. Mivart, "Notes on the Crania and Dentition of the *Lemuridæ*," in *Proc. Zool. Soc.* 1864 (pp. 611-648) and 1867 (pp. 960-975); Mivart and Murie, "On the Anatomy of the *Lemuroidea*," in *Trans. Zool. Soc.* 1872, vol. vii. pp. 1-113; W. Turner, "On the Placentation of the Lemurs," in *Phil. Trans.* vol. clxvi. pp. 569-587; F. Pollen and D. C. Van Dam, *Recherches sur la Faune de Madagascar*, 2me parte, "Mammifères," 1868. For the fossil types see M. Schlosser, "Die Affen., Lemuren, etc., des Europäischen Tertiärs," in *Beitr. Pal. Œstr-Ungar*, 1888.

Suborder ANTHROPOIDEA.

This suborder includes the whole of the remaining members of the Primates, namely, those animals commonly known as Marmosets, Monkeys, Baboons, and Apes, together with Man himself. The characters by which the Anthropoidea are distinguished as a whole from the Lemuroidea may be summarised as follows. Skull with the orbit separated from the temporal fossa by a vertical plate of bone joining the postorbital bar, and the lachrymal foramen situated within the margin of the orbit. Pollex sometimes rudimentary or absent; second digit of manus always well developed, and that of the pes usually with a flattened nail (not so in *Hapalidæ*). The cerebral hemispheres of the brain either completely or almost completely cover the cerebellum, and are much convoluted. Uterus not bicornuate. The placenta is deciduate and discoidal; and the allantois is small. There are never abdominal mammæ. As additional points of distinction from the Lemuroidea, it may be mentioned that the anterior cornu of the hyoid is shorter than the posterior; the inner pair of upper incisors are in contact in the middle line; and the transverse portion of the colon extends uninterruptedly across the abdomen.

The Anthropoidea may be divided into the five families—*Hapalidæ*, *Cebidæ*, *Cercopithecidæ*, *Simiidæ*, and *Hominidæ*, of which the first and second are confined to the New, and the third and fourth to the Old World.

In noticing some of the salient features in the external and internal structure of the Anthropoidea it will be found convenient to allude to all the members of the first four families as Apes, in contradistinction to Man. In respect to relative size the extremes are found in the Gorilla on the one hand and *Hapale* on the other; the difference in this respect between these two forms being greater than that between Man and a Squirrel. The relative proportions between the limbs and the body, and also between the fore and hind limbs, are subject to great variation. Thus in *Hylobates* and *Ateles* both pairs of limbs are much elongated; in the former case the pectoral being much longer than the pelvic pair (Fig. 335). In other cases, as in the Orang (Fig. 354), while the arms are very long, the legs are short; but in the subfamily *Cercopithecinæ* both pairs are short and subequal. Only in the *Hapalidæ* and some of the *Cebidæ* are the legs proportionately as long as in Man.

The tail is as much as three times the length of the body in *Ateles;* while in the *Simiidæ* it is totally absent. In the majority of genera it is long in all the species; but in some cases, as in *Macacus*, it may be either long, short, or absent in the different species of a single genus.

Equally marked variations occur in the shape of the head. Thus in *Ateles* it is rounded; while in the Orang it is elevated vertically; in *Chrysothrix* it is produced posteriorly; and in the Baboons (*Cynocephalus*) it is characterised by the great production of the muzzle and the terminal position of the nostrils, whereby a characteristic Dog-like form is assumed. The eyes are always directed forwards, and are never more separated from one another than in Man, although, as in *Chrysothrix*, they may be closer together. They are of very large size in *Nyctipithecus*, while in the Baboons they are relatively small in proportion to the size of the head. The ears are invariably well developed, and are usually pointed at their postero-superior angle. Those of man are characterised by the soft depending portion known as the "lobule," of which there is a rudiment in the Gorilla. In the majority of Apes the nose is but very slightly prominent; but it attains an extraordinary development in *Nasalis larvatus*, and is scarcely less remarkable in *Semnopithecus roxellanæ* (Fig. 349). Among the Gibbons the Hoolock has a distinctly aquiline nose. The nostrils are terminal in the true Baboons; and while in all the Old World Apes they are approximated, in those of the New World they are separated by a broad septum. With the exception of the Orang, the lips of the Apes are thin.

The pollex makes a nearer approach in form to the human thumb in the Chimpanzee than in any other Ape. Man differs from all the Apes in having the hallux frequently longer instead of shorter than the other digits of the foot. The hallux of the Orang is peculiar in having no nail, but in other cases the nail is flat; the nails of the other digits of the Apes are never quite flat, and in some of the *Cebidæ* they are decidedly compressed laterally, while in the *Hapalidæ* they assume the form of sharp and curved claws.

All the Apes have the greater part of the body well clothed with hair. In the Gibbons and the *Cercopithecidæ* the buttocks have naked ischiatic callosities, which attain their greatest development in *Cynocephalus* and its allies. The male of the Orang has a well-developed beard, and in *Cercopithecus diana* there is long hair on the cheeks and chin, while in *Macacus silenus* the face is surrounded by a fringe of long hair, separated by an interval on the forehead. Long hair is found on the head in *Hapale œdipus* and in some species of *Semnopithecus;* while in the Bonnet Monkey (*Macacus sinicus*) it radiates in all directions from a central point on the vertex. Long hair clothes the shoulders in *Cynocephalus hamadryas* and *Hapale humeralifer;* and the end of the tail has a tuft in two species of *Cynocephalus* and in *Macacus sinicus.* Many of the African *Colobi* and some species of the Howlers have very long hair on the flanks; and in *Pithecia* this development of hair extends to the greater part of the body and the tail, *P. satanas* also having a long beard. In all the lower Apes the hairs on the arm and fore-arm are directed towards the hand quite down to the wrist; and the same arrangement obtains in *Hylobates.* In the other *Simiidæ,* however (as in man), the points of the hairs of the arm and fore-arm converge at the elbow. Darwin's explanation of this peculiarity is that these Apes are accustomed to sit with the arms bent, so that the rain is thus enabled to run off at the elbow.

In one species of *Hapale* the hair is of a silky texture, and in the South American *Eriodes* and *Macacus tibetanus* (as in all the mammals inhabiting the arid and severe climate of Tibet) it becomes woolly.

The development of very brilliant colours on the naked parts of the body, such as the face, sexual organs, and ischiatic callosities is a marked feature of many of the *Cercopithecidæ* and some other Apes.

With the exception of the long tail found in most forms, the general structure of the skeleton of the Apes is very similar to that of man, but there are marked differences in the form of the jaws and of the innominate bones. The proportion of the facial to the cranial region of the skull varies with the shape of the head, of which brief mention has already been made; the greatest development of the facial portion being in the Baboons. Curiously enough, some of the lower American Monkeys, and more especially *Chrysothrix,* have the greatest relative development of the cranial part of the skull of all the Apes; this character being, however, one common to all the smaller representatives of particular groups, and obviously necessary to provide the requisite amount of brain-space. In the convexity of the frontal region of the skull the American forms, and more especially *Pithecia,* make the nearest approximation to man, and the same is true with regard to the occipital production, which is most developed in *Chrysothrix*. Most of the *Simiidæ* exhibit, however, a distinct convexity of the occiput, and thereby differ markedly from the *Cercopithecidæ*, in which this region is flat. The rotundity of the cranium is obscured in the larger Apes, such as the Orang (Fig. 353) and Gorilla, by the development of prominent bony ridges for muscular attachment; these attaining their maximum in the males of the species last named, where the sagittal crests and the supraorbital ridges are very prominent. The mastoid process is always smaller in the Apes than in Man, and as it diminishes in size the petrosal tends to assume an inflated or bullate condition. The orbits in shape are most like that of Man in the Gorilla; and, in accordance with the size of the eyes, they are of enormous dimensions in *Nyctipithecus*.

The angle formed by the plane of the foramen magnum with that of the basicranial axis is subject to variation according to the degree of convexity of the occiput, but is generally smaller than in Man, although larger in *Chrysothrix*. There is an external bony meatus auditorius in Man, the *Simiidæ*, and the *Cercopithecidæ*, but none in the *Cebidæ* and *Hapalidæ*.

The premaxillæ of the Apes are always large; and, except in the Chimpanzee, the premaxillo-maxillary suture persists until after the permanent dentition has been developed. The nasals are smaller and flatter than in Man, but are largest in *Mycetes*. The two rami of the mandible are invariably completely ankylosed at the symphysis in the adult. The Siamang (*Hylobates syndactylus*) is the only ape in which the mandibular symphysis shows a slight projection in front corresponding to the human chin. In *Mycetes* the angle of the mandible attains an enormous development (Fig. 338) to protect the huge inflated basihyal. The frontal sinuses, though present in the *Simiidæ*, are generally replaced in the *Cercopithecidæ* by a coarse diploë, but they are present in the *Cebidæ* and *Hapalidæ*, being especially large in *Cebus*. In fully adult individuals the cranial sutures become obliterated, the inter-nasal suture disappearing at an early age in the *Simiidæ* and most of the *Cercopithecidæ*. As in many Carnivora, the tentorium, or membrane separating the cerebrum from the cerebellum, may become ossified in some of the American forms.

The number of the teeth in the Old World Apes is invariably the same as in Man, namely $i\ \frac{2}{2}$, $c\ \frac{1}{1}$, $p\ \frac{2}{2}$, $m\ \frac{3}{3}$, total 32; but in the *Cebidæ* the cheek-teeth are $p\ \frac{3}{3}$, $m\ \frac{3}{3}$, and in the *Hapalidæ* $p\ \frac{3}{3}$, $m\ \frac{2}{2}$. It is probable that the two pairs of incisors correspond to the first and third of the typical series of three. In all Apes the dental series is interrupted by a diastema, and the canines of the males are large. Man alone has an uninterrupted dental series of a horse-shoe-form, without prominent canines.

According to recent researches the Chimpanzee and some of the other *Simiidæ* exhibit a more or less close approximation to the sigmoid curvature of the vertebral column which is so characteristic of Man, and there is also some approach to it in the Baboons. The number of dorsal vertebræ in the Apes may vary from eleven, as in some species of *Cercopithecus* and *Macacus*, to fourteen in certain forms of *Hylobates*, and to fifteen in *Nyctipithecus*. The *Cebidæ* generally have thirteen; and the same number obtains in the Chimpanzee and Gorilla, while the Orang resembles Man in having but twelve. The lumbar vertebræ show a range in number of from four to seven. In the *Simiidæ* there are four or five of

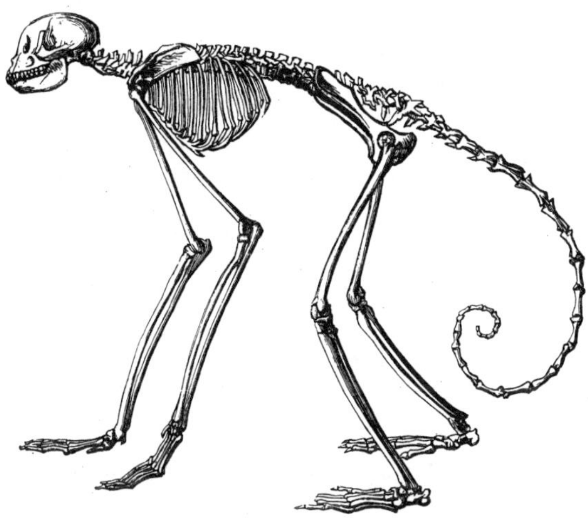

FIG. 335.—Skeleton of the Black-handed Spider Monkey (*Ateles geoffroyi*). From De Blainville.

these vertebræ, the length of the lumbar region being shorter in this family than in the other Apes, with the exception of *Ateles*. The shortness of the lumbar region in the last-named genus is compensated by the relative length of the dorsal region, as is shown in Fig. 335.

The sacrum is longest in the *Simiidæ* and Man, its greatest absolute length occurring in the Gorilla, and the relative greatest length being found in *Hylobates*. The *Simiidæ* never have less than five, and may have six sacral vertebræ; while in the lower forms there are generally only two or three, although occasionally four in some of the American forms. The Orang and some of the Baboons make the nearest approximation to Man in the marked angle formed at the junction of the sacrum with the lumbar vertebræ. Except in the *Simiidæ* and *Macacus inuus*, the number of caudal vertebræ in the Apes always exceeds four, but they may be reduced to five in the Mandrill (*Cynocephalus maimon*). In *Macacus* and *Uacaria* the shortness of the tail is attained by the small size of the vertebræ themselves, the number of which may be from fifteen to seventeen. Other forms usually have from twenty to thirty-three caudals, the latter number occurring in *Ateles* (Fig. 335), where the tail is relatively the longest. The tail is, however, absolutely longest in *Semnopithecus*, *Colobus*, and their allies, the length being partly due to the size of the component vertebræ. Chevron bones are present in all forms having a distinct tail; and, together with other processes for muscular attachment, attain their greatest development in *Ateles*.

The vertebral processes known as metapophyses and anapophyses, which are generally inconspicuous in Man, and are but small in the *Simiidæ*, attain a large development in the lower forms. The metapophyses generally commence in the eighth or ninth dorsal, and continue to the anterior caudals, where they gradually merge in the prezygapophyses. The anapophyses, which are most developed in the *Cebidæ*, project outwards and backwards from one vertebra to embrace the prezygapophyses of the succeeding one. They occur generally in the same region as the metapophyses, but usually cease at the penultimate lumbar, although in some cases they can be traced on to the posterior cervicals and anterior caudals, in the latter region passing into the transverse processes.

In most Apes the sternum is narrow, and consists of a more or less enlarged manubrium, followed by a chain of subequal and antero-posteriorly elongated bones, from three to six in number In the *Simiidæ* alone is there a broad sternum, or one consisting of a manubrium, followed by a single bone only, as in *Hylobates*. The Orang presents a peculiarity, in that the sternum long remains made up of ossifications arranged in pairs, side by side, successively. The true ribs are seven in number on each side in the highest forms, but in *Hylobates* there are sometimes eight. In *Ateles* there are sometimes nine pairs. In *Hapale* the number varies from six to eight, and it is seven or eight in the other genera. The angles of the ribs are never so marked as in Man; although most marked in *Hylobates*. *Pithecia* is distinguished by the greater relative breadth of the ribs. In no Ape is the thorax half as broad again as it is deep from back to breast; but in the *Simiidæ* its transverse diameter exceeds its depth by from about one-fourth to a little under one-third of the latter. In *Ateles*, and sometimes in *Mycetes*, the thorax is wider than deep, but in all the rest it is deeper than wide.

In regard to the appendicular skeleton it may be observed that the Gorilla and Orang make the nearest approach to Man in the form of the scapula; and that the supraspinous fossa of this bone is largest in *Gorilla* and *Mycetes*, being remarkably small in *Simia*.

The *Cebidæ* have a distinct suprascapular notch which is often converted by a bar of bone into a foramen; this bar in *Mycetes* giving rise to a peculiar flat process. The acromion and coracoid processes are most developed in the *Simiidæ* and *Ateles*.

The relative length of the fore and hind limbs has been already briefly touched upon. The humerus closely resembles that of Man throughout the suborder; the nearest approximation occurring in the *Simiidæ*. As in the Lemuroidea, this bone never has an entepicondylar foramen, but in many of the American forms it has a supracondylar perforation. The radius and ulna, like the tibia and fibula, are always perfectly distinct throughout their length; and the hand can be pronated and supinated upon the forearm. Man, the Gorilla, and the Chimpanzee differ from other forms in having no os centrale in the carpus.

The brain of Apes is always much smaller in absolute dimensions than in Man. Thus, according to Professor Mivart,[1] "the cranial capacity is never less than 55 cubic inches in any normal human subject, while in the Orang and Chimpanzee it is but 26 and 27½ cubic inches respectively. The relative size of the brain varies inversely with the size of the whole body, but this is the case in warm-blooded vertebrates generally. The extreme length of the cerebrum never exceeds, as it does in Man, two and a quarter times the length of the basicranial axis. The proportion borne by the brain to its nerves is less in the Apes than in Man, as also is that borne by the cerebrum to the cerebellum. In general structure and form the brain of Apes greatly resembles that of Man. Each half of the cerebrum contains a triradiate lateral ventricle, and though in some *Cercopithecidæ* the posterior cornu is relatively shorter than in man, it again becomes elongated in the *Cebidæ*, and in many of the latter it is actually longer relatively than it is in man. The posterior lobes of the cerebrum are almost always so much developed as to cover over the cerebellum, the only exceptions being the strangely different forms *Mycetes* and *Hylobates syndactylus*. In the latter the cerebellum is slightly uncovered, but it is so considerably in the former. In *Chrysothrix* the posterior lobes are much more largely developed relatively than they are in man. The cerebrum has almost always a convoluted external surface. In this group, however, as in mammals generally, a much-convoluted cerebrum is correlated with a considerable absolute bulk of body. Thus in *Hapale* (and there only) we find the cerebrum quite smooth, the only groove being that which represents the Sylvian fissure. In *Simia* and *Gorilla* and *Anthropopithecus*, on the contrary, it is very richly convoluted. A hippocampus minor is present in all Apes, and in some of the *Cebidæ* it is much larger relatively than it is in Man, and is absolutely larger than the hippocampus major. Of all

1 Article APE, *Encyclopædia Britannica*, ninth edition.

Apes, the Orang has a brain which is most like that of Man; indeed, it may be said to be like Man's in all respects, save that it is much inferior in size and weight, and that the cerebrum is more symmetrically convoluted and less complicated with secondary and tertiary convolutions. If the brain of *Simia* be compared with that of *Gorilla* and *Anthropopithecus*, we find the height of the cerebrum in front greater in proportion in the former than in the latter; also the bridging convolutions, though small, are still distinguishable, while they are absent in the Chimpanzee. Nevertheless this character cannot be of much importance, since it reappears in *Ateles*, while two kinds of the genus *Cebus* (so closely allied as to have been sometimes treated as one species) differ strangely from each other in this respect. The corpus callosum, in Apes generally, does not extend so far back as in Man, and it is very short in *Pithecia*. In the Orang and Chimpanzee there are, as in Man, two corpora albicantia, while in the lower Monkeys there is but one. The vermis of the cerebellum is larger in the *Cebidæ* than in the *Simiidæ* and *Cercopithecidæ*. In all Apes below the *Simiidæ* each lateral lobe of the cerebellum gives off a small lobule, which is received into a special fossa of the petrous bone. Certain prominences of the medulla oblongata, termed corpora trapezoidea, which are found in lower mammals, begin to make their appearance in the *Cebidæ*."

The organs connected with the functions of alimentation, circulation, and excretion, as well as the muscles, conform generally to the type obtaining in Man, of which full description will be found in works on human anatomy. The tongue is longer in Apes than in Man; and a uvula is generally present, although rudimentary in the *Cebidæ*. The peculiar sacculation of the stomach in the subfamily *Semnopithecinæ* has been already mentioned; this sacculation is most developed at the cardiac extremity, where it somewhat resembles a colon spirally coiled. In *Hylobates* the stomach is very like that of Man, differing only in the more elongated and distinct pylorus. *Pithecia* has a more globular stomach, while in *Hapale* the cardiac and pyloric apertures are approximated. The intestine of Apes is devoid of valvulæ conniventes, and it is only in Man and the *Simiidæ* that the colon is furnished with a vermiform appendage. The colon varies from a fully sacculated form in *Hylobates* to a smooth one in *Cebus*.

The liver of Apes is subject to a considerable amount of variation. In the *Simiidæ* it comes more or less close to the human type; that of the Orangs being usually divided only into two principal lobes by the umbilical vein, and showing no trace of lateral fissures. In the Gorilla these fissures are present, so as to produce right and left lateral and central lobes. *Hylobates* has a liver (Fig. 352) which perhaps is nearer to the human than that of any of the other *Simiidæ*. In the *Cercopithecidæ* the liver differs from that of the *Simiidæ* by the deeply cleft lateral fissures, and has a comparatively small and pointed caudate lobe. The enormous size of the stomach in *Colobus* causes the liver to be very narrow, and pushed to the left side. The liver of the *Cebidæ* (Fig. 336) and *Hapalidæ*, in addition to the deeply cleft lateral fissures, is characterised by the great size and quadrangular form of the caudate lobe (*c*), which attains its maximum development in *Ateles*. The gall-bladder is always present.

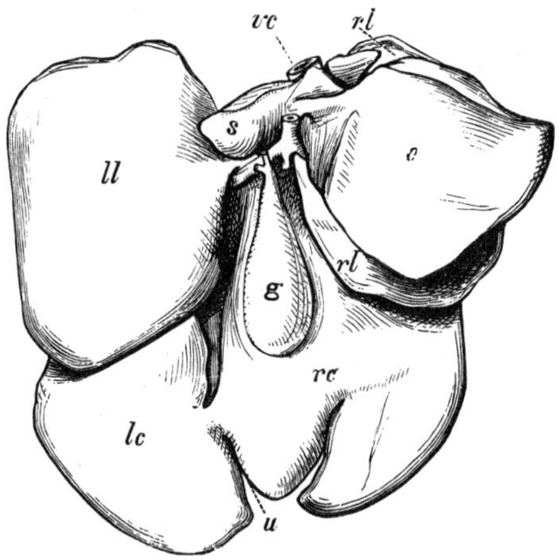

FIG. 336.—Under surface of the liver of the Black-handed Spider Monkey (*Ateles melanochir*). *u*, Umbilical fissure; *vc*, vena cava; *ll*, left lateral lobe; *lc*, left central lobe; *rc*, right central lobe; *rl*, right lateral lobe; *s*, Spigelian lobe; *c*, caudate lobe; *g*, gall-bladder.

The larynx is in many Apes furnished with sac-like appendages, which are variable in different species as regards number, size, and situation. They may be dilatations of the laryngeal ventricle, as in *Simia*, *Gorilla*, and *Anthropopithecus*, or they may open above the false vocal chords so as to be extensions of the thyro-hyoid membrane, as in *Hylobates*. There may be but a single median opening in the front part of that membrane at the base of the epiglottis, as in the *Cercopithecidæ*. There may be a single median opening at the back of the trachea, just below the cricoid cartilage, as in *Ateles*; there may be but a single sac, or there may be five, as sometimes in *Mycetes*. These may be enormous, meeting in the middle line in front and extending down to the axillæ, as in the Gorilla and Orang. A sac may occupy the cavity of the expanded body of the hyoid, as in *Mycetes*.

The hyoid has its basilar part generally somewhat more convex and enlarged than in Man; but in *Mycetes* it becomes greatly enlarged and deeply excavated, so as to form a great bony bladder-like structure. The posterior cronua of the hyoid (thyro-hyals) are never entirely absent, but the anterior or lesser cornua may be so, as in *Mycetes*. The anterior cornua never exceed the posterior cornua in length; but they may be (*e.g.* in *Cercopithecus*) more largely developed relatively than in Man, and may even be jointed, as in *Lagothrix*.

The lungs have generally the form of those of man; but the right lung may have four lobes, as in *Hylobates*. The great arterial trunks in *Simia*, *Gorilla*, and *Anthropopithecus* are arranged as in

Man. In *Hylobates* and the lower Apes, however, the left carotid artery may take its origin from the innominate artery.

In regard to their distribution in time the earliest record that we as yet have of the occurrence of Apes is in the Middle Miocene of Europe, where forms are met with apparently so closely allied to some of the higher existing types that it is evident we must look much farther back before we can get any clue to the origin of the suborder. Since all the known fossil Old World Apes are referable to the *Simiidæ* or *Cercopithecidæ*, and no representatives of these families have been obtained from the Tertiaries of America, it would appear that the distinction of the Apes of the Old World from those of the New is of very old standing.

At the present day Apes are mainly confined to tropical and subtropical regions. In the Old World *Macacus inuus* is found as far north as Gibraltar, *M. tibetanus* and *Semnopithecus roxellanæ* inhabit western Tibet, while in Japan we have *M. speciosus*. In the New World one species of *Ateles* is known to occur as far north as latitude 19° in Southern Mexico, and may range a few degrees higher. To the southward species are found near the Cape, in Timor, and the Malay Archipelago; while in America they range in Brazil and Paraguay to about latitude 30°. The Tibetan species are found at a very high elevation; and in the outer Himalaya the Langurs (*Semnopithecus*) may be seen in winter and spring leaping from bough to bough of snow-covered pines.

Apes are very abundant in the Ethiopian and Oriental regions, as well as in that part of America which extends from Panama to Southern Brazil. Ceylon, Borneo, Sumatra, and Java may be mentioned as islands where Ape-life attains great development; but they are unknown in Madagascar and the West Indian Islands, and of course in the Australasian region.

We have already alluded to the circumstance that while the *Simiidæ* and *Cercopithecidæ* are exclusively confined to the Old World, the *Cebidæ* and *Hapalidæ* are equally restricted to the New, and we may accordingly proceed to notice a few points in relation to generic distribution. Of the larger *Simiidæ* the Gorilla and Chimpanzee are confined to Equatorial Africa, and the Orang to Malayana; but there is evidence of the former existence of a species of Chimpanzee (*Anthropopithecus*) and not improbably of an Orang (*Simia*) in Northern India. The Gibbons (*Hylobates*) are now exclusively Oriental. Europe has only *Macacus inuus* of Gibraltar, also found in Africa north of the Sahara, and therefore strictly Palæarctic in distribution. The Ethiopian region includes in the *Cercopithecidæ* the genus *Colobus* (the African analogue of *Semnopithecus*), *Cercopithecus*, and the Baboons (*Cynocephalus*, etc.) The Baboons range, however, into Arabia and Syria, and also existed during the Pliocene epoch in Northern India. *Semnopithecus* and *Macacus* are very characteristic of the Oriental region; but, as already mentioned, outlying species extend into various parts of the Palæarctic region. *Macacus* has indeed a very wide distribution, extending from Gibraltar and North Africa to Japan. The allied *Cynopithecus*, represented only by *C. niger* of Celebes, approximates to the Baboons; while the one species of *Nasalis* is peculiar to Borneo. Remains of *Semnopithecus* and *Macacus* occur in the Tertiaries of India and Europe, which also yield allied extinct types noticed in the sequel.

In America, north of Panama, the genera known to be represented are *Chrysothrix*, *Nyctipithecus*, *Cebus*, *Ateles*, *Mycetes* and *Hapale* in Veragua; *Nyctipithecus*, *Cebus*, *Ateles*, and *Mycetes* in Costa Rica and Nicaragua; *Ateles* and *Mycetes* in Guatemala; and *Ateles* in Southern Mexico. Brazil is the headquarters of the American Apes; but different portions of that vast region have a somewhat distinct Ape fauna. Thus the genus *Eriodes* appears in South-Eastern Brazil to represent the species of *Ateles* inhabiting the more northern and western parts of the empire. Southwards, the genera *Cebus*, *Mycetes*, *Chrysothrix*, and *Callithrix* extend farthest; but they do not probably all extend to the farthest limit yet known, namely 30° S. The species found farthest south are *Mycetes caraya*, *Cebus fatuellus*, and *Callithrix personatus*.

Family HAPALIDÆ.

Dentition: $i\ \frac{2}{2}$, $c\ \frac{1}{1}$, $p\ \frac{3}{3}$, $m\ \frac{2}{2}$; total 32. No bony external auditory meatus, a broad internarial septum, and no cheek-pouches. Tail non-prehensile; no ischiatic callosities. Pollex not opposable; a long, curved, and pointed claw to all the digits except the hallux.

This family, which includes the smallest representatives of the suborder, commonly known as Marmosets, is confined to the New World. In addition to the diagnostic characters given above, it may be mentioned that the pollex is elongated and the hallux very small, while the pectoral limbs are not longer than the pelvic pair; and the tail is long and more or less thickly covered with elongated hairs.

The dentition of the Marmosets sufficiently distinguishes them from all other members of the suborder, although they are evidently nearly allied to the *Cebidæ*. The small size of the hallux, and the total incapacity of the pollex to oppose itself in the least degree to the other digits, as well as the presence of claws on all the digits of the manus, are, however, equally characteristic features. These animals (Fig. 337) are not larger than Squirrels, and are of active arboreal habits, living in small companies, and adding insects to the ordinary fruit diet. Frequently, as in the figured species, the head is furnished on either side with a long tuft of hair projecting outwards and backwards. They give birth to as many as three young ones at a time, and thereby differ from all other members of the suborder, in which one is the normal number. They are divided into two genera, according to the proportionate size of the lower canine to the incisors; but some species present an intermediate condition, so as to render this distinction of somewhat doubtful value.

Hapale.[1]—Lower canine not longer than the incisors. A number of species have been described, among which may be mentioned

FIG. 337.—The Golden Marmoset (*Midas chrysoleucas*). From *Proc. Zool. Soc.* 1868, pl. 24.

H. jacchus, *H. albicollis*, *H. aurita*, and *H. humeralifer*. Remains of species of this genus have been found in the cavern-deposits of Brazil.

Midas.[2]—Lower canine considerably longer than the incisors. No less than twenty-four species of this genus have been named, among which the Silky Marmoset (*M. rosalia*) of Columbia, the Pinche Monkey (*M. œdipus*) of South-Eastern Brazil, and the Golden Marmoset (*M. chrysoleucas*, Fig. 337) are well-known types.

[1] Illiger, *Prodromus Syst. Mamm. et Avium*, p. 71 (1811).
[2] Geoffroy, *Ann. du Muséum*, vol. xix. p. 120 (1812).

Family CEBIDÆ.

Dentition: $i\ \frac{2}{2}$, $c\ \frac{1}{1}$, $p\ \frac{3}{3}$, $m\ \frac{3}{3}$; total 36. Tail frequently prehensile; digits with nails; other characters as in the *Hapalidæ*.

The members of this American family are at once distinguished by the dental formula, which is numerically higher than in any other Apes. The various species range over the whole of tropical America, but are most abundant in the dense forest regions of Brazil, where they live a completely arboreal life, to which the prehensile tails of many of them are so specially adapted. They are in most respects closely allied to the *Hapalidæ*, but the pollex diverges somewhat from the plane of the other digits; while the retention of the third molar is a very distinctive feature. None of the species attain the dimensions of the larger *Cercopithecidæ* of the Old World. The genera are usually arranged in five subfamilies.

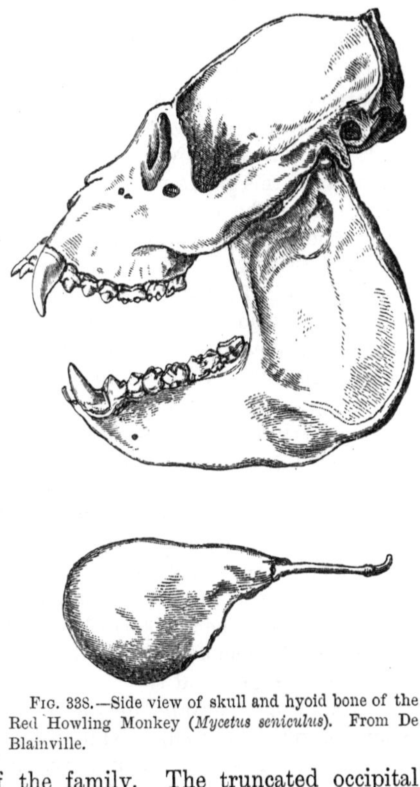

FIG. 338.—Side view of skull and hyoid bone of the Red Howling Monkey (*Mycetus seniculus*). From De Blainville.

Subfamily **Mycetinæ**.—Lower incisors vertical; hyoid bones enormously inflated; tail long and prehensile, naked beneath at the end; pollex well developed.

Mycetes.[1]—The sole representatives of this subfamily are the well-known Howling Monkeys, all of which are included in the genus *Mycetes*. They are of more bulky build, and have more produced muzzles than the other members of the family. The truncated occipital region, and the extraordinary development of the rami of the mandible, especially of their angular and ascending portions, are the chief peculiarities by which the skulls (Fig. 338) of the members of this genus are characterised. The last named character, which is more marked in the male than in the female sex,

[1] Illiger, *Prodromus Syst. Mamm. et Avium*, p. 70 (1811).

is related to the enormous size of the vocal organs, which the rami of the mandible enclose and protect. The inflated hyoid bone, which forms a deep cup, is shown in the figure. The Howlers are subject to great individual and sexual variation of colours, so that the discrimination of species from local races is difficult. In one species the male is black and the female straw-coloured; and several of the species have bright red or golden hair on the flanks. In disposition these creatures are sluggish and stupid, but their chief characteristic is their prodigious power of voice. Mr. Bates, in his *Naturalist on the Amazons*, observes that "when Howlers are seen in the forest there are generally three or four of them mounted on the topmost branches of a tree. It does not appear that their harrowing roar is emitted from sudden alarm; at least it was not so in captive individuals. It is probable, however, that the noise serves to intimidate their enemies."

Several species have been described, the Red Howler (*M. seniculus*) and the Ursine Howler (*M. ursinus*) being well-known forms. Remains of this genus probably referable to existing types are found fossilised in the cavern-deposits of Brazil. An allied fossil form from the South American Pleistocene has been described as *Protopithecus*.

Subfamily **Pitheciinæ**.—Lower incisors inclined forward at their summits; hyoid bone normal; tail long or short, non-prehensile; pollex well developed. Two genera are included in this subfamily, readily distinguished by the length of the tail.

Pithecia.[1]—The Sakis, as the representatives of this genus are commonly termed, are readily characterised by the length of the tail; the angle of the mandible is expanded, although less so than in *Mycetes*. A number of species have been described, the Black Saki (*P. satanas*) of the Lower Amazons, being one of the best known. While some species, like *P. hirsuta*, have long hair covering the whole of the head, body, and tail, in others only the head, or the cheeks and chin, are so clothed.

Uacaria.[2]—The Uakari Monkeys differ from all the other *Cebidæ* by their short Baboon-like tail. The Bald Uakari (*U. calva*) of the Rio Negro, and the closely allied *U. rubicunda* of the Upper Amazons, are remarkable for their scarlet face, which forms a striking contrast to the long, silky, whitish hair covering the body. According to Mr. Bates, the Uakaris live in forests which are inundated during a great part of the year, and never descend to the ground; they appear to be rare and of local distribution. The third species, *U. melanocephala*, differs considerably from both the others. The cæcum of *U. calva*, according to Mr. F. E. Beddard, measures

[1] Geoffroy, *Ann. du Muséum*, vol. xix. p. 115 (1812).

[2] Gray, *Proc. Zool. Soc.* 1849, p. 9. Amended from *Ouakaria*: Syn. *Brachyurus*; Spix, *Sim. et Vesp. Brasil*, p. 11 (1823). Preoccupied by Fischer, 1814.

upwards of "10 inches along the greater curvature; it is separated from the colon by a very marked constriction; it is not sacculated, and when fully distended with air is curved on itself into a little less than a circle; it is furnished with a well-developed median frenum carrying blood-vessels." A similar type of cæcum is also found in *Callithrix* and *Pithecia*.

Subfamily **Nyctipithecinæ**.—Lower incisors vertical; hyoid normal; tail long, non-prehensile; pollex well developed.

Three genera are included in this subfamily, the species being partly insectivorous.

Callithrix.[1]—Head small, depressed, and not elongated; nares

FIG. 339.—The Moloch Teetee (*Callithrix moloch*). From *Archives du Muséum*, vol. iv. pl. 3.

widely separated; canines small; angle of mandible expanded as in *Pithecia*; tail with long hair.

This genus comprises several small species, mostly from Brazil and the Amazons, and commonly known as Teetees, one of the best-known species (*C. moloch*, Fig. 339) being represented in the accompanying woodcut. The smaller eyes and the more widely separated nostrils distinguish them from *Nyctipithecus*; while the

[1] Geoffroy, *Ann. du Muséum*, vol. xix. p. 112 (1812).

small canines and the bushy tail readily mark their distinction from *Chrysothrix.* Remains of *Callithrix* have been found in the Brazilian caves.

Chrysothrix.[1]—Head greatly elongated; orbits large and closely approximated; canines well developed; tail with comparatively short hair.

The small Squirrel Monkeys, of which four species have been

FIG. 340.—The Lemurine Douroucouli (*Nyctipithecus lemurinus*). From *Archives du Muséum*, vol. iv. pl. 2.

described, are characterised by the great backward projection of the occipital region of the skull, and by orbits approximating in size to those of the next genus.

Nyctipithecus.[2]—Head rounded; orbits very large, separated by a narrow septum; nares somewhat approximated.

The Dourouculis (Fig. 340), as the members of this genus are called, are of nocturnal habits, in association with which the eyes are of enormous dimensions, as in the Lemuroid genus *Loris.* The following account of two species of this genus is taken from Mr.

[1] Kaup, *Thierreich,* vol. i. p. 51 (1835).

[2] Spix, *Sim. et Vesp. Brasil,* p. 25 (1823).

Bates's *Naturalist on the Amazons:* "They sleep all day long in hollow trees, and come forth to prey on insects and eat fruit only in the night. They are of small size, the body being about a foot long, and the tail 14 inches, and are thickly clothed with soft gray and brown fur, similar in substance to that of the Rabbit. Their physiognomy reminds one of the Owl or Tiger-Cat; the face is rounded and encircled by a ruff of whitish fur; the muzzle is not at all prominent; the mouth and chin are small; the ears are very short, scarcely appearing above the hair of the head; and the eyes are large and yellowish in colour, imparting the staring expression of nocturnal birds of prey. The forehead is whitish, and decorated with black stripes, which in one of the species (*N. trivirgatus*) continue to the crown, and in the other (*N. felinus*) meet on the top of the forehead. *N. trivirgatus* was first described by Humboldt, who discovered it on the banks of the Cassiquiare, near the head-quarters of the Rio Negro."

Subfamily **Cebinæ**.—Lower incisors vertical; hyoid bone normal; tail long and prehensile; pollex present or absent.

This subfamily includes the typical members of the family, which are arranged in four genera.

Ateles.[1]—Form slender; limbs very long; fur not woolly; pollex absent; tail naked beneath distally; nails not much laterally compressed and pointed.

This genus includes the well-known Spider Monkeys (Fig. 341), which by their long limbs and tail are admirably adapted to a purely arboreal life, although they lack the active and agile habits of the Old World Gibbons. The tail with the under surface of its extremity naked affords the most completely prehensile type of this organ, and can sustain the weight of the whole body. Objects are not unfrequently grasped by it and brought within reach of the hand or mouth. Owing to the absence of the pollex the power of grasping is very imperfect in the hand. At least fourteen species of this genus have been described, among the best-known being *A. melanochir* (Fig. 341), *A. paniscus* of Guiana, *A. geoffroyi* of Central America, *A. ater* of Eastern Peru, and *A. hybridus* of Colombia.

Eriodes.[2]—Form slender; limbs very long; fur woolly; inter-nasal septum narrower than usual in the family; pollex rudimentary; tail naked beneath distally; nails exceedingly compressed laterally, and pointed.

This genus is represented by three species from South-East Brazil, which, while closely allied to the true Spider Monkeys, differ by their woolly hair, the narrow internasal septum, the presence of a rudimentary pollex, and the great compression of the

[1] Geoffroy, *Ann. du Muséum*, vol. vii. p. 260 (1806).

[2] I. Geoffroy, *Dict. Class.* vol. xv. p. 443 (1829).

nails. The species are *E. arachnoides*, *E. hemidactylus*, and *E. hypoxanthus*.

Lagothrix.[1]—Form rather robust; limbs moderate; fur woolly; pollex well developed; tail distally naked beneath.

The Woolly Monkeys differ from the preceding genera by the presence of a well-developed pollex. They resemble *Eriodes* in their fur and compressed nails, but differ in the more widely

FIG. 341.—The Black-handed Spider Monkey (*Ateles melanochir*). From *Proc. Zool. Soc.* 1871, pl. 15.

separated nares. The tail resembles that of the preceding genera. Speaking of these Monkeys Mr. Bates observes that "the Barrigudos are very bulky animals, whilst the Spider Monkeys are remarkable for the slenderness of their bodies and limbs. I obtained specimens of what have been considered two species, one (*L. olivaceus?*) having the head clothed with gray, the other (*L. humboldti*, Fig. 342) with black fur. They both live together in the same places,

[1] Geoffroy, *Ann. du Muséum*, vol. xix. p. 106 (1812).

and are probably only differently coloured individuals of one and the same species. I sent home a very large male of one of these kinds, which measured 27 inches in length of trunk, the tail being 26 inches long; it was the largest monkey I saw in America, with the exception of a black Howler, whose body was 28 inches in height. The skin of the face in the Barrigudo is black and wrinkled, the forehead is low, with the eyebrows projecting. . . . In the forests the Barrigudo is not a very active animal; it lives exclusively on fruits, and is much persecuted by the Indians on

FIG. 342.—Humboldt's Lagothrix (*Lagothrix humboldti*). From *Proc. Zool. Soc.* 1863, pl. 31.

account of the excellence of its flesh as food." Five species are usually recognised, viz. *L. canus*, *L. humboldti*, *L. castelnaui*, *L. tschudii*, and *L. geoffroyi*.

Cebus.[1]—Form rather robust; limbs moderate; fur not woolly; pollex well developed; tail not naked beneath distally.

This, the typical, genus includes the Sapajous or Capuchins (Fig. 343), which are so commonly seen in this country in captivity, being the favourite Monkeys of itinerant musicians. They are smaller and stouter in build than the Spider Monkeys, from which they are readily distinguished by the well-developed pollex, and the absence of a naked under surface to the extremity of the tail.

[1] Erxleben, *Syst. Règne Animal*, p. 44 (1777).

At least twenty species have been described (*C. fatuellus*, *C. lunatus*, *C. capucinus*, *C. albifrons*, *C. hypoleucus*, etc.), but it is probable that some of these are not entitled to stand, since there is a large

Fig. 343.—The White-cheeked Sapajou (*Cebus lunatus*). From *Proc. Zool. Soc.* 1865, pl. 45.

amount of individual variation. Fossil remains of species of *Cebus* have been described from the Pleistocene cavern-deposits of Brazil.

Family Cercopithecidæ.

Dentition: $i\ \frac{2}{2}$, $c\ \frac{1}{1}$, $p\ \frac{2}{2}$, $m\ \frac{3}{3}$; total 32. Crowns of molars elongated antero-posteriorly, with the tubercles forming a pair of imperfect transverse ridges, and the last lower molar usually with a hind talon. A bony external auditory meatus. A narrow internarial septum. Tail non-prehensile. Ischiatic callosities present. Cheek-pouches present or absent. Pollex, when present, opposable. Pelvic limbs never much longer than pectoral. Sternum narrow. Cæcum without vermiform appendage.

This family includes all the Old World Apes, with the exception of the *Simiidæ*, and may be divided into the subfamilies *Cercopithecinæ* and *Semnopithecinæ*.

Subfamily **Cercopithecinæ**.—Pelvic and pectoral limbs approximately equal; tail variable; cheek-pouches present; stomach simple.

This subfamily comprises the African Baboons, the common Indian Monkeys constituting the genus *Macacus*, together with the African *Cercopithecus* and *Cercocebus* and a few allied types.

Cynocephalus.[1]—Muzzle much elongated (Fig. 344), with the nostrils terminal; ischial callosities very large; tail more or less short; muzzle swollen by enlargement of the maxillæ. Now confined to Africa and Arabia.

This genus comprises the typical Baboons, and we may select the well-known Mandrill (*C. maimon*), of tropical West Africa, as a good illustrative example. It may be mentioned in passing that the name Mandrill appears to have been first introduced into English literature by William Smith in his *New Voyage to Guinea*,

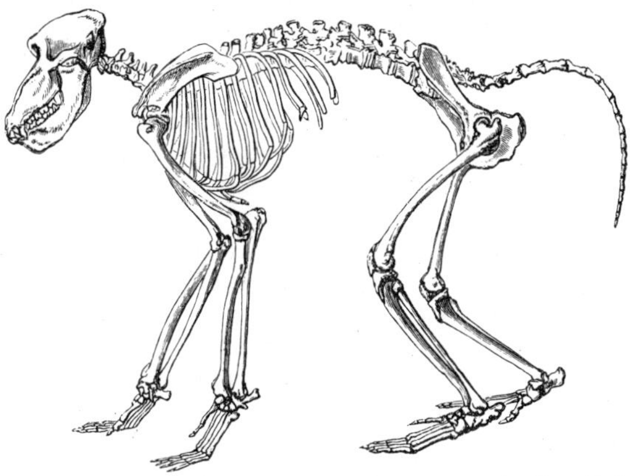

FIG. 344.—Skeleton of the Chacma Baboon (*Cynocephalus porcarius*). From De Blainville.

published in 1744, wherein he mentions among the animals of Sierra Leone one "called by the white men in this country Mandrill," but adds, "why it is so called I know not."[2] Smith gives sufficiently accurate details to show that his animal is not that now called Mandrill, but the Chimpanzee. Buffon, however, while quoting Smith's description, transferred the name to the very

1 Lacépède, "Nouv. tabl. méth." (1799) in *Mém. de l'Institut*, vol. iii. p. 490 (1801).

2 "'Mandrill' seems to signify a 'man-like Ape,' the word 'Drill' or 'Dril' having been anciently employed in England to denote an Ape or Baboon. Thus in the fifth edition of Blount's '*Glossographia*, or a dictionary interpreting the hard words of whatsoever language now used in our refined English tongue . . . very useful for all such as desire to understand what they read,' published in 1681, I find 'Dril, a stonecutter's tool wherewith he bores little holes in marble, etc. Also a large overgrown Ape and Baboon, so called.' 'Drill' is used in the same sense in Charlton's *Onomasticon Zoicon*, 1668. The singular etymology of the word given by Buffon seems hardly a probable one."—Huxley's *Man's Place in Nature*, p. 10, 1863.

different species now under consideration, and to that it has been attached ever since.

The Baboons generally are distinguished from most other Monkeys by the comparative equality of the length of their limbs, which with the structure of the vertebral column adapts them rather for quadrupedal progression on the ground than for climbing among the branches of trees; and some of them, like the South African Chacma (*C. porcarius*), of which the skeleton is shown in Fig. 344, live habitually among rocks, and are much less completely frugivorous than other Apes. They are also remarkable for the great size of their face and jaws as compared with the part of the skull which encloses the brain. The Mandrill, in addition to these characters, is distinguished by the heaviness of its body, stoutness and strength of its limbs, and exceeding shortness of its tail, which is a mere stump, not 2 inches long, and usually carried erect. It is, moreover, remarkable for the prominence of its brow ridges, beneath which the small and closely approximated eyes are deeply sunk; the immense size of the canine teeth; the great development of a pair of oval bony prominences on the maxillary bones in front of the orbits, rising on each side of the median line of the face, and covered by a longitudinally ribbed naked skin; and more especially for the extraordinarily vivid colouring of some parts of the skin. The body generally is covered with a full soft coating of hair of a light olive-brown above and silvery-gray beneath, and the chin is furnished underneath with a small pointed yellow beard. The hair of the forehead and temples is directed upwards so as to meet in a point on the crown, which gives the head a triangular appearance. The ears are naked and of a bluish-black colour. The hands and feet are naked and black. A large space around the greatly developed ischial callosities, as well as the upper part of the insides of the thighs, are naked and of a crimson colour, shading off on the sides to lilac or blue, which, depending not upon pigment but upon injection of the superficial blood-vessels, varies in intensity according to the condition of the animal—increasing under excitement, fading during sickness, and disappearing after death. But it is in the face that the most remarkable disposition of vivid hues occur, more resembling those of a brilliantly coloured flower than what might be expected in the cutaneous covering of a mammal. The cheek-prominences are of an intense blue, the effect of which is heightened by deeply sunk longitudinal furrows of a darker tint, while the central line and termination of the nose are a bright scarlet. Notwithstanding the beauty of these colours in themselves, the whole combination, with the form and expression of features, quite justifies Cuvier's assertion that "il serait difficile de se figurer un être plus hideux que le Mandrill."

It is only to fully adult males that this description applies. The female is of much smaller size, and of more slender make; and, though the general tone of the hairy parts of the body is the same, the prominences, furrows, and colouring of the face are very much less marked. The young males have black faces. At the age of three the blue of the cheeks begins to appear, but it is not until they are about five, when they cut their great canine teeth, that they acquire the characteristic red of the end of the nose.

The Mandrills, especially the old males, are remarkable for the ferocity of their disposition, as well as for other disagreeable qualities, which are fully described in Cuvier's account of the animal in

FIG. 345.—The Yellow Baboon (*Cynocephalus babuin*). From *Archives du Muséum*, vol. ii. pl. 34.

La Ménagerie du Muséum d'Histoire Naturelle (1801), but when young they can easily be tamed. Like the rest of the Baboons, they appear to be rather indiscriminate eaters, feeding upon fruit, roots, reptiles, insects, scorpions, etc., and inhabit open rocky ground rather than forests. Not much is known of the Mandrill's habits in the wild state, nor of the exact limits of its geographical distribution. The specimens brought to Europe all come from the west coast of tropical Africa, from Guinea to the Gaboon.

An allied species, the Drill (*C. leucophæus*), which resembles the Mandrill in size, general proportions, and shortness of tail, but wants the bright colouring of the face which makes that animal so remarkable, inhabits the same district. Other well-known species are the Yellow Baboon (*C. babuin*), of West Africa (Fig. 345); the Arabian Baboon (*C. hamadryas*), of Arabia and Abyssinia; and the Anubis Baboon (*C. anubis*), of West Africa.

It is very noteworthy from a distributional point of view, as showing the former intimate connection between the faunas of the Oriental and Ethiopian regions, that fossil remains of Baboons have been found in the Pleistocene cavern-deposits of Madras, and also in the older Pliocene beds of the Siwalik Hills in Northern India; the two species from the latter deposits having been described as *C. subhimalayanus* and *C. falconeri*.

Theropithecus.[1]—Distinguished from *Cynocephalus* by the nostrils not being terminal, but situated as in *Macacus*. This genus is represented by the Abyssinian Gelada (*T. gelada*) and the allied *T. obscurus*.

Cynopithecus.[2]—The Black Ape of Celebes (*C. niger*) forms a

FIG. 346.—The Tibetan Macaque (*Macacus tibetanus*). From Milne-Edwards, *Recherches des Mammifères*, pl. 34.

connecting link between the Baboons and the genus *Macacus*; the skull differing from that of the latter in the development of longitudinal ridges on the sides of the upper surface of the maxillæ, as in some of the species of *Cynocephalus*. The muzzle is also more produced than in *Macacus*.

Macacus.[3]—Muzzle considerably produced; nostrils not terminal; cheek-pouches and ischial callosities well developed; tail long, short, or absent; a distinct talon to the third lower molar.

With the exception of the Barbary Ape (*M. inuus*) of Northern Africa and Gibraltar, the Macaques are now exclusively Asiatic, one species (Fig. 346) occurring in Tibet, and another (*M. speciosus*)

[1] I. Geoffroy, *Arch. du Muséum*, vol. ii. p. 576 (1841).
[2] I. Geoffroy, *Voyage de Belanger*, p. 66 (1834).
[3] Lacépède, *Mém. de l'Institut*, vol. iii. p. 450 (1801). Amended.

being found in Japan. All these Monkeys are of stout build, and it is chiefly by the greater production of the muzzle, the larger ischiatic callosities, and the frequent shortness of the tail that they are distinguished from the under-mentioned African genera. The transition from the longer-tailed to the short-tailed forms is so complete that the proposed generic separation of the latter as *Inuus* is impracticable. In *M. inuus* the tail is wanting; in *M. tibetanus* (Fig. 346) and *M. nemestrinus* of Tenasserim it is short; in the common Bengal Monkey (*M. rhesus*) it is about one-half the length of the head and body, while in *M. cynomolgus* and its allies it is still longer. In the Indian Lion-tailed Monkey (*M. silenus*) it is tufted at the end.

The following summary of the habits of the Macaques is taken from Mr. W. T. Blanford's *Mammals of British India*: "The species of the present genus resemble each other in their habits; they are found in flocks, often of considerable size, and generally composed of individuals of both sexes and of all ages. They are active animals, though less rapid in their movements, whether on trees or on the ground, than the *Semnopitheci.* Their food is varied, most of the species, if not all, eating insects as well as seeds, fruits, etc., and one kind feeding partly on crustacea. They have occasionally been known to devour lizards, and, it is said, frogs also. All have the habit of cramming food into their cheek-pouches for mastication at leisure—a practice that must be familiar to any one who has fed monkeys in confinement. The voice and gestures of all the species are similar, and differ entirely from those of both the Gibbons and *Semnopitheci.* . . . The majority of the species are very docile when young. They thrive well, and several of them have bred in confinement. The period of gestation is about seven months, only a single young one, as a rule, being produced at a birth. They become adult at the age of four or five years, but breed earlier."

The Common Indian *M. rhesus* is found in the Himalaya at an elevation of over 8000 feet.

Fossil remains of *Macacus* are found in India in the Pleistocene of Madras and the Pliocene of the Punjab; and they also occur in the Pliocene of France and Italy, those from the latter deposits having been incorrectly separated as *Aulaxinuus.* Part of the jaw of a Monkey from the Pleistocene of Essex has been described as *Macacus pliocenus*, and is very interesting as showing the presence of Apes in Europe at that late period.

Cercocebus.[1]—An African genus agreeing with *Macacus* in the presence of a hind talon to the third lower molar, but with the other characters of *Cercopithecus.* The species of this genus are known as Mangabeys, or White-eyelid Monkeys, and include

[1] Geoffroy, *Ann. du Muséum*, vol. xix. p. 97 (1812).

C. collaris, *C. fuliginosus*, *C. æthiops*, and *C. albigena*; all being from West Africa.

Cercopithecus.[1]—Muzzle more or less short; ischial callosities moderate; tail long; no talon to third lower molar. Build more slender than in *Macacus*. Confined to Africa.

The members of this and the last genus include those Monkeys which in their comparative slender build and length of tail make the nearest approach to the next subfamily. There are numerous species, among which the Green Monkey (*C. callitrichus*), the Grivet (*C. griseo-viridis*), the Vervet (*C. lalandi*), the Pluto Monkey (*C. pluto*, Fig. 347), the Patas (*C. ruber*), the Diana Monkey (*C. diana*), and the Mona Monkey (*C. mona*) are well-known types.

FIG. 347.—The Pluto Monkey (*Cercopithecus pluto*). From Gray, *Proc. Zool. Soc.* 1848, p. 57.

Subfamily **Semnopithecinæ**.[2] — Pelvic limbs longer than the pectoral; tail very long; no cheek-pouches; stomach sacculated. Build slender.

This subfamily is represented by three genera, of which one is African and two are Asiatic. Mr. W. T. Blanford, in his *Mammals of British India*, observes that "the members of this subfamily are readily distinguished by their slender form, and by the absence of cheek-pouches. They are more purely herbivorous than the Macaque Monkeys, and a considerable portion of their food consists of leaves and young shoots. In consequence probably of the nature of their food, these Monkeys are more delicate than the species of *Macacus*, and are thus less easily kept in captivity. They are consequently far less well represented in European museums, and have been less studied by European naturalists. Very little is known of their general life-history or of their feeding habits."

[1] Erxleben, *Syst. Regnc. Animal*, p. 22 (1777).

[2] Or *Colobinæ*.

Their digestive organs are much modified, the stomach attaining an extraordinary complexity, which may be described as follows. An ordinary stomach must be supposed to be immensely elongated, and gradually tapering from the cardiac end to a very prolonged, narrow, pyloric extremity. Then two longitudinal muscular bands, corresponding in situation to the greater and lesser curvatures of an ordinary stomach—the former commencing just below the fundus, and the latter at the cardiac orifice, and both proceeding towards the pylorus—are developed, so as to pucker up the cavity into a number of pouches, exactly on the same principle as the human colon is puckered up by its three longitudinal bands. These pouches are largest and most strongly marked at the œsophageal end, and becoming less and less distinct, quite cease several inches before the pylorus is reached, the last part of the organ being a simple smooth-walled tube. The fundus, or cardiac end of the stomach, is formed by a single large sac, slightly constricted on its under surface by the prolongation of the inferior longitudinal band, or that corresponding to the great curvature. The œsophagus enters into the upper part of the left, or pyloric end of this sac, or rather at the point of junction between it and the second (also a very large) sacculus. Furthermore, the whole of this elongated sacculated organ is, by the brevity, as it were, of the lesser curvature, coiled upon itself in an irregularly spiral manner, so that when *in situ* the pylorus comes to be placed very near the œsophageal entrance.

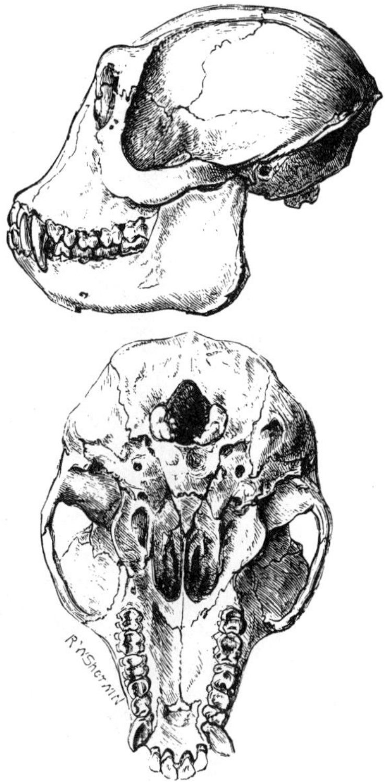

FIG. 348.—Lateral view of the skull and palatal aspect of the cranium of *Semnopithecus nemæus*. (From De Blainville.)

Nasalis.[1]—Skull resembling that of the *Cercopithecinæ* in that the lower border of the nasal bones extends considerably below the lower border of the orbits, whereas in the other *Semnopithecinæ* the aperture of the nares extends upwards between the orbits. Nose

[1] Geoffroy, *Ann. du Muséum*, vol. xix. p. 90 (1812).

produced into a large proboscis. Other characters as in *Semnopithecus.*

This genus includes only the Proboscis Monkey (*N. larvatus*) of Borneo, remarkable for the great prolongation of the nose in the adult. In young animals the nose is relatively much shorter, and

FIG. 349.—*Semnopithecus roxellanæ.* (From Milne-Edwards, *Recherches des Mammifères*, pl. 36.)

bent upwards after the manner of that of *Semnopithecus roxellanæ* (Fig. 349).

Semnopithecus.[1]—Pollex small; narial aperture extending upwards between the orbits. Now confined to Asia.

This genus is characteristic of South-Eastern Asia from the Himalaya southwards, the Oriental region being its head-quarters. The development of the muzzle is less than in the Macaques, and the facial angle is higher, but it does not appear that this indicates greater intellectual capacity. The outlying *S. roxellanæ*[2] (Fig. 349),

[1] F. Cuvier, *Hist. Nat. des Mammifères* (1821), "Semno-pithèque."

[2] Separated generically by some writers as *Rhinopithecus.*

of the highlands of Eastern Tibet and Kansu, is remarkable for the peculiar upturned nose, in which respect, as already mentioned, it recalls the young of *Nasalis larvatus*. The genus is represented in India and Burma by no less than fourteen species, of which the common Indian Langur, or Hanuman Monkey (*S. entellus*) and the larger Himalayan Langur (*S. schistaceus*) are two of the best known. In the former the length of the head and body is about 24, and that of the tail 38 inches in adult males. This monkey, owing to the veneration in which it is held by the Hindus, is a great pest in many parts of India, frequently pilfering grain from the shops in the native bazaars. According to Mr. Blanford, it "is usually found in smaller or larger communities, composed of individuals of both sexes and of all ages, the youngest clinging to their mothers and being carried by them, especially when alarmed. An old male is occasionally found solitary, as with so many other mammals. . . . Apart from villages, the high trees on the banks of streams or of tanks, and, in parts of Central India, rocky hills are the favourite haunts of these monkeys. Whether on trees, on rocks, or on the ground, they are exceedingly active." The closely allied *S. schistaceus* attains a larger average size, full grown males attaining a length of 30 inches, the tail measuring 36 inches. In the spring and winter this species may be observed in the Kashmir Himalaya leaping among the snow-laden trees of the forest. In a fossil state *Semnopithecus* occurs in the Pleistocene and Pliocene of India, and it has also been recorded from the Pliocene of France and Italy.

Colobus.[1]—This African genus differs from *Semnopithecus* in that the pollex is absent or reduced to a small tubercle, which may or may not carry a nail. About eleven species have been described, some of which are remarkable for the beautiful mantle of long silky hair which hangs down from each side of the body, and for their tufted tails. In *C. guereza* from Abyssinia these are white, and the rest of the body and limbs black. Others (as *C. satanas*) are entirely black. The skins of the long-haired species are largely imported into Europe for the manufacture of ladies' muffs, etc.

Extinct Genera.—Certain types of Apes from the European Tertiaries indicate genera referable to the *Cercopithecidæ*, but distinct from any of those now living. Of these *Mesopithecus*,[2] from the Lower Pliocene Pikermi beds of Attica, is known by almost complete skeletons, and resembles *Macacus* in the shortness and stoutness of the limbs, but agrees with *Semnopithecus* in the characters of the skull and teeth. An allied Monkey from the Lower Pliocene of Perpignan, in France, differs from *Mesopithecus pentelici* by its superior size, proportionately more produced muzzle, and larger

[1] Illiger, *Prodromus Syst. Mamm. et Avium*, p. 69 (1811).
[2] Wagner, *Gelehrte Anzeigen*, vol. viii. No. 38, p. 310 (1839).

hind talon to the last lower molar; it has been described under the name of *Dolichopithecus*.[1]

The genus *Oreopithecus*[2] was founded upon the remains of an Ape from the Middle Miocene of Monte Bamboli, in Tuscany, of somewhat larger size than a Gibbon, and apparently presenting characters connecting the *Cercopithecidæ* and *Simiidæ*. According to Dr. Ristori,[3] it resembles the former, especially *Cynocephalus* and *Semnopithecus*, in the long dental series and the elongation of the last molars; but in the shortness of the face, rounding of the chin, and the diagonal arrangement of the molar tubercles, it approximates to the *Simiidæ*, of which it may have been an ancestral type.

Family SIMIIDÆ.

Crowns of molars relatively wide, with the angles more or less rounded off, the tubercles not forming tranverse ridges, and the last lower molar without a hind talon. No tail. No cheek-pouches. Ischiatic callosities, if present, small. Pectoral limbs much longer than pelvic. Sternum broad. Cæcum with vermiform appendage. Centrale of carpus sometimes absent. Other characters as in *Cercopithecidæ*.

This family contains the true Anthropoid Old World Apes, namely the Gibbons, Orangs, Chimpanzees, and Gorillas, which are the most highly organised of all the Apes, and thus make the nearest approach to Man.

Hylobates.[4]—Skull not produced at the vertex; body and limbs slender, the pectoral limbs being so elongated that the hands reach the ground when walking upright; hallux well developed; a centrale in the carpus; and small ischiatic callosities. Size smaller than in the following genera, the height of the largest species (*H. syndactylus*) not much exceeding 3 feet. Now confined to Asia.

The Gibbons, or Long-armed Apes (Figs. 350, 351), are readily distinguished from the remaining members of the family by the characters given above, as well as by the circumstance that they are the only Apes which habitually walk in an upright position. It is in these animals that we meet with the last traces of the ischial callosities so largely developed in the *Cercopithecidæ*. The species are now restricted to South-Eastern Asia, being especially abundant in the Malay Archipelago and adjacent regions.

The largest species is the Sumatran Siamang (*H. syndactylus*), which attains a height of 3 feet, and has been generically separated by some writers as *Siamanga*. It is remarkable as

[1] Depéret, *Comptes Rendus*, vol. cix. p. 982 (1889); see also *Mém. Soc. Géol. France*, "Palæontologie," vol. i. (1890). [2] Gervais, *Comptes Rendus*, vol. lxxiv. p. 1217 (1872). [3] Scimmie Fossili Italiane, *Boll. Comm. Geol.* 1890.

[4] Illiger, *Prodromus Syst. Mamm. et Avium*, p. 67 (1811).

having a better developed chin and wider sternum than any other Ape, and differs from the other members of the genus by the circumstance that the second and third digits of the pes are united by skin as far as their last joints. Exclusive of this species, the Gibbons differ but little from one another in size and general conformation, and since the colour of individuals undoubtedly referable to a single species is remarkably variable, there is much uncertainty about the number of species, and much confusion in the nomenclature. Among well-marked species we may mention the Hoolock (*H. hoolock*), ranging from the South of Assam through Sylhet and Cachar to the Irawadi Valley near Bhamo, the White-handed Gibbon (*H. lar*, Fig. 350), which is found in Tenasserim and throughout Malayana, the Dun-coloured Gibbon (*H. entelloides*, Fig. 351) of Malayana, and the Tufted Gibbon (*H. pileatus*) of Siam and Cambogia.

FIG. 350.—The White-handed Gibbon (*Hylobates lar*). From Blanford, *Mammals of British India*, p. 8.

The following account of the habits of the Gibbons is taken from Mr. W. T. Blanford's *Mammals of British India*. "Gibbons are thoroughly arboreal, and Hoolocks are almost, if not entirely, confined to hill-forest. They move chiefly by means of their long arms, by which they swing themselves for prodigious distances from branch to branch and from tree to tree. They descend hillsides at a surprising pace, their descent being accomplished by grasping bamboos or branches that bend beneath their weight, and allow them to drop until they can seize the ends of other bamboos or branches lower on the slope, and take another mighty swing downwards. They also ascend with great rapidity, swinging themselves from tree to tree. When walking on the ground the Hoolock rests on its hind feet alone, with the sole flat on the ground, and the great toe widely separated from the other digits. The arms are usually held upwards, sometimes horizontally, their great length giving the animal a very peculiar aspect. Gibbons walk rather

FIG. 351.—The Dun-coloured Gibbon (*Hylobates entelloides*). From *Archiv. du Muséum*, vol. ii. pl. 29.

quickly, with a waddling gait, and can easily be overtaken by men when on the ground. The food of these Apes consists of fruit, leaves, young shoots, spiders (of which they are very fond), insects, birds' eggs, and almost certainly of young birds, if not of any birds they can capture. Anderson found that small birds were killed and devoured by Hoolocks in confinement with a method and eagerness that showed this prey to be the natural food of the Apes. The Hoolock drinks with its lips, putting its head down to the water as Monkeys do. All species of *Hylobates* have a powerful voice, and the common name of the Hoolock is taken from its peculiar double call, which is repeated several times. At a distance the sound much resembles a human voice; it is a peculiar wailing note, audible from afar, and in the countries inhabited by these animals is one of the most familiar forest sounds. The calls commence at daybreak, and are continued till 9 or 10 A.M., several of the flock joining in the cry, like hounds giving tongue. After 9 or 10 o'clock in the morning the animals feed or rest, and remain silent throughout the middle of the day, but recommence calling towards evening, though to a less extent than in the earlier part of the day."

The skull of the Gibbons, although agreeing with that of other Apes in its prognathism, presents a somewhat human appearance, and the molar teeth are also very like diminutive human molars. In the anterior inward inclination of the two series of cheek-teeth and the inward position of the upper premolars the Gibbons make an approach to the human type unknown in other Apes.

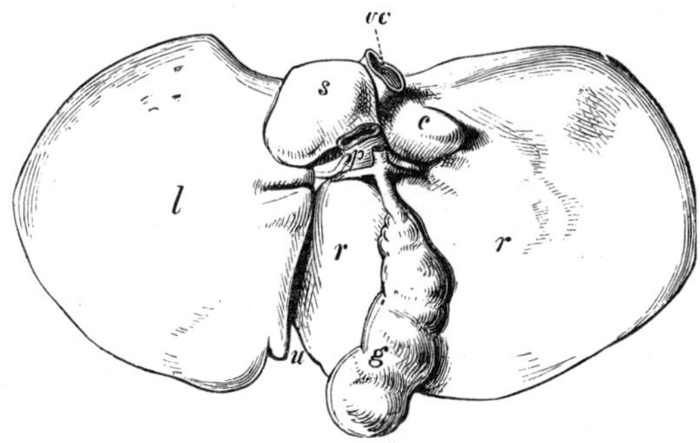

FIG. 352.—Under surface of the liver of *Hylobates lar*. *u*, Umbilical fissure; *p*, portal fissure; *vc*, vena cava; *l*, left lobe; *r*, right lobe; *s*, Spigelian lobe; *c*, caudate lobe; *g*, gall-bladder.

The figure of the liver of one species of this genus is introduced to show the general absence of lateral fissures and the small size of the caudate lobe (*c*) characteristic of the liver of all the *Simiidæ*, except *Gorilla* (see p. 706), as well as that of Man. Another specimen of the liver of the same species showed scarcely any trace of a caudate lobe.

A fossil Ape from the Middle Miocene of France, originally described as *Pliopithecus*, indicates an extinct Gibbon which does not appear to be generically separable from *Hylobates*.

Simia.[1]—Skull (Fig. 353) produced at the vertex; body and limbs massive; the pectoral limbs reaching to the ankle; a centrale in the carpus; hallux very small; sixteen dorso-lumbar vertebræ, and twelve pairs of ribs; no ischiatic callosities. Oriental.

This genus includes the large red-haired Apes from Sumatra and Borneo commonly known as Orangs, or Orang-Utans,[2] of which there is probably only a single species (*S. satyrus*). These animals

[1] Linn. *Syst. Nat.* 12th ed. vol. i. p. 34 (1766).

[2] A Malay word, signifying "Man of the Woods."

inhabit the swampy forests near the coasts; and the males attain a height of about 4 feet 4 inches. The body is very bulky and the legs exceedingly short, but the arms are very long, reaching in the erect posture down to the ankles. The Orang walks resting on the knuckles of the hands and the outer sides of the feet, with the soles of the latter turned mainly inwards, as in Fig. 354. Its movements appear to be slow and deliberate, and in those specimens which have been kept in captivity in this country the demeanour is languid and melancholy, although this is far from being the case with those shown in the more congenial climate of the Zoological Gardens at Calcutta. The habits of these animals are arboreal, and they build a kind of shelter or nest of boughs and leaves; their food appears to consist mainly of fruits, and is exclusively of a vegetable nature. The whole of the body is clothed with long hair of a reddish-brown colour, and full-grown males have a well-developed beard; the males not unfrequently also develop a large warty protuberance, formed of fibro-cellular tissue, on either side of the face. The hands are long, and are characterised by the small size of the pollex, which does not reach to the end of the metacarpal of the index finger. The feet have a similar structure, the hallux only reaching to the middle of the proximal phalange of the adjacent toe, and being often destitute not only of a nail, but likewise of the terminal phalange. The presence of a centrale in the carpus is a feature in which *Simia* agrees with *Hylobates* and the lower Apes, and differs from the two following genera and Man. With very rare exceptions the number of dorso-lumbar vertebræ is sixteen, of which twelve carry ribs, and therefore belong to the dorsal series, while the remaining four are lumbar. The distinction between the last lumbar and the first sacral vertebræ is clearly marked in young skeletons by the additional pleurapophysial ossifications (sacral ribs) in the transverse processes of the latter. Thus though *Simia* presents a closer resemblance to Man than does *Anthropopithecus* in the number of ribs, it differs in the more

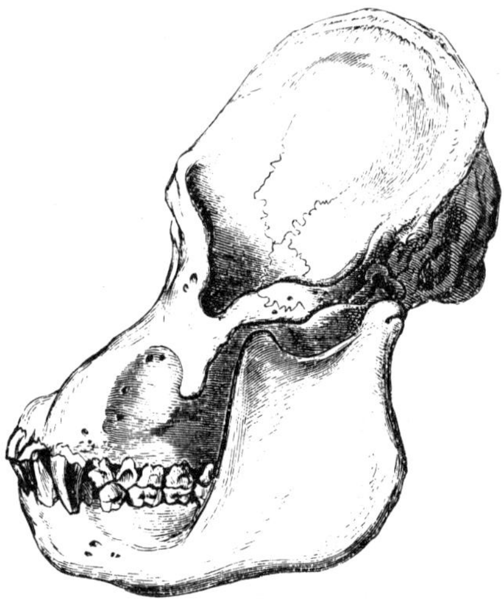

FIG. 353.—Side view of the skull of adult Orang (*Simia satyrus*). From *Trans. Zool. Soc.* vol. i. pl. 53.

important characters of that of the whole series of trunk-vertebræ.[1] The hemispheres of the brain are much convoluted; the whole brain being more human-like than in any other Ape. The larynx is remarkable for having a prolongation from each ventricle, which in the adult become of enormous dimensions, and unite in front of

FIG. 354.—The Orang-Utan (*Simia satyrus*). From Mr. Wolf's sketch at the Zoological Gardens.

the trachea to form one large sac extending downwards between the muscles to the axilla.

The skull of the Orang (Fig. 353) is characterised by its highly vaulted cranial portion, which is comparatively short (brachycephalic). The sagittal crest is well developed on the vertex, and has a highly convex contour; the superciliary ridges are but moderately developed, and do not stand out in the prominent manner so characteristic of the Gorilla. The aperture of the nares in the skull is more pear-shaped than in the two following genera.

The canines of the male Orang attain a great development;

[1] One skeleton in the Museum of the Royal College of Surgeons has five lumbar vertebræ, and has thus given rise to the statement that the number of vertebræ in the Orang is the same as in Man.

and the molars are characterised by the complex structure of their cusps and the numerous rugosities on the crown surface. The outer border of the upper premolars is placed in the same line as that of the molars.

The broken canine tooth of a large Anthropoid Ape from the Lower Pliocene of the Siwalik Hills probably indicates the existence at that period of a species of *Simia* in Northern India.

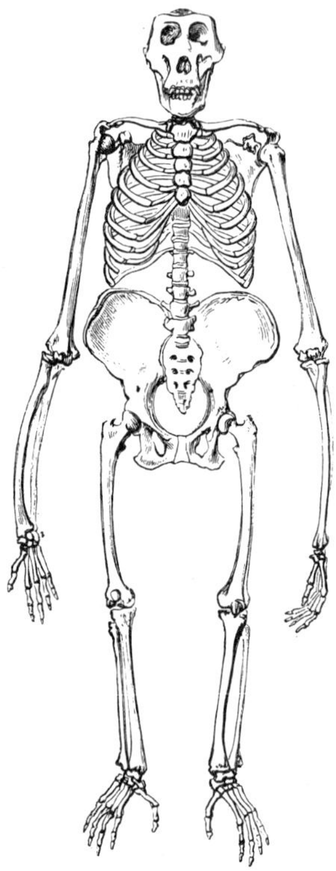

FIG. 355.—Skeleton of the Gorilla. (From De Blainville.)

Gorilla.[1]—Skull not produced at the vertex; body and limbs massive, the pectoral limb not reaching below the middle of the lower leg (Fig. 355); no centrale in the carpus; hallux well developed; seventeen dorso-lumbar vertebræ, of which thirteen carry ribs; no ischiatic callosities. Male much larger than female, and with very strongly marked cranial ridges, which are wanting in the latter. Mandibular symphysis long. Ethiopian.

The well-known Gorilla (Fig. 356), of which there seems to be only one species (*G. savagei*), is found in Western Equatorial Africa, chiefly or entirely in the district enclosed by the Cameroon and Congo rivers. It is the largest of all the Apes, its bulk considerably exceeding that of man, although from the shortness of the legs it appears never to attain a greater height than $5\frac{1}{2}$ feet. The first introduction of this animal to the notice of zoologists was made in 1847 by Dr. Thomas Savage, but it was not fully known till many years later.

The skin of the Gorilla is entirely black, the hair being blackish, but turning more or less gray in old individuals. The arms reach down as far as the middle of the lower leg; while the pollex extends only a short distance beyond the base of the first phalange of the index finger, and the hallux reaches nearly as far as the distal extremity of the corresponding digit of the foot. The digits of both the hand and foot are united together by integument as far as the distal extremities of the first phalanges. The larynx has very capacious air-sacs, which meet in front of the trachea and communicate with the ventricles; and in advanced age these sacs

[1] I. Geoffroy, *Comptes Rendus*, vol. xxxiv. p. 84 (1852).

may extend to the axilla. The ears are relatively small. The skull is of an elongated or dolichocephalic type; that of the adult male being characterised by the enormous development of the supraorbital ridges, which form a kind of penthouse over the eyes, and contribute to the peculiarly ferocious appearance of the animal. The sagittal crest is also very large. The canine teeth of the male are very large, and are inclined outwards in both jaws. In the

FIG. 356.—The Gorilla (*Gorilla savagei*). From *Trans. Zool. Soc.* vol. iv. pl. 43.

cheek-teeth the upper premolars are of considerable antero-posterior extent, with their outer border placed in the same line as that of the molars; and the third upper molar is larger than either of the others.

The posterior cervical vertebræ are characterised by the great height of their neural spines, which thus form a strong basis for the powerful cervical muscles supporting the massive skull. In some instances the fourth lumbar vertebra becomes ankylosed to the sacrum, as is occasionally found to be the case in some of the lower human races.

In the absence of a centrale to the carpus, and also in the

number of the dorso-lumbar vertebræ, the present and following genus resemble man; although they both differ in having thirteen in place of twelve pairs of ribs.

The brain of the Gorilla, according to Dr. Hartmann, resembles that of the Orang in the complexity of its convolutions, and is thereby distinguished from that of the Chimpanzee. In form it is of the long oval characteristic of Man; the brain of the Chimpanzee and Orang being more rounded.

Gorillas live in family parties in the depths of the dense forests of Western Equatorial Africa, seeking their food during the day, while at night it is said that the female and young ascend a tree at the foot of which the male sleeps. They walk with the backs of their closed hands and the flat soles of the feet placed on the ground. Although there has been much exaggeration on this point, it appears certain that the male Gorilla is an extremely ferocious and dangerous animal when brought to bay, but the statements as to its making unprovoked assaults on men do not appear authentic. They utter deep guttural sounds, which on some occasions may be described as grunts and at others as a roar.

Anthropopithecus.[1]—One of the most important differences of this genus from the preceding is the absence of any marked disparity between the two sexes, either in the size or the conformation of the skull, although the male can always be distinguished by the larger size of the canine teeth. The mandibular symphysis is also much shorter. Differences in the characters of the teeth are described below. The genus is confined at the present day to the Ethiopian region.

The Chimpanzees (Fig. 357) inhabit Western and Central Equatorial Africa; and there has been much discussion whether they should all be included under one specific name (*A. troglodytes*), or whether there are really two or more species. A female specimen now living in the London Zoological Gardens, characterised among other distinctive features by the nearly bald head, clearly indicates, however, a second species, which probably corresponds to the imperfectly defined *A. calvus* of Du Chaillu.

The region inhabited by the Chimpanzees extends from the Gambia to the Benguela, reaching as far inland as 28° E. long. The Common Chimpanzee is a smaller animal than the Gorilla, its height not exceeding 5 feet. In colour it is darker than the latter, and the ears are relatively larger. In the upright position

[1] De Blainville, *Leçons Orales* (1839). The Chimpanzees have been very generally described under the name of *Troglodytes*, but since this name is preoccupied for a genus of birds, it is incumbent to follow the strict rule, and adopt the name *Anthropopithecus*, although both the present witers have elsewhere expressed the opposite opinion.

the arms reach only a short distance below the knee, in which respect the Chimpanzee is more human-like than any of the other Apes. The face is furnished with distinct whiskers, eyebrows, and eyelashes. The pollex reaches nearly or quite to the base of the first phalange of the index finger, and the hallux to the base of the second phalange of the corresponding digit of the foot. The laryngeal sacs are as largely developed as in the Gorilla.

Although the skull of the Chimpanzee has distinct superciliary

FIG. 357.—The Chimpanzee (*Anthropopithecus troglodytes*). From Mr. Wolf's drawing of a young individual in the Zoological Society's Gardens.

ridges, yet the high bony crests of the calvarium of the male Gorilla are wanting, and the whole coronal region of the skull is more rounded and far less rugged.

The canine teeth of the male Chimpanzee are relatively much smaller than in the Gorilla and Orang. The upper molars are characterised by the third one being smaller than either of the other two, as well as by the presence of an indistinct cingulum on their inner surfaces. The upper premolars differ from those of the other genera of the family by the shortness of their antero-posterior diameter, and also by the larger size of their external as compared with their internal cusps; while the outer border of these teeth is placed internally to that of the upper molars. In all these respects the teeth of the Chimpanzee make a decided approximation to the human type.

Many young individuals of the Chimpanzee have been brought to Europe, but they appear to succumb sooner or later to the effects of an unsuitable climate. All these examples show that the disposition of this Ape is gentle, lively, and intelligent, and in all respects markedly opposite to that of the Orang. In a wild state these Apes are essentially forest-dwellers, and are more arboreal in their habits than the Gorilla. They live either in families, or in small parties of several families. Frequently at least they construct a kind of nest in the trees as a sleeping-place; the male being said to sleep on a forked branch below the level of this nest. In walking the Chimpanzee usually supports himself on the backs of his closed fingers, and either on the soles of the feet or on the closed toes.

From a distributional point of view the discovery of a fossil Ape in the Pliocene of the Punjab, apparently closely allied to the Chimpanzee, is of great interest. This determination rests upon the evidence of an imperfect palate originally described under the name of *Palæopithecus*, but subsequently referred to the present genus. The teeth of this jaw present all the essential characters of those of the Chimpanzee, but the two series of cheek-teeth have a slight anterior convergence, the premolars are shorter in the antero-posterior direction than is usually the case in that species, and the outer incisor is relatively narrower than in the latter. In these features the extinct *A. sivalensis* makes a nearer approximation to the human type than is the case with its living congeners.

Dryopithecus.[1]—The extinct *Dryopithecus* of the Middle Miocene of France is represented by a single species of the approximate size of the Chimpanzee, and appears to be the most generalised member of the family. According to the recent observations of Professor Gaudry,[2] while it resembles the Gorilla in that the two series of lower cheek-teeth diverge anteriorly and the penultimate premolar is larger than the last of that series, it differs in having a much longer and narrower mandibular symphysis, and thus indicates a transition to the *Cercopithecidæ*. A gradual transition in the form of the mandible may, indeed, be traced from *Dryopithecus*, through *Gorilla*, to *Anthropopithecus;* the latter having a short and wide symphysis, with the two series of cheek-teeth slightly converging anteriorly, and the penultimate premolar being not larger than the last. In all these specialised characters the jaw of the Chimpanzee

[1] Lartet, *Comptes Rendus*, vol. xliii. p. 219 (1856).

[2] *Mém. Soc. Géol. France*, "Palæontologie," vol. i. Mém. No. 1 (1890).

approximates to that of Man, in which the symphysis is still further shortened and widened, and the anterior convergence of the cheek-teeth so much increased as to produce a horse-shoe-like form in the whole dental series.

Family HOMINIDÆ.

In the *Systema Naturæ* of Linnæus Man was separated only generically from the Apes, but in the next great work which exercised a wide-spread influence over the progress of zoological science, the *Règne Animal* of Cuvier, he forms a distinct order under the name of Bimana, the Monkeys and Lemurs being associated together as Quadrumana. This has been the prevailing arrangement in the zoological systems of the present century, though in the classification of Owen his position is still farther removed from that of the Monkeys, as in it the genus *Homo* forms one of the four primary divisions or subclasses of the Mammalia, called Archencephala, the Quadrumana being united with the Carnivora, Ungulata, and others in another division called Gyrencephala. On the other hand, the tendency of most modern systematists, for reasons which have been fully stated by Professor Huxley,[1] is to revert towards the Linnæan position.

Considering solely the facts of Man's bodily structure, it can be clearly demonstrated that the points in which he differs from the Ape most nearly resembling him are not of greater importance than those by which that Ape differs from other universally acknowledged members of the group; and therefore, in any natural system, if Man is to be made a subject of zoological classification upon the same principles as those applied elsewhere, he must be included in the order which comprises the Monkeys. We say upon the same principles as are applied elsewhere, since zoological classification has never taken into consideration the psychological characteristics which distinguish the subjects of its investigations, but only their tangible and physical structure, otherwise endless confusion would result, at all events with our very imperfect knowledge of animal psychology. The essential attributes which distinguish Man, and give him a perfectly isolated position among living creatures, are not to be found in his bodily structure, and should therefore either be left entirely out of consideration, or have such weight given to them as would remove him completely out of the region of zoological classification. To profess to classify Man as if he were one of the animals (as in all points of the structure and functions of his organs he undoubtedly is), to place him in the class Mammalia, and then

[1] *Man's Place in Nature*, 1863, and *Anatomy of Vertebrated Animals*, 1871. See also the more recent investigations of Broca into the comparative structure of Man and the higher Apes, published mostly in the *Revue d'Anthropologie*.

to allow other considerations to influence the judgment as to the particular position he should occupy in the class, is most illogical.

Man, therefore, considered from a zoological point of view, must be included in the order Primates, even if the Lemurs be removed from it, since his structural affinities with the Monkeys are far closer than are those of the so-called "Half-Apes." We may, without treading upon debatable ground, go farther, and say that the differences between Man and the Anthropoid Apes are really not so marked as those which separate the latter from the American Monkeys. This being admitted, perhaps the best exposition relating to the present condition of the order will be to regard Man as representing a fifth family of the Anthropoidea, which should be known as the *Hominidæ*. In thus ranking Man as one of the five principal families or sections of the suborder it should, however, be observed that this course does not in the least degree imply that such families are precisely equivalent to one another, or that the intervals by which they are separated are of equal importance; all that we commit ourselves to being that they are five perfectly distinct groups, all branches from a common stem, and in the present state of nature not united by any intermediate types.

The distinctions between the *Hominidæ* and *Simiidæ* are chiefly relative, being greater size of brain and of brain-case as compared with the facial portion of the skull, smaller development of the canine teeth of the males, complete adaptation of the structure of the vertebral column to the vertical position, greater length of the lower as compared with the upper extremities, and greater length of the hallux or great toe, with almost complete absence of the power of bringing it in opposition to the other four toes. The last feature together with the small size of the canine teeth are perhaps the most marked and easily defined distinctions that can be drawn between the two groups.

Man is universally admitted to form a single genus, *Homo* of Linnæus, but a question of considerable importance in treating of him from a zoological point of view, and one which has been a subject of much controversy, is whether all men should be considered as belonging to a single or to several species. This question is perhaps of less importance now than formerly, when those who maintained a plurality of species associated with the hypothesis plurality of origin. One of the strongest arguments against the view that the various races of Man represent more than one species is that none of those who have maintained it have been able to agree as to how many distinct specific modifications can be defined, almost every number from three to twenty or more having been advocated by different authors. If the distinguishing characters of the so-called species had been so marked, there could not be such a remarkable diversity of opinion upon them. Again, the two facts

—(1) that, however different the extremes of any two races may be in appearance (and it must be admitted that, as advocated by many polygenists, the differences are greater than many which are considered specific among other animals), every intermediate gradation can be found through which the one passes into the other, and (2) that all races are fertile *inter se*—are quite conclusive in favour of considering Man as representing a single species in the ordinary sense in which the word is now used, and of treating of all his various modifications as varieties or races.

The great problem at the root of all zoology, the discovery of a natural classification which shall be an expression of our knowledge of the real relationship or consanguinity of different forms, is also applicable to the study of the races of Man. When we can satisfactorily prove that any two of the known groups of mankind are descended from the same common stock, a point is gained. The more such points we have acquired the more nearly shall we be able to picture to ourselves, not only the present, but also the past distribution of the races of Man upon the earth, and the mode and order in which they have been derived from one another. But the difficulties in the way of applying zoological principles to the classification of Man are vastly greater than in the case of most animals. When groups of animals become so far differentiated from each other as to represent separate species, they remain isolated; they may break up into further subdivisions—in fact, it is only by further subdivision that new species can be formed; but it is of the very essence of species, as now universally understood by naturalists, that they cannot recombine, and so give rise to new forms. With the varieties of Man it is otherwise. They have never so far separated as to answer to the physiological definition of species. All races, as said above, are fertile with one another, though perhaps in different degrees. Hence new varieties have constantly been formed, not only by the segmentation of portions of one of the old stocks, but also by various combinations of those already established.

Without entering into the difficult question of the method of Man's first appearance upon the world, we must assume for it vast antiquity,—at all events as measured by any historical standard. Of this there is now ample proof. During the long time Man existed in a savage state—a time compared to which the dawn of our historical period is as yesterday—he was influenced by the operation of those natural laws which have produced the variations seen in other regions of organic nature. The first Men may very probably have been all alike; but when spread over the face of the earth and subjected to all kinds of diverse external conditions, —climate, food, competition with members of their own species or with wild animals,—racial differences began slowly to be developed through the potency of various kinds of selection acting upon the slight variations which appeared in individuals in obedience to the tendency planted in all living things. These differences manifested themselves externally in the colour of the skin, the colour, quality, and distribution of the hair, the form of the head and features, and the proportions of the limbs, as well as in the general stature.

Geographical position must have been one of the main elements in determining the formation and permanence of races. Groups of Men isolated from their fellows for long periods, such as those living on small islands, to which their ancestors may have been accidentally drifted, would naturally, in course of time, develop a new type of features, of skull, of complexion, or hair. A slight set in one direction in any of these characters would constantly tend to intensify itself, and so new races would be formed. In the same way different intellectual or moral qualities would be gradually developed or transmitted in different groups of Men. The longer a race thus formed remained isolated the more strongly impressed and the more permanent would its characteristics become, and less liable to be changed or lost when the surrounding circumstances were altered or under a moderate amount of intermixture from other races—the more "true," in fact, would it be. On the other hand, on large continental tracts, where no mountain ranges or other natural barriers form obstacles to free intercourse between tribe and tribe, there would always be a tendency towards uniformity, from the amalgamation of races brought into close relation by war or by commerce. Smaller or feebler races would be destroyed or absorbed by others impelled by superabundant population or other causes to spread beyond their original limits; or sometimes the conquering race would itself disappear by absorption into the conquered.

Thus for untold ages the history of Man has presented a shifting kaleidoscopic scene: new races gradually becoming differentiated out of the old elements, and, after dwelling a while upon the earth, becoming either suddenly annihilated or gradually merged into new combinations; a constant destruction and reconstruction; a constant tendency to separation and differentiation, and a tendency to combine again into a common uniformity—the two tendencies acting against and modifying each other. The history of these processes in former times, except in so far as they may be inferred from the present state of things, is a difficult study, owing to the scarcity of evidence. If we had any approach to a complete palæontological record, the history of Man could be reconstructed; but nothing of the kind is forthcoming. Evidence of the anatomical characters of Man as he lived on the earth during the time when the most striking racial characteristics were being developed, during the long ante-historic period in which the Negro, the Mongolian, and the Caucasian were being gradually fashioned into their respective types, is entirely wanting, or if any exists it is at present safely buried in the earth, perhaps to be revealed at some unexpected time and in some unforeseen manner. Even the materials from which a history of the modifications of the human species as known to our generation must be constructed are rapidly passing away, since the age in which we live is an age in which, in a far greater degree than any previous one, the destruction of races, both by annihilation and absorption, is going on. Owing to the rapid extension of maritime discovery and commerce, changes such as have never been witnessed before are now taking place in the ethnology of the world—changes especially affecting the island populations among which, more than elsewhere, the solution of many of these problems may be looked for. The subject is, however, attracting the attention of observers of all countries to a greater degree than it ever has before, and such progress has been made in perfecting the methods of investigation of racial characteristics that we are beginning to learn what lines of research are profitable and what are barren, so that we may hope the time is not far distant when we may get some clear insight into the knowledge of the natural classification and relationships of the races of Man.

The following is a brief summary of the principal results which appear to have been attained up to the present time by the study of this somewhat difficult subject.[1]

The most ordinary observation is sufficient to demonstrate the fact that certain groups of men are strongly marked from others by definite characters common to all members of the group, and transmitted regularly to their descendants by the laws of inheritance. Thus the Chinaman and the Negro, the native of Patagonia and the Andaman Islander, are as structurally distinct from each other as are many of the so-called species of any natural group of animals. Indeed, it may be said with truth that their differences are even greater than those which mark the groups called genera by many naturalists of the present day. Nevertheless the difficulty of parcelling out all the individuals composing the human species into certain definite groups, and of saying of each man that he belongs to one or other of such groups, is insuperable. No such classification has ever been, or, indeed, can ever be obtained. There is not one of the most characteristic and most extreme forms, like those just named, from which transitions cannot be traced by almost imperceptible gradations to any of the other equally characteristic and equally extreme forms. Indeed, a large proportion of mankind

[1] "On the Classification of the Varieties of the Human Species," by W. H. Flower, *Journal of the Anthropological Institute of Great Britain and Ireland*, May 1885.

is made up, not of extreme or typical, but of more or less generalised or intermediate forms, the relative numbers of which are continually increasing, as the long-existing isolation of nations and races breaks down under the ever-extending intercommunication characteristic of the period in which we live.

The difficulties of framing a natural classification of Man, or one really representing the relationship of the various minor groups to each other, are well exemplified by a study of the numerous attempts which have been made from the time of Linnæus and Blumenbach onwards. Even in the first step of establishing certain primary groups of equivalent rank there has been no accord. Thus four primitive types were sketched out by Linnæus—the European, Asiatic, African, and American. These were expanded into five by Blumenbach by the addition of the Malay,[1] and reduced by Cuvier to three by the suppression of the last two. Many later writers have largely increased the number of these so-called primary divisions, but the conclusion, so often arrived at by various anthropologists, and so often abandoned for some more complex system, that the primitive man, whatever he may have been, has in the course of ages divaricated into three extreme types, represented by the Caucasian of Europe, the Mongolian of Asia, and the Ethiopian of Africa, and that all existing individuals of the species can be ranged around these types, or somewhere or other between them, seems, on the whole, to give the clearest view of the facts of the case. Large numbers are doubtless the descendants of direct crosses in varying proportions between well-established extreme forms; for, notwithstanding opposite views formerly held by some authors on this subject, there is now abundant evidence of the wholesale production of new races in this way. Others may be the descendants of the primitive stock before the strongly marked existing distinctions had taken place, and therefore present, though from a different cause from the last, equally generalised characters. In these cases it can only be by most carefully examining and balancing all characters, however minute, and finding out in what direction the preponderance lies, that a place can be assigned to them. It cannot be too often insisted on that the various groups of mankind, owing to their probable unity of origin, the great variability of individuals, and the possibility of all degrees of intermixture of races at remote or recent periods of the history of the species, have so much in common that it is extremely difficult to find distinctive characters capable of strict definition by which they may be differentiated. It is more by the preponderance of certain characters in a large number of members of a group, than by the exclusive or even constant possession of these characters

[1] The Malay of Blumenbach was a strange conglomeration of the then little known Australian, Papuan, and true Malay types.

in each of its members, that the group as a whole must be characterised.

Bearing these principles in mind, we may endeavour to formulate, as far as they have as yet been worked out, the distinctive features of the typical members of each of the three great divisions, and then show into what subordinate groups each of them seems to be divided.

We begin with the Ethiopian, Negroid, or Melanian, or "black" type. It is characterised by a dark, often nearly black, complexion; black hair, of a kind called "frizzly" or, incorrectly, "woolly," *i.e.* each hair is closely rolled up on itself, a condition always associated with a more or less flattened or elliptical transverse section; a moderate or scanty development of beard; an almost invariably dolichocephalic skull; small and moderately retreating jugal bones (mesopic face); a very broad and flat nose, platyrhine in the skeleton; moderate or low orbits; prominent eyes; thick, everted lips; prognathous jaws; large teeth (macrodont); a narrow pelvis (index in the male 90 to 100); a long forearm (humero-radial index 80); and certain other proportions of the body and limbs which are being gradually worked out and reduced to numerical expression as material for so doing accumulates.

The most characteristic examples of the second great type, the Mongolian or Xanthous, or "yellow," have a yellow or brownish complexion; black coarse straight hair, without any tendency to curl, and nearly round in section, on all other parts of the surface except the scalp scanty and late in appearing; a skull of variable form, mostly mesocephalic (though extremes both of dolichocephalism and brachycephalism are found in certain groups of this type); a broad and flat face, with prominent, anteriorly-projecting jugal bones (platyopic face); nose small, mesorhine or leptorhine; orbits high and round, with very little development of glabella or supraciliary ridges; eyes sunken, and with the aperture between the lids narrow; in the most typical members of the group with a vertical fold of skin over the inner canthus, and with the outer angle slightly elevated; jaws mesognathous; teeth of moderate size (mesodont). The proportions of the limbs and form of the pelvis have yet to be worked out, the results at present obtained showing great diversity among different individuals of what appear to be well-marked races of the group, but this is perhaps due to the insufficient number of individuals as yet examined with accuracy.

The last type, which, for want of a better name, we must still call by the misleading one that has the priority, Caucasian, or "white," has usually a light-complexioned skin (although in some, in so far aberrant cases, it is as dark as in the Negroes); hair fair or black, soft, straight, or wavy, in section intermediate between the flattened and cylindrical form; beard fully developed; form of cranium variable, mostly mesocephalic; jugal bones retreating; face narrow and projecting in the middle line (pro-opic); orbits moderate; nose narrow and prominent (leptorhine); jaws orthognathous; teeth small (microdont); pelvis broad (pelvic index of male 80); forearm short, relatively to humerus (humero-radial index 74).

In endeavouring to subdivide into minor groups the numerous and variously-modified individuals which cluster around one or other of these great types—a process quite necessary for many practical or descriptive purposes—the distinctions afforded by the study of physical characters are often so slight that it becomes necessary to take other considerations into account, among which geographical distribution and language hold an important place.

I. The Ethiopian or Negroid races may be primarily arranged as follows:—

A. African or Typical Negroes.—Inhabitants of all the central portion of the African continent, from the Atlantic on the west to the Indian Ocean on the east, greatly mixed all along their northern frontier with Hamitic and Semitic Melanochroi, a mixture which, taking place in various proportions and under varied conditions, has given rise to many of the numerous races and tribes inhabiting the Sudan.

A branch of the African Negroes are the Bantu—distinguished chiefly, if not entirely, by the structure of their language. Physically indistinguishable from the other negroes with whom they come in contact in the Equatorial regions of Africa, the Southern Bantu, or Kaffirs, as they are generally called, show a marked modification of type, being lighter in colour, having a larger cranial capacity, less marked prognathism, and smaller teeth. Some of these changes are probably due to crossing with other races.

B. The Negrillos—diminutive sub-brachycephalic tribes, inhabiting the dense forests of Central and Western Equatorial Africa—represent a distinct section of the Negro race. They form the only exceptions to the general dolichocephaly of the African branch of the Negroid division, and when found in a pure state are the smallest of all known human races, averaging scarcely more than 4 feet in height. The colour of their skin is yellowish rather than black.

C. The Bushmen (Bosjesmen, men of the woods, of the Dutch colonists of South Africa) constitute a very distinct modification of the Negro type. The hair shows the extreme of the frizzly character; being shorter and less abundant than that of the ordinary Negro, it has the appearance of growing in separate tufts, which coil up together into rounded balls compared to "pepper-corns." In their yellow complexion, wide cheek-bones, and peculiar form of the eyes they so much resemble some of the

Mongolian races that anthropologists have been inclined to trace affinities to or admixture with them, although the character of the hair makes such a supposition almost inadmissible. The width of the cheek-bones and the narrowness of the forehead and chin give a lozenge shape to the front view of the face. The forehead is prominent and straight; the nose extremely flat and broad, more so than in any other race; the lips prominent and thick, although the jaws are less prognathous than in the true Negro races. The cranium has many special characters by which it can be easily distinguished from that of any other race. The average height of the males is about 4 feet 8 inches. There is every reason to believe that the Bushmen represent the earliest race of which we have any knowledge inhabiting the southern part of the African continent, but that long before the advent of Europeans upon the scene they had been invaded from the north by Negro tribes, who, being superior in size, strength, and civilisation, had taken possession of the greater part of their territories, and, mingling freely with the aborigines, had produced the mixed race called Hottentots, who retained the culture and settled pastoral habits of the Negroes, with many of the physical features of the Bushmen. These in their turn, encroached upon by the Kaffirs from the north and by Europeans from the south, are now greatly diminished, and threatened with the same fate which will surely soon befall the scanty remnant of the early inhabitants who still retain their primitive type.

D. Oceanic Negroes or Melanesians.—These include the Papuans of New Guinea and the majority of the inhabitants of the islands of the Western Pacific, and form also a substratum of the population, greatly mixed with other races, of regions extending far beyond the present centre of their area of distribution.

They are represented, in what may be called a hypertypical form, by the extremely dolichocephalic Kai Colos, or mountaineers of the interior of the Fiji Islands, although the coast population of the same group has lost the distinctive characters by crossing. In many parts of New Guinea and the great chain of islands extending eastwards and southwards ending with New Caledonia they are found in a more or less pure condition, especially in the interior and more inaccessible portions of the islands, almost each of which shows special modifications of the type recognisable in details of structure. Taken altogether, their chief physical distinction from the African Negroes lies in the fact that the glabella and supra-orbital ridges are generally well developed in the males, whereas in Africans this region is usually smooth and flat. The nose also, especially in the northern part of their geographical range, New Guinea, and the neighbouring islands, is narrower (often mesorhine) and prominent. The cranium is generally higher and narrower.

It is, however, possible to find African and Melanesian skulls quite alike in essential characters.

The now extinct inhabitants of Tasmania were probably pure, but aberrant, members of the Melanesian group, which had undergone a modification from the original type, not by mixture with other races, but in consequence of long isolation, during which special characters had been gradually developed. Lying completely out of the track of all civilisation and commerce, even of the most primitive kind, they were little liable to be subject to the influence of any other race; and there is in fact nothing among their characters which could be accounted for in the way above suggested, as they were intensely, even exaggeratedly, Negroid in the form of nose, projection of mouth, and size of teeth, typically so in character of hair, and aberrant chiefly in the width of the skull in the parietal region. A cross with any of the Polynesian or Malay races sufficiently strong to produce this would, in all probability, have also left some traces on other parts of their organisation.

On the other hand, in many parts of the Melanesian region there are distinct evidences of large admixture with Negrito, Malay, and Polynesian elements in varying proportions, producing numerous physical modifications. In many of the inhabitants of the great island of New Guinea itself and of the islands lying around it this mixture can be traced. In the people of Micronesia in the north and New Zealand in the south, although the Melanesian element is present, it is completely overlaid by the Polynesian, but there are probably few, if any, of the islands of the Pacific in which it does not form some factor in the composite character of the natives.

The inhabitants of the continent of Australia have long been a puzzle to ethnologists. Of Negroid complexion, features, and skeletal characters, yet without the characteristic frizzly hair, their position has been one of great difficulty to determine. They have, in fact, been a stumbling-block in the way of every system proposed. The solution, supported by many considerations too lengthy to enter into here, appears to lie in the supposition that they are not a distinct race at all, that is, not a homogeneous group formed by the gradual modification of one of the primitive stocks, but rather a cross between two already-formed branches of these stocks. According to this view, Australia was originally peopled with frizzly-haired Melanesians, such as those who still do, or did before the European invasion, dwell in the smaller islands which surround the north, east, and southern portions of the continent, but that a strong infusion of some other race, probably a low form of Caucasian Melanochroi, such as that which still inhabits the interior of the southern parts of India, has spread throughout the land from the north-west, and produced a modification of the physical characters, especially of the hair. This influence did not extend across Bass's Straits into Tasmania, where, as just said, the Melanesian element remained in its purity. It is more strongly marked in the northern and central parts of Australia than on many portions of the southern and western coasts, where the lowness of type and more curly hair, sometimes closely approaching to frizzly, show a stronger retention of the Melanesian element. If the evidence should prove sufficiently strong to establish this view of the origin of the Australian natives, it will no longer be correct to speak of a primitive Australian, or even Australoid, race or type, or look for traces of the former existence of such a race anywhere out of their own land. Absolute proof of the origin of any race is, however, very difficult, if not impossible, to obtain, and there is nothing to exclude the possibility of the Australians being mainly the direct descendants of a very primitive human type, from which the frizzly-haired Negroes may be an offset. This character of hair is probably a specialisation, for it seems very unlikely that it was the attribute of the common ancestors of the human race.

E. The fourth branch of the Negroid race consists of the diminutive round-headed people called Negritos, still found in a pure or unmixed state in the Andaman Islands, and forming a substratum of the population, though now greatly mixed with invading races, especially Malays, in the Philippines, and many of the islands of the Indo-Malayan Archipelago, and of some parts of the southern portion of the mainland of Asia. They also contribute to the varied population of New Guinea, where they appear to merge into the taller, longer-headed, and longer-nosed Melanesians proper. They show in a very marked manner some of the most striking anatomical peculiarities of the Negro race, such as the frizzly hair, the proportions of the limbs, especially the humero-radial index, and the form of the pelvis; but they differ in many cranial and facial characters, both from the African Negroes on the one hand, and the typical Oceanic Negroes, or Melanesians, on the other, and thus form a very distinct and well-characterised group. Wherever they are still found they are obviously holding their own with difficulty, if not actually disappearing, and there is much about their condition of civilisation and the situations in which they occur to induce us to look upon them, as in the case of the Negrillos of Central and the Bushmen of South Africa, as the remains of a population which occupied the land before the incoming of the present dominant races.

II. The principal groups that can be arranged round the Mongolian type are as follows:—

A. The Eskimo appear to be a branch of the typical North

Asiatic Mongols, who in their wanderings northwards and eastwards across the American continent, where they have been isolated almost as perfectly as an island population would be, hemmed in on one side by the eternal Polar ice, and on the other by hostile tribes of American Indians, with which they rarely, if ever, mingled, have gradually developed characters, most of which are strongly-expressed modifications of those seen in their allies who still remain on the western side of Behring Strait. It has also been shown that these special characteristics gradually increase from west to east, and are seen in their greatest perfection in the inhabitants of Greenland, at all events in those where no crossing with the Danes has taken place. A typical Eskimo skull presents a combination of characters by which it can be at once distinguished from that of any other of the groups of mankind. Such scanty remains as have yet been discovered of the earliest inhabitants of Europe do not present any structural affinities to this type, and there is therefore no justification for the supposition that they belonged to the same race, although it is not unlikely that similar external conditions may have led them to adopt similar modes of life.

B. The typical Mongolian races constitute the present population of Northern and Central Asia. They are not very distinctly, but still conveniently for descriptive purposes, divided into a Northern and a Southern group.

a. The members of the former, Mongolo-Altaic or Sibiric group, are united by the affinities of their language. These people, from the cradle of their race in the great plateau of Central Asia, have at various times poured out their hordes upon the lands lying to the west, and thence penetrated almost to the heart of Europe. The Lapps, Finns, the Magyars, and the Turks are each the descendants of one of these waves of incursion, but they have for so many generations intermingled with the peoples through whom they have passed in their migrations, or whom they have found in the countries in which they have ultimately settled, that their original physical characters have been completely modified. Even the Lapps, that diminutive tribe of nomads inhabiting the most northern parts of Europe, supposed to be of Mongolian descent, show so little of the special attributes of that branch that it is difficult to assign them a place in it in a classification based upon physical characters. The Japanese are said by their language to be allied rather to the Northern than to the following branch of the Mongolian stock.

b. The southern Mongolian or Sinitic group, divided from the former chiefly by language and habits of life, includes the greater part of the population of China, Tibet, Burma, and Siam.

C. The next great division of Mongoloid people is the Malay, forming the bulk of the population of the Indo-Malayan Archipelago and (mixed with the Negro) of Madagascar, subtypical it is true, but to which an easy transition can be traced from the most characteristic members of the type.

D. The brown Polynesians, Malayo-Polynesians, Mahoris, Sawaioris, or Kanakas, as they have been variously called, seen in their greatest purity in the Samoan, Tongan, and Eastern Polynesian Islands, are still more modified, and possess less of the characteristic Mongolian features; but yet it is difficult to place them anywhere else in the system. The large infusion of the Melanesian element throughout the Pacific must never be forgotten in accounting for the characters of the people now inhabiting the islands—an element in many respects so diametrically opposite to the Mongolian that it would materially alter the characters, especially of the hair and beard, which has been with many authors a stumbling-block to the affiliation of the Polynesian with the Mongolian stock. This mixture is physically a fine one, and in some proportions produces a combination, as seen, for instance, in the Maories of New Zealand, which in all definable characters approaches quite as near, or nearer, to the Caucasian type than to either of the stocks from which it may be presumably derived. This resemblance has led some ethnologists to infer a real extension of the Caucasian element at some very early period into the Pacific Islands, and to look upon their inhabitants as the product of a mingling of all the three great types of men. Though this is a very plausible theory, it rests on little actual proof, since the combination of Mongolo-Malayan and Melanesian characters in different degrees, together with the local variations certain to arise in communities so isolated from each other and exposed to such varied conditions as the inhabitants of the Pacific Islands, would probably account for all the modifications observed among them.

E. The native population (before the changes wrought by the European conquest) of the great continent of America, excluding the Eskimo, present, considering the vast extent of the country they inhabit and the great differences of climate and other surrounding conditions, a remarkable similarity of essential characters with much diversity of detail.

The construction of the numerous American languages, of which as many as twelve hundred have been distinguished, is said to point to unity of origin, as, though widely different in many respects, they are all, or nearly all, constructed on the same general grammatical principle—that called *polysynthesis*—which differs from that of the languages of any of the Old World nations. The mental characteristics of all the American tribes have much that is in common; and the very different stages of culture to which they had attained at the time of the conquest, as that of the Incas and Aztecs and the hunting or fishing tribes of the north and south, which have been quoted as evidence of diversities of race, were not greater than those between different nations of Europe, as Gauls and Germans on the one hand, and Greeks and Romans on the other, in the time of Julius Cæsar. Yet all these were Aryans, and in treating the Americans as one race it is not intended to imply that they are more closely allied than the different Aryan peoples of Europe and Asia. The best argument that can be used for the unity of the American race—using the word in a broad sense—is the great difficulty of forming any natural divisions in it founded upon physical characters. Thus there is no difference throughout the whole continent in the important character of the hair, this being always straight and lank, long and abundant on the scalp, but sparse elsewhere. The colour of the skin, notwithstanding the enormous differences of climate under which many members of the group exist, varies but little. It is true that in the features and cranium certain special modifications prevail in different districts, but the same forms reappear at widely separated parts of the continent. Thus skulls almost undistinguishable from one another may be met with from Vancouver's Island, from Peru, and from Patagonia.

Naturalists who have admitted but three primary types of the human species have always found a difficulty with the Americans, hesitating between placing them with the Mongolian or so-called "yellow" races, or elevating them to the rank of a primary group. Cuvier, indeed, does not seem to have been able to settle this point to his own satisfaction, and leaves it an open question. Although the large majority of Americans have in the special form of the nasal bones, leading to the characteristic high bridge of the nose of the living face, in the well-developed superciliary ridge and retreating forehead, characters which distinguish them from the typical Asiatic Mongol, yet in many other respects they resemble them so closely that, while still admitting the difficulties of the case, we are inclined to include them as aberrant members of the Mongolian type.[1] It is, however, quite open to any one adopting the Negro, Mongolian, and Caucasian groups as primary divisions to place the Americans apart as a fourth.

Now that the high antiquity of man in America—perhaps as high as that which he has in Europe—has been discovered, the puzzling problem, from which part of the Old World the people of America have sprung, has lost its significance. It is, indeed, quite as likely that the people of Asia may have been derived from America as the reverse. However this may be, the population of America, except at the extreme north, was, before the time of Columbus, practically isolated from the rest of the world. Such visits as those of the early Norsemen to the coasts of Greenland,

[1] No one can have seen a group of Botocudos from Brazil or of natives of Tierra del Fuego without being struck by their markedly Mongolian external characteristics.

Labrador, and Nova Scotia, or the purely accidental stranding of a canoe containing survivors of a voyage across the Pacific or the Atlantic, can have had little appreciable effect upon the characteristics of the people. It is difficult, therefore, to look upon the anomalous and special characters of the American people as the effects of crossing, as was suggested in the case of the Australians —a consideration which gives more weight to the view of treating them as a distinct primary division.

III. The Caucasian, Eurafrican, or white division, includes the two groups called by Professor Huxley Xanthochroi and Melanochroi, which, though differing in colour of eyes and hair, agree so closely in all other anatomical characters, so far, at all events, as has at present been demonstrated, that it seems preferable to consider them as modifications of one great type than as primary divisions of the species. Whatever their origin may have been, they are now intimately blended, though in different proportions, throughout the whole of the region of the earth they inhabit; and it is to the rapid extension of both branches of this race that the great changes now taking place in the ethnology of the world are mainly due.

A. The Xanthochroi, or blonde type, with fair hair, eyes, and complexion, chiefly inhabit Northern Europe (Scandinavia, Scotland, and North Germany), but, although much mixed with the next group, they also extend as far as Northern Africa and Afghanistan. Their mixture with Mongoloid people has given rise to the Lapps, Finns, and some of the tribes of Northern Siberia.

B. Melanochroi, with black hair and eyes, and skin of almost all shades from white to black. They comprise the great majority of the inhabitants of Southern Europe, Northern Africa, and South-West Asia, and consist mainly of the Aryan, Semitic, and Hamitic families. The Dravidians of India, the Veddahs of Ceylon, and probably the Ainos of Japan, and the Maoutze of China, also belong to this race, which may have contributed something to the mixed character of some tribes of Indo-China and the Polynesian Islands, and, as before said, have given at least the characters of the hair to the otherwise Negroid inhabitants of Australia. In Southern India they are largely mixed with a Negrito element, and in Africa, where their habitat becomes coterminous with that of the Negroes, numerous cross-races have sprung up between them all along the frontier line. The ancient Egyptians were nearly pure Melanochroi, though often showing in their features traces of their frequent intermarriages with their Ethiopian neighbours to the south. The Copts and fellahs of modern Egypt are their little-changed descendants.

In offering this scheme of classification of the varieties of the human species, it is not suggested that it is one universally accepted by anthropologists, or that it is likely to be final. Whatever care be bestowed upon the arrangement of already acquired details, or whatever judgment be shown in their due subordination one to another, the acquisition of new knowledge may at any time call for a complete or partial rearrangement of the system. The difficulties which encompass the subject have, indeed, been already indicated, and will be found abundantly illustrated in the writings of those authors who have specially devoted themselves to its elucidation.

Bibliography.—P. Topinard, *Éléments d'Anthropologie Générale*, 1885 ; A. de Quatrefages, *Histoire Générale des Races Humaines* (1. *Questions Générales*, 1887 ; 2. *Classification des Races Humaines*, 1889) ; Quatrefages and Hamy, *Crania Ethnicæ* (1873-1879) ; D. G. Brinton, *Races and Peoples*, 1890.

INDEX

THE END

Printed by R. & R. CLARK, *Edinburgh*

BIOLOGISTS AND THEIR WORLD

An Arno Press Collection

Adler, Kraig, editor. **Early Herpetological Studies and Surveys in the Eastern United States.** 1978

Adler, Kraig, editor. **Herpetological Explorations of the Great American West.** Two vols. 1978

Agassiz, [Jean] Louis [Rodolphe]. **Contributions to the Natural History of the United States of America.** Four vols. in two. 1857/1860/1862

Allard, Dean Conrad, Jr. **Spencer Fullerton Baird and the U.S. Fish Commission.** 1978

Altum, Bernard. **Der Vogel Und Sein Leben.** 1868

Azara, Don Felix de. **Apuntamientos Para La Historia Natural De Los Quadrúpedos Del Paragüay Y Rio De La Plata.** Two vols. in one. 1802

Baer, Karl Ernst v[on]. **Reden Gehalten in Wissenschaftlichen Versammlungen und Kleinere Aufsaetze Vermischten Inhalts.** Three vols. in two. 1864/1876/1873

Barrett-Hamilton, Gerald E. H. and Martin A. C. Hinton. **A History of British Mammals.** Two vols. 1910-1921

Boulenger, G[eorge] A[lbert]. **The Tailless Batrachians of Europe.** Two parts in one. 1897/1898

Brocchi, [Paul]. **Mission Scientifique Au Mexique Et Dans L'Amérique Centrale,...** Étude Des Batraciens De L'Amérique Centrale. 1882

Buffon, [Georges L. L.]. **The History of Singing Birds.** Translated from the French. 1791

Buffon, [Georges L. L.]. **The Natural History of Oviparous Quadrupeds and Serpents.** Translated by Robert Kerr. Four vols. in two. 1802

The Cabinet of Natural History and American Rural Sports. Three vols. in one. 1830/1832/1833

Candolle, A[ugustin] P[yramus] de and K[urt] Sprengel. **Elements of the Philosophy of Plants.** Translated from the German. 1821

Cassin, John. **United States Exploring Expedition During the Years 1838, 1839, 1840, 1841, 1842 Under the Command of Charles Wilkes, U. S. N.:** Mammalogy and Ornithology. Two vols. 1858

Chapman, Frank M. **Essays in South American Ornithogeography**. Edited by Keir B. Sterling. 1978

Cope, Edward D[rinker]. **The Vertebrata of the Tertiary Formations of the West.** One vol. in two. 1883

Cuvier, [Georges]. **The Class Mammalia.** Five vols. 1827

Donovan, E[dward]. **The Natural History of British Fishes.** Five vols. in two. 1802/1803/1804/1806/1808

Duméril, Auguste [Henri André], [Marie-Firmin] Bocourt and [François] Mocquard. **Mission Scientifique Au Mexique Et Dans L'Amérique Centrale,...** Étude Sur Les Reptiles. Two vols. 1870-1909

Flower, William Henry and Richard Lydekker. **An Introduction to the Study of Mammals, Living and Extinct.** 1891

Forbush, Edward Howe. **Birds of Massachusetts and Other New England States.** Three vols. 1925/1927/1929

Girard, Charles. **United States Exploring Expedition During the Years 1838, 1839, 1840, 1841, 1842 Under the Command of Charles Wilkes, U. S. N.:** Herpetology. Two vols. 1858

Grinnell, Joseph. An Account of the Mammals and Birds of the Lower Colorado Valley. 1914

Howard, H[enry] Eliot. **Territory in Bird Life.** 1920

Hume, Edgar Erskine. **Ornithologists of the United States Army Medical Corps.** 1942

Huxley, Leonard. **Life and Letters of Sir Joseph Dalton Hooker.** Two vols. 1918

LeConte, John L. and George H. Horn. **Classification of the Coleoptera of North America.** 1883

Linsley, E. Gorton, editor. **Beetles From the Early Russian Explorations of the West Coast of North America, 1815-1857.** 1978

Linsley, E. Gorton, editor. **The Principal Contributions of Henry Walter Bates to a Knowledge of the Butterflies and Longicorn Beetles of the Amazon Valley.** 1978

Murray, Andrew. **The Geographical Distribution of Mammals.** 1866

Osborn, Henry Fairfield. **Cope:** Master Naturalist. 1931

Peale, Titian R. **United States Exploring Expedition During the Years 1838, 1839, 1840, 1841, 1842 Under the Command of Charles Wilkes, U. S. N.:** Mammalogy and Ornithology. 1848

Phillips, John C. **American Game Mammals and Birds.** 1930

Rafinesque: Autobiography and Lives. 1978

Ray, John. **Synopsis Methodica Animalium Quadrupedum et Serpentini Generis.** 1693

Ray, John. **Synopsis Methodica Avium & Piscium.** [Edited by William Derham]. 1713

Richardson, John. **Fauna Boreali-Americana;** Or the Zoology of the Northern Parts of British America, Part Third: The Fish. 1836

Richardson, John, William Swainson and William Kirby. **Fauna Boreali-Americana;** Or the Zoology of the Northern Parts of British America, Part Four: Insecta, 1837

Riley, Charles V[alentine]. **[Nine] Annual Report[s] on the Noxious, Beneficial and Other Insects of the State of Missouri, 1869-1877** *and* **General Index and Supplement.** Ten vols. in three. 1869-1877/1881

Say, Thomas. **The Complete Writings of Thomas Say on the Entomology of North America.** Edited by John L. LeConte. Two vols. 1859

Schuchert, Charles and Clara Mae LeVene. **O. C. Marsh:** Pioneer in Paleontology. 1940

Sclater, William Lutley and Philip Lutley Sclater. **The Geography of Mammals.** 1899

Seton, Ernest Thompson. **Trail of an Artist-Naturalist:** The Autobiography of Ernest Thompson Seton. 1940

Smith, James Edward. **A Selection of the Correspondence of Linnaeus and Other Naturalists, From the Original Manuscripts.** Two vols. 1821

Stuckey, Ronald L., editor. **Development of Botany in Selected Regions of North America Before 1900.** 1978

Stuckey, Ronald L., editor. **Essays on North American Plant Geography From the Nineteenth Century.** 1978

Stuckey, Ronald L., editor. **Scientific Publications of Charles Wilkins Short.** 1978

Swammerdam, John. **The Book of Nature.** 1758

Wadland, John Henry. **Ernest Thompson Seton:** Man in Nature and the Progressive Era, 1880-1915. 1978

Waterhouse, G[eorge] R[obert]. **A Natural History of the Mammalia.** Two vols. 1846/1848

Weiss, Harry B. and Grace M. Ziegler. **Thomas Say:** Early American Naturalist. 1931

Wheeler, Geo[rge] M., [editor]. **[Reports Upon Insects Collected During Geographical and Geological Explorations and Surveys West of the One Hundredth Meridian During the Years 1872, 1873, and 1874].** 1875

Willughby, Francis. **De Historia Piscium** *and* **Icthyographia Ad Amplisimum Virum Dnum., Samuelem Pepys, Presidem Soc. Reg.** Two vols. in one. 1686/1685

Youmans, William Jay, editor. **Pioneers of Science in America.** Revised edition. 1896

DURHAM UNIVERSITY LIBRARY
3 0104 01313814 9